Advances in Manufacturing T

Advances in Manufacturing Technology VIII

Proceedings of the Tenth National Conference
on Manufacturing Research
Loughborough University of Technology
5-7 September 1994

Edited by

Keith Case and Steven T. Newman

Loughborough University of Technology
Loughborough, Leicestershire

Taylor & Francis
Publishers since 1798

file copyright

UK   Taylor & Francis Ltd, 4 John St, London WC1N 2ET

USA Taylor & Francis Inc., 1900 Frost Road, Suite 101, Bristol, PA 19007

A catalogue record for this book is available from the British Library

ISBN 0-7484-0254-3

Printed in Great Britain by Burgess Science Press, Basingstoke, on paper which has a specified pH value on final paper manufacture of not less than 7.5 and therefore 'acid free'.

Contents

Acknowledgements

Academic aspects of the conference, particularly the refereeing of papers has been undertaken by a Scientific Committee consisting of the editors and Christopher Backhouse, Leslie Davis, John Edwards, Allan Hodgson, Roy Jones, Paul Leaney, John Middle, Geoffrey Modlen, David Whalley, Bob Young and Bob Wood. Organisation and Administration has been provided by the Department of Manufacturing Engineering and in particular the editors would like to thank Tessa Clark and Jo Mason for their invaluable efforts.

Preface

The Consortium is an independent body and was established at a meeting held at Loughborough University of Technology on 17 February 1978. Its main aim is to promote manufacturing engineering education, training and research. To achieve this the Consortium maintains a close liaison with those Government Departments and other bodies concerned with the initial and continuing education and training of professional engineers and responds to appropriate consultative and discussion documents and other initiatives. It organises and supports national manufacturing engineering education conferences and symposia. The Institution of Electrical Engineers, with whom there is a close working arrangement, undertakes the secretarial duties.

The Consortium consists of those university departments or sections whose first priority is to manufacturing engineering and who have a direct responsibility for running honours degree courses in the field of manufacturing engineering.

COMED decided in 1984 that a national forum was needed in which the latest research work in the field of manufacturing engineering and manufacturing management could be disseminated and discussed. The result of this initiative was that an annual series of these national conferences on manufacturing research (NCMR) was started; the first NCMR was held at the University of Nottingham in 1985 with subsequent conferences at Napier College (Edinburgh), the University of Nottingham, Sheffield Polytechnic, Huddersfield Polytechnic, Strathclyde University, Hatfield Polytechnic, the University of Central England and the University of Bath. The Tenth Conference, of which this is the published proceedings, was held at Loughborough University of Technology, 5–7 September 1994.

Dr Keith Case and Dr Steven Newman, Editors and Conference Chairmen, Department of Manufacturing Engineering, Loughborough University of Technology, Loughborough, Leicestershire, LE11 3TU.

Keynote address

REQUIREMENTS AND TRENDS IN MANUFACTURING SYSTEMS

Professor Richard Weston

Pro-Vice-Chancellor Research
Professor of Flexible Automation
Loughborough University of Technology
Loughborough, Leics, LE11 3TU

The nature of typical manufacturing systems has changed significantly over the last few decades. Some of the most marked of these changes are identified, as are factors which have induced them. By so doing the paper seeks to provide a broad requirement specification, in respect of which many of the conference papers can be classified.

Introduction to the Conference

The 10th National Conference on Manufacturing Research promises to be an exciting and enlightening event, judged by the breadth and quality of papers and authors assembled by the organising team. Hence Loughborough University of Technology (LUT) is particularly delighted to host this event, being a University which prides itself on the *relevance* of its research and its *industrial partnerships*. At LUT, "*manufacturing*" constitutes a major interdisciplinary research theme; a recent audit showing that we have over 300 active researchers in the arena working with a total budget in excess of £10M per annum. Hence, personally, I was delighted to be asked to start the ball rolling. What I aim to do is outline emerging requirements and trends in manufacturing which will set a backcloth for the work of us all.

Observable Trends - Globalisation, Increased Complexity and Increased Rate of Change

"He who does not think far ahead is certain to meet troubles close at hand" Confucius, born 551BC.

The last century has seen many dramatic changes which have impacted on the methods, processes, tools and organisational forms adopted by manufacturers. Though the impact of such changes has been far from uniform, with many present day examples of companies persisting, albeit often highly successfully, in using approaches which would not have been out of place one hundred years ago. Despite variances there are clearly observable trends. For example, the centre of gravity of employment in technically developed nations has moved during the last four or five decades from primary (agriculture), through secondary (manufacturing) to tertiary (service) industries. Of course this does not necessarily imply that agriculture or manufacturing is becoming less important than service industries but such trends, coupled with enormous observable variability certainly illustrates the occurrence of natural selection processes.

Very few manufacturing companies still enjoy a protected market niche. Increasingly competition is on a global scale, with so-called third world countries of a few decades back becoming market leaders and even dramatically reshaping those and related markets.

Fuelled by dramatically improved communication and transport systems, virtual manufacturing enterprises have arisen which can be shaped around key business processes to optimise their responsiveness to customer need. Often timescales have dramatically collapsed giving rise to more highly responsive businesses. An ethos of "have it customised to my particular requirements now and pay for it later" has been engendered in customers, manufacturers and governments alike. The result may be instability through positive feedback, despite natural non-linearities and constraints. Certainly many manufacturing companies are in a state of constant turmoil, reorganising themselves regularly; for example first believing big is beautiful, then devolving responsibility to smaller units, then seeking to rationalise, and so on.

Clearly the use of the latest technology (particularly through the guise of *automation* and *IT based integration*) can lead to better products, more quickly and cheaply. Particularly technology can provide a means of quickly responding to customer need and can facilitate metamorphosis of an enterprise. However, the use of this technology has significantly increased the complexity of solutions so that although well designed solutions can provide very significant operational benefit the technology can itself present major hurdles in the event of unforeseen and medium-to-long-term changing needs, Weston(1993).

The philosopher will observe conflicting implications of increased *system complexity* and *rates of change*, yet both are implicit properties of a highly competitive manufacturing enterprise. What is more, external factors can greatly influence internal ones. For example, characteristic properties of product markets and supply chains will directly influence the optimal choice of manufacturing methods, processes and organisational forms. Similarly financial and labour markets can have a marked impact on best practice, particularly in respect to the adoption of high tech solutions.

Considerable hype followed the widespread industrial use in the early 1970s of CNC machines and robots, which was accentuated during the 1980s by the accelerated industrial and commercial adoption of networks, databases and intelligent systems. However, for the foreseeable future it is clear that seldom will technology be used to completely replace people, rather it will be used to semi-automate their tasks (as depicted by Figure 1), this by providing computational tools, machinery and mechanisms to help

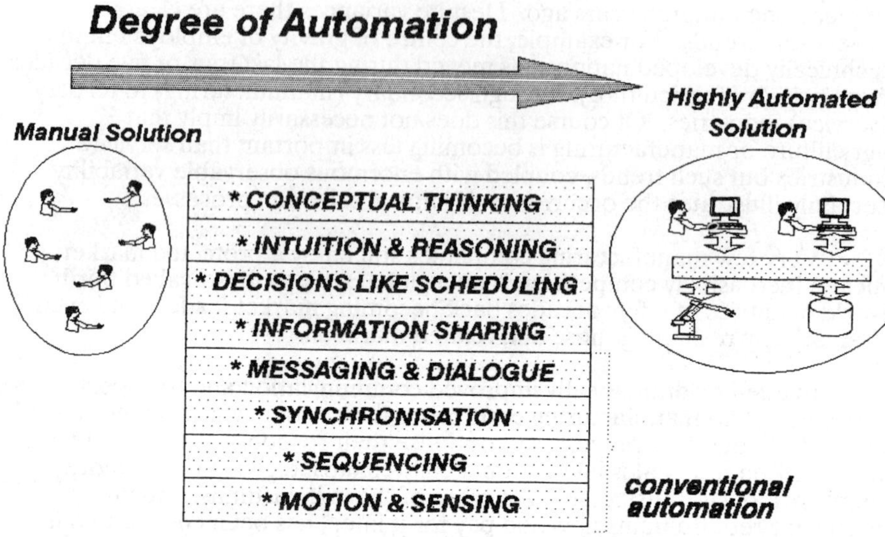

Degree of Automation

Manual Solution

* CONCEPTUAL THINKING
* INTUITION & REASONING
* DECISIONS LIKE SCHEDULING
* INFORMATION SHARING
* MESSAGING & DIALOGUE
* SYNCHRONISATION
* SEQUENCING
* MOTION & SENSING

Highly Automated Solution

conventional automation

Figure 1 Activities which could be automated

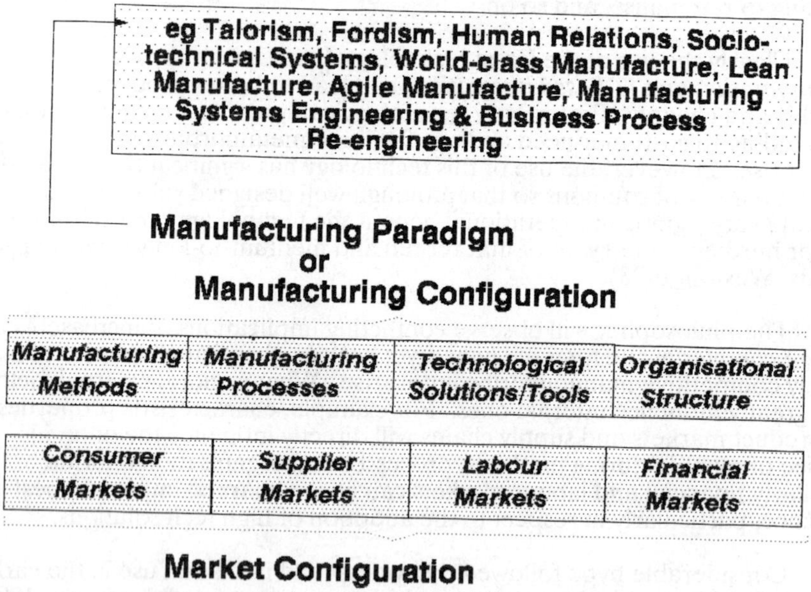

eg Talorism, Fordism, Human Relations, Socio-technical Systems, World-class Manufacture, Lean Manufacture, Agile Manufacture, Manufacturing Systems Engineering & Business Process Re-engineering

Manufacturing Paradigm
or
Manufacturing Configuration

Manufacturing Methods	Manufacturing Processes	Technological Solutions/Tools	Organisational Structure

Consumer Markets	Supplier Markets	Labour Markets	Financial Markets

Market Configuration

Figure 2 Need to identify and adopt
an appropriate Manufacturing Paradigm

them make better and faster decisions and carry out more appropriate and timely actions. Hence we should look for more effective combinations of technology and people when seeking a solution to a set of requirements.

Very interesting interrelationships exist between the 'manufacturing paradigm' adopted by leading manufacturers during this century and characteristics of the markets within which they operate, Buchanan(1994). Such exemplar companies have recognised particular properties of their markets and evolved an appropriate manufacturing paradigm to realise a competitive edge. Figure 2 highlights key factors involved here. This raises many very important questions; particularly the question *how can best practice (or exemplar solutions) be recognised and adopted?* which is one which faces most manufacturers, whereas the research community will be actively answering the question *what should be the next set of manufacturing paradigms?* there will always be a need for different manufacturing paradigms and different ways of realising any given paradigm; this in pursuit of competitive differentiation. Clearly, in different markets vastly different solutions may be required, even though we might anticipate increased uniformity as the impact of globalisation reaches all parts of the globe.

Dealing with Complexity, Globalisation and Change

It is a requirement of the manufacturing research community to formalise ways of dealing with increased complexity, globalisation and increased rates of change.

It is not sufficient to hope that competitive solutions will naturally emerge.

"Hope is a waiting dream" Aristotle.

Computer, management and behavioural scientists continue to evolve enabling technology to help us do this. Central to our task will be the modelling of manufacturing systems, so that we turn the art of manufacturing into a widely adopted science. The levels of complexity typically involved dictate a need to model manufacturing systems from a variety of perspectives and with different levels of abstraction, formalism and completeness. Where we cannot model things in an analytical way we will need to seek less formal description methods. We will also need reference models of good practice, to guide and support senior managers, system designers, system constructors and industrial practitioners, who collectively have responsibility for the various life-cycle phases of an enterprise. Also required will be tools (many of which will be computerised) to empower such persons, thereby enabling them to carry out their tasks more effectively. Also required will be computational infrastructures which support the management of change and provide operational mechanisms which underpin the concurrent interworking of groups of people, software systems and automated machines; where each may be located at different geographical locations around the globe. As appropriate such an arrangement will traverse conventional customer, manufacturer, supplier product chains.

In manufacturing systems of the future, models and tools will be used to determine requirements from a strategic perspective and to help implement changes specified, guided by a best practice manufacturing paradigm (as depicted by Figure 3). The infrastructure will aid implementation and management of change in an operational sense, Clements, Coutts and Weston (1993).

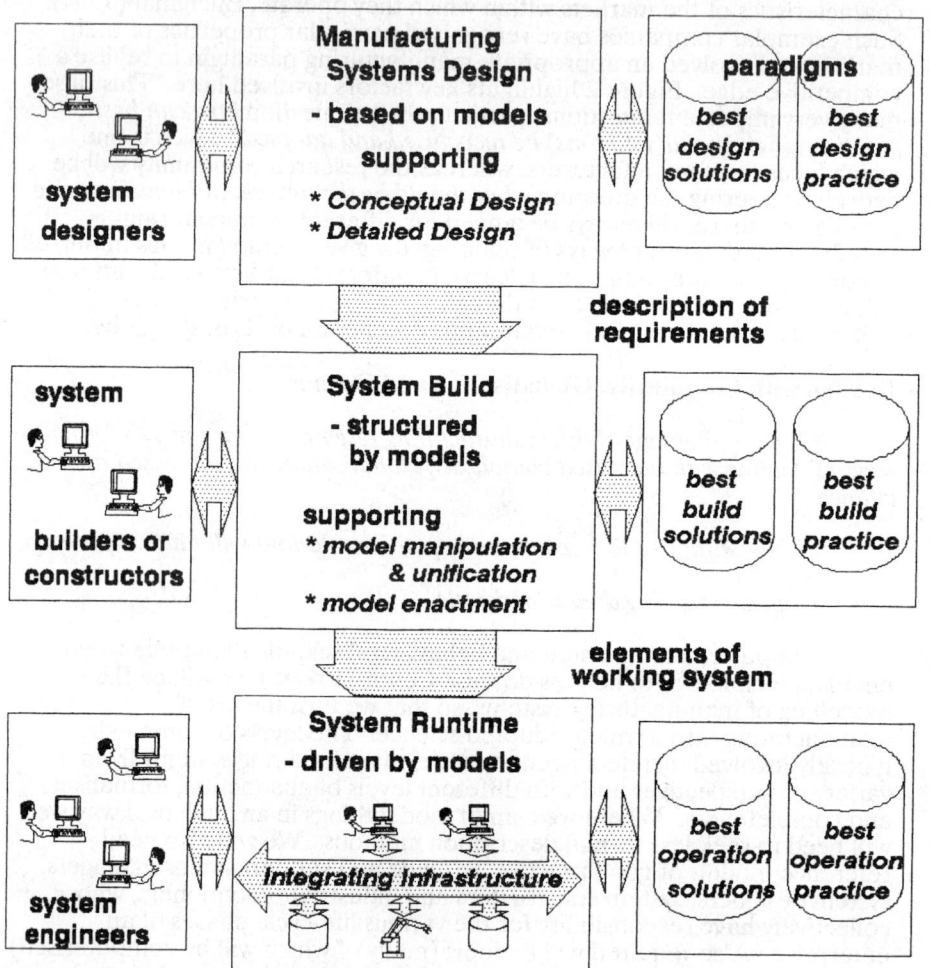

Figure 3 Model Driven Approach to Creating Integrated Manufacturing Systems

Perhaps a clear note of caution. Any solutions derived by the research community will need *ultimately* to be delivered to industry in a tractable form and must be capable of dealing with inherent constraints, including the need to build on and advance any legacy of existing personnel and installed systems. This will require closer partnerships between researchers, vendors and users of solutions.

References

Weston, R.H. 1993, *Steps towards enterprise-wide integration: a definition of need and first-generation open solutions*, Int. J. Prod. Res., Vol. 31, No. 9, 2235-2254. (Taylor & Francis, London).
Buchanan, D. 1994, *Theories of Change*, Loughborough University Business School Research Series, Paper. 1994:5 ISBN 185901 0350.
Clements, P., Coutts, I.A. and Weston, R.H. 1993, A Life-cycle Support Environment Comprising Open Systems Manufacturing Modelling Methods and the CIM-BIOSYS Infrastructural Tools, *Proc. of the Symposium on Manufacturing Applicaton Programming Language Environment (MAPLE) Conference*, Ottawa, Canada, 181-195.

Manufacturing management

ENVIRONMENTAL MANAGEMENT - STANDARDS, CERTIFICATION AND REGULATION

Philip Sayer

Napier University
Edinburgh, EH10 5DT

Industrial organisations are forever facing new challenges in both technology and management systems. Today, the new challenge facing industry is the requirement to improve their environmental performance, with: public pressure for "green" measures, the increasing cost of waste disposal, new legislation and regulation requiring organisations to obtain process authorisations and measure, control and reduce emissions. To do so effectively requires a formal environmental management structure, an appropriate environmental policy, a planned programme of action with quantifiable objectives and targets, and operational control procedures. This paper reviews the development and implementation of standards currently available and progress towards third party certification of environmental management systems.

Introduction and History

Pressure has grown over a number of years for both industry and the public at large to clean up their environmental performance. The public have been encouraged to purchase "environmentally friendly" products, such as: CFC free aerosols, phosphate free washing powders and organic produce. However, the claims of manufacturers about the environmental soundness of their products have been proved in many cases to be highly exaggerated and often at best misleading. These problems are insignificant though when one considers the environmental damage created by most manufacturing operations. It is this that has promoted the introduction of extensive environment legislation and regulations to control and ultimately reduce the environmental effects of manufacturing and service operations.

In order to ensure that organisations fulfil all the requirements of the regulators, their financiers and insurers, it is essential that they implement rigorous management responsibility and systems to control their environmental impact. The Eco-Management and Audit Scheme requires such rigorous management and sets the scene for environmental improvement through the appropriate use of an environmental management system. It is to this end that the British Standard BS7750 Environmental Management Systems has been developed.

Eco-Management and Audit Schemes (EMAS)

The Eco-Management and Audit Scheme Regulation (no. 1836/93) finally came into force in all 12 member states on 29 June 1993. The scheme has been established in the words of Article 1 of the regulation "for evaluation and improvement of the environmental performance of industrial activities and the provision of the relevant information to the public".

This statement sums up the ethos of the regulation and its operation in practice; requiring those organisations who volunteer for the scheme to undertake an environmental review and as a result implement an environmental management system to control such effects as may be discovered. Finally the organisation will, at specified intervals, audit its environmental performance and prepare an environmental statement which, once verified by an independent verifier, will be published.

The scheme is open to any industrial organisation, specifically: manufacturing (defined in the Council Regulation no. 3037/90), electricity, gas, steam and hot water production, and the recycling, treatment, destruction or disposal of solid or liquid waste. Organisations can, however, only be registered on a site by site basis with effect from 10 April 1995.

BS7750 Environmental Management Systems

This British Standard was first published on 16 March 1992 after a short development period. Publication was followed by a pilot programme to assess the suitability of the standard in controlling and improving an organisation's environmental effects. The findings of this pilot study have been incorporated into the latest issue of BS7750 published in February of this year. It is intended that BS7750 should be fully compatible with the Eco-Management and Audit Schemes management system (Annex 1 of the EMAS regulation).

There are a number of key differences between the Eco-Management and Audit Scheme however, principally that BS7750 is open to any organisation, not just those outlined above, and that the Eco-Management and Audit Scheme requires an environmental review and the publication of a verified environmental statement. It is therefore inappropriate to suggest that a straight forward adoption of BS7750 will fulfil the requirements of the Eco-Management and Audit Scheme.

BS7750 and BS5750 - Joint Systems Development

The Environmental Management Standard has been developed from the British Standards Institutions experience of the Quality Systems standard BS5750. The two documents therefore have the same "look and feel". For those organisations who already have a certified Quality System the implementation of BS7750 within the current Quality System should be a relatively straight forward matter, and indeed this is the experience of Wavin.

It has been suggested by a number of organisations that without an existing management system, such as BS5750, the implementation of BS7750 would be difficult to achieve. In some respects this opinion is valid since BS7750 does not include in its clauses the extensive management structure required by BS5750. This is to say that BS7750 relates closely to the requirements of environmental management, whilst BS5750 acknowledges that the quality is an organisation wide issue and therefore requires a management system that reflects this.

Developing an Environmental Management System

The development of an Environmental Management System is straight forward providing that it is approached in a logical and systematic manner (Figure 1). The following details the steps which should be undertaken to develop and implement an appropriate management system.

Organisation and Personnel

In order that a consistent approach is taken in the development of the system, it is important that a suitably qualified and motivated team is identified at an early stage in order to drive the implementation of the system. The leader of this team must be at an appropriate level in the organisation to develop and implement policy.

Preparatory Environmental Review (PER)

The second stage in the development of a system is the PER identifying an organisation's environmental impact (effects) or potential impact as a result of its activities, products or services. As a minimum this should include a review of:

- current legislation and regulation effecting the organisation;
- the site, plant and equipment;
- previous accidents or incidents;
- existing management documentation.

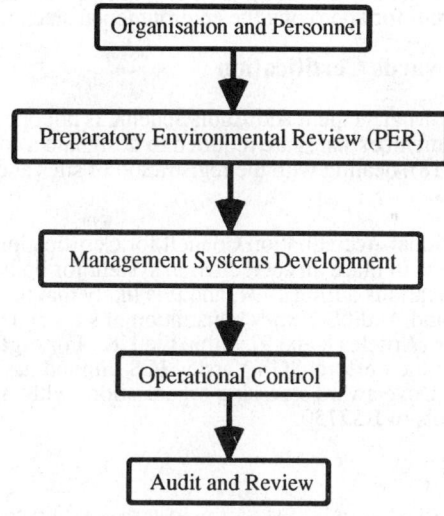

Figure 1

The PER is not a requirement of BS7750, though it is felt that this is a vital first step in the assessment of an organisations environmental effects, and is a useful tool in the assessment of performance against initial goals.

Management System Development

The results of the Preparatory Environmental Review will be used in developing an organisation's Environmental Policy, Objectives and Targets and Environmental Management Programme. This is also a suitable time to develop the Management System (Documentation Control, Records and Training), Management Review and Audit Systems necessary to maintain an Environmental Management System.

If the Eco-Management and Audit Scheme is to be adopted then this would be an appropriate time to develop the necessary provision for the development and publication of the verified environmental statement.

Operational Control

The final stage in the development of an Environmental Management System is the preparation of procedures to control and monitor the organisation's effect or potential effects. A phased development of operational control procedures is most appropriate since this reduces the impact on the organisation. Where a phased approach is opted for, those areas singled out as having a significant environmental effect should be procedurised first, followed by those areas identified as having an environmental or potential environmental effect in ranked order.

Audit and Review

Following the development and implementation of the Environmental Management System it is necessary to audit the procedures to ensure compliance with the chosen standard. Following audits it is necessary to undertake a management review to assess the organisation's compliance and performance in relation to its stated policy, objectives and targets, and current legislative and regulatory requirements. For those organisations implementing the Eco-Management and Audit Scheme this would also be the forum for approving the environmental statement, prior to verification.

Progress Towards Certification

The Eco-Management and Audit Scheme is not operational until 10 April 1995. However, the member states are required to designate a competent body by 29 June 1994 (Article 18), dealing with the registration of sites and the publication of a list of registered sites.

The National Accreditation Council for Certification Bodies (NACCB) is expected to have in place an accreditation system for bodies wishing to undertake BS7750 certifications during 1994, and it is likely that they will also operate the Eco-Management and Audit Schemes registration of sites and publication of the list of registered sites (Articles 8 and 9) within the UK. For organisations wishing early recognition for their efforts SGS Yarsley ICS Limited have been offering certification to their Green Dove award assessing organisations with management systems meeting the requirements of BS7750.

Conclusions

The publication of the revised British Standard BS7750 and the coming into force of the Eco-Management and Audit Scheme are significant milestones towards improved environmental performance within Europe. There has of course been criticism of both of these systems in that they are inappropriate or lack real bite in terms of performance improvement. However, the proof of the pudding is in the eating, and as with other management standards such as BS5750, the key to real success is the development and implementation of appropriate systems. As far as the Eco-Management and Audit Scheme goes, the requirement for the publication of an environmental statement is significant in that an independent verifier assesses the statement against the environmental performance of the organisation in much the same way as company accounts are audited. This should ensure that the scheme gains credibility and respect from both commerce and the public at large.

The adoption of an international standard by the International Standards Organisation (ISO) may be decided when the main ISO subcommittee meets in Australia in May 1994. The British Standard BS7750 is being considered for adoption, though the Americans are objecting that the standard is too prescriptive, and believe that both BS7750 and the Eco-Management and Audit Scheme are "regional" initiatives.

On a more optimistic note Article 13 of the Eco-Management and Audit Scheme makes provision for the Community to actively promote the scheme to small and medium sized enterprises (SME's). It is felt that this Article will lead to funding for those SME's who wish to register for the scheme, though details of such provision are unlikely to arise until nearer April 1995.

Finally, consideration of the likely benefits of implementing an environmental management system is appropriate. These will of course vary depending upon the system implemented, but at least should:

-        reduce the risk of unwanted environmental incidents;
-        ensure compliance with environmental regulators and legislation;
-        improve the public image of the organisation;

and at best:

-        motivate staff;
-        improve profitability.

References

Annon 1994, *BS7750 Environmental management systems,* (British Standards Institution)

Hillary R. 1994, *Linking BS7750 and the Eco-Management and Audit Scheme,* (British Standards Institution)

Rothery B. 1993, *Implementing the Environmental Management Standard and the EC Eco-management scheme,* (Gower)

Gilbert M. J. 1993, *Achieving Environmental Management Standards,* (Pitman Publishing)

Annon 1993, Pitfalls discovered in LA eco-management and audit: Environment Business, 6/10/93, (Information for Industry), 4

Annon 1993, Companies concerned about eco-management and audit scheme: Environment Business, 3/11/93, (Information for Industry), 5

Annon 1994, EMAS undermined by plans for international standard: Environment Business, 23/02/94, (Information for Industry), 4

ed. Peckham A. C. 1994, *Fact file series: Environmental Management; 1. An Introduction to BS7750,* (Institute of Environmental Managers, Centre for Environment & Business in Scotland).

MATERIAL SUPPLIER, COMPONENT SUPPLIER AND OEM INFORMATION FLOW EFFECTS

David Newlands and Phil. Southey

Coventry University
Coventry, Warks, CV1 5FB

Following interviews with material suppliers and part producers, three information exchange flow models became apparent. A serial model and two triangular models are presented. It is proposed that all three organisations may be transplants or indigenous to markets they serve. Supplier development, working in relationships, benchmarking, quality, kaizen improvement and lean production initiatives have been driven from OEMs to their component, sub-assembly and system suppliers. This working paper identifies the materials section of the chain, the authors believe to be not yet fully affected by these sweeping changes. Communication and logistic control differences between the models are proposed. These models are presented for discussion, based on initial research of the plastics industry, but may equally be applicable to other material supply chains.

Introduction

Driving Forces for Change

Japanese Original Equipment Manufacturers (OEMs), in manufacturing sectors, pioneered fundamental changes in product quality, production tools, stock reduction with increased flexibility together with a range of manufacturing techniques and management philosophies. Advances such as these lead to increased market share. Observing Japanese steady market share growth performance, organisations based in other regions changed strategies, reducing support function and production costs, whilst reducing delivery lead times of quality products through simultaneous initiatives. Competitive pressures increased, forcing management to review their organisations and supply chains, to find potential improvement areas.

OEM driven supplier development initiatives were introduced into transplant assembly plants, such as those of Nissan, Toyota and Honda. Cloned from Japanese original plant models, but modified to suite the local requirements. Component manufacturers and assemblers have been OEM targets for partnerships or longer term working relationships, forming first, second and third tier supplier layers.

Three Organisational Interface Models

The authors' findings produced three models (Newlands and Southey 1994). They were given the names serial, conventional or traditional flow, material supplier active and pro-active supply chain. Some international multinational enterprises (MNEs, from Dunning, 1993) use more than one model. The term OEM here is used to represent both well studied supply chains such as automotive and electrical, but also food package users or [can] 'fillers'.

Model 1. Serial or Traditional Communication and Logistic Routes Within the Supply Chain

Material flows from one company or process to another serially. In the serial model, negotiations are open to tender, or organizations are invited to submit proposals and quotations "in series". Information including delivery schedules, cost/price negotiation, quality, design and payment, are all serial. Inventory is kept at to safe guard against stock outs, poor quality, insufficient and late delivery. See Figure 1 (adapted from Newlands and Southey 1994). Information about component operating conditions and material specifications is created by the OEM and then shared with component moulders and suppliers.

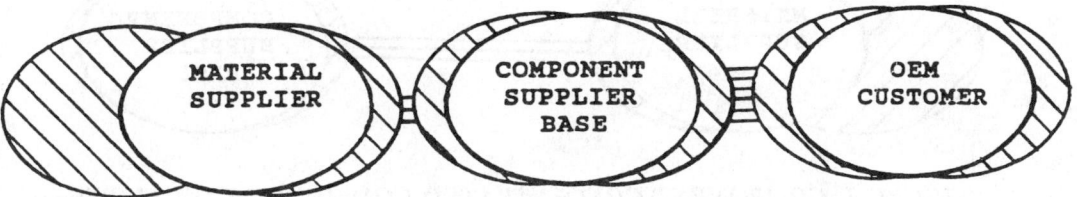

Figure 1. SERIES COMMUNICATION AND MATERIAL FLOWS

Material supplier choice is delegated to component suppliers. OEMs expectations of direct suppliers increase periodically. This leads the suppliers to select material suppliers who are willing to work in partnership or harmony. The selection of material suppliers is delegated to the component suppliers. This obligates material suppliers to sell direct to moulders. Contracts were short term between OEM and component suppliers, thus material suppliers are not assured of long term contracts either and must therefore maintain a high marketing profile with many prospects. The initial stages of supplier development concentrated on first tier suppliers of complete assembles. Negotiations with material suppliers direct by OEMs were not common, due to the practice of passing on any cost increased to the end user.

Model 2. Active Material Supplier Supply Chain Relationships

Within collectivism cultures, OEM organisations generate sibling strength bonds with suppliers. These suppliers are in close, direct, communication with the OEM and negotiate for, and on behalf of, the OEM on technical issues with part producers. Supplier responsibilities include specification creation and scheduling details with lower tier suppliers and material suppliers. Material supplier selection is carried out by OEMs, using their greater purchasing leverage to drive down prices and ensure availability. Monitoring or benchmarking of material suppliers is carried out, however, delivery schedules and other specification arrangements are left to moulders. A weak information transfer route is established between assembler and material supplier. This communication route will temper specifications with available materials, availability and guarantee of supply, quality and cost (world wide price and cost stability as part of the OEMs' efficiency initiatives). Consequently, a weaker bonds exist between suppliers and the material suppliers. Communication and bonding intensities between raw material producers and OEM appear low, principally due to OEMs' trust in tier supplier judgement, leading to the tiered supplier/material supplier total relationship ownership. See Figure 2.

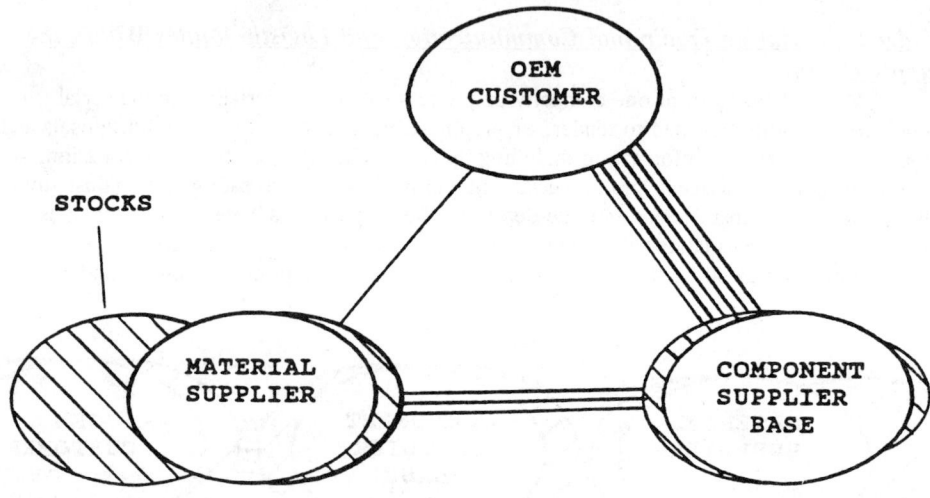

FIGURE 2. COMPONENT CONCENTRATED COMMUNICATION FLOWS

Model 3. Pro-active Material Supplier Supply Chain Communication Developments

Active supplier base reduction polices using selection procedures and capability development, are being successfully employed in various assembly industry sectors. Automotive, electro' technical and food industries have progressively reduced their supplier bases. OEMs are now becoming more willing to negotiate with selected strategic suppliers, further down the supply chain.

Raw material producers coordinate with OEMs and the component suppliers (material forming or processing), on material specifications requirements, technical developments in material mixes and process implications of the various concoctions. Stock sites are reduced in volume and material is only converted on a make to order basis. Stock is pushed down the supply chain, back to the material supplier and to its most basic but available form, (see figure 3). Material supply and modification activities to form components regarded as black box activities, are carried out with or without OEM direct consultation by both parties in order to meet business objectives. Working in a partnering relationship, each organisation considers both of the others to be 'on their side'. Hence the sales accounts become easier to manage, with staff to staff stability increased.

Material supplier representatives approach OEMs and 'fillers', etc, directly in order to learn what material performance specifications are required for the end product. Raw material sources are increasingly specified by the OEM on a world wide supply basis. World wide single material supplier sourcing contracts are made on behalf of component suppliers. Economies of this scale reduce prices, while global availability factors and quality levels, guarantee that product is sold at one price.

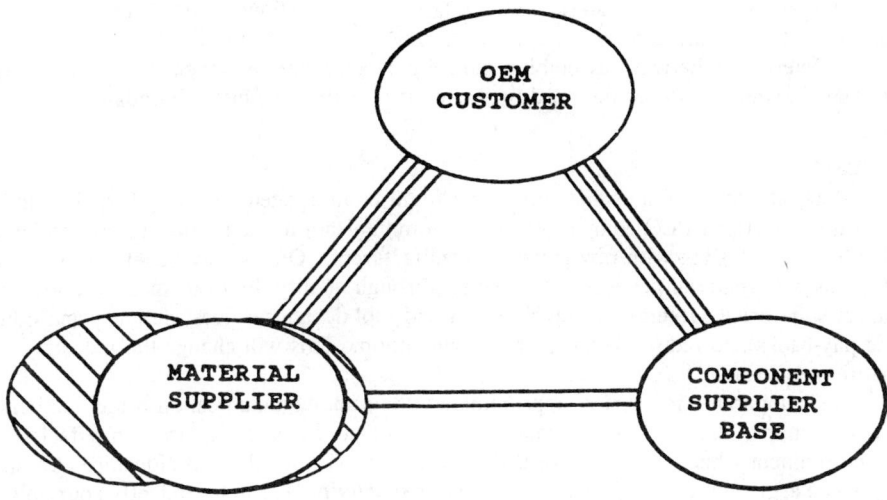

FIGURE 3. EQUAL EMPHASIS COMMUNICATION FLOW

Exchange rate variations tend to differentiate material prices in the various locations. This has come about by material producers promoting their materials on cost and processing advantages, technical specifications and services. Profit levels are ensured through renumeration packages which take into account the following factors:-value added contribution, customer relationships and satisfaction, production technology/knowledge/efficiency improvements, complementary services, alternative cost of skill development, supplier commitment factors, local operating/employment costs, living costs, relocation/support costs. (adapted from Coulson-Thomas, 1992).

Environmental Issues

BS7750, together with other environmental/market forces are forcing OEMs to re-examine material suppliers. IBM for example, realises that its products are only as 'green' as the least green supplier's process which produced components for the product. "... it isn't enough to operate in an environmentally sensitive way; companies must also ensure that raw materials and components come from equally responsible sources. Every part of the business has an impact on the environment." (Gillett, 1992)

Stock Issues

Stock, any where in the supply chain has value. Work done to materials increases the value of the materials. OEM initiatives to push stock back to suppliers, leaving enough components for a specified number or time period, with JIT deliveries does not remove stock from the supply chain. JIT manufacture is becoming more common. JIT material deliveries from suppliers reduces on site inventory to a minimum. Small material users are delivered to less frequently than is desirable.

The use of nominated carriers to carry out milk rounds, picking up daily required raw material quantities from suppliers, delivering to moulders, picking up mouldings to drop off at higher tier assembler plants and so on back to the OEM assembly plant, is similar to, and an extension of cellular manufacturing. Nissan Manufacturing U.K., utilize Ryder in this manner. (Smith 1994)

The removal of stock materials forces fast reactions, allows more flexible production. Manufacturing rates are controlled and coordinated centrally by OEM/first tier suppliers. Interfacing between assembler and reliable suppliers, together with coordinated logistics by assembler reduces the need to retain large WIP and finished goods levels.

Conclusions

Material Process Plants and suppliers should be integrated into supply chains, in order to improve U.K. PLC competitiveness. Communication route numbers grow slightly for OEMs, when dealing with raw material manufacturers. OEMs may benefit from direct negotiations with material producers/suppliers, (through quality, build to specification, world or market wide price agreements, development and tool design services, delivery and other service pay-backs), to reduce Total Cost of Acquisition. This will change the industry competition equation.

For comparable industries, approximately the same communication route numbers exist within supply chains. 'Middle men' finished component manufacturers may feel safer in an environment where, they control and own the material supplier relationship. An equal and opposite argument is: where suppliers are not specifying and independently sourcing, are placed in the difficult position (between supplier and customer), of being relatively easy to replace. Each party may, whilst working together feel that the other two organisations are 'on their side'. Customer/supplier development can occur from various sources, in both directions: OEM with component supplier, material supplier with component supplier, material supplier with OEM.

Financial measurement and control can be increased when the OEM knows the purchase price paid by the component supplier for materials. If several component suppliers are used to source similar components, monitoring and comparisons between them will produce performance benchmarks. Component supplier replacement ease may well improve tier supplier quality and service performance levels should they decide to continue in that business. Quality procedures have been recognised and certified in some economic regions, but a levelling process is under way. Transplanted and subsidiary plants work together in the way the original plants do, but incorporate locally demanded changes. These changes hybrid the relationships, creating learning situations. The issue of stock piling in more basic materials for bespoking, within the supply chain, should be addressed. More research into, and capability development of, these relationships within the supply chain in all industries, needs to be carried out.

Acknowledgements

This research into the area of Supplier Development and Benchmarking is currently being supported by the Economic and Social Research Council, (ESRC).

References

Coulson-Thomas, C. (1992) *"Creating the Global Company"* p278 McGraw Hill.
Dunning J H. (1993) *"The Globalization of Business"* Routledge, New York.
Gillett, J. (1992) "The green line", *IBM United Kingdom Annual Review 1992, p38-39.*
Newlands D.J, and Southey P. (1994) "Which Information System Is The Most Advantageous? How Should OEMs Relate to Material Suppliers?", *3RD Conference, p363-372, IPSERA Glamorgan*, South Wales.
Smith, K. (1994) "Problems of Just-In-Time Distribution And A Solution!", *3RD Conference, p535-543, IPSERA Glamorgan*, South Wales.

MANUFACTURING WITH EXTREME SEASONAL DEMAND

Dr Glyn B. Williams and Markus Westhues

University of Birmingham
School of Manufacturing and Mechanical Engineering
Edgbaston
Birmingham B15 2TT

Seasonal variations can be identified in the demand patterns of many products. Agricultural equipment for example is often only employed during short periods. Accordingly, delivery data for such products exhibits pronounced seasonal peaks. Typically, a similar set of problems occurs. Pure make to order strategies can not be applied, making to a forecast is an alternative. Consequently high inventory levels seem to be unavoidable. In order to gain a competitive advantage, the challenge is to develop a strategy that deemphasises the effects of seasonality. This paper reviews some techniques applied in industry. The approaches are grouped into four categories: i) improving the seasonality of production, ii) reducing the seasonality of demand, iii) improving the predictability of the markets and accuracy of forecasts, iv) combining products with different seasonality.

A few examples of seasonality and seasonal industries

Seasonal variations can be identified in the demand patterns of almost every product, from cars to television sets. This paper tries to focus on all those industries that face extreme seasonality.

A wide variety of annually reoccurring events might determine typical highs (or lows) in a company's sales. Sales of jewellery, pens and toys are high in the few weeks before Christmas. In some industries, trade fairs are an occasion where a large percentage of annual orders come in. Other products sell best at the beginning of the summer vacation. Every year retailers offer "back-to-school" packages. Whether the product is ski equipment, barbecues, electric cooling fans or suntan lotions, the fact is that sales to the consumer are highly precarious, subject as they are to the vagaries of the weather (Gabb, 1991).

The seasons dictate a strict schedule for agriculture and related industries such as food processing or the agricultural machinery industry. Farm equipment, especially harvesting machinery (i.e. combine harvesters), can only be employed during the short harvesting period. Some sugar refining plants only have a three month period of full operation (Benders, 1991). Demand for farm inputs like fertilisers or herbicides varies greatly over the year (Bloemen and Maes, 1992). As an additional burden, the uncertainties of the weather often make it hard or impossible to make reliable forecasts and schedules in industries related to agriculture.

What problems can be caused by seasonality?

For most industrial companies that have to cope with seasonal demand patterns, a similar set of problems occur. High inventory levels seem to be unavoidable, particularly for specialised manufacturers. In order to utilise plant and equipment during the off season months when demand is low, many manufacturers are building up extensive finished goods inventories according to their expectations and forecasts.

Seasonality often poses a major challenge for a company's logistics. As pure make-to-order strategies are usually not feasible, make-to-stock is one alternative during the off-season months. With the season approaching, a firm will often find differences in between what they have in their stores and what is actually required. A seasonal firm will then have to provide flexibility. Either they are able to produce the remaining actual demand within a short period of time or they rework and change units taken from the stocks before delivery.

The dependence on forecasts rather than on real orders can be identified as another major problem for seasonal industries. Because of forecasting inaccuracies, many vagaries exist throughout the year. A forecast for a seasonal product can generally be expected to be less accurate than a forecast for a product with constant demand. This is so for two reasons. Firstly, the time horizon of twelve months is comparatively long, especially as often there is no continuous inflow of data. Secondly, uncertainties because of the weather are difficult to account for. Though variances have to be expected in every forecasting cycle, seasonal industries suffer from the one year long time lag that makes it impossible to respond continuously and quickly.

Methods and techniques to overcome seasonalities

A seasonal demand pattern should not be considered a problem, but a challenge. After all it is better to have seasonal demand than to have declining or scarce demand. The challenge for a company is to develop an appropriate strategy that matches seasonality in the best possible way. Coping with seasonality better than others means to gain competitive advantage. Only if seasonality is ignored, the characteristics (i.e. high inventory level) might bring about problems and threats for a company.

This paper summarises some methods and techniques that are applied in different industries to approach seasonality. The right strategy for a specific product or company consists of a combination of these options. Most options can be grouped in four categories:

1. Improving seasonality of production
2. Reducing seasonality of demand
3. Improving predictability of demand, accuracy of forecasts
4. Combining different seasonalities of two products to achieve more even pattern

Below, these categories will be elaborated in more detail.

1. Improving seasonality of production

One approach to seasonalize production is to vary the weekly working hours according to demand. Several companies have introduced a concept of "annual hours", which will be illustrated by another example. Bomford Turner (a manufacturer of grass-cutting equipment) introduced annual hours - that is employees are contracted to work a given number of hours a year but on a flexible basis so that they receive a constant weekly wage regardless of the hours worked in a particular week (Cheesright, 1993). That way, a company can typically expect to reduce expensive overtime at the peak-season. Other companies using similar concepts are Frigoscandia frozen foods (Charlton, 1991) and Crosslee, a manufacturer of tumble dryers (Tighe, 1993).

Another step to achieve more seasonality in a firm's labour availability is to motivate the workforce to take holidays off-season. Arrangements could be that every employee who does not take a day off during the peak-season will have one additional holiday at any other time of the year. Also, longer downtimes because of maintenance or the installation of new machinery can be allocated off-season. The same applies to operator training etc.

Benders (1991) provides a detailed example for a model applied in a sugar refining plant, which only operates for a few months per year and where the workforce was split into core (full-time) and a peripheral (seasonal) employees. The core employees perform what firms regard as their most important and unique activities. They tend to be male, full-time, permanent, with long job tenures, and to deploy skills which the firm cannot easily recruit outside. The peripheral workers, by contrast, conduct what are seen as routine and mechanical activities. They are more likely to be female, part-time, possibly temporary, with shorter job tenures and deploying skills which are readily available on the labour market... Functional flexibility, or the ability of firms to reorganize jobs, is to a great extent provided by the core workforce, while numerical flexibility (defined as the ability of firms to adjust the number of workers, or the level of hours worked, in line with changes in the level of demand for them) is provided by the peripheral workforce but also by externalizing labour.

The models above lead to a more flexible and seasonal labour force. They enable a firm to accumulate labour when production peaks and to reduce the production rate when demand is low. As a result, a firm would not use the full technical capacity of plant and equipment. If production is capital intensive, including expensive machining processes, a manufacturer is more likely to aim for maximum use of machinery. If production is labour intensive, flexible and seasonal labour models are more likely to be applied.

For example, a manufacturer offers a range of lawn-mowers, some of which have an expensive combustion engine, others a less expensive electric motor and there is also a push-type model without engine. If possible, the company tries to produce the work-intensive push-type model first. This way the inventories are reduced, as expensive purchased parts are not stored or only for a short period of time.

For a lawn-mower manufacturer, it makes sense to produce and assemble the mowing units all year round in order to make full use of plant and equipment. However, purchased parts, like combustion engines and electric motors, are added to the product just-in-time before delivery. This way, inventories resulting from purchased parts can be reduced.

The insufficient accuracy of forecasts has to be considered when setting up a master production schedule. During the off season months, a company should produce models that have the highest probability of being sold. Other models with questionable demand should be delayed until later when better and more accurate forecasts are available.

Different models of a product line usually have some parts and components in common. Without knowing exactly what model-mix is actually going to be sold, a manufacturer could still predict the demand for these common components and prefabricate. With the season approaching and forecasts becoming more accurate, these components can then be fitted in the assembly line. The idea is to produce components with predictable demand first and to assemble them later as soon as orders come in.

A manufacturer of patio furniture offers all products in a wide range of colours. As the painting is done at the end of the process, the manufacturer is building up temporary stores of unpainted furniture. The painting is done as soon as orders come in.

2. Reducing seasonality of demand

Small examples where products with seasonal demand are increasingly used all year round can be found. Bicycle-carriers for cars that carry skis as well are such an example, or garden tractors that can also be used for snow-ploughing.

Especially for some consumer products, advertising during the off-season months is another effort to deseasonalize demand (Advertising for ice-cream is increasingly placed in winter months). Sales of many food products can be levelled out by offering them frozen, canned or dried. Compare sales of fresh and canned tomatoes. Demand of heating energy is extremely seasonal for (not storable) electricity or gas, less seasonal if it is provided by (storable) coal or oil.

Agricultural machinery sales are usually supported by financial packages that offer attractive conditions during the winter months. Every autumn and winter, manufacturers of combines offer discounts and reduced interest leasing contracts to customers that order their machines early. The salesforce should be motivated to bring in orders during the off-season months. A salesman should receive a higher premium for a barbecue-set sold in January than for a unit sold in June. Sales to countries like Australia or Southern Africa can help as well.

3. Improving predictability of markets and accuracy of forecasts
If every product is tailor-made and manufactured according to customer specifications, no forecasting system can be accurate enough to predict all details of the units to be sold. If, on the other hand, the company offers only one standard product, forecasting would only be needed for the quantity. Here again, a compromise would have to be made.

Because of the length of the time horizon and numerous uncertainties, the manufacturer should develop close links to the dealers. No delay should be accepted. Exact knowledge about orders, sales and possible left-over stocks is crucial to improve the accuracy of a forecast.

4. Combining products with different seasonalities
Many manufacturers of sports-equipment are active both in winter as well as in summer sports. Rossignol for example offers not only skis, but also tennis rackets. Any diversification would typically be a compromise between beneficial similarities on the one hand and distinct differences to aim for an opposed seasonality on the other hand. A manufacturer of garden-tractors and lawn-mowers tried to diversify into the growing market of snow-mobiles. Its limited success in that latter market was partly due to the fact that these two products have different customers and are sold in different areas. Different types of co-operation might be feasible. These include the exchange or joint use of plant and equipment, labour or resources. Even the acquisition of other manufacturers or product lines with opposed seasonality might be considered.

References
Benders, J. 1991, Core and Peripheral employees in a Seasonal Firm, *Journal of General Management,* **Vol.17,** Nr.1, Autumn 1991, 70-79
Bloemen, R. and Maes, J. 1992, A DSS for optimizing the aggregate planning at Monsanto Antwerp, *European Journal of Operational Research,* **Vol. 61,** 30-40
Charlton, D. 1991,Rewards of Annual Hours, *Management Services,* **Nov.1991,** 16-19
Cheesright, P. 1993, The cutting edge, *The FT Engineering Review,* **14.Sep.93,** 16
Gabb, A. 1991, Braving the elements, *Manageent Today,* **January 1991,** 60-63
Tighe, Ch. 1993, A line that suits tumble dryers, *Ingenuity: The FT Engineering Review,* **14.Sep.1993,** 11

BUSINESS INTEGRATION METHODS: A REQUIREMENTS SPECIFICATION

Umit S Bititci

University of Strathclyde
Glasgow, G1 1XJ

This paper describes the work done as part of a research and development programme focusing on business integration. More specifically the paper provides an in-depth analysis of the state of the art with respect to business integration and goes on to introduce a reference model for business integration and details a requirements specification for methods which would facilitate business integration. The paper concludes with a brief discussion on the significance of the work described and makes references to related work which is currently under way.

Introduction

Today worldwide competition between manufacturing enterprises is increasing significantly. Facing these challenges a manufacturing business can improve its competitive position through integration. In this context integration is defined as:-

> "*All functions of a business working together to achieve a common goal, using clearly defined objectives, benchmarks, disciplines, controls and systems in a flexible, efficient and effective way to maximize value added and to minimize waste.*"

The literature in this area indicates that the background to business integration is technology. The availability of technology has driven the integration effort which led to the concept of *Information Integrated Enterprise*. At present there are a number of tools, techniques and methods available to facilitate the integration of information systems within an organisation. These include I.CAM methodology (CAM.I, 1983), Information Engineering Methodology (Inmon, 1988), GRAI Methodology (Doumeingts, 1989), Strathclyde Integration Method (Carrie and Bititci, 1990). On the other hand there is considerable emphasis being placed on streamlining of business activities for improvement of business performance. Work in this area, including TQM, focuses on the business needs and order winning criteria.

It is now clear that technology is no longer the constraint for achieving business integration but methodologies are required to facilitate the creation of an integrated business platform. This conclusion is strongly supported by Self (1992) and Hodgson & Waterlow (1992).

Background

In recognition of the need for a practical methodology to facilitate business wide integration a broad literature survey was carried out. The purpose of this literature survey was to identify the current best practice with respect to business integration and improvement. The results of this survey may be summarised in the following thirteen points.

- Shared vision and business values
- Common understanding of business objectives
- Commitment at all levels
- Facilitation oriented leadership
- Focus on internal and external customers
- Systematic approach to human resource development
- Continuous education and training to improve skills and business awareness
- Participation and involvement by employees at all levels
- Accountability and ownership
- Performance measurement and benchmarking
- Effective communications
- Focus on business processes
- Recognition of the criticality of the information flows within a business

Having identified these points, existing practice to achieve business wide integration was surveyed in detail. This included in depth studies of companies who focused on business integration over the past three to five years as well as in depth interviews with some of the leading consultancies on their approach to achieving business integration. This review resulted in the conclusion that; Existing methods are either analysis or design oriented and none of the existing methods focus on implementation. That is they provide support up to the specification/design or action planning stage but they fail to facilitate the implementation process. Furthermore, none of the existing methods completely address the thirteen points identified through the broad literature survey. Some address higher level issues, some focus only on operational detail, some take prescribed solutions and attempt to implement these and so on.

Based on the discussion thus far, it should become clear that different methods take different viewpoints and emphasis on different aspects of integration. Of course this may be expected as the subject area is very broad and covers all aspects of a business. However, the main reason behind this level of variety is the lack of a reference model or a standard which defines key aspects of business integration. As there is no specific reference model, there is no specific methodology which would facilitate the achievement of this model. With these points in mind the specific research issues were identified as the development of:-

- a reference model for business integration
- a methodology to facilitate the achievement of this reference model.
- an audit method to monitor an organisation's progress towards integration.

A Reference Model For Business Integration

In the context of this work, a reference model is best explained through an example. The field of quality, more specifically TQM, contains a number of reference models which outlines the performance areas a business must satisfy to be a Total Quality business. The two most widely used reference models are the Malcolm Baldrige Model and the EFQM (European Foundation for Quality Management) Model.

Based on the structure of the Malcolm Baldrige and the EFQM Models, the objectives for business integration as defined earlier in this paper and the thirteen points identified through the broad literature survey, a reference model for business integration has been developed. This reference model consists of six performance areas under the general headings of Drivers and Enablers. The six areas of performance are:-

Drivers: -   Shared Vision
                  -     Shared Strategy and Common Goals

Enablers: -   Awareness of Roles and Responsibilities
                  -     Integrated Processes
                  -     Conforming Working Practices
                  -     Integrated Performance Measures

The objective and performance criteria for each one of the above performance areas is detailed in table 1.

Business Integration Methods: A Requirements Specification

Having developed a reference model for business integration the next key step was the development of a methodology which would facilitate the achievement of this reference model. With this objective in mind a requirements specification was compiled. The requirements specification has been categorised under three headings. These are: Scope, Functional and Application.

The functional requirement from such a methodology is the provision of procedures, tools, techniques and guidelines to facilitate the achievement of the performance criteria specified within the reference model.

The requirements specified under the general heading of Scope attempt to define the boundaries of such a methodology. The requirements with respect to the Scope of a methodology are:-

- Incremental approach   - Independency from technology
- Provision for systems implementation   - Hard and Soft Systems Orientation

The application requirements reflect the requirements of potential users, ie senior management in small, medium and large enterprises. The requirements with respect to the Application of the methodology are:-

- Low Cost   - Simplicity
- Structured and Planned Approach   - Early Benefits

Discussion and Closing Remarks

The work described in this paper has been conducted as part of a larger programme of research focusing on development of tools, techniques and methods to facilitate measurable improvements in business performance through integration. This paper presents the conclusions of the literature survey, details of the reference model which has been developed and a requirements specification for methodologies to facilitate business wide integration.

In addition to the work presented in this paper, a methodology, Total-I, has been developed which provides a comprehensive set of tools, techniques, guidelines and procedures and facilitates the achievement of the performance criteria stipulated in the reference model. This methodology has been rigorously applied in a number of manufacturing and service organisations. The results of these have been reported in other publications (Bititci, 1993).

At present this work is focusing on the development of an audit method to measure the level of integration within an organisation. It is envisaged that this audit method will make use of the audit points (performance criteria) specified within the reference model to create a scoring system. The method will also provide supporting procedures and guidelines for auditing to ensure that audits are conducted with minimum room for interpretation. The objective behind the auditing method is to provide the management with a tool for benchmarking, monitoring and planning their company's progress towards integration.

References

Bititci U S. 1993, *Total-I: A Methodology For Business integration*, Proceedings of ICCIM'93, Singapore.

CAM-I Architects Manual. 1983, *ICAM Definition Method*, Computer Integrated Manufacturing International Inc., Texas, USA.

Carrie A S and Bititci U S. 1990, *Tools for Integrated Manufacture*, Proc. of MATADOR'90.

Doumeingts G. 1989, *GRAI Approach to Designing and Controlling Advanced Manufacturing Systems in a CIM Environment*, Advanced Information Technology for Industrial Material Flow Systems, NATO ASI Series, Volume 53, Springer Verlag.

Hodgson A and Waterlow G. 1992, *Integrating the Engineering and Production Functions in a Manufacturing Company*, Computing and Control Engineering Journal.

Inmon, W H. 1988, *Information Engineering for the Practitioner: Putting Theory in to Practice*, Prentice Hall.

Malcolm Baldridge National Quality Award, 1994 Award Criteria, American Society for Quality Conrol.

Self A. 1992, *Factory 2000 - A Response to Order Winning Criteria*, Computing and Control Engineering Journal.

Table 1 - A Reference Model for Business Integration

Shared Vision

Objective of this performance area is to ensure that the business direction is defined and this direction is known and understood by employees at all levels of the organisation. To achieve this the management should ensure that:-

a. The current position of the company is defined with respect to competitors and customers, financial performance, internal strengths and weaknesses, external threats and opportunities.

b. A long term, realistic, vision is defined.

c. The key competitive criteria which would enable the company to achieve its vision is defined.

d. Items a-c above is understood by employees at all levels and employees can relate their own job functions to the achievement of this vision.

Shared Strategy and Common Goals

Objective of this performance area is to ensure that the company has defined a plan which details how it is going to progress from its current position towards its vision in terms of long and short term goals, and that employees at all levels understand this plan. To achieve this the management should ensure that:-

a. The vision is expressed in terms of measurable short and long term goals.

b. A strategy for achieving the vision has been formulated and expressed in measurable terms in the form of short and long term targets and priorities.

c. Items a and b above are understood by employees at all levels and employees can relate their own job functions to the long and short term strategic targets.

Awareness of Roles and Responsibilities

Objective of this performance area is to ensure that all roles and responsibilities are defined and that there is a general awareness of all roles and responsibilities across the business. To achieve this the management should ensure that:-

a. All strategic and operational processes within the business are identified.

b. All process responsibilities are allocated to business functions.

c. All functions clearly understand their process responsibilities.

d. All functions demonstrate an awareness of other functions roles and responsibilities.

Integrated Processes

Objective of this performance area is to ensure that the objectives, interfaces and the procedures for each process is clearly understood by the personnel responsible for those processes and that these objectives, procedures and interfaces are agreed with the customers and suppliers of those processes. To achieve this the management should ensure that:-

a. The objective of each process is defined.

b. The inputs, outputs and controls (ie. policies, constraints, etc) of each process is clearly defined.

c. The customers (ie the destination of outputs) and suppliers (ie the source of inputs) of each process is defined.

d. The procedure (ie. the actions necessary for converting the inputs into the outputs) for each process is defined.

e. The customers and suppliers agree with the objectives, inputs, outputs and the procedures for each process.

f. All personnel responsible for the process understand a-d above.

b. Conduct regular reviews to ensure that the processes remain integrated.

Conforming Working Practices

Objective of this performance area is to ensure that the working practice conforms to the processes as described in the previous performance area. To ensure this the management should:-

a. Conduct regular audits

b. Take corrective actions as necessary (eg training, retraining, redeployment, revision of the processes, etc)

Integrated Performance Measures

Objective of this performance area is to ensure that the goals and targets set at the strategic levels are deployed to individual functions and processes within the organisation through an integrated set of performance measures. To ensure this the management should:-

a. Define the performance criteria of each process.

b. Identify indicators which objectively measure the ability of these processes to fulfil the performance criteria.

c. Establish an understanding of the relationships between the process level performance measures and strategic measures and goals.

d. Establish the relative priorities at process levels with respect to strategic priorities.

e. Encourage the use of the integrated performance measurement tool as a management tool.

A NATURAL SYSTEMS APPROACH TO MANUFACTURING STRATEGY

Dr R Maull, Ms S Whittle

School of Computing, University of Plymouth, PL4 8AA
Sheffield Business School, Sheffield, S17 4AB

This paper questions the fundamental assumptions behind most modern manufacturing strategy literature. It will begin by briefly reviewing the content and process literature, providing a framework summarising process and content issues where there is general agreement. The main part of the paper focuses on expressing an emergent theory of the process aspects of manufacturing strategy. Existing approaches are based strongly on a rational systems perspective assuming that organisations are purposefully designed for the pursuit of explicit objectives. This paper synthesises an approach to manufacturing strategy based on the natural systems model emphasising the importance of unplanned and spontaneous processes.

Introduction

In 1969 Skinner (1969) identified the absence of manufacturing in the corporate strategy making process. Since then a considerable amount of effort has been expended on researching the general area of manufacturing and a wide number of publications have occurred on manufacturing strategy as summarised by Neely (1993). These papers can be broadly divided into two types; those that focus on what should be addressed in a strategic debate - the content literature - or how strategies should be developed and implemented - the process literature.

Content Literature

Substantial agreement exists in published work on the appropriate content of manufacturing strategy. According to Leong, Snyder and Ward (1990) the most important areas can be divided into two broad categories; decision areas that are of long term importance in manufacturing and competitive priorities based on corporate and/or business unit goals.

There is general agreement on categories of decision areas for manufacturing based on the original eight of Hayes and Wheelwright (1984); capacity, facilities, technology, vertical integration, workforce, quality, production and organisation. To these are often added a ninth and tenth; the fit between products and processes and their life cycles and performance measurement.

Competitive priorities may be defined as a consistent set of goals for manufacturing. A general model would include the following; quality, delivery performance, (dependability and speed of delivery), cost, flexibility (product mix and volume).

The Process Literature

The literature describing the process of manufacturing strategy formulation suggests a hierarchical rational systems model in which corporate strategy drives business strategy. This in turn drives manufacturing and other functional strategies. This dominant view of strategy is that it should be top-down.

In the top-down model a strategy making process occurs within an environment that consists of markets (competitors, customers and financial) and stakeholders. Corporate and business level strategies determine an appropriate pattern of functional strategies for each business unit. The functional strategies themselves consist of the process of strategy formulation and implementation. The Leong et al. (1990) model recognises the role of capabilities and the reality that they may well develop from unplanned activities rather than through any strategic goal. The collective effort and capabilities of the manufacturing business's functional areas serve to produce a service enhanced product (the bundle of goods and services available to the customer when purchasing the product). Next, internal performance measures are included to reflect the degree to which the service-enhanced product meets the strategic objective. Finally, the market itself, provides an external performance measure of the service enhanced product and hence strategy (Hill 1985).

The strength of this model and the concepts is also reflected in the current interest in policy deployment (Ernst and Young 1992). The key elements in policy deployment are that top management develops and communicates a vision, which is then deployed through the development and execution of annual policy statements. Implementation of the policy involves the execution of the plans where everyone determines what specific actions will be taken in the very near term to accomplish the plans. Monthly audits and an annual audit by senior executives provide monitoring processes - a strongly hierarchical rational model.

Critique of Model

There are a number of criticisms that may be made of this model. Firstly, the model may not be so hierarchical and rational in practice. One issue central to the theoretical basis for strategy formulation is the rational/natural systems assumptions underpinning strategy formulation. This was summarised by Scott (1981) as;

"The rational system model assumes that organisations are purposefully designed for the pursuit of explicit objectives. The Natural systems model emphasises the importance of unplanned and spontaneous processes. Organisations are viewed as coalitions of participants who lack consensus on goals, sharing only a common interest in the survival of the organisation".

Recently, Pettigrew and Whipp (1991) have provided an analysis which refutes many aspects of the rational model, including the issue of the *pursuit of explicit objectives*. For example, they point out that many theorists use a Porterian (Porter 1980) model to understand a firms competitive environment. Such analyses often concentrate on the central themes of the value chain and generic strategies. Yet on closer inspection Porters work (1987) recognises that

"no tool can remove the need for creativity in selecting the right strategy. Great companies don't imitate competitors, they act differently ..... we do need to discover how the process of creativity takes place".

Pettigrew and Whipp point out that a firm does not assess its environment through a single act or a given point, instead strategy arises through a sequence of actions through time. That process is open to many influences for no matter what methods, techniques a firm uses they are used by someone - *what matters is who uses them and how they are used*. It is not simply that those who assess the firms environment are influenced by a greater or lesser degree by their own personal values, the rational model assumes that there exists external to the firm an absolute objective environment which awaits inspection, though by the very act of assessment and their subsequent behaviour managers are in part creating their own environment.

The building of theories on the basis of the rational model ignores the historical, cultural and personal dimensions of organisational life and as a result is a poor base from which to derive a model of strategy making in firms. Many organisational theorists, (for example Cooper, Hayes and Wolf 1981) had long ago concluded that decisions are events that occur when problems and solutions come together in a "fortuitous confluence" rather than on the basis of a logical application of reasoning.

The natural systems model stresses the importance of the unplanned and spontaneous, where problems and decisions come together. Tom Peters in his recent book "Liberation Management" (Peters 1993) devotes whole sections to the disorganisation and de-centralisation necessary for the fast changing 90s. According to Cooper (Cooper et al 1981). this may be termed the "technology of foolishness". This technology which is derived from learning theory has two elements. Firstly, it suggests a relaxation of strictures of imitation, coercion and rationalisation and secondly, it requires the suppression of imperatives towards consistency. Cooper et al. propose a framework which encourages "playfulness" (for example, exploring alternative visions of the future). The importance of play is that it forces rethinking and recombination of elements of behaviour into different elements thus opening up different ways of looking at problems - a highly appropriate behaviourial model entirely in tune with Peters "crazy ways for crazy days".

Another criticism of the rational model is its failure to take sufficient account of the extent to which the organisation can manipulate its external environment through the development of internal capabilities. Stalk, Evans and Shulman (1992) define capabilities based competition as the ability to emphasise behaviour ie *the organisational practices and business processes in which capabilities are rooted*. Bartmess and Cerny (1993) define capabilities as *"a company's proficiency in the business processes that allow it to distinguish itself continually in ways that are important to its customer"*. Such a capabilities focus identify a set of strengths in core processes which enables companies to compete in entirely different set of competitive environments - these may be termed the critical capabilities that are difficult to develop, these include;

•   Complexity - critical competencies tend to be developed in business processes that are highly complex and involve patient organisational learning.

•   Organisational differences - critical capabilities involve processes which nearly always cut across traditional organisational functional boundaries and frequently involve external groups.

•   Team culture - critical capabilities depend on the way individuals and organisations have learned to work together, rather than on specific organisational or individual skills.

The notion in the Leong et al (1990) composite model that capabilities necessarily follows strategy is open to question. Capabilities emerge and develop within organisations, Miller and Hayslip (1989) argue that manufacturers should balance the use of strategy formulation and implementation with capability building to achieve competitive advantage. This implies that the process model should be revised to reflect this parallel development. In addition, the capabilities model focuses on processes not functions: the model also needs to be re-defined to reflect business process strategy development, not functionally based strategy development.

Alternative planning modes

Cognitive limitations and environmental uncertainty serve to limit top managements ability to separate strategy formulation and implementation. Consequently, at the opposite end of the spectrum from the traditional rational, hierarchical model of planning, several authors have incorporated the theme of involvement into their strategy making typologies, for example in Ansoff's (1987) organic and Mintzberg and Waters' (1985) unconnected modes, strategy making is mostly unmanaged and strategy is the result of serendipity. These modes posit high levels of independent action by organisational actors.

In support of a strategy making mode based around the middle ground of interaction between top down and bottom up strategy making Hart (1992) proposes a transactive mode of planning. Cross functional communication amongst organisational members is central to this mode, with feedback and learning the crucial components. This transactive mode is reflected in many companies recent efforts to foster employee involvement, customer focus and TQM. This mode usually involves the creation of "process based" communication channels for involving customers and other key customers. Companies currently actively pursuing this mode of planning include Motorola, Xerox and

Ford. Initiatives such as the Deming prize, the EQA (European Quality Award) and the Baldridge award are based on a firms ability to demonstrate strong organisational learning capability fostered by transactive relationships among suppliers, customers and employees.

Conclusions

This paper argues that it is insufficient for companies to regard the creation of knowledge and judgements of their external competitive world as simply a rational, technique based exercise. Rather there is the need for organisation to become open learning systems - the assessment of the competitive environment does not remain the preserve of a single function nor the sole responsibility of one senior manager. Strategy creation emerges from the way a company acquires, interprets and processes information about its environment.

References

Ansoff I 1987 The emerging paradigm of strategic behaviour *Strategic Management Journal* **8**, 501-515

Bartmess A and Cerny K 1993 Seeding plants for a global harvest *Mckinsey Quarterly* 1993 **2**, 107-120

Cooper D J, Hayes D and Wolf F 1981 Accounting in organised anarchies: Understanding and designing accounting systems in ambiguous situations. *Accounting, Organisations and society* **6**, 3

Ernst and Young Quality improvement consulting group 1992 *Total quality a managers guide for the 1990's* Kogan-Page

Hart S L 1992 An integrative framework for strategy making processes *Academy of management review* **17**, 327-351

Hill, T 1985 *Manufacturing strategy*, Macmillan

Leong G K Snyder D L and Ward P T 1990 Research in the Process and Content of Manufacturing Strategy *Omega* **18**, 109-122

Miller J G and Hayslip W 1989 Implementing Manufacturing Strategy Planning *Planning Review*

Mintzberg H and Waters J 1985 Of strategies deliberate and emergent *Strategic Management Journal* **6**, 257-262

Neely A, 1993 Production/Operations Management: Research process and content during the 1980s. *International journal of Operations and production management* **13**, 1

Pettigrew A and Whipp R, 1991 *Managing for Competitive Success* Blackwell, Oxford,

Porter M 1987 The man who put the cash cows out to grass. *Financial Times* 20 March, 18

Porter M E, 1980 *Competitive Strategy* New York: The Free Press,

Scott, W R, 1981 Developments in Organisational Theory 1960-1980, *American Behaviourial Scientist*, **24**, 3

Skinner W, 1969 Manufacturing- Missing Link in Corporate Strategy, *Harvard Business Review*. May-June

Stalk, G, Evans, P and Shulman, L, E. 1992 Competing Capabilities: The New rules of Corporate Strategy. *Harvard Business Review*, March-April

SIMULATING PRICE QUOTATIONS FOR SMALL SIZED MANUFACTURING ORGANISATIONS

Mr Pete Cameron-MacDonald and Dr Martin Moltrecht

Napier University, Department of Mechanical,
Manufacturing and Software Engineering.
10, Colinton Road, Edinburgh EH10 5DT

This paper will discuss how small/medium sized manufacturers are coming to terms with the fall of the Berlin wall and the effect that has had on manufacturing companies. There has been a marked increase in the number of small/medium compaies over the past few years.. However these companies do not have a great deal of market knowledge and also the use of cost accounting in small/medium companies is negligible. The availability of new computing techniques is becoming more widespread but computing knowledge is lagging behind. It is apparent in East Germany that the smaller specialised companies cannot afford the cost of employing the larger software houses for their turnkey solutions, however they do need specialist software for manufacturing applications.

1.0 Introduction

The problem confronting East German companies is how to overcome these barriers. This paper will show a possible solution which encompasses Computer Integrated Business philosophies and the use of cost calculation and price quotation simulation. Inherent in this is an open integrated data processing function which will aid management to improve on their response time to customer enquiries. The model created is able to show the bread-even analysis points, cost analysis and price quotations and also includes the capability to show the data in graphical. The simulation is designed to aid management when there are changes in the manufacturing processes re: costs, overheads, elements of risk and profit.

2.0 Background to small/medium size companies

In the UK small/medium size manufacturing companies have 200 employees or less however in Germany and Continental Europe there is a different definition in that a small company has less than 50 employees and a medium sized company between 50 and 300 employees.

Since the fall of the wall there has been an increase in the number of small/medium companies in former East Germany and a reduction in the number of large manufacturing companies. (Reference: Table 1)

Table 1: New small/medium firms in Berlin
(only manufacturing firms)

Employment Size in	09/91		09/92	
	west	east	west	east
1 - 19	1,226	163	1,215	211
20 - 49	549	97	521	139
50 - 99	242	62	242	59
100 - 199	171	50	160	45
200 - 499	104	51	94	30
500 --->	57	37	54	23

This is a familiar trend, that can be seen all over Europe, in that smaller, leaner companies are replacing larger less cost effective ones.

3.0 Definition of the problems encountered by small companies

In general, small manufacturing firms in East Germany have five main problem Areas:

(1)Most of these companies do not have a great deal of **market knowledge** and the use of **cost accounting** is negligible.

(2)The availability of new computing techniques is becoming more wide spread but the **computing knowledge** is lagging behind.

(3)It is apparent that the smaller companies cannot afford **the cost of employing** the larger software houses for their turnkey solutions.
These companies have specialised manufacturing applications and do require specialist software.

(4)If small firms are to work efficiently, they must install a **computer system**. Such systems are expensive, however one solution for small firms is using **standard programs** which can reduce the costs.

(5)The **management systems** in small firms are different to large firms. In small firms these systems must be more flexible and adapt to change faster. Therefore to overcome these problems a management information system for small/medium manufacturing companies has to be developed taking into account the points raised above.

4.0 The development of the Management Information Systems for small companies

Small manufacturing companies have comprehensive specialist departmental groupings and management activities, this gives rise to an increase in movement of information between departments and creates a complex information flow. This is due to increasing the number of products and the growth of specialist departmental groupings (functions) companies require for production processes. These functions and products are interdependently connected and this interdependence has two criteria: Data and decision taken re production.

Table 2 shows an example of a prefabrication company, with 50 employees, who manufacture furniture components for only 6 products (order types).Table 2 shows the management activities to be carried out by the Production Engineer at his work station in the different specialist departmental groupings.

Table 2: Management activities in different specialist departmental groupings

Spec.dep.group. Man.act.	production	purchasing	marketing	.....
planing	x			...
organisation	x			...
controlling	x	x	x	...
coordenation	x	x	x	...
....	.....	....	......	...

Due to the increasing number of interdependent functions and the large number of interdependent objects (products and services , performance) there is a major problem in the information flow especially in small and medium size firms. Because the number of employees is low and the efficiency of the tasks to be performed is high there is an increase in competitive pressure. These complexities are shown in the rectangle in Figure 1.

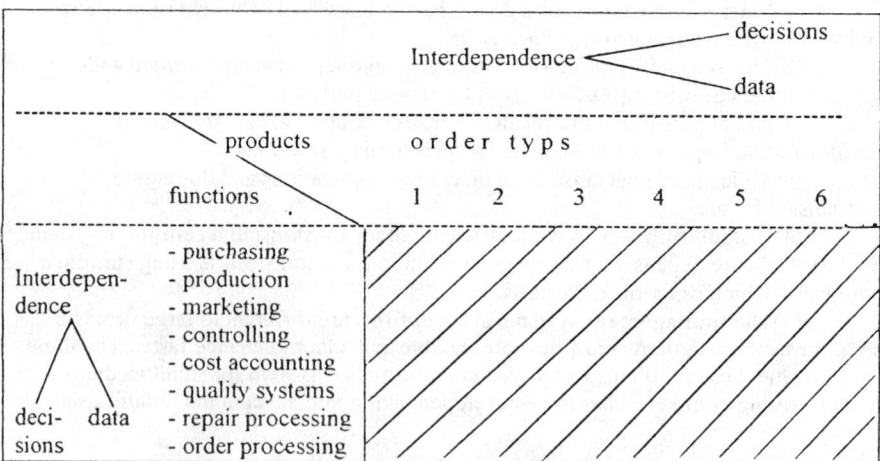

Figure 1. increasing complexity of manufacturing processes

J o b s of Production Engineer	A c t i v i t i e s
(1) production control	
- parts lists, material schedules, works order	production planning
(2) Materials control	
- estimate material	purchasing
- actual material consumption comparison material consumption	materials control
- profitabilities	controlling
- intermediate stock	production planning
(3) Cost analyses independent from plants	
- direct costs in order types production cost/cost increase coefficient	cost accounting
- direct total costs	cost accounting
(4) price quotation when a prompt answer has to be given to a customer	
- price quotation/price simulation	marketing
- break-even analyses	controlling

Figure 2. Jobs to be done at an Production Engineers Work Station

Figure 2 shows an example of interdependence of information at a Production Engineer's Work Station in a prefabricated furniture factory making furniture components.

5.0 What can small firms do in this case ?

Due to the high level of data integration and multi functional decision making required, the managers concerned must have a good working knowledge of business management, as well as Engineering and Computer skills.

To meet this demand specialist training is required in the area of CIB for managenent of small/medium Manufacturing companies. An example of this simple Computer Integrated Business philosophy is shown in figure 3. The example is a factory application where material processing and direct materials costs are very high.

A specialist solution is by integrating marketing and production control processes.

Cost calculation/ Price quotation simulation incl. Break-even analyses						
Cost Accounting for Small Manufacturers						
Calculation	Cost	ORDER TYPES				
items	summry	OT 1	OT2	OT3	OT4	OT5
(1)	(2)	(3)	(4)	(5)	(6)	(7)
Direct Materials	1300	10	11,37	13,14	4,06	4,91
+ Direct Labour		0,5	1,15	0,25	0,03	0,15
= Production cost		10,5	12,52	13,39	4,09	5,06
+ Overhead cost	1150	8,85	10,06	11,62	3,59	4,34
88,46%						
= Total cost per unit (sq. m.)		19,35	22,58	25,01	7,68	9,40

Please put in OT	ot1				
Price simulation ot1					
Cost per unit	Total quantity	Total cost	Price simulation	P/L TDM	P/L (%)
(1)	(2)	(3)	(4)	(5)	(6)
19,35	2000	39	19,00	-0,69	-1,79
19,35	2000	39	20,00	1,31	3,38
19,35	2000	39	21,00	3,31	8,55
19,35	2000	39	22,00	5,31	13,72
19,35	2000	39	23,00	7,31	18,89
19,35	2000	39	24,00	9,31	24,06
19,35	2000	39	25,00	11,31	29,22

Figure 3. Cost calculation/ Price quotation simulation

6.0. Conclusions

This simulation model permits small companies to research the market to obtain optimum prices for materials and services for existing products, as well as new ones, this can help to establish a high profit margin due to fast accurate decisions regarding production. The model also gives the break-even point in each case and it is possible to calculate a competitive price which allows the company to intruduce the product into the market at a low profit margin to obtain a share of the market and attract new customers.

The Production Engineer can be use this model, to make simulations of cost forecasts and price quotations for new markets and new products and is be able to create management activities so that costs could be reduced.

References

Moltrecht, M. 1994, *Production Engineers Workstation*, Eigenverlag., Berlin

Evans,J. and Anderson,D. et al 1990, *Applied Production and Operations Management* (West, St Paul USA) 41-46, 76-84.

Askin, R. and Stainbridge C. 1993, *Modeling and Analysis of Manufacturing Systems*, (Wiley) 95-117.

Regional Statistical Office 1984, *Small Manufacturing Company Data* Berlin .

EFFICIENCY MEASUREMENT IN COMPANIES ENGAGED IN ADVANCED MANUFACTURE

M A Fowler, J C Beamish and D S Spurgeon

Division of Manufacturing Systems
School of Engineering, University of Hertfordshire
College Lane, Hatfield, Herts. AL10 9AB

Companies engaged in manufacture must compete in the international market-place while producing to world-class standards. Firms using Advanced Manufacturing Technologies (AMT) must respond to demand with reference to delivery, price, quality and flexibility in the face of decreasing product life-cycles and smaller lot sizes. This paper reviews the usefulness of current efficiency measures to companies using AMT and refers to research currently taking place at the University of Hertfordshire to develop appropriate measures of efficiency in such companies.

Advanced Manufacturing Technology

Definitions of AMT vary in degrees but common factors remain. Namely the application of information technology to manufacturing operations that allow a common information processing base with the potential to be combined into more comprehensive production systems. Benefits to be gained from introduction have been related to; high quality, accuracy, repeatability, increased machine utilization, reduced manufacturing lead-times, reduced inventory, lower labour costs and quicker response to new product design. However, the difficulties relating to the measurement of efficiency of these systems might be reflected in the still considerable confusion about the true definition and meaning of "cost-effective AMT". Furthermore, the reality of introducing AMT within companies already engaged in traditional manufacture compounds the need for re-examination.

Standard Efficiency Measurement

Measures of efficiency have been developed in order that management might exercise timely control of production and to enhance and ease decision making. Historically, measures developed in industry have taken money or the most expensive factor inputs as the index to be monitored. Ratios of inputs to outputs have been given in value terms with managers combining a number of suitable measures in order to achieve optimum advantage. The benefit of such measures lies in their use of available operational and costing data whilst managers in standard manufacturing operations can achieve a degree of measurement flexibility.In technology-led industries a movement towards higher indirect costs, reduced direct labour costs and higher capital turnover in companies operating under restrictive market pressures has reduced the significance of these conventional indexes.In contrast, the need for quantitative knowledge related to manufacturing is a key factor. Adequate performance measures are needed to drive the production systems and the relevancy of standard measurement systems which are static and often independent of original influence is questionable. Many companies are failing to identify factors relating to unavailable or inadequate cost information. Timely management information is needed to monitor and exercise prudent business decisions. On examination, existing measurement and control methods can be related to systems measuring inventory, production, quality, costing, budgeting and capital expenditure. All the areas identified use either productivity or cost accounting measures to indicate efficiency,however, most suffer from the narrowness of the approach taken. Whilst the combination of measures may relate to control across areas, in practice each covers only a small area of a company's operations, control being exercised by ensuring that individual components are kept within a predetermined plan. Measures are required to highlight inefficiencies in order that remedial action can be taken. They must also be sufficiently plausible to enable decision making and refined enough to allow for production and operations control. The lack of such suitable measures in AMT has led to discussion centred on productivity and accounting techniques, each related in turn to the difficulties often encountered when justifyingnew technologies in financial terms.

Productivity

Productivity measurement generally relies on taking "output" as the governing index. Over time, this has developed as a comparable measure and is often expressed in monetary terms. The data used is taken from existing accounting and production systems, with the input indexes commonly chosen on the grounds of proportional influence related to cost, classically, direct labour. However, particular factors effecting performance in high technology manufacture have already been observed to differ from those traditionally considered, most obviously, direct labour. Partial productivity measures, whilst remaining the dominant approach to measurement, have questionable usefulness.The productivity paradox as discussed by Skinner (86) has also become more widely accepted and industry has experienced that attempts to alter inputs to standard measures, such as machine usage, have raised productivity and then lowered it again through overproduction. (Ohno 82)

Accounting

Many managers now believe that conventional accounting systems are costly to administer, fail to control costs and usually yield erroneous data and that within AMT, the data resulting from standard measures can be misleading. As with productivity measures, calls for change have varied from manipulating existing individual inputs to the development of total systems capable of considering factors across the board. Brimson (88) discusses cost management systems, Roth and Morse (83) recommend that quality costs be included in the organizational accounts (88), Seed (84) attempts to replace direct labour and overheads using a conversion cost whilst the management of overheads, a particularly problematic area in AMT, is addressed using activity based costing. (Cooper and Kaplan 88) However, all these suggested measures still relate to historical financial activity rather than the benefit managers might expect from efficiency measures that help record and control present performance. In addition factors relating specifically to the introduction of technology, other than the general recognition that short-run marginal costs can approach long-run marginal costs and that we must now somehow account for the growing influence of overhead, have not been identified. These inadequacies might prove vital if the more non value-added activity that is needed to support production, the greater the down stream costs that are incurred. The unsuitability of conventional measures lies in their being based on the steady production of mature products with known characteristics and stable technologies and attempts to reflect the growing importance of overhead as a large part of total cost can lump many important inputs together. To devise measures for AMT, attempt at delineation must be made and the importance of individual factors in relation to outcome understood.

Total systems

The philosophy of Total Systems has led to the discussion and development of a number of optimization models that seek to integrate individual measures. Mohanty and Rastogi (86) propose the development of a measure to show Total Productivity whilst Son and Park (87) discuss the possibility of quantifiable measures that might monitor efficiency in three key areas, Total Quality, Total Flexibility and Total Productivity. They suggest that this could be regarded as an integral manufacturing performance measure. Sink (85) goes on to identify a Multi-factor Approach, attempting to record variances over time in important performance areas including profit, productivity, price and quality. However, all these models ignore those factors that have been consider unquantifiable using standard measures and suggest only that these factors can only be included by looking at them separately and then allowing them influence on the decision making process. Attempts to account for all factors using total systems have so far concentrated on the continued use of existing productivity and cost accounting models with occasional attempts to include factors from other areas that have been felt to now exert more influence as a result of the introduction of the technology. It should also be recognized that the import of various factors will change over time and with circumstance.

Capital justification

The area of justification was a natural choice for investigation when related to the high cost of AMT. The situation in industry has developed to the point where some managers have turned down proposals developing manufacturing technology on the grounds that conventional techniques have shown the development to be unjustifiable. Research at the University of Hertfordshire (UH) indicates that elsewhere development has proceeded solely as an "act of faith" (Fowler et al 93). It has been said that many times a justifiable proposal has been rejected because the preparation of the financial analysis has prohibited the inclusion of important "intangible" factors related to AMT. Discussion has centred around attempts to include many so called indefinable benefits common to AMT. Undoubtedly, existing models fail to take account of all the factors that result from the introduction of AMT. Investment optimization models have been developed that attempt to determine the effect of individual investment allocation on other financial results within the organization. However, as with other proposed efficiency measures these proposals are either built around existing models or attempt to include all possible factors quantifiably. A need exists for rigorous and effective control over the derivatives used in investment appraisal related directly to AMT.

Research to develop measures of efficiency

Advances in process technology are interdependent on advances in performance. Concurrently, effective management of assets in AMT will increasingly determine competitive success. The conception might be that progress in AMT comes principally from improved technical performance, but progress requires that wider standards are established, constraints imposed and performance monitored. These can only be achieved if all factors effecting the efficiency of the system are identified and their relevant influences incorporated in the measurement. Research at UH has concentrated on existing inefficiencies in companies engaged in AMT manufacture, in particular Surface Mount Technology (SMT) manufacture. SMT comprises a technologically advanced area of electronics and its management in manufacture embodies all the vital component of AMT namely; IT dependency, high-value products, small-batch, large variety, fast moving technology and capital intensive etc. By profiling a sample of companies representative of size and stage of development in managing AMT, a model representing existing key areas related to efficiency was developed.

In developing the model care was taken to ensure that inefficiency played a meaningful role in each of the areas identified. Inefficiency was operationally defined as a falling short of expectation in any case and could be defined from the point of view and perception of the companies reviewed and could also be related to expected outcomes in operational or financial terms. The fact that each company may vary in their operational/ financial/ profitability criteia for success was not considered, each being free to judge. A preliminary examination of the companies selected was used to compile an exhaustive list of possible inefficiencies in order to construct a detailed questionnaire from which categories can be developed in order that a causal model can be constructed. The model will be used to explore those ways in which the identified areas might depart from overall efficiency, the degree to which there is a departure, the component types of inefficiency and the structural relationships between components. The aim of the model is to isolate factors effecting efficiency and to define relationships with other identifiable variables eg. organizational size, length of experience in the technology etc.

Future research

The need to re-examine current efficiency measures in relation to AMT is inescapable. Standard measures do not work with the result that companies are forced to make decisions without reference to all the facts. In operations, those same companies may be operating within wide margins of inefficiency due to a lack of suitable measures related to monitoring and control. Measures capable of reflecting operation accurately not only have to be developed but they must also be confidently related to operations at all the stages of development of an AMT. The above approach is an attempt to apply empirical methods as a first step in isolating common and related factors. When proven, such a model would allow specific measures to be developed related to a number of identifiers thus allowing managers to concentrate on those areas of inefficiency most likely to occur at a given time or stage of development.

References

Brimson J A and Berliner C. 1988, *Cost management for today's advanced manufacturing*, Harvard Business School Press, Boston, Mass.
Cooper R and Kaplan R 1988, *Measure costs right : Make the right decisions*, Harvard Business Review, Sept-Oct.
Fowler et al 1993, *Beyond the financial* frontier, Manufacturing Engineer, IEEE, June.
Mohanti RP and Rastogi JC 1986, *An action research approach to productivity measurement,* Int. Jnl. of Operations and Production Management.
Ohno T 1982, *Workplace management*, Productivity Press. Cambs. Mass.
Roth HP and Morse WJ 1983, *Let's help measure and report quality costs*, Management Accounting. Aug.
Seed AH 1984, *Cost accounting in the age of robotics*, Management Accounting. Oct.
Sink S 1985, *Productivity management: Planning, measurement and evaluation, control and improvement.* Wiley. NY.
Skinner 1988, *The productivity paradox*, Business Week. June 16.
Son YK and Park 1987, *Economic measure of productivity, quality and flexibility in advance manufacturing systems,* Jnl. of Manufacturing Systems.

EXPERIENCES OF COMPETITIVE BENCHMARKING RESEARCH IN UK MANUFACTURING

Mrs Gwyn Groves and Mr James Douglas

School of Industrial & Manufacturing Science
Cranfield University, Cranfield, Bedford MK43 0AL

Benchmarking is a useful approach that can be adopted by companies wishing to acquire competitive advantage. Identifying competitors' superior performance, and the practices which have led to this, will help companies to develop their own winning strategies. This paper draws upon experiences of competitive benchmarking research in contrasting sectors of UK manufacturing. Particular reference is made to clothing manufacture, electronic equipment manufacture and engine assembly plants. Issues covered include the selection and design of meaningful measures of practice and performance. Lessons learned about good competitive benchmarking practice are included, such as obtaining cooperation from participants, and assuring them of confidentiality.

The Use of Benchmarking

Benchmarking is the process by which companies compare performance. Comparison can be made either internally (between different parts of one company) or externally.

External benchmarking enables companies to compare their performance relative to other companies, and is of three main types. These are
- competitive. Comparison with competitors in the same industry.
- functional. Comparison with companies having similar operations, processes etc. but who are not competitors.
- generic. Comparison with excellent performers from other industries in order to identify transferable practices which can be applied in order to improve performance.

The name "benchmarking" has become very fashionable in the '90s but, as with so many fashionable ideas, other versions have been around for a long time. Companies operating in competitive environments, and those wishing to enter new markets or to develop new products, have often undertaken comparative performance measurement in the past. This comparison may have been called "industrial market research" or even on occasion "industrial espionage" rather than benchmarking.

A great deal of "benchmarking by another name" has been carried out in the accounting area. Financial benchmarking in the form of ratio analysis is a well developed technique which can be used to compare the financial position of companies with others operating in the same industry. There are severe limitations to this type of comparison. Published accounts often contain different accounting practices, and the level of information on subsidiaries and small/medium enterprises can be very limited. Where financial reporting is on a corporate basis, comparison of the performance of specific functional units may not be possible. Some of these, and other objections, can be overcome if companies are visited and accurate and comparable data is collected and verified, as exemplified by research carried out in the UK clothing and electronic equipment manufacturing sectors (Groves and Hamblin, 1989 and 1990). However, benchmarking which is based solely on financial measures will be unlikely to reveal the underlying reasons for a company's relative performance. Groves and Hamblin (1990) uncovered only a proportion of the links between company financial practices and financial performance in electronic equipment manufacture, and concluded that there were other major performance drivers involved. These would include factors such as company strategy and management style, as well as other non-financial operational practices.

Recent approaches to benchmarking such as Camp (1989) focus on the selection of meaningful performance measures and identifying company practices which lead to superior performance.

Competitive Benchmarking

There can be a danger in focusing solely on competitive benchmarking to identify best practice, in that it limits the scope for improvement to ideas and practices which are already being exploited in the industry under study. However, in many cases useful and comparable performance measures are best obtained by benchmarking performance against direct competitors, for example productivity levels for a particular industry sector or sub-sector. Once relative performance has been established then generic benchmarking can be carried out, in order to identify the best practices which can be applied to improve performance.

It is important to have clear aims and objectives before setting out on a benchmarking exercise, so that the appropriate methodology is chosen. If for example, a manufacturing company making pharmaceutical products was concerned with its customer service and customer image, it is just as likely to find examples of good practice in customer management from a machine tool manufacturer as from another manufacturer of pharmaceutical products.

On the other hand, if specific operational issues such as plant productivity really are the prime concern, and no reliable means of comparison with non-competitors can be identified, then competitive benchmarking will be appropriate.

Two fundamental problems arise when faced with developing a competitive benchmarking exercise. These are
- the need to define a set of meaningful practice and performance measures
- acquisition of reliable information, which usually requires active cooperation from competitors.

Obtaining Cooperation

Generally, published information will not furnish enough useful data that is accurate, timely and sufficiently detailed. Hence information must be obtained directly from competitive companies. However, whilst most companies want to know what their competitors are up to, most companies do not want their competitors to know what **they** are up to.

In almost all cases, useful feedback must be provided to participating companies so that they can judge their own performance relative to others. At the same time this feedback must avoid identifying particular performance levels with specific companies. In general, companies will not want to be identifiable by competitors, or by other groups such as customers. On the one hand poor performance may be embarrassing to a company, and on the other hand excellent performance may be seen as a source of competitive advantage and therefore viewed as commercially sensitive.

Douglas (1993) encountered particularly severe problems in a study of UK engine assembly plant performance. The total sample of comparable plants in this case was very small, and hence there was a very marked need to convince companies of the benefits of cooperation. The need to have a code of conduct agreed to by all of the participants was indicated. The code of conduct used on this occasion can be summarised as follows:
- Information given will be treated with strict confidentiality.
- Names of companies participating will not be communicated to other participants or externally unless prior permission has been granted.
- All participating companies will be given feedback. This will summarise their performance relative to other companies surveyed. The other companies will not be identified.
- Information requirements will be cleared in advance of plant visits.
- Participants are under no obligation to provide strategically sensitive data.

With reference to the last point listed above, it is recommended that where participants have felt unable to provide data, they are not provided with this comparable information for other companies. Unless this practice is adopted, some participants may be tempted to provide only limited, non-sensitive information about their own operations in the hope that they will gain full information about competitors' practices and performance.

The involvement of a detached third party can be critical to the success of competitive benchmarking. Research should generally be carried out by a person, or organisation, to whom participants can trust the collection, analysis, and dissemination of sensitive information.

The authors have experience of competitive benchmarking in three contrasting sectors of UK manufacturing, namely clothing manufacture, electronic equipment manufacture, and engine assembly. In all cases, one of the greatest challenges was in gaining participation from companies. Participation is much more likely if senior managers are approached and become involved at an early stage. They must be persuaded that the benefits of participation will outweigh any perceived risks.

Selection and design of comparative practice and performance measures

The purpose of carrying out a competitive benchmarking exercise is to enable companies to compare their performance with competitors in the same industry. Clearly, the more similar companies are, the simpler the process of comparison. An extreme example can be found in the electronics industry, where some multi-national companies have assembly plants in a number of countries which are virtually "clones" of the parent. Internal benchmarking for a company of this type will be fairly straightforward, and differences in performance will be relatively simple to identify and address.

In a competitive benchmarking situation, there is a need to find similar companies, but there are always going to be differences between companies which will need to be taken into account. These differences could include size, market characteristics, product differences, production variables such as level of automation, integration, workforce and management characteristics etc.

Indeed, some companies feel their operation to be unique, and can be hard to convince that they can be compared with others. When studying UK clothing manufacture many examples of this were found (Groves and Hamblin 1989). Shirt manufacturers, for example, were defensive about being compared with other shirt manufacturers because there might be detailed differences in the style such as the construction of a collar, and this would influence the work content in a garment. This kind of difference between companies, at the micro level, will always be present. The selection of useful and meaningful performance measures is therefore critical.

It is important to use a number of performance measures, rather than just one overall measure, in order to make useful comparisons between companies. Historically, garment manufacturers have focused management attention on the control of work content ("standard minutes"). Their management practices, investment, factory design and so on have largely ignored opportunities for improvements in other cost areas. A few pence saved on work content is less significant than the potential savings to be made by reducing factory throughput time e.g. in inventory, inspection, handling and recording, or by lowering risk by later commitment of fabrics to styles, customers and uncontrollable variables such as weather and fashion. In designing a competitive benchmarking study for the clothing industry then, it would be important to focus managers' attention away from the traditional "work content per garment" measure which is at the heart of so many costing systems and investment decisions in this industry.

A contrasting industry sector, engine assembly, poses a similar problem. Assembly plant productivity measures tend to rely heavily on "hours per engine", dating from the days when this was a highly labour-intensive industry. The engine manufacturing process is too complex to be meaningfully reduced to one overall measure of performance. Further, any approach to measuring performance must also take into account differences with respect to factors such as engine size, complexity, variability, amount of in-house machining, automation level etc.

In the electronics industry, equipment manufacturers vary greatly in the amount of design, sub-assembly, test and other functions they are concerned with. Performance measures designed for this industry need to take these different practices into account.

Once a set of useful measures has been derived, it is also important that these are normalised so that individual companies cannot be identified. Often this involves the use of ratios rather than absolute values - for example expressing measures in terms of "per full-time employee", or "per product" or "per m^2 of production floor space", as appropriate.

Above all, it must be remembered that the measurement of performance alone will not help a company to improve competitive performance. In order to take action which will improve performance, the underlying practices must be identified, understood, and changed where necessary. Research into the clothing industry (Groves and Hamblin, 1989) found that companies had invested heavily in computer systems, and that this was in reality linked with poor company performance. Closer examination revealed that many large systems had been installed in order to manage large inventories resulting from traditional manufacturing with high levels of work-in-progress, complex production control, multiple data recording, complex payment and costing methods. In this industry and others, the measurement and comparison of working methods, organisational structures and other management practices can be vital. The rate at which companies adopt competitive management practices will determine whether or not they survive.

References

Camp, R. 1989, *Benchmarking: The Search for Industry Best Practices that Lead to Superior Performance*, 2nd edn (ASQC Quality Press, Milwaukee, Wisconsin)
Douglas, J. 1993, *Benchmarking of UK Engine Manufacturing Plant Performance*, MSc Thesis, Cranfield.
Groves, G. and Hamblin, D. 1989, *The Effectiveness of AMT Investment in UK Clothing Manufacture* (British Clothing Industry Association, London)
Groves, G. and Hamblin, D. 1990, *The Effectiveness of AMT Investment in UK Electronic Equipment Manufacture*. Report of SERC Grant GR/E/22794, Cranfield.

A SOFT SYSTEMS APPROACH TO MANUFACTURING REDESIGN

Mr Adam Weaver, Dr Roger Maull, Dr Stephen Childe, Mr Jan Bennett

School of Computing, University of Plymouth, Drake Circus, Plymouth, PL4 8AA

The aim of this paper is to examine whether a "hard" systems model-based approach to the redesign of manufacturing companies will produce the radical improvements required to compete in today's global market place.

A working definition of a model is established and the theoretical preconceptions that are employed to create models of processes are discussed. The paper then proceeds to describe the deficiencies in both "hard" and "soft" systems approaches and how BPR methodology that integrates both "hard" and "soft" model-based approaches will encourage more radically improved business processes.

Introduction

In today's global market place many multi-national organisations have implemented successful Total Quality Management programmes resulting in a set of incremental improvements and a change in organisational culture. These organisations are now looking towards the more radical approaches of Business Process Re-engineering (BPR) and Business Redesign (Johansonn, 1993) to keep ahead of their competitors.

BPR differs from other approaches to business regeneration by explicitly recognising that many business activities cut across both internal and external organisational boundaries. Hammer (1990) states that BPR methods, which strive to "break away from the old rules about how we organise and conduct business," offer organisations the only hope of achieving radical performance improvements. He states that re-engineering cannot be accomplished in small or cautious steps but must be viewed as an "all-or-nothing proposition."

Kaplan and Murdock (1991) have identified several benefits to thinking of an organisation in terms of its core processes. They maintain that the adoption of such a viewpoint helps a firm to link its strategic goals to its key processes. These include the complete chain of organisational activities independent of departments, geography, cultures. Most importantly they also embrace suppliers and customers.

The process-oriented viewpoint emphasises cross functional performance rather than encouraging departmental optimisation and the consequent system-wide sub-optimisation. It also encourages the firm to focus on business results, particularly where total lead times are an issue. Process focus also offers an organisation the opportunity to re-engineer the process or radically reduce the number of activities it takes to carry out a process, often through the application of IT. This in turn provides opportunities to reduce the cost base and/or improve service levels.

The process focused approach concentrates first on identifying the business processes, then analyzing and re-engineering each process. Many current BPR methodologies (Harrington(1992), Harrison & Platt(1993)) are designed to enable organisations to:

- Define business processes and their internal or external customers;
- Model and analyze the processes that support these products and services;
- Highlight opportunities for both radical and incremental business improvements through the identification and removal of waste and inefficiency;
- Implement improvements through a combination of IT and good working practices;
- Establish mechanisms to ensure continuous improvement of the redesigned processes.

The model-based tools advocated by these authors to facilitate the analysis and redesign of the business processes are used in what could be described in as a "hard systems approach" (Checkland, 1981). The "hard systems approach" is frequently used by most organisations, systems analysts and engineers since they are all "analytical and detail oriented in their problem solving" (Hammer, 1993).

This paper examines whether the hard systems approach which is reductionistic in nature and assumes a well-defined problem situation and objective is the most appropriate approach to be used when attempting a radical improvement through the use of a BPR methodology.

Initially a number of theoretical issues concerning the use of models as part of any approach are discussed. The deficiencies with both hard and soft systems approaches with respect to the objectives of a BPR methodology are examined and finally it is proposed that a "softer" approach is required to enable the organisations to gain radical improvements from the application of a BPR methodology.

Models and their application within a BPR methodology

"The way forward lies in gaining an increased understanding of theory and of its relationship to practice. We need to be more aware of our theoretical preconceptions and the way these affect attempts to change the real world." Jackson(1982)

In the modelling of any system or process there is always a tendency for the analyst to forget the theoretical preconceptions that exist when the initial decision is made to produce a model. These preconceptions need to be taken into account when analyzing and designing a process by a model-based approach. To examine the preconceptions it is worthwhile establishing a definition of a model that will be used throughout the paper.

The definition that will be used in this paper to describe what a model is, its objective and how it models a subject and combines the simplicity a definition used by Meredith (1993) and Dubin's (1969) emphasis on boundaries and two goals of science which are to "predict and understand" Dubin(1969).

> "A model is a bounded representation or abstraction of reality. It describes, replicates, or reflects a real event, object or system, with the objective of understanding or predicting."

The term "process" and "system" are assumed to be interchangeable within this paper since both a "process" and a "system" are;

> "a set of elements connected together which form a whole, this showing properties which are properties of the whole, rather than properties of its component parts" Checkland (1981)

The important element of the definition is that a model is a bounded representation or abstraction of a system. It is therefore an enabler or aid to understanding and prediction, but with limitations. The importance of the limitations are emphasised by Meredith (1993) thus;

> "The primary difficulty in using models to analyze situations is obtaining adequate simplification, while maintaining sufficient realism."

The limitations are a result of the subjective decisions that are made by the modeller or in as Ashby (1965) describes a "list of variables nominated by an observer".

Not only must the analyst be aware of the limitations of the model but also their preconceptions as Cleland and King(1983) state;

> "The kind of model used depends on the knowledge, experience, profession and life-world an individual is embedded in; it is not necessarily determined by the subject to be modelled"

The use of models within a BPR methodology whether to gain an understanding of the existing processes or to redesign or to predict performance of a process must be done with the knowledge that the models are an analyst's abstraction of reality. The analyst must also recognise that the models are influenced by their experience and perspective and may constrain a radical approach which could exist external to the analysts perspective. According to Askin & Standridge (1993), the value of a model "lies in its substitutability for the real system for achieving its intended purpose". A process model used within a BPR methodology should therefore encourage the possibility of radical redesign otherwise it does not give maximum value.

Deficiencies in Hard and soft systems approaches

The objective of a BPR methodology is a radical improvement in the performance of the business processes within an organisation. To gain this radical improvement Hammer (1990) tells us that we need to "break away from the old rules". The problem situation that the BPR methodology must address is that the existing processes within the organisation are not producing the desired performance.

A business process is a complex socio-technical system, the elements within the process that interact can be divided into physical designed systems such as machine tools in the case of a manufacturing element in a business process and human activity systems such as the managers and operators within a business process. The emergent properties of a business process can be essential to the strategic aims of the business.

The "order-flow" process within a manufacturing company has a starting point is the customer's order and its end point is the delivery of goods to the customer is a socio-technical system that can have problems that are ill-defined and stochastic in nature. BPR methodologies are used to provide a radical improvement in the emergent properties of business processes. In the case of the "order-flow" process these could include the time taken from receipt of order to delivery and cost of processing the order.

A hard systems approach to BPR used by so many organisations would start from the perspective that the problem situation or process was well-defined and deterministic in nature. It would then proceed to reduce the process into components, the components could then analyzed and modelled by using accurate data, and a solution could be constructed from the improved components. Hard systems approaches are based on rationality described in Trought (1993) as a type of rationality that;

"... assumes that within the system there is access to all the relevant data which is also accurate, and that all alternatives of choice and consequences are unambiguous."

In comparison a soft systems approach to BPR would start from the perspective that the problem situation or process is ill-defined and the process is stochastic in nature. It would then attempt to create "rich" picture of the problem situation and the systems involved. Conceptual models would be developed and compared with reality. Feasible and democratic changes would be agreed upon through debate and the changes implemented. By using the soft systems approach, the properties that are only a result of interactions between all elements in a process (emergent properties) can be considered more easily.

Considering the objective of radical improvement in the emergent properties of the business processes within an organisation a soft systems approach provides a holistic approach that encourages open debate and creative thought to break down the old rules. The soft systems approach also develops a set of conceptual models of the relevant systems for the problem situation which;

"is envisaged ... will lead to improved conceptualisation of user requirements." Dooner (1991)

Although hard systems approaches may assume incorrectly at times that a business process is a well defined problem, accurate data is available and it is deterministic in nature, it has been successfully used for many years in addressing the problems of physical designed systems.

At the higher levels of abstraction the problem situation or process is less well-defined it is therefore suggested that a BPR methodology should initially use a soft systems approach for analysis and design of processes. A soft systems approach would consider the emergent properties of the whole process and where a single change could result in a more radical change of the emergent properties of the process.

At the lower levels of abstraction where the problem situation is less ambiguous and more accurate data is available (Trought, 1993) a hard systems approach should be used.

The output from the soft systems phase applied to the higher levels of a process is used as the input to the hard systems phase at the lower levels of detail. The two approaches can therefore be regard as "complementary" and form part of a single BPR methodology.

Conclusion

The paper has described the increasing interest in Business Process Re-engineering and its objectives of radically improving an organisation's performance using a process perspective. It has been identified that a model-based approach is dependent on the analysts experience and preconceptions and could therefore reduce the possible number of radical alternatives considered in the analysis and redesign.

A structure for a BPR methodology has been proposed to facilitate radical improvements in the performance of process. It should use a soft systems approach at the higher level process analysis and redesign. To complement the soft systems approach, a hard systems approach at the lower levels of process abstraction is required especially where the process being analyzed has sub systems that are physical designed system at the lower levels such as the "order-flow" process within a manufacturing organisation.

References

Ashby W R 1965, *Design for a brain,* (Chapman & Hall)

Askin R G & Standridge C R 1993, *Modelling and Analysis of Manufacturing Systems* (Wiley)

Checkland P 1981, *Systems thinking, systems practice,* (Wiley)

Cleland D & King W R 1983, *Systems Analysis and Project Management*, (McGraw-Hill)

Dooner M 1991, Conceptual Modelling of Manufacturing Flexibility, *International Journal of Computer Integrated Manufacturing*, **4**, No.3 135-144

Dubin R 1969, *Theory Building*, (The Free Press)

Hammer M 1990, Re-engineering work: Don't Automate, Obliterate, *Harvard Business Review*, July-August

Hammer M & Champy J 1993, *Re-engineering the Corporation*, (Harper Collins)

Harrington H J 1992, *Business Process Improvement*, (McGraw-Hill)

Harrison D B & Platt M D 1993, A methodology for re-engineering businesses, *Planning Review*, March-April

Jackson M C 1982, The nature of Soft Systems Thinking: The work of Churchman, Ackoff and Checkland, *Journal of Applied Systems Analysis*, **9**

Johansonn H J, McHugh P, Pendlebury A J, Wheeler W A 1993, *Business Process Re-engineering*, (Wiley)

Kaplan R B and Murdock L 1991, Rethinking the Corporation: Core Process Redesign. *The McKinsey Quarterly*, **2**

Meredith J 1993, Theory building through conceptual models, *International Journal of Operations and Production Management*, **13**, No.5

Trought B 1993, Model Collapse: the people effect, *Proc. 9th NCMR*, Bath , UK

An analysis of Activity Based Cost Management software.

D Steeple , C. Winters

*Coventry University, School of Engineering,
Priory St, Coventry, CV1 5FB*

This paper summarises the findings of research that analyses the use of "off
the shelf" Activity Based Cost Management (ABCM) software. It assesses
the degree to which organisations have found the use of ABCM software to
be beneficial and outlines possible areas for improvement. It also reports on
the authors' evaluation of a number of ABCM computer software packages,
and analyses the current development of fully integrated manufacturing and
financial software systems, incorporating ABCM that can provide an holistic
view of all information and go beyond a mere general ledger system.

Introduction

Activity Based Cost Management (ABCM) is seen as a suitable system for
identifying the causes and consumption of costs, so as to overcome the deficiencies of
traditional costing systems (Steeple and Winters, 1993). In addition to providing strategic
cost information to enable organisational managers to assess the long term profitability of
activities and processes within their particular organisation, this technique focuses
attention on the underlying causes of cost, profit and resource consumption throughout the
organisation (Johnson, 1988).

The long-term success of an ABCM project is to some extent governed by the
effective processing of data within an ABCM package. The identification of software
system structure is vital for an organisation undertaking an assessment of ABCM as a
management information system and also to assess the practical implications that can be
gleaned by implementing the technique. This analysis is made more complex once an
awareness is gained of integrated financial and manufacturing software systems,
incorporating ABCM.

While the introduction of updated and enhanced ABCM packages is beneficial, it
is always important to recognise that such systems will only provide cost information, not
instantaneous solutions for the business. It is the skill and commitment of the management
team which will develop the business and address the problems identified by the ABCM

system, however as an aid to management decision making a computer based ABCM model can be invaluable (Steeple and Winters, 1993).

ABCM Software

The authors' evaluation of ABCM software is restricted to two commercially available packages, which in their view have something to offer many organisations at a significant but affordable cost. To describe these ABCM packages and the author's evaluation, they are designated Product East (PE) and Product West (PW).

Product West (PW)

PW is a stand alone product, whose system structure can be classified into three main areas for analysis which are as follows: -

i) Activity costing analysis.

The main area for system design this section will take information from traditional financial systems primarily the general ledger and provide the primary data source for departmental cost through general ledger account classification and quantification. Once activities are defined departmental costs can be allocated to them.

Activities can be assigned as support and as such their respective cost can be reallocated to other departments on an objective basis, relating to the consumption of the activity. Likewise activities which are direct in nature, and therefore do not service other departments are allocated down to products.

While the activity type definition is important it is always important to "drive down" costs to either products or customers. Therefore the most common activity types utilised will be product, customer, and reallocate, while the business definition is useful for allocation of research and development and financial activities and departments.

Having identified the structure of the system in terms of department, activity definition, and the utilisation of primary data (i.e. general ledger codes and values), it is necessary to undertake the cost object analysis.

ii) Cost Object Analysis.

When all activities have been analysed the cost of activities needs to be assigned to cost objects. The cost drivers utilised to do this are user specified. PW will allow the user to define a customer or product hierarchy for detailed analysis by product or customer segment.

iii) Profitability analysis.

After product and customer costs have been calculated a further analysis is undertaken of profitability. This is undertaken through the introduction of customer revenues and product values. A conjoint analysis of products and customers is useful to identify the most profitable mix.

System functionality and flexibility.

While the systems functionality is undeniable, the flexibility is often questioned. The flexibility is provided by identification of activity type and breakdown of cost object into a functional hierarchical structure. The onus of developing flexibility into a system is no longer the purpose of the system designer; as required by more unstructured packages. The system designer merely identifies the options required from the flexible operating structure provided by the original software supplier (OSS).

Product East (PE)

PE operates within a relational database, and its modular operating structure can be classified as follows: -

1) Master File Development.

This module enables the dimensions of the analysis to be defined, and is used to divide up costs, and is considered to be a route map for understanding costs. The module allows for the classification of costs, activities, cost drivers, functions, resources, and product lines. Resource classification allows for the summarising of activities according to the principal elements of the profit and loss account. defining product lines allows for the preparation of a product group analysis of costs, which forms the basis for profitability analysis.

The dimensions can be used in any way to split the business needs. They need to be defined carefully as the real power of PE lies in its analysis potential, and a little thought at the outset will ensure that its full potential is realised.

2) Interdepartmental Distribution.

Some departments within an organisation provide activities which cannot be directly related to specific products, these activities are utilised by other departments. This module provides a method akin to a direct charging mechanism for the allocation of service department costs. Within the module activity pools for distribution are allocated costs, appropriate bases are selected for the distribution, and then redistribution is carried out.

3) Activity Based Rate Development.

The development of Activity Based cost rates is the heart of ABC. Within this module departments can be selected and cost values assigned to particular activities, therefore developing rates for particular activities.

4) Product Costing.

In the above modules activities within departments have had costs assigned to them. Within this module materials are introduced into the system, and products can therefore be defined as a combination of a specific set of resources involving both activities and materials, using appropriate driver rates. this module can deal with either a single - or multi - level bill of materials.

Reporting Procedures

Within PE there are options for graphical and tabular analysis. This provides a detailed information on product costs, driver rates, activity rates and a comparative cost analysis. A graphical representation is always more visual in its impact, reports in tabular form will still be necessary, in particular to aid reconciliation with the accounts.

Fully Integrated Financial and Manufacturing Software Systems.

The introduction of an ABCM module into an integrated software system designed to cater for the financial, distribution and manufacturing requirements of today's modern company's produces many substantial benefits. Benefits are achievable through the acknowledgement that the general ledger, bill of materials and other such modules can be directly interfaced with the ABCM module. Therefore the data structure does not have to be changed to be imported into the ABCM module and data interchange is minimised. The ABCM module operates in both a stand alone and integrated mode. There are unlimited overhead cost elements and automatic cost roll up and stock re - evaluation if required.

It must be recognised that the current stand alone systems developed by companies in the market of ABCM are more appropriate to the technique as they have been developed within a competitive environment and the features which are contained within those packages have been designed to meet accurately the needs of the user after assessing requirements from previous implementations and modifying the software accordingly. A company offering a fully integrated Manufacturing, Financial and distribution system cannot develop an ABCM module to the same level therefore many of the features that are required by ABCM users are not in the system, such analyses involve customer lines and business resources. Such a detailed cost breakdown is vital for identifying strategic analyses for use in the organisation, however the use of an ABCM module is a good introduction in the use of the technique and for identifying a requirement for a more detailed stand alone system. If such systems are to be marketable then substantial development needs to take place to ensure their effectiveness.

Practical Utilisation of ABCM Systems

The practical utilisation of ABCM software was analysed through a questionnaire and follow up interviews in both British based and international organisations. Forty organisations responded to the questionnaire, a rate of return of 16.6%.

The software section of the questionnaire dealt primarily with reasons for selection of package, differentiation between packages, benefits and limitations of the chosen package, and an overall satisfaction rating for the software. The initial results are as follows:

Software Selection.

50% of respondents indicated that software purchased was based on recommendation. Other characteristics indicated for purchase were, Functionality (22.5%), System Structure (27 %), Cost (15%), and Reporting Facilities (15%).

Overall satisfaction was analysed on the basis of a 10 point scale. 64.7% of recommendation respondents of 17 usable responses proved to be more than satisfied with their software package by indicating a value of > 6 on the scale; 35.3% of respondents indicating selection through recommendation were unsatisfied with their package (a score of <5).

On the scale of satisfaction companies utilising consultants as part of the project team, tended to be less than satisfied with their software, an interpretation of this is that the utilisation of consultants tends either to lead to more benefits being required before satisfaction is acknowledged, or that the software selected was inappropriate for the organisation.

Software Differentiation.

The majority of responses for product differentiation were based on cost. However this differentiation was not compared with PW (outlined previously), which has become a major player in the ABCM software market. Companies utilising PW indicated differentiation on the basis of functionality, flexibility, and the services provided by the Original Software Supplier (OSS).

Overall Satisfaction.

The overall satisfaction levels outlined previously are subjective. Further analysis needs to be undertaken to determine how organisations have interpreted satisfaction, and how the software package utilised satisfied those requirements.

Conclusions

The development of stand alone ABCM systems is increasing rapidly, and even the most expensive ABCM package can be obtained at reasonable cost. This brings forward doubts about the suitability of integrated ABCM modules within manufacturing, distribution, and finance packages, which are comparatively more expensive, and are combined with poor system structure and flexibility.

Clarification is still required on practical software issues. However the results outlined, certainly suggest that the current demands on stand alone systems are being adequately met. It is imperative to recognise that although the introduction of ABCM and associated software can be beneficial , it only allows organisations to compete on equal terms, it is how organisations play the game that will determine their success.

Acknowledgements

This work is currently being supported by the Economic and Social Research Council (ESRC).

References

Steeple, D. and Winters, C. 1993, The Development of Activity Based Costing. *Proceedings 10th Irish Manufacturing Committee, Annual Conference 1993, Galway, Ireland.*
Johnson H.T, 1988 "Activity based information: A Blueprint for World Class management accounting" *Management Accounting (USA)*, June 1988.

SURVEY OF PRODUCTIVITY MEASUREMENT TECHNIQUES & APPLICATION TO 5 MAJOR INDUSTRIES IN JORDAN

Dr Tarik Al-Shemmeri and Bashar Al-Kloub

Staffordshire University, School of Engineering
Beaconside, Stafford, Staffs, ST18 OAD UK

The productivity of five major Jordanian industries was studied and analysed. The Multifactor Productivity Measurement Model (MFPMM) was applied to study the relationship between productivity, price recovery and profitability. The study revealed that the subject companies have increased their profits through price recovery and not productivity. Fluctuation in partial and total productivities for each Company was also analysed during the period of study (1986-1988).

Introduction

Performance measurement is a critical component in the general management process. Reliable measurement systems constitute a sound basis for continuous monitoring and control of organisational performance.

Research and resources have been directed towards the productivity measurement, evaluation, planning and improvement (the four phases of productivity cycle); with the objective of effective and efficient utilisation of resources.

Productivity is one of the best tools used to minimise inflation, create employment opportunities, increase competitiveness, provide for increasing capital investment, and improve the quality of life in general, Sumanth (1985).

In Jordan, most companies do not have an explicit, goal oriented, on-going program for productivity measurement. In order to derive tangible results, the function of productivity measurement should assume an integral place in company activities. It has to be planned, monitored and directed, Al-Kloub (1991).

Many studies show that a decrease in productivity in Jordan took place in recent years, for example Malkawi (1987), Hammour (1988), and Malkawi (1989). The growth of gross domestic product (GDP) during the last 20 years is mainly due to capital investment, more than any other factor. In addition, the difficulty of acquiring new resources due to current financial conditions leads to the conclusion that an important way to increase the GDP is by the efficient use of existing resources, Al-Kloub (1991).

In recent years variations in the basic definition of productivity observed. Three basic definitions of productivity appear in he literature, Sumanth (1985), described them stating their merits: partial productivity (the ratio of output to one class of input), total productivity (ratio of total output to the sum of all input factors) and total factor productivity (ratio of net output to sum of Capital and labour input). Sink (1985) described in detail four techniques for productivity measurement which are:

1. The Normative Productivity Measurement Methodology (NPMM)
This is a structure participative approach utilising the Nominal Group Technique (NGT) to develop a decentralised measurement and evaluation system. Once the productivity measures are identified, it becomes the task of the group to operationlize and implement the productivity measurement system.

2. The Multifactor Productivity Measurement Model (MFPMM)
This technique is similar to the total productivity model of Sumanth (1985)but it is capable of blending the major inputs of a particular organisational system together and relate the resulting aggregate input to the total output of the same system. MFPMM can be utilised to measure productivity change in labour, material, energy and capital. It, also, measures the corresponding effect each one has on profitability. It can provide additional insight into the individual factors that are most significantly affecting profits.

3. The Multicriteria Performance/Productivity Measurement Technique MCP/PMT
Once the criteria against which productivity is to be evaluated have been chosen a mechanism for aggregating this vector of criteria needs to be developed. The approach is to develop a prioritised set of productivity/performance measures utilising the nominal group technique.

A list of heterogeneous measures is prepared, then aggregating or evaluating performance against these criteria in an integrated fashion is sought. The next step is the development of a "utility" curve for each of the priority measures which converts all the uncommon measures into some common measure. A ranking and rating process is to be executed so as to weight the relative importance of each productivity measure. The final step is to integrate the performance (utility) graphs with the criteria weighting which will allow the development of one indicator that will indicate the overall performance of the organisation.

4. Surrogate Approaches
A surrogate productivity measurement technique does not measure productivity but measures other indicators that are highly correlated with productivity.

Methodology, Analysis and Results

The MFPMM was implemented on five major Jordanian natural resources industries for the following reasons:
1-It tracks changes in productivity over time.
2-It tracks changes in price recovery over time.
3-It tracks changes in profitability over time.
4-It enables analysis of the relationship between productivity, price recovery, and profitability.
5-It shows the dinar (Jordanian dinar, JD) impact of above changes.
6-It allows users to ask "What if questions".
7-It provides partial factor productivity ratios.
8-It allows users to analyse cost drivers.
9-Methods of specifying dominant factors and potential factors and productivity improvement factors are possible.
10-It is accounting-system based.
11-Diagnostic in an objective sense.

This model is capable of blending the major inputs of a particular organisational system together and relate the resulting aggregate input to the total output of the same system. MFPMM can be utilised to measure productivity change in labour, material, energy capital and total productivity. It also measures the corresponding effect each input has on profitability and can provide additional insight into the individual factors that are most significantly affecting profits.

The MFPMM is based on that profitability is a function of productivity and price recovery; that is, profit growth of an organisational system in generated from productivity improvement and/or from price recovery. The data for MFPMM are periodic data for quantity, price and value of each output and input of the system. It compares data for one period with data for another period. This comparison forms the basis of analysis. The ratios and indexes generated from the model include weighted change ratios, cost/revenue ratios, productivity ratio, weighted performance indexes and total dinar effect on profits.

Weighted change ratios depict the percentage increase (or decrease) of an item from the base to current period. Both price and quantity weighted change ratios are generated by the model to show the percentage changes from period to period. Cost/revenue ratios reflect the percentages of reported revenue consumed by a particular input in a given period. The most common method of productivity improvement is cost reduction and those ratios show exactly where cost reduction will pay the biggest dividends. Productivity ratios depict absolute productivity values in the base and current period. These absolute values used in calculating the price weighted productivity indexes which show increase or decrease in productivity for the overall system as well as for each component. Weighted performance indexes are output over input change ratios from period 1 to period 2. The final set of indexes are dinar effects on profits. These indexes indicate the impact (in dinars) which is caused by changes in productivity, price recovery and profitability. The ratios and indexes, along or together, provide management with information about their systems of outputs and inputs. The ratios and indexes identify areas that need improvement and they also identify areas that are operating at acceptable levels.

To provide the hierarchical structure of the measurement model, inputs and outputs of the system are disaggregated by class, type, level and sublevel.

The data base of the model contains quantity, unit value, and value information about inputs and outputs at the lowest level defined. The definition of the outputs and inputs for this study is as follows, Al-Kloub (1991):

Inputs: 1-Labour
2-Capital (both fixed and working)
3-Material
4-Energy
5-Other expenses

Outputs: 1-Finished
2-Partial units produced
3-Dividends from securities
4-Other income

During the period 1980-1986 the industrial sector averaged about 18% of GDP, (about 60% came from the five industries under study), year 1986 was chosen as a base year, and data were collected for the following 3 years (1986, 1987 and 1988).

The data required was taken mainly from a accounting departments of the companies. This data was not easy to collect since most of it is considered confidential, because it includes prices of outputs and inputs. Another problem was that outputs, or inputs for some companies were too many (e.g. 136 different outputs for one of the companies).

A special questionnaire was designed (based on the requirement of the MFPMM) to collect data regarding inputs and outputs for these companies.

The spreadsheet program Lotus 123 was used as a modelling tool which simplified the data base construction and model building effort considerably. Data transfer from the low levels to high levels of the hierarchical model was also performed by the aid of this software.

The results of analysis of applying the model on one company will be presented as an example: there were 14 different outputs and 47 different inputs for the company, the results shown in Table 1 are aggregated outputs and inputs, the following results are deduced from analysis:

A)For the years 1987 & 1988 the only input causing profit is "materials". In view of the fact that only 6.4% more materials were used in 1987 productivity of material was increased from 1.39 to 1.42 but in 1988 productivity decreased to 1.02 (partial productivities). The reasons for productivity of materials changes is not clear. This, however, is not within the scope of this study. It remains an area for further research.

B)For the year 1987 the only source of profit was "material". All other inputs were causing losses in the following descending order: 1. Capital, 2. Energy, 3. Labour, 4. Other Expenses, costs.

In order to increase productivity the cost of capital, energy, labour and other expenses need to be reduced. The following questions need to be investigated to find the best way to increase productivity:

1-How can the cost of capital be reduced?

2-How can the cost of energy be reduced?

3-How can the cost of Labour be reduced?

4-How can the cost of other expenses be reduced?

C) For the year 1988 the causes of loss were the same but in a different descending order1-Other expenses, 2. Labour, 3. Energy, 4. Capital.

it is not evident why this change occurred, but still there is a need to reduce costs in order to raise productivity .

D) The relationship between productivity, price recovery, profitability was studied. This company has increased it profitability through price recovery and not productivity which is something rarely maintained in the long run.

E) Dominance of material productivity on total productivity is clear. Primary analysis points out the extensive fluctuations of productivity indexes throughout the analysis periods.

References

Al-Kloub, B. 1991, *Productivity Measurement and Productivity Incentives Relationship*, MSc Theses, Jordan University, Industrial Engineering Department.

Hammour, A. 1988, *Industrial Sector in Jordan*, Industrial Development Magazine, Vol 1,

No. 5, 17-23.

Malkawi,, M. 1989, *Productivity and Technological change*, Dirasat Journal, No. 17, 55-60

Malkawi, M. 1987, *Productivity in Industrial Enterprise Employing Twenty workers and More*, Dirasat Journal No 1, 51-60.

Sink, S. 1985. *Productivity Management, Planning, Measurement, Evaluation, Control and Improvement,*

Sumanth, T. 1985, *Productivity measurement, Evaluation, Planning and Improvement.*

TABLE 1: RESULTS OF APPLYING THE MFPMM TO THE JORDAN FERTILIZERS COMPANY FOR THE YEARS 1987 AND 1988 (BASE YEAR 1986)

	Cols 7-9			Cols 10-11		Cols 12-13		Cols 14-16			Cols 17-19		
	Weighted Change Ratios			Cost/Revenue Ratios		Productivity Ratios		Weighted Performance Indexes			JD Effects on Profits		
	Q output	P Price	V Reven.	Period 1	Period 2	Period 1	Period 2						
RESULTS 1987	Col 7	Col 8	Col 9	Col 10	Col 11	Col 12	Col 13	Col 14	Col 15	Col 16	Col 17	Col 18	Col 19
Outputs	1.0900	0.9380	1.0220										
Inputs													
Labour	1.5100	1.0000	1.5100	0.0480	0.0723	20.4400	14.7500	0.7210	0.9380	0.6770	-821345	-132392	-953675
Material	1.0600	0.9100	0.9720	0.7190	0.6835	1.3900	1.4250	1.0240	1.0260	1.0520	748952.4	694952.3	1443904
Energy	1.4360	1.1580	1.6630	0.0840	0.1375	11.8300	8.9850	0.7590	0.8100	0.6150	-1166889	-995853	-2162743
Capital	1.4750	1.0000	1.4750	0.1400	0.2030	7.1200	5.2620	0.7390	0.9380	0.6930	-2160777	-379738	-2540515
Others	1.3680	1.0000	1.3700	0.0516	0.0690	19.3800	15.4300	0.7960	0.9380	0.7470	-573593	-139571	-713164
Total Inputs	1.1850	0.9620	1.1400	1.0443	1.1649	0.9600	0.8810	0.9196	0.9750	0.8960	-3973653	-952540	-4926193
RESULTS 1988													
Outputs	1.1070	1.2130	1.3420										
Inputs													
Labour	2.0850	1.0000	2.0860	0.0480	0.0760	20.4400	10.8450	0.5310	1.2130	0.6430	-1912922	460276	-1452646
Material	1.1550	0.9400	1.0820	0.7190	0.5795	1.3900	1.3300	0.9580	1.2940	1.2400	-1384382	8851998	7467615
Energy	1.5000	1.5100	1.7270	0.0840	0.1087	11.8300	8.7300	0.7370	1.0540	0.7770	-1329773	31227.7	-1298545
Capital	1.4170	1.0000	1.4170	0.1400	0.1482	7.1200	5.5600	0.7810	1.2130	0.9470	-1738514	1320827	-417686
Others	4.2400	1.0000	4.2400	0.0516	0.1630	19.3800	5.0590	0.2610	1.2130	0.3170	-6453886	485464.1	-5968422
Total Inputs	1.4140	0.9770	1.3820	1.0443	1.0754	0.9600	0.7490	0.7830	1.2410	0.9710	-1.3E+07	11149793	-1669685

INNOVATION AND IMPLEMENTATION OF TECHNOLOGY IN MATURE INDUSTRIES

Dr K. Ridgway and Mr I. McCarthy
Mr R. Scaife
Mr J. Murphy
Mr S. Dawson

University of Sheffield, Sheffield
Chapmans Agricultural Ltd, Sheffield
Evenseek Ltd, Sheffield
Allform Tools Ltd, Sheffield

This paper will examine a definition of innovation and discuss the relevance of the definition to small and medium sized manufacturing companies in Sheffield. Using Teaching Company Schemes as case studies the paper presents a brief analysis of the business strategy adopted by the various companies and describes the innovative steps taken.

Introduction: defining innovation

The Oxford dictionary defines innovation as "the introduction of a new process or way of doing things." Drucker in his book "*Managing for Results*" uses a broader definition of innovation. He defines innovation as "The design and development of something new, as yet unknown and not in existence, which will establish a new economic configuration out of the old, known existing elements. It will give these elements an entirely new economic dimension. It is the missing link between having a number of disconnected elements each marginally effective and an integrated system of great power."

It is this second definition which is most applicable to the mature traditional industries. The manufacturing processes are well established and represent the known existing elements. The Teaching Company Associates have been involved in developing the new economic configurations. This has involved the identification of the companies core competencies and their development to make the Company more effective. It has also involved the identification of the factors which can give the companies a competitive edge and the introduction of appropriate new technologies to enhance and integrate their manufacturing systems.

This paper will describe the work carried out at the Teaching Companies and discuss the definition of innovation pertinent to the mature industries.

Business Strategy

To examine the impact of innovation on SMEs it is useful to examine the relevance of the strategic model proposed by Porter (1991).

STRATEGIC ADVANTAGE

		Uniqueness perceived by customer	Low cost position
STRATEGIC TARGET	Industry wide	Differentiation	Overall cost Leadership
	Particular Segment	Focus	Focus

Porter identifies three generic strategies which can be used to out perform the competition:
Overall Cost Leadership
Differentiation
Focus

To obtain an overall cost leadership position a company needs to maintain a relatively large market share. The traditional industries rely on a high level of manual labour. This can make them uncompetitive against companies which manufacture in areas with low labour costs. Cost leadership is not usually a feasible position for a small company manufacturing traditional products.

To differentiate the Company needs to distinguish the products manufactured from the competition. The aim is to gain an industry wide reputation for producing a superior product which can be sold at a premium price.

A focus strategy is a strong position for a small company. In this case the company aims to achieve lower costs and/or differentiation by serving a specific segment of the market effectively. A Company developing a focus strategy has a good relationship with its customers and aims to meet their specific requirements.

Chapmans Agricultural Ltd

Chapmans Agricultural Ltd manufacture agricultural wear parts such as plough blades points and scrapers. A typical product is cut from a 6-10mm steel sheet and heated, cropped, punched and bent before quenching in oil. The product is shot blasted, cleaned and painted before packing and dispatch. Traditional profitability exercises look to reduce labour costs by automating handling processes. In this case the area where most improvement can be gained is the design and manufacture of press tooling. This is one of the areas where an innovative approach is being used.

A two dimensional PC based CAD system has been introduced. As many tools are combinations or modifications to existing designs, the CAD system reduces the time to produce accurate drawings of tool sets. Where possible tool designs have been standardised and set up times have been reduced. A CNC electron discharge machine and CNC lathe have been introduced to manufacture punches and dies in house.

Work is now being carried out to examine ways of reducing the time to manufacture press tools. Tools are usually manufactured from samples provided by the customers. Although CAD can be used to produce engineering drawings for the tool room a major problem is the transfer of surface positional data to the CAD system. With only a sample to work to it is difficult and time consuming for the draughtsman to calculate the co-ordinates of a complex surface. An innovation which is currently being developed involves the collection of surface data using a probe fitted to the spindle of a CNC machining centre. The memory within the CNC system is insufficient to hold the amount of surface data collected. To overcome this problem the data from the probe passes through the CNC machine control system into a PC. In the PC it is manipulated before drip feeding back to the machine tool during the subsequent machining process. The introduction of probing will represent the "development of something new" which establishes the new economic configuration out of the "known existing elements". The introduction of probing will provide the link between the sample and the finished press tool which will reduce the manufacturing lead time.

A recent development has involved the introduction of water quenching. Water quenching is faster than oil quenching and the hardness of the steel is increased. Unfortunately the rapid quenching causes distortion which reduces the quality of the product. To overcome this problem the components are clamped before quenching and the Company is able to produce superior products which can be differentiated from the competition. Again the introduction of clamp quenching represents the "development of something new" which gives the Company competitive edge.

In both cases the innovations introduced have enabled the Company to pursue the business strategy and differentiate the products manufactured. Through the innovations the Company is able to produce a higher quality product and reduce manufacturing lead time.

Evenseek Ltd

Evenseek manufacture leaf springs for the automotive and railway rolling stock markets. The Company realised that they needed to segment the market and focus their attention on the replacement leaf spring market. Thus a core area of their business is the supply of replacement leaf springs to commercial vehicle operators. In this market segment the factor which enables the company to differentiate from the competition is due date delivery and response to customer demand.

A low level of technology is used in the manufacture of leaf springs and the Company uses a high level of manual labour. The inherent flexibility of the manual labour can be used to advantage as the operators can change quickly from job to job. The main area for improvement was the factory layout, reducing throughput time and availability of components such as bushes and fasteners. To meet market demand the Company needs to deliver from stock or within a very short time. A computerised stock control system (based on an IBM PC and database) has been introduced to monitor the level of finished stock.

On the supply side the steel stocks held are governed by the minimum delivery quantities of the steel supplier. There is little need to computerise the levels of raw material stored. Components such as bushes and fastenings are low cost and a simple manual two bin stock control system is sufficient.

The new factory layout and simple stock control systems represent the new elements which are enabling the Company to compete in the market segment identified. The Company identified response to customer demand as the order winning criteria and new technologies have been introduced to enable them to pursue this strategy.

Allform Tools Ltd

Allform Tools Ltd manufacture cutting tools for the aerospace industry. The key business variables are the technology incorporated in the cutting tool and the quality of the tool produced. The market for specialist aerospace cutting tools is small but the customers are willing to pay premium prices for tools designed to meet a particular machining requirement. To meet this demand the Company need to manufacture small batches or single tools. Producing small batches increases the number of machine set-ups and ultimately manufacturing costs. Many of the tools are manufactured on multi axis CNC tool grinding machines which take up to three hours to set up.

The priority for Allform was to reduce the set-up time to acceptable limits. In 1993 a new multi axis CNC grinding machine came onto the market which could change between tool sets within three minutes. This technological step was such that the Company had to make a strategic decision to purchase one of the new machines. The ability to manufacture small batches has enabled the Company to pursue the high value aerospace market.

The ability to change tooling highlighted a new market opportunity. If the Company could change tools quickly they could consider introducing a re-sharpening service. The main problem associated with re-sharpening is the difficulty of locating previously machined surfaces. This was solved when the Company introduced concentricity software which enabled the operator to load the tool and identify datum surfaces using a touch sensitive probe. The touch sensitive probe and concentricity software have provided the Company with the opportunity to exploit a niche market.

In each case the technology has enabled the Company to introduce a focused business strategy and produce high value tools to a specific market.

Footprint Tools Ltd

Footprint Tools Ltd manufacture hand tools which are supplied to the competitive DIY market. Part of their Teaching Company Programme involved a detailed investigation into the hand tool market and the changing distribution trends. The study highlighted the problems of low cost imports and confirmed that the Company should focus on the provision of quality hand tools to the professional user.

The Company introduced manufacturing cells to manufacture hand planes and wood products and significantly increased output and reduced lead time. Critical to the successful introduction of the plane cell was the realisation that the production flow through the plane shop should balance. This was achieved through the purchase and installation of a second hand milling machine. The introduction of manufacturing cells represents the unknown element which has enabled the Company to develop a new economic configuration.

Summary and Conclusions

The case studies demonstrate that the definition proposed by Drucker is applicable to small and medium sized manufacturing companies. The cases considered demonstrate that the innovation does not need to incorporate an entirely new technology. The technology may be well proven, yet new to the Company adopting it.

The case studies demonstrate the following:

i) Small to medium sized companies operating in traditional industries need to adopt a focused strategy. They need to identify specific market segments and differentiate their product on the basis of technology, quality or response to customer demand.

ii) The new technology or technological innovation should be introduced as part of the business strategy. When a clear business strategy has been developed the elements required to improve the performance of the system can be identified more clearly.

iii) The innovation introduced does not need to be a new invention or involve a large investment. In traditional manufacturing companies an innovation may be the link which converts known elements into a new or improved system.

Acknowledgements

The authors wish to thank the Billy Ibberson Trust who support research in industrial change at the University of Sheffield.

References
Drucker P.F., 1964, Managing for Results, Heinemann Professional Publishing, UK, pp138-139.
Porter M.E., 1980, Competitive Strategy, The Free Press, New York, pp34-39.

INSTALLING MRP & CONTINUOUS IMPROVEMENT
A BOTTOM UP IMPLEMENTATION

Steve Webb* and Harry L Cather**

* Business Analyst
General Motors Europe
Germany

**Senior Lecturer
University of Brighton
Brighton, BN2 4GJ

There are many problems associated with introducing a change in technology. The introduction of a MRP (Material Requirement Planning) system is especially problematic because many Production & Material Control systems are riddled with errors. This paper describes the introduction of a Continuous Improvement (CI) scheme to deal with the aftermath of installing a new MRP system after repeated "hard" management methods had failed. The CI programme turned the staff from their positions of isolation and self-interest towards mutual support and company-wide solutions. These groups were successful, even after several rounds of redundancies that saw the workforce reduced by 50%

THE FOUR DIMENSIONS OF CI

Drawing together information from both published literature and personal experience across a wide field of Human Relations Management, Total Quality Management, Project Management and Continuous Improvement it was noted that some 75% of CI installations were failures. It was observed that the best chance of success came from having four distinct dimensions biased towards CI in an organisation:

Philosophy, Infrastructure, Processes & Tools

The **Philosophy** level is the most abstract, but it determines the commitment of top managers to the key principles behind CI. It requires a genuine change in attitude from that normal within UK management. The movement from relying solely on technology driven large innovations accompanied by maintaining the new standards to also embracing the creeping small but continuous improvement concept is often difficult for "men of action" managers.

Simply preaching CI is not sufficient unless management show clearly by its day-to-day actions that they embrace the concept. For example, to aid problem solving CI requires a no-blame atmosphere and many managers cannot bring themselves to act this way. In addition it means spreading some of the decision making powers i.e. empowering people at all levels within the organisation to make decisions and act on them.

Along with the philosophy must go a re-examination of the company's whole **infrastructure** from its management structure to its method of communication. Tall rigid structures tend to inhibit CI, especially where cross-discipline approaches are called for.

In addition to how people are formally organised and managed, the **process** of CI also requires careful nurturing to deliver the desired effect. Suggestion schemes by themselves are insufficient and often hinder rather than encourage the small self improvements which are so much part of CI. There should be a portfolio of methods used to raise problems which reduce productivity and other opportunities for improvement.

Methodologies which can guide employees through the problem can be provided by a set of simple **tools** which can easily be taught to all employees.

The following case describes the events following a MRP installation in which the CI process was successful where normal management driven project teams were not.

MRP INSTALLATION.

An MRP system however is completely computer based and relies on the data contained therein being of a high degree of accuracy. Existing information such as inventory records, bills of material (product structures), routings, leadtimes, etc. often is fed into the MRP database complete with their errors. The output of the MRP system is therefore wrong and this calls into doubt the integrity of the whole system, and staff naturally fail to trust this output. The system is then blamed for the problem rather than the personnel who are responsible for the data accuracy.

There are three key elements making up an MRPII system. The software, the hardware and the people operating the system. Of the three, by far the most important element to get right is the people element, it is too often not properly addressed initially.

The company had just completed the purchase of a PC mounted MRPII production control system. The company had been an MRP user for almost a decade at this time, the previous system was Mainframe based. One main problem of the old system was that much of the data was incorrect. When bringing the new system on stream, it was decided because of the time factor and against advice, that the data existing on the old MRP system would be transferred over to the new system - yes the new system was a reflection of the corrupted old system.

In the early days of using the new software, there was a tendency for people to blame operational problems on "the system". This reference to the software and the hardware was a convenient way of diverting the focus from procedural problems and data integrity to technical problems. Over the course of the first half-year of use it became more and more apparent that the technical side of the MRP system was in fact running correctly. The operational and procedural side however was not. As a result a MRP task team was set up.

The meetings of the task group were structured as formal meetings with the agenda set and controlled by management. The meetings were often made difficult by the *normal* habit of people attempting to apportion blame to others, hence protecting departmental and sectional interests by many of the participants. Management did attempt to involve staff in determining root causes by exhortation of the obvious need, but without much success at the end of the day.

Data Integrity Studies.

The earliest activities undertaken by the team were related to problems of Data Integrity. Any MRP system is highly dependant on the integrity of data in two dimensions: the Static/rule data, and the Dynamic process data.

Of the two, it is essential that the former static/rule data is correct. This data acts on the dynamic data and if incorrect it will create as a result incorrect dynamic data. The data contained in the PC system was however exactly the same inaccurate data as in the previous system.

A report was prepared proposing clear accountability and responsibility for source data throughout the user base of the company. It was proposed to install improvements through the use of specific techniques to "Cause" data integrity. These techniques included: Mistake Proofing Data-Entry through the use of Templates, Checksheets, Rounding Disciplines, Cyclical Counting, Root Cause Investigations and of course a thorough training and education investment. These were not implemented because it was felt that they were too complex to operate and too costly to install. The problems with the "system" continued.

The Class "A" MRPII Users Project.

It became clear that to get the new system into place, a further concerted effort to use the MRP system as a tool to improve the business would be required. It was decided by management that the goal would be to achieve Class "A" MRP status within 2 years. (Class "A" status is awarded by the recognised guru's of MRPII the Oliver Wight Organisation. Based on a 260 Point Check-list, the Oliver Wight organisation will make an assessment of a company and, if it is achieving all of the points noted in the check-list which are essential to the correct operational use of MRPII, then it will award an MRP User Classification - "A" status being the top grade.)

The objective presented a major project and careful planning of this project was essential to its sustained momentum. A PC networked, on-line project management system was established. This created an accessible bulletin board for the vast amount of work which would have to be undertaken to achieve Class A.

The objective effectively called for all existing procedures to be replaced with streamlined procedures which reflected the requirements of the Class A goal. This was to be achieved with minimum disruption to operations and without additional capital expenditure. As a result the goal of the project was already in jeopardy because Oliver Wight insist on evidence of an investment in training for staff.

The project was launched and dates for initial review meetings were established. However before these review meetings could take place the project was suspended temporarily due to considerable pressures on management caused by a severe Working Capital trauma. This resulted in the announcement of redundancies in the workforce and the Class "A" project was further suspended whilst industrial relations problems were allowed to settle. The suspension was to prove permanent as the project was sidelined to allow more immediate fixes to be addressed.

The problems, of course, still existed in the system.

CONTINUOUS IMPROVEMENT ACTIVITIES.

The Concept of Continuous Improvement had been made in a presentation to all Supervisors and Managers at the Monthly Supervisors meeting, where it attracted an initial scepticism. However, it as the MRP task teams starting losing steam that the concept was implemented by the IT manager calling together interested people to form an improvement team.

The initial team was small and informal and met outside the normal working time. As they progressed the team grew and management recognised their commitment by bringing the meetings back inside company time thereby formally accrediting the team members' roles to include the CI activity.

The key task was defined as the reduction of component shortages at every stage of production.

The team felt that they required a formal method for the resolution of problems and as a result the CEDAC Method was introduced. CEDAC stands for Cause and Effect Diagrams with the addition of Cards. CEDAC was developed in the Sumitomo Corporation of Japan in the 1970's. and is a complete eight step methodology for solving problems ; the eight steps are :

* Decide what needs to be improved.
* Decide how to measure improvement.
* Set goals and targets.
* Gather and integrate information about causes of the problem, facts and opinions.
* Generate ideas possible solutions and proposals for improvement.
* Implement new ideas and changes - take improvement action.
* Establish standards to prevent problem reoccurring.
* Develop ways to ensure adherence to all new procedures.

The CEDAC methodology uses large charts (see figure 1) to track both the Cause and Effect side of a problem. Each problem uses a CEDAC chart to structure its progress whilst being solved.

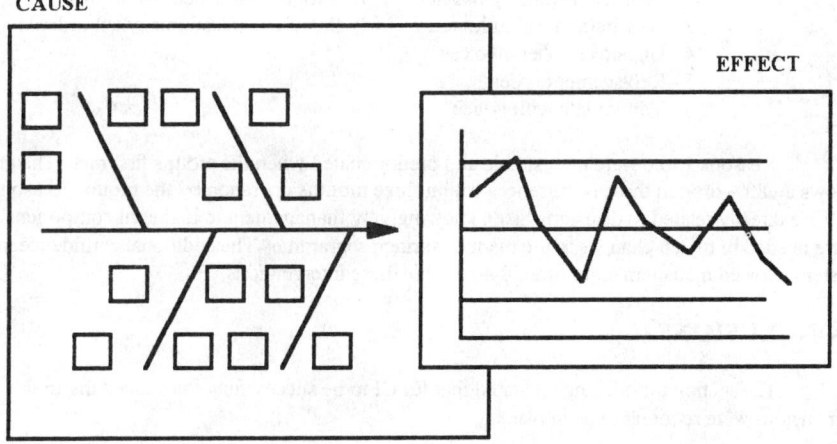

CAUSE   EFFECT

FIGURE 1 - CEDAC DIAGRAM

The CEDAC diagram is in two, inter-related parts:

The traditional fishbone Causes diagram is modified by using cards to hold full thoughts on the areas to be investigated. These cards can be moved around, modified or replaced as the investigation progresses.

The Effect side plots the changes on a day to day basis hence tracking progress

To begin problem solving in earnest the main group was sub divided into four project groups such as the "Shop Capacity group", and the "Wrong Parts Group". Each group operated its own CEDAC diagram.

An example of the tangible effects of the operation of the CI groups can be seen in figure 2 below. This shows the changes achieved in the reasons for shortages on the shop floor from undesirable "uncontrollable" events towards ones that could be readily foreseen.

FIGURE 2: SHORTAGE OF PURCHASED ITEMS

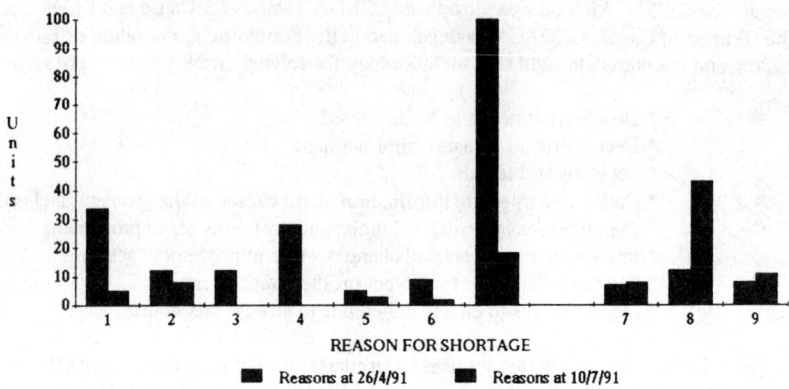

REASON FOR SHORTAGE

■ Reasons at 26/4/91   ■ Reasons at 10/7/91

Where: 1 - Vendor late without notice   7 - Unplanned spares requirement
          2 - Incorrect inventory balance   8 - Programme scheduled in
          3 - New item in Bill of Material   9 - Insufficient leadtime when ordered
          4 - Unpaid supplier invoices
          5 - Replacement order
          6 - Vendor late with notice

Reasons 1 to 6 were undesirable and predominated when the groups first met. The graph shows an 80% drop in these occurrences within three months operation by the group. Reasons 7 to 9 are directly related to decisions taken knowingly by management to highlight components that were needed to match changes in the manufacturing programme. The additional confidence in the system allowed management to more than double these interventions.

CONCLUSIONS

The section introducing CI stated that for CI to be successfully introduced the four dimensions were required to be in place:

Philosophy, Infrastructure, Processes & Tools

In this case, we could see them all in place, even at the top management level who may have appeared from the description of events as following rather than leading the process. The main problem was not that management did not embrace the philosophy but that the other staff did not believe that management had. The fact that the IT manager was perceived not to be in the normal command chain and hence not part of the *usual* blame sequence enabled the groups to start operating. The groups commitment to CI was re-inforced by management's re-action to the group implementation and was initially emphasised by the decision to operate the groups during normal working time without them being formally managed.

The Continuous Improvement groups remain in operation and the following quote by the Chairman of the group demonstrates the on-going objectives.

"It was considered that the most significant achievement was that the team
and its sub-groups were continuing to meet and to focus attention on the many issues
which, by the application of the Continuous Improvement philosophy, will lead to
the achievement of the teams objectives."

A FRAMEWORK FOR TECHNOLOGY MANAGEMENT EDUCATION AND RESEARCH

Kazem Chaharbaghi

Cranfield University
Cranfield, Bedford, MK43 0AL

This paper is based on the supposition that in order to identify and address present and future learning challenges it is necessary to examine the historical development of organisations. This examination is first presented. A framework is then described the purpose of which is to facilitate learners, industry and academia in meeting present and future learning challenges.

Introduction

The view that technology is a key strategic factor in achieving competitive advantage is so widely accepted in industry and literature that it has become an axiom. However, management of technology represents the greatest challenge to any organisation as the process of technological change requires investment in both capital and time, is unpredictable in nature and can encompass changes in strategies, culture, skills, working practices, organisation structures, supply chains and trading alliances. Management of technology is the ability to achieve an advantage by exploiting technology in products, services and processes accompanied by the necessary changes in internal and external structures. Such an ability requires a wide range of expertise spanning the development of technology through to managing the change process itself. The growing importance of technology management has led an increasing number of higher education institutions to develop and deliver education and research programmes in this field. In the development of these programmes it is important to take an integrated approach in which the technology, engineering and management issues are brought together as a coherent whole. The learning objectives should be aimed at developing strategic thinkers who are technical rather than theorists with business minds capable of understanding the complexity of the process of technological change. The programme delivery mechanism should maximise the learning skills of learners so that they can cope with the fast changing competitive rules as opposed to developing knowledge and skills about a deterministic world. This paper proposes a framework for education and research in management of technology which is based on an industry-academic interface. It is shown that such an interface will be mutually beneficial to industry, academia and learners.

The Learning Revolution

In order to explore new learning challenges and the ways in which they can be addressed it is necessary to examine the future in the light of the past and present. In the past organisations were confined to industries and markets with well-defined boundaries and had the strategic aim of achieving economies of scale by producing in long runs products with relatively long life cycles and reducing costs by constantly improving the process employed. These considerations led to the creation of functional areas within organisations where the key assumption was that high performance is achieved by focusing individuals on one task. The learning curve concept was used to support this assumption with the explanation that individuals learn to perform their tasks faster and better over time. Using this mindset, individuals within organisations were developed to acquire specialised knowledge together with skills necessary to solve problems within their functional areas. As organisations grew so did the need for better communication and control. This led to the introduction of more and more management layers and as a result functional hierarchies coupled with a top-down policy framework became an organisation norm within different industries.

The emergence of functional hierarchies and the need for specialised knowledge and skills together with the necessity for better communication, co-operation and co-ordination between different functions within an organisation resulted in the 1970s and 1980s becoming a period of rapid expansion in the number of programmes offered by higher education institutions particularly in the fields of technology and management. These programmes can be grouped in two major categories: specialised and non-specialised programmes. The aim of the former has been to maximise the knowledge and skills of individuals whose interests lie in a particular functional area. The aim of the non-specialised programmes (e.g. general management programmes) has been to provide an understanding of different functions and wider business issues while improving *inter alia* communication and team-working skills through the use of virtual worlds (e.g. case studies which are representative of real life situations in the past).

There is now a general cognizance that functional hierarchies hinder organisation learning and hence their ability to innovate (Peters, 1992). For their survival, organisations are having to learn faster in order to become leaner and exploit a wider range of business opportunities. The increase in the level of knowledge together with its widespread dissemination and application has resulted in a rapid rise in the level of competition and vice versa. The accelerating dynamics of competition has resulted in the following major trends emerging over the last decade:
- The borders between different competitive environments are eroding while competition is shifting towards a global scale.
- Organisations are becoming flatter and are more dominated by technology.
- The command and control structures for the supervision of people and systems are increasingly being supplanted by leadership, motivation and support mechanisms.
- There is a greater emphasis on multi-functional teams and cross-functional business processes.

- Organisations are continuously restructuring themselves as their critical success factors are constantly being challenged.
- There is a growing obsolescence of knowledge which has to be addressed by relearning.

The future will represent a logical extension of the past where organisations will need to cope with a greater level of uncertainty. Their survival will increasingly depend on their ability to anticipate and develop future opportunities and manage effectively and efficiently the necessary knowledge gaining, dissemination and application processes in order to exploit such opportunities. As a result of the accelerating dynamics of competition the world is experiencing a learning revolution which has major implications for industry, academia and society at large. Industry must increase its competitive capacity by continuously improving the quality of its knowledge-base. Industry therefore needs a new breed of highly innovative individuals who are able to create new market opportunities and manage technology to develop superior offerings that dominate such opportunities. The learning revolution is therefore moving the centre of gravity for career opportunities towards knowledge workers who have maximised their learning skills which enable them to identify and fulfil their own learning needs, communicate ideas, shape and direct their own work and work with people (Chaharbaghi, 1991 and Drucker, 1992). The greatest challenge for academia is how to best expand the breadth and relevance of their knowledge-base in order to help already highly educated, achieving and keen learners to relearn. The framework presented in figure 1 aims to address these challenges through a joint effort between industry and academia and problem-based learning. The elements of this framework are detailed below.

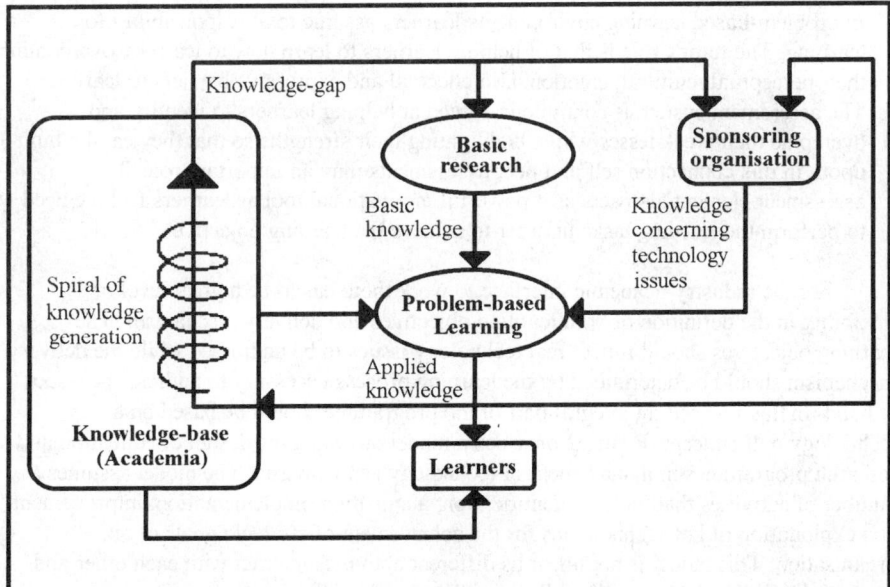

Figure 1. Development, dissemination and application of knowledge for advantage.

Industry-academic links: the best of both worlds

The process of technological change represents a learning activity the driving force behind which are problems or issues that need to be addressed and previous knowledge. This highlights the significance of greater links between industry and academia. Industry can improve its competitive capacity by providing a platform upon which the knowledge-base of academia can be used to address major strategic technology issues and problems. Industry-academic links will result in a spiral of knowledge generation which develops a cumulative understanding of technology management issues and this in turn helps academia and industry to identify knowledge gaps which can be used to explore new frontiers of research and opportunities for new industrial applications. In this way the relevance of research including its applicability and timeliness is ensured. Industry-academic links also develop a fertile ground for the application of problem-based learning which has been growing in importance over the years (Chaharbaghi, 1992). This represents a total learning philosophy the aim of which is to create a supportive environment so that deep learning is fostered and learners further develop their learning skills. The features of a problem-based learning programme include:

- There are three distinct learning objectives: knowledge, skills and attitudinal. Knowledge refers to the principles involved but does not by itself enable learners to perform a task competently. Skills are abilities to demonstrate while attitudes refer to values or preferences of learners as represented by their actual behaviour.
- The programme is designed around a real problem, the solution to which requires the achievement of the learning objectives.
- The concept of learning teams lies at the heart of the learning process, enabling learners to share their knowledge and experiences as well as negotiating and making sense of new ideas.
- In problem-based learning environments learners assume total responsibility for learning. The tutor's role is that of helping learners to learn how to learn by overcoming their perceptual, cultural, emotional, intellectual and expressive barriers to learning.
- The assessment system is continuous, aimed at helping learners to identify and overcome their weaknesses whilst highlighting their strengths so that they can be built upon. In this connection self and peer assessments play an important role. The assessment element also acts as a powerful motivational tool as learners feel the need to perform the learning tasks in order to achieve the learning objectives.

For the industry-academic interface to work there has to be a great level of flexibility in the definition of both learning objectives and delivery mechanism. The learning objectives should reflect real technology issues to be addressed while the delivery mechanism should be determined by the learning process necessary to address the issues at hand. In this respect, an integral part of the programme should be based on a technology pull concept. Figure 2 provides a model on which the design of education and research programmes in management of technology can be based. The model assumes a number of activities that include identification, acquisition, implementation, improvement and exploitation of key technologies for the achievement of strategic goals of an organisation. This model is not linear as different activities interact with each other and one activity does not necessarily follow another sequentially.

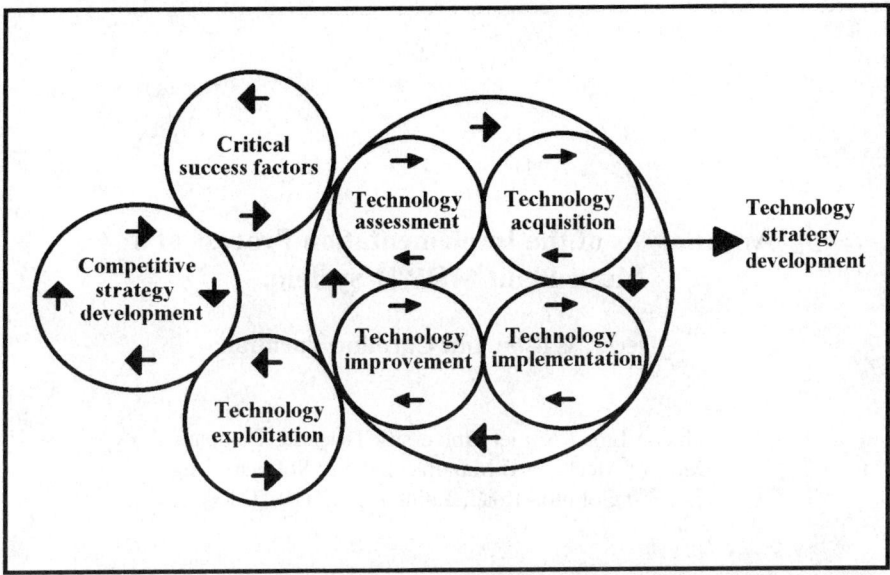

Figure 2. A model for strategic use and management of technology.

The framework presented above means that academia should follow the lead of industry and move away from a highly specialist mindset towards more multi-disciplinary research and education processes. This is because technology management issues and problems are becoming more and more complex as the level of uncertainty grows and therefore require many individuals with different backgrounds. No single individual can provide a total solution for any given problem as the expanse of knowledge required is potentially very large.

Conclusion

The learning revolution that the world is experiencing demands industry and society to continuously learn and relearn. This requires a greater partnership between industry and academia and a much more flexible framework in which the content and process of education and research programmes are defined. Such a framework should accommodate the elements of technology pull to match the needs of industry while encompassing the necessary knowledge push from academia. Problem-based learning is a powerful mechanism for this purpose as it culminates in the effective development, dissemination and application of knowledge for the advantage of learners, industry and academia.

References

Chaharbaghi, K. 1991, Learning the Ropes, *Manufacturing Engineer*, **70** , 36-38.
Chaharbaghi, K. 1992, *Problem-based Learning: An effective Approach to Professional Education*, (Cranfield University, Cranfield).
Drucker, P. F. 1992, *Managing for the Future*, (Butterworth-Heinemann, Oxford).
Peters, T. 1992, *Liberation Management*, (MacMillan, London).

An Analysis of the Implementation Process of a Multi-Plant MRP II System.

Derek Wilson and Caroline Turnbull

Ethicon Ltd. / Napier University Teaching Company.
Dept. of Mechanical Manufacturing & Software Eng.
10 Colinton Road, Edinburgh EH10 5DT

Manufacturers today have operations that encompass multiple manufacturing facilities and complex supply chains. These operations require information systems that can support planning and execution over multiple facilities and boundaries. Systems must be flexible enough to accommodate the vast differences in not only the network itself, but in each company's approach to planning and executing material requirements within these networks. Multi-plant MRP II systems allow manufacturers both simplification and flexibility in addressing multi-plant requirements.

This paper examines the initial implementation of a European multi-plant MRP II system at the manufacturing plant based in Scotland. The paper outlines the project plan, background methodology, implementation process used, problems encountered during the implementation are highlighted and recommendations offered for the future.

Introduction

The company is involved in the design, manufacture, marketing and sales of a comprehensive range of surgical wound closure sutures, more commonly known as needle and threads found in surgeries and operating theatres.

In order to improve customer service levels, the company has committed resources to develop a European Logistics Centre (ELC) in Brussels. The centre will be responsible for the central logistics control for manufacturing and distribution for the three facilities in Scotland, France and Germany (see figure 1).

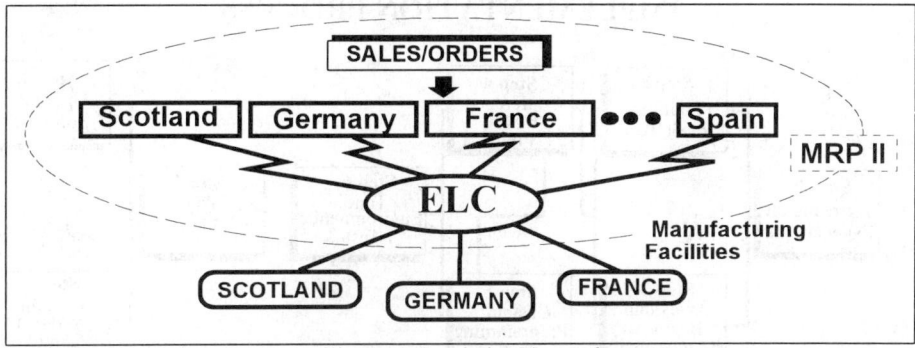

Figure 1. Logistics Network

It is intended that the ELC facility carries out central planning functions for manufacturing at each of the manufacturing sites allowing for a certain amount of rationalisation, both in terms of product range, stock and resources. They will be ultimately responsible for long term capacity projections, labour requirements and negotiations for major raw material supplies. The logistics centre will co-ordinate the efforts of the three manufacturing companies, providing a focus for customer service. To control the logistics operations within the ELC, the company is currently installing a multi-plant MRP II (Manufacturing Resource Planning) computer software system. This will enable ELC to plan production for each of the sites allowing full visibility of the work in progress, stock, resources, constraints, etc. with ELC making planning decisions based on the available information and updating customers on the position of orders.

Background Methodology

MRP II systems enable the co-ordination of all the functions in a company to develop an achievable business plan. MRP II was the next logical step in the progression of materials planning and control. It extended the scope of MRP (Materials Requirements Planning) by integrating all other functional departments' computing requirements through a common database. MRP II systems are seen as an effective means of collecting and manipulating important business data from all the functional departments into useful management information. The integrated nature of such systems eliminates duplication of effort and allows immediate access to information, which further fuels their competitive advantage. Multi-plant MRP II takes this development a stage further by linking the manufacturing operations of multiple sites through a central control centre. Added complication of multiple language, multi-currency and the ability to perform centralised one time MRP/MPS runs for multiple plants while retaining the ability to perform individual plant scheduling are easily handled.

Implementation Process

The implementation of the MRP II software was based around the methodology recommended by the software vendor and involved both pilot and phased introduction of modules running in parallel with the existing system. It was intended to use a modular approach as a means to revising the plan for future implementation of other modules, if required. Fig. 2 outlines the implementation process.

IMPLEMENTATION PROCESS

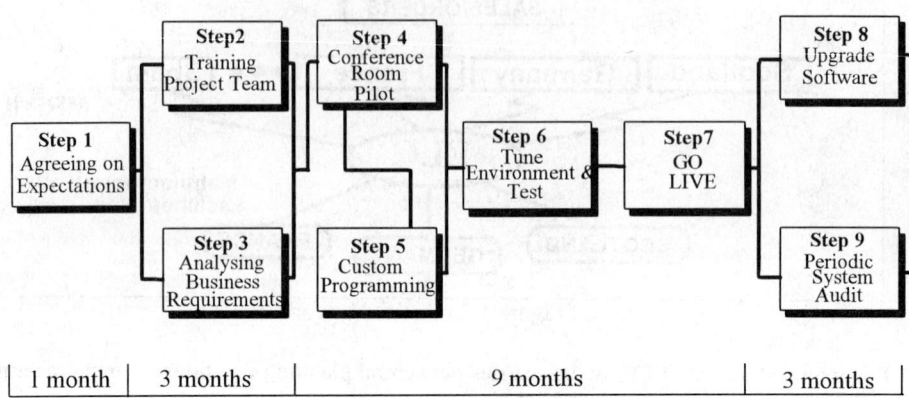

TIME SCALES

Figure 2. Multi-Plant Implementation Process

The Nine Step Implementation Methodology

STEP 1 - Agreeing on Expectations

The most important part of any successful project is to define the proper expectations of what the organisation wants to attain throughout the overall project. This phase refines the implementation plan and establishes the project team and lines of communication needed to help the implementation of the project. This step also includes the appointment of a project steering committee to monitor progress, recommend changes and resolve issues where necessary.

STEP 2 - Training Project Team

This step is to identify and arrange the training of the project team through the detailed functions of the software.

STEP 3 - Analysing Business Requirements

This stage allows the members of the software vendors project team to become better acquainted with details of the business objectives.

STEP 4 - Conference Room Pilot

The project team conducts a pilot demonstration using the MRP II software and the data to emulate the business functions. Software applications are evaluated for their performance with respect to the organisation's business needs.

STEP 5 - Custom Programming

This stage is the formal development of the modifications that need to be made to the standard software to meet the specific functionality.

STEP 6 - Tune the Environment and Test

The project team performs systems acceptance tests to make sure the software applications implemented are performing according to specifications. Additionally, during this step other tasks are accomplished to prepare the system for production:
- test inputted data
- develop end-user training requirements
- develop end-user training schedule
- test all written documentation

STEP 7 - Go Live

This step involves a series of activities to prepare the system for production. Items included in this are:

-complete end-user training
- finalise system profiles and security
- save existing data to 'test' library
- create production library with all system data
- run live data transactions through production system
- compare new system with old system data
- discontinue operation of old system

STEP 8 - Upgrade to latest Version of Software

This is a post-implementation stage to bring the new system into production taking note of software enhancements or correction to any of the existing software.

STEP 9 - Periodic System Audits

As in step 8, this stage is a post-implementation activity that assesses the performance of the software application installed, making sure the software is performing to specification.

Problems Encountered

There are a number of factors that affected the overall implementation of the project that can be attributed directly to the multi-plant aspect of the project.

The co-ordination and logistics of bringing together project team members from the three manufacturing facilities, agreeing on location for project team base, common holidays and also breaking down the language/communication barriers, all contributed to slowing down the implementation process initially.

Data formats, field lengths, unit measurements and a general lack of standardisation all meant that a lot of time and effort had been spent in agreeing on common parameters. Variations in performance measures including customer delivery targets, stock levels and variances again added time in agreeing on a common set of goals.

Many project team members were still required to carry out their daily responsibilities as well as participate in the project. This meant there was a lack of time and resource available to the overall implementation.

Lack of expertise in some functions within the core project team e.g. Quality Assurance, meant that decisions were not able to made without further lengthy consultation

Recommendations

Top-down commitment, in the form of an Executive sponsor, must be gained to show organisational commitment to the project, someone who, with sufficient power, can effect the success of the project and also give effective leadership in steering the local team in developing objectives, plans and business requirements.

In order to increase the success rates of future implementations of MRP II, there are a six key elements that should be considered. These elements have evolved during the implementation process and have contributed to the success of the project to date.

The six elements are:

1. European Management Steering Committee

Business directors meet on a regular basis (at least monthly) to monitor the course of the implementation process. They are responsible for overseeing the project implementation and monitoring its position relative to the project plan, objectives and time scales.

2. Full Time Project Managers / Core Team / Local Team

Key resource to both the steering committee and the implementation sites. Their roles are to determine the implementation support and tools needed to accomplish the organisation's objectives. They are responsible for ensuring their own companies are fully represented in the activities being carried out and also co-ordinating the efforts required from their own companies to satisfy the project's objectives.

3. Software Consultant

Consultants main role is to serve as catalyst for change and actively participate in the program design and implementation.

4. Education

People must have a working knowledge of the concepts and application principals of MRP II. On-site education and training programs must be aggressively pursued to include all people in the organisation. The uses of in-house trainers are effective and can be extremely cost efficient.

5. Performance Management

The use of regular feedback and positive reinforcement for progress and successes provides visibility and builds enthusiasm throughout the organisation.

6. Performance Monitoring

Knowledge about current levels of performance and the status of the business as plans are implemented is critical. Project plans must be developed and reviewed regularly. Implementation programs must establish two or three measures for each key area and be visible from the shop floor to top management.

Conclusions

The selection and phased involvement of those people on the multi-plant MRP II implementation project is critical. The level of resource and commitment required throughout the project, in terms of technical effort, must not be underestimated. Successful implementation is largely determined by the degree of commitment, the early involvement of people and not least a thorough understanding of the use of MRP II.

References

Cooper, R.B. and Zmud R.W., 1989, Materials Requirements Planning System Infusion. OMEGA International Journal of Management Science, Vol 17, No.5.

Cox, J.F. and Clark, S.L., 1984, Problems in Implementing and Operating a Manufacturing Resource Planning Information System. Journal of Management Information Systems, Vol 1, No.1.

Luscombe, M. 1993, MRP II: Integrating the Business- a practical guide for managers. (Butterworth-Heinemann)

Wight,O., Manufacturing Resource Planning: MRPII, (Oliver Wight LTD., NY)

EDUCATING MRP USERS
A PRE-REQUISITE FOR SUCCESS

H L Cather* and M Montekio**

*Department of Mechanical & Manufacturing Engineering
University of Brighton
Brighton, BN2 4GJ

**Centro de Investigacion en Computacion Educativa
Universidad Anahuac
Mexico

The success of installing any change in the methods or system of working can be problematic. Where it is a major and strategically important change such as the installation of a Material Requirement Planning (MRP) system where many people and departments are involved the problems can be very severe. So much so that the installation time is extended, only partial installation takes place or even a complete failure occurs. The key to a successful installation in these cases is a sound common understanding by those involved and the key to this lies in education. This education should be capable of being absorbed quickly by a large amount of people working at their own rate. Computer Aided Learning can deliver in these circumstances as the package described in this paper shows. The package delivers an inactive, graphic based introduction to the concepts involved in MRP.

EDUCATION'S ROLE IN MRP

In his book on Manufacturing Resource Planning, O.W. Wright devotes a chapter to education's importance in the introduction of Material Requirements Planning (MRP) systems. He quotes a case where 20% of the cost of installing a successful MRP system was accounted for by education. His conclusion is that the role of education in the success, or failure of a MRP installation is not understood and therefore too often not enough is invested in this aspect.

There is ample evidence from other cases on installation to support this view in that the major problems associated with MRP introduction is not technical issues but organisation ones. The resistance to change is but one aspect that can be controlled by better education beforehand. All the personnel involved in a MRP system operation require to fully understand the system in order to get the most out of it.

The typical approach to training people is an open common environment such as a lecture room. This has distinct disadvantages in an organisational sense because the degree, or absence, of each participant's personal knowledge and differing learning ability is exposed to others.

Learning Environment

It was this disadvantage that has lead many organizations to move towards a degree of self paced learning. This initially was via linear book exercises then into a branching mode where reinforcement of problematic points could be directed towards each participant's particular speed and comprehension.

The advent of the computer, especially the cheap personal PC, attracted educator's towards developing appropriate software packages designed around a particular aspect of an organisation's operations.

The fact that MRP is itself a computer based system reinforces the need to develop appropriate learning packages to aid personnel to understand the basic operation of the system so that they can more readily appreciate their own role in the system's operation. The package developed at the University is such.

THE PACKAGE DESIGN

Design Specification

The specification selected for the package's design was:

It must be visually attractive

It must be highly interactive

It must have help facilities

Above all else, of course, it must fully cover the subject in a clear, logical progression.

Software Tools

Many software tools were required to meet the packages remit. The initial program suite was developed on the Macintosh computer and a PC-DOS version was developed from this later. The packages used were:

Stratavision 3D as a CAD working environment to create models for the product and its assemblies and components.

Photoshop was used to edit and colour the images produced and transfer file formats.

Infini-D was used to create the animation sequence.

ResEdit was used to define objects such as mouse buttons, icons and menu bars.

Authorware Professional was used for the integration of all the graphic resources, animations, interactive control routines, graphics and text displays, variable definitions and control, and to compile the program and produce run time editions.

The files once created and proved out were compressed using DiskDoubler as a file compression utility to produce the portable run time package.

User Interface

The user interface is completely graphical. Buttons are provided to select in menus, move onto the next screen and to access pull-down help including an on-line glossary of terms used and access to reference material. Three dimensional graphics make the interactive sessions realistic and easy to understand.

Most interactions are mouse driven with text input only used when this is necessary. This includes click-on-and-drag icons and the rotation of 3D views used in some of the interactive exercises.

The system allows the user to make errors which are trapped and, depending on the position in the program, the user is either informed that a mistake has been made or is directed to what, or where, the correct entry should be.

Strong use of colour has been made to highlight important points in the instructions, error and help messages and different instructions.

The user's progress is recorded and shown constantly on the screen for reference.

THE TUTORIAL

On entry to the tutorial, the user has to identify himself by name. This allows the system to create a file recording the user's progress and ensures the user is one from a pre-recorded list. Once cleared the user is presented with the main menu to access one of the six parts of the tutorial. Once each part has been completed, the menu carries a tick and "DONE" against that part.

Instruction

The first part of the tutorial contains facts and explanations about the user interface. The user is introduced to the button and uses this method to progress through an explanation of the tutorial itself, the use of the menu bar, glossary, etc. This part concludes with a simple, interactive session on dragging objects around the screen.

What is MRP ?

This part briefly introduces the user into the concept of MRP itself. It concludes with an interactive introduction to the concept of lead time and points towards the following exercise-

Exercise 1

This is a fully interactive graphic based exercise explaining the concepts of a product structure and how the lead times of individual assembles and components have consequences on the assembly process. A skateboard is used as the example product.

The user has to estimate the number of parts used therein and can rotate the product to see from all views the components used. The user then has to dis-assemble the skate board, by clicking on each component following prompts.

After the dis-assembly is completed the parts (sub- assemblies and components) are listed in a table indicating if they are bought out or internally produced. The table also contains information on the lead times for each part. This is then reproduced as a product tree (Figure 1) denoting levels of assembly with the quantity of each part used in its parent. The product tree is retained as a snapshot which can be accessed later by the user.

The exercise ends with the user having to construct a time based graph (figure 2) showing overall lead times for each part and hence the accumulative lead time for the skateboard itself. The icons are dragged into position by the user. This plan once complete forms a snapshot which the user can access in later exercises.

Exercise 2

The second exercise is carried out by using an on screen MRP worksheet. The user is then guided through its completion using the same model as in exercise two to produce one lot of skate boards (figure 3).

The user therefore has to work out at each level the recognised sequence of determining Gross demand, offsetting available stock to find Net Demand, using lead time to determine when the associated lower level parts must be available and their gross demands.

The exercise concludes by describing the normal output from a commercial system such as Planned Order Release/receipt.

Exercise 3

This exercise deals with the differing lot sizing techniques such as Lot-for-lot, Economic Batch Quantity, Periodic Review, Part-period Balancing and Safety Stock to show the effect on overall inventory costs by the variations in ordering and holding cost against a variety of demand profiles. This is done by leading the user through a series of exercises, showing the cumulative ordering and holding costs for different techniques.

MRP Test

The final part of the tutorial is a simple test using the concepts and techniques worked through in the three examples and using the on screen MRP sheet. The Product Structure and Leadtime Chart are available to the user as help facilities.

Help Options

One of the pull down options available to the user at all times during the tutorial is a glossary of terms used in MRP. This lists alphabetically 35 MRP terms with a brief description on each. In addition there are the references mentioned above - the product tree and the graph of overall leadtime.

CONCLUSION

The package produced serves an important function in being a user driven free standing tutorial presented in a highly interactive manner to lead the user through the basic concepts of MRP. It can be tailored to suit any situation and expended if necessary into areas such as Capacity Planning and Market Forecasting if required.

As an example of the ratio of development against learning times, the package represents over five hundred man hours of programming and in use it took between four and five hours to work through the first time by a variety of users.

References

Wright, O W, 1984, *Manufacturing Resource Planning: MRP II*,
Oliver Wright Ltd. Publications Inc., New Work

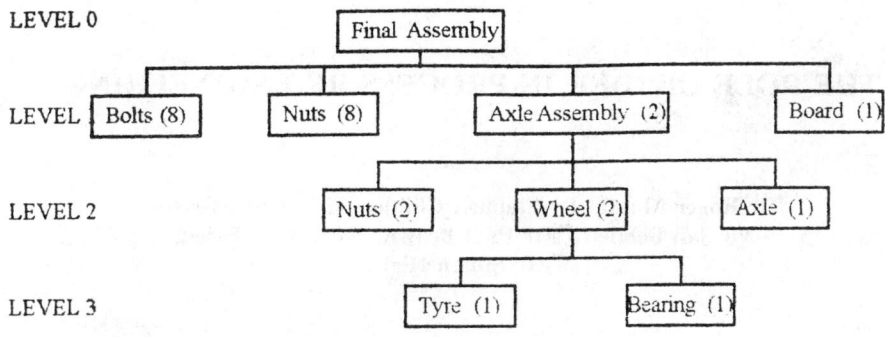

LEVEL 0

LEVEL 1

LEVEL 2

LEVEL 3

FIGURE 1 - PRODUCT TREE

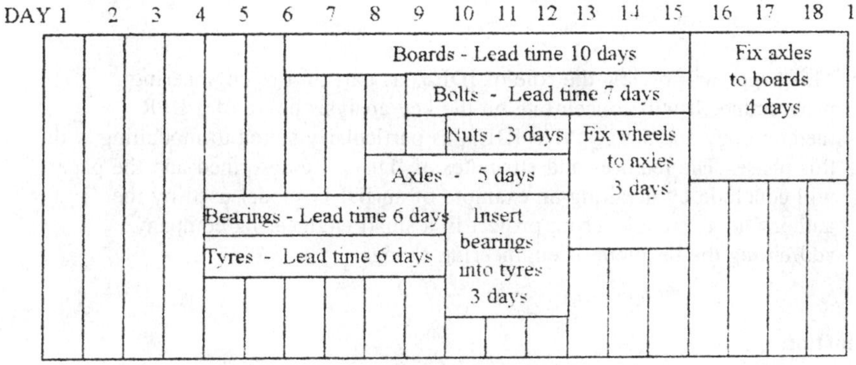

DAY 1   2   3   4   5   6   7   8   9   10   11   12   13   14   15   16   17   18   19   20

FIGURE 2 - TIME BASED ASSEMBLY CHART

In each part or assembly block, you will find Lot Sizing information, and in some cases the Safety Stock quantity. Follow the conditions for each block when writting the required quantities in the worksheet cells. This is an example of how this conditions may look:

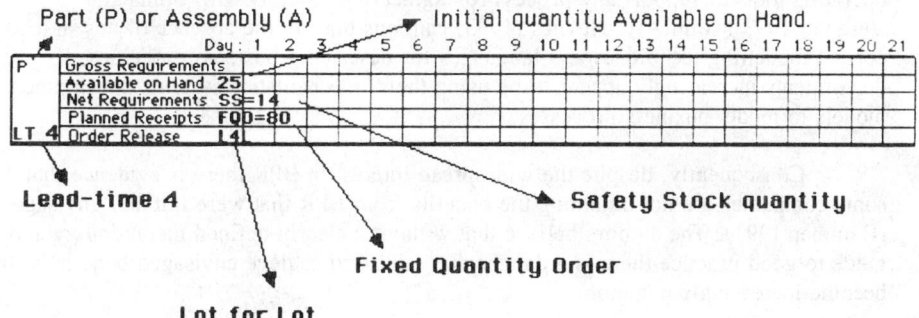

FIGURE 3 - MRP WORKSHEET

THE ROLE OF IDEF$_0$ IN PROCESS RE-ENGINEERING

*Dr Roger Maull, *Dr Stephen Childe, *Mr Adam Weaver
**Mr Jan Bennett, *Mr Paul Brown, *Mr Jim O'Brien,
**Mr Simon High.

*School of Computing, University of Plymouth, Plymouth PL4 8AA

**Teaching Company Centre, University of Plymouth, Plymouth

This paper will review the role of IDEF$_0$ as part of a re-engineering programme. It will concentrate on the key analysis phase of a BPR methodology and suggest that IDEF$_0$ is particularly suited to modelling within this phase. The features and attributes of IDEF$_0$ are described and the paper will conclude by detailing an example of an IDEF$_0$ model used by the authors' in a re-engineering project in a small engineering company addressing the problem of engineering change.

Introduction

Business Process Re-engineering (BPR) is becoming a key enabler of the 1990's for companies seeking to achieve competitive advantage. BPR offers the opportunity for sustained competitive advantage through radical reductions in lead time and cost and substantial service level improvements. Increasing attention is being paid to BPR by many manufacturing companies including Lucas, IBM, ABB, BAe, HP and Rank Xerox.

Despite the widespread interest there is a lack of conceptual models and operating tools to support any process re-engineering (Bartezzaghi, Spina and Verganti 1993). Similarly, Heynes (1993) cautions that, in the absence of any agreed, correct modelling techniques and languages for describing business processes, IS departments increasingly appear to be using their "mechanistic" systems development models to model business processes.

Consequently, despite the widespread interest in BPR there is evidence,that some companies are not obtaining the benefits from BPR that were initially envisaged (Hammer 1991). The authors believe that without a clearly defined methodology and guide to good practice there is a danger that failure to achieve envisaged benefits will become increasingly common.

BPR Methodology

A number of authors (Davenport and Short 1990, Kaplan and Murdock 1991) and companies for example, IBM (Snowden 1991) have proposed, in very general terms, the stages of a BPR methodology. The authors have distilled from these what they believe to be a good composite BPR methodology consisting of five phases. These are:

Phase 1   Create/Identify corporate, manufacturing and IT strategies
Phase 2   Identify key process(es) and performance measures
Phase 3   Analyse existing process(es)
Phase 4   Re-design process(es)
Phase 5   Monitor and continuously improve new process(es)

For those wishing to read an excellent overview of the entire BPR process, the authors have produced a working paper defining each stage in considerable detail. This paper will continue by focusing upon the development of methods for what the authors regard as the key phase of the methodology - phase three.

Phase 3 Analyse Existing Processes

This phase defines key business processes and identifies possible opportunities for re-engineering by comparing corporate objectives and business drivers within the defined processes.

The first activity is to carry out a key process profile. This profile attempts to understand process flow in terms of activities/tasks/steps performed, cycle times for products/services produced, individual task timings, redundant tasks or steps, delays and work volumes

In our view, in order to provide a basis for incremental and radical change it is necessary that some comprehensive effort be made to analyse existing processes. This may best be achieved through the development of a process model. A number of possible modelling tools exist which could be used at this stage. The most widely used techniques include flow charting (Oakland 1989) Role Activity Diagrams (Ould 1993) and IDEF$_0$ (Le Clair 1982). There is insufficient space to provide an analysis of each of these methods, this paper will now concentrate on describing the most widely used technique - IDEF$_0$ and its application to analyse a process in a manufacturing company.

ICAM Definition Method

IDEF$_0$ consists of three to six boxes. Three is felt to be a reasonable minimum (a diagram of two can usually be incorporated into a higher level diagram) and six a maximum because of individual cognitive limitations. The graphical language of IDEF$_0$ uses boxes and arrows coupled together in a simple syntax. Boxes on a diagram represent activities. The arrows that connect to a box represent real objects or information needed or produced by the activity. The side of the box at which an arrow enters or leaves shows the arrow's role as an input, a control or an output.

The strength of $IDEF_0$ is that it is a tool designed for modelling processes and in our view it is relatively easy to use (though more difficult than flow charting). It uses a structured set of guidelines based around hierarchical decomposition, with excellent guidance on abstraction at higher levels, if used well this ensures good communication and a systems perspective. It is also becoming the defacto standard modelling tool for business process modelling.

The main weaknesses in using $IDEF_0$ are that some users claim it is too complex to use and that it is not possible to produce a detailed software specification directly from the $IDEF_0$ diagrams, thus its use in linking stages three and four is very limited.

Case study

The application of $IDEF_0$ is illustrated in Figure 1. Here we can see an example of $IDEF_0$ applied in a small engineering company based in Plymouth. The sub-process that the authors analysed was engineering change. There are six key activities A11..A16. The first activity is to filter the engineering change proposal, the key control on the filtering process are the company policies on acceptable engineering change requests. The marked drawings are then used to input a hypothetical effectivity date into the CAPM system. At this stage the effectivity date is always 1999 ie some future date. The drawing and engineering notices are then used by the draughtsman to produce the changed drawings which are then evaluated by supply for a true effectivity date. Supply will assess their stock levels and if, for example, they have a large stock of material affected by the drawing change they will request that the effectivity date be pushed out as far as possible. The feedback loop is to the product engineer who has to interface with the customer to identify whether the proposed effectivity date is acceptable. This activity produces an effectivity in date for the new part and also an effectivity out date for the existing part which is entered on the Bill of Materials. The final activity is where the drawing and engineering notices are appraised and signed off for implementation by the product engineer.

Discussion and Conclusions

The $IDEF_0$ models are useful in identifying areas for improvement in three main ways. Firstly, they act as a means of understanding the process. The $IDEF_0$ models developed of the process were the first time that the process had been modelled in such detailed manner. Secondly, because of the hierarchical nature of $IDEF_0$ the models are useful in communicating this understanding of the process to senior executives. In essence, because $IDEF_0$ insists on consistency amongst levels yet allows for abstraction of terms, the models can be shown to strategic meetings where radical re-engineering decisions are made. Thirdly, the models allow an analysis of the process to take place. The team are currently engaged on developing a specification of a methodology for BPR which will take the $IDEF_0$ models and indicate areas for radical and incremental improvement.

References

Bartezzaghi E, Spina G and Verganti R, Modelling the lead-time of the operations processes, in Johnston R and Slack N D C, (Eds), Service Operations, Operations Management Association UK, 1993, pp117-124

Davenport T H and Short J E, The new industrial engineering: information technology and Business Process Redesign, *Sloane Management Review*, Summer 1990

Hammer M in Shocking to the Core, *Management Today* August 1991

Heynes C, Presentation at British Computer Society BPR Conference, London, 29 June 1993

Kaplan R B and Murdock L, Rethinking the Corporation: Core Process Redesign, *The McKinsey Quarterly*, 1991 No 2

Le Clair S, *IDEF The method, architecture the means to improved manufacturing productivity*, SME technical papers, Society of Manufacturing Engineers 1982

Oakland J S, *Total Quality Management*, Butterworth 1989

Ould M, Process Modelling with RADs *IOPENER* The Newsletter of the IOPT club. Volume 2, Number 1, 1993.

Snowden D, Business process management and TQM, *Proc. 4th International Conf. on TQM*, IFS Publications, June 1991

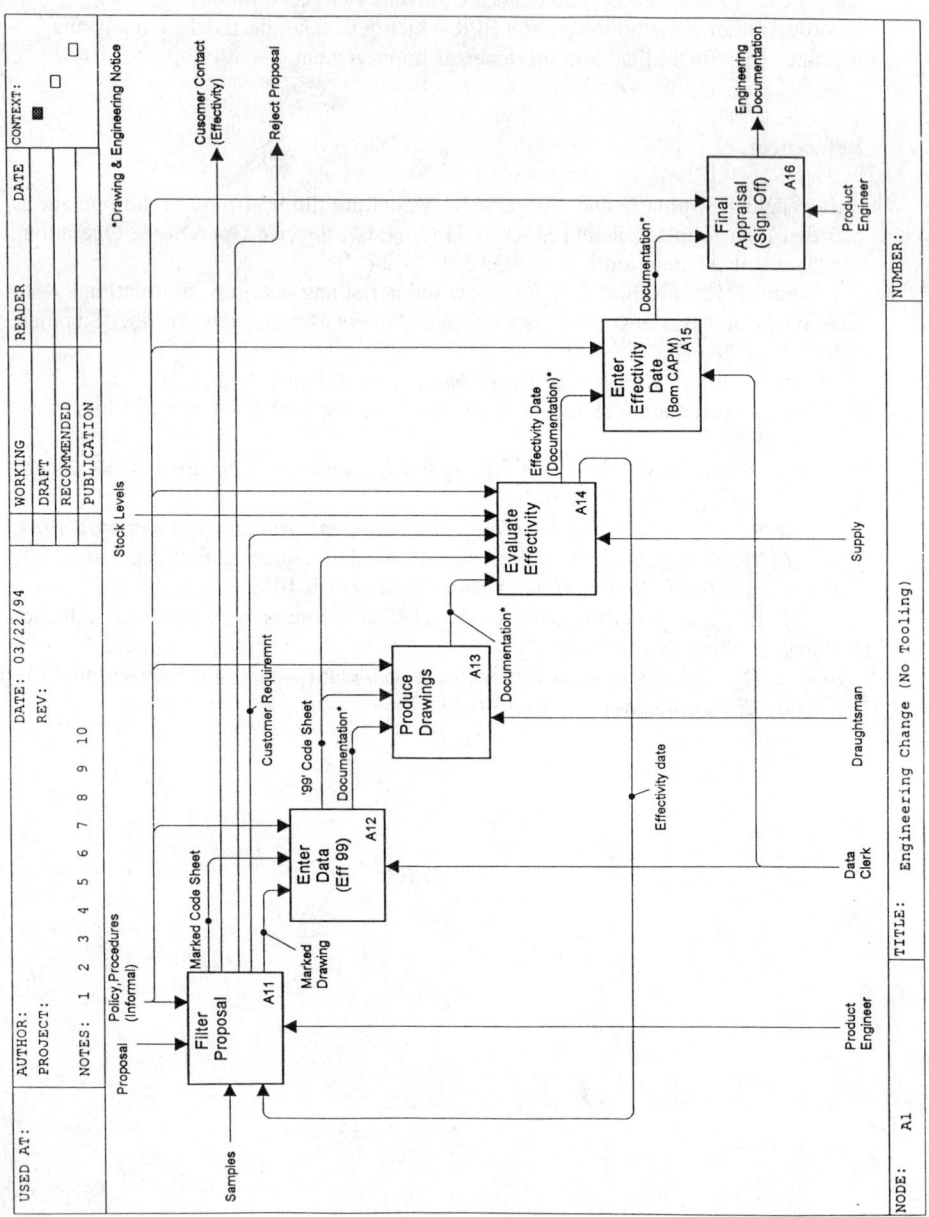

MANUFACTURING STRATEGY: A SURVEY OF BEST PRACTICE

William J. Bailey, Grant MacKerron, Martin Askey

Napier University, M.M.S.E
10 Colinton Road, Edinburgh, EH10 5DT

This paper investigates the potential contributions to the strategic aims
of a company which can be made by its choice of manufacturing
strategy. The key literature is reviewed, and the process, context and
content, including the key strategic elements of quality, cost, flexibility,
delivery and innovativeness are investigated. Evidence is drawn from
case studies of several international manufacturing companies, and this
is used to illustrate and expand upon the theoretical models found in the
literature.

Introduction

This papar defines manufacturing strategy, and examines its formulation,
context, and content. This is achieved by a review of the literature, followed by case
study examples of electrical goods manufacturers, undertaken by a Napier University
research team.

There are a number of different views on what manufacturing strategy actually
consists of. For example Hayes and Wheelwright (1979) suggest that a business'
manufacturing strategy is the "pattern of structural and infrastructural decisions...a
sequence of decisions that, over time, enable a business unit to achieve a desired
manufacturing structure, infrastructure, and set of specific capabilities". Gupta *et al*
(1991) proposes the view that a manufacturing strategy is how manufacturing 'supports
the overall business objectives through the appropriate design and utilisation of
manufacturing resources and capacities'. Hill (1985) points out that "The purpose of
thinking and managing strategically is not to just improve operational performance or
to defend market share. It is to gain competitive advantage and it implies an attempt to
mobilise manufacturing capability to help gain this competitive edge".

The Process of Manufacturing Strategy Formulation

The subject of how manufacturing strategy is actually formulated has been the
subject of much discussion, for example Garvin (1993), although less empirical
research. In discussing this issue one must distinguish between manufacturing strategy
as the written result of a formalised strategy making process, the view taken by

Wheelwright(1981), who suggests a manufacturing strategy should be drawn up by senior management and explicitly integrate specialised objectives into a consistent and coherent programme; or alternatively manufacturing strategy can be seen as the pattern of management decisions relating to manufacturing and its effect on the long-term competitive stance of the company. It is the authors' view that, as many companies have only informal manufacturing strategy making mechanisms; and as formal statements of strategic intent do not necessarily result in actual changes in behaviour, it is equally valid to view manufacturing strategy as actual patterns of behavioural change, and to investigate the strategy making process as the actions and events which brought about these changes, which may or may not include a formal strategy making process. Thus a key determinant in the actual formulation of manufacturing strategy is the context in which it occurs: i.e. the structures and relationships within the firm.

Manufacturing Strategy in Context

Skinner (1969) suggested that the main reasons why manufacturing strategy does often not support the competitive priorities of the business, are that top management suffers from a sense of 'personal inadequacy' in dealing with manufacturing operations, as they are seen as highly detailed and requiring specialist, expert attention; and also there is a lack of top management awareness of the strategic importance involved in manufacturing decision making, specifically in deciding upon the trade-offs in deciding manufacturing priorities. This problem is compounded by manufacturing's domination by technical experts, be they industrial engineers or computer specialists, who are given responsibility for taking strategically important manufacturing decisions, but are excluded from the wider strategy debate, at least until it is too late for them to have any real opportunity to shape business strategy in line with manufacturing competencies and opportunities. Meredith and Vineyard (1993) suggested that manufacturing's influence in strategic decisions was reduced as their firm's performance got worse, and was also reduced in times of high environmental uncertainty. These problems, it has been suggested, can be reduced by using a formal strategy making process of the type suggested by Hill (1985).

Manufacturing Strategy Content

Wheelwright (1981) identified 8 major types of decisions, which the manufacturing function is normally responsible for, capacity, facilities, vertical integration, production technologies and processes, work force, quality control and product assurance, production planning and materials control and organisation. He said that capacity, facilities, vertical integration and process technology are normally viewed as having long-term strategic implications, and so need top-management control, but goes on to say that the others dimensions, while being dependent on specific situations and requiring detailed operational knowledge, do have a potentially vital cumulative strategic effect.

Order Winning and Qualifying Criteria

New (1992) suggests that there are 4 'competitive edge' criteria : Delivery (lead time and reliability); quality (capability and consistency); flexibility (design and volume); and price. Wheelwright also identifies four competitive priorities, which are dependent on a firm's manufacturing strategy : cost, quality, flexibility and dependability. Meredith and Vineyard (1993) consider that manufacturing's mission consists of five basic competitive priorities : cost, quality, delivery (speed and dependability), flexibility, and innovativeness (both product and process). Hill (1985) distinguishes between 'order-winning' and 'qualifying' criteria. Qualifying criteria must be provided to allow a company to enter or remain in a market, but doing them particularly well will not increase orders. This is the role of order winning criteria, which should be identified as part of the business' competitive strategy, and given priority in its manufacturing strategy. New uses the alternative terminology of 'competitive-edge' and 'market hygiene' factors.

Strategic Trade-offs

Much discussion (e.g. Skinner [1969], New [1992], Corbett [1993]]) has concentrated upon the various trade-offs, for example cost v. quality, which, some say, must be made in a firm's manufacturing strategy. Against this traditional view a number of people, notably Schonberger (1986), have put forward the ideal of World Class Manufacturing, which claims that, with the introduction of new manufacturing management techniques such as Total Quality Control and Just-in-Time manufacturing, these trade-offs do not need to be made: In fact reducing, for example, lead times increases flexibility and dependability of delivery; improving quality by removing the causes of faults reduces costs, due to reduction in wastage, so in fact higher quality can result in lower cost - competences build in a cummulative way. Whilst it is true that the nature and dynamics of the trade-offs have changed considerably, it must be noted that no one can be perfect in all areas, so improvement priorities still need to be made, based on strategic criteria.

Manufacturing Technology and Manufacturing Strategy

Hayes and Wheelwright[1978] suggest that, just as products often go through distinct stages throughout their life-cycles, so too do manufacturing processes. They identify four stages in the process life-cycle: Jumbled flow (job shop), disconnected line flow (batch), connected line flow (assembly line), and continuous flow. They posit that different management tasks are involved at each stage, and that the combination of the stages of the product and process in their life-cycles largely determines the business' distinctive competence, i.e. the activities the company can do better than its competitors. In a study of three companies over extended periods of time, Meredith and Vineyard (1993) attempt to analyse the role of manufacturing technology, specifically the introduction of FMS systems, in the companies' business strategy. They found that the choice of this manufacturing technology was primarily governed by the desire to 'modernise' and to improve the image of their manufacturing operations, rather than any explicit strategic realignment of their operations competencies with regards to cost, quality, delivery, flexibility, or innovativeness.

Case Study Examples

Case A : A major Japanese manufacturer in the electrical consumer goods market is attempting to reduce costs and decrease manufacturing lead-times by introducing a 'common build policy', whereby the common stages of different models are now being built together in the same batch, to simplify production, thereby reducing cost and increasing speed of manufacture. They are also considering the introduction of a unit which could be suitable for use anywhere in Europe, with minor adjustments or small changes of components. This would increase unit costs in terms of materials, but would lead to a huge simplification of the manufacturing task, resulting in increased speed of manufacture, flexibility, and possibly overall cost reductions. This shows how the competitive position of a company is dependent upon how the decisions made relating to product design and manufacturing strategy can reinforce each other, or can have negative effects if mis-aligned. These decisions must be taken within a coherent, integrated framework, which must be linked with the business strategy of the firm, and manufacturing must have strategic influence in decisions relating to product design. It also demonstrates the value of simplifying the manufacturing task as much as possible. This approach has been put forward by a number of writers, including Schonberger (1982), who suggests "simplify and goods will flow like water", and recommends the use of just-in-time and TQM techniques. As several writers, notably Skinner (1974), have suggested simplifying the task of manufacturing further, by reducing the number of products, volumes, and processes dealt with in a single factory, and where necessary building separate factories or moving people and equipment to make plants-within-a-plant. Although this plant had only a single product type, and a limited number of processes, it was not truly focused in that it produced a very large range of models, which were of widely different complexity and were produced in very different volumes.

Case B : A multinational manufacturer of hi-tech electronic equipment is concentrating large amounts of effort on quality improvement measures, including fixing quality targets which cannot possibly be met with existing suppliers and processes, despite the fact that the quality of their finished goods is regarded as exemplary by the marketplace, and cost and delivery are seen as more of a weakness relative to the competition. This demonstrates that order winning criteria need to be objectively analysed, and manufacturing priorities should be aligned appropriately. It also demonstrates a lack of understanding of the difference between order winning and qualifying criteria, as defined by Hill (1985).

The same factory built a high-tech factory in the UK, primarily as a showpiece, to improve image, despite the fact, that in the manufacturing manager's view, the same goods could be produced to the same quality with substantially lower capital equipment costs. This example supports the findings of Meredith and Vineyard (1993), who found that many companies did not use the introduction of new technology to improve their manufacturing competence in the key competitive areas of cost, quality, delivery, flexibility and innovativeness. The state-of-art factory does, however, have benefits in these areas, notably in terms of flexibility - and of the plants products can be assembled in any volume, with no substantial loss of efficiency, and in speed. It is the authors' view that this potential is not fully capitalised on by the firm in question, as the supporting systems are not integrated to allow them to offer these facilities to their customers; the main problem being slow internal communication, and manufacturing being remote from the ultimate customer. These problems are not seen as critical at present, as the company does have significant competitive advantages in its primary product technology, giving an edge in performance, although the competition is catching up, and cost and delivery are seen as becoming increasingly important.

Case C : A successful American manufacturer of scientific equipment has a policy of transferring managers between functions, so they get to understand the operational choices which have to be made in several departments, thereby improving communications and aiding the formulation of departmental policies which support each other, rather than pulling in different directions. For example the current manufacturing manager started as an accountant, and the marketing manager has worked for a period in manufacturing. The same company also has a policy of splitting divisions in two when they get over a certain size. This enables each division to keep focused on a limited product range, identified by Skinner (1974) as being vital in allowing the plants to concentrate on the limited number of manufacturing tasks which are essential in providing competitive advantage in a particular market. They also have extremely good communications between the divisions which means they do not miss out on the advantages due to the sharing of knowledge of new developments in different markets and countries.

    The same plant identified speed of delivery as being the critical order-winning factor, so has launched a major programme to try to reduce their lead-time. They realised that while the actual manufacturing lead time was reasonable, major delays were being caused as time was wasted processing the order in marketing then in manufacturing. To reduce these problems they have reorganised, and adapted their systems. This has involved moving order administration out of manufacturing and into marketing, which has had the effect of removing one layer of schedulers from the process, and has in fact improved the communications between the two departments.

They are also in the process of linking their manufacturing computer system with other systems, thereby allowing manufacturing to respond faster to demand. This example demonstrates the benefits of analysing the important competitive factors, as proposed by many writers on manufacturing strategy, including Hill (1985), and translating these into operational programmes which encompass all the relevant parts of the organisation, not just the manufacturing department or the marketing department.

The company's strategic planning mechanisms are also in line with some of the formal strategic planning frameworks put forward in the literature. For example Garvin (1993) suggested a framework the main elements of which include : Describe the business strategy; identify and rank the strategic priorities of the business; disaggregate the top two or three priorities; review existing manufacturing policies for consistency with the disagregated priorities; if priorities and policies are not aligned, develop programmes to ensure a better fit [Garvin calls these strategic manufacturing initiatives or SMIs]; then develop more SMIs to improve operations in line with the priorities.

Conclusions

It is the view of the authors that manufacturing operations have a critical effect on the strategic competitive stance of a manufacturing company. Thus it is imperative that the strategic implications of manufacturing decisions need to be well thought out, and integrated into a coherent business strategy. Formal manufacturing strategy making mechanisms can be a vital element in achieving this goal, depending upon the structure of the organisation, the nature of the business, and external economic and market factors; but the critical elements are open communications between manufacturing, other departments and senior management, the use of suitable control mechanisms, and a detailed top management understanding of the firm's manufacturing competencies and imperatives. The case studies demonstrate that potential competitive advantage can be lost if a firm does not view its operations strategically, or if only certain aspects are considered.

References

Hill, T. 1985, *Manufacturing Strategy*, (Macmillan, London).
Corbett, C. and Van Wassenhove, L. 1993, Competence and Competitiveness in Manufacturing Strategy, *California Management Review*, **4**, 107-122.
Garvin, D.A. 1993, Manufacturing Strategic Planning, *California Management Review*, **4**, pp.95-107.
Skinner, W. 1969, Manufacturing-missing link in corporate strategy, *Harvard Business Review*, May/June, pp.136-45.
Skinner, W. 1974, The Focused Factory, *Harvard Business Review*, May/June, pp.113-21.
Schonberger, R.J. 1982, *Japanese Manufacturing Techniques,* (Free Press, N.Y.).
New, C. 1992, World-class Manufacturing versus Strategic Trade-offs, *International Journal of Operations and Production Management*, **6**, pp.19-31.
Meredith, J. 1987, The Strategic Advantages of New Manufacturing Technologies
Hayes, R.H. and Wheelwright, S.C. 1979, Link manufacturing process and prodcut life cycles, *Harvard Business Review*, Jan/Feb, pp.15-22.
Meredith, J. and Vineyard, M. 1993, A Longitudinal Study of the Role of Manufacturing Technology in Business Strategy, *International Journal of Operations and Production Management*, **12**, pp.4-24.
Wheelwright, S.C. 1981, Japan-where operations really are strategic, *Harvard Business Review*, July/August, pp. 67-74.
Gupta, P. Lonial, S.C., Mangold, W.G. 1991, An Examination of the Relationship between Manufacturing Strategy and Marketing Objectives, *International Journal of Operations and Production Management*, **10**, pp.33-43.
Kotha, S. and Orne, D. 1989, Generic Manufacturing Strategies: A Conceptual Synthesis, *Strategic Management Journal,* Vol.10 No.3, pp.211-32.

IMPLEMENTING OPT, APPLYING THE PRINCIPLES AND USING THE SOFTWARE.

Steve Martin and Sharon Cox

School of Engineering
Coventry University
Coventry, CV1 5FB.

This paper uses the case study of an original equipment manufacturer (OEM) of plant and equipment for the construction industry, who in introducing Optimised Production Technology (OPT) scheduling software to plan and control both the procurement function and the manufacturing activity, has recognised the need to establish the ideas of OPT in the workplace prior to software implementation. This paper briefly describes the philosophy of OPT. It details the changes that have been made to the manufacturing system as part of the improvement process. Also described are the developments which have led the company to rethink the use of the software to perform the manufacturing control function, and instead to consider relying on the simpler 'pull' replenishment approach for the manufacture of parts, sub-assemblies, and for assembly.

Introduction

OPT is a manufacturing management strategy which has been developed in more recent years. It has been less widely adopted than either Manufacturing Resource Planning (MRP II), or Just-In-Time (JIT), however it shares many of the features of both the alternative strategies. In particular it shares with JIT the requirement to synchronise the manufacturing activities and to concentrate on issues such as quality, leadtimes, lot sizing and machine set-up times to ensure the optimisation of the manufacturing resource. In common with the use of MRP II to manage more complex manufacturing systems, the adoption of an OPT strategy involving software support requires a large, and often complex data base of product and machine information to enable meaningful scheduling calculations to take place. More recent research has suggested that OPT given the many complementary characteristics that it shares is potentially the most beneficial of all three manufacturing strategies, Lawrence (1990).

OPT, a brief description.

Manufacturing according to OPT is for making money, this activity is measured in terms of throughput, the rate at which money is generated by the system through sales; inventory, the money that the system has invested in purchasing items that it intends to sell; and operating expense, the money the system spends to turn inventory into throughput. Simultaneously increasing throughput and decreasing both inventory and operating expense will move the organisation towards the aim of making money, Harrison (1985); Goldratt & Cox (1984).

The root of OPT is constraint management, ensuring that all of the activities support the optimisation of throughput; that is the rate at which the money is generated through the constraint resource. At its heart, OPT has a scheduling philosophy that recognises that all the manufacturing activity, if it is not to be wasteful must be synchronised, and if this is so, will be paced by the performance of the constraint. This scheduling philosophy and a number of operating principles or rules are incorporated into a proprietary software package. However for organisations for whom the manufacturing management function is not regarded as complex enough to require the creation and manipulation of a large database, with the need for complicated calculations to develop realistic schedules, the principles can be utilised alone to bring about improvements in business performance, Jacobs (1984); Booth (1988).

There is a requirement therefore, independent of whether a software solution is implemented, to ensure that in the workplace the principles of OPT are recognised and readily applied if the system is to support the improvements sought, this will entail changes on the shopfloor to physical layout, policies and procedures which the total workforce will be involved in.

OPT, the implementation.

A previous paper, Martin (1993), identified three distinct areas of challenge for this organisations implementation. These are briefly; the technical challenge of mating the OPT software with the existing MRP II data base and identifying the manufacturing resources in such a way that they can be modelled in a meaningful manner, to provide for the identification of constraints during planning and allow for alternative choices of action to be evaluated. The people challenge of the need to ensure that the philosophy of OPT is appreciated by the workforce and to gain their commitment to making changes within the workplace, often cutting across existing and long standing practice and procedures to affirm that information about activities on the shopfloor are as realistic as is possible. Also creating the opportunity to make ongoing improvements. The final challenge is perceived as the need to develop cost accounting procedures which are able to provide the necessary link between shopfloor activity, which is product focussed and numbers driven, and the organisational performance, measured as you would expect in profit & loss and balance sheet terms. It is beyond the scope of this paper to detail the response to the accounting challenge, this paper will confine itself to addressing aspects of the technical and people challenges identified.

The manufacturing system, the existing situation.

Having previously identified the compaction products as the area to concentrate on, and with a product team in place which included the disciplines of purchasing, engineering, and marketing & sales in addition to manufacturing, objectives were identified to improve the output from the line by some 40 per cent, as a reduction in the time taken to complete the assembly of each unit through the synchronisation of production activities, and to reduce the inventory held in the area to the equivalent of 1 weeks build requirement. Analysis of an activity sample, (Fig 1) undertaken in the build area revealed that non-productive activity accounted for only 17 per cent of the daily activity of the assembly team. From this it was apparent that if an improvement

	Roll Subs.	H/F Tank	Wtr Tank	Frm Assy	Hydr. Fit	Eng Subs.	Assy Final	Test	Notice Board	Walk Thro'	Crane Opn	Abs
Tot	15	1	1	18	6	10	11	8	1	1	1	11
%	18	1	1	21	7	12	13	10	1	1	1	13

Figure 1. Activity Sample.

of the magnitude necessary was to be realised, it would require not just the elimination of all of the non-productive activity, but more fundamentally would require a change to the current method of assembly.

A further analysis of average inventory levels of the major sub-assemblies and bought-out components held within the build area revealed that a reduction of 50 per cent by value would be necessary if the objective established for inventory was to be achieved.

The manufacturing system, planning the improvements.

Reducing the inventory resident within the unit build area has released much wanted space and allowed for not only the re-layout of the area into a flow-line, (Fig 2) and the re-assignment of tasks between the assembly operators, thereby improving the balance of the workload and creating greater synchronisation of activities, it has also provided additional benefits. The first of which is to enable the incorporation of the total build of all models of tandem compactors, where previously only the roller sub-assembly and engine preparation activities were carried out in this area for the two smaller capacity units, assembly for these being completed elsewhere in the plant. concentrating on the non-productive activities has led to the proposal to introduce two changes that will contribute to the overall objectives. One to provide a detailed parts line-side storage administered by the parts supplier, which will eliminate the need for the storekeeping function and the necessity for the assembly operator to requisition these generally small, low value items. Secondly the kitting of hydraulic hoses per machine with all the necessary nuts, 'o' rings and olives to complete the assembly. The supplier will deliver in these kits to a given schedule. This will reduce considerably the amount of time spent selecting the correct hose type and fittings for any particular assembly.

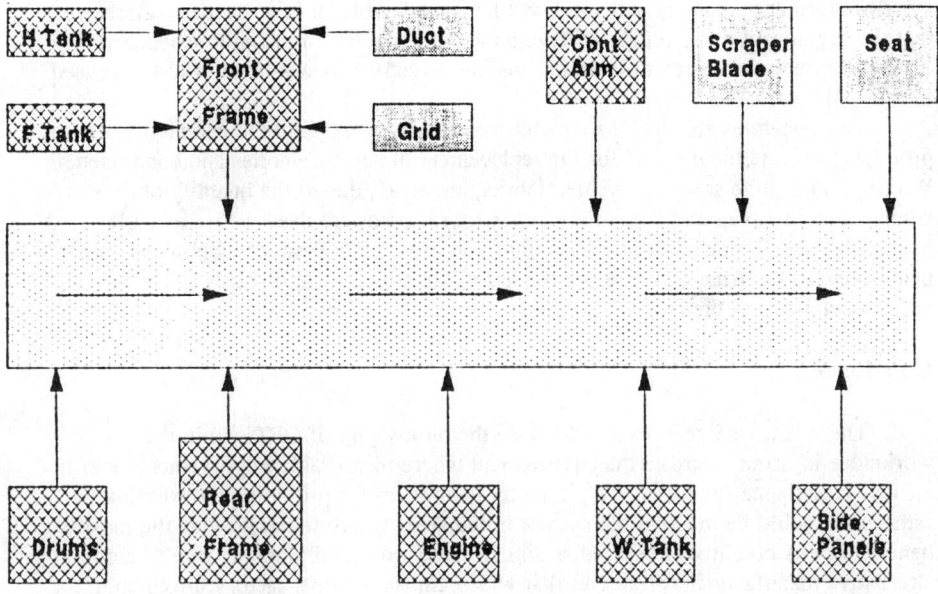

Figure 2. Flow of major sub-assemblies to line build.

OPT, creating the schedule.

The OPT software was originally purchased for the purpose of creating schedules for the bought-out parts, with the intention of later expanding this to include a shopfloor scheduling function. The purchasing software has now been fully developed, but in doing so many of the original procedures have been changed. One of the aspects of using this software that has now become apparent is the necessity to make use of all the data associated with the bill of materials, routing files, inventory information and the order requirements or build programme and download these from the MRP II system into the OPT software. This is an extremely time consuming routine, but is required prior to the scheduling run. Only after this has been completed can the purchasing staff begin to develop the 'best' schedule. The time involved means that scheduling by necessity has become a monthly occurrence.

Scheduling the shopfloor activity.

It is anticipated that the generation of schedules for the shopfloor will place an equal demand on the system in the preparation of data and the time taken. Additionally to maintain the flexibility required would mean that ideally these schedules should be created on a weekly basis, something that will, under the present arrangements be difficult to achieve. With labour recognised as a constraint, many of the manufacturing areas will require scheduling in order to achieve optimum throughput. This suggests that the amount of work involved in tracking shopfloor activity and in keeping the system information current and valid will be considerable. It is no surprise therefore that the organisation has sought to develop a simpler solution of 'pull' schedules, Kanbans based upon cards or the more visible inventory squares similar to those used in many JIT systems.

The system established throughout the factory relies upon the completion of products to create the demand for the replacement of sub-assemblies and components. Within the machine shop and welded fabrication areas, due to the quantity of components involved and the variety experienced, cards are used within a single kanban system. Throughout the assembly build area, with greater model focus, it has been found to be beneficial to adopt the use of squares instead of cards to control the pace of activity and the level of inventory generated.

Conclusion.

There is a very real need to advance the philosophy of OPT within the workplace in order to create the environment where measurable improvements will be achieved. Reliance should not be placed on the system for providing the solution but rather use should be made of the system in the areas where it can provide the most benefit, in this case it will be used as an extremely powerful planning tool to model alternative manufacturing strategies that will optimise the total factory throughput. Where a simpler but no less effective method of shop floor scheduling can be realised, which will compliment the use of OPT, this should be evaluated and utilised if it offers advantages. OPT like any other system should be developed and applied to maximum advantage for the organisation, even if this leads to a current redundancy of certain features.

References.

Booth, J. 1988,Leaving traditions behind: the case of York International, the challenge of JIT for small businesses, *IFS*.

Harrison, M.C. 1985, The concepts of optimised production technology OPT-the way forward, *BPICS Control, June-July*.

Goldratt, E.M. & Cox, J. 1984, The Goal-beating the competition, *(Aldershot: Gower)*.

Jacobs, R. 1984, OPT uncovered: many production and scheduling concepts can be applied with or without the software. *Industrial Engineering, October*.

Martin, S. 1993, OPT-the challenges identified for successful implementation, *National Conference for Manufacturing Research (NCMR9), Bath, 7th-9th September*.

THE REQUIREMENTS OF INTEGRATED PERFORMANCE MEASUREMENT SYSTEMS

D.F Kehoe and P. D. Patterson

Department of Industrial Studies
University of Liverpool
Liverpool, L69 3BX

This paper describes the framework established by a collaborative research programme for identifying the requirements for developing integrated performance measurement systems. Previous research has shown that the performance measurement systems employed within a manufacturing company are a key contributor to manufacturing integration. The dimensions to this integration include Structural, Informational and Behavioural perspectives. This paper outlines a framework in which performance measurement systems can be audited and more integrated systems developed. The second element of the paper discusses a complimentary research programme working in the area of supply chain performance measurement. The need for integration of measures between the various elements of the supply chain will be presented in the light of the above research.

Introduction

Manufacturing Systems researchers have seen a number of issues emerge during the 1990's which are critical pre-requisites of world-class performance. These have included:

- Manufacturing/Supply Chain Integration
- Business Process Re-Engineering
- Benchmarking
- Total Quality Management.

Manufacturing industry has enthusiastically embraced such philosophies only to find that they have failed to deliver the promised step-change in business performance.

In many areas of manufacturing systems development there exists a basic lack of framework for understanding both the techniques involved and the current status of the company. Most organisations are unable to fully understand the relationship between manufacturing initiatives and business performance due to the complexities of the dynamic nature of manufacturing systems; the large number of variables involved; the need to consider both cultural and technological issues.

Previous research has shown the importance of providing manufacturing systems engineers with:

-        a mechanism for understanding the current position of the company [Mann and Kehoe 1994]
-        decompositional methods for handling the increasing level of detail and complexity [Kehoe, Little and Lyons 1993(a)]
-        an effective performance measurement system to support the proposed manufacturing development [Kehoe, Little and Lyons 1993(b)]

The aim of this paper is to identify some of the manufacturing research issues associated with the analysis and design of integrated performance measurement systems and to identify suitable approaches.

The 'Dimensions' of Integrated Performance Measurement Systems

In developing a framework for evaluating performance measurement systems, a set of criteria needs to be established which reflect the overall aims of integration. Previous research [Kehoe et al 1993(b)] has identified three critical dimensions:

Structural   the extent to which the performance measures deployed through the organisation reflect the overall business objectives

Informational the ability of the organisation's information systems to provide accurate and appropriate data to support the provision and use of the performance measures

Behavioural   the effect on the behaviour of individuals and groups of the performance measures

In order to design integrated performance measurement systems analysis skills are required in each of these domains. Techniques such as Functional Decomposition [Kehoe et al 1993(b)] are required to systematically decompose strategic objectives down to increasing levels of detail whilst maintaining the integrity of the overall system. Methods are then required to evaluate the timing, accuracy and availability of the information required and finally tools are needed which can assess the human behaviourial factors such as culture, response and motivation.

Through having an effective methodology for the design of integrated performance measurement systems, the manufacturing manager can then utilise the considerable amount of research emanating from the accountancy profession on the linking of non-financial to financial measures [Nair 1990].

Audit Methods

To facilitate the design of improved performance measurement systems it must first be recognised that most manufacturing companies are in a brownfield rather than a greenfield situation and a transitional change model is implicit. This approach therefore requires both an audit method for analysing/evaluating the current system of performance measurement and also reference models which provide a mechanism for focusing improvements.

In order to provide a coherent method for auditing performance measurement systems an overall systemic framework is required. The viable systems model shown in Figure 1 illustrates the usefulness of the VSM approach [Beers 1979] in evaluating simultaneously each of the three dimensions of strategy (systems 5 and 3), information (systems 4, 3, and 2) and behaviour (systems 3 and 1).

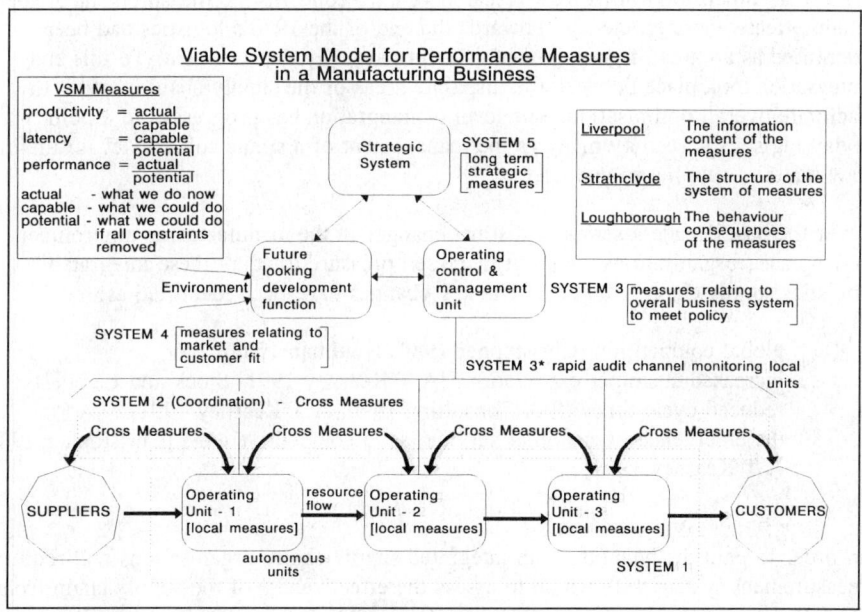

Figure 1 The VSM with 5 sub-systems each with inter linked measures

Reference Models

Having developed an effective analysis method, the output from this analysis (if correctly formatted) can be mapped onto appropriate reference models to enable both gap analysis and benchmarking.

Previous collaborative research has shown the important research role in providing manufacturing industry with reference models extracted from a range of businesses and business sectors. The usefulness of reference modelling has been illustrated using:

-        a configurations approach to identifying an organisation's behaviour patterns and culture
-        process maps which decompose to detail the expected information flows between specified business functions
-        GRAI grids to identify decisional inconsistencies between different operational levels within the organisation

Through the integration of the analysis method and the reference models an appropriate methodology for the design of effective performance measurement systems can be developed.

Supply Chain Integration

The final dimension to this joint research venture concerns the measurement of supply chain effectiveness/efficiency. Towards the end of the 1970's logistics had been identified as an area of great potential for productivity improvement. To this end integration took place between the disparate areas of the supply chain in order to facilitate overall optimisation. The level of integration has progressed to a point where today logistics is seen by many as the management of a single cohesive chain linking raw material supplier to end user.

Over the past decade a series of distinct changes in the manufacturing environment have placed organisations under an increased pressure to adopt these integrated logistics/supply chain structures. The key changes have been identified as:

-        global competition [Christopher 1988, Houlihan 1988]
-        increased customer expectations [A T Kearney 1991, Stock and Lambert 1987]
-        reduced cycle times [P A Consulting 1990, A T Kearney 1991]
-        the emergence of customer service as a competitive edge [Christopher 1988, A T Kearney 1991]

In order to gain the benefits of an integrated supply chain, organisations will require a measurement system with which to assess the effectiveness of the supply chain from a customer service perspective, whilst maintaining efficiencies from an overall business perspective.

In addition to the structural, informational and behavioural dimensions discussed earlier, management will therefore require a framework with which to assess the overall performance of the supply chain whilst understanding the elemental contribution of the functional activities. Just as the structural element of the research seeks to establish measurement system consistency from a strategic to operational level, the supply chain element seeks to maintain consistency throughout the chain. In doing so this element of the measurement system must also reconcile the customer measures of success with those internal business measures which are applicable from a management perspective.

In order to negate the functional boundaries which have hindered previous measurement system design [Skinner 1986, Spijkerman 1988] this research will examine the supply chain from a process perspective and employ functional decomposition techniques in order to reconcile the detail activities/measures.

Conclusions

- The lack of a comprehensive framework for assessing business performance has impaired the successful application of recent manufacturing initiatives.

- The critical dimensions of such an integrated framework have been identified as: Structural, Informational and Behavioural.

- The additional dimension of horizontal integration between functional elements must also be considered in order to promote supply chain integration.

- The framework should consist of an audit method for evaluating the current system of performance measurement and reference models which provide a mechanism for focusing improvement.

References

Beers, *The Heart of the Enterprise*, John Wiley & Sons, 1979
Christopher, Martin, "Global Logistics for World Markets", *IFS Executive Briefing in Logistics*, IFS, 1988
Houlihan, J., "Exploiting the International Supply Chain" ,*IFS Executive Briefing in Logistics*, IFS, 1988
Kearney, A.T., *Achieving Quality and Productivity in the Logistics Process*, CLM, 1991
Kehoe, D.F., Little, D. and Lyons, A.C.(a), "Strategic Planning For Information Systems Enhancement", *Journal of Integrated Manufacturing Systems*, Vol. 4 No. 2 1993, pp 29-36
Kehoe, D.F., Little, D. and Lyons, A.C.(b), "A Composite Methodology For Improved Manufacturing Systems Integration", *SERC ACME Conference Sheffield*, 1993, pp 125-134
Mann, R.S. and Kehoe, D.F., "The Quality Improvement Activities of Total Quality Management", *Quality World Technical Supplement, Journal of the Institute of Quality Assurance*, March 1994, pp 43-56
Nair, C.J.M. et al, "Do Financial and Non-financial Measures Have to Agree?", *Management Accountancy*, Feb 1990, pp 23-36
P A Consulting, *Manufacturing into the Late 1990s*, HMSO, 1990
Skinner, Wickham, "The Productivity Paradox", *Harvard Business Review*, No. 4 1986 Spijkerman, G., "The Control and Measurement of Logistics Performance", *IFS Executive Briefing in Logistics*, IFS, 1988
Stock, James R., Lambert, Douglas M., *Strategic Logistics Management*, Irwin, 1987

PERFORMANCE MEASURES IN SMALL MANUFACTURING ENTERPRISES : Are Firms Measuring What Matters?

C.J Addy, J. Pearce and J. Bennett

Plymouth Teaching Company Centre
University of Plymouth, Devon, PL4 8DE

This paper discusses the nature and extent of performance measures used within small manufacturing companies based on a number of case studies carried out within small manufacturing firms in the South West. It is suggested that characteristics particular to small firms such as ownership structure, limited resource and skills availability, management style and culture influence the set of performance measures used. The way in which small firms pursue competitive advantage is also discussed and the paper concludes by offering a framework for use in the identification of an appropriate set of performance measures.

Introduction

In 1993, CIMA published the findings of a study on *Performance Measurement In The Manufacturing Sector*. One of the key findings was that there appears not to be an optimal mix of both financial and non-financial performance indicators applicable to all manufacturers and that each organisation must find a balance of measures which it views as appropriate for its operational activities and which focuses the organisation on the development and maintenance of competitive advantage. In the process of producing annual accounts, an activity which all companies must carry out, the performance of the company for a particular accounting period is expressed in financial terms. This level of performance measurement, however, does not provide the detailed information about operational performance required by managers to monitor and control a companies activities on a day-to-day basis. Further, beyond the requirements of shareholders, this top level financial performance measurement does not provide a direct link between the companies activities and the extent to which the requirements of other stakeholders are being met.

Likewise, many of the other traditional performance measures used within manufacturing companies as identified in the CIMA study, fail to capture and represent the new production methods currently being employed in even the relatively small firms, for example, flexible manufacturing systems (FMS). The ability to collect, interpret and act upon information about those organisation's activities which create a competitive advantage is becoming increasingly crucial as markets shrink and competition increases, both locally and globally. The performance measures of the key activities which create and sustain advantage must be accurate, timely and, above all, relevant.

Competitive Advantage and Small Firms

Much of the seminal literature on competitive behaviour (Porter 1985, for example) focuses on large, corporate and very often U.S. organisations. Brytting (1990) notes that many of the traditional theories and models used to explain and prescribe organisational behaviour do not function well enough to capture the kind of "complex, capricious and turbulent events often found in small firms". There is now a considerable body of research (Bamberger 1989, Oakey 1984, Schroeder et al 1989, Meredith 1987, for example) which recognises that small firms compete in different ways to large organisations.

In the report published by The Institute of Electrical Engineers (1993), *Small and Medium Size Enterprises: A Recipe for Success*, several factors which influence the competitive position of small firms are identified. Key aspects include limited availability of finance, management and labour skills, and difficulty in acquiring and implementing new technology. Given these limiting factors, it is crucial that any attempt made to improve a small firms competitive position is understood in terms of impact across the whole organisation and does actually contribute to competitive advantage.

One common and fundamental characteristic of small firms is the centralisation of ownership and control. The owner/manager is often the key to the organisations success or failure and the wisdom he/she develops in the process of initiating and growing the organisation is often irreplaceable and is in itself a rich source of competitive advantage. Many of the decisions made by this key individual are instinctive, based on an in-depth knowledge and understanding of the particular market. Operational performance of the organisation is also frequently intuitively gauged rather than measured on a formal and precise basis.

Two Small Manufacturers in the South West

From a number of case studies carried out in small manufacturing firms in the South West, two particular organisations best illustrate some of the issues involved in performance measurement and small firms. Company A. manufactures and markets a range of standard and special purpose cutting tools. The company employs 18 staff and has an annual turnover of £700k , 80% of which is export. The level of manufacturing technology the company employs is relatively advanced including several CNC machine tools, electronic gauging and full CADCAM facilities. Their strategy is to build on their present success by reducing lead times even further and offering delivery of a typical £2,000 cutting tool, anywhere in the world, within 10 working days of receiving the customer specification.

Company B, offers a design and manufacture service for materials handling systems to the food industry, complimented by on-site installation and maintenance. The company has customers throughout the UK, although most are food processors in the South West where company B is estimated to hold 50% market share. The company employs 14 staff and current turnover is £600k p.a. Company B has grown substantially over the past few years and turnover is forecast to increase by 50% over the next 2 years. This growth is based on the strategy of delivering high quality service, responding rapidly to their customer's needs and new market penetration into the servicing of food processing equipment and own product development in the area of food handling equipment. Company B is in the process of updating much of its production equipment, and there is now a recognised need to update the manufacturing and maintenance systems and formalise procedures in order to avoid the risk that continued growth could bring a fall in quality and customer service.

Table 1. briefly outlines the types of performance measures used internally for companies A and B respectively, in order of importance to the organisation. The owner/manager of each company, however, was quick to point out that the measurement is informal in that it is on an ad hoc basis and not reported in any particular format or at pre-defined time intervals.

Table 1.

COMPANY A	COMPANY B
Lead Time	Cash Flow Projection
Schedule Adherence	Creditor Days
Capacity Utilisation	Debtor Days
Supplier Lead Time	On-time Delivery
Supplier Product Quality	Customer Complaints
% Rework/reject Level	Orders on Hand
Direct Labour	Price
Indirect Labour	Product Quality
Inventory	Manufacturing Lead Time
% Yield on Production	% Rework/reject Level
Set-up Time	Schedule Adherence

Both companies A and B recognise the competitive actions in which they must engage in order to build and maintain their competitive advantage. As with many small firms in the manufacturing sector, both organisations perceive the closeness of relationship with the customer to be of fundamental importance and much effort is placed in `courting' the customer, particularly by the firms owner/manager. Hence, by developing a close working relationship (often reinforced by social contact), the owner/manager has direct and continual feedback from the customer. The owner/manager can, therefore, continually monitor performance in terms of customer satisfaction, albeit on an informal basis, and has direct information on the aspects of those firm's operations which are of key importance to the customer. Eccles (1991) notes the importance of efforts to generate measures of customer satisfaction, commenting that "What quality was for the 1980's, customer satisfaction will be for the 1990's". It would appear, therefore, that the inherent nature of the small firm naturally facilitates the measurement of customer satisfaction. This natural process of informal performance measurement automatically reinforces competitive advantage.

The performance measures identified in Table 1. reflect those aspects of each company's operations which the owner/manager feels to be of greatest importance both in terms of customer requirements and maintaining their own personal control and influence over the organisation. Much of what happens is by instinct or accident. The need is evident for a single tool for identifying how changes can help small firms create competitive advantage.

Linking Performance Measures To Competitive Advantage

One of the characteristics which differentiates small firms from larger companies is the power of a whole range of stakeholders. In an attempt to analyse how a small firm copes with the pressures and constraints imposed by its powerful stakeholders, the following framework has been generated.

The operations of a business can be considered as a series of trading activities with each of its stakeholders. Stakeholders can be broadly categorised into Traders and Modifiers;

Traders include - customers, suppliers, employees, banks, shareholders etc.
Modifiers include - competitors, Government, pressure groups etc. (those stakeholders which seek to influence and change the nature of the trading relationship between the business and its Traders).

Traders will trade with the organisation which provides the best mix of product, price and service, requiring:

-Value: perceived benefits gained for a given price
-Responsiveness: closeness of offered product and/or service to the traders requirements
-Security: likelihood of the transaction being considered successful in the long and short term

The organisation can, therefore, gain competitive advantage over its competitors if;
- it offers more benefits at a lower cost
- it has greater flexibility to respond to changing stakeholder demands
- it minimises the risk of failure by having better control over its operations

The action of Modifiers tends to negate the existing forms of competitive advantage. The organisation must, therefore, continually find new ways of creating advantage (or at worst preventing disadvantage) through innovation. In anticipation of Modifier actions, the organisation must also seek to secure barriers around its existing competitive advantage by attempting to control its external environment. To cope with the actions of Modifiers, the organisation must:

- continually strive to improve the benefits offered for a given cost
- have the flexibility to offer new packages of benefits
- have the flexibility to utilise/trade with alternative sources, channels and stakeholders
- create and maintain control over aspects of the business system by raising barriers to entry, developing alliances with stakeholders, patents, long-term contracts, etc.

Considering both the requirements of the Traders and the actions of the Modifiers as outlined above, the competitive actions the organisation must engage in are; to maximise value, quality and flexibility, minimise cost, and exercise control measures over the external environment.

Table 2.

COMPANY	COSTS	BENEFITS	QUALITY	TRADING FLEXIBILITY	EXTERNAL CONTROL
A	Reduce design time from 1 day to 10 mins.	Lead times reduced to 50% of competitor's. Performance data direct to customer. Request and specify on spot.	Standardisation and removal of paper based system. Direct link from customer requirement to machining.	Customer specifies prod characteristics.	Create and control channel by installing direct link in large customers and near distributors
B	More accurate and timely monitoring and reporting of costs.	Progress reporting to customer.	Early warning of non-conformance. More professional quality of trading.		

Table 2 above identifies the ways in which the introduction of new technology has contributed to improved performance in both companies A and B by focusing on the competitive actions identified previously. Company A developed a system for right-first-time production, linking CAD facilities installed in the company's largest customer sites to an advanced turn-mill centre through computer-generated NC programs.

Company B has introduced project management software to assist the process of forming and monitoring customer proposals. The software has also enabled to company to formalise organisational systems, reducing the risk of failing quality and customer service as the organisation continues to grow.

References

Bamberger, I. Developing Competitive Advantage in Small and Medium Size Firms, *Long Range Planning,* **vol.22**, No.5, 1989, pp.80-88,

Brytting, T. Spontaneity and Systematic Planning in Small Firms - A Grounded Theory Approach, *International Small Business Journal 9,* **1**, 1990, pp.45-63.

Eccles, R.G. The Performance Measurement Manifesto, *Harvard Business Review,* Jan-Feb, 1991, pp.131-137.

Levy, J. (ed), *Small and Medium Sized Manufacturing Enterprises; A Recipe for Success,* The Institute of Electrical Engineers, July 1993.

Meredith, J. The Strategic Advantages of New Manufacturing Technologies for Small Firms, *Strategic Management Journal,* **vol 8**, 1987, pp.249-258.

Oakey, R. *High Technology Small Firms,* Frances Pinter Publishers, London, 1984.

Porter, M.E. *Competitive Advantage: Creating and Sustaining Superior Performance,* Free Press, New York, 1985.

Schroeder, D.M, Gopinath, C., Congden, S.W. New Technology and the Small Manufacturer: Panacea or Plague?, *Journal of Small Business Management,* **vol.27**, No.4, Oct.1989, pp.45-60.

THE RELATIONSHIP BETWEEN MANUFACTURING AND HUMAN RESOURCES IN NEW PRODUCTION SYSTEMS

Peter Carr, Gwyn Groves, John Kay

School of Industrial and Manufacturing Science
Cranfield University, Cranfield, Bedford, MK43 0AL

Most implementations of manufacturing improvement concentrate on either manufacturing systems or human resources elements. Many initiatives to develop manufacturing systems, such as 'Just in Time', focus mainly on the changes to manufacturing techniques which are required, ignoring the people requirements of these systems. Meanwhile, human resources initiatives, such as 'Empowerment', seek to maximise the contribution of people to manufacturing, without taking into account manufacturing systems developments. The inter-relationship of these factors is a critical factor in the success which is achieved in New Production Systems. This paper examines the nature of the relationship between these factors and outlines a method for charting this.

Introduction

Excuses for the failure to properly involve people in new manufacturing systems in western industry proliferate both industry and the academic community. Few dispute the importance of this involvement, yet it remains the main factor in the failure to realise the benefits of new methods of production. Both new manufacturing system 'engineers' and human resources 'professionals' are able to argue for manufacturing systems developments and people management activities which will result in improved company performance, but there has been a failure to integrate implementation and operational activities.

The new manufacturing systems which are referred to in this paper are those stemming from Japan and known as Just In Time or Lean Production and featuring low stocks, short leadtimes, reduced set up times, flexibility of labour, improved maintenance and, when combined with Total Quality Management techniques, improved product and process reliability. In order to operate successfully, these systems require a new role for employees. There are four reasons for this.

First, new manufacturing systems rely on a rapid response to fluctuations in demand on processes and products. Reliance on managerial action to activate this response is likely to result in sub-optimal performance. Second, new manufacturing systems are more vulnerable to immediate production problems, due to the lower levels of stock which are held. Employees must act to solve these problems as they arise. Again, reliance on managerial action to activate this response is likely to result in sub-optimal performance. Third, these systems are far more vulnerable to disruption by employees who have a grievance to express. While strike action is relatively rare in the U.K., other forms of employee expression of dissatisfaction (e.g. go - slows, overtime bans, etc.) are also important. Fourth, the continual improvement of new manufacturing systems and the products they make will aid implementation and further development. All of these factors require a radical improvement of employee involvement.

Human resource management activities are intended to develop appropriate employee action to support the business objectives. Various approaches are taken to this, with a wide ranging debate in the literature as to the normative effects. However, the Price Waterhouse Cranfield Survey (1991) of European human resource management practices suggests that there is little integration with manufacturing systems developments, either at strategy level or in practice.

In this paper the suggested ideal position is referred to as New Production Systems - the full integration of new manufacturing systems and human resources activity. The charting method which is outlined in this paper, allows the relationship between new manufacturing systems and human resources activities to be studied.

The Charting Method

The new charting method is based on a two axis chart. The Y axis represents new manufacturing systems activity within the organisation being studied. The X axis represents progress with the implementation of human resources activities. The bottom left corner of the chart represents a traditional manufacturing environment, while the top right corner represents a New Production Systems environment (see Figure 1).

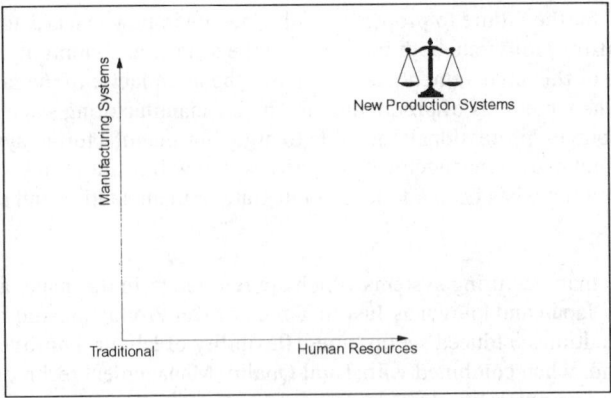

Figure 1. The blank chart.

Identification of the positions which are plotted on the chart is possible in a variety of ways. The authors have used study teams, surveying to identify company perceptions of change and complemented this with study of company performance data. Positions have also been identified as part of discussion with company management teams, a valuable activity to assist planning of company development.

Relative Positions

The most basic application of the chart is the depiction of the relative position of individual companies or parts of companies. Figure 2 shows 2 companies, A and B. Company A has focused predominately on new manufacturing systems changes, with little attention to people aspects of improvement. They are therefore plotted in the top left hand corner of the chart. Company B has concentrated on human resources changes and engaged in little manufacturing systems change.

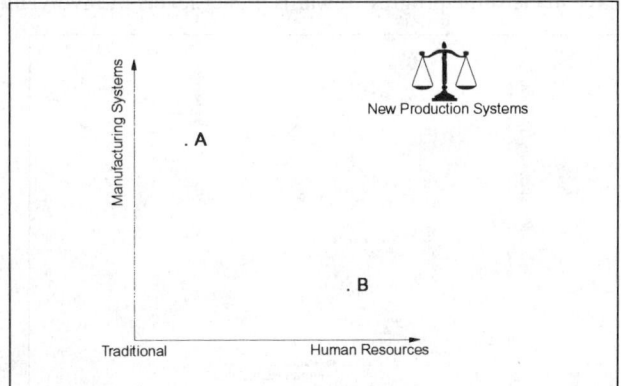

Figure 2. The relative position of each company can be shown.

The Area Under Each Point

The area under each point on the chart can be used to show the extent of change which has been undertaken in each company. This is shown in Figure 3. Both companies

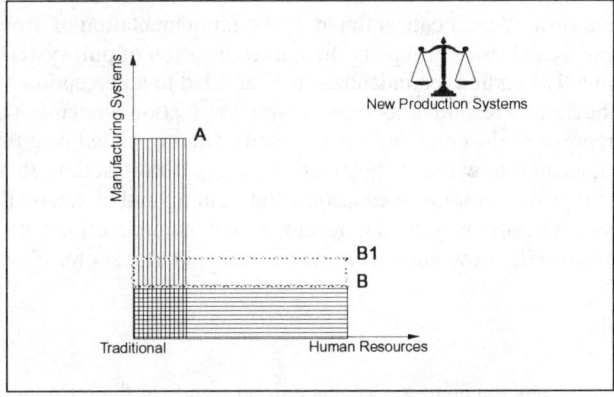

Figure 3. The area represents the progress made.

have made change towards New Production Systems, with a similar area appearing under each point. While extensive empirical research would be necessary to establish a clear relationship between area and benefits gained, it is suggested that the area under each point does provide an indication of the benefits which might be expected by the company. Figure 3 also shows the relative effect of change from a company's current position. If the area under each point represents progress made, an increase in manufacturing systems activity by company B is likely to have more improvement benefits than in company A.

Charting Past Progress

The chart can also be used to show the progress over time. The effect of individual changes can be shown, or the 'net' position for particular time periods. This is shown in Figure 4. At the beginning of the time period being represented, the company was thought to have made some progress towards higher levels of involvement.

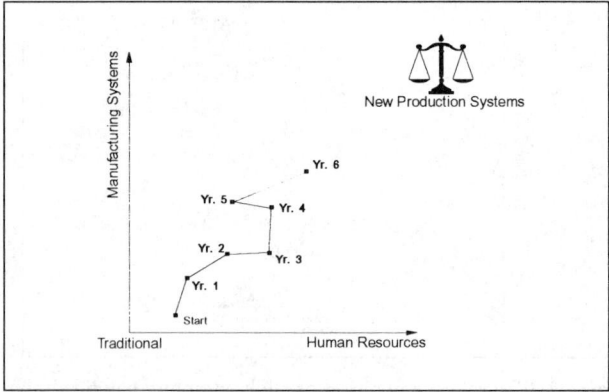

Figure 4. The path taken can be traced.

At the end of Year 1 of its improvement programme more emphasis had been put on manufacturing systems (e.g. the introduction of shopfloor data capture to improve production scheduling). Year 2 featured the introduction of teambriefing and some effort to simplify process flow. Year 3 concentrated on the implementation of problem solving groups, while Year 4 used these groups in the implementation of pull systems of process flow. A round of badly handled redundancies in Year 5 led to a perception of a deterioration in the human resources aspects of New Production Systems. The dotted line in Figure 4 represents the company's future plans. Having undertaken this study of their activities, they could now more effectively decide on future action. In Year 6 they plan to introduce improved employee education and training and to extend the use of Kanbans. The relative positions of human resources and manufacturing systems activities on the chart will allow judgements to be made about resource allocation.

Current Trends

It is possible to use the chart to suggest current trends in the introduction of New Production Systems in Western industry. It is suggested that 2 trends can be discerned.

The first trend may be described as a 'Directed' approach. In this trend there has been a concentration on new manufacturing systems. This is likely to have been led by manufacturing management and it has neglected the human resource aspects of change. The second trend may be described as an 'Involved' approach. Here, human resources activity has dominated. It is suggested that current industrial practices result in companies tending towards one of these extremes. Rehder (1994) provides evidence of this in his discussion of the Saturn (General Motors) and Uddevalla (Volvo) experiments.

Compatibility

The critical factor in the implementation of New Production Systems is the effective combination of new manufacturing systems and human resources activities. Figure 5 illustrates the problem. While sub-optimal improvement paths are common, companies which appear in the shaded centre of the chart are uncommon, if not non-existent. Issues arise as to the appropriate combination of activity which would allow a company to move into this integrated position. It is clear that some combinations are unlikely to be successful. For example, individual performance appraisal and reward systems are unlikely to be compatible with new manufacturing systems which require improved group working.

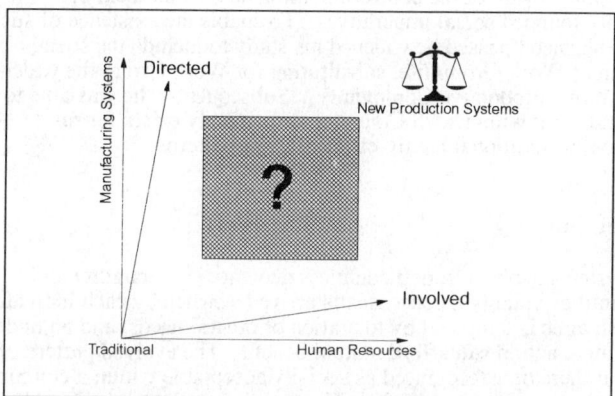

Figure 5. Integration is the critical issue.

Similarly, the charting method illustrates the weakness of prescriptive approaches to N.P.S. Different companies will require different activities, depending on the point at which they appear on the chart. The charting method should facilitate an improved approach to the implementation of effective, integrated N.P.S.

References

The Price Waterhouse Cranfield Project, 1991, Report
Rehder, R.R. 1994, Saturn, Uddevalla and the Japanese Lean Systems: Paradoxical Prototypes For The Twenty-first Century, The International Journal of Human Resource Management **5:1**, February 1994, 1-31.

WORKPOINT MANAGEMENT

R. Pringle

9 Seafield Court, Ardrossan, KA22 8NS

During the 1970's while studying the operation of Work Groups the writer became aware of the controlling influence within them of historically founded social imperatives. To enable the existence of such forces to be encompassed he widened his study to include the social dimension of Work Group life, substituting for Work Group the wider-reaching term 'Productive Community'. Subsequently, he was able to demonstrate that within such Communities capacity exists for re-acquisition of traditional highly-efficient work patterns.

1.   Introduction

Social systems evolve through countless generations of conflict and confrontation until eventually a consensus is arrived at whereby each individual's pursuit of selfish aims is tempered by toleration of others' needs, and an understanding of how co-operative action can afford mutual benefit. The evolved pattern of behaviour is that thereafter recognised as socially acceptable within a community and enforced by religious and civil authority. This process applied equally well at the workplace, and the system of operation in the craft shop prior to the Industrial Revolution by virtue of its genesis met completely the human needs of its members.

2.   Differentiating Between Work And Activity

To set the scene for modern developments in work management we can do no better than quote Dahrendorf (1982) : 'Work is crystallised human action, hardened into highly defined posts in factories, offices, organisations. ...As a general medium of life such work is a modern invention; for most the lines between living and working were blurred in pre-industrial circumstances. To understand life in those times a distinction is in order which is as old as it is important: the distinction between work and activity. Work is human action which is heteronomous, imposed by external needs, be they needs of survival or of power. Activity, on the other hand, is human action which is freely chosen, which offers opportunities for self expression, which carries its satisfaction within itself, which is autonomous.'

3. The Pre-Industrial Revolution Activity Culture

Within this culture, evolved to stability over numberless generations, we find some familiar, some unfamiliar elements.

a) Pride in possession of, and fulfilment through the exercise of individual work skills.

b) An easy familiarity with the 'minimum energy for maximum achievement' self-controlled work pace.

c) A sound grasp of what comprises fair return for effort.

But these are individual attributes; those relating to the Work Group as a whole, the archetypal Productive Community in the Activity Culture, are of even greater significance.

d) A sense of place in the group hierarchy related to the level of individual work skill possessed.

e) A strong drive to ensure 'fair-shares' of work and reward in the group.

f) Recognition of the position at the top of the group hierarchy of the master craftsman - the Master.

4. High-Productivity Management - Back To Basics

Life for the worker in contemporary industry is full of performance-inhibiting stress, while through a return to the historically-evolved pre-industrial revolution work management system access can be regained to a great fund of repressed productive energy.

In modern management terms what is needed is:

a) Task Groups of 3-15 workers under the leadership of a Master (cf. the German 'Meister') managing his team as an autonomous 'Business Area'.

b) Within each Meister's area installation of a work management system designed to meet and fulfil workers energy releasing needs, viz:
c)

Hygiene Needs

1. SECURITY of place in the undifferentiated social/work-life of the community.

2. EQUITY of return for contribution to the community effort.

Motivational Needs

3. EXPOSURE to the whole range of community problems.

4. FREEDOM OF ACTION i.e. self-pacing control over application of one's own skills and effort.

5.   Non-Conformist Management

A useful way of viewing Tayloristic 'Scientific Management' is to recognise that by enforcing a purely instrumental outlook towards work upon manual workers, by imposing a new management system that did not conform with that that was their inheritance a situation of great stress was created. However, while heavily suppressed workers' drive to satisfy their basic needs, to obtain some intrinsic satisfaction from work, persisted through the imposition of ever tighter controls in an attempt to contain all activity within the confines of each company's formal management system. Under sustained attack, but never wholly suppressed, in each company an all-powerful 'informal' system of control facilitating a degree of satisfaction of workers' basic needs emerged to regulate workplace activity.

Effectively, therefore, what we have been witnessing as pragmatic (often symptom-chasing) work improvement ideas have been introduced in recent years is a move towards 'conformist' management, i.e. a style that will embrace the principles and utilise the great productive capacity restrained by ill-considered 'non-conformism'. In the process what has not been recognised is that only temporary advantage can be obtained from new systems which only fulfil one or more basic needs. Only when a system can satisfy all four needs can the long process of recovery, orthogenesis, be set in motion. When even one of the four is missing, or is subsequently denied then collapse of the project and anguish among the participants, workers and managers alike, will ensue.

Such is the picture presented by many pioneering projects, as the following case-histories will illustrate. In the first, an example will be given to demonstrate how, after enforced conversion to instrumentalism if equity is subsequently denied then the submerged informal control system will surface to crush a company's 'formal' system, an event as common as it is unappreciated.

6.   Case Histories

6.1   Bonus Working In Engineering

In 1950 the major portion of our engineering workforce were working at Standard rate and earning a bonus of one third extra onto their weekly pay. By 1980 it was common to find shops where much under Standard was the norm and where 'stint' working (deliberately restricted performance) could be observed. The source of this problem dates from a post-1950 policy of many employers in the industry to 'claw-back' inflation-driven rises in basic pay by freezing the datum from which bonus earned was calculated at each successive pay round. This was a particularly ill-founded policy, for a reason overlooked or simply not understood at the time.

At the turn of the century when Bedaux established his performance rating scale he adopted as datum the tireless working pace of the Activity Culture. Terming this NORMAL he then extended his scale one third upwards to a point he termed STANDARD i.e. the one third more intensive pace that factory managers sought from their workforce. Many years later, when establishing a new rating scale, the British Standards Institute, observing that STANDARD was then the normal pace in factories saw the NORMAL point on the scale as an anomaly and dropped it. Thereafter it became universally accepted that STANDARD was the datum upon which payment systems could be based, and companies' formal control systems were designed about this 'fact'.

This assumption was, however, entirely false since the traditional NORMAL, the tireless 'minimum energy for maximum achievement' pace, the pace with the solid biological foundation, remained the true pivot, and each company's hidden but all-powerful informal control system came into operation to maintain equity by enforcing compensatory rises in bonus percentage. When, eventually, figures of near 100% bonus were reached and exceeded, and managements foolishly capped payments, then the equity-maintaining mechanism brought about 'stint' working.

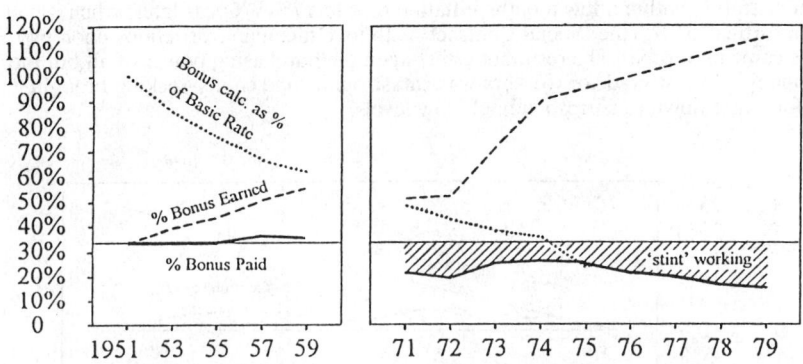

Graph 1 : Equity maintenance and bonus-capped 'stint' working

This equity-maintenance mechanism, being a universal feature of economic life, finds national and international expression, driving and maintaining a 'black economy' in most countries. In Britain each 'pay pause' during a time of inflation is, because of this mechanism, simply a direct attack upon the labour productivity of those affected.

6.2   Job Enrichment

Publication of the Two-Factor theory (Herzberg, 1966) sensitised progressive managements to the value of widening the scope of their manual employees work. In Britain during the late 1960's a leading company suffering poor labour efficiency following a descent to ineffectiveness of its bonus working system (in the way described above) implemented a far-reaching programme of manual workers' Job Enrichment to replace it.

Joint teams of workers and managers analysed all the various jobs, and re-wrote job descriptions to enrich each worker's job by including elements of self-management. The subsequent presentation of new job descriptions to a Job Assessment panel, for regrading (and therefore higher payment), was made a very formal affair before witnesses. Rapidly acquiring the characteristics of a 'coming of age' ceremony the process generated a high level of expectation and commitment to success among all manual staff.

When within only a few years these expectations came to nought, and instead of recovering labour productivity fell further, the company's managers were disappointed but able to move on to another new development. Among the workers, however, the effect was very different. Having had their expectations of escape from rigorous control raised so high and then dashed gave rise to scenes of greater anguish than ever previously witnessed in industry by the writer. While the sources of failure were hard to understand by managers (and impossible by angry and acutely distressed workers) within the context of this work they can be surfaced for examination.

To recover the situation created by the old 'drifted' bonus scheme a considerable across-the-board pay rise had been awarded. Following implementation of the new flat-rate payment system it was decided some 'draw-back' could be made during the annual pay round. Thus, although the lowest grades of manual worker received in full the proper return for the 3% annual increase in labour productivity the new system targeted (and at first produced) the higher grades did not.

Then, after a few years the (Heath) government fell, and under the new administration within a few months inflation rose to 17%. A year later, when it had risen further to 26% the 'Social Contract' with its Draconian restrictions upon pay rises came into force. The resultant effect upon the purchasing power of all but the company's lowest grade of workers was catastrophic, and equity-seeking brought performance down to correspondingly low levels.

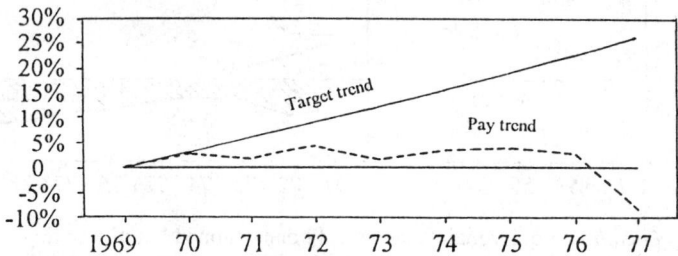

Graph II : Effect on Senior Grade's pay of 'draw-back'
and Social Contract pay restraint

More subtle but more immediately damaging, however, had been the emergence from within the management and technical staff of covert resistance to the process of orthogenic recovery at the workplace. Sharing the 'Officers and Troops' view of society of the educated and cultured minority most companies' Technical and Managerial component find self-regulating Productive Communities at the workplace virtually impossible to accept.

7.   Summary

It was in Britain's traditional craft shops that the Industrial Revolution was born and in its early manufactories that it flowered. If we would but recognise the old wisdom and go back to the basic work management of the time then high productivity performance can be regained.

References

Dahrendorf, R. 1982, *'On Britain'* (BBC, London) **181-182**
Herzberg, F. 1966, *'Work and the Nature of Man'* (Staples Press, London) **71-91**
Pringle, R. 1987, *'New Concepts in High-Productivity Management'* Unpublished M.Phil thesis, University of Strathclyde, U.K.

MODEL COLLAPSE 2: THE CULTURAL IMPERATIVE

Brian Trought

*Management Research International
59 Belton Lane, Great Gonerby, Grantham,
Lincolnshire, NG31 8NA*

The central point which is being made in this paper is that models or part-models of manufacturing systems tend to collapse, not so much because of inherent relationships in the models, but because of factors and relationships in the cultures of manufacturing companies, in which the models are embedded. Over a period of time few models escape this collapse. A common danger which hastens model disintegration is the destructive, incompatibility between the functioning of a model and the functioning of human behaviour in a company's culture. It is argued here that it is necessary for a specific company culture to be configured appropriately for any models of systems operation to thrive in that company. In effect, this means that a model and a culture need to be compatible and each designed with the other in mind.

The research base

The data on which this paper is based has been collected by observation and recording of the work activities of a sample of manufacturing managers in U.K. companies. Additionally, interactive process analysis (Bales 1951), a critical incident technique (Flanagan 1954) and statistical significance testing, have been employed. The work has been continuous for nineteen years and is still proceeding. About thirty managers comprise the sample. It has become apparent through observations and statistical analysis that control of a systems model after implementation within a company, is extremely elusive. In complementary fashion, research suggests that a company culture in which a model is embedded, is of primary importance to a model's success. Several theoretical aspects of the research base for this paper were addressed in Trought (1993b) and therefore the present paper can be considered as a continuation of Trought (1993b).

Models of manufacturing systems

Engineers appear to have a penchant for systems models, be they manual or computer operated. Still to be seen, and in use in many small companies, are wall-mounted Gantt charts - preogative of order movement vested in the foreman. Alternatively, computer operated models, such as MRP 2, need no comment as to their widespread use. Computer simulations of, for example, the management of projects are likewise widespread. Arguably, we can take the idea of a model a stage further and apply them to what perhaps are better described as organising philosophies or principles. Currently used examples of these philosophies or principles in manufacturing systems are Just-in-time product-flow, concurrent engineering facilities and cell working. The underlying theme of most of these model applications, particularly if they stemmed from academia, appears to be the use of a hard systems methodology (HSM). This methodology implies:

'....that systems are goal-seeking, operate by rules of cause and effect and that systems data and information are accurate.' (Trought 1993b, pg.181)

It is not difficult to demonstrate empirically, that in practice, hard system models do not deliver these features. The detailed arguments which support this statement were set out in Trought (1993b, pg.182). Although it is argued in this paper that the main causes of model collapse are cultural and behavioural, one should not lose sight of the fact that some models may also possess inherent, structural factors which tend to generate collapse. For example, Bowden et.al. (1991) note that:

'MRP uses estimates of lead times which by definition are not accurate.' (pg.251)

The behavioural environment

Uncertainty

The uncertainty which pervades manufacturing systems and their environments in the U.K. has a direct impact on the implementation and control of systems models. It makes implementation and control much more difficult if not impossible. This uncertainty is largely human based and appears to be a function of weak work disciplines combined with, what Hofstede (1980) calls, the '....tolerance for ambiguity.' of U.K. managers and employees. More recently Hofstede (1991) confirmed the cultural tolerance of people in the U.K. for a high level of uncertainty, by demonstrating the comparatively low score achieved by the U.K. sample on his Uncertainty Avoidance Index. Uncertainty avoidance is defined by Hofstede (1991) as:

'....the extent to which the members of a culture feel threatened by uncertain or unknown situations.' (pg.113)

He goes on to explain that this feeling is:

'....expressed through nervous stress and in a need for predictability.' (pg.113)

The inference is that, the behavioural characteristics which allows tolerance for uncertainty in our systems, does not provide the best incentive to combat and reduce this uncertainty. Hence likewise, imperfectly working models of systems can also be tolerated. Observational data records a whole range of other features which stimulate manufacturing uncertainty, or, in the words of Tom Peters, writing in the Independent on Sunday, 13 March 1994: 'the mess is the message.'

Managerial lack of control

One such feature was raised in an early study by Carlson (1951) who drew attention to the reactive nature of managerial activities. He said:

'I always thought of a chief executive as a conductor of an orchestra, standing aloof on his platform. Now I am in some respects inclined to see him as a puppet in a puppet-show with hundreds of people pulling the strings and forcing him to act in one way or another.' (pg.52)

Since 1951 this view of managerial activity has been reinforced many times by researchers. Managerial behaviour and activities are mainly reactive and rarely proactive. This adds to the uncertainty. Examples of uncertain features which cause managerial reacton are: incorrect lead times, cost data and stock figures; uncertain quality schedules, customer requirements and reasons for scrap, (Trought 1988 and 1993a). Compounding the lack of control which managers experience is the incidence of fraud on, and theft from, companies, (Trought 1993a). It is endemic and substantial and in some cases it rises to the status of an acceptable perk for the job. However, it does make the job of managing much harder than it need be, and more uncertain.

Non-rational decision making

That decision making in organizations does not have a rational basis is well supported by research, examples of which are (Nutt 1984) and (Trought 1990). Furthermore, Trought (1990) using a statistical significance test, showed that routine and non-routine manufacturing decisions were made in a similar, non-rational manner. This non-rationality often conflicts with the cause and effect rationality of models in action. This can be shown to lead to managers and others, in companies, becoming uncertain as to likely outcomes of decisions which have been made. In fact, unsuspected outcomes of decisions taken are not uncommon.

Work overlap

The uncertain working environment of people in manufacturing systems leads directly to the 'firefighting' style of most managers in these systems, and also to the extensive work overlap between managers and non-managers, (Trought 1991). Working in these systems is a very pragmatic process. Trought (1991) showed in one research project that 59% of discrete managerial activities were overlapping with, and similar to, activities performed by other system workers. For example, managers drove fork-lift trucks, repaired machinery, transported components and progress chased urgent orders. These were all jobs which other workers were also employed to do. Why? The short answer given by many managers is that they do what needs to be done. They 'firefight' on very short timescales to keep the product flowing and so highlight the uncertainty, but at the same time try to minimise it.

Empowerment

Empowerment of workers by upper management has become a flavour of the year. However, in the hands of U.K. management it perhaps more frequently leads to delegated work, so that layers of supervision, inspection and middle management can be dispensed with - a cost cutting exercise. The original intention of empowering workers was that authority, responsibility and decision making were also devolved with the work. Workers, it was suggested, were the best people to organize their own work area and product-flow. Rarely has this degree of delegation happened in the U.K. Worker participation is not a popular concept in the manufacturing environment, where, a manager's right to manage is an oft stated, self-evident managerial criteria. Do workers want to be empowered? is a pertinent question to ask. Who benefits from empowerment? is another important question. Finally, have workers often received an increase in pay to compensate for the 'enriched' jobs? In the way that work has been re-distributed, mostly downwards in a still retained hierarchy, leaves many workers with little interest in empowerment. This probably leads, in many cases, to insufficient labour to get the managerial work done in manufacturing systems - a recipe for more uncertainty. It does not have to be like this, but to change the work culture radically in a company can be a massive undertaking. Just how massive is recorded by Semler (1993) in his fascinating, blow by blow account of changing the culture of one company - the company which he owns in Brazil. To the present time, he and many others, have devoted some twelve years to the changes and estimate that it is about 30% finished. Whether empowerment leads to greater or less uncertainty, probably, also depends to some extent on the retention or elimination of class barriers in our companies. It is this feature, class barriers, which completes my review of what observational studies suggest is the behavioural environment for U.K. manufacturing company cultures.

Class barriers

Stewart (1991) drew attention to:

'....the persistence of class divisions in British society, and of a low degree of trust between management and other employees.' (pg.203)

Similarly, Trought (1992) argued that the class barrier:

'....goes to the heart of the divide in U.K. manufacturing systems. It has a cultural base. Whilst upper management seek to maintain this difference - this status quo - in the U.K., the fully productive potential of cooperative, fear-free manufacturing systems is unlikely to be realised. There is a penalty to be paid for working out our culture. (pg.349)

Surely, these class divisions prevent the mobilisation of the whole workforce in an objective-orientated direction? This then is another facet of the empowerment failure and a compounding of behavioural uncertainty in which systems models are embedded.

The uneasy fit between model and company culture

In Trought (1993b) it was argued that:

'....the crux of why a HSM is usually a poor analytical and decision making tool, lies broadly in the fact that it is individual people both inside and outside the system who act, not the system.' (pg.182)

So, although a model may not appear to contain inherent factors which speed collapse, the problematic surrounding uncertainties, endemic in current company cultures, do not provide an adequate basis for the model to function as intended. People acting out their perceptions of situations, invariably nullify the work of system designers. In the U.K., humans in and around manufacturing systems, and that includes managers, rarely appear to have the joint ability and commitment to allow or make models function satisfactorily within the prevailing cultures.

References

Bowden, R. Duggan, J. and Browne, J. 1991, The Design and Implementation of a Factory Co-ordination System. In Pridham, M. and O'Brien, C. (eds.), *Production Research; Approaching the 21st. Century. Selected papers from those presented at the Tenth International Conference on Production Research,* held at the University of Nottingham, U.K., August 1989. (Taylor and Francis, London)

Carlson, S. 1951, *Executive Behaviour: A Study of the Work Load and Working Methods of Managing Directors.* Strombergs, Stockholm, Sweden.

Flanagan, J.C. 1954, The Critical Incident Technique, *Psychological Bulletin,* **51,** No.4, 327-358.

Hofstede, G. 1980, *Cultural Consequences: International Differences in Work Related Values,* (Sage, Beverly Hills, U.S.A.)

Hofstede, G. 1991, *Cultures and Organizations: software of the mind,* (McGraw-Hill,

Book Co, London)

Nutt, P.C. 1984, Types of Organizational Decision Processes, *Administrative Science Quarterly,* 2 9, September, Cornell University, 414-450.

Semler, R. 1993, *Maverick!,* (Century Publishers, an imprint of Random House Ltd. London)

Stewart, R. 1991, *Managing Today and Tomorrow,* (Macmillan Academic and Professional Ltd. Basingstoke)

Trought, B. 1988, Controlling Factory 2000: the managerial perspective, *Proceedings of the International Conference on Factory 2000: integrating information and material flow,* Churchill College, Cambridge, August-September, (Institution of Electronic and Radio Engineers, London)

Trought, B. 1990, Manufacturing Decision-Making: Irrationality Rules - OK! In Carrie, A. and Simpson, I. (eds.), *Advances in Manufacturing Technology 5; Proceedings of the Sixth National Conference on Production Research,* (Strathclyde University, Glasgow), September.

Trought, B. 1991, Actual Managerial Jobs in and Around Advanced Manufacturing Systems, In Hollier, R.H, Boaden, R.J. and New, S.J. (eds.), *International Operations: Crossing Borders in Manufacturing and Service, Proceedings of the Seventh Annual Conference of the Operations Management Association (U.K.),* UMIST, U.K. (Elsevier Science Publishers B.V. Amsterdam, The Netherlands), June.

Trought, B. 1992, Management by Fear, *Proceedings of the Eighth National Conference for Manufacturing Research,* (University of Central England in Birmingham, U.K.), September.

Trought, B. 1993a, Total Manufacturing Systems: the operational reality, In Kochar, A.K. (ed.), *Proceedings of the Thirtieth International Matador Conference,* UMIST, U.K. (The Macmillan Press Ltd. Basingstoke), March-April.

Trought, B. 1993b, Model Collapse: the people effect, In Bramley, A. and Mileham, A. (eds.), *Proceedings of the Nineth National Conference on Manufacturing Research,* (University of Bath), September.

AN SME'S APPROACH TO HARNESSING THE WORKFORCE

K Wheeler* and H L Cather**

*Chairman/Managing Director (Retired)
MD Norcon (Norris) Ltd
Burgess Hill
West Sussex

**Senior Lecturer
Department of Mechanical & Manufacturing Engineering
University of Brighton
BrightonBN2 4GJ

Since the Industrial Revolution and Adam Smith's treaty on the economics of specialisation in processing, organisations have concentrated on separating the management function from the processing function and de-skilling the workforce. Present market conditions forces organisations to become leaner by removing layers of management and enlisting the talents of its workforce heavily. This paper describes a manufacturing organisation which has re-engineered itself along the lines suggested by Eli Goldratt in "The Goal" and by enplacing a simple flowchart empowered its workforce.

MAN MANAGEMENT - HISTORY

Before the Industrial Revolution, work was carried out by individual craftsmen, such as saddlers, blacksmiths, cobbler, etc., who supplier a small local market. With the Industrial Revolution and the associated infrastructure improvement of canals then railways, work became substantially factory based with mechanisation replacing, or changing the basic skills. Management, however, because of the prevailing conditions where labour was cheap and markets plentiful tended to ignore the finer points of man management.

In the eighteenth century, the noted economist Adam Smith noted that as demand rose the scope for specialisation within the production process increased. This specialisation along with the greater application of mechanical energy through machinery drove costs down substantially during the initial stages of the industrial revolution. Thereafter increased specialisation became the focus to increase productivity.

Towards the end of the nineteenth century, F W Taylor increased the momentum towards specialisation. His technique of Work Study emphasised the need to break work down into its smallest element and ensuring these elements were carried out in an efficient manner. Taylor also emphasised the separation of the planning & control function from the doing operations.

Henry Ford discovered that to achieve the high output and low costs he desired he had to standardise his product and its components and to simplify, de-skill and re-organise the work content by breaking it down into every small elements because of the low technical skill of his workforce. He quickly found that both the productivity and quality of his workforce outdid that achieved by the all-embracing craftsmen which were being used elsewhere to produce the motor car.

Gradually the theory of division of labour spread throughout the larger organisations who were encouraged to continue sub-dividing work as economy of scale with slow changing products became possible. This specialisation spread into management resulting in separate functions and a hierarchical management structure.

In the nineteen-twenties, another movement arose which attempted to encourage the greater use of the potential of the individuals employed in the organisation. There have been many observed experiments, such as the Hawthorne Investigations and those carried out by the Tavistock Institute, which appeared to prove that employees were being under-utilised by their organisations. People such as Douglas McGregor & Frederick Hertzberg pointed out that the framework of reference which were applied by many managers was faulty. The so called Humanist movement however failed to convince many of the existing upper management that there was a need to, or even a possibility of, harnessing their workforce's full potential.

Today one of the latest "techniques" of good management is to re-engineer the organisation. This necessitates delegating down to the lowest point all pertinent decision making and empowering the workforce to commit the organisation to suppliers and customers, exactly what the humanists recommended.

THE CORE BUSINESS PROCESS

Over the last twenty years, Norcon has replaced the functional structure of its organisation with a lean management structure built around a core business process. This has enabled substantial levels of productivity and returns to be achieved. It has been fundamental that the whole organisation of product design, procurement, manufacturing, distribution and marketing were built around and subordinated to this core business concept. The main problem came back to the classic control problem (Figure 1):

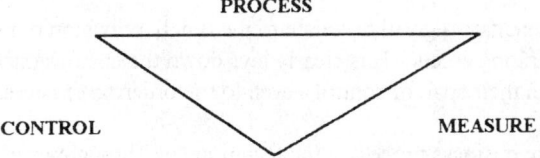

FIGURE 1 - Three elements of a control problem

The classic organisation structure appeared to spend more energy avoiding blame than solving problems and had to change. Through trials of various organising and controlling methods, the business structure has developed to a small, multi-disciplined team with a maximum of outsourcing of basic components feeding into the final processing and assembly.

Three algorithms have been developed to optimise the return (Figure 2):

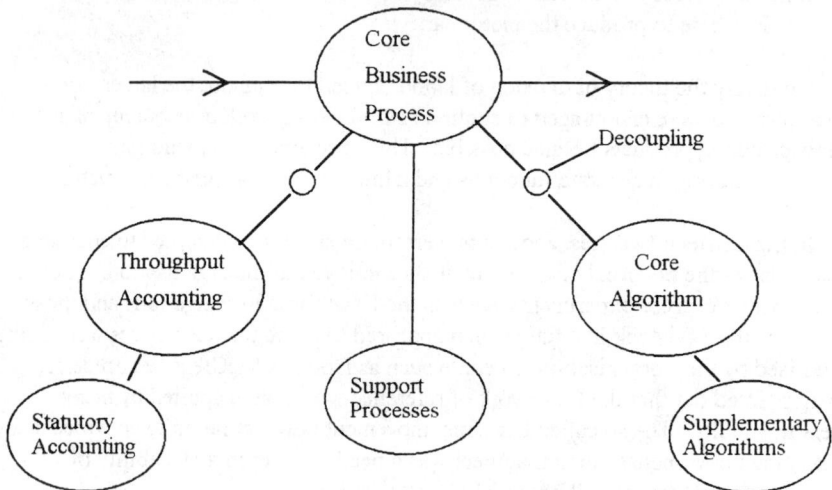

FIGURE 2 - The three algorithms

The core process works in real-time. It is the basic process which is carried out by the organisation and is designed to maximise throughput in key resources, i.e. bottlenecks.

The core algorithm is represented by a flowchart (Figure 4) which illustrates the quality audit trail required by BS 5750 documentation. This flowchart is displayed freely throughout the organisation - including the shop floor. By adhering to the flowchart, everyone knows s/he is contributing to the basic business process.

Every process is then converted into a "pull" demand mode tied back to sales with allowances made for the bottle-neck resources. This is important within the organisation as it supplies a large product mix using a small range of components.

The supplementary algorithms relate to the timely provision of resources and maintenance of operating assets. This clearly lays down the empowerment of employees to take action within their span of control - even to the ordering of necessary materials.

Allied to the business process is the technique of Throughput Accounting in a similar manner to that suggested by Goldratt (1989). This states that within an organisation the true profit can only be expressed as:

Sales (throughput) - Expenses = Profit .... Equation (1)

In the short term, the processing costs (expenses less material) can be taken as constant and all work-in-progress is treated as having no value except that of a raw material. The basic profit maximisation therefore comes down to movement of material through the process and into sales. For statutory accounts, the material content of the product is backflushed through the inventory records.

A simple control used within Norcon is the key ratio of material purchases to sales which is used as a breakeven analysis:

If we take equation (1) and break it down into its component parts, we have at the breakeven point:

Sales = Expenses = Processing + materials, or

Sales - materials = processing

But processing is taken as a constant, therefore

Sales (S) - materials (m) = transform function (kS)

Dividing both sides by S, we have

Breakeven co-efficient (k) = (1 - (material / sales)) ....Equation (2)

We can therefore determine profitability by comparing the simple ratio of material : sales against our normal process transform coefficient. A graph (Figure 3) is produced showing this coefficient against different operating patterns.

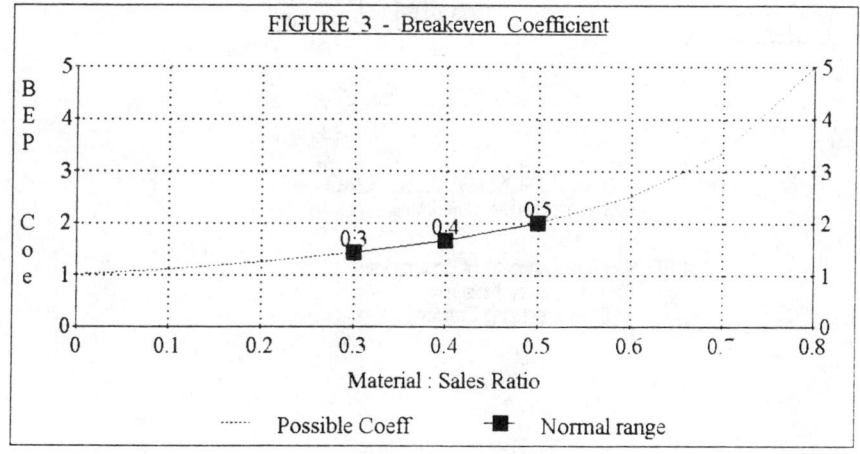

Therefore at Norcon they have shown that through a careful blend of training and organisation coupled to a empowerment of the workforce they have made a reactive and profitable business possible.

Reference

E Goldratt & J Cox, 1989, *The Goal*, Revised edition, (Gower Press)

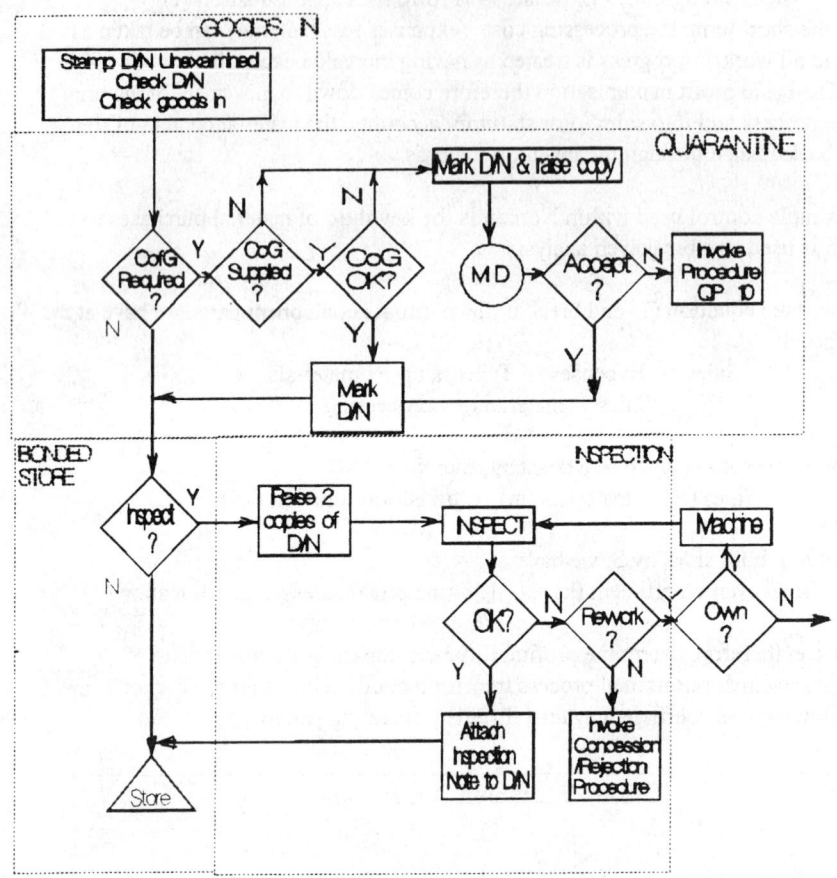

FIGURE 4 – EXTRACT FROM PROCESS FLOWCHART
(showing goods inwards section)

NOTE: CofG = Certificate of conformance
DN = Delivery Note
MD = Managing Director

THE EFFECT OF TECHNOLOGY ON ORGANISATIONAL STRUCTURE, SYSTEMS AND CULTURE

Trevor R. Pye

The School of Engineering
Coventry University
Coventry, CV1 5FB

How do differing levels and types of technology, create a culture within the organisation, that needs to be addressed as an integral part of the Business Process Redesign? Many differing theoretical approaches have been suggested in proposing cultural and organisational responses to situations such as BPR. Does the mass adoption of Information Technology that supports BPR affect these proposals? In creating an organisation that can respond to the new challenges that BPR presents decisions need to be made concerning hierarchical levels, spans of control, decision making and managerial relationships. Does the technology being adopted affect these decisions?

The Development of Organisations

The study of organisations and their behaviour seems to be as old as civilisation itself. With the coming of the Industrial Revolution the effect of change was noticeably increased. From this time onwards the study of organisations has continued, with all facets of the organisation being researched. Much of this research can be seen as an attempt to understand and rationalise the way in which people, when grouped together, respond to technological change. Systems Engineering or Business Process Re-design can be seen as the latest attempt to reconcile the organisation with the current technology.

Whilst it may be difficult to directly associate technological advances to organisational changes, Chandler (1962) and Scott (1970), have recognised the link. They have advanced the concept that the structure of an organisation develops through distinct stages as the company responds to the needs of the market. The structure of the organisation may also be configured by factors other than those relating to the product or process that the organisation is engaged in.Mintzberg (1979) suggests that the internal elements of the organisation may, by virtue of their existence, mould the shape of that structure.

135

The technology of the industry, and the nature of the technical control system being employed, can have considerable impact in deciding how many people one individual can be responsible for within their personal capacity. The classical theorists had, as the cornerstone of their views, the notion that spans of control should be limited. The study by Woodward (1965) suggested that the technology employed by organisations may override this basis concept.

The need to compete with the Japanese predominantly, has raised the need for questions to be asked about culture. The basic premise is; are they more successful than us because of the way they are rather than what they do?

The work of Parnaby (1987), in Lucas Industries has identified three elements necessary if change is to be managed successfully. The work on a Systems Engineering approach to manufacturing finds that the responses to these stimuli can be categorised as :-

Structure - Change the shape and function of the organisation
Systems   - Increase the technological content of the task
Culture   - Ignore it

These three factors can be used to examine the impact of technology on organisations.

Structure

The suggestion that organisations change from simple to complex structures needs to be examined by understanding the nature of these options. What are the choices available and where might they be of use in adapting to change. Once organisational structures develop beyond the stage of little or no formality, typified by the 'one man show', the way in which activities are grouped together needs to be considered. The need for the structure to respond to the influences of economies of scale, division of labour and job specialisation is generally seen as a choice between organisational models. These are generally seen as Functional, Product, and Matrix organisational structures.

Spans of control, and the number of levels in the hierarchy, are traditionally seen as being a measure of the effectiveness of the chosen model. The traditional view of span of control relates to the number of direct subordinates, this clearly needs to be modified to consider the type of interaction that exists between the manager and those who work for him. This interaction will clearly depend upon the training and expertise of the subordinates and the part of the organisation in which they are engaged. The technology of the industry, and the nature of the technical control system being employed, can have considerable impact in deciding how many people one individual can be responsible for within their personal capacity.

The span of control discovered by Woodward (1965), varied in relationship to the increase in technology involved in the process. That is, the largest span of control was experienced in the organisation that employed mass-production technology.

he median figures found for each category were :-
   Group 1 : 1 supervisor per 20-30 operators.
   Group 2 : 1 supervisor per 40-50 operators.
   Group 3 : 1 supervisor per 10-20 operators.

ystems

Whilst much organisational theory could be considered the responses of organisation, respective of the level of technology employed, Perrow (1971) and Woodward (1965) imply that nowledge technology and production technology do affect structure. The work of Burns and talker (1961), J.D. Thompson (1967) and Lawrence and Lorsch (1967), investigated the impact of ie environment on organisations. Within the environment, they recognised the power of :chnology to shape the organisation introducing or relying on that technology. This approach is to iew technology in a broad way. In this context it refers to the 'standardised means for attaining a redetermined objective or results'. Is this definition by Ellul (1965) a more constructive view to ike of technology in our post micro-computer society.

The environmental approach to discussing the impact of technology on the organisation ttempts to differentiate corporations in terms of Environmental Change and Environmental egmentation. Environmental change reflects the variability of the market and/or the technology ith which the organisation is involved. The other consideration that needs to be recognised is the ariability in the components of the task, Thompson (1967) describes this as Environmental egmentation.

The work of Lawence and Lorch (1967) has taken the notion that technology affects the tructure by examining the relationship within the organisation. They employ two central concepts) examine how organisations achieve effective communications without sacrificing the epartmental differences necessary for success in their differing fields. Differentiation is described s the attitudes and practices between managers in different departments, whilst Integration is the :ate of collaboration in existence to achieve the unity of purpose necessary to combat the nvironment.

ulture

:ulture is described as a 'set of ways of thinking about, and reacting to, events and problems'. :learly the history of an organisation will be a major determinate in formulating the mind sets that ill frame any decision making process. Handy (1989) suggests, that the culture of the rganisation will be the major factor in determining their success, as they respond to changing work atterns brought about by the increasing technological content of their work. Culture is difficult to efine at a national level and even more difficult at an organisational level. Handy (1978) attempts) use a cultural definition of management to come to terms with the increasing and changing :chnological nature of work. Using the notion of four management cultures, he sets out a ramework in which individuals can exist within organisations, these cultures are described as ower, Role, Task and Person.

Technology in Organisations

Structure

The theories relating technology to organisational characteristics need to be examined against firms involved in differing levels of technology. Firms within the aero-space and retail sectors would appear to the outsider to be involved in differing levels of technologies. An aero-engine manufacturer, in terms of span of control and organisational levels is a group 2 category as described by Woodward. Retail DIY firms are not easily categorised in the Woodward terminology as the investigation was conducted in manufacturing organisations. The nature of the business, with the emphasis on customer relationships, places them more readily in Group 1 where the emphasis is on the individuality of the product. Neither firms however, exhibit the organisational shape expected from the Woodward analysis. Our aero-engine manufacturer had spans of control of between 10-25 operators working for each foreman. It was noticeable that the maintenance areas had smaller spans of control than the assembly and manufacturing sections. The move to manufacturing cells was impacting on the structure, with smaller teams being organised by team leaders who themselves reported to a reduced number of foremen. Within the administrative and technical functions the spans of control were considerably smaller with figures of between 1 and 10 being the norm.

Within the DIY organisation there were variations in Spans of Control between the different disciplines. The two areas reviewed, the retail stores and the computer systems, showed quite different numbers of people reporting to individual managers. The Management Information Services had a clearly defined structure, with a maximum of eight people working with an individual supervisor or manager. This was at the operational end of the structure, with programmer/analysts being responsible to a section supervisor. Higher up the organisation, the span of control was less, with the norm being between two and four direct reporters. In the Retail Stores the spans of control are more difficult to estimate as the effect of part-time working clouds the issue. On the organisation charts the numbers reporting to each manager or supervisor was clearly shown, but in reality both the supervisors and the stores staff could be part-time.

Systems

In terms of market pressures the two companies can be seen to be operating within similar environments based on the Thompson definition. Both are responding to markets characterised by a high degree of environmental change and a low degree of environmental segmentation. In this situation both have responded with increasingly sophisticated computer systems, backed up with direction from the top of the organisation and devolved working practices at the operational end.

Within the retail trade, EPOS (Electronic Point of Sale) technology is now widely used to further enhance this flow of information from the consumer, as is the facility to electronically mail suppliers through the JET (Joint Electronic Trading) system. With the addition of a stores management system, MLP (Mechanising Layout Plans), decisions relating to pricing structures, stores layout and promotions can be made with increased sophistication. The same drive in the manufacturing sector is increasing the adoption of MRP II to provide on line real time systems to integrate manufacturing, engineering, shop control and inventory data.

Culture

Both organisations showed a preference for Role/Task culture. Role cultures are typified by bureaucratic structures enforced by systems and rules, which are enforced by assigned rights and responsibilities. All aerospace activities are highly regulated both by internal need and external direction. Increasingly the working relations within the retail organisation are being regulated by the EPOS , JET and MLP computerised management control systems. These systems require strict adherence to time frames and consistent data patterns. In return they dictate the actions to be carried out in the organisation. Organisations that portray the Task Culture, put achievement and effectiveness above authority and procedures. Success is seen as being concerned with the continuous and successful solution of problems. Responsibility and authority are placed where the qualifications and expertise lie, and they may shift rapidly to respond to changing priorities. The organisation is based around a network of loosely based task forces where talent and creativity are the requirements for team membership. Power in the organisation does not lie at the top nor the centre of the structure but at the interfaces of the teams. This approach represents the vision and resultant structures being developed in pursuit of Systems Engineering principles.

Conclusions

Woodward uses the term technology to represent the types of machines being used. With the increased use of CNC, and DNC in manufacturing, and Micro and Mini Commuters for management control, this definition becomes less practical. As already suggested, a systems approach which encompasses the notions of Ellul is perhaps more appropriate. In this consideration, our two companies exhibit remarkable similarities. Their Structure, Systems and Culture stand the comparison, and the perceived differences in levels of technology appear to have little effect. If we adopt the wider view of technology this would be expected. Both companies are attempting to manage a logistical chain within their corporate objectives of providing customer satisfaction. Given the technology available, a similar set of procedures are required. This in turn requires a similar structure, similar systems and, if they are to be successful, a culture that supports their objectives.

References

Chandler A.D. (1962) Strategy and Structure : MIT Press, Cambridge Massachusetts.
Scott D.R. (1970) Strategies of Corporate Development : Harvard University Press, Boston, MA
Mintzberg H. (1979) The Structuring of Organisations : Prentice Hall, New Jersey.
Woodward J. (1965) Industrial Organisation, Theory and Practice : Oxford University Press
Parnaby J. (1987) The 5th. Manufacturing Forum, London May 1987
Perrow C.W. (1971) Organisational Analysis A Sociological View, Wadsworth, Belmont, CA
Burns T. and Stalker G.M. (1961) The Management of Innovation : Tavistock, London.
Thompson J.D. (1967) Organisations in Action : McGraw-Hill, New York.
Lawrence P.R. and Lorsch J.W. (1967) Organisation and Environment : Harvard University Press
Handy C. (1978) The Gods of Management : Pan, London.
Handy C. (1984) The Future of Work : Blackwell, London.
Ellul J. (1965) Technological Society : Cape, London.

GROUP TECHNOLOGY AND CELLULAR PRODUCTION

Prof John L Burbidge

Cranfield University
School of industrial & Manufacturing Science
Cranfield Bedford MK43 0AL

Group Technology (GT) evolved in the period covering the 1950s to the end of the 1970s. Starting with the work of Professor Mitrofanof of Leningrad University, it developed into a Group production method of organisation for manufacturing companies.

Some time in the 1970s an attempt was made to change the term "Group Technology" (GT) into "Cellular Production" (CP) and the term "Group" into "Cell". Although CP was initially a synonym for GT, the two terms evolved in different ways and now have very different meanings. This paper submits that both ideas are valuable and proposes a reconciliation between them.

1. Group Technology

The term "Group Technology" was introduced by Professor Mitrofanov of Leningrad University as the title for his research into the relationship between component shape and processing methods. [1] Among his findings was the fact that it is possible to sequence operations on lathes so that the same or similar set-ups can be used for operations on several different parts, one after the other, giving major savings in set-up time. By adding other machine types to the lathes, Groups were formed which could complete "families" of parts. It has since been found that if "Production Flow Analysis" (PFA) is used to plan the groups, GT can be substituted with advantage for Process organisation, in any case where automated Continuous Line Flow (CLF) is not appropriate. [2]

Group Technology can be defined today as: "A method of organisation for Manufacturing, in which the machine tools and other processing facilities are totally divided into Groups which complete all the parts or assemblies they make, through the major processing stage at which they operate, without backflow between groups at following stages, and without cross-flow between Groups at the same stage. The Group machines are laid out together in their own special area, and are manned by

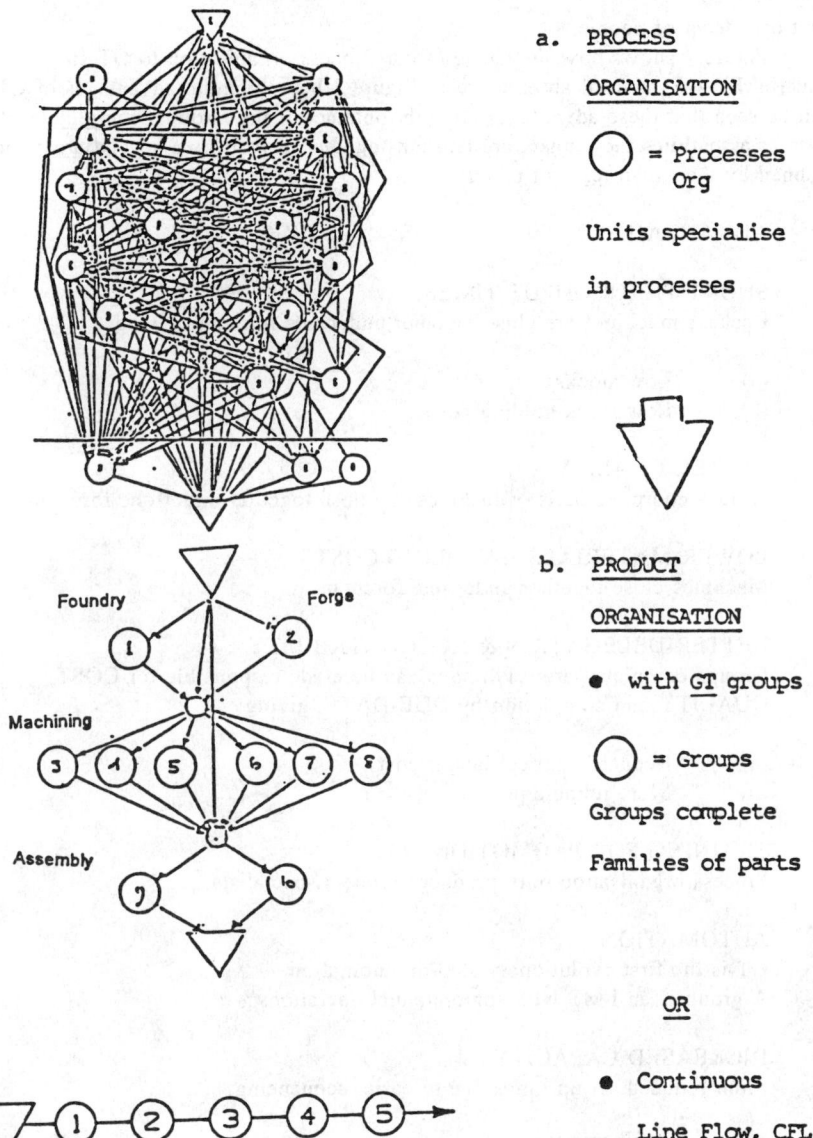

a. PROCESS
ORGANISATION

◯ = Processes
Org

Units specialise

in processes

b. PRODUCT
ORGANISATION

● with GT groups

◯ = Groups

Groups complete

Families of parts

OR

● Continuous

Line Flow, CFL

◯ = Stations

Figure 1   The Change to Group Technology GT

their own team of people."

Figure 1 shows how the change from Process organisation to GT simplifies the
Material Flow System of an enterprise. Figure 2 lists the main advantages of GT. It
will be seen that these advantages are only obtained if the Groups: complete all the
parts or assemblies they make; are laid out together in their own special area; and are
manned by their own team of people.

No.   ADVANTAGES

1   SHORT THROUGHPUT TIMES
       Because machines are close together under one foreman, giving:

       (a)   Low stocks
       (b)   Low stock holding costs

2   BETTER QUALITY
       Groups complete parts. Machines are close together under one foreman

3   LOWER MATERIALS HANDLING COSTS
       Machines close together under one foreman

4   BETTER DELEGATION & ACCOUNTABILITY
       Groups complete parts. Foremen can be made responsible for COST,
       QUALITY and completion by DUE-DATE, giving:

       (a)   Reduced indirect labour cost
       (b)   More reliable production

5   TRAINING FOR PROMOTION
       Process organisation only produces process specialists

6   AUTOMATION
       GT is the first evolutionary step in automation.
       A group is an FMS with some manual operations.

7   INCREASED CAPACITY
       From reduced set-up times due to easier sequencing

8   JOB SATISFACTION
       From finishing parts, and from team work etc

Figure 2 ADVANTAGES OF GT

It should be noted however, that these advantages are not automatic. GT
makes them possible but further action must be taken to achieve them. For example,
GT makes it possible to obtain improved accountability but no savings are achieved
until the responsibility and authority to manage the Groups are transferred to the
Group teams.

A second point: with GT some savings are quickly achievable with the completion of the first Group e.g. reduced throughput time, stocks, stock holding costs and materials handling cost. The total potential savings can only be obtained however, when a total change to GT has been achieved, and when the production control; process planning; wage payment; accounting; purchasing and other operating systems have been changed to suit.

A final point: to many the greatest contribution of GT is that it increases job satisfaction and makes it possible for the workers in industry to take a significant part in decision making. It is probably true that such changes are only possible if they start with the team structure of GT organisation. There is little hope of gaining these advantages with process organisation.

The primary conditions of GT (completing parts; layout together in one place, and a team of people) are not negotiable. Without these conditions, the savings in Figure 2 cannot be achieved.

2.   Cellular Production

When the term Cellular Production (CP) was introduced its sponsors saw it as a synonym for Group Technology. It has evolved with time however, to describe much smaller units than GT groups composed of a few machines which are normally used together but do not complete all the parts they make. Cells in this sense, formed by random selection from the total plant list, are never likely to combine to form larger groups which complete all the parts they make. That this is now the general meaning of the term Cellular Production can be seen in many published research papers on Cellular Production from the USA and other countries.

Because cells do not complete all the parts they make, they cannot achieve the savings listed in Figure 2, for GT groups.

In a recent research paper entitled "Financial Pathos : Cellular manufacturing gone wrong", published by Ingersoll Engineers [3], they show that the first few cells can only be expected to produce modest fractional savings such as:-

> ½ an operator
> ¼ of a fork truck
> 1/17 of a warehouse
> 1/12 of a foreman
> 200 sq ft of floor space (dispersed)
> etc

None of these savings is bankable. "Most benefits as a practical matter, are not achievable until near the end".

If a company follows a policy of introducing cells, it will eventually reach a state where it can cash in on some accumulated fractional savings but it is still unlikely to achieve the major system change savings listed in Figure 2, because cells formed at random are most unlikely to combine into larger organisational units, which complete parts.

Cells composed of a small number of machine tools and other processing facilities, laid out together to form single work centres, tend to simplify material flow systems by reducing the numbers of operations and of different work centres on which work must be scheduled. If there are many cells, some savings can be achieved in throughput time and in costs. However, the major system change savings listed in

Figure 2 can only be achieved if cells are formed inside GT groups, which complete all the parts they make.

3.   Finding GT Groups and Families

Most of the early work in planning GT groups and families was based on "classification and coding", (C & C). In other words Groups were found by analysing the information given in component drawings to find sets of parts which were similar in shape. It was reasoned that if a set of parts all had the same or similar shape, one set of machines and other processing facilities could be used to make them all.

This approach was unsuccessful, partly because: it only found "Families" of similar parts, it didn't find the Groups of machines needed to make them; it didn't find a total division into Groups and Families; and it generally required an investment in new plant to make even a limited division possible.

Production Flow Analysis (PFA), which has replaced C & C, is based on the information in process routes. It can always find a total division into both Groups and Families and with PFA it has never yet been necessary to buy additional machines to make the change to GT possible. [4]

PFA consists of the succession of sub-techniques illustrated in Figure 3. Providing that FFA finds a division into primary processing stages (or departments), which complete all the parts they make, it is always possible using "Group Analysis" (GA), to divide these departments into Groups, which complete all the parts they make and are provided with all the machine tools and other processing facilities they need to do so.

"Group Analysis" can be seen as a method based on the resolution of a matrix on a diagonal. The matrix shows the parts and the machines used to make them. A very simple matrix is shown in Figure 4. GA simplifies the much larger matrices found in practice by combining parts to form a smaller number of "modules", or mini-matrices based on a series of key machines. These modules are combined to form Groups. As a final step, any exceptions, where the machine shown in the process route for an operation on a part, is not in the Group where it is to be made, are eliminated by reallocating the operations to other similar machines, which are in the group. [5]

It is the high level of flexibility in the process routes, which makes it very unlikely that we will ever find a case where the change from Process organisation to GT is impossible or unprofitable.

4.   Cells in GT Groups

The fourth sub-technique of PFA is Line Analysis (LA). This analyses the way in which materials flow between the machines and other work centres in a Group, to provide the information needed for plant layout.

Among other information, LA indicates the existence of any sub-groups of facilities, which can be integrated and laid out together (as cells), in order to make close-scheduling possible, or so that one or a few operators can operate several different machines.

An example is illustrated in Figure 5. Each machine in the Group was given a two letter code number. Each part made in the Group was given an "Operation Route No" (ORN) code showing the machines used to make it, in usage sequence. The

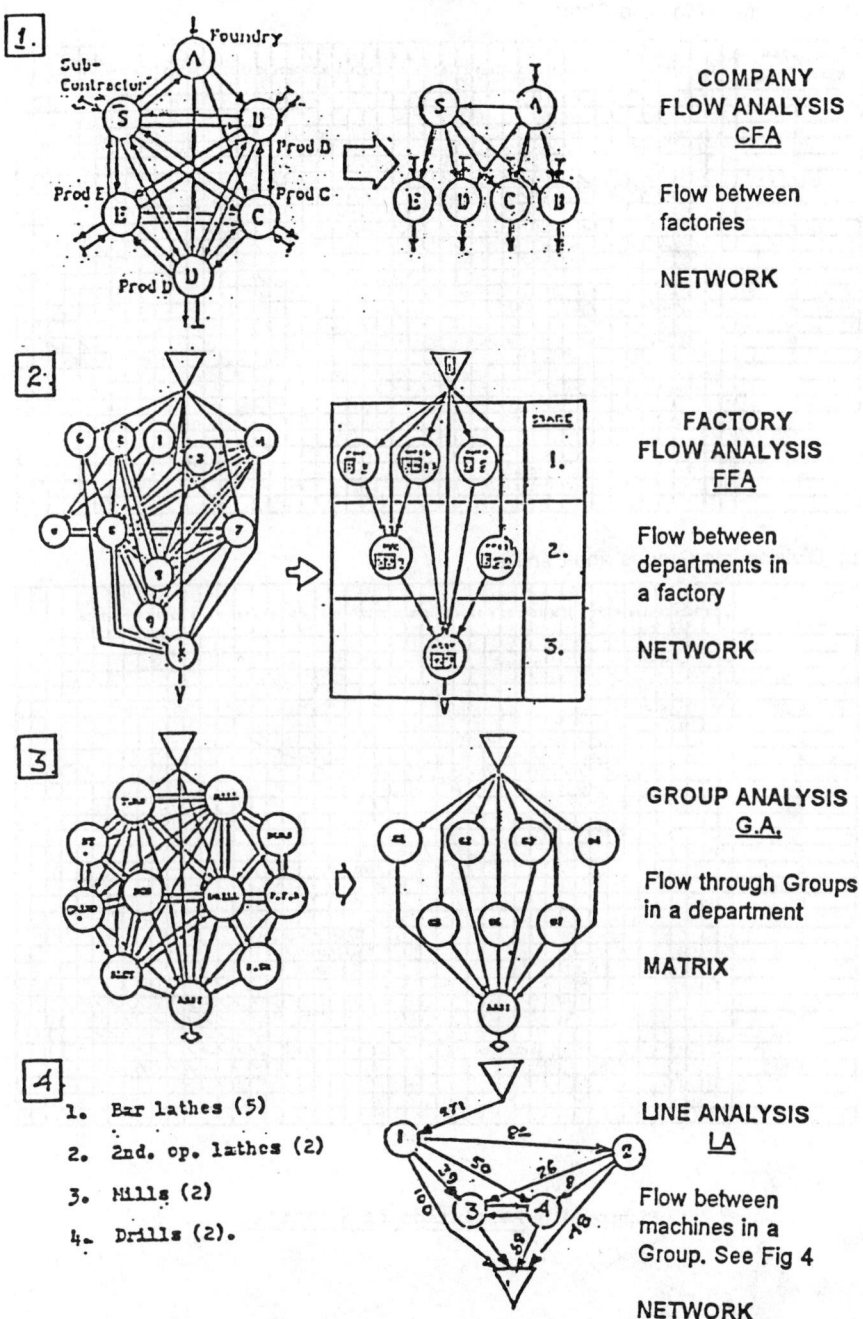

COMPANY
FLOW ANALYSIS
CFA

Flow between
factories

NETWORK

FACTORY
FLOW ANALYSIS
FFA

Flow between
departments in
a factory

NETWORK

GROUP ANALYSIS
G.A.

Flow through Groups
in a department

MATRIX

1. Bar lathes (5)
2. 2nd. op. lathes (2)
3. Mills (2)
4. Drills (2).

LINE ANALYSIS
LA

Flow between
machines in a
Group. See Fig 4

NETWORK

Figure 3   Sub-techniques of FFA

a) Component Machine Chart

b) Division into Groups and Families

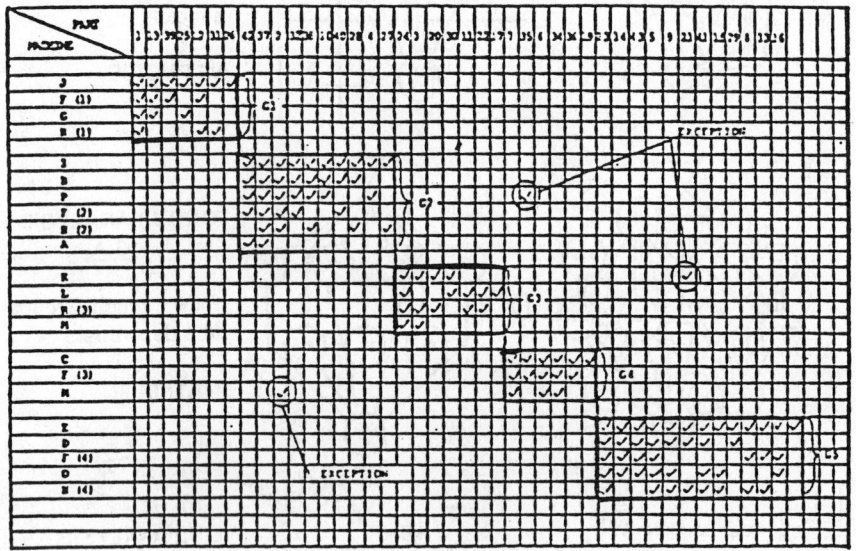

Figure 4   Resolution of a Matrix

Item	ORN.																	FREQ.
1	CQ	CO																1 ᶜʳ
2	CQ	(CS)	CT															3
3	CQ																	9
4	CQ	CS																5
5	CQ	CT																5
6	CQ	CT	CS															2
7	CQ	CT	CS	CR														2
8	CQ	(CY)	CZ	CW														1
9	CQ	CZ	CW	CO	CV													6
10	CN	CQ	CS															1
11	CN	CQ	CT	CQ	CR	CS												1 x
12	CN	CQ	CT	CR	CS													1
13	CN	CQ	CT	CX														1
14	CN	CQ	CT	CZ	CO	(CT)	(CT)											4 x
15	CN	CQ	CT	(CX)	CQ	CO	(CT)	(CT)										1 x
16	CN	CQ	CT	CZ	CT	CT												1
17	CN	CQ	(CR)	CT														1
18	CN	CP	CT	CZ	CO	(CT)												2 x
19	CN	CP	CT	CZ	CO	(CT)	(CT)											3 x
20	CN	CP	CT	CZ	CY	CR	CS											1 x
21	CN	CP	CT	CS														1
22	CP	CQ	CT															1
23	CP	(CS)	CZ	CW	CO	CU												3
24	CP	CZ	CW	CO	CV													1
25	CT																	1

Key: CQ = CNC lathe.   CS = Rad drill.   CW = De-burr.   CO = Nitride.
      CN = face & ctr.   CR = Mill   CX = Gear cut.
      CP = CNG lathe.   CY = Spline.   CU = Hone.
      CT = cyl. grind.   CZ = Gear cut.   CV = Hone   ◯ = Exceptions

Figure 5 Divide Group into Cells

ORNs for all parts made in the Group are summarised in the figure.

It will be obvious that 5 of these machines (CQ, CP, CN, CT and CZ) are normally used in the same sequence. If they are laid out close together in this sequence, close scheduling is possible. Parts can be transferred between machines as single units giving low throughput times and low stocks. This sub-group of machine was called Cell 1.

A nitriding furnace (CO), two honing machines (CU and CV), and a deburring bench (CW) were normally used one after the other in different combinations. They were called Cell 2. A third cell - Cell 3 - contained four machines, (CS, CR, CY and CX), mainly used for finishing operations.

Cells 2 and 3 had the advantage that they supported the more efficient use of flexible labour. Cell 1 had this advantage and also the additional advantage of close-scheduling. The cells produced their own special savings and because they were formed inside a GT group which completed all the parts it made, the Group also achieved the main GT advantages listed in Figure 2.

Conclusions

"Group Technology" is a form of organisation in which "Groups" of machines and people complete all the parts they make at the major processing stage at which they operate. eg component processing stage, followed by assembly stage.

"Cellular Production" forms smaller sub-groups of machines known as "cells". Cells integrate some machines to form a smaller number of larger work centres.

GT can achieve major benefits but to achieve them, groups must complete all the parts they make, without back-flow or cross-flow between groups. The groups must be laid out together in one place and each group must have its own team of workers.

Cells can achieve some savings on their own but there is a danger that if they are formed at random, they may make it impossible to form GT groups which complete parts and thus make it impossible to achieve the much greater savings possible with GT.

It is not difficult, or expensive, to plan the division into Groups and Families. Ideally, GT groups should be planned first, and cells should be seen as elements to be planned during Group Layout.

References

1.   Mitrofanof, S.P. 1966, *Scientific Principles of Group Technology*, National Lending Library for Science and Technology, Boston Spa, Yorkshire, UK.

2.   Burbidge, J.L. 1989, *Production Flow Analysis*, (Oxford University Press, Oxford, UK).

3.   Ingersoll Engineers 1994, *Financial Pathos*, Bourton on Dunsmore, Rugby, CV22 9SD, UK.

4.   Burbidge, J.L. 1992, Change to GT Process Org is obsolete, *IJPR*, **30**, No 5, (Taylor & Francis, UK).

5.   Burbidge, J.L. 1992, Production Flow Analysis for Planning GT, *Journal of Operations Management*, **10**, No 1, USA.

WHO SHOULD IMPLEMENT AN FMS, HOW AND WHY? - A GUIDE FOR FMS PROJECT MANAGERS

*S.M. Ngum and **Dr K.F. Gill

**AMT Group, Manufacturing Systems Engineering, University of Hertfordshire,
Hatfield AL10 9AB
**Dept. of Mechanical Engineering, University of Leeds, Leeds LS2 9JT.*

As UK manufacturing industries struggle to escape the tight grip of the recession, conventional manufacturing techniques and their inadequacies will, and must, come under strict examination. Advanced integrated systems of the FMS type will, to some companies, be the step forward, and, in many respects, a significant leap from traditional approaches. The rigours of present investment appraisal techniques and the antagonism known to exist between engineers and accountants are bound to intensify. This paper is a guide for project managers contemplating Flexible Manufacturing Systems (FMS) implementation. It examines FMS-related issues highlighting why, and how companies should set about tackling the challenges that will lie ahead, to reap its rewards.

Introduction

FMS has, to most companies involved, resulted in significant improvements in competitiveness through lower manufacturing and operating costs, increased responsiveness and better quality products Ngum(1989). Until recently, the application of concepts associated with FMS, showed an expected upward trend, indicating the desire by companies to consolidate their competitive positions. Unfortunately, the beginning of the 1990s saw, in the UK and most industrialised countries, a slump in imports and exports of manufactured goods. In the UK, this manifested itself in reductions in manpower and factory closures. The difficulties in a recession, is the reduction in a company's skill-base, capital reserves and effects in the event of a recovery. FMS and related forms of automation could, for some of these companies, be the step forward as they struggle to regain their competitive positions.

Who should? - FMS as a manufacturing strategy

The conventional manufacturing environment is characterised by inefficiencies linkable to operational (high WIP, multiple handling, layers of supervision); system (utilisation, layouts); and personnel (poor industrial climate, labour standards) factors. The cumulative effect is one of low resource utilisation, increased scraps/re-works, and inability to meet deliveries. Integrated systems show, as indicated by recent figures Ngum(1989), enormous potential in minimising these undesirable high costs-low productivity factors. FMS as a significant, and logical, leap from traditional manufacturing approaches, is also capital intensive. Hence the need for the question - who should implement FMS? Each concern must evaluate the degree to which the aforementioned factors influence inefficiencies and the scope for improvements. FMSs are more suitable for mid-volume/mid-variety production and may prove not to be the most cost-effective route, particularly if frequent modifications or new product introductions are not envisaged.

Approaching FMS implementation and feasibility studies.

From an economic viewpoint, an FMS is required to be: responsiveness to market changes and flexible enough to cater for part variety - adaptability (parts mix, new product and production flexibility); expandable, with limited alteration - (expansion flexibility); capable of re-routing parts in the event of breakdowns (routing flexibility); serviceable and reliable. Because of these requirements, its design entails a thorough analysis of numbers, sizes, weights and geometries of parts, tooling and operational sequences. Without these, configuration(s) of machine tools cannot be established nor economically designed. A top-down approach based on a detailed examination of the company's development strategy is essential. This will ensure compatibility with strategic business objectives, determine resource requirements and levels of flexibility. With these established, consideration must be given to software and hardware requirements. Ease and simplicity should be chosen in preference to complexity and the element of human resources and effective management should not be treated lightly - they hold the key to the success of such a venture.

A sound feasibility study at the outset of the proposal is critical and a vital step towards establishing its practicability and benefits. In addition, it is a useful means of putting the resources, cost/benefits in their true perspective. As Chesi and Semararo (1992) argue, this process is important from the buyer's viewpoint in that several possible solutions can be evaluated to find out that which will assure the best return on investment (ROI); while from a supplier's perspective, a good solution in the shortest time will be the objective. Fig.1 is a suggested concurrent planning procedure for such a study and emphasises on the requirement to:

• Detailly examine the manufacturing strategy to include volume/variety of parts involved, the product life (product flexibility) and impact on final system design. For the most part, the level of flexibility required will be high for products with shorter lives. Table 1 details factors and impact when examining the range of components to be processed.

• Decide on layout to determine space requirements and constraints. Consideration should be given to storage, load/unload space buffer stores for decoupling machines if required and how these affect existing equipment. These would largely depend on the variation between part families, sizes, weights, complexity and desired manpower.

• Determine control strategy in relation to the supervisory control systems, software requirements, management information system (MIS), priority and emergency procedures.

• Analyse components for possible grouping into part families based on design or manufacturing similarities. Assign parts to dedicated fixtures and standardise products to minimise complexities and to facilitate automatic handling or machining.

• Analyse system within the conceptual framework of the business objectives and justify its performance. High costs, non-productive times must be addressed. Information feedback, for example, would assists in the evaluation of tools, machines, system and manpower utilisation. Present accounting techniques make it imperative to critically determine initial capital outlay, working capital requirements, and forecasts of revenue in comparison to existing methods.

The essence is clearly to gain financial acceptance often based on quantifiable evaluations using one or more of the measures in table 2: These quantification measures tend not to reveal the true benefits, as in the case of the intangibles in table 3.
- A detailed study will establish and assess the risks/consequences of failure. It must be remembered that new/existing machine tools may need extensive modifications; that complexity might increase; and that the proposed level of flexibility might be achieved only with some of the most expensive pieces of equipment. The study must be looked upon as an iterative process subject to refinements.

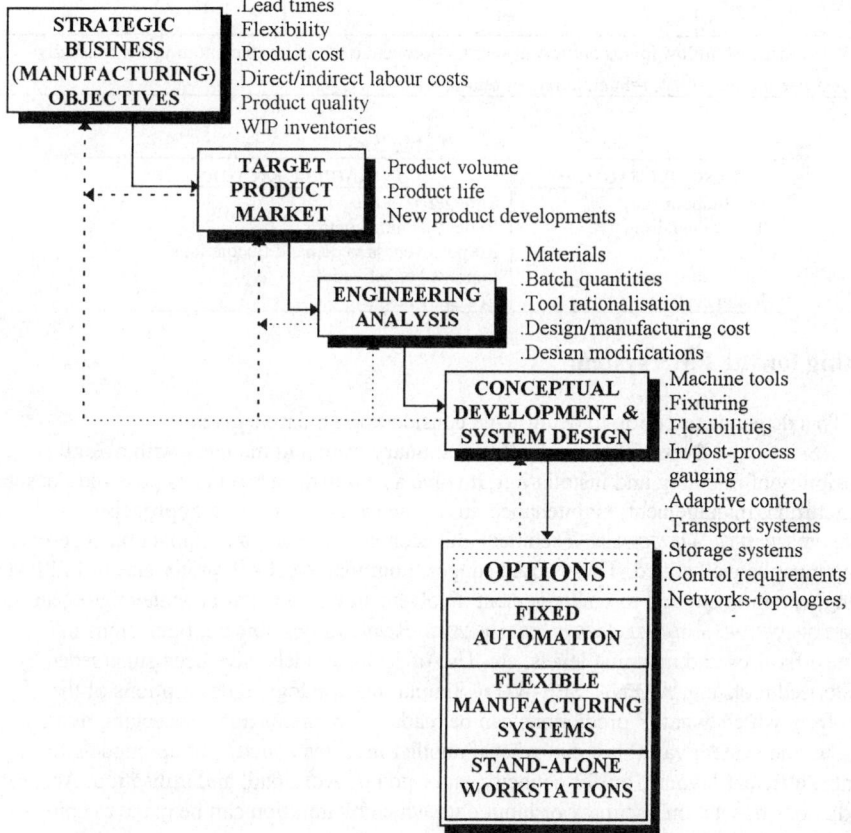

Figure 1. Feasibility Study Planning Model - A Concurrent Approach

Table 1

FACTORS	ATTRIBUTE	IMPACT
Volume	High	More machines required
Variety	Large	High level of flexibility
Part Size	Large	Larger working envelope of machines
Part Weight	Heavy	Better handling systems AGVs, Robots, Shuttle carts
Part Material	Hard	High spindle Horsepower, Tooling
Tolerance	Tight	Machine design, tooling, Inspection, fixturing

Table 2

MEASURE	ASSESSMENT	ACCEPTANCE CONDITION
Payback	Time to recover initial capital outlay.	Shortest time usually in years
Return On Investment (ROI)	$ROI = (^{Net\ profit}/_{Capital\ employed})*100$	>0
Internal Rate of Return (IRR)	$IRR = \sum_{t=1}^{T} CF_t/(1+i)^t$	$IRR \geq r$
Net-present-value (NPV)	$NPV = \sum_{t=1}^{T} CF_t/(1+r)^t$	$NPV \leq 0$

KEY: CF_t=net cash inflow minus outflow at year t; T=economic life of investment in years; r=usually interest rate on capital(%); i=Internal rate of return(%).

Table 3

TANGIBLE FACTORS	INTANGIBLE FACTORS
Direct labour	Stronger skill base
Material handling	Better planning/control
Scraps/re-works	Responsiveness to demand fluctuations
WIP/Lead times	Improved effectiveness
Design costs	Company image

Planning for the FMS system

This demands attention to detail. Due consideration must be given to:

[I]. TEAM SELECTION: Dependent on a multi-disciplinary team and manager with overall responsibility of planning and installation. Involve as many departments as possible - design, manufacturing, management, maintenance, in a concurrent engineering approach.

[II]. DEFINE STRATEGIC OBJECTIVES: Technical and economic factors, and impact on corporate strategy must be well stated. These could suggest upgradable cells in preference to full FMS. Training, commitment and top management involvement will be vital in achieving objectives.

[III]. DETERMINE AND SIMULATE ALTERNATIVE DESIGNS: Relate to possible configurations of machines; fixtures and manning levels, etc. Use of scaled models have been superseded by computerised tools e.g. ARENA, SEE-WHY. Simulations are logical descriptions of the system from which 'what if' predictions can be made. They are means of checking dynamic behaviour and system variables, e.g. what if another machine added? Set up models to determine efficient layouts, buffer capacity, entry points, workload, and utilisation. Accuracy of predictions depend on accuracy of input data which by iteration can be made to compare favourably with the eventual FMS.

[IV]. ANALYSE DESIGN IN RELATION TO THE PRODUCTION SYSTEM: Its introduction requires an analysis of how it integrates with current set-up and what changes will be required. Optimisation requires fine balancing so that all systems operate efficiently and in harmony.

[V]. SELECT VENDOR: This should be based on ability to act as main contractor for the project and provision of maintenance support; interfaces for various subsystems e.g. CAD, MRP, guarantee compatibility if a step-by-step implementation is preferred; competence of his team and willingness to liaise with in-house team, and system cost.

[VI]. DETERMINE MANPOWER AND SKILL REQUIREMENTS: Managing the project, installing and operating it requires substantial levels of skills. They do not benefit from intuition, and flexibility of operators and corrections at machine level are unacceptable. Personnel must be trained to develop multi-functional skills required to run the system efficiently.

Installing, operating and analysing the benefits

Once decided, the team must begin preparing plans for stoppages; acceptance testing procedures; re-configure departments for efficient integration; reinforce floors for heavy machines; and prepare plans for unforeseen contingencies e.g. resignations of key personnel.

The operational phase entails adjustments to reach full operation. Schemes should be set up to collect operational data (cycle; set-up; quality; breakdowns; costs,etc.). With such information deviations from targets and forecasts of future requirements can be assessed more objectively and used to influence future expansion and investment policies.

Provided planning, engineering, installation, commissioning and management, are suitably catered for, considerable business gains are possible. These benefits can be classified as time, cost, market and technology related as summarised in table 4.

Summary

FMS is a logical advancement from the CNC machine shop. However, a poorly designed FMS system can only function at a fraction of its full potential. This paper has highlighted some key issues of FMS design but it cannot be over-emphasised that optimum flexibility and functionality lies in efficient design of the control software. Emphasis must be on modularity, compatible with the company's business systems for better interaction with all databases. Unimpaired data flow, and the coordination of the activities of subsystems will largely depend upon its quality. The issue of intelligent control will be the subject of a subsequent publication.

Table 4

BENEFIT	TIME	COST	TECHNOLOGY	MARKET
Lead times	■			
System availability	■			
Set-up times	■			
Re-routing flexibility	■			
Direct/indirect costs		■		
Tooling requirements		■		
Inventory		■		
Floor-space requirements		■		
Better quality products		■		
Equipment utilisation		■		
Expandability			■	
Integration			■	
Competitive position			■	
Image			■	
Responsiveness				■
Throughput				■
Small batches				■
Adaptability				■

References

Brandon, J.A. 1992, Strategic perspectives for implementation of FMS, Robotics and CIM, *Proceedings of the Twenty-ninth International MATADOR Conference.*

Chesi, F. and Semeraro, Q. 1992, A new approach to the automatic design of FMS through a simple but effective forecast of production needs, *Proceedings of the Twenty-ninth International MATADOR Conference.*

Ngum, S.M. 1989, The reliability of Flexible Manufacturing Systems (FMS) in the UK, MSc Dissertation, Department of Engineering, University of Warwick.

ADVANCED MANUFACTURING PERFORMANCE : THE CAUSAL LINKS EXPLORED

David Hamblin and Albert Lettman

*Putteridge Bury Faculty of Management, University of Luton,
Luton England LU2 8LE*

Investment in Advanced Manufacturing Technology may be opportunity driven (perceived opportunity to improve performance) or performance driven (cash available from previous success). Granger-causality methods have been used to identify which of these models operate in the UK metal component industry. It has been determined that both models operate depending on the performance variable studied. The methodology has widespread application in correlation problems.

Introduction

Investments in advanced manufacturing technology (AMT) have been seen as the answer to improving the competitiveness of UK manufacturing industry. The investment performance of the UK manufacturing sector has been inferior to that achieved by other industrial countries (CBI, 1991). Schonberger (1986) however argues that substantial gains have been achieved without extensive investments and raises the question whether higher investments are required for further productivity. He concludes that unless investment is managed properly, this will detract rather than improve financial performance. Similarly, Beatty (1992) argues that technology must be managed effectively and its use and introduction is crucial in an organisation's performance.

Research however has failed to demonstrate a consistent causal relationship between investment in AMT and its impact on organisational performance. Improved performance has not universally been the case in practice. A survey by New and Myers (1986) established that companies which have been looking for a quick financial payoff from the implementation of AMT have been disappointed. They suggests that the perceived payoff from AMT investment has either been low or non existent. This paper will explore the causal links between investment and performance essential for managers and practitioners for managing AMT effectively. Managers need to determine the strength and direction of causal relationships between investment and organisational performance in order to make better resource allocation and investment decisions. The results from a historical study of the metal components industry, funded by the Department of Trade & Industry, are reported

We have hypothesised that companies follow one of two investment policies. Either investing when opportunity arises to enhance performance, primarily capacity and effectiveness (opportunity driven) or, investing when cash is available (performance driven). One of the aims of this paper is to report the empirical work done to confirm or refute this hypothesis by using cross sectional longitudinal analysis. This approach is different from the views of academics who postulate simple correlation, or those whose main proposition is that bicausality exists between performance and investment. Hence whether or not managers distinguish between investment based on performance, or performance based on investment, the financial performance of the organisation would be the same.

The paper is organised as follows. First, a synopsis of the literature on investment and performance is presented. Then the data and methodology are introduced. Next the empirical results are presented along with a discussion. The discussion that follows reports the findings of the research team to date in the metal components industry, UK wide. Our results indicate that investment allowed the companies to perform substantially better but the reverse causality was also found to be true implying that both performance driven and opportunity driven models operate.

Examining links between Investment and Performance: The Literature

Various authors have investigated the links between AMT investments and their subsequent impact on organisational performance. Table 1 presents some of the research carried out and reviews the propositions of the study concerned, the methods used and the main conclusions.

The paper extends the work of Granger (69) in the field of Econometrics to the field of Manufacturing Management. Moreover, the development of econometric tools such as PANEL DATA has allowed the analysis of multi-respondent observations throughout time, as well as across time.

Table 1 Summary of selected literature review

Author, Year	Proposition	Method	Results
Pike et al, 87	How manufacturing practices have changed with Respect to AMT	Mail survey, 47 UK manufacturing firms	Difficult to quantify AMT benefits
Swann & O'Keefe, 90	Quantifying benefits of AMT and Investment decision process	Literature review	AMT benefits depend on organisation's strategies and their link with AMT
Hamblin & Nugent, 90	Examined differences between manufacturing practice and performance across industry sectors with respect to AMT usage and adaptation	Panel Analysis, interviews, in house studies, quantitative data collected for 250 firms in 5 industries	Various drivers of performance and investment with respect to AMT adaptation

Data and Methodology

Thirty-nine of the organisations reported by Hamblin & Nugent (90) in the study " The effectiveness of AMT investment in UK metal components industry " were selected. In order to measure investment in AMT, three measures of investment were utilised. These are:

1. Total investment spending as a proportion of value added (SPVA)
      (Amount spent by company in ALL investments excluding land and buildings)
2. Investment in machinery as a proportion of value added (MCSP)
      (Amount spent by company in conventional machinery, excluding AMT)
3. and finally investment in AMT as a proportion of value added[*1]

To measure the impact of organisational performance, 2 types of performance variables were used:
1. Return on sales ROS (profitability variable) - (pre-tax profits divided by turnover)
2. Total factor productivity TFP (productivity variable) - (value added divided by the sum of employment costs and capital usage).
The findings of other studies suggest that these variables are adequate measures of investment and performance (Hamblin, 90; Peters and Waterman, 82).

All organisations utilise inputs measured by investment. In the case of three areas of investment outlined above, these relate specifically to plant and machinery. These investments must be utilised effectively to impact on an organisation's performance. The key issue is how well the organisation is achieving its performance goals measured by ROS and TFP as a result of the investment. There may in addition be growth goals, which are outside the scope of the analysis presented here.

The data set reported above was made up of time series data from 1979 to 1988. Therefore we were in the unique position of carrying out both cross sectional and longitudinal analysis. To this end the method used to analyse the data was panel data analysis (Hamblin & Lettman, 94). To test the model of opportunity driven and performance driven investment, the notion of Weiner Granger causality was employed (Granger, 69).

The estimation of Granger's causality involved the following steps

1. Fitting the regression equation

$$1.1) \ P_{it} = b_{it} + a\Sigma \ I_{it-n} + E_{it}$$
$$1.2) \ I_{it} = b'_{it} + a\Sigma \ P_{it-n} + E'_{it}$$
where P is the performance time series and I is the investment time series; the individual effect is b_{it} , which is taken to be constant over time, t, and specific to the individual cross-sectional unit , i, and E_{it}, E'_{it} have a random distribution across t & i.

The two investment policies can be summarised in equation 1.1 and 1.2 as:-.

---

[1]AMT investment breaks down into three main areas
            a. CNC machine tools (plus handling and measuring equipment)
            b. Computer aided design (CAD/CAE/CAM)
            c. Computer systems for management and control

a. when opportunities arise to enhance performance, primarily capacity and
   effectiveness (i.e. opportunity driven); this formulation as shown in equation 1.1
b. when cash is generated (performance driven); equation 1.2

2. Specifying the lag lengths

In this study an arbitrary approach to the issue of lag specification is employed. Various
other work suggests that the impact of investment on performance is realised within the first
3 years, even AMT investments were expected to have a payback period of under 4 years
(Primrose & Leonard, 1986) This became the basis upon which our lag specification was
made. Extended lag specifications cannot in any case be estimated without very long time
series of data. For practical reasons, therefore, a maximum lag of 2 years was used.

3. Diagnostics

There are various approaches used in tackling the issue of causality; time series data allows
the use of the vector auto regressive method. Due to the very nature of our data set being
both cross sectional and longitudinal we chose to use the method of panel data study. Greene
(1993) suggests the analysis of panel or longitudinal data is the subject of the most active
and innovative bodies of literature in Econometrics. The Lagrange multiplier test and
diagnostic testing by the use of the F statistic are employed in order to obtain results on
causality.

Empirical Results

From the bi-causal hypothesis that performance allows investment and that investment
generates performance, six different pairs of model formulations were tested using panel
data software - LIMDEP. Of the six pairs, three were found to demonstrate causality of the
type under study. When the measure of performance was return on sales (ROS), the
resulting data analysis implied that performance allows investment (measured by SPVA).
Table 2 below shows this.

Table 2 Summary of Empirical results between investment and performance

Variables	Causal relationship	Variables
Performance		Investment
ROS	----------------->	SPVA
TFP	<-----------------	AMSP and MCSP

Notes: Causal relationships are defined as follows ------> implies right hand variable "causes" the
left-hand variable and vice-versa.

One of the opportunity driven estimations found for the Metal Component sector is:-

$$TFP_{it} = a_{it} + .32*MCSP_{it} + 1.3*MCSP_{it-1} - .12*MCSP_{it-2}$$
$$\quad\quad\quad (.46) \quad\quad\quad (.41)*** \quad\quad\quad (.38)$$

and Adjusted $R^2 = 0.87$ for the fixed effect model
where $*** = p < 0.001$ (significant at the .01% level)
and $a_{it} = $ individual observation for companies.

Manufacturing spending in the year preceding investment in conventional equipment is
significant at the .01% level, with a explanatory power of 87 %. This high level of
explanation relates to the use of the company specific constant, because there is a
substantial stability of the TFP for a particular unit over time. The explanatory power of the

three spending terms alone is less than 20%, but the one year lag spending nevertheless has a highly significant effect on TFP. Opportunity driven investment companies are either adding increased value or are reducing their capital/employment costs and their investment in AMT and conventional equipment results in improved performance. The improvement flows through to TFP in the year after investment. The effect on ROS is not significant. As described in Hamblin, 1990, there are impacts of investment on other overheads which offset the gains in TFP.

However, performance driven investment is demonstrated by the effect of ROS on total spending. Having made a profit, companies will spend it. Further analysis suggests that companies who invest as a result of good performance do so because of their perception of the structure of the businesses/ markets in which they operate; frequently there is perceived opportunity for future growth.

Conclusion

This paper has presented empirical research into the causal links between AMT investment and organisational performance, using Panel Data Analysis. The results show that there is a relationship between investment and performance for the metal components industry. Where the manufacturing strategy is to add value or reduce capital/ employment costs, then it is investment that will cause improved performance. However, the previous performance of the firm has a significant impact on investment. Cross industry comparisons will strengthen the hypothesised relationships within other sectors, and this is the objective of the next strand of our research.

Acknowledgement
The authors gratefully acknowledge the contribution of Dr E Nugent, Cranfield University, in the assimilation of data for this analysis

References
Beatty, C A. 1992 " Implementing Advanced Manufacturing Technologies: Rule of the Road ", Sloan Management Review, Summer 1992.
CBI, 1991, Competing with the Worlds Best,
Granger, C W J, 1969 "Investigating Causal Relations by Econometric Models and Cross-Spectral Methods," Econometrica, Vol. 37, No 3, July.
Greene WH, 1993, Econometric Analysis, Macmillan, New York
Hamblin, DJ, 1990, The Effectiveness of AMT investment decisions in the UK process machinery industry, Adv Manuf Eng, 2, 3, July.
Hamblin & Lettman, 1994, 1st European Operations Management Association
Hamblin & Nugent, 1990, "The Effect of AMT Investment in the Metal Components Industry, Cranfield University Press
New & Myers, 1986, Managing manufacturing operations in the UK, BIM
Peters T J & Waterman R H 1982, In Search of Excellence: Lessons From America's Best Run Companies, Harper & Row, New York.
Pike, R, Sharp, J., and Price, D, 1988, " AMT Investment in the larger UK firms", International Journal of Operations & Production Management, Vol. 9 No. 2.
Primrose, PL & Leonard R, 1986, The financial evaluation and economic application of advanced manufacturing technology, Proc Inst Mech Eng, 200, B1 pp27-32
Schonberger R. 1986, World Class Manufacturing, Macmillan
Strotz & Wold 1960, "Recursive v Non-recursive Systems:an Attempt at Synthesis", Econometrica,28 pp417-427
Swann, K., and O'Keefe, W D. 1990, " Advanced Manufacturing Technology: Investment Decision Process Pt2, Management Decision, 28,3, 27-34.

INTRODUCING FOCUSED & CELLULAR MANUFACTURING IN A TRADITIONAL MANUFACTURING COMPANY.

Ian McCarthy and Dr Keith Ridgway

Dept of Mechanical and Process Engineering,
University of Sheffield,
Sheffield S1 4DU

Manufacturing is a dynamic and complex process and managing this complexity is key to achieving effectiveness. This complexity evolves from the size and interrelationships of the elements (machines, materials and people) which constitute the manufacturing system. This paper details the process of change that took place at Footprint Tools Ltd using a Teaching Company Scheme (TCS) as the transfer medium and Teaching Company Associates (TCA's) as the transfer co-ordinators. The aims of the scheme were to develop a marketing orientation and to restructure the manufacturing system to produce a focus which would achieve the business and operational goals.

Introduction

In the context of change and technology transfer this paper will concentrate on:

(i)   Identifying the problem. (Current business situation versus desired business situation)

(ii)   Identifying an appropriate manufacturing solution. (Focused manufacturing by means of cellular manufacturing)

(iii)   Implementing the solution. (New attitudes, structures, and working practices)

The TCS was formed between the University of Sheffield and Footprint Tools Ltd. Footprint Tools was established in 1875, and is now one of the largest privately owned hand tool manufacturers in the United Kingdom. The range of tools produced by Footprint include: masonry tools, plumbing tools, woodworking tools and engineering tools.

Concept of Focused Manufacturing

Focused manufacturing (FM) is a state and condition which a company must attain to achieve the operational goals of a focused business. Skinner (1974) explains FM with the following definitions: (i) Learning to focus each plant on a limited, concise, manageable set of product, technologies, volumes and markets; (ii) learning to structure basic manufacturing policies and supporting services so that they focus on one explicit manufacturing task instead of on many, conflicting, implicit tasks.

Concept Cellular Manufacturing

The concepts of cellular manufacturing are more widely known by the synonym of Group Technology. The concept of Group Technology was first published by Mitrofanov (1959) and later transferred into English (1966). Group Technology has developed rapidly in the last 30 years and has changed radically. Cellular Manufacturing is considered to be an holistic application of Group Technology, which has a broader impact on the manufacturing environment. *"A cellular system is not just a simple arrangement of similar components, tooling and grouping of machines, but is a total manufacturing system which involves the complex interaction of market, management and men. It is a new approach to batch production affecting not only the production aspect of the company but the entire organisation."* Schaffer et al (1992)

Cellular manufacturing is the term used in this paper as it closely identifies with the manufacturing units (cells) which are to be established and reflects that this change affects all elements of a manufacturing organisation (especially people), and is not just the application of algorithms. A conceptual evolution in the development of Group Technology/cellular manufacturing was the development in the UK of Production Flow Analysis (PFA). This technique was introduced by Burbidge (1989). Production Flow Analysis is a series of techniques which analyse the existing situation and then develop recommendations for the formation of cells.

Process of Change

Current Business Situation

In order to develop competitive strength there existed a need for change and therefore a TCS was initiated to develop and implement solutions. A new business and manufacturing approach was needed to satisfy customer and market demands. A marketing orientation was developed where the management and workers of Footprint no longer regarded the Company as a hand tool manufacturer, but rather as a business that served the needs of hand tool users and buyers. The development of such an attitude enabled Footprint to develop realistic and effective strategies and goals. These goals would be attained by focusing the entire manufacturing system on the output of the system: products.

Developing A Focus

Focused manufacturing can only exist within a focused business, that is the company must have a knowledge of its corporate objectives, in terms of the markets in which it operates and intends to serve. This was achieved by developing a marketing orientation. The introduction of a marketing orientation passed through three distinct phases. Each phase required a detailed understanding of the cultural and human aspects relevant to Footprint Tools.

(i)   Understanding Existing Orientations
        An orientation can best be described as "the manner in which business is conducted". Orientations can include conducting business with an emphasis on cost, capacity, or even a mixture of several different factors. To identify the existing orientations a series of interviews were conducted with senior management. The dominant orientations were found to be those emphasising the product, manufacturing capacity and was linked to the internal consideration of economies of scale. The predominate orientations were production orientated.

(ii) Identifying Marketing Effectiveness
        The orientations listed, highlighted the values and attitudes of the Company and the management team. The development of a marketing orientation would be built on these values. To identify the Company's current marketing effectiveness, attributes such as: customer philosophy, marketing organisation, marketing information, strategic orientation and operational efficiency were analysed. The purpose at this stage was to find out and communicate to senior executives the perceived level of marketing effectiveness within the Company. The information also served as useful evidence of the need to introduce a marketing orientation.

(iii) Developing a Marketing Orientation / Business Focus
        To develop a marketing orientation it was necessary to change the way in which managers perceived their business world. This process was based on a fundamental re-education of the management team. The important factors at this stage were developing skills and changing attitudes, rather than simply transferring knowledge. The development of a marketing orientation is not only the installation of a marketing function within the organisation, but rather a process of change that enables Directors through to operators, to perceive Footprint as not just a hand tool manufacturer, but rather as a business that serves the needs of hand tools users. This change in business orientation provided the direction and goals required for focused manufacturing (Figure 1).

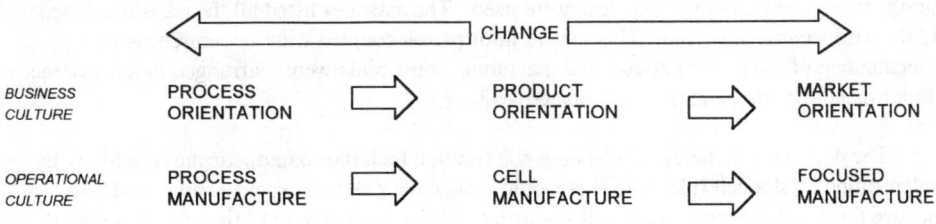

Figure 1. Culture Changes

Achieving Cellular Manufacturing
        To introduce cellular manufacturing required a process of restructuring, rationalising and amalgamating the existing system. Figure 2 shows the generic stages of change and continuous improvement.

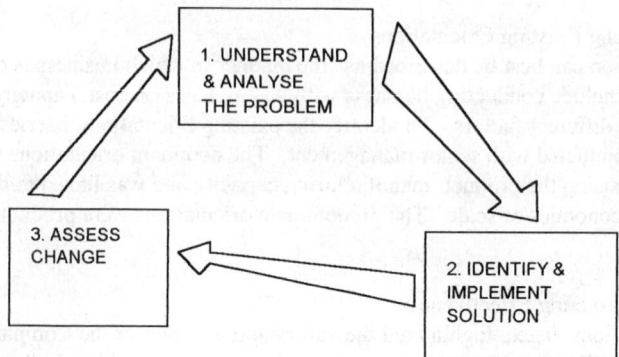

Figure 2. Generic Stages of Change

Analysis
    Within the context of cellular manufacturing, classification and coding techniques are available to perform the analysis. These techniques apply the principles of GT (usually by means of computer algorithms) and are normally supported by the technique; Production Flow Analysis, Burbidge (1989). Classification and coding systems are not essential for forming groups, but do allow the analysis to be carried out in a systematic manner. Production Flow Analysis and its sub techniques; Factory Flow Analysis (FFA), Group Analysis (GA)and Line Analysis (LA) were used to examine the manufacturing system and material flows for the company as a whole, as a focused product cell and finally for individual machines. Focused cells were identified for the following products and parts: carpenters planes, joiners tools, tinsnips and wooden parts.

Identify and Implement Solutions
    The analysis stage gathered information on the existing situation so as to create a picture of the problem. The next stage was to identify appropriate solutions or technology which could be successfully transferred to the organisation. Within the context of cellular manufacturing, cell grouping matrices were used. The matrices listed all the parts machined and the work centres involved. This matrix information coupled with a comprehensive understanding of the process routes and machining constraints were rearranged to achieve the cellular manufacturing and business objectives.

    The decision was made to introduce the solution (cellular manufacturing) gradually by implementing a pilot cell (planes). This would reduce the risk and expense and would be a test case for cellular manufacturing at Footprint. The analysis phase of the project revealed obstacles which could have prohibited the successful introduction of cellular manufacturing. This resistance was primarily generated from the expectations and motivations of people in terms of participating in a change programme.

Change Review
    The review phase is standard for any project and is an assessment of the innovation. This is an important stage of the change and learning process as it reveals the benefits and problems associated with change. The initial change was a plane product cell with a linked cell layout and multi-machine operation. Piecework payment was removed and new scheduling practices such Optimised Production Technology, Goldratt (1984), and Period Batch Control, Burbidge (1989) were introduced. Factors such as competitiveness and performance benefits (lead times, quality, cost etc.) were measured. The benefits obtained

from the pilot cell were clear and substantial and therefore strong recommendations were made to install cellular manufacturing practices in other areas of the company (joiners tools, wooden parts and tinsnips). The benefits obtained from the plane cell included:

1. Lead Times have been reduced from 13 weeks to 6 weeks. Production statistics indicated that customers were receiving satisfactory orders at a 98.4% yield.

2. Quality has improved with the use of alternative parts suppliers and the removal of piecework.

3. Shop floor inventory has been reduced by 30% due to the 45% reduction in transfer batch size.

4. Output was increased by 36%, compared with the previous year.

5. Increased job satisfaction has been achieved by involving workers in decision making, having personalised work relationships, and a variety in tasks.

Conclusion

Conclusions are drawn from this study which suggest a model for the successful transfer of technology. There is no systematic method which ensures success but rather a generic approach which analyses the existing situation and problems, and develops a solution which will overcome the forces against change and achieve the project goals. This can be related to the various stages in the introduction of cellular manufacture as illustrated in figure 3.

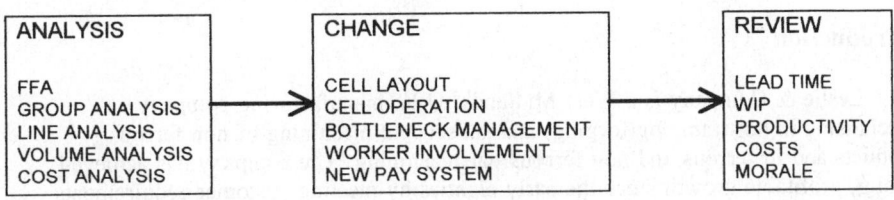

Figure 3.

References

Burbidge, JL 1989, *Production Flow Analysis For Planning Group Technology*, Clarendon Press .

Goldratt, E 1984, *The Goal*, Gower Publishing

Mitrofanov, S P. 1966, *Scientific principles of Group Technology*, National Lending Library for Science and Technology.

Schaffer, H et al, 1992, The Consideration of Human Factors in the Design of Group Technology Manufacturing Systems", Int'l Conf on Flexible Automation and Information Management 92.

Skinner, W. 1974, The Focused Factory, *Harvard Business Review*, May - June 113 - 121

REFINING THE CELLULAR STRUCTURE: A CASE STUDY IN FACTORY RE-LOCATION

R. Duffill, * D.G.M.Brooks, * J. Rowley

*School of Engineering, Coventry University,
Coventry. CV15FB. and *Leslie & Co.
Waterloo Rd. Birmingham. B258LD*

This paper is a case study about factory relocation. Early in 1994, Leslie & Company, a successful hot brass stamping and machining company, moved from Coventry to Birmingham. The opportunity was taken to refine the existing cellular manufacturing system and to provide flexible factory services so that new cellular formations could be created, as needed, at short notice. Training to support the factory move was provided.

Introduction

Leslie & Company is a West Midlands, U.K. manufacturing company specialising in hot stamping(forging) and subsequent machining of non ferrous products and in ferrous and non-ferrous bar machining. The company has achieved strong, profitable growth since the early eighties by meeting customer requirements for high quality and rapid response with no minimum quantity.

Underpinning the recent competitive development of the company has been the change from process based departments to cellular factory organization. However, the Coventry premises occupied before the move imposed severe restraints on the adoption of the desired cellular system. The move to a larger factory in Birmingham was therefore the ideal opportunity to review the future competitive requirements of the company.

Cellular organization

The cellular architecture of the Coventry factory (Brooks & Duffill, 1992) was based around the 'runner', 'repeater' and 'stranger' categories of product, where these stand for:-

runner - the same product made regularly
repeater - the same product made periodically
stranger - prototype or very low volume jobbing product

The runner category consists of a product which is made in three high volume variants and a number of low volume patterns, all of which follow the same process route. This was the most well defined cell in the Coventry factory.

The repeater category covers all other proven products and these were handled in the Coventry factory by a combination of traditional process based layout and cellular layout. That is, the billet sawing, stamping, flash clipping and vibratory finishing processes resided in their own separate departments, whereas the machining operations were done in a generic grouping of different machining processes arranged to suit typical product flow and manned by a dedicated cell team. In addition, two 'U' shaped, one piece at a time, Nagare cells were used to machine complex forgings in medium volume.(Brooks & Duffill, 1992)

The stranger category encompasses all prototype work. The new product introduction process is a key aspect of the company's business since successful handling of this stage leads to the probability of long term orders. The problem is that producing prototypes conflicts with day to day manufacturing priorities. It has not so far been possible to set up a dedicated machine group and cell team to completely isolate prototypes from the rest of manufacturing. However the concept of a product introduction team of key personnel who spend a managed part of their time on prototypes has been found helpful.

Key competitive requirements

It was intended that the new factory layout would contribute to the company's competitiveness by refining the cellular structure. The objectives were to:-
1. Arrange all equipment in well defined cellular groupings reflecting natural material flow routes and shortest possible flow distances.
2. Try to reduce the interdependence of cells by , for example, providing additional equipment to avoid shared use of resources.
3. Provide flexible factory services, air, power, etc. to enable rapid re-positioning into new cellular groupings as the volume/variety characteristics of products change over their life cycles and to monitor and cost energy demand in each cell.
4. Provide an environment supportive to continuous improvement and enabling the highest standards of housekeeping.
5. Contain each production grouping into the minimum area in line with lean production concepts. (Brooks, 1994)
6. Provide training to assist understanding by all personnel on the competitive benefits to be gained.

Planning the move

From the exchange of contracts for the new premises, to the proposed initial occupation date, was a period of about three months during which a detailed planning process took place. The premises had previously been used as a distribution centre and there were certain fundamental repairs and modifications to the basic fabric of the building to be carried out, which were independent of the proposed factory layout, so these were initiated immediately. Meanwhile the planning process commenced.

The new plant layout was developed using an iterative refining process as follows.

1. Resources were allocated to characterize each item of production equipment in terms of its physical dimensions, operator access, maintenance requirements, services needed and connection points.
2. The relevant details were entered into a spreadsheet for storage and to assist sorting into groups.
3. The information obtained was used to create a detailed plant profile record for each piece of equipment. Each record was printed on a single sheet of A4 and included a dimensioned plan and side elevation with any foundation requirements.
4. Initial layout proposals for each cell were developed with the participation of cell management and personnel. This stage used plant profile records drawn on gummed paper to develop trial layouts.
5. The resulting layout proposals were then redrawn accurately on to a factory floor plan using the Autocad software. This typically revealed conflicts with other cell layouts which had to be resolved.
6. Further consultation then took place with stages 4, 5 and 6 being repeated as many times as necessary to achieve an agreed solution.

The new factory layout

All the key competitive objectives listed earlier were met by the new layout, although some compromise was necessary in that environmental considerations dictated that stamping presses should be shielded from the rest of the factory by a sound damping wall and this inevitably compromised the clarity of cell definition.

There is now virtually no need to share resources between the cells and short, logical material flow routes have been established.

A further significant aspect of the new layout is the provision for future flexibility. It is recognised that the current cellular architecture will only be appropriate for a limited time. It is therefore very important that the factory be capable of being rapidly re-configured to deal with new products or changes in product volume. To this end, the provision of factory services received detailed attention at the planning stage.

Factory services

The services required in the factory are electricity, gas and water. For all of these a zoning logic was applied. The zoning concept has implications both for flexibility and for efficient energy management. The idea is that the factory is divided into a number of areas, the services for each of which can be completely isolated from the rest. Energy usage can also be monitored separately for each zone. This means that, for example, a part of the factory which is not currently used can have all services turned off, thus saving energy, but can be rapidly revitalised if necessary. For example, factory zones could be turned off even for short periods, such as lunch breaks. This would yield considerable savings over a year.

The electricity and compressed air supplies were of course relatively easy to zone. It is more difficult to gain instantaneous control over heating and several options were adopted. Fast acting radiant heaters were extensively used for production areas together with recovery of waste heat from compressor units. A conventional piped hot water, gas fired system was used for offices.

It was essential to provide a large number of connection points to tap into air and electricity supplies, to provide for future layout changes. This was done by installing service loops for compressed air and electrical busbars from which local connections could be dropped down to floor level, as required. The overhead services were positioned by 'layering' the pipework and busbar circuits onto the cellular layout drawing using the facilities of the Autocad system. Twelve different layers were needed for the full factory installation.

Training requirements

The majority of the Coventry workforce were expected to transfer to the Birmingham factory. Improvement groups based on cell teams were already a feature of the Coventry factory but it was intended to take the opportunity to further develop the continuous improvement culture, and, to dramatically raise standards of housekeeping in the new premises.

Since success in these aspirations would depend on the response of employees, a programme of induction training was delivered to all personnel.
The programme covered specific induction topics relevant to the relocation, but, in addition, the opportunity was taken to improve employee knowledge on aspects of competitive manufacturing such as total quality, cellular manufacturing, lean production and just in time.

The training was supported by the publication and distribution of an employee handbook.

Concluding remarks

The planning phase was all important for the achievement of the key competitive requirements. Although it cannot be claimed that such a complex exercise went 'without a hitch', it is certainly true that translating the broad vision of the new factory into detailed, phased stages allowed the move to take place with minimum disruption. Following the three month planning period, there was a further four months of dual site operation before production was fully transferred. It is estimated that production was only about 10% below normal during the changeover period.

Inevitably, this paper has focused on only a small part of the total activity involved in a major factory relocation. The development of the detailed specification and schedule for the factory services, for instance, was a major project in its own right. However, it is hoped that this paper has given some insight into the problems, methodologies and solutions used in this case.

References

Brooks, D.G.M. & Duffill, R. 1992, *Towards World Class at Leslie & Co.* Proceedings of the Eighth National Conference for Manufacturing Research, (University of Central England), 178-182.
Brooks, D.G.M. 1994, *The Competitive Edge in Manufacturing Industry*, Technovation Vol.14, (Elsevier Advanced Technology)

DESIGN OF CELLULAR MANUFACTURING SYSTEMS USING GRAI METHODOLOGY

Mr Kah-Fei Ho and Dr Keith Ridgway

University of Sheffield
Sheffield, S1 3JD

This paper describes the use of the GRAI method for the analysis and design of production planning and control systems prior to the introduction of cellular manufacture.

Introduction

Cellular manufacturing, through the application of group technology, has become increasingly popular with many advocates of the system describing the advantages that can be achieved. These include reductions in: material handling, tooling costs, set-up time, expediting, work-in-process, part make span, and improvements in human relations and operator expertise. However, there are associated disadvantages which are increased capital investment and lower machine utilisation. Before cellular manufacturing system can be introduced the design, operation and control of the complete manufacturing system needs to be examined closely. Besides the physical reorganisation of the machines and equipment into cells, where one family of products or parts can be manufactured, it is also necessary to introduce an effective production planning and control system which can operate within the manufacturing constraints and capacities.

There are numerous publications (Wemmerlov and Hyer, 1987) reporting on the configuration techniques for organising products and machines into cellular manufacturing, but virtually none on the analysis and design methodology for effective production planning and control (Sinha and Hollier, 1984), Wemmerlov and Hyer, 1987) in that environment. Wemmerlov (1988) presented various production planning and control procedures for cellular manufacture, but no overall design framework incorporating them. Moreover, Wemmerlov and Hyer (1989) and Slomp and Gaalman (1993) surveyed cellular manufacturing users and suggested that in practice the production planning and control system were not analysed and designed using a suitable methodology.

The complexities of decision making in the cellular manufacturing environment, are similar to those at a decision centre defined by the GRAI (Graphical Results and Activities Interrelated) method. The method was therefore proposed as an appropriate tool for the analysis and design of the production planning and control system in a major UK electric motor manufacturer. Doumeingts (1985) presented a GRAI model with a hierarchy of decentralised decision centres which is an analogy to the inter-related cells making consistent and coherent decision. GRAI has been applied to various situations as reported by Doumeingts et al (1987), Ridgway and Downey (1991), and Ridgway (1992), but not particularly in cellular manufacture.

Problem Definition

The UK company studied is a major electric motor manufacturer, producing motors from a few watts up to 650 Kilowatts. There are numerous variations of motor designs and sizes to cater for different areas of application e.g. corrosive environment, hazardous conditions, clean room environment. The company is in the process of implementing cellular manufacture throughout the production facility. It has grouped the electric motors into five major categories (GP1, GP2, GP3, GP4 and GP5). Presently, it lacks an effective production planning and control system to coherently co-ordinate all the necessary operations from the incoming customer orders to the final products.

Framework for Production Planning and Control in Cellular manufacturing

The objective was to integrate the different levels of the decision hierarchy in the proposed co-ordinated and concerted cellular manufacturing system as shown in Figure 1.

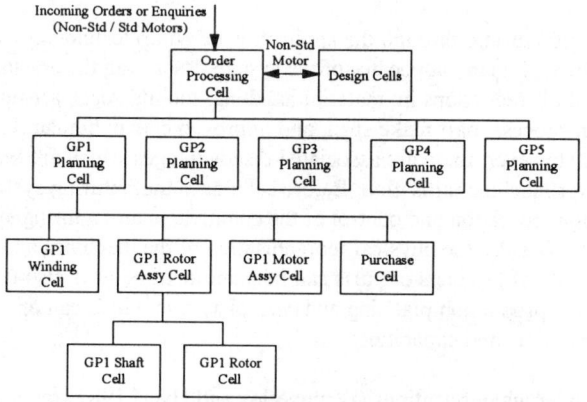

Figure 1: Hierarchical Structure of Related Cells

By having a helicopter view of the production planning and control, a GRAI grid is constructed as shown in Figure 2. The important cells (decision centres) are identified by their relative position in the row (horizon and review period) and column (cell functions). The horizon defines the production lead time of the electric motor or its main components, and the review period the time bucket for assessing the status of the job after being released towards completion. Both horizon and review period will be evaluated and reduced as the manufacturing effectiveness of the cells improves. The single and double arrows show the interaction of important information and decision flows respectively within the cellular manufacturing system. This top-down approach ensures that any inconsistencies are eliminated.

Function \ Time	External Information	Order Processing Cell	Design Cells (Electrical / Mechanical)	Purchase Cell	Control Mechanism for product families (GP1,GP2,GP3,GP4,GP5)					Internal Information
					Planning Cell	Main Production Cells				
						Winding	Rotor Assy	Motor Assy		
H = 30 D P = 5 D	Non-Std Motor Orders / Enquiries	Group and Check Orders	Technical Specification and evaluation of relevant design							
H = 10 D P = 2 D	Std Motor Order / Enquiries	Group and Check Orders		Required Special Accessories	Complete Motor Schedule					
H = 5 D P = 1 D				Possible Material Shortages		Winding Schedule	Rotor Assy Schedule	Motor Assy Schedule		Design Specification → Store

Legends: D -- Day, H -- Horizon, P -- Review Period

——► Important Information ⇒ Transmission of Decision Frame

Figure 2: GRAI grid: Proposed Cellular Manufacturing System

Bottom-up analysis, using GRAI nets, is used to examine each cell in detail. The operation of the order processing cell for non-standard motor with a horizon of 30 days and review period of 5 days is shown in Figure 3. This is similar to a standard motor with a horizon of 10 days and review period of 2 days. The incoming customer orders are classified into the five major categories before determining the viability of each order according to the technical (electrical and mechanical) specifications. For non-standard motors, technical assistance is sought from the common design cell, from which a detailed design specification is forwarded to the relevant main production cells. It is noted that the status of the relevant planning cell is consulted before any commitment is made to the customers. After accepting a feasible order, the customer details will be entered into the order entry system, confirming the delivery date and price to the customer and motor due date to the relevant planning cells.

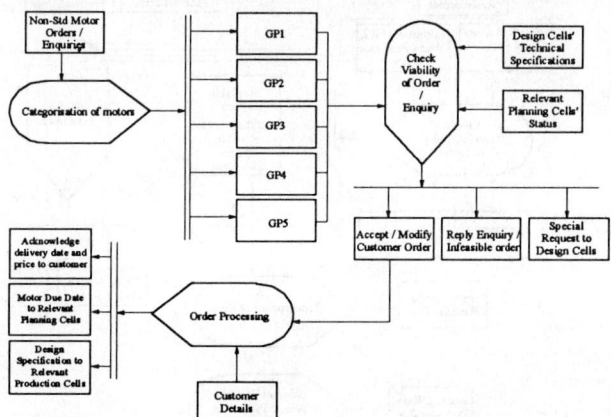

Figure 3: Ordering Processing Cell for non-standard motor

Figures 4 and 5 show the planning and winding cells for the GP1 group of electric motors. The planning cell is the brain that oversees the smooth inter-dependence of operations between the main production cells; that is winding, rotor assembly and motor assembly. It synchronises the production plans using material requirement planning (MRP) techniques for the main production cells. It also plans the purchase of special parts and/or accessories for non-standard motor. Re-planning procedures are initiated if any constraints and/or capacity problems occur in any of the main production cells. In the worst case the feedback system instructs the order processing cell to inform the customer of any delayed delivery. The winding cell matches the new production orders from the planning cell and the requirements from other sources and plans within its constraints and capacity i.e. priority, labour, parts availability, and produces a current feasible schedule. If there are problems in meeting the demand for some wound packs this is fed back to the planning cell for action. The algorithm within the scheduler is still under development. The capabilities of the rotor assembly and motor assembly cells are similar to the winding cell.

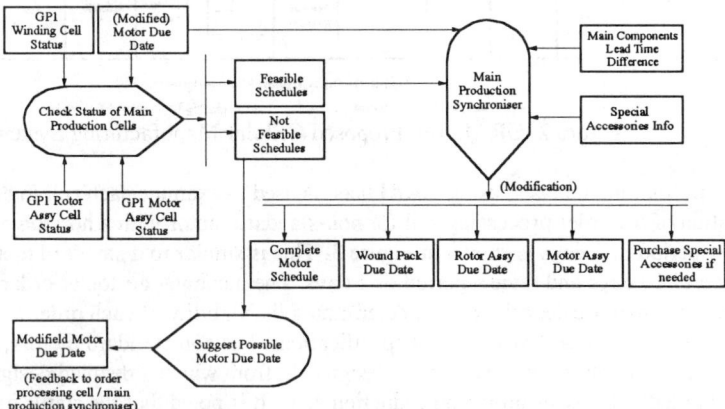

Figure 4: GRAI net: GP1 Planning Cell

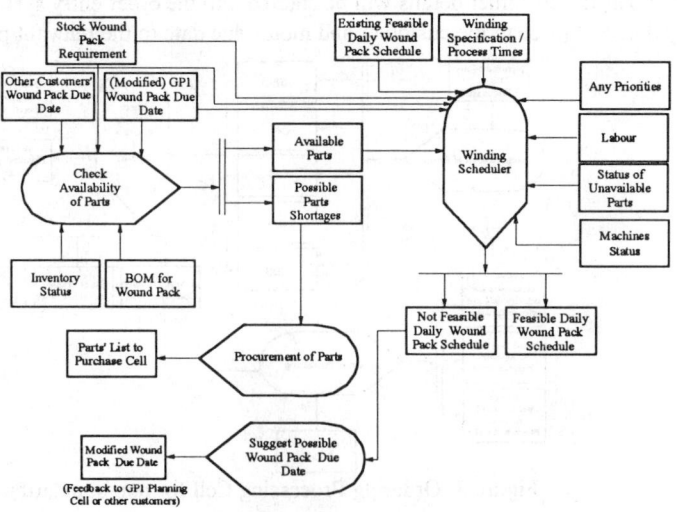

Figure 5: GRAI net: GP1 Winding Cell

The production planning and control for GP2, GP3, GP4 and GP5 motors will be achieved in a similar manner to that used for the GP1 case. The above framework provides the possibility of satisfying the following criteria when the whole system is operational:

- provide synchronisation between the production schedules of the main production cells.
- provide automatic updating of production schedules at each cell without human intervention.
- undertake automatic evaluation of capacity and constraints for each cell.
- provide dynamic feedback of production cell status.
- provide dynamic 'what-if' capability.
- provide traceability of job order relative to customer order.
- possibly reduce delivery lead time to customer due to dynamic nature of information on manufacturing resources.
- responsive to new incoming customer orders or enquiries.

Conclusions

A case study has been used to demonstrate the use of the GRAI method in the analysis and design of cellular manufacturing systems. It is critical to have constant evaluation, verification, and modification of the GRAI model with the production control staff to achieve the desired model for the Company. It was found to be a useful tool that assists the development of a detailed understanding of the distribution (decentralisation) of the main decision making functions which occur when cellular manufacturing is introduced. As a result of this research, it is hoped to establish a generic model which can be adapted to different production management systems.

References

Doumeingts, G. 1985, How to decentralize decisions through GRAI model in production management, *Computers in Industry*, 6, 501-514.

Doumeingts, G., Dumora, E., Chabanas, M. and Huet, J. F. 1987, Use of GRAI method for the design of an advanced manufacturing system, *Proc. 6th Int. Conf. Flexible Manufacturing Systems*, 341-358.

Ridgway, K. and Downey, S. 1991, A graphical technique for planning the introduction of AMT: the GRAI method, In E. H. Robson, H. M Ryan and Wilcock (eds.), *Integration and Management of Technology for Manufacturing*, 55-65.

Ridgway, K. 1992, Analysis of decision centres and information flow in project management, *International Journal of Project Management*, 10(3), 145-152.

Sinha, R. K. and Hollier, R. H. 1984, A review of production control problems in cellular manufacture, *International Journal of Production Research*, 22(5), 773-789.

Slomp, J., Molleman, E. and Gaalman, G. J. C. 1993, Production and operations management aspects of cellular manufacturing - a survey of users, In I. A. Pappas and I. P. Tatsiopoulos (eds.), *Advances in Production Management Systems (B-13)*, 553-560.

Wemmerlov, U. and Hyer, N. L. 1987, Research issues in cellular manufacturing, *International Journal of Production Research*, 25(3), 413-431.

Wemmerlov, U. 1988, *Production Planning and Control Procedures for Cellular Manufacturing Systems: Concepts and Practice*, (Production & Inventory Control Society).

Wemmerlov, U. and Hyer, N. L. 1989, Cellular manufacturing in the U.S. industry: a survey of users, *International Journal of Production Research*, 27(9), 1511-1530.

TOOLROOM ORGANISATION FOR WORLD CLASS MANUFACTURE

Mr Terry Oliver & Mr Mark Hopkins

University of Portsmouth
School of Systems Engineering
Portsmouth, Hampshire PO1 3DJ

This paper deals with the restructuring of a traditional toolroom to facilitate a transition from isolated skill centre to integrated skill base. The changes discussed are concentrated around a central concept of improved communication and empowerment. These ideas are used together with systems such as TPM, QOS & ABM to create a homogeneous and balanced structure capable of self sustaining continuous improvement in a commercial context. Throughout the reorganisation of manpower and resource, World Class Objectives and a JIT manufacturing policy are cited as the key drivers.

Introduction

The company concerned are volume manufacturers of plastic and metal fastening systems for the Automotive industry. The demands of World Class Manufacturing as defined by the company combined with the specific processes employed in production, mean that tooling plays a crucial role in both the total product quality and intrinsic manufacturing cost. In developing a new tooling operation the primary considerations lie in achieving a structure in which design for manufacture can be effectively implemented at the earliest stage, backed by empirical data from process performance. This not only offers the shortest leadtime to realisation of Customer demand, but also ensures that tooling is functionally and commercially capable throughout it's life.

The traditional role of the toolroom and associated tool design has been split between the manufacture of new tools and repair and maintenance of existing tooling. These basic requirements were split again between tools producing metal components by stamping and forming, and those producing plastic components by injection moulding. This gave rise to four distinct areas of activity in the Toolroom, each with different priorities and demands on available capacity.

New Tooling

In the manufacture of new tooling, key requirements were defined as reduction of both leadtime and tool cost. Additionally the demands for consistency, optimum cycle time and quick setting were to be considered. The company had identified the need to reduce all

New Tool prices by 20%. It was also felt that a reduction of 50% in current leadtime was vital if the overall business strategy was to be supported.

Although there are differences between the manufacturing routes of the two tool types, the primary skills are similar and the equipment utilised identical. The nominal manufacturing route at the outset was primarily via traditional manual machines, although the company did have limited CNC milling capabilities.

The underlying concept for the improved New Tooling operation, was that it should be fed 'Kits' of parts from the CNC areas and provide the final finishing, assembly and production proving of new tools.

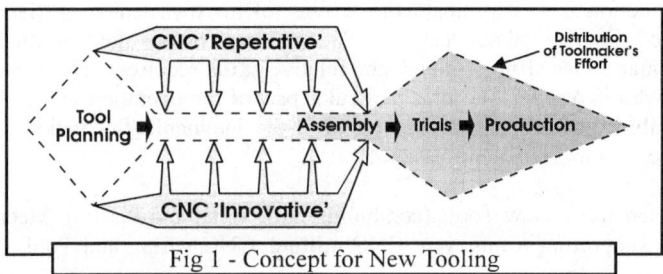

Fig 1 - Concept for New Tooling

Tool Repair

The requirements for repair and maintenance were defined largely in terms of tool availability and repair cost. A fundamental need was identified as the accurate communication of problems and their causes back to Tool design. This not only converts the production environment into a valuable source of development and improvement information, but also effectively targets and validates productivity improvements.

Tool repair was found to be the most distinctly polarised area in terms of Metal and Plastic production tooling. The patterns of working were significantly different between the two groups, as were the labour skill requirements.

Metal tooling requires a comparatively high level of day to day maintenance, with a large number of potential failure modes derived from every day activity. Additionally the metal tooling is split between four different configurations as determined by production machine types. The variation in tool types combined with some tooling exceeding 40yrs in age meant that maintenance requirements varied enormously depending on the production shop loadings. This problem was further aggravated by the fact that tools had traditionally been prepared before a run, rather than being stored in a ready to run condition. In order to effectively run a JIT production facility the conversion of tooling to be stored ready to run and the elimination of maintenance variation were given high priority.

Plastic tools require significantly less day to day attention, with problems tending to be very much more focused. Also plastic moulds tend to be broadly similar in their basic design. Mould tools ran an effective system based on storage in a ready to run condition.

New Structure

Fundamental to the re-structuring programme was the need to improve communications and flow of information. This included improved planning, promoting stronger links with tool design and with production and improving access to management to facilitate change. This combined with the Company's commitment to employee empowerment led to the concept of splitting the toolroom into small and highly focused teams able to control and shape the development of their own areas. A move to 'Cell' based manufacture in the company was also envisaged so development of small 'transplantable' teams complete with procedure and systems was considered, particularly for the repair teams.

In considering the new structure a further need was to effectively integrate existing CNC equipment with the other toolroom functions. Effective integration was seen as one of the cornerstones of both leadtime and cost reduction.

The new structure is flatter than the original, with each team being supervised by a team leader whose function is to coordinate the activity of his own team and liaise with other team leaders to optimise labour usage. The precise composition of the team varies according to demand placed upon it and the relative skills requires. This structure creates one individual who is responsible for a particular part of the operation and is expert in that field. The flexibility of labour allocation allows the inevitable fluctuations in Customer demand to be accommodated more easily.

The teams selected were; New Tools (combining both Metal and Plastic), Metal Repair and Plastic Repair. Supporting teams were CNC milling, CNC wiring and Tool Stores.

Development of the teams was tackled initially by the definition and analysis of working procedures. This gave the opportunity to identify required communication links and to begin to specify new equipment to meet the team needs. The ongoing improvement of these team 'owned' procedures and subsequent definition of specifications is a major part of the process of defining a new layout.

Key measures were identified for each of the teams and the data is logged and displayed on a regular basis. As more data is collected the teams will be involved in analysing data and devising action plans to tackle improvements in their areas. The measurement and analysis will be conducted within the framework of the Company's Quality Operating System (QOS) and TPM programmes.

In addition to establishing a framework for continuous measurement, analysis and improvement, the combination of procedures and measures will provide some of the basic data required for Activity Based Management. The move towards this system of business management will provide a further method of analysis and will complement the information collected by the toolroom teams. In particular the use of value analysis stemming from both ABM itself and activities defined in procedures, is expected to eliminate cost from the process. This approach is likely to yield savings in both cost and leadtime, whilst providing clear justification for investment against defined and quantified activities.

Re-Layout

The need to improve the layout of the Toolroom and to upgrade facilities had been recognised for some time. Contribution to the new layout is seen as a crucial part of team development and ownership. The physical change alone is likely to yield benefits from an improved working environment, however the more detailed involvement with each team means that a much greater level of consultation is possible.

An important factor in developing a new culture for the Toolroom is integration with the rest of the production areas. As stated previously communication is seen as crucial to effective performance. The breaking down of traditional barriers between trades and functions is at the root of this requirement. Having tackled the internal barriers through reorganisation, the re-layout can also tackle some of the more physical constraints such as the removal of walls between Toolroom and Production. A further consideration to the layout is the company's intention to move towards 'cell' based manufacture, this will result in the repair teams eventually being distributed out into the production process.

In developing the new layout consideration has been given to the flow of work in the shop and in particular to the flow through CNC equipment. This has led to the centralisation of CNC milling and CNC wiring and it is hoped that multi-skilling in the area will lead to the eventual merger of the two small teams into a single 4 man team, capable of operating all equipment. This flexibility combined with greater consideration for CNC at the initial planning stages of new tools will build the foundation for increasing CNC usage in the future.

In general the process of re-layout will be an opportunity for each of the teams to determine their own working procedures and to specify equipment and environment to suit their own specific needs.

Results

Early results have been very promising with the teams developing well and beginning to take on the challenges of autonomy. The main problems have been in changing the culture to get Teams Leaders to understand that they can make their own decisions and can implement the ideas of their teams.

The largest immediate area of improvement has been in the area of CNC milling, where the teams have doubled utilisation, changed suppliers to improve cutter performance and are in the process of respecifying work holding methods to improve cycle times. Further improvements are anticipated, as shop planning changes to place CNC in the first line of manufacture rather than the fringes as it had been. This is a very significant change for a Toolroom which currently sees an unbreakable link between craftsmanship and manual machines. Proving the capability of the CNC mills to the more traditionally minded Toolmakers is now a priority.

Analysis of value added to tooling using the techniques of ABM is in it's early stages at the moment, but even preliminary work has shown that there is likely to be massive scope for reduction of both manufacturing cost and leadtime.

Key measures (such as hours booked per operation) are well established and the suspicion they initially provoked has largely been overcome by feeding the information back as a matter of course. Making the data available and actively encouraging toolmakers to be data literate is proving to be a major advantage. As shown by experience in other companies, the measurement and understanding of tasks by the person performing the task is leading to useful and constructive ideas.

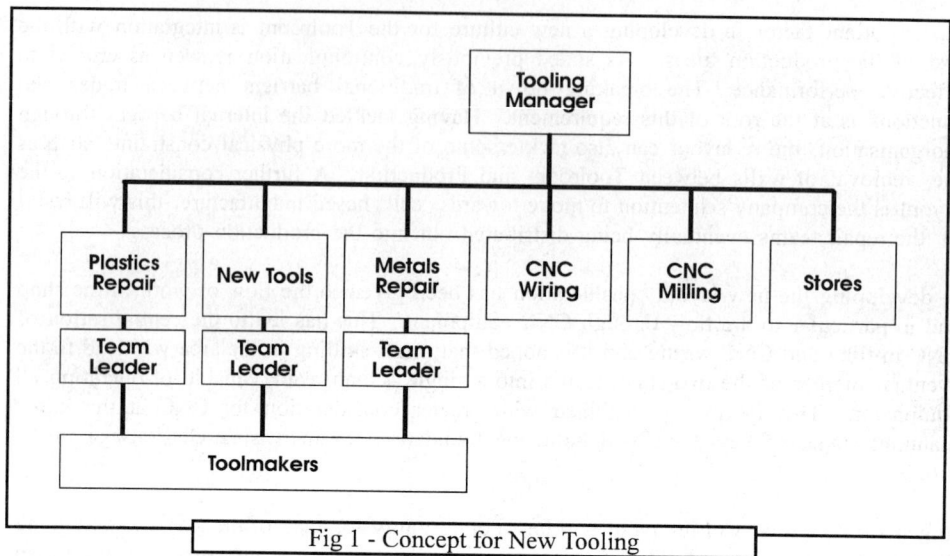

Fig 1 - Concept for New Tooling

The future appears very positive for the company, with orders and consequently new tool demand rising. The major challenges now lie in changing the remaining culture to allow new methods of tool manufacture and maintenance to be adopted.

Conclusion

The development of the Toolroom via the root of small locally responsible teams has been on the whole successful and scope exists to see even greater improvements. A complete change to more flexible ways of working is essentially governed by a perceived loss of status for the Toolmaker, who now becomes just an assembler. However it is the broadening of the Toolmaker's role to include advise on design for manufacture and reduction in set-up times which is hoped to reduce these fears.

References
Schonberger R. J. 1986, World Class Manufacturing, (Free Press)
Ichiro M. 1992, The Shift to JIT, (Productivity Press)
Drury C. 1989, Activity Based Costing, (Management Accounting)

HOW QFD CAN HELP DESIGN CELLULAR MANUFACTURING SYSTEMS

Tim Nimmons, Prof. John Kay and Dr Gareth Williams

Cranfield University
Cranfield, Bedford MK43 0AL

Planning and implementing cellular manufacturing (CM) successfully in practice requires more than grouping parts and machines. This paper argues that the focus of current research is too narrow and that a more holistic approach is required. It is proposed that Quality Function Deployment (QFD) could improve CM planning by translating strategic goals into features of the CM system, and by facilitating an understanding of the interdependency between those features. A way of applying QFD to this purpose is described.

Introduction

Cellular manufacturing (CM) has been described as a hybrid production system. In other words, it combines elements of traditional production systems (jobbing, batch and flow manufacturing) to achieve the favourable qualities of each without their respective disadvantages (Hill 1985; Bennett 1986). Burbidge (1979, 1989) describes the desirable features of Group Technology (which he prefers as a synonym for CM) as follows: 1) each cell should have its own team of workers who stay together as far as possible; 2) a cell should complete its own special set of products; 3) each cell should directly control all the machines and equipment necessary to complete the products in its family; 4) a cell's machines and equipment should be laid out in one special area reserved for the group (The layout should minimise material handling and maximise social contact within the cell.); 5) the number of workers in the cell should be small enough to obtain social cohesion; 6) tasks and output targets should be given to the cell as a whole and not separately to its individual workers; 7) the cell should regulate its own operations (at least its own dispatching, and should be responsible for its own quality); 8) material flow within the cell should be continuous, not intermittent or completing one stage on all products before going on to the next stage. While recognising that not all these will be desirable or feasible in all situations, he recommends that successful CM requires the presence of a majority of these features.

Significant benefits are claimed for CM across many dimensions of performance, and it is becoming accepted as one of the most widely applicable approaches to the organisation of production systems. A survey by Ingersoll Engineers (1990) reported that nearly two-thirds of the companies contacted were either using CM or were planning to. There are many prominent success stories, such as Schlumberger (Stoner, Tice and Ashton 1989), Northern Telecom (Taheri 1990), and Lucas and British Aerospace (Kellock 1992). CM is also associated with the recent triumphs of Japanese manufacturing techniques (Schonberger 1982, 1986). Several surveys (Dale and Wiley 1980; Wemmerlöv and Hyer 1989; Ingersoll Engineers 1990) report dramatic improvements in performance upon implementing CM. Unfortunately, the range of improvements reported, shows that good results are not guaranteed; see Table 1.

Table 1. Benefits of Cellular Manufacturing (Wemmerlöv and Hyer 1989).

Type of Benefit	Average	Minimum	Maximum
Throughput time reduction	45.6%	5%	90%
WIP reduction	41.4%	8%	80%
Improved operator satisfaction	34.4%	15%	50%
Set-up time reduction	32%	2%	95%
Improved part quality	29.6%	5%	90%

The research described in this paper considers the contribution made by the design process to the performance of the resultant cellular manufacturing system.

Design of Cellular Manufacturing Systems

Although the concept of CM is more than thirty years old, we do not understand it very well. We have only a vague understanding of how most design decisions affect the CM system's final performance. Black (1983) states, "Few rules and virtually no theory, exists for designing cellular manufacturing systems." This is still very much the case today.

Burbidge (1979) concluded from his observations of several CM implementations that simply forming cells on the shop floor would provide few benefits. He argues that the cellular organisation provides opportunities to radically change the way manufacturing is managed, and that most benefits are obtained from identifying and exploiting these opportunities. Despite this, the majority of research about CM has focused on the grouping of parts and machines. Wemmerlöv and Hyer (1986) documented over seventy contributions to this issue, and since then there has been a continuous stream of new methods. The International Journal of Production Research alone, has published over fifty new part-machine grouping procedures since 1987. Brandon and Schäfer (1992) suggest that this stream of research is often more concerned with the technicalities of the grouping procedure than striving directly to design better CM systems. It is not surprising then that so few of the methods that have been developed are used in industry. Parnaby (1986) has developed a high level, Systems Engineering based, process for designing CM systems. The process, which has demonstrated its utility in many practical CM implementation projects, defines a series of broad steps from the identification of objectives through to the development and evaluation of a solution. While the evaluation of the final solution can be related to the strategic objectives, the link between those goals and the individual design decisions is not made explicit, and no mechanism is provided for dealing with the interdependency among design decisions. Wu (1991) developed a similar process with a stage for conceptual design, where several key design decisions can be evaluated together, for compatibility and their potential for satisfying the strategic objectives, before tackling them in greater detail.

Quality Function Deployment

Introduction

Quality Function Deployment (QFD) is a practical design tool that has proven useful in helping design products that satisfy customer requirements. The following section will introduce QFD and show how it can be used to support the design of manufacturing systems.

Dr Akao developed QFD a quality planning tool in the late sixties (Akao, 1990). He defines it as "converting the customers demands into "quality characteristics" and developing a design quality for the finished product by systematically deploying the relationships between the demands and the characteristics, starting with the quality of each functional component and extending the deployment to the quality of each part and process." Burn (1990) describes QFD as a clear pictorial system for logically analysing the interrelationships between customer requirements and product design parameters. He claims that this a powerful tool to help companies understand, anticipate and prioritise customer needs in their totality, and to incorporate them effectively into the product and service they provide: critical design parameters can be identified and reasoned judgements can be made concerning design trade-offs. Akao (1990) has developed a comprehensive methodology for QFD. He makes it clear however, that this should not be followed religiously, but should be used by companies for reference and explanation to develop individual approaches to QFD that suit their specific circumstances. A less complex process is commonly described in western literature. Sullivan (1986), Hauser and Clausing (1988), Burn (1990), and ASI (1992), all describe similar procedures to the one outlined below.

Overview of the QFD Process

The basic mechanism used by QFD at each stage, to develop an understanding of the relationships between a set of requirements and the means for their satisfaction, is a matrix. Requirements are prioritised and listed along the vertical axis of the matrix, and a set of measures to achieve them are developed along the horizontal axis. Each location in the matrix relates a requirement with a mean and can be assigned a value to indicate the strength of the relationship. Measures that have a significant effect on the total

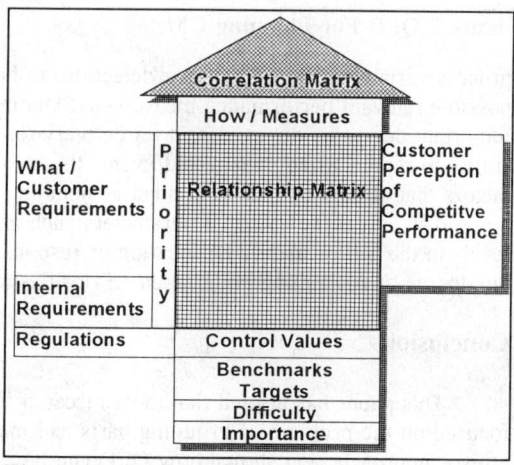

Figure 1. Basic QFD Matrix

requirement can be identified for improvement and target values set. Critical measures (new, high impact or difficult to achieve) can be cascaded as the requirements of a subsequent QFD matrix. Figure 1, shows a QFD matrix with an area for competitive assessment and correlation matrix to trace the interdependency between means. Typically, for product design, a four matrix cascade is advocated. First, customer requirements are translated into product characteristics, which in turn require certain part characteristics. Process parameters that will produce the required part characteristics can then be determined, and finally operating and control requirements for the processes can be derived.

QFD Applied to CM system Design

Black (1991) comments, that the design of manufacturing systems is a neglected aspect of Concurrent Engineering, which has focused on the relationship between product design and process design. Though Burn (1990) claims that QFD is not limited to designing physical products, most of the literature refers to deploying product quality requirements as described above. Note for example, that this cascade does not consider the direct impact of an assembly process on customer satisfaction. Furthermore, a recent survey (Rose, 1994) reports that most companies claiming to use QFD are only working with the first matrix.

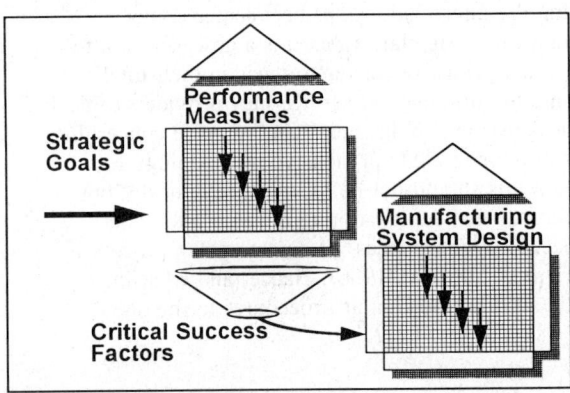

It is proposed that the mechanism of QFD can be used to translate strategic goals into features of the CM system. Use of the basic matrix remains as it would for product design but the inputs and outputs change, as does the cascade, see Figure 2. For example, a company may have such key objectives as quick and dependable delivery lead time. Working through the matrix might show, planning lead time, setup time, queuing time,

Figure 2.QFD For Planning CM

process variability, speed of error detection, to be the most influential of the many possible relevant performance measures. Planning lead time may be particularly important due to its direct impact on delivery lead time and because it affects the ability to replan when unexpected events threaten customer delivery. Focusing on factors that affect key measures such as planning lead time, the second matrix would be used to assess the effect of parameters such as: product part count, number of levels in the bill of materials, location of responsibility for planning and control, and the degree of replanning undertaken. Priorities and targets could then be determined.

Conclusions

This paper has argued that most research into the planning of CM is overly focused on the problem of grouping parts and machines. It has called for a more holistic approach, and shown how QFD can be applied to this problem. A project is currently being developed to test this procedure with a manufacturing company that already has significant knowledge about planning CM. Their experience of designing CM systems will provide a useful benchmark against which the utility of the QFD approach can be measured. This work is also the subject of an ACME proposal to develop the procedure into a usable tool.

References
Akao, Y. 1990, *Quality Function Deployment* (Productivity Press, Portland)

American Supplier Institute, 1992, *Quality Function Deployment: Practitioners Manual*, (ASI, Milton Keynes)

Bennett, D. J. 1986, *Production Systems Design* (Butterworths, London)

Black, T. J. 1983, Cellular Manufacturing Systems Reduce Setup Time, Make Small Lot Production Economical, *Industrial Engineering*, **November**, 36-48.

Black, T. J. 1991, The Design of Manufacturing Systems: Axiomatic Approach, *PED-Vol 53, Design, Analysis and Control of Manufacturing Cells*, (ASME), 1-13.

Brandon, J. A. and Schäfer, H. 1992, Strategies for Exploiting Emerging Programming Philosophies to Re-invigorate Group Technology, *Proceedings of 29th MaTaDoR Conference*, 187-194.

Burbidge, J. L. 1979, *Group Technology in the Engineering Industry* (Mechanical Engineering Publications, London)

Burbidge, J. L. 1989, *Production Flow Analysis* (Clarendon Press, Oxford)

Burn, G. R. 1990, Quality Function Deployment. In B. G. Dale and J. J. Plunkett, *Managing Quality*, (Philip Allen, New York) 66-88.

Dale, B. G. and Wiley, P. C. T. 1980, Group Technology: Predicting its Benefits, *Work Study*, **February**, 15-24.

Hauser J. R.and Clausing, D. 1988, The House of Quality, *Harvard Business Review*, **May-June**, 63-73.

Hill, T. 1985, *Manufacturing Strategy* (Macmillan, London)

Ingersoll Engineers 1990, *Competitive Manufacturing The Quiet Revolution* (Ingersoll Engineers Ltd, Rugby)

Kellock, B. 1992, Unlocking the Rewards of Cellular Manufacture, *Machinery and Production Engineering*, **21st February**, 32-45.

Parnaby, J. 1986, Competitiveness Through Systems Engineering, *Proceedings of the Cambridge Manufacturing Forum*, (Institution of Mechanical Engineers) 1-15.

Rose, E. 1994, Working Paper (CIM Institute, Cranfield University)

Schonberger, R. 1982, *Japanese Manufacturing Techniques* (Free Press, New York)

Schonberger, R. 1886, *World Class Manufacturing* (Free Press, New York)

Sullivan, L. P. 1986, Quality Function Deployment, *Quality Progress*, **June**, 39-50.

Stoner, D.L. Tice, K. J. and Ashton, J. E. 1989, Simple and Effective Cellular Approach to a Job Shop Machine Shop, *Manufacturing Review*, **2.2**, 119-128.

Taheri, J. 1990, Northern Telecom Tackles Successful Implementation of Cellular Manufacturing, *Industrial Engineering*, **October**, 38-43.

Wemmerlöv, U. and Hyer, N. L. 1986, Procedures for the Part Family/Machine Group Identification Problem in Cellular Manufacturing, *Journal of Operations Management*, **6.2** 125-147.

Wemmerlöv, U. and Hyer, N. L. 1989, Cellular Manufacturing Practices, *Manufacturing Engineering*, **March** 79-82.

Wu, B. 1991, *Manufacturing Systems Design and Analysis* (Chapman and Hall, London)

ASSESSING AND IMPROVING MANUFACTURING FLEXIBILITY

Ms Nicola Bateman and Dr David Stockton

De Montfort University
Leicester LE1 9BH

The purpose of this paper is to propose a viable tool which can help
manufacturers improve their flexibility. This is achieved by initially
outlining a model for flexibility which deals with a manufacturing system's
ability to change between different products. The model is then developed
to include a database tool which can assist manufacturers in identifying
which areas of their manufacturing system inhibit flexibility. Once a
manufacturer has identified their flexibility "bottlenecks" alternatives can
be considered. These alternatives can be similarly assessed and the
benefits compared with the current system.

1. Introduction

Flexibility in manufacturing is perceived by many companies as important, as it
enables them to deal with dynamic markets and changing environments. Implementing
flexibility in manufacturing systems has however, proved difficult. This is demonstrated
by Jaikumar (1986) who showed the use of high technology hardware, such as FMS's,
has not provided a solution to the flexibility problems of many companies, and Hill and
Chambers (1991) who identifed that companies do not evaluate the type of flexibility they
need before investing in new manufacturing systems.

This paper attempts to simplify flexibility concepts by proposing a model which
looks purely at the outputs of a manufacturing system i.e. the products it makes. This
model incorporates the concepts of manufacturing flexibility range and response as
outlined by Slack (1991).

2. Model and Theory

Shown in Figure 1 is a model of the range of products which a company makes.
Area β represents the products that the company *currently* makes and area α represents
the products the company *could* make. Area β is finite and composed of discrete

elements, with each element representing a specific product, and area α has a finite boundary although it consists potentially of an infinite number of different products. The area outside α is the universe of all potential products "U" which is infinite and unbounded. A similar model was proposed by Dooner (1991) who describes an application domain which represents the possible states of the manufacturing system, thus focusing internally on the manufacturing system rather than as Figure 1 shows, the products it can produce.

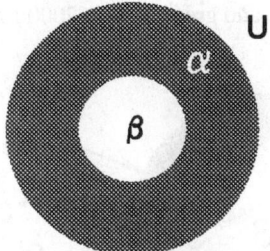

Figure 1 Model of Product Ranges

Examining this model in terms of Slack's (1991) definitions of flexibility range and response, range can be identified in Figure 1 as the size of areas α and β. Response, however, must be divided into three types, ie:

Response 1:- Assuming the manufacturing system can only make one product at a time, a type 1 reponse is the time it takes to move around area β i.e. the amount of time it takes to change from manufacturing one product to another.

Response 2:- The effort it takes to enlarge area β. This can also be expressed as moving products from area α to β. i.e. the time and cost of introducing a new product to manufacturing.

Response 3:- The effort it takes to enlarge area α. i.e. increasing the manufacturing capabilities of the system.

If the boundaries between areas α and β and the universe of product "U" are considered, it is possible to identify what products are in area β - those products the company actually makes. Defining the edges of area α, however, can be less clear. To actually introduce a new product may for example simply require programming a CNC lathe or writing a process plan. Some products, in order to produce them, will however, require a low level of investment such as a jig. Other products are a stage beyond this, and may require, for example, more expensive production machinery to be installed in the manufacturing system. Thus the boundary between area U and α is not easily defined.

To deal with this viewpoint, the model shown in Figure 1 has been modified to that shown in Figure 2 with a number of concentric bands for area α, each band represents a higher level of investment in either money or time or both. Thus:

α_A represents the range of products that can be manufactured simply through changes in the information system, such as process plans or computer programs.

α_B represents the range of products which can be manufactured through simple hard ware changes such as jigs and fixtures.

α_C represents the range of products which would need a medium cost investment in the manufacturing system.

α_D represents the range of products which would require a high level of investment to adapt the manufacturing system to enable production.

The actual levels of investment in terms of pounds and hours of time invested, will vary between industries. In the case study of a simple FMS the ranges were; up to £250 for α_B , between £250 and £2000 for α_C and greater than £2000 for α_D.

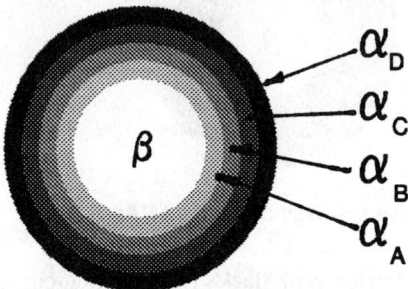

Figure 2 Modified Product Mix Model

3.Methodology for Developing Database Structure

Translating the model outlined above into a useful tool can accomplished by representing the model on a computer. In order to do this it is necessary to establish the range of each of the α areas in terms of a physical reference scheme, such schemes are commonly used for categorising products in group technology. It should be noted that different industries and technologies will have to define which characteristics are relevant to them.

In the case study FMS we used shape, size, materials, operations and routes to define the α areas. Each of these types of physical attribute is termed a "characteristic".

Characteristics then are assigned a number of criteria which defines a numerical range or highlights one category of the characteristic. For instance, for the shape characteristic the criteria may be, rotational, flat, irregular or extrusion. To relate this to the manufacturing system, the capabilities of each part of the system is assessed in terms of each characteristic's criteria for all characteristics which are appropriate. Each part of the system can typically be expressed as a machine or subsystem. This builds up into a three dimensional database: characteristics, criteria and machines or subsystems, being the three dimensions.

The assessment is performed in terms of the categories α_A , α_B , α_C and α_D. It is designated an A if the machine has the capability to perform the characteristic to the required criteria level, a B if it would require a small degree of investment, C for a larger investment and D the largest. An example showing the size characteristics for the FMS system, (called system 1), is shown in table 1.

The methodology outlined above only considers responses 2 and 3. Modelling response 1 can be done using some of the principles of probability and flexibility in Chryssolouris and Lee (1992).

Table 1 Showing "Size" Capabilities of System 1

Char: size	Criteria	-	Dimension	d in mm
Machine or subsystem	0-40	40-80	80-200	>200
CNC lathe	A	A	C	D
CNC miller	A	A	C	C
AGV	A	A	A	A
Robot R1	A	B	D	D
Robot R2	A	B	D	D
Robot R3	A	B	D	D
Robot R4	A	B	D	D
Carrousels	A	B	B	B
Software	A	A	A	A
O/p station	A	B	B	B
I/p station	A	B	B	B

4. Applications

The need to build manufacturing systems which can easily incorporate new products is a challenge facing manufacturing engineers. The problem is, that at the time of designing the manufacturing system, the exact form of these new products is not known, hence the need for flexibility.

The database outlined above provides an ability to map the capabilities and potential capabilities of a manufacturing system. Hence manufacturing engineers can evaluate alternative system's abilities to respond to future changes. This can be performed for all the product types covered by the range of characterisitcs, or as outlined below one or two charecterisitcs for which there is an anticipated market demand. For example if an alternative System 2 were proposed which replaced the robots 1-3 by a larger robot with a larger working envelope and a gripper with a wider size range, this would modify the robotic aspects of Table 1 to that shown in Table 2. Note the larger robot is denoted as RL.

Table 2 Showing Robotics "Size" Capabilities of System 2

Char: size	Criteria	-	Dimension	d in mm
Machine or subsystem	0-40	40-80	80-200	>200
R4	A	B	D	D
RL	A	A	B	B

It can be seen from the data that products with dimensions between 40 and 80 mm in System 1 will require changes to R1, R2, R3, R4, the carousel, input station and output station. System 2 will require changes to R4 the carousel, input station and output station. Thus in terms of cost the difference between these two systems for introducing products up to 80mm in size areshown in table 3

Table 3 Cost Comparison Of Developing Systems 1 and 2

	System 1	System 2
R1	250	250
R2	250	-
R3	250	-
R4	250	-
RL	-	-
Carousel	250	250
I/p station	250	250
O/p station	250	250
TOTAL	1750	1000

Table 3 demonstrates that System 2 is more flexible in terms of sizes up to 80 mm that System 1. The system designer can also consider other characteristics such as routing flexibility which may be restricted by the use of a larger robot. The database could also be used to identify "bottlenecks " i.e. those machine which greatly restrict the product range of the system. These bottlenecks can then be replaced with improved alternatives.

5. Conclusions

The model outlined above is a simple method of modelling the complex subject of flexibility. It looks at the outputs of the system rather than the internal system structure, thereby focusing upon what really matters to manufacturers i.e. products. The methodology described in 4 represents a simple and robust technique for mapping the capabilities of the system. Different types of process technology can use this system by identifying appropriate "characteristics".

6. References
Chryssolouris G. and Lee M. (1992) An Assessment of Flexibility in Manufacturing Systems, *Manufacturing Review* **vol 5 no 2 June 1992**

Dooner M. 1991, Conceptual modelling of manufacturing flexibility, *Computer Integrated Manufacturing* **vol 4 No 3** 135 - 144.

Hill T. and Chambers S. 1991, Flexibility - a Manufacturing Conundrum. *International Journal of Operations and Production Management* **vol 11 No 2** 5-13

Jaikumar R., Postindustrial Manufacturing. *Harvard Business Review* **Nov/Dec 1986**

Slack N. 1991, *The Manufacturing Advantage* 1st Edn (Mercury)

Stockton D.and Bateman N. 1993, Flexibility in Manufacturing. *9th NMCR Bath Univ.* Sept 7-9th 1993

CELLULAR MANUFACTURING: HOW TO ACHIEVE COMPETITIVE QUALITY THROUGH JOB DESIGN

Jane Goodyer and Dr Stuart Spraggett

Coventry University
Coventry, West Mids., CV1 5FB

Cellular manufacturing is emerging as one of the major tools being used in the drive to manufacturing competitiveness. The Ingersoll Engineers survey (1990) shows that the approach has delivered many benefits. Yet despite general satisfaction with the approach many installations have failed to yield excellent quality early in the lives of the cells. This survey shows that the main benefits of cellular manufacturing are from devolution of control and management to the lowest possible level in the organisation. The dismantling of traditional centralised control functions means that cell design routes must be capable of delivering designs for peoples jobs in the new situation. The paper will describe our early approaches to the cell design process which will allow the design information for jobs in the cell to be produced in a rational way.

Introduction

The popularity of cellular manufacture reflects the unpredictable nature of modern business. Flexibility, quick response and freedom to alter direction are survival issues. According to Ingersoll Engineers (1993), 73% of UK companies employ cellular manufacturing. It allows companies to modularise their activities and devolve power to lower levels in the organisation. This application is sometimes described as a 'business-within-business' approach.

Many companies are following the trend towards product focussed cells, with the equipment, multi-skilled people and support functions needed to produce a finished product, sub-assembly or component.

The recent survey by Goodyer and Higgins (1993) showed that companies' main objectives in introducing cellular manufacturing are:

- On time delivery
- Improved customer response
- Reduced inventory
- Improved quality

These are the 'hard' business drivers relevant to industrial production. They relate mainly to improved market response and better logistics and engineering. Quality improvement is one of the main objectives for introducing cellular manufacture, and one of the key issues to tackle when gaining competitive advantage.

Cellular Manufacturing Results

The Ingersoll survey (1990), research, Wemmerlov and Hyer (1989), and numerous case studies, Lloyd and Mason (1993) and Caulkin (1988) have shown that companies are meeting their initial objectives, and that significant improvements have been made, particularly, in delivery times and reduced inventories.
Typically these are:

- Reduced inventories by 30%
- Reduced lead times by 50%

Yet, even though the great successes lie in the operational areas of businesses (as illustrated) substantial quality improvements are generally eluding many cell installations. Too often world competitive levels of quality have been achieved by subsequent painful revisions to the design of the cell and the way it is operated.

In addition to these 'hard' issues, cellular manufacturing is also playing a large part in people issues. Generally, cells are creating a new form of employee, one who has various skills, whose tasks may include: machine operating, quality control, scheduling, maintenance, ... the list is forever increasing. The evidence is that the effect for those who work in the cells is that their jobs are now more demanding, interesting and rewarding.

The Ingersoll (1990) and Goodyer and Higgins (1993) surveys have highlighted the importance of preparing people to understand and accept the new way of working and to have the necessary skills. Typical replies highlights the importance people have on the success of a cell:

- "cellular manufacturing is as much about people as machines"
- "on those areas where the cell approach is beginning to work it is due to the shop floor worker"
- "employees must be allowed involvement in cell and process design or they will not own the cells"

It can be deduced that generally the 'hard' business objectives are the drivers for cellular manufacture, but it is the 'soft' people issues which will ensure success.

Cellular Manufacture - The Shortfalls and their Cure

As previously discussed cell manufacturing is extremely popular and is gaining benefits in many business areas, but its shortfall is quality. Initial research has highlighted that the reasons for this quality shortfall are that the design process of the cell is insufficiently focused on defining and achieving specific business benefits.

Many tools for the design and management of products and processes for high quality have been introduced over the last few years. However there has been no effective overall design methodology for use in a 'Simultaneous Engineering' mode to design cells and their products.

The main issues which companies have found to be difficult are:

- Breakdown of whole-product and whole factory specifications to specifications for components and cells.
- Design of jobs to help people to create positively good quality.
- Design of process controls for high variety short batch manufacture.
- Establish 'pulse-point' measurement and reporting techniques for quality measurement in a system.

Currently work has been undertaken at Coventry University to generate a methodology whereby product, process and jobs are designed concurrently, refer to Figure 1. This will enable customer wants and needs to be the driver throughout the cell design process, resulting in world competitive quality early in the lives of the cells.

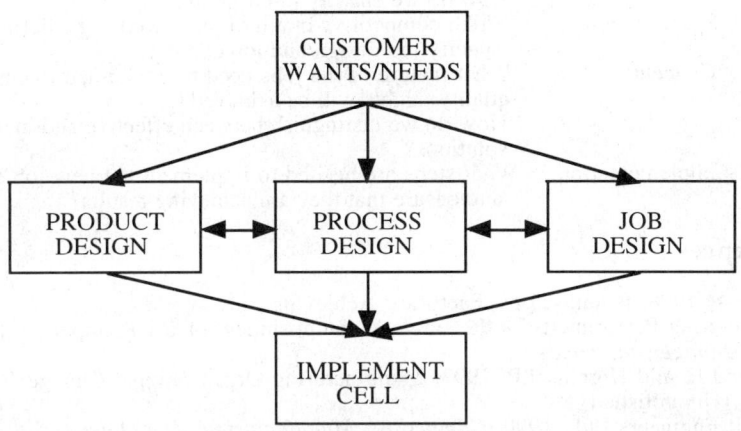

Figure 1. Concurrent Cell Design

Job Design

As previously mentioned, people have the most significant effect on the success or failure of a cell. Generally, many installations have reported many great results of people working in cells. For example cells can create:

- Ownership
- Autonomy
- Flexibility

These results are often 'hit or miss'. Some installations plan the skills that are needed, using skill matrices and training the skill shortfalls, others provide training workshops on problem solving techniques and team building. Yet, many installations do not have a structured approach when designing jobs, they are usually evolutions of previous jobs, leaning towards free-play. The problems which can arise are:

- There is a 'grey' area about peoples' responsibilities (cell boundaries / support functions are ill-defined). Usually cell operatives' responsibilities are left very open: they need to be as flexible as possible.
- Usually there is no learning capability built into jobs.
- Generally existing job designs are applied, with little change, to the cell.
- There is no system for evaluating the effectiveness of the job design.
- Job designs are usually constructed after the cell layout is designed.

In order to address these issues as an integral part of the methodology, the objectives must include:

'to design the total human task and the individual tasks needed in a cell to achieve its quality objectives'.

At each stage of the methodology the job design process will define what is needed to achieve the overall objective. The questions which will need to be answered are:

Measurement:	What quality measures refer to the job performance and range of skills needed ?
	How do we classify job designs?
Specification:	Which competitive quality issues need to be defined to specify job design and job content?
Concept:	What feature of the jobs need to be defined to ensure quality targets will be achieved?
	How do we distinguish between effective and ineffective solutions?
Implementation:	What steps are needed to implement the new job designs and ensure that they are achieving results.

References

Caulkin, S. 1988, Britains' Best Factories: Achieving truly World Class Manufacturing Performance is the single accomplishment of 6 UK Super plants, *Management Today*.

Goodyer J.E. and Higgins J.P. 1993, Manufacturing Organisational Change Survey, (Unpublished).

Ingersoll Engineers Ltd., 1990, *Competitive Manufacturing: The Quiet Revolution*, (Ingersoll Engineers, Rugby).

Ingersoll Engineers Ltd., 1993, *The Quiet Revolution Continues*, (Ingersoll Engineers, Rugby).

Lloyd, M.R. and Mason, M. 1993, Large Electrical Machines: Powering a business renaissance for the twenty-first century, *Proceedings of the Institution of Mechanical Engineers*, **207**,

Wemmerlov, U. and Hyer, N.L. 1989, Cellular Manufacturing Processes, *Manufacturing Engineering*, 79-82.

APPLYING PERFORMANCE INDEX TECHNIQUE TO MONITOR FLEXIBLE MANUFACTURING CELLS

Mr.D. Stuart McLachlan, Mr.Terry Shaw and Mr.Nick Brigham

Department of Mechanical Engineering, University of Teesside
Middlesbrough, Cleveland, TS1 3BA

The Performance Index (PI) technique is described enabling sensory information gathered from key areas of flexible process machines within a manufacturing cell, to be analysed and presented at the Human Computer Interface (HCI) in real time. Human cognition is greatly improved by use of a PI which monitors and predicts quality sensitive areas of the processed product as well as tooling and machine control areas. Applications in flexible assembly and machine tools are demonstrated and discussed.

Introduction

In the application of flexible robotic assembly, mating of parts can be quite complex and therefore using active compliance and Feed-Transport-Mate (FTM) methodology from Selke (1988), sensory data is broken down into its generic components suitable for monitoring the process of assembly. The PI method makes use of statistical process control (SPC) and provides an economical method of analysis, data storage, and fulfils product liability requirements. Case studies demonstrate a real time HCI which can intelligently inform current status, process predictions with suitable diagnostics.

The PI Method

Initially the method overcomes the mutually exclusive goals of gathering large amounts of data to analyse in real time and the ability to store this data to comply with product liability laws, E.E.C. council directive 85/374/EEC (1985). Parameterising the sensory trace to a straight line reduces manipulated data and enables SPC monitoring, as discussed by McLachlan, Bannister, et al, (1992). Applying regression analysis allows the monitoring of this straight line relationship and a correlation coefficient monitors the spread of data, experienced as varying cell dependencies and system dependant noise. Since calculating the parameters is by summation, real time analysis is performed with minimal computing compared with alternative methods.

The HCI now automatically interprets the numerous possible SPC charts of simultaneous short duration subspaces. HCI presentation can cause poor cognition as discussed by Patrick (1987), and therefore the human operative is presented with a single **performance index** (PI) number between 0 and 100%.

PI Assessment on Parameter Control Charts

Applying SPC control charts to each parameter, the PI value corresponds to the position of the point relative to the mean and the control limits. As the point moves away from the mean, its chances of occurrence correspondingly reduce. Action limits are set at $\pm 4s$ (standard deviations) and shut down of the machine takes place at/or before this point. The PI assessment variance of the point about the mean is summarised in Figure 1 where the diagram shows that the PI varies from 100 at $\pm 2s$, and reduces linearly to 0 at $\pm 4s$.

Initially the moving average and range is found from the sub sample of the current value and the previous four values found in the history array. These points are then compared with their associated control limits and the PI assessment found using the formula given in Figure 1. The PI value is also influenced by chart trends such that a deteriorating trend of 7 consecutive points reduces the PI value to 30%.

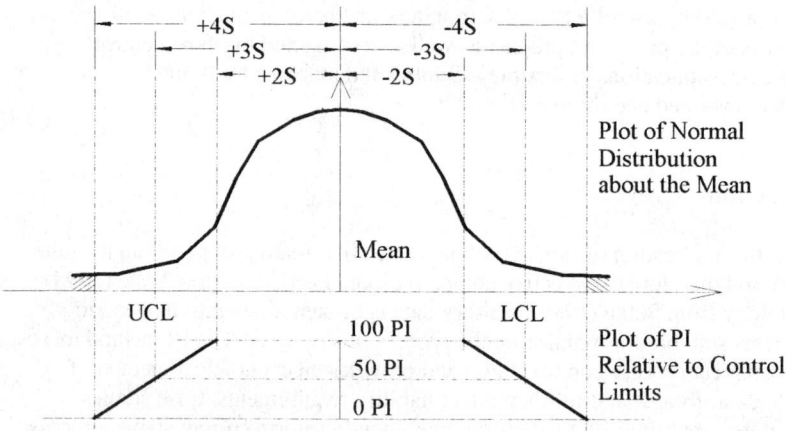

Figure 1. Assessing PI Relative to the Control Limits

The main HCI presentation is in the form of a bar graph varying from green at 100% to red at 0%, displaying the **worst performance index** (WPI) which indicates the SPC chart interpretation of the offending parameter.

Case Study: Britax Lamp Assembly (see SERC contract No.GR/F 70037)

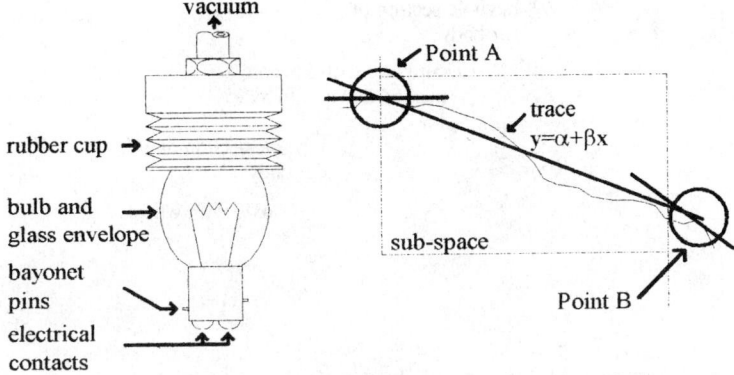

Figure 2. Plain Bayonet Bulb with sensory trace Sub-Space

Part of a car lamp assembly from Britax (PMG) Ltd. is shown in Figure 2. Due to the varying tolerances of the glass envelope of the plain bayonet bulb, a vacuum cup gripper is used. The sensory trace should obey Hook's Law whilst compressing the bulb holder spring, but suffers from **system dependencies** due to compression of the rubber vacuum cup, rubber material cavities and the compliance of the tooling and robot. Other **system dependencies** have been catalogued by McManus and McLachlan, et al, (1992).

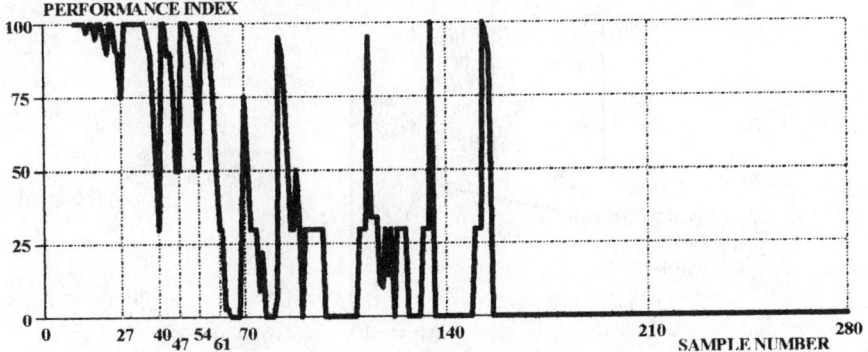

Figure 3. Graph of Worst Performance Index

Algorithms and software by McLachlan (1993) enables us to focus a subspace on the **quality sensitive** area of the trace. Sampling 280 assemblies using a deteriorating force sensor, the regression analysis intercept parameter a gave the earliest discontinuity at sample number 98 followed by the gradient β parameter several samples later. The correlation coefficient *r* indicating spasmodic breakdown of the force sensor beyond sample 150, where the force data becomes more randomised with less association to the modelled straight line. Plotting SPC parameter control charts, the α parameter gave earliest warnings of non-control above sample number 60, however, Figure 3 shows the benefits of a moving average method employed in calculating the PI with an earlier prediction of problems given on sample 40.

Case Study: Hornby Model Car (see SERC contract No.GR/F 70037)

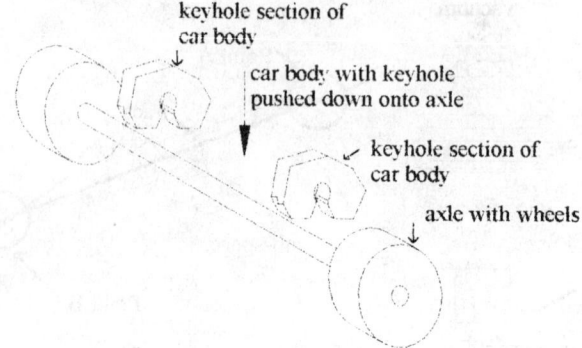

Figure 4. Mating Front Wheel Axle Parts of Car Body

A study on a "Scalextric Walter Wolf" racing car model produced by Hornby, shows in Figure 4, a "double click fit" due to the keyhole moulding in the body shell and the non-instantaneous fitting of both sides of the axle. PI analysis creates **quality sensitive area** subspaces shown in Figure 5 such that if the moulding conditions change gradually during the process, cracking during insertion will result in predictable changes of the α and β parameters.

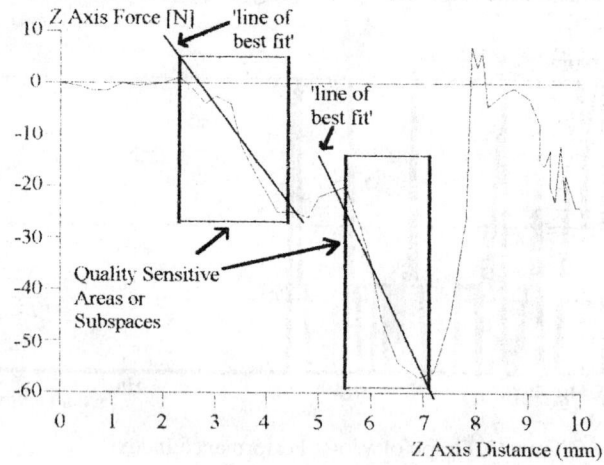

Figure 5. Axle Assembly: Quality Sensitive Sub Spaces

Case Study: Monitoring Single Point Lathe Tool

On-line tool wear monitoring is achieved by sampling piecewise every 5mm tool travel whereby the β parameter is relevant to tool wear. Since magnitude of force varies to cut diameter, machine diagnostics requires further analysis. Figure 6 shows tool wear PI early warning during diameter 5 much earlier than observed overheating on diameter 7 and tool 'crumbling' during diameter 8. This method overcomes cyclic forces due to 'crabbing' of the slides and would also enable less downtime compared with 'tool life' methods.

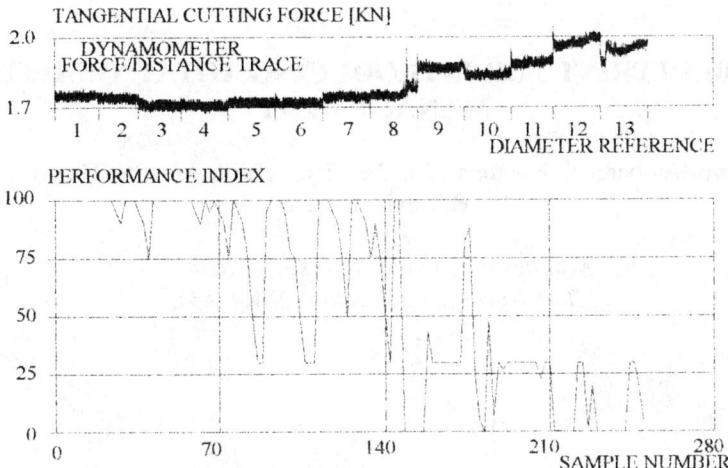

Figure 6. Graph of β Performance Index

Discussions

The PI concept for improved human cognition within a HCI has been presented which uses established SPC techniques with automatic interpretation of the control charts. The predictive nature of the PI method appears to be reliable with work on flexible cell diagnostics within generic subspaces ongoing. Other applications include hand operated assembly fixtures where sensing the force and movement of the mating part would enable detailed SPC to be accomplished.

References

E.E.C. Council Directive 85/374/EEC, Regulations and Administrative Provisions of the Member States Concerning Liability of Defective Products, Official Journal of the EEC, No.L210/29, 1985.

McLachlan, D.S., Bannister, B.R., Joyce, A.D. and McManus, D. 1992,*Process Control Parameters in Flexible Assembly*, 11th IASTED Int. Conf., MIC 92, Innsbruck, Austria, pages 439-440.

McLachlan, D.S. 1993, *Flexible Assembly Cell Manipulator*, submitted Ph.D. Thesis, September 1993, University of Hull.

McManus, D., McLachlan, D.S., Selke, K.K.W., and Joyce, A.D. 1992, *Parameter Relationships for Interactive Flexible Assembly Cell Control*, 8th Int. Conf. Computer Aided Production Engineering, University of Edinburgh, pages 235-240.

Patrick, J. 1987, *Information at the Human-Machine Interface*, New Technology and Human Error, John Wiley & Sons Ltd. pages 341-345.

Selke, K.K.W. 1988, *Intelligent Assembly in Flexible Automation*, Ph.D. Thesis, University of Hull.

SERC contract No.GR/F 70037, *Assembly Cell Control based on Generic Processes*, funded by the ACME directorate.

A BLUEPRINT FOR INTRODUCING TOTAL QUALITY MANAGEMENT

Abby Ghobadian, Jonathan Liu, David Bamford, Annette Thompson
& David Gallear

Middlesex University Business School
The Burroughs, London NW4 4BT

This paper is based on the collaboration between Betts Plastics, a high
volume plastic components manufacturer, and Middlesex University
Business School (MUBS). The aim is to develop a workable
methodology for the introduction and sustainment of a "Continuous
Improvement" programme at Betts. The approach adopted was
developed at MUBS, then modified to fit Bett's requirements. It was
decided to pilot the model in one area, 'Containers', before applying it
in the company as a whole. In accordance with the proposed
methodology, a "steering committee" was established to determine
priorities and plan implementation. This process produced an
implementation blueprint. The paper gives an overview of the
programme, then goes on to describe the progress made in certain areas.

Introduction

Betts Plastics (part of the Courtaulds group) manufactures high volume plastic
components for the packaging industry. The main manufacturing process is injection
moulding. The products include: closures, thin walled containers and pharmaceutical
products. The company has recently introduced a 'Continuous Improvement'
programme, with the help of Middlesex University Business School, as part of a
Teaching Company Scheme that commenced on 1 January 1993. Two teaching
company associates (TCAs) are facilitating the project. The project at Betts is now
roughly fourteen months old. It was decided to pilot the approach in one area, the
Containers moulding shop, before applying it in the company as a whole.

The Middlesex Approach

The implementation methodology was developed at Middlesex University
Business School (Ghobadian, 1993). The fishbone diagram in figure 1 depicts the
five key elements of the Middlesex approach: **market focus**; **process focus**; **people**

focus; communication and measurement; and management process. This methodology, unlike those proposed by Oakland (Oakland, 1993), Juran, Crosby, and Deming (Ghobadian and Speller, 1994) is neither sequential or prescriptive. The improvement initiative under each major heading is identified by the host organisation. A spreadsheet is then used to develop the detailed implementation plan. For each improvement initiative the implementation team needs to address the following :

(a)   Description of the existing position;
(b)   Strengths of the existing position;
(c)   Weaknesses of the existing position;
(d)   Description of the desired position;
(e)   Actions required to attain desired position;
(f)   Financial resources requirement;
(g)   Human resources requirement;
(h)   Scheduled completion date;
(i)   Responsible person;
(j)   Priority;
(k)   Other areas affected by the programme; and,
(l)   Nature and consequence of the impact.

The process is integrated, and in the long run all five elements and their sub-elements need to be addressed.

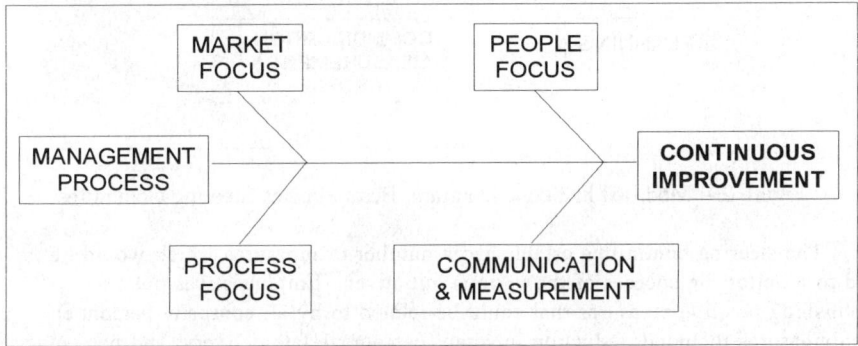

Figure 1 - The Salient Elements of Continuous Improvement

Application to Betts - Establishing the Blueprint

Shortly after the start of the Teaching Company programme, the Managing Director at Betts set up a steering committee to oversee the company wide implementation of continuous improvement. This committee considered the approach proposed by Middlesex and re-defined the five key elements in line with the operational characteristics of the company (figure 2). Sub-elements were discussed in detail and added. This resulted in the development of a definitive document to guide the initial work within 'Containers' operations. The following projects were identified as having a high priority :-

a.   Initial **audit** of the Containers operations.
b.   Introduce all Betts personnel to the concept of continuous improvement via **presentations**.
c.   Employ **brainstorming sessions** to gather suggestions for improvement projects.
d.   Perform a **communication audit** to establish the current state of the documentation system.
e.   Perform a **supplier review**.
f.   Use **news sheets** to communicate the continuous improvement message.
g.   Arrange **visits to customers** for Betts employees, to help establish stronger links.
h.   Introduce **machine monitoring**.
i.   Introduce **scrap graphs**.
j.   Perform a **customer survey** to ascertain customer perceptions.
k.   Form **continuous improvement** teams.

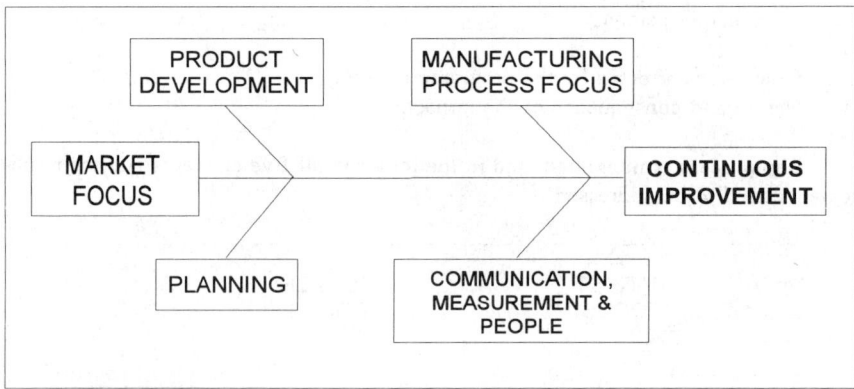

Figure 2 - Modified Fishbone Diagram, Betts Plastics Steering Committee

The steering committee established a number of measures which would be used to monitor the success of the various initiatives. Emphasis was put on establishing positive measures that could be related to by all company personnel. Such measures included; reduction in scrap; percentage late delivery; and percentage first-time quality.

To achieve the objectives as set out in the blueprint, five project groups, were created; Communication, Measurement and People; Product Development; Manufacturing Process Focus; Market Focus; and Planning. The original remit of these project groups was to analyze what initiatives were underway within their area, and ensure that those initiatives considered essential were completed as fast as possible. The ongoing remit is to instigate new investigations and initiatives within their area to continue the process of change. Prior to any action taking place, each group leader presents the outline and objectives of a new initiative to the steering committee. The group leader decides who will be on each project group, either as a permanent or temporary member.

Specific Initiatives

Presentations - Communication, Measurement and People

To introduce the idea of continuous improvement and the concepts it involves, a presentation was developed and made to all shifts. This provided the ideal opportunity to explain to all employees the ideas behind continuous improvement, the company's approach, and some of the tools and techniques that could be used. Informing people in this manner helped to remove some of the fear and resistance to change. The communication programme helped to lay the foundations, ensuring that everyone had a basic understanding of continuous improvement that could be strengthened as the implementation programme progressed.

Customer Survey - Market Focus

Continuous improvement puts a high emphasis on customer satisfaction and enhanced value. With this in mind it was considered essential to carry out a customer survey. The survey's aims were to identify key order winners, and how Betts performs in the case of each order winner, and compared to its main competitors. A questionnaire was developed, (i) to identify the importance attached by customers to various criteria that would affect the purchase decision; and (ii) to measure how well Betts performed against each of these criterion. Personal interviews were held with the customer's purchasing management representative, using the questionnaire for structure. The survey was carried out for the market sector served by Containers, and used as a pilot study to assess the validity and benefits of carrying out similar surveys for the remaining market sectors of the business. The response rate was 60%. This was felt to be adequate for a pilot study although a larger sample would have been preferable. The data collected proved extremely useful. However, it would have been more beneficial if interviewees had been more forthcoming with information as regards Betts' competitors. Despite some inadequacies, this pilot study demonstrated the necessity of such surveys. An independent consultant will be used to carry out the remaining surveys. It is hoped this will overcome any problems of access to sensitive information.

News Sheets / Whiteboards - Communication, Measurement and People

Improving communication is fundamental to continuous improvement. Communication at Betts was considered to be less than adequate. To address this, two new forms of communication were introduced. The first was a white board system to enable direct communication between production managers and production personnel. This system allowed a free flow of information between the two parties, providing a mechanism whereby the personnel could ask for responses on matters they wished to know more about. To further promote the continuous improvement project, a series of news sheets have been published. These took the form of local sheets, specific to a particular area, issued as and when required, and site sheets, which are issued monthly and used to ensure that everyone knows what is happening in all areas. Both types are put out with the aim of publicising achievements and progress. They also provide a vehicle for recognising those people who are responsible for improvements. Reaction so far has been very positive. This leads to greater understanding of how people are involved, and people become an integral part of the drive to improve communication.

Scrap Measurement - Communication, Measurement & People

'Scrap' is a highly visible indicator of problems within the process. It, therefore, seemed appropriate to use scrap as a simple performance measure. The scrap level was routinely collected and thus, the information was readily available. However, the information was not utilised to either take corrective action or bring the issue to the notice of operatives or provide feedback. Hence, the data was presented graphically and displayed on noticeboards. With monthly updates, operators were able to see for themselves how the scrap levels changed over a period of time. This initiative will now focus on tracing the peaks on the graphs back to specific causes.

Machine Monitoring - Manufacturing Process

Monitoring problem areas of the manufacturing process followed as a direct result of the brainstorming sessions. During these sessions specific areas responsible for producing high amounts of scrap were identified. Having identified the problem areas, the objective of the monitoring sessions was to establish the cause, and obtain some measurement of the significance and severity of the situation. The most successful example is that of vending machine cups, which after moulding, did not fall from the tool into the collection bin correctly, thus resulting in large amounts of otherwise good quality cups becoming contaminated or damaged. The period of monitoring provided sufficient data to allow a cost to be applied to the problem - £880 per week (based on 24 hour production). This made people take notice and action to correct the situation was immediate. An air manifold has been fitted at a cost of £470, which although not completely curing the problem, has improved the situation dramatically.

Conclusions

The continuous improvement process at Betts is at its initial stages. The Management recognises the need for change and has decided to embrace the principles of "Total Quality Management (TQM)" as the catalyst for change and the introduction of a "Continuous Improvement" culture. It is often said that TQM is a never ending journey. It is widely accepted that TQM is a continuous programme and not a one-off project. However, a TQM programme is made of many projects / journeys. Management at Betts has recognised this reality and decided to systematically plan the implementation of the "Continuous Improvement" programme rather than allow it to mushroom in all directions, without clear direction.

After fourteen months of the programme the beginnings of a "culture change" are discernable. The top management's commitment and understanding is an essential ingredient of change. The top management at Betts is committed to change.

References
Ghobadian, A. Dec 1993, 'Integrating Operations Strategy and Quality Improvement - The Way Ahead', Inaugural Professorial Lecture, Middlesex University.
Ghobadian, A. and Speller, S., 'Gurus of Quality : A Framework for Comparison', pending publication in *Total Quality Management*, **5**(3), June 1994.
Oakland, J.S. 1993, *Total Quality Management* (Butterworth Heinemann. 1993).

THE SELECTION AND DESIGN OF PPC SYSTEMS

J F Hill and I C Brook

University of Warwick
Coventry, CV4 7AL
and
University of East London
Dagenham, Essex. RM8 2AS

A number of attempts have been made to systemise the processes of selecting and designing appropriate PPC systems to meet the demands of particular manufacturing environments. Previous work has tended to focus on partial analysis of the manufacturing situation. This paper discusses the findings of a benchmarking study and presents a checklist of factors as the basis of an analytical framework for manufacturing managers to participate in the specification and implementation of appropriate PPC systems for their particular manufacturing situations.

Introduction

The selection and design of Production Planning and Control (PPC) systems has been a source of interest to both academics and industrialists [Prabhu (1983), Plenert and Best (1986), Schofield and Bonney (1981)] for many years. The nature of the problems inherent in these activities has evolved from decisions about which manual planning and control techniques to apply to a particular situation into the need for a series of decisions which today reflect the increased knowledge about the manufacturing environment, the influence which PPC systems can have on the performance of the manufacturing system and the range of new tools and technologies available [Hill (1993), Doumeingts (1991), Kochhar and McGarrie (1992)]. This paper presents the findings of a benchmarking study into "best practice" at a sample of users of MRPII systems and then proposes a checklist of factors against which manufacturing managers can assess the effectiveness of current PPC systems, identify opportunities for improvement and monitor the implementation of changes to the PPC system. It is considered that the checklist presented here has the advantage of not being system-specific.

Benchmarking Study

Manufacturing consultants were asked to identify a sample of companies with experience of successful implementations of MRPII systems in order to identify common factors contributing to success. This proved to be difficult because many of the consultants approached claimed that their clients would not be prepared to reveal the sources of their competitive advantage. Despite this, a sample of six user companies was identified. The study was conducted by interviews and the findings were compared with a company (Company A) in the course of implementing an MRPI system in order to investigate the similarities. Some findings of the study are summarised in Tables 1 and 2.

Table 1: Comparison of Survey Companies' Manufacturing Environments

Company	Production System Classification	Main PPC task	Main performance measure
A	Medium Batch	Ordering	Order backlog
B	Large Batch	Inter-plant scheduling	Schedule compliance
C	Medium Batch	Materials procurement	Inventory level Gross MRP vs. actual demand
D	High volume batch/flow	Plant scheduling	Output vs. std. rate. Delivery Performance
E	Low volume batch & flow	Plant scheduling	Delivery Performance Order backlog
F	High volume batch	Plant scheduling	Delivery Performance
G	High volume flow	Plant scheduling	Delivery lead time.

Companies in the survey fell into two groups: those genuinely successful in their MRPII implementations and those that stopped at MRPI. Successful companies required large amounts of effort, modification and selective introduction of standard MRPII modules. None of the Class A MRPII systems identified fit the standard MRPII structure. All have tasks performed by either manual, stand-alone or stand-alone/linked computer systems that were considered to be too expensive of time and resources to obtain via a conventional MRPII system. Shopfloor control and capacity planning were the two tasks most likely to be performed by non-MRPII systems. The add-on systems were integrated with the MRPII system to enhance timeliness, provide faster simulation capabilities and improve the level of production resource utilisation. These survey findings lend weight to the view that standard PPC system solutions are not universally applicable.

Table 2: Comparison of Survey Companies' Implementation Procedures

Company	Implementation timescale	Implementation approach	Start point	ABCD class.
A	ongoing	function by function	Sales Order Processing	D
B	NA	NA	NA	C
C	24 months	pilot on small product range.	Sales Order Processing	C
D	4 years and 6 months	function by function	Sales Order Processing	A
E	18 months	pilot on small product range.	from a corrupted MRPI system	A
F	11 mth study plus 30 mths.	function by function	Sales Order Processing	A
G	study plus 30 months	function by function	Master Scheduling	A
Recommended procedure	18 months inclusive	pilot then by function	Sales Order Processing	

Analysis of the survey data indicated that the most successful companies had the following attributes:

1. main performance measures related to the order-winning criteria of the markets served.
2. selective use of standard MRPII modules and special modules or stand-alone systems where appropriate.
3. sustained commitment to the implementation process over long periods.
4. a comprehensive feasibility study prior to implementation.
5. implementation by a project team led by PPC staff.

Table 3: Checklist for Judging PPC System Success

Factor	Criteria for total success	Criteria for total failure
1. Performance on due date, inventory, capacity	Better than sector average	Worse than sector average
2. Use of other performance criteria for PPC system	Yes, developed by PPC staff	None used or considered
3. Use of performance targets	Yes, with good results	None
4. Written policies and procedures for system	Fully documented	Not available or incomplete
5. Assignment of tasks & responsibilities	Fully documented	Not available or incomplete
6. User commitment and understanding of system	Yes, with required knowledge	No, without reqd. knowledge
7. Integration of all data requirements for the manufacturing system	All possible data sharing provided	Some data not shared where beneficial
8. Data accuracy and timeliness	Accuracy >98% Availability as reqd. for use	Accuracy <98% Alternative system sought
9. Data duplication	None	Multiple non-linked files
10. Data provision for PPC	Files for design process, time standards, buying	Basic data not provided
11. Existence of parallel systems	None, systems integrated	Informal & hidden systems
12. Ability to replan capacity & materials	Replanning takes <30 mins.	Replanning takes >30 mins.
13. System used to monitor Suppliers/customer orders	Yes	No
14. System used to make business plans	Used for strategic planning	Not used for strategic planning
15. Philosophy of continuous PPC system review	Yes, in writing	None

PPC System Checklist

The findings of the Benchmarking Survey and analysis of the manufacturing situation at Company A led to the development of the PPC system checklist in Table 3. The checklist considers four aspects of PPC: performance, systems, data and management. Criteria for total success and total failure are defined for factors in each category with the user responsible for scoring the level to which success is achieved. A suitable scoring system gives points on a scale of 0 to 5 for factors in the first three categories and 0 to 10 points for management factors, the inherent weighting giving due regard to management's responsibility to ensure the PPC system can cope with its environment, has its integrity maintained and is used by managers to formulate and test the viability of manufacturing strategies. As with Bill Belt's scheme (1981, 1988), the maximum score is 100. One major difference is that the factors in this checklist are not optional for certain situations or certain PPC systems.

Discussion

Although suffering from the same deficiencies as Prabhu's attempt to use factor analysis in the selection and design of PPC systems, the checklist does have a useful role as a system audit tool and as a means of identifying essential PPC system attributes at the feasibility stage of an implementation project. The checklist can be applied to both manual and computer-based systems, although it has an underlying bias towards use of a common database.

The evidence from the benchmarking survey indicates that other companies could benefit from use of such a broadly applicable diagnostic tool to aid the design and maintenance of PPC systems.

References

Belt B. 1981, Checklist of Classification Criteria for MRPII systems. (Oliver Wight Associates).

Belt B. 1988, "Class A" Manufacturing Education Services in the European Market 1992. (Oliver Wight Companies).

Doumeingts G. et al. 1992, GRAI Integrated Methodology: A Methodology for Designing CIM Systems (Version 1.0).
(University of Bordeaux LAP/GRAI).

Hill T. 1993, Manufacturing Strategy (2nd edn). Macmillan.

Kochhar A. and McGarrie B. 1992, Identification of the Requirements of Manufacturing Control Systems: A Key Characteristics Approach.
In Integrated Manufacturing Systems 3, No. 4.

Plenert G. and Best T. D. 1986, MRP, JIT, OPT: What's Best?
In Production & Inventory Management 1/2nd Quarter.

Prabhu V.B. 1983, Towards a Methodology for Determining Appropriate PPC Systems in Small Companies. PhD thesis. (Open University).

Schofield N.A. and Bonney M.C. 1981, The Use of Manufacturing Profiles to Predict the Applicability of Computer Systems.
In Int. Jnl. of Operations and Production Management 1, No. 2.

A MODEL OF THE PRODUCTION MANAGEMENT SYSTEMS IN A MAKE TO STOCK COMPANY.

**Mr Ian McCarthy, Mr Aslam Chodari and
Dr Keith Ridgway**

*University of Sheffield,
Sheffield, S1 3JD*

This paper describes the use of the GRAI method to construct a conceptual model of the production management systems in a 'Make to Stock' Company. The work described is the first phase of a research programme to develop a generic model of a production management system suitable for the 'Make to Stock' environment.

Introduction

This paper details one area of a research project which aims to develop a methodology for business process re-engineering (BPR), (Hammer and Champy, 1993) of Production management Systems (PMS), whilst considering the company's business and market strategies and organisational structure. The research area discussed applies a classification technique to PMS types, so that reference models can be established using GRAI decision grids. Once the system has been classified, and an ideal reference model has been constructed, a company's existing system can be related and analysed against the appropriate ideal model.

This paper describes the application of the GRAI technique in a 'Make To Stock' company. The emphasis of the application was to analyse the decisional activities of a PMS and associated information and materials flow. The main objectives of the application were:

i)   Develop a graphical model for a Make to Stock PMS.
ii)   Examine decision hierarchy, information flows and activities for the PMS.
iii)   Assess GRAI grids and nets to identify inconsistencies and develop recommendations for improving the structure and operation of the PMS.
iv)   Refine and systemise the method of applying GRAI to other PMS types.

Review of GRAI Method

The GRAI method was developed in the late 1970's by the GRAI Laboratory at the University of Bordeaux I in France. The laboratory gets its name from its principal area of research; <u>G</u>roupe de <u>R</u>echerché en <u>A</u>utomatisation <u>I</u>ntegee. However, the method is named after a tool which produces a graphical model; Graphs with Results and Actions Inter-related, Grislain and Pun (1979). It is a method which was developed explicitly for analysing PMS types and is therefore a suitable technique for the research project. The analysis does not simulate the systems activities, but examines the structure of the decision centres and the flow of information within this structure. The theory of the GRAI method is that decisions start and terminate events within a PMS, and these events will determine the performance and operating characteristics of the system. As a PMS is a dynamic system, serving both internal and external customers, decisions will only be appropriate for given time states (Horizons). Before the decision Horizon is reached, decisions can be adapted. This is reflected by a time interval known as the Period, and represents the frequency with which decisions are reviewed. Using this philosophy a GRAI model is produced. The model consists of two tools, GRAI grid and GRAI net.

GRAI Grid

This represents the macro structure of the system showing the links between the main decision centres. It comprises columns, representing the various elements (To buy, to make, etc.) in the PMS and rows representing the hierarchy level of the decision centre. Each level has a corresponding Horizon and review Period. The relationship and structure of the decision centres are indicated by two symbols as described by Doumeingts (1985).

========= >   A double arrow indicates a frame between two or more decision centres. A decision frame specifies all the resources required to execute the decision. Decision links always lead from a higher level to a lower level.

------------- >   A single arrow represents transmission of information between the decision centres. Information flows can be in any direction.

GRAI Nets

Nets examine the micro structure of the system by analysing the components within the decision centres. A GRAI net is an illustrative portrait of a decision's activities and their interrelationships as shown in figure 3.

GRAI Application

In using the GRAI method it is important to define the boundaries of the PMS to be examined. For instance some organisations would consider the sales function to be an activity of the PMS, while others would consider sales to be an internal customer. In order to identify the decisions within a PMS and their structure, the management responsible must be consulted. This group of people is termed the synthesis group, Ridgway (1992). The synthesis group provide the information which determines the hierarchy and links between decisions. This leads to the generation of the GRAI grid (top down analysis). A bottom up analysis is then performed to scrutinise the decision centres and construct GRAI nets.

Application Of GRAI to Presto Engineers Cutting Tools

Presto Engineers Cutting Tools is a 'Make to Stock' company manufacturing high quality cutting tools. It has more than 400 employees and exports some 30% of the total production to over 90 countries. The Company manufacture 24000 stock lines

which comprise high speed steel twist drills, reamers taps and dies, tool bits, milling cutters, solid carbide twist drill and milling cutters, carbon steel taps and dies and a range of pipe threading tools and associated threading equipments. The company comprises of five distinct departments, which are co-ordinated by the managing director of the company. These departments are:
- Finance and administration
- Sales (marketing and export)
- Production
- Technical control
- Planning

Each department splits in to several sub departments or sections which are co-ordinated by the head of department.

The study

The synthesis group comprised the heads of the departments and heads of sections responsible for the operation of the PMS. The interviews related to their responsibilities, functions and planning. The first phase of analysis is the top down analysis which examines the overall production management system and produces an actual picture of the existing system in the shape of a GRAI grid. After interviews, the following PMS functions were identified:

To buy: This function involves purchasing materials for products, machines and general equipments.

To plan: The planning function determines the production activities, the sales forecast and production forecast are used to produce production plans.

To make: This function converts the raw material into finish products. Several process are involved, including heat treatment, machining and inspection.

To develop: This function produces technical information such as product drawings and N.C computer programs.

To sell: This function receives orders from customers, generates sales forecasts for future planning and manages the information about markets for both home and export.

According to GRAI rules each grid must have two function columns, for representing internal and external information. These functions represent the direct production involvement. The horizons, review periods and decision centres are also identified. With the identification of the functions, horizons, review periods and activity centres relevant to the production management system of the company, it was possible to construct an initial GRAI grid of the existing production management system. The GRAI grid for the existing production management system is shown in figure 1.

Bottom Up Analysis (Construction Of GRAI Nets)

This phase of the analysis involved detailed interviews with the synthesis group to evaluate the decision centres identified in the top down analysis in greater detail. Using the information from the interviews and the GRAI grid constructed in the top down analysis a GRAI net was produced for each decision centre. To analyse the decision centre identified, it was necessary to determine:

a) The various activities performed and decision taken at a given decision centre.

b) The relationship between the activities and decisions.

c) The information required to perform a particular activity or make a specific decision.

The GRAI nets constructed for the each decision centre represent the different activities performed in the production management system.

Results

After detailed analysis of the GRAI grid and nets several inconsistencies were found. The lack of a master production schedule for the system resulted in the assumption that everything was required on the day of the planning run. A lack of capacity information meant that the system assumed output could be achieved on that day. The sales forecast did not appear to be used by any area of the production management system and there was no horizon to the computerised production management system. The analysis indicated that there was insufficient information transmission from the internal and external sources. In addition there was no evidence of a systematic collection of information or review of existing information. To eliminate the inconsistencies identified a modified GRAI grid (figure 2) and GRAI nets (figure 3) were constructed.

Conclusion and Recommendation

The GRAI method was used to analyse and design the production management system for a 'Make to Stock' company. The GRAI grid and GRAI nets were constructed and a number of inconsistencies were identified. Using GRAI rules a modified GRAI grid and nets were constructed. The study demonstrates that the GRAI method could be used to develop a generic model of a 'Make to Stock' production management system. It is anticipated that rules governing the design and organisation of the production management system can formulated, and that similar categories of 'Make to Stock' companies should confirm to these rules. That is, the management decisions within the PMS should adhere to a similar cohesive structure.

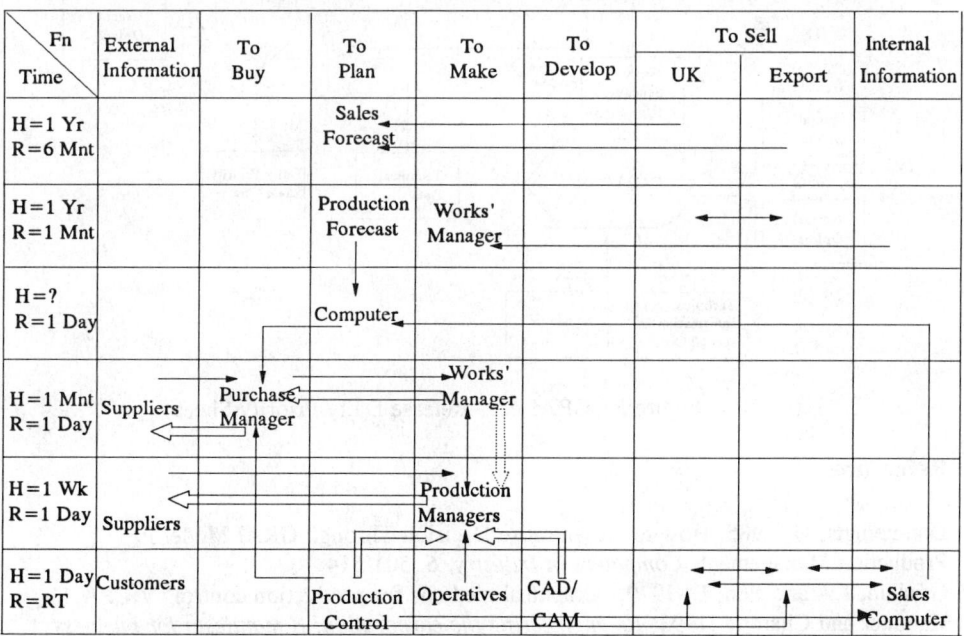

Figure 1. GRAI grid of Presto Engineering Cutting Tool

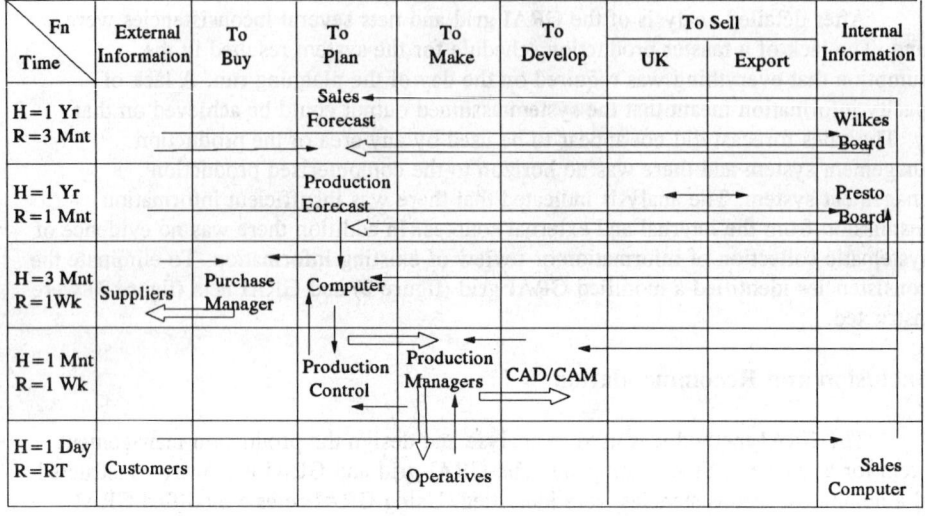

| Fn ＼ Time | External Information | To Buy | To Plan | To Make | To Develop | To Sell | | Internal Information |
						UK	Export	
H=1 Yr R=3 Mnt			Sales Forecast					Wilkes Board
H=1 Yr R=1 Mnt			Production Forecast					Presto Board
H=3 Mnt R=1Wk	Suppliers	Purchase Manager	Computer					
H=1 Mnt R=1 Wk			Production Control	Production Managers	CAD/CAM			
H=1 Day R=RT	Customers			Operatives				Sales Computer

Figure 2. Modified GRAI grid

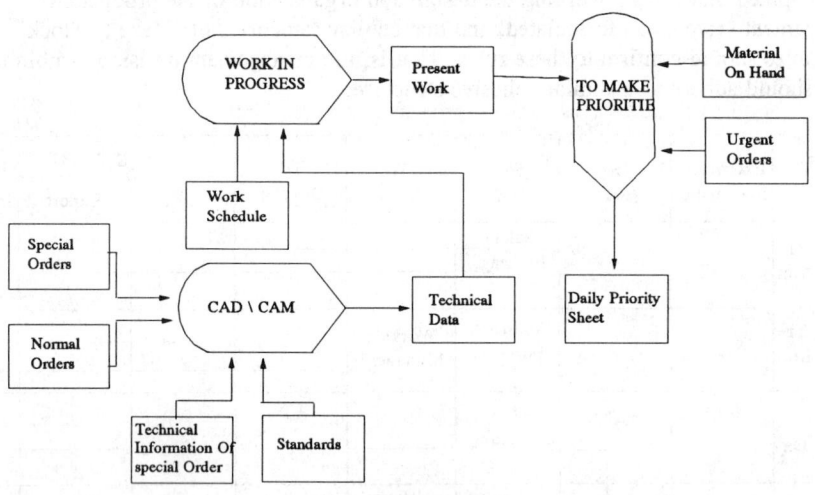

Figure 3. GRAI net:- Release Daily Priority Sheet

References

Doumeingts, G. 1985, How to Decentralise Decision Through GRAI Model in Production Management, *Computers in Industry,* **6,** 501-514.

Grislain, J.A and Pun, L. 1979, 'Graphical methods for production control' VICPA, Hammer and Champy ,1993, *Re engineering the corporation: A manifesto for business revolution.* (Harper Collins Inc)

Ridgway, K. 1992, Analysis of decision centres and information flow in project management. *International journal of project management,* **10** (3) 145-152.

THE USE OF SEMANTIC DATABASES IN MANUFACTURING DSS FOR SMALLER MANUFACTURING ENTERPRISES

D.N.Halsall, A.P. Muhlemann, D.H.R. Price and F.P. Wheeler

University of Bradford Management Centre,
Emm Lane, Bradford, West Yorkshire,
BD9 4LJ, UK

Smaller manufacturing enterprises encounter significant difficulties in carrying out effectively planning and scheduling. User acceptance of a DSS in this area is greatly facilitated by transparency and clarity of the underlying structure. This paper outlines the role of semantic models in achieving this objective.

Introduction

Existing Research

Studies (Muhlemann et al. 1990, Afferson et al. 1992) have been carried out over a number of years which have concerned the decision support needs of smaller manufacturing enterprises (SMEs) and ways of satisfying these needs. A focus of this work has been to produce a decision support system (DSS) which would satisfy the requirements of a broad range of SMEs, thus reducing the cost of individual systems. The approach adopted was to identify a small set of core production management areas (Sharp et al. 1990) and to develop for each area, using a control loop based data model (Price et al. 1992), a template to support the appropriate decision making. These templates were linked to form the generic system which was capable of providing most of the decision support required. Because the system was developed using a fourth generation language (4GL) and a relational database management system (RDBMS), it was possible to modify it to meet many company specific needs.

There were three aims for the design of the DSS. Firstly, that the system should support rather than attempt to replace the decision making of managers. Secondly, that the system should recognise and assist with the problems many SMEs experienced in specifying their requirements of a DSS. Thirdly, that the system should, as far as possible, reflect good production management practice rather than be based solely on a single methodology (e.g. MRP, OPT). During the design of the DSS, particular efforts were made to produce as flexible a system as possible. This entailed: a) careful design of the underlying database to maximise its flexibility, b) development of the system in a suitable 4GL, which would enable users to generate reports for themselves. This flexibility is considered a key characteristic of the DSS, without which the developments discussed below would be very difficult to carry out.

Survey

A survey of a group of companies, which were expected to show good production planning and control policies, was conducted in 1992 (Halsall et al. 1994), which concentrated on the companies' experience of scheduling systems. A few significant findings of the survey are summarised below. When asked about the stability of the production environment, substantial proportions of the respondents reported that: in many cases, customer requirements were not fully known before jobs were started and that many delivery dates were changed after jobs had been started. When reporting on the performance of the scheduling systems, many of the companies reported a frequent need to adjust schedules. Only 7% of the companies had not needed to modify their scheduling systems which had been installed for a mean period of two years. Overall, there was a clear impression that even companies expected to represent good practice were experiencing significant difficulties with their scheduling systems.

Scheduling Philosophy

In order to provide more support to managers engaged in real scheduling tasks, it was decided to enhance the scheduling facilities in the DSS. It was quickly recognised that it would be difficult to incorporate fully the new scheduling module in the DSS, and that a better arrangement was for the new module to operate outside the 4GL/RDBMS environment of the DSS and to call for data as required. The role of the DSS was to provide transaction processing facilities and to act as a source of static data about the manufacturing system. As a result of the careful design of the DSS database, many of the tables were "recognisable" objects and, consequently, an object orientated language (Smalltalk) proved to be a convenient and powerful medium in which to develop the new module.

As the new module was required to operate under the volatile conditions experienced in many SMEs, a hierarchical structure was chosen for the module (Bowers and Jarvis 1992) with successively shorter time horizons at successively lower levels. Feedback has been provided between levels, so that if a plan proves to be infeasible at a given level in the hierarchy, then it can be passed up a level for the replanning necessary to resolve conflicts to be carried out. In order to enhance the "transparency" of the module, the structure of the hierarchy in the module has been designed to mirror the planning processes in the organisation in which it is to be used.

The rapidly changing priorities in many SMEs cannot be easily captured by an automated system, consequently it is important to incorporate managerial expertise in conflict resolution. The scheduling module links individual manufacturing operations to finished goods and customer orders with the aid of a tree structure for each product which incorporates the information from the Bill of Materials with that from the technological route and has been termed the Bill of Production (Halsall et al.1993). Thus, given a conflict for resources, a manager can identify the customer orders giving rise to the conflict and use his knowledge of priorities in the resolution of the conflict. In order to assist the manager in interacting with the system, a visual interactive planning board has provided the vehicle for interaction.

Development

Background

A task of production management is not only to plan how to use the resources available to best satisfy the company's objectives and customers' needs, but to adjust the plans as problems are encountered. It is common to find in SMEs that a high degree of flexibility is required in all areas of the business to remain competitive. Because the production environment is frequently complex, even in small manufacturing companies with simple products, experienced staff have been found in the majority of cases to out-perform computerized decision systems. The problem can be considered in two parts, firstly where staff have an expert knowledge of the production environment and the decision support system is attempting to match or better this knowledge. Secondly where the production environment is changing or new situations are encountered and the consequences of various actions have to be worked through.

Management Role in Production Planning

In the first case, which is usually at the operational level, where the production environment could be described as well understood, many of the hard objectives and constraints of a company can be stated and used by a DSS in determining the best possible course of action at any particular time. A typical example of this may be minimising the number of jobs delivered late. But the majority of production environments are complex in nature resulting in many soft constraints which also have to be taken into account, for example which customers are viewed as the most important. These constraints are very difficult to embody in a DSS. Also it was recognised in early work on visual interactive modeling (Bell and O'Keefe 1987) that in many cases planning staff negotiate with production staff while developing plans. From this it can be seen that it is very important to have clearly defined roles for both the DSS and experienced staff. It is also important for the DSS not to restrict the abilities of staff by limiting their access to data or preventing courses of action.

In the second case, which is usually in response to changing market requirements and to maintain a competitive position, it is frequently necessary for companies to improve their products and production methods. Because a body of expert knowledge is not usually available the consequences of various actions have to be worked through. Due to human cognitive limits this can take some time and may not necessarily lead to an optimum solution. Few small companies have excess capacity or pilot plants to permit testing out new methods or products to build up staff experience. Because of the financial implications of any stoppage or disruption to the production environment, changes are usually introduced in an incremental fashion to minimize the risk and give time for staff to gain experience in the new systems or products. Many methods proposed in the literature have never been tried out in a real situation because management are not convinced any potential gain is worth the risk of failure. Computer simulation of the production environment is a tool which can be used in these situations but has only found limited applications in small companies due to high cost and need for the users to have a good grasp of the method.

Semantic data models

We are proposing a semantic data model (Hull and King 1987, Peckham and Maryanski 1988) as an extension of the current work on production planning. The aim is to allow users, with appropriate authority and security controls, to manipulate a semantic data model. This is motivated by two observations. Firstly, that users hold significant information about the manufacturing environment in their heads but that they have difficulty including this in a conventional (i.e., relational, hierarchical or plex) database. Even when this information is contained in a conventional database it is difficult to extract it and therefore difficult to use it. This is because traditional database management systems are based upon techniques that encourage the efficient storage and retrieval of data (Ullman 1988). They have simple, fixed data structures based on records. This means that they cannot support directly such information as relationships, data abstraction, constraints or inheritance.

Secondly, at any time information is known at different levels of abstraction. In the production environments of interest to us, the level of abstraction maps to the time available for action, which usually maps to the level of managerial responsibility. This means that lower levels of abstraction correspond to less time to act. In other words, at lower levels of the hierarchy information must be specified in more detail and this information is more relevant to lower managerial levels. Therefore, in our area of application, the fact that semantic models permit information to the viewed at different levels of abstraction is particularly relevant. The higher levels of abstraction will generally correspond to higher levels of authority and to decisions that have longer lead times. Since semantic schemas can be considered at various levels of detail, it is possible to isolate information that is appropriate to the type of decision. Furthermore, if the system supports derived schema components it should be possible to compute individual arbitrary views of the data. In our case, a decision-support environment has been built to offer the user the means to extract and combine manufacturing data from a relational data base to make scheduling decisions. This environment uses the object-oriented paradigm. A semantic data model is applicable when there is a need to isolate information about links between objects from information about object behaviour. Therefore the semantic model will allow the implied structural properties of objects to be revealed. Furthermore, the abstractions of a semantic model will enable the structural aspects of the database to be open to manipulation. This is particularly important for the execution of such operations as structural queries, see Figure 1.

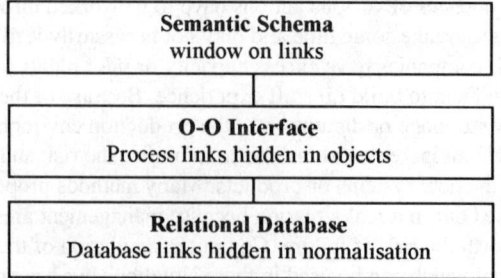

Figure 1. Schematic representation of the role of the semantic model

Summary and Conclusions

Our aim is to provide users with the facility to extract and to add semantic information to an existing relational database and decision-support system. The purpose is to make use of the person's understanding of information rather than to focus on the physical representation of the data, as occurs in a purely relational representation. The benefit of semantic data modelling is the ability to express complex data interrelationships. Whereas a traditional relational model has only two constructs that allow such relationships to be specified , semantic models generally have several. We plan to use a hypermedia interface to communicate the structural content of the database. Combined to this will be the facility to create linkages in the database using personal semantic knowledge. An interactive hypermedia interface should allow users to personalise the database by creating, at will, connections between segments of information. These connections will link concepts that have meaningful associations for the person. Instead of applying conventional queries, these structural queries involve active navigation, such that the user moves across connected information by a process of browsing. This development platform should overcome many of the SMEs difficulties in using decision support for scheduling.

References

Afferson, M., Andrews, J.K., Muhlemann, A.P., Price, D.H.R. and Sharp J.A. 1992, Generic manufacturing information systems development via template prototyping. European Journal of Information Systems, **1,** 379-386.

Bell, P.C. and O'Keefe, R.M. 1987, Visual interactive simulation - History, recent developments, and major issues, Simulation, **49**, 109-116.

Bowers, M.R. and Jarvis, J.P. 1992, A hierarchical production planning model. Decision Sciences, **23**, 144-159.

Halsall, D.N., Muhlemann, A.P. and Price, D.H.R. 1993, A production planning and resource scheduling model for small manufacturing enterprise. In A Bramley and T Mileham (eds.), *Advances in manufacturing technology VII,* (Bath University Press, Bath), 161-165.

Halsall, D.N., Muhlemann, A.P. and Price, D.H.R. 1994, A review of production planning and scheduling in smaller manufacturing companies in the UK. Production Planning and Control, to appear.

Hull, R. and King, R. 1987, Semantic database modeling. ACM Computing Surveys, **19**, 201-260.

Muhlemann, A.P., Price, D.H.R., Sharp, J.A., Afferson M. and Andrews J.K. 1990, Information systems for use by production managers in smaller manufacturing enterprises. Proceedings of the Institution of Mechanical Engineers (Part B), **204**, 191-196.

Peckham, J. and Maryanski, F. 1988, Semantic data models, ACM Computing Surveys, **20**, 153-189.

Price, D.H.R., Muhlemann, A.P. and Sharp, J.A. 1992, A taxonomy for supporting the development of computer based production planning and control systems. European Journal of Operational Research, **61**, 41-47.

Sharp, J.A., Muhlemann, A.P., Price, D.H.R., Afferson M. and Andrews, J.K. 1990, Defining production management core applications for smaller businesses. Computers and Industrial Engineering, **18**, 191-199.

Ullman, J.D. 1988, *Principles of Database and Knowledge-Base Systems.* Computer Science Press, Rockville, USA.

COMMUNICATION: THE KEY TO LOCAL PRODUCTION CONTROL

W F Garner and R K Haagensen

Coventry University, Coventry CV1 5BP and
Rolls Royce Plc, Derby D2 8BJ

Manufacturing industry is constantly reminded of the need to achieve
'world class' standards, not only in design and technical capability but in
cost, quality and delivery performance. With competition strong in all
these areas, it is surprising to find delivery dates are often achieved by
expediting the most urgent order only.

This paper reports on the development of a local production control
system, introduced into a turbine blade cell, Rolls Royce Plc, Coventry.

Introduction

Success of manufacturing industry depends upon the ability to compete and win
orders in a world market place. 'World class Manufacturing' is a common phrase,
but the achievement of this is not widespread. Product design and product capability
are tangible standards which are easily measured and improved upon. Price and
quality are criteria which are monitored avidly by management and are subjected to
strident controls. Delivery performance is a further criteria which although considered
important by the manufacturer is often left in the hands of fate. Raw material is
launched into the factory at the correct time, allowing the proposed leadtime for
manufacture. However an all too familiar scenario of expediting and firefighting
follows as the impending delivery date becomes closer. Urgent, very urgent and
wanted yesterday are common methods of prioritising work. Delivery dates are
achieved through queue jumping, persuasion, coercion and brute force.

MRP systems are used extensively by businesses as the controlling mechanism
for the company. However the complexity of the whole company often becomes too
unwieldy and does not allow the MRP system to work effectively at a local level
controlling the movement of the work in progress.

New philosophies have introduced cellular manufacturing as the way forward;
dividing the 'whole' into small autonomous units, which are less complex in
structure. Each cell has different characteristics; volume, variety, technical content
etc. and the local production control system should reflect this.

This paper shows how a local production control system was developed to meet
a cell's characteristics and then integrated upwards into the controlling MRP system.

The Turbine Blade Cell

The high volume schedule of the turbine blade enabled the cell to be set up to manufacture one product type in batches of 100. Each operation of the manufacturing process had a dedicated machine, which eliminated the need for changeovers (see fig 1)

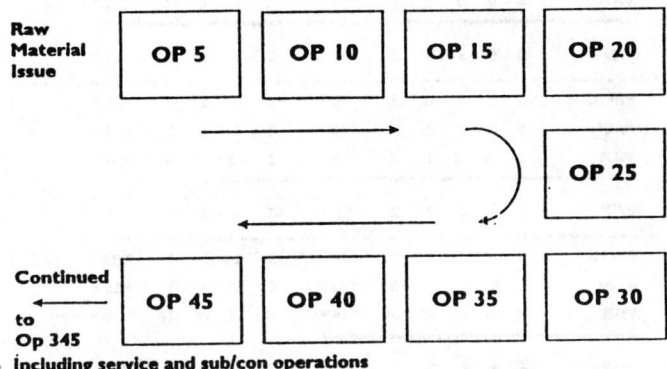

Figure 1 Schematic layout of Turbine cell

The capacity of the cell was matched to the schedule requirement by adjusting the number of operators employed in the cell. The operators were flexible and multi-skilled and moved to the machine where work waited with the highest priority. There was therefore a need to communicate and direct the operator to the correct machine.

Local Production Control

The initial schedule requirement was imposed on the cell from the company MRP system. This was subsequently translated to two levels of local control;

i)   Weekly - Line of Balance

ii)   Daily - Mark up boards

From the monthly MRP schedule a Line of Balance (LOB) was generated. The LOB identified key stages in the manufacturing process with leadtimes estimated in weeks, the total being equal to that quoted on the MRP system. A weekly forecast of output was then applied to the final stage of the LOB and was cascaded through each of the five stages (see fig 2) creating intermediate requirements at each stage. Batch number 24 was due to stores Accounting Period (AP)3 week 4, completion of stage 4 AP 3 week 3, completion of stage 3 AP 2 week 2, completion stage 2 AP 2 week 1 and completion stage 1 AP 1 week 1.

The LOB document was updated weekly and used as a management tool to monitor the progress of the cell. Any lateness is quickly highlighted by an imbalance of the achievement lines (see fig 3) and actions can be taken to resolve this.

The information from the LOB was communicated to the shopfloor for daily use via the 'markup boards' (see fig 4).

The operation sequence is defined on the y-axis, with the batch numbers on the x axis. The batch number is the number given to the batch when all batches are ranked in order of operations completed, 1 is the batch furthest advanced in the manufacturing process. This information is collected daily by a cell administrator who uses a database to rank the information and then allocate the batch numbers. This

DESCRIPTION			AP 1						AP 2					AP 3			
		YTD	1	2	3	4	*CUM*	1	2	3	4	*CUM*	1	2	3	4	*CUM*
STAGE 5	PROG	0	0	0	2	3	*5*	0	1	2	6	*14*	2	3	2	3	*24*
	ACH	0	0	0	2	2	*4*	3	2	1	2	*12*					
2 WK STAGE	VAR	0	0	0	0	-1	*-1*	2	3	2	-2	*-2*					
	WIP	0	0	3	4	6	*6*	3	3	3	3	*3*					
STAGE 4	PROG	0	2	3	0	1	*6*	2	6	2	3	*19*	2	3	2	3	*29*
	ACH	0	0	3	3	4	*10*	0	2	1	2	*15*					
3 WK STAGE	VAR	0	-2	-2	1	4	*4*	2	-2	-3	-4	*-4*					
	WIP	3	3	3	3	2	*2*	2	1	2	1	*1*					
STAGE 3	PROG	5	1	2	6	2	*16*	3	2	3	2	*26*	3	2	3	0	*34*
	ACH	3	0	3	3	3	*12*	0	1	2	1	*16*					
2 WK STAGE	VAR	-2	-3	-2	-5	-4	*-4*	-7	-8	-9	-10	*-10*					
	WIP	4	4	6	3	1	*1*	4	4	5	5	*5*					
STAGE 2	PROG	8	6	2	3	2	*21*	3	2	3	2	*31*	3	0	0	0	*34*
	ACH	7	0	5	0	1	*13*	3	1	3	1	*21*					
4 WK STAGE	VAR	-1	-7	-4	-7	-8	*-8*	-8	-9	-9	-10	*-10*					
	WIP	9	9	5	8	10	*10*	9	11	13	16	*16*					
STAGE 1	PROG	21	3	2	3	2	*31*	3	0	0	0	*34*	0	3	3	3	*43*
	ACH	16	0	1	3	3	*23*	2	3	5	4	*37*					
1 WK STAGE	VAR	-5	-8	-9	-9	-8	*-8*	-9	-6	-1	3	*3*					
	WIP	0	0	1	1	2	*2*	4	4	3	0	*0*					

Key:	AP	Accounting Period	YTD	Year To Date	CUM	Cumulative
	PROG	Program	ACH	Achieved	VAR	Variance

Figure 2 Example of a Line of Balance Document

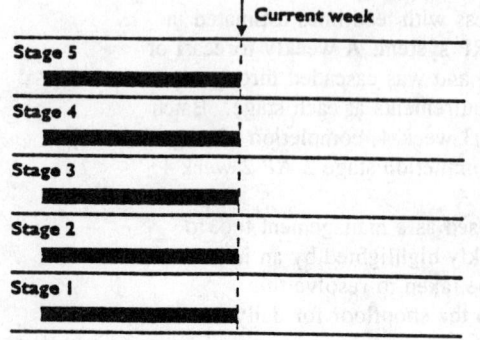

Figure 3a) Balanced LOB

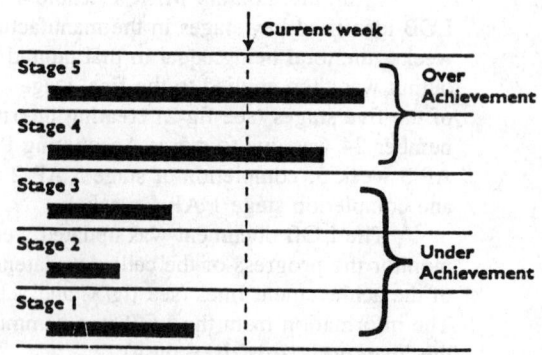

Figure 3b) Imbalanced LOB

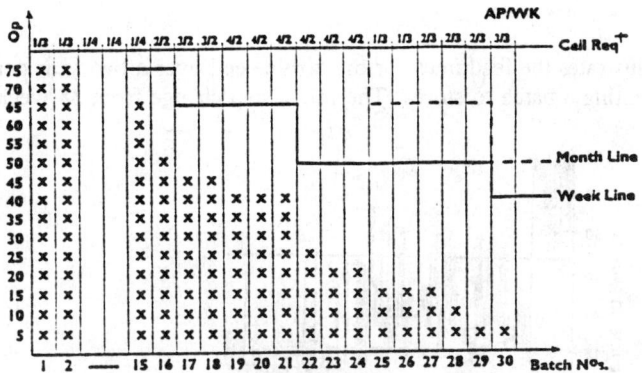

Figure 4 Mark Up Board

may seem a long and arduous task, but in reality took approximately 90 minutes and had the benefits that all batches were audited each day, any problems were discovered immediately and work was moved in and out of the service areas quickly. Once sorted the information was used to update the markup boards to show the latest position of each batch.

The communication of the cell requirement could now be conveyed using the simple instruction "batch number 29 must be at operation 45 by Friday". This instruction could be passed from one shift to the next, from cell manager to cell operator and was not open to misinterpretation. The achievement of the task was monitored on a daily basis via the markup boards with batches 1 - 28 being automatically progressed up to at least operation 45. The instruction was extended by using a stepped line on the board to show the end of the week requirement for all batches (see fig 4). A different colour line was used to show the end of month requirement. These lines were drawn at the key stages identified on the LOB and the targets were set to match those of the LOB. An important aspect of the control system was that the batch numbers quoted on the LOB correlated directly to those used on the markup boards. Therefore "batch 27" had the same meaning whichever monitor was used.

With the information updated cell leaders and cell operators could quickly see which operations had the highest priority. The product requirement had been communicated from the company MRP system and was now simply translated to the batch furthest from the 'end of week' line taking priority. Visibility was also given to the month ahead and enabled resources and work in progress to be managed effectively.

The system was extended to the process areas eg. heat treatment where a section of the Markup board was displayed and updated showing batches moving through the operations preceding the treatment. Again the service department was given its task by quoting batch numbers to be achieved. Previously the service areas would work on batches waiting in their departments, now armed with information showing the position of work coming to them, they could take a proactive role, planning capacity in advance and sometimes expediting close batches to fulfil their own need and to operate their resources at a more efficient level.

The local production control system worked on the principle of communication. The company MRP system generated the customer schedule. This was translated into a cell requirement by the LOB. The LOB was updated by information used daily on the shopfloor and used the same language, namely batch numbers. The targets set by the LOB were communicated directly to the shopfloor by using the markup boards and progress could be seen against these targets on a daily basis.

Results

Figure 5 illustrates the leadtimes for the turbine cell over a two year period. Each cross representing a batch to stores. The initial step change from 38 weeks to 20 weeks

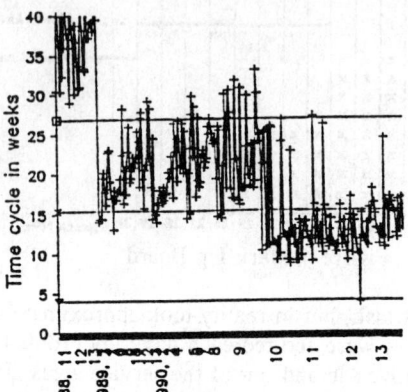

Figure 5 Leadtimes for the Turbine Blade Cell

was due to the machines being moved from a process layout to a cellular layout. The second step change from 20 weeks to 12 weeks was caused by the introduction of the local production control system.

The local production control system had a positive effect on the cell and the reduction in leadtime was a tangible saving. Further unquantifiable savings were also made in the general management of the cell. The simple communication of the requirement enabled all members of the cell to work consistently to a common instruction. Priorities were no longer changed every day and the visibility of work targets using the markup board highlighted problems in advance.

The common database of information provided by the cell administrator, relieved managers of collecting information independently and eliminated discussions on the accuracy of the data, allowing efforts to be concentrated on achieving the targets.

Senior management could monitor progress through the weekly LOB meetings or by walking around the boards when required. The module leader was no longer required to resolve the day to day rows over priorities but could consider the long term strategy for the module.

The use of different colours each week to update the markup board gave a clear history of the movement of work through the cell. Constraint operations were identified and estimated leadtimes through the key stages could be verified or refined.

Continuous improvement will change the system in the future. The combination of data collection and updating the board and LOB into one single transaction is an initial area for improvement. However care should be taken as communication becomes more effective through computers and charts, that people in the cell cease the real art of communication, talking.

Conclusions

The local production control system introduced into the turbine blade cell was simple, visible and designed to meet the cell charateristics. A markup board communicated the product schedule requirement and weekly targets to the shopfloor, resulting in a 40% reduction in leadtime.

AN APPLICATION OF FUZZY PROGRAMMING IN THE STANDARDIZATION OF QUALITY PARAMETERS

Mr Shunian Yang*, **Dr Valentin Vitanov and Mr Nick Bennett**

Department of Mechanical & Manufacturing Engineering
University of Portsmouth, Anglesea Building
Anglesea Road, Portsmouth, PO1 3DJ

This paper considers a fuzzy logic approach for setting an actual standard for a quality parameter of a product. Because the requests for setting this standard, proposed by the decision-makers of a factory, are abstract and difficult to quantify, a fuzzy decision-making model for the standard is developed in the paper. The authors have given the membership functions of the objective and constraint of the model. The essence of the task to set the standard is to find the intersection of the objective set and the constraint set by means of fuzzy programming. The programming results are given and discussed.

Introduction

In order to obtain an optimum result of standardization, many scholars have performed extensive study on the method of setting or revising standard scientifically and given mathematical programming methods such as function-cost method, minimum cost method, least life period method, etc., Shen Jingning(1983), for the optimisation of parameter of the standardization object. The objective functions of these methods are respectively expressed in the form of currency directly or indirectly, therefore, a common shortcoming exists   in these methods, i.e. all the mathematical models of the standardization objects do not reflect on the uncertain factors that effect the optimum result of standardization such as some factors concerned with sociology, psychology, etc. which cannot be expressed in the form of currency.

In the standardization process some uncertain, even fuzzy factors are often involved and they result in difficulty to establish a mathematical model for the standardization object. For example, the customer's appraisal of some performance of a product will have influence on setting a quality standard for this product. But the extent of customer's satisfaction in this performance could not be simply described as 'Yes' or 'No' by means of the classical set theory while setting the standard. Thus, a fuzzy approach would be useful on this occasion.

*Mr Shunian Yang is a Visiting Scholar from the Huazhong University of Science & Technology, China

In the following a fuzzy programming model for setting an actual standard for a quality parameter of a product is introduced. Since the work concerned here deals with the internal quality standard of a factory, only one programming method is chiefly described and the names of the factory and its product are neglected.

Requirements of setting the standard

The engineering product decision-makers of the factory gave the requirements of setting a standard for the quality parameter, i.e. (1) implementation of this standard must suit the production ability of their factory; as well as (2) satisfy the demand of customers for the quality reflected by the parameter.

Here the words 'suit' and 'satisfy' involve fuzzy concept obviously, therefore a fuzzy decision-making model would be established to setting the standard.

Objective, constraint and analysis

It is not emphasised which is important among the two requirements mentioned above, thus they could be on an equal status in a decision-making model. Hence a symmetric model of fuzzy programming would be used. Owing to their playing the same role in a decision-making model, we let the first requirement be as an objective and the second be as a constraint and let x indicate the value of the quality parameter.

Objective
After testing the x values of a thousand of these products and analyzing them statistically, we have the mean value $\bar{x}$ of x obeying approximately a normal distribution with 2 as an estimation of its mean value and 0.14 as an estimation of the population mean square deviation, namely the density function of the approximate normal distribution is

$$f(\bar{x}) = \frac{1}{0.14\sqrt{2\pi}} \cdot e^{-\frac{(\bar{x}-2)^2}{2\times0.14^2}}$$

Figure 1 shows the histogram of $\bar{x}$ and its density function curve.

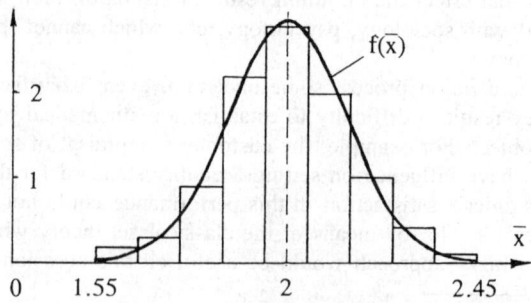

Figure 1. Histogram and density function curve of $\bar{x}$

As the confidence interval of $\bar{x}$ under a confidence level of 95% is [1.73, 2.27], we consider the average production ability of the quality parameter x in this factory fluctuates between 1.73 and 2.27. However from this interval which value is taken to be the assigned value of the quality standard that fits better the production ability of this factory is a fuzzy objective.

Hence it is necessary to describe this objective set $\underset{\sim}{B}$ of $\bar{x}$ by its membership function $\mu_{\underset{\sim}{B}}(\bar{x})$ according to $f(\bar{x})$, He Zhongxong (1985).

$$\mu_{\underset{\sim}{B}}(\overline{x}) = \frac{f(\overline{x}) - m}{M - m}$$

where M, m are respectively the upper and lower absolute bounds of $f(\bar{x})$, and we consider the domain of $\bar{x}$ is [1.73, 2.27] in the view of engineering application, thus

$$M = \max f(\overline{x}) = 2.85, \quad 1.73 \leq \overline{x} \leq 2.27$$

$$m = \min f(\overline{x}) = 0.44, \quad 1.73 \leq \overline{x} \leq 2.27$$

Then we have the membership function of $\underset{\sim}{B}$

$$\mu_{\underset{\sim}{B}}(\overline{x}) = \frac{f(\overline{x}) - 0.44}{2.41}, \quad 1.73 \leq \overline{x} \leq 2.27$$

Constraint

The quality standard set here must meet the demand of customers of various kinds, which come from different social stratum and have different professions, different attitudes toward value, different preferences of this parameter x of the product, etc. therefore satisfying the demand of customers is a fuzzy constraint for setting the standard.

On the basis of extensive market investigation, by using statistical method, He Zhongxong (1985), we find out the membership function $\mu_{\underset{\sim}{C}}(\bar{x})$ of fuzzy constraint set $\underset{\sim}{C}$ of parameter x which satisfies the demand of the customers, and

$$\mu_{\underset{\sim}{C}}(\overline{x}) = e^{-20(\overline{x}-2.1)^2}, \quad \overline{x} \geq 1.5$$

Fuzzy programming and decision

From above, fuzzy constraint set $\underset{\sim}{C}$ and objective set $\underset{\sim}{B}$ are given. Now, we want to find a $\bar{x}^*$ satisfying the demand of customers as well as suiting the production ability of the factory, i.e to find the point being constrained by $\underset{\sim}{C}$ and suiting the objective $\underset{\sim}{B}$. This is a problem of finding the symmetric model of fuzzy programming. We have the strategy set here, Terano, Asai & Sugeno (1992)

$$\underset{\sim}{D} = \underset{\sim}{C} \cap \underset{\sim}{B}$$

whose membership function is

$$\mu_D(\overline{x}) = \mu_C(\overline{x}) \wedge \mu_B(\overline{x})$$

as shown in figure 2.

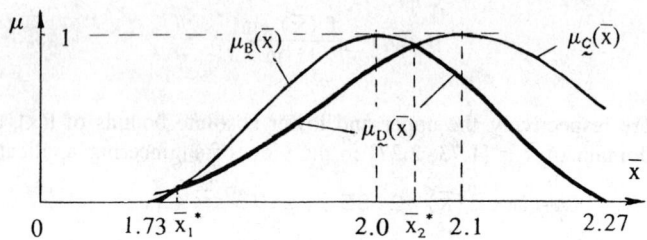

Figure 2. $\underset{\sim}{D} = \underset{\sim}{C} \cap \underset{\sim}{B}$

According to the requirements for setting the standard, our decision principle is to select proper $\overline{x}^*$, such that $\overline{x}^*$ satisfies

$$\mu_D(\overline{x}^*) = \max_{\overline{x} \in R} \mu_D(\overline{x})$$

then

$$\mu_D(\overline{x}) = \mu_C(\overline{x}) \wedge \mu_B(\overline{x})$$

$$= \begin{cases} \dfrac{f(\overline{x}) - 0.44}{2.41} & 1.73 \leq \overline{x} \leq \overline{x}_1^* \\[2mm] e^{-20(\overline{x}-2.1)^2} & \overline{x}_1^* \leq \overline{x} \leq \overline{x}_2^* \\[2mm] \dfrac{f(\overline{x}) - 0.44}{2.41} & \overline{x}_2^* \leq \overline{x} \leq 2.27 \end{cases}$$

The $\overline{x}^*$ is determined by

$$\frac{f(\overline{x}) - 0.44}{2.41} = e^{-20(\overline{x}-2.1)^2}$$

Solving

$$\overline{x}_1^* \approx 1.765 \quad , \quad \overline{x}_2^* \approx 2.045$$

Since $\mu_D(\overline{x}_1^*) \approx 0.11$ and $\mu_D(\overline{x}_2^*) \approx 0.94$, we consider 2.045 as the allowable value of the internal quality standard for the parameter x. This decision has met the requirements proposed by the decision-makers of the factory at a membership degree of 0.94, thus they have adopted our result in the quality standard of their factory.

Discussion

According to the decision-making method proposed here, one, several or countless solutions may be obtained. There are only two solutions here and one of them has a membership degree of 0.94 so that our decision could be made successfully. If the membership degrees of the solutions are small or not big enough (here the words 'small' and 'big enough' are concerned with fuzzy concept.), or several solutions have a same membership degree, a further decision should be made to select a better one among them, then a fuzzy multidecision problem will be involved.

In addition, multiple objectives and constraints are more often proposed while setting a quality standard. On this occasion a strategy set could be given as follows

$$D = C_1 \cap C_2 \cdots \cap C_n \cap B_1 \cap B_2 \cdots \cap B_m$$

where C_i (i=1,2,...,n) are fuzzy constraints and B_j (j=1,2,...,m) are fuzzy objectives.

References

He Zhongxong 1985, *Fuzzy Mathematics and its Application*,second edn (Tianjin Science and Technology Press, Tianjin)
Shen Jingming 1983, *Technical Economics in Mechanical Industry*, (Mechanical Industry Press, Beijing)
Terano, T., Asai, K. and Sugeno, M. 1992, *Fuzzy systems Theory and its Applications*, (Academic Press, Inc. Harcourt Brace Jovanovich, San Diego)

TOTAL QUALITY MANAGEMENT IN SMALL AND MEDIUM SIZED ENTERPRISES

Ms Susan Goh and Dr Keith Ridgway

University of Sheffield
Sheffield, S1 3JD.

This paper describes a survey, carried out with thirty small and medium sized manufacturing enterprises in Sheffield to examine the level of TQM implementation and their attitudes to TQM. The paper identifies five key elements of TQM and describes how a questionnaire and structured interview were used to assess the performance of the companies against the five elements. The results are presented and key issues are discussed.

Introduction

This paper describes the first stage of a project, carried out at the University of Sheffield, to design a framework to assist SMEs in the implementation of Total Quality Management (TQM). The first phase of the project involved a survey of 30 Sheffield based SMEs to determine their current state of TQM implementation and evaluate their attitudes and perception of TQM. The survey was based on a structured interview with the senior manager responsible for quality issues.

The survey indicated that many SMEs take the attainment of BS5750 certification as an end point. Few companies realised that TQM had implications for profitability and growth and there was evidence that the attainment of BS5750 was inhibiting innovation.

Components of the Survey

To carry out the survey it was necessary to develop a framework that could be used to compare the companies examined. The framework was developed around the basic concepts of TQM, based on the works of Oakland (1989, Flood (1993) and Dale (1990) and the quality gurus, Deming (1986), Crosby (1980), (1984), Feigenbaum (1991) and Ishikawa (1985). The framework identified five key concepts that the authors have defined as the "5 Pillars" because all five are vital for the establishment of a full TQM company. They form the basis of the whole TQM philosophy. The "5 Pillars" supporting the TQM system are identified as:

i)   Management Commitment (Commitment to Quality). This is essential to successfully implement TQM as resources and management leadership are required.

Management needs to establish a sound quality policy. This policy will state the company's corporate policy, its objective(s), its mission and vision on the quality of the company's products and on its commitment to its customers, together with arrangements for implementation. The contents of the policy must be communicated and understood at all levels of the organisation. Employee commitment and participation are vital. Employee morale, motivation and dedication to improving efficiency, effectiveness and quality must be enhanced by management.

ii)   Customer focus. This ensures that the company will continue to survive and grow. The following are key sources of information on customer requirements: Customer surveys and trials, working with key customers, competitor analysis, customer complaints and compliments, trade surveys and trials.

iii)   Quality Costs. Costs are incurred ensuring that products and services meet the customer's requirements. The huge and non-productive costs associated with poor quality and non-conformance are avoidable through the implementation of TQM. According to Crosby (1980) quality costs average 25 to 30 per cent of annual sales. Unnecessary costs can be eliminated by identifying the causes of non-conformance and failure through the implementation of prevention activities, which is another important aspect of TQM.

iv)   Quality Systems. These are an important aspect of TQM. Although it does not meet the TQM objective of delighting the customer, it is an important step that can be expanded into a TQM system. However, for SMEs, there are obstacles in adopting BS5750 because of the bureaucracy and costs associated with certification to BS5750.

v)   Continuous Improvement. (Increasing Growth and Profitability). This is a continual search for excellence and customer satisfaction. Both of these escalate and evolve into ever higher standards and greater expectations so that any company wanting to rank among the market leaders must actively engage in this Pillar of TQM to improve growth and productivity. The search for excellence involves a constant review and improvement of all management and planning activities, and processes in the company. The standard is the best practices in the industry and relevant practices outside the industry. One technique of achieving this is bench marking with other companies to provide reference points for comparison. Any business aspect can be bench marked, such as, customer satisfaction, financial performance, market share, distribution, design and management practice.

Innovation is another essential characteristic of a market leader in a competitive environment. Innovation is not limited to new products and processes. It can apply to all aspects of the company's business operations and management. The focus is on new markets, new products and processes, new ways of doing things leading to improvements, growth and profitability.

Development of Survey Questionnaire

The layout of the survey is similar to Crosby's (1980) Maturity Grid that indicates the 5 stages on the path to TQM, starting from "Uncertainty" to "Certainty". In this survey, the company's performance is rated against each of the fore-mentioned Pillars which have a rating scale from 1 to 5. The highest rating of 5 is given when the company has reached TQM status for any of the 5 Pillars. The average rating of 3 is given to a company that has quality system coupled with a focus on customer requirements. A questionnaire was also developed and used in the survey to determine whether the company had a well-communicated quality policy, the quality programs or processes employed in the company, whether the company was BS5750 certified, whether TQM implementation was a part of future improvement plans, and how much non-conformance to quality was costing the company annually.

Summary of Survey Results

The companies were grouped into the following categories to facilitate analysis. The number of companies per category is shown below:

Table 1: Survey Results

Categories	Current Status	
	No. of Co.	% of Total
1.0 No Plan for BS5750	1	3.3
2.0 Planning for BS5750	11	36.7
3.0 Has Implemented BS5750	11	36.7
4.0 Planning for TQM	6	20.0
5.0 No Plan for TQM	1	3.3
Total	30	100

As shown, 96.7% of the companies are taking some action to improve quality. However, only 20% of the companies are planning for TQM. This is a low percentage that can be improved upon if the 73.4% of BS5750 companies can be encouraged to advance to TQM.

The majority of the companies interviewed do not have any future plans to continue to TQM status. Most companies felt that TQM was not applicable to their company's operations, their common misgiving being that TQM is only for large companies and multi-national corporations. These companies on the other hand have between 7 to 250 employees. For the 11 BS5750 certified companies, it is unfortunate that most of them see this as an end in their Company's quality drive.

Companies surveyed, just recovering from the costs of implementation and consultancy fees, despite financial assistance under the DTI's Enterprise Scheme, and from the newly imposed load of paperwork associated with BS5750, are apprehensive of getting involved in yet another round of campaigns and employee talks and meetings. Most of the companies stated that widespread resistance was encountered, especially from employees, towards the increased amount of paperwork associated with maintaining BS5750.

Of the eleven BS5750 certified companies, five have had BS5750 for between 1 to 2 years and none of these are planning for the implementation TQM. Although three companies have had BS5750 for at least 5 years, they are not keen to embark upon the TQM route.

A further reason cited by the companies for not considering TQM was the current economic climate in the UK. Many of the companies are uncertain about their future, and are thus unwilling to venture into yet another quality program. BS5750 has been implemented as a response to customer pressure. As the customers are not demanding TQM the companies are reluctant to invest. The general feeling was that the cost of getting BS5750 certification was investment enough.

Further analysis examines the strengths and weaknesses of the weighted average rating of the 30 Sheffield companies against each TQM Pillar, as shown in Table 2.

Table 2: Weighted Average Rating per Pillar.

	TQM Pillar	Wt Ave
Strongest	Commitment to Quality	4.38
	Quality Systems	4.02
	Customer Focus	3.99
	Increasing Growth & Profitability	3.90
Weakest	Quality Costs	3.49

The main reasons for the deviation of the average rating per Pillar from the ideal rating of 5 are:

Commitment to Quality: With 50% of the companies having 50 employees or less, the level of top management commitment is high as they are directly involved in the day-to-day operations of the company. For these small companies, the MD is usually working together with his men on the shop floor and any problems encountered are reported directly to him. This direct communication of problems in production processes with top management is similar to the cultural change attained in TQM. Although many companies do have quality policies, these policies are not actively communicated to all employees, hence the importance of the quality policy is not felt by the employees.

Quality Systems: Fourteen (47%) companies have no more than 20% of their workforce trained in quality awareness or any other quality practices relevant to the company's operations. Management felt that their workforce is sufficiently skilled to perform their jobs and there is no pressing need for training programs to be implemented.

Customer Focus: Only two companies conduct customer surveys to determine their expectations and future needs. Surveys are not conducted to determine customer satisfaction with the company's products. The companies mainly rely on feedback through customer complaints or compliments with the sales team serving the dual function of being the 'customer contact department'. The companies have concentrated their efforts on establishing good customer relationships. The response time to customer complaints was generally very good. Only 40% of the companies provide formalised training for their customer contact employees. The majority of the companies either conduct only limited comparison with their competitors or just monitor trends within their own company. The companies tend to be very inward looking and are content as long as they out-perform their neighbouring competitors. They have not realised the importance of bench marking and the advantages it holds for their company.

Increasing Growth and Profitability: Only eight companies have continuous improvement programs in place, and are actively monitoring and striving to continuously improve their product quality. Four companies have no plans to improve upon any aspect of their company's product quality. Only 10% are actively seeking improvement opportunities.

Quality Costs: Only five companies ensure that detailed cost of quality records are kept and that these costs are well communicated to the workforce. Six companies confirmed that do not keep any quality cost records.

Summary and Future Research

The main points identified from the results of the survey are:

1. The majority of companies viewed BS5750 as the end-point in their quality drive.
2. Top management in most of the companies feel that TQM is inappropriate, viewing it as being applicable only to large companies.
3. Companies need to apply Creative Technologies and Innovation to their operations. It was evident that the cost of implementing BS5750 had reduced the investment in new product and process innovations.
4. Companies were unaware of the importance of keeping detailed, accurate cost of quality records, of the factors involved in the calculations of the cost of quality, and that full cost of quality records comprised Prevention, Internal Appraisal, Internal Failure and External Failure costs.
5. Companies which have been BS5750 certified are apprehensive of embarking on yet another quality program. The current economic climate has resulted in some companies shelving plans for TQM implementation because of financial constraints.
6. The need for a formal system to determine customer satisfaction with the company's products and their future requirements has been neglected by almost all the companies.

To address theseproblems a further project has been established in which large organisations, in the Sheffield area, will act as 'Uncles' to local SMEs and assist them to implement TQM. This process will be monitored by the University research team and developed into a framework for TQM implementation in SMEs.

Acknowledgements

The authors wish to thank the various SMEs surveyed for their help during the project. The authors also wish to thank the Billy Ibberson Trust who fund research into Industrial Change at the University of Sheffield.

References

Dale B. G. and Plunkett J. J. (Eds.), 1990, *Managing Quality*, Philip Allan, Hemel Hempstead, UK.
Hakes C. (Ed.), 1991, *Total Quality Management,* Chapman and Hall, London.
Oakland J. S., 1989, *Total Quality Management*, Nichols Publishing Company, New York.
Flood R. L., 1993, *Beyond TQM*, Wiley, Chichester, UK.
Deming W. E., 1986, *Out of the Crisis*, MIT, Cambridge, Massachusetts.
Crosby P. B., 1980, *Quality is Free,* Mentor, New York.
Crosby P. B., 1984, *Quality Without Tears,* McGraw-Hill, New York.
Feigenbaum A. V., 1991, *Total Quality Control*, McGraw-Hill, New York.
Ishikawa K., 1985, *What is Total Quality Control?,* Prentice-Hall, New York.

BENEFIT, IMPACT AND SHORTCOMINGS OF THE FOUR MAJOR QUALITY AWARDS

Professor Abby Ghobadian, Hong Woo and Jonathan Liu

Middlesex University Business School
Centre for Interdisciplinary Strategic Management Research
The Burroughs, London, NW4 4BT

Globalisation of markets and competition has changed the competitive priorities. Total quality in many markets is a condition of survival. The need to improve competitiveness has resulted in the birth of a number of transnational and national quality awards. These awards recognize the importance of management process, customer satisfaction, people, and quality in the attainment of superior competitive positions. This paper describes, compares, and highlights the key strengths and weaknesses of the following four major national and transnational quality awards: the Deming Application Prize; the European Quality Award (EQA); the Malcolm Baldrige Quality Award (Baldrige Award); the Australian Quality Award (AQA). In addition, the key requirements of each award, their underlying assumptions, and their impact are discussed.

Introduction

The economic success of Japan has drawn the attention of Western managers to Japanese management techniques and strategies (Treveor 1986). The fact that Japanese corporations used , "superior quality" to: capture; hold; and build market share has not escaped the attention of Western managers. Today, most chief executives, in the West, recognise that in all markets, "quality" is a basic requirement for continuous existence, and in some markets, differential quality is an important source of competitive advantage.

In 1951, the "Deming Prize" was introduced in Japan to help the development of quality culture among Japanese corporations. This was instrumental in spreading "quality methods" throughout the Japanese industry. Almost three decades later USA, Australia, and Europe recognised the value of an annual quality prize and established their own awards. The broad aims of these awards are:

(a)   to increase awareness and interest in the "quality of offerings" and "quality management" as an ever more important element of competitiveness;

(b)   to encourage co-operation between organisations on a wide range of non-commercially sensitive issues;

(c)   to stimulate sharing and dissemination of information on successful quality strategies and on benefits derived from implementing these strategies;

(d)   to promote understanding of the requirements for "quality excellence" and "quality management"; and

(e)   to spur organisations to introduce "quality management" improvement process.

In this paper the strengths, shortcoming, and the impact of: the Deming Prize; the EQA; the Baldrige Award; and the AQA are discussed.

Benefits

The quality awards provide a universal framework for evaluating and establishing the position of a company with regards to key actions and processes that influence the quality of the final offerings and competitiveness. Garvin (1991) stated that the Baldrige Award "not only codifies the principle of quality management in clear and accessible language but, it also provides companies with a comprehensive framework for assessing their progress towards the new paradigm of management and such commonly acknowledged goals as customer satisfaction and increased employee involvement". These sentiments equally apply to the other awards.

The Awards are particularly useful for smaller companies. Small businesses are vital to economic growth and creation of new jobs. Thus, they must be encouraged to strive to improve the quality of their offerings. Yet most small companies lack knowledge of how to implement total quality and cannot afford to engage expensive consultants. The awards provide a framework for the implementation of Total Quality and this is particularly useful to small companies.

There is a substantial difference between the requirements of the awards and those of ISO 9000 or BS 5750. The standards are an audit of procedures, the awards are concerned with wider issues, for example, leadership and results. The standards represent a minimum common denominator. Thus, it is highly improbable that any company can win one of the quality awards without first being able to satisfy the requirements of ISO 9000 or BS 5750 or their equivalents.

The awards have captured the attention of top management of many businesses. For example, it is claimed that the Baldrige Award has influenced the behaviour of US businesses more than any other award and applying for, and, winning the award has become an obsession for many US businesses (Sunday & Liberty 1992). It is generally perceived that winning one of the quality award's helps to enhance the organisations' competitiveness and position as the leader in its field. Ron Burnstein, an executive of NRMA, one of the AQA winners is on the record as saying that "winning the quality award offers significant promotional opportunities particularly to organisations that use service quality to achieve a marketing edge" (Ghobadian and Woo 1993). Winning the award will help to raise the profile of the organisation as well as creating publicity for it. Thus, from top management's point of view, the lure of the award is simple enough - publicity (Crainer 1994). Moreover, because the awards have become a symbol of quality excellence, customers often judge the company by their efforts to apply for the Award. Motorola, for example, requires all its suppliers to make the effort to apply for the Baldrige Award.

Gaining market advantage and publicity is not a sole reason for applying. There are also internal reasons. Ron Burnstein summarised these as (Ghobadian and Woo 1993):

(a) genuine recognition that a company will benefit from the process of preparing a submission and going through the award entry;

(b) quality improvement is a long-term process and involves a great deal of effort by everyone in the organisation, the award is a form of recognition for this effort, and improves staff morale and motivation;

(c) self appraisal helped the organisation to identify its strengths, weaknesses and plan to remedy the latter.

Impact

A major aim of the awards reviewed was to raise quality consciousness among industrialists and the general public. One way of determining the success of the awards in this respect is to measure the number of organisations requesting information and the number of organisations applying for the award. In the case of Baldrige only 12,000 organisations requested information in 1988. This figure had increased to 175,000 by 1992. These figures do not provide the full story, because these awards have been adopted at lower level. Most states in the USA offer a quality award based on Baldrige, the best known of these is the Excelsior award open to for-profit and non-profit organisations in New York State. In January 1994, the British Quality Foundation launched a new quality award based on the EQA model. The mission of the foundation is to promote total quality in the UK. The DTI is backing the Foundation with a grant of £500,000. In addition, 70 organisations are supporting the Foundation by providing donations. Another aim of these awards was to facilitate sharing of experiences and to encourage corporative behaviour. For example, Xerox, as a Baldrige award winner, has made 2500 presentations to more than 250,000 customers (Rickard, Debate, 1992). This benefits the customers and Xerox. Xerox will have the opportunity to listen and learn from their customers.

The awards help organisations to establish a benchmark from which all future progress can be measured as well as providing the businesses with improvement ideas. The evidence suggests that some companies apply for no other reason than to obtain feedback from the examiners. At the cost of about four thousand US dollars, these companies can expect feedback from experts concerning: their current performance; what is expected of them; and most importantly, how they can improve (Stratton 1990). Thousands of businesses and business units are increasingly using the criteria propagated by these awards to benchmark their quality programs and quality efforts (Sunday & Liberty 1992). Our experience in UK reinforces this finding. Many organisation in UK use EQA's model for self appraisal.

Winning an award spurs winners to greater efforts. Ikuro Kusaba, Professor Emeritus, Nagoya Institute of Technology, found that enterprises which are awarded the Deming Prize, usually exert greater effort to enhance quality control in the years after winning the Prize. Receipt of the Deming Prize enables companies to integrate the wills and activities of all employees, strengthen corporate structure, and upgrade the quality of their products and services, and this, in turn, has a positive effect on profits. This, then, has greatly contributed to the betterment of Japan's industry as a whole. In a recent article, J.M. Juran found that Baldrige Award winners have achieved significant breakthroughs within a few years.

Shortcomings of the Awards

The amount of effort involved in the preparation of an application and the size of investment required is sited as one of the major disadvantages of the awards. This may account for the large difference between the actual applications and the number of requests for information. To meet the award's requirements, it is necessary to: gather vast amounts of internal and external information; analyse the data; and expend substantial amounts of managerial effort. For example, Corning devoted over 14,000 man-hours toward its award submission in 1989. Westinghouse's Thermo-King Division was estimated to have spent almost $500,000 on internal labour and an equivalent expenditure on outside technical assistance in putting their application together. Florida Power & Light, the first winner of the Deming Prize for Overseas Companies spent $850,000 on fees to Japanese consultants to help them improve their quality systems to the point that they could meet the Prize's standards. The amount of effort and investment required raises an important question - are small businesses and non-multinationals able to take part seriously in the process? This is particularly worrying as small business are vital to the economic growth of most nations and must be encouraged to strive for improved quality. Small businesses will have a difficult task affording the long-term investment in total quality, and cannot be expected to expend limited managerial capability and resources on awards (Cooper 1992). On the other hand, a number of winners appear to have invested little effort or money. For example, Globe Metallurgical have spent almost nothing on their application process for the Baldrige Award (Cole 1991).

Poor performance of past winners, for example, Cadillac, Federal Express, Wallace Company and Motorola have led pundits to question the value of these awards in terms of gauging the quality of the offerings and profit potential of an organization. There were special reasons for the poor performance of these organisations immediately after winning the award. The profitability of an organisation is affected by external as well as internal circumstances. Moreover, the awards do not solely focus on product or service quality. In fact, the awards primarily focus on management systems and processes. The argument put forward by the advocates of the awards is that all companies are susceptible to economic downturn, shifts in technology, and changes in fashion, but companies that embrace total quality and receive a high score are more robust and capable of recovering from these external set backs.

Baldrige, the AQA and, to a lesser extent, the EQA are criticised because they lack sufficient focus on results. The antagonists of the awards argue that they are too process-oriented and place too much emphasis on TQM as a "check the box" activity and not as a path to sustainable results. They argue that more emphasis must be placed on results over time and not simply reinforce the culture of "just do it". The other criticisms include:

(1)   award criteria are static and not dynamic;
(2)   supplicants nominate themselves and are not nominated by customers;
(3)   the EQA, Baldrige and Deming and, to a lesser extent, the AQA fail to define quality and thus, fail to help organisations to reach a common understanding;
(4)   awards encourage a home grown approach to quality and this will not help them to achieve world class performance;
(5)   companies may focus on winning the award rather than opportunities for self examination, learning, and improvement; and
(6)   pursuing the award distracts key executives from running the business.

Conclusions

The awards are based on a perceived model of "total quality management". They do not focus solely on either product or service perfection or traditional quality control methods, but consider a wide range of management principles and factors that influence the quality of the final offerings. The models upon which the awards are based implicitly recognize that the quality of the "final offerings" is the end result of integrated processes and employees' effort. The quality awards appear to have had a positive impact. For example, Baldrige award participants point to lessons learnt in preparing the application and new policies instituted. They state that the process of preparing the application forced them to take a long hard look at the way the company was run. Using the award application as a self-audit tool, companies can assess how well their quality programs are doing and determine what steps to take next. Those taking part in the competition report instigating change and introducing improvement programmes in a variety of areas.

References
Cole, R. E. 1991, Comparing the Baldrige and the Deming, *Journal for Quality and Participation*, 1991, 94-104.
Cooper, G., E., Debate 1992, *Harvard Business Review*, 1992, 138-139.
Crainer, 1994, *The Times*, 24 February 1994.
Garvin, D.A. 1991, Does the Baldrige Award Really Work, *Harvard Business Review*, 1991, 80-93.
Ghobadian, A. and Woo, H. 1993, Quality at NRMA, Unpublished Case, Centre for Interdisciplinary Strategic Research, Middlesex University Business School.
Rickard, N., E., Debate 1992, *Harvard Business Review*, 1992, 146-147.
Stratton B. 1990, Making the Malcolm Baldrige Award Better, *Quality Progress*, 1990, 30.
Sunday, J. L., and Liberty, L. 1992, Benchmarking the Baldrige Award, *Quality Progress*, 1992, 75-77.
Treveor, M. 1986, Quality Control - learning from the Japanese, *Long Range Planning*, **19**, 1986, 46-53.

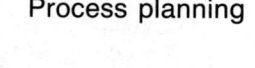

Process planning

STATE-OF -THE ART REVIEW OF CAPP: ITS IMPACT AND ITS APPLICATION IN CIM

Dr. A Ajmal and Dr. H G Zhang

School of Engineering , Systems and Design
South Bank University
London SE1 0AA

Process planning plays a key role in CIM and has emerged as a strategic link between CAD and CAM. This paper presents a comprehensive summary on the development and research trends in CAPP, highlighting areas which are likely to be applicable in the near future. The advances in the feature recognition, neural network and object oriented programming techniques are listed and major CAPP systems that have been described in the open literature are analysed. Suggestions for future research are also presented.

Introduction

The efficient processing and distribution of information is vital to the development of a truly CIM system, and automated process planning aims in providing the strategic integration link between CAD and CAM. According to Sirihari (1987) Computer-aided technique was first used in process planning in 1960's and since then several methodologies, and concepts have been used to automatic process planning. These include application of expert systems (Ajmal 1989) and relevant application.

This paper presents a review of the development and application of CAPP systems. Recent trends and development in the application of neural networks , feature definition and recognition, automatic tolerancing and dimensioning, and object oriented programming in CAPP have also outlined. The paper concludes with suggestions for future developments in process planning.

Feature extracting and feature-based design for process planning

Feature recognition

Extracting information from CAD is critical for integrating CAD and CAPP, and various efforts and many methodologies have been proposed and the major ones are as follows:
- Syntactic pattern recognition

This approach uses the theory of formal language and a feature is represented by some primitives which have semantic association with geometric entities. Another methodology   state transition diagrams, uses the relationship between adjacent primitives to describe a matching feature instead of grammar syntactic pattern recognition. These approaches are limited to 2-D and $2\frac{1}{2}$-D primitives and mainly focus on the rotational part families (Rong-Kwei Li, 1988).

- Volume decomposition

    These techniques uses several approaches , e.g. destructive solid geometry (DSG) , delta volume , edge extension, half space, and face extension (Hendric, 1990). Recently, Kim (1992) presented alternating sum of volume (ASVP) for the recognition of form feature based on B-rep solid modelling. The system implemented on Sun workstation with PROLOG, has been used to convert ASVP decompositions into form features and classify the recognised features. This is aimed towards constructing a meaningful feature volume for process planning.

- Expert system (ES) approach

    This approach assumes that the human process planner performs feature recognition based on some pre-defined notions of features. These ES rules of feature pattern and faces of part description are loaded to the knowledge base of an expert system (Wang , 1987). According to Wang, Ming-Tzong (1990) implementation of such an ES based feature recognizer is quite simple. However, in real implementations because of the exhaustive backward chaining search strategy, it faces combinatorial and computationally problems.

- Extracting from CSG solid modelling

    Constructive solid geometry (CSG) representation has the problem of non-uniqueness and redundancy (Lee and Fu, 1987). Shpitalni and Fischer (1994) used the approach on 2D design and suggest that it can be used for 3D design as well.

- Graph based approach

    Joshi(1988) developed a feature recognition methodology based on graph-based representation and heuristic in 1988. In this approach, the input B-rep is transformed to a attributed adjacency graph (AAG) and a manufacturing cavity heuristic is applied to convert the whole AAG graph to sub-graphs. Floriani (1989) uses another graph based algorithm for extracting shape features of an object from a relational boundary model.

Feature based Design

Li (1991) developed a prototype system embedded in AUTOCAD and use pre-defined standard primitive and feature commands. Gupta, etal (1992) presented a feature based CAPP system for prismatic parts.   Cranford (1993) outlined a feature based process planning for sheet metal part. The system provides a library of standard low-level features such as plate, bend, hole and spur and has the flexibility for user to define their own features. Duan (1993) outlined the Feature solid modelling tool (FSMT). which uses a unified method to create a feature.

Other methodologies and their applications

Neural network and its applications

Neural networks have become a very popular field of research in various fields of neurobiology , computer engineering/science, signal processing, optics, and physics (Kung, 1993), and could be applied in CAPP (Knapp and Wang 1992). However, because of the potential advantages, it would play an very important role in CAPP. An artificial neural network is an abstract simulation of real nervous systems that contain a collection of neurone units communicating with each other via axon connections. There are various types of neural network models (Kung ,1993) and Fig.1 is back-propagation network based on a three-layer network.

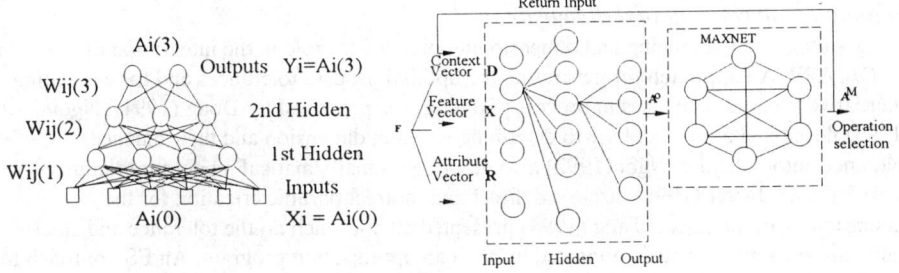

Fig 1 The multiple-layer network   **Fig. 2 Network operation for operation selection**

Prabhakar and Henderson (1992) presented an automatic feature recognition system based on neural networks.

Traditional process planning knowledge uses ES such as rules, and is time consuming, costly and error-prone. Furthermore, the ES cannot adapt to change in manufacturing practice and technology and it cannot generalise from the past experience to handle new, unforeseen cases (Knapp and Wang 1992). Neural networks have potential advantages to overcome these limitations. Knapp, etal outlined an approach as shown in Fig.2. The first is back-propagation networks and other is MAXNET. Chen(1994) also presented a system for solving setup generation and feature sequencing using back-propagation neural network. Kaparthi (1991) also use this approach for automatic coding system.

Conjugative concept

Zhao (1992) presented a system based on the conjugative concept as shown in Fig.3. and the core of the system is an innovatory definition which is used to establish the relationships between design and manufacturing information

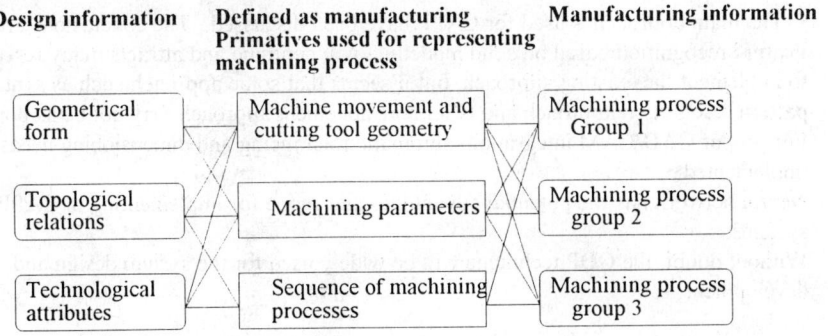

Fig. 3 Conjugative model

Machine feature extracting from STEP/PDES

STEP and PDES both are standards data exchange for product data developed in mid-1980's (1993) and provides an avenue to exchange data between CAD/CAM/CAE/CIM systems. Shah (1991) outlined an experimental investigation of the STEP form-feature information model.

Automatic tolerancing and dimensioning

Automatic tolerancing and dimensioning play a vital role in the integration of CAD/CAPP. An approach for generating the optimal process tolerances and for evaluating alternative process sequences in process planning was presented by Dong (1991). Ngoi (1993) developed a methodology to determine working dimension and their balance tolerance automatically. Ngoi (1992) also described a mathematical model for tolerance chart balance. Juster (1992) discussed direct and indirect parameterization for the dimensioning of an object. Tang (1994) presented an approach on the tolerance and stock removals are simultaneously allocated through an optimisation program. An ES approach to generate alternative process plans for machined parts with tolerance requirement has been reported by Abdon (1993). Ping Ji (1993) presented a tree-theoretic representation for a tolerance chart. Very little has been publised on the following, related to tolerance problems, and needs futher investigation: (i) Extraction of surfaces and tolerance information from CAD database ; (ii) Implementation of dimensional and geometrical tolerances verification in CAD; (iii) Design of rule based ES incorporating tolerance and; (iv) Using neural networks to implement tolerancing and dimensioning.

Object oriented programming (OOP) in CAPP

OOP is a revolutionary change in programming (Coad and Nicola 1993). The object oriented database has many advantages over the relational database (Hughes,1991). OOP applications in CAPP are still in their infancy. Yut and Chang (1994) presented a object-oriented process planning system, the primary objective of the architecture is modularity, where this is achieved mainly through polymorphism. Wang (1988) presented a C++ layered hierarchy based object-oriented feature-based CAD/CAPP/CAM integration framework. It is based on an object oriented feature-based part model which represents a workpiece as an organised set of main shape features attached with some standardised features.

Conclusions and future works

The main approaches used for CAPP have been discussed, The conclusions are:
- Feature recognition based on solid modelling will continue and attracts many researcher to implement the existing approach, but it seems that some approach such as syntactic pattern recognition approach and CSG rearrangement approach may not be as popular;
- For proper CAD/CAM integration, automatic tolerancing and dimensioning needs to be implemented;
- Neural network has the potential advantages and scope for implementing in CAPP systems;
- Without doubt, the OOP technique will be widely used for the system design and development;

- The implementation and application of standard product data exchange scheme such as STEP will continue;
- Feature-based design is an attractive research field. but the key point will focus on the flexibility of the system and user interface .

References

Srihari. K , 1987, *Proc. World Productivity Forum, Int. Ind. Enginnering Conf., USA*, 438-444

Ajmal. A, 1989, *Proc. 4th. Int. Conf. on CAD/CAm and Robertics*, 527-536

Li, Rong-Kwei , 1988, *Int. J. Prod. Res.*, **26**, **9**, 1451-1475

Wang, Ming-Tzong , 1990, *A geometric reasoning methodology for manufacturing feature extraction from a 3-D CAD model :Ph.D thesis, Purdue University*

Hendric, V J, 1990, *Automatic resoning of machinable feature in solid model :Ph.D thesis , The university of rochester*

Kim,Y.S., 1992, *Computer-aided design*, **24**, **9**, 461-476

Henderson, M R and David, C. 1984, *Computer in industry* ,**5** ,329-339

Shpitalni,M and Fischer,A, 1994, *Computer aided design*, **26**, **1**, 46-58

Joshi,S and Chang,T,C , 1988, *Computer aided design* , **20**, **2**, 58-66

Leila De Floriani, 1989, *Proc. of 3rd ACM Symposim on computational geometry*, 100-109

Li Rong Kwei , et al , 1991, *Int. J. Prod. Res.*, **29**, **1**, 133-154

Gupta,S.K. and Rao,P.N and N.K. Tewari , 1992 , *Int. J. Adv. Manuf. Technol.* ,**7**, 306-313

Cranford,R, 1993, *Int. J. Computer integrated manufacturing* , **6**, **1-2** , 113-118

Li ,Rong Kwei , Ying-Mei Tu and Tao H Yang, 1993, *Int. J. Prod. Res.*, **31**, **7**, 1521-1540

Ping Ji, 1993, *Int. J. Prod. Res.*, **31**, **5**, 1023-1033

Duan,W, Zhou,J and Lai,K , 1993, *Computer-aided design*, **25**, **1**, 29-38

Kung,S.Y 1993 *Digital neural networks:PTR Prentice Hall* , ISBN 0-13-612326-0

Knapp,G M and Wang , H.P1992 ,*Journal of intellegent manufacturing*,3, 333-344

Prabhakar,S and Hendson,M.R.,1992, *Computer-aided design*, **24, 7**, 381-392

Chen,C L P and LeClair,S.R., 1994, *Computer-aided design*, **26**, **1** , 59-75

Kaparthi,S and Suresh,N.C.,1991, *Int. J. Prod. Res.*, **29**, **9**, 1771-1784

Zhang,H.C and Huq,M.E., 1992, *Int. J. Prod. Res.*, **30**, **9**, 2111-2135

Zhao Z , 1992, *Generative process planning by conjugative coding design and manufacture infomation : PhD thesis , Staffordshire University*

Qiao,L H, Zhang,C. Lium,T.H. Wang ,H.P and Fisher ,G.W,1993, *Computer in industry*, **21**, 11-22

Shah ,J Jand Mathew , 1991, *Computer-aided design*, **23, 4** , 283-296

Dong,Z and Hu,W , 1991, *Computer in industry* ,**17** , 19-32

Ngoi,B K A and Teck, O. C,1993, *Int.J.Prod. Res*,**31**, **2**,453-469

Ngoi ,B K A,1992, *Int. J Adv. Manuf Technol* ,**7** , 187-192

Juster,N.P. 1992, *Computer-aided design*, **24, 1**, 3-17

Tang,G R et al , 1994, *Int. J. Prod. Res.*, **32**, **1**,23-35

Absou,G and Cheng ,R, 1993, *Int. J. Prod. Res.*, **31**, **2**, 393-411

Ping Ji, 1993, *Int. J. Prod. Res.*, **31**, **5**, 1023-1033

Coad,P and Nicola ,J,1993, *Object-oriented programming: Yourdon Press*

Hughes ,J.G. 1991, *Object oriented database :Prentice Hall* , ISBN 0-13-629882-6

YUT,G A and Chang ,T.C. 1994, *Int. J. Prod. Res.*, **32, 3**, 545-562

Patel,R.M and Mcleod,A.J, 1988, *Computer-aided engineering journal*, **Oct.** 181-183

Lee,Y.C and Fu ,K.S,1987 *IEEE Computer and graphics applications*, **7** , 20-23

Wang ,H.P, 1987, *Computer in industry*, **8** , 293-309

PROCESS CAPABILITY MODELS FOR EQUIPMENT SELECTION

Nabil N Z Gindy*, Tsvetan M Ratchev**, Keith Case**

Department of Manufacturing Engineering and Operations Management, University of Nottingham, Nottingham, NG7 2RD

**Department of Manufacturing Engineering, Loughborough University of Technology, Loughborough, LE11 3TU*

Due to the increased complexity of modern manufacturing facilities and the increased demands for product variability and system flexibility there is a need for coherent formal representation of the basic knowledge domains supporting manufacturing applications such as equipment selection. The paper presents integrated framework for equipment selection based upon describing process capability at generic, machine tool and manufacturing system levels. The decision making process is designed as a sequence of steps for transforming component design information into processing requirements which are mapped into specific physical machines organised as a processing system.

1. Introduction

Equipment selection is a complex and time consuming process which requires a number of important decisions to be made and a large number of inter-related factors to be taken into account (Kochhar and Pegler, 1991). Many of the difficulties faced when attempting to formalise the decision making logic for equipment selection arise from the need to transform information, at several levels, between closely inter-related domains of components, manufacturing processes and manufacturing resources used for their production. Knowledge integration to facilitate transformations between the three domains is one of the crucial issues to be addressed during development of equipment selection systems.

The concept of using features for CAD/CAM applications has demonstrated significant potential during the recent years (Shah, 1992). Features, treated as generic entities for describing component attributes, act as basic information carriers during component design and machine selection. By attaching geometric as well as technological attributes to the features, main-stream activities such as process selection can be directed at a feature level.

Several approaches have been used to describe manufacturing processes and their capabilities (Alting, 1982; Eversheim, Marczinski and Cremner, 1991) Manufacturing processes were traditionally closely associated with specific types of machine tools used for their execution. The advent of multi-axis machining centres and the general increase in capabilities of machine tools, however, has reduced the relevance of such close linking and has underlined the need of machine independent process representation which can capture the exclusive and the repeated processing capabilities of different machine tools.

The approach adopted in this work is to relate process capability to three fundamental levels. The first is "form generating schema" to be used for describing process knowledge at a level that is independent of the machine tool and machining facility used for process execution. The second level of abstraction is the "operation" level that is used to attach machine specific attributes to manufacturing processes. The third is the "resource element" level which is used to complete the loop by relating both the form generating schema and operation levels to the way in which machine tools are organised as a manufacturing system.

The integrated model used as decision making support system for equipment selection is shown in Figure 1. The following sections of this paper give a brief overview of the model entities and equipment selection methodology adopted in this work.

2. Component Data Model

The input to the system is a feature based component description. Component features are considered as geometric entities which have significance in the context of component design and manufacture. Examples of such features are hole, step, pocket, etc., each of which is uniquely classified and described within a component model.

Features are treated as volumes enveloped by entry/exit and depth boundaries. Feature geometry is described by deciding on its external approach directions, i.e. the number of imaginary faces included in feature definition, its form variation with respect to its depth axis. Component form features are organised into a hierarchical structure for their definition and classification which includes categories, classes and sub-classes (Gindy, 1989).

Component connectivity which describe the relationships between adjacent features is represented by two types of links: external access directions for relating individual features to the basic component directions; and inheritance links that relate adjacent features, with some features becoming parents to other features (Gindy, Huang and Ratchev, 1993). At the component level, the technological relationships (geometrical and positional tolerances) between features are also recorded.

3 Process Capability Models

Process capability knowledge used for equipment selection is described at three basic levels: form generating schemas, machining operations and resource elements.

3.1 Form Generating Schemas

Form generating schema (FGS) is defined as technologically meaningful combination of tool of specific geometry, set of relative motions between a part and the tool, and nominal levels of technological output that can be associated with using that combination of tool and relative motions (see Figure 2). Form generating schemas are machine independent, but can be used to provide a generalised description of machine tool capabilities.

The links between the form generating schemas and the component features are provided through feature transition diagrams (FTD) - collections of form generating schemas capable of producing specific feature geometry and various levels of technological output. Feature transition diagrams reflect company specific knowledge on how features are machined.

Process selection is based on finding the set of terminal nodes in a feature transition diagram which match the feature requirements in terms of surface finish and tolerances. All feasible, equally weighted processing sequences are then selected through backtracking.

3.2 Machining operations

A form generating schema acts as the root for defining various machining operations. Operation inherits the attributes of the FGS it originated from and the division of motions between a part and the cutting tool from the machine tool used for its

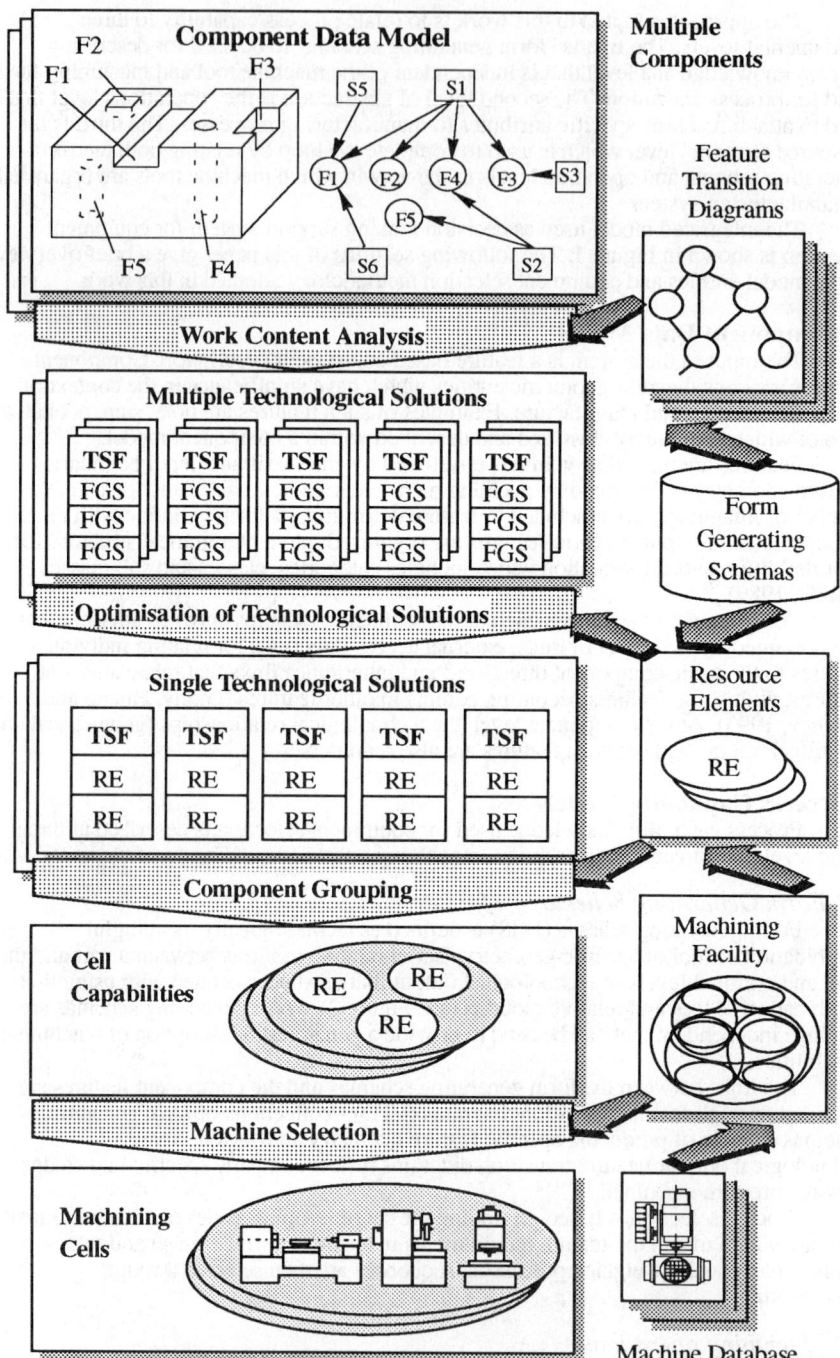

Figure 1. Equipment selection system - an overview.

execution. A machine tool performing an operation also provides the specific levels of technological output to be attached to machining operations.

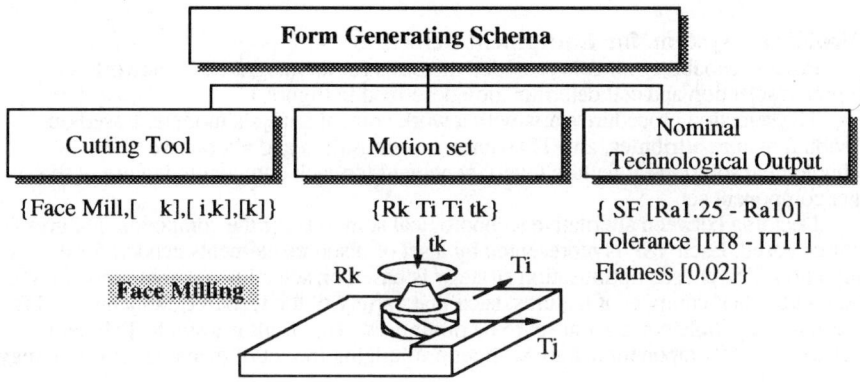

Figure 2. Form generating schema - an example.

A typical example are the machining operations originated from FGS "drilling" when performed on a lathe and on a drilling machine - their tool and motion set are identical with the only difference being that on a lathe component is rotating and the tool is translating while on a drilling machine both motions are given to the tool.

3.3 Resource Elements

Many manufacturing facilities contain identical machines and several machine tools with overlapping capabilities in terms of form generation and technological output. During equipment selection, however, a methodology is needed for comparing machine capabilities to provide a basis for deciding between alternatives before a final selection is made.

The set of machine tools defining a manufacturing facility are described using a set of resource elements (RE). Each RE represents a collection of form generating schemas which define uniquely the exclusive machine tool capability boundary and the shared boundaries between machine tools (see Figure 3).

Resource elements are machining facility specific and capture information relating to the distribution (commonality and uniqueness) of form generating schemas among the machine tools included in the machining facility.

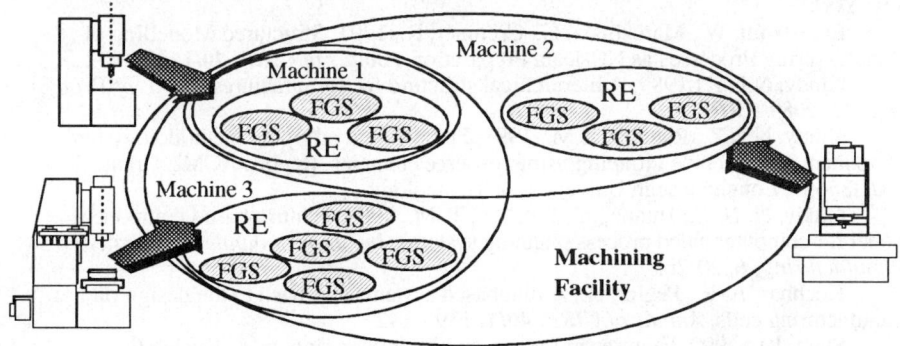

Figure 3. Machining facility represented using resource elements.

Although an RE may be attached to several machine tools, a form generating schema can belong only to one RE. A machine tool has to be capable of carrying out all the form generating schemas of the REs associated with it.

4. Prototype System for Equipment Selection

Process capability models provided the basis for an integrated framework for equipment selection and cell determination described in Figure 1.

The selection procedure starts with a work content analysis module. Based on individual feature attributes, an FTD database is consulted and alternative TSFs (technological solutions at feature level) generated for each component feature of the target component set.

Deciding between alternative technological solutions for the component features is resource based. Each TSF is represented by a set of resource elements needed for its completion. The chosen optimisation strategy is based on selecting a combination of TSFs for the individual component features, such that variety of the resource elements used for processing the whole component set to be minimised. The result is a single TSF being attached to each component feature without pre-judging the set of machine tools that may be finally selected.

The capability requirements for each machining cell are first defined by component grouping using resource elements as basic grouping primitive. Grouping approach is based on fuzzy clustering with cluster validation (Gindy, Ratchev, 1993). The cell capability is defined by the sets of REs required by each component group.

The final step of the selection process is machine specification in terms of transforming the capability requirements of the designed machining cells into physical machines. The selection is carried out based exclusively on capability requirements without considering planning and scheduling issues at this stage.

5. Conclusions

The concept of resource elements is proving useful in providing a common basis for decision making throughout the machine selection and cell determination process. Resource elements provide flexibility in comparing the capabilities of individual machine tools and describing the processing capabilities of manufacturing facilities/cells.

Adopting modelling approach to integrating the information requirements of an application such as equipment selection has many advantages. The developed models help simplify the decision making logic and facilitate development of structured and modular systems which are easy to update and maintain.

6. References

Alting, L., 1982, *Manufacturing Engineering Processes*, (Marcel Dekker Inc., New York).

Eversheim, W., Marczinski, G., Cremner, R., 1991, Structured Modelling of Manufacturing Processes as NC-Data Preparation, *Annals of CIRP*, **40/1**, 429-432.

Gindy, N.N.Z., 1989, A hierarchical structure for form features, *Int. J. of Prod. Res.*, **27**, 2089-2103.

Gindy, N.N.Z., Ratchev, T.M., 1993, Fuzzy based grouping methodology for component and machine grouping using resource elements, Report, ACME Grant GR/G35657, Loughborough University of Technology.

Gindy, N. N. Z., Huang, X., Ratchev, T. M., 1993, Feature-based component model for computer aided process planning systems, *Int. J. of Computer Integrated Manufacturing*, **6**, 20-26.

Kochhar, A. K., Pegler, H., A rule based systems approach to the design of manufacturing cells, *Annals of CIRP*, **40/1**, 139 - 142.

Shah, J.J., 1992, Features in Design and Manufacturing, in A. Kusiak (ed.), *Intelligent Design and Manufacturing*, (J Wiley & Sons, New York), 39-72.

A FEATURE BASED APPROACH TO DETAIL AND ASSEMBLY COSTING IN THE AEROSPACE INDUSTRY

[a]T M Hill, [b]R J Forster and [a]G Smith

[a]University of the West of England
Frenchay, Bristol, BS16 1QY

[b]British Aerospace Airbus Ltd
Filton, Bristol, BS99 7AR

The introduction of Design for Manufacture (DFM) practices at the early stages of the design process has been shown to generate cost effective design and rationalisation of mechanical design function. This paper outlines the development and implementation of individual DFM systems at British Aerospace Airbus Ltd and the University of the West of England. The three level approach to DFM by these organisations is exemplified by KBS enhancements to conventional CAD systems, architectural developments within an Intelligent CAD environment and new research into feature based DFM/CAD system development.

Introduction

It has long been realised that cost assessment should occur at the earliest possible time in the design process, preferably at the concept stage. Yet detailed information is rarely available at this stage and influence in the past has been limited to advising on good design practice in purely functional terms, on material selection and on generic cost avoidance practices. As detailed design follows schematic representations, the potential for cost reduction decreases but level of detailed information available for cost/manufacture analysis substantially increases. Cost analysis is more accurate at this level but is still subject to influences at the manufacturing planning stage. The planning stage is able to call on very specific costing methods and data and will generate accurate costs uniquely associated with the plant available to manufacture the product. The computational expense of invoking process planning systems, if they are available, during the design phase is prohibitive and burdensome to the designer. Hence some other means of encouraging the designer to consider the manufacturing and cost of the design during the design phase must be sought.

One popular means of reducing design costs is to employ concurrent engineering teams operating in an interactive mode. The deployment of functional

249

teams which include expert design, planning, cost and manufacturing engineers undoubtedly achieve dramatic results but flag inadequacies in our ability to represent their skills in software form. An alternative, the approach adopted here, is to replace these domain experts with knowledge based systems. What follows are descriptions of several attempts at BAe and UWE to do just this in an industry where designers have been burdened by the need to meet the exacting functional requirements of their products. The opportunity to trade function for cost of manufacture is still far more limited than with most other industries, but commercial reality has forced the industry to address this problem with some urgency. This paper will outline the evolutionary development of DFM systems at both UWE and British Aerospace Airbus Ltd. Figure 1 gives an outline of the various developments within the two organisations.

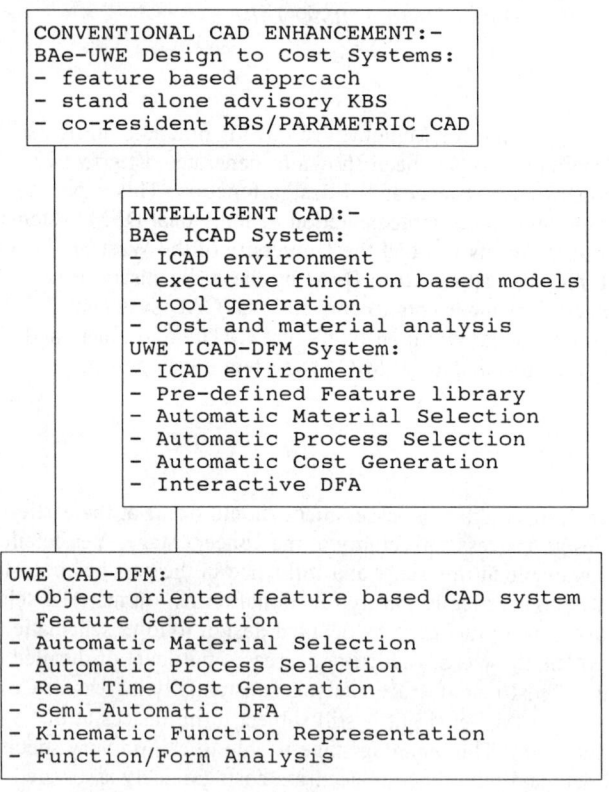

```
CONVENTIONAL CAD ENHANCEMENT:-
BAe-UWE Design to Cost Systems:
- feature based apprcach
- stand alone advisory KBS
- co-resident KBS/PARAMETRIC_CAD
```

```
INTELLIGENT CAD:-
BAe ICAD System:
- ICAD environment
- executive function based models
- tool generation
- cost and material analysis
UWE ICAD-DFM System:
- ICAD environment
- Pre-defined Feature library
- Automatic Material Selection
- Automatic Process Selection
- Automatic Cost Generation
- Interactive DFA
```

```
UWE CAD-DFM:
- Object oriented feature based CAD system
- Feature Generation
- Automatic Material Selection
- Automatic Process Selection
- Real Time Cost Generation
- Semi-Automatic DFA
- Kinematic Function Representation
- Function/Form Analysis
```

Figure 1 Design for Manufacturing Systems

Enhancements to CAD Systems

Initial attempts to provide cost related advice during the design phase have resulted in the development of stand alone advisory knowledge based systems which classified form features and assigned pre-defined costs to those features, Mirza (1994). The costs were computed from simple multiplicative correlating equations which were derived firstly by identifying the primary cost driving features. The cost driving features were identified by conducting a extensive knowledge elicitation exercise with design, planning and manufacturing personnel. Correlating coefficients, attached to the feature parameters, were determined from multiple regression analysis using historic cost data from the company database. This type of scheme requires many correlating equations, one for each class of manufacturing process, and is not easily adapted to handle assembly operations. In its present form it operates a bottom-up approach and attempts to influence detail design through design simplification and feature attribute optimisation. Trials with such a system has predicted costs on families of parts with average errors in the region of ten percent. The system is useful for indicating which features to limit or indeed avoid and for comparing alternative designs. In this role it operates in an advisory capacity as a stand alone design aid. To be effective it requires continuous maintenance since its accuracy requires the cost drivers and their attendant correlating coefficients to match current manufacturing processes and working practices. In a large organisation this constitutes a major effort since the knowledge required to maintain the system is held by a disparate range of manufacturing personnel. Its inherent flaws are thus its static nature and lack of integration with the CAD system. The KBS could, in practice, be based any of a number of proprietry shells but it has been found that object oriented frame based systems (eg KAPPA) provide the best framework for such schemes.

Later developments in this area have improved the flexibility of the system by replacing the correlating equations with more complex estimating procedures and heuristics in order to generate costs. Interactive graphics have also been added to these systems to improve the user interface.

An enhancement to the above system has been achieved in a experimental system, Forster (1994), by integrating a KBS (KEE) with a parametric feature based solid modeller (PRO-ENGINEER). This co-resident system attempted to provide on-line cost related advice to·the designer who was able to request DFM analysis whilst the components were being created. The communications between the two systems were difficult to set up and a three layer communications protocol was devised for that purpose. The system was designed to work in a design by features mode with a set of user designed features.

Knowledge Based CAD Systems

Recent additions to the CAD market have prompted yet another approach. Knowledge based CAD systems (eg ICAD, WISDOM) now provide the means to create 'total product models' which can capture functional and manufacturing related knowledge and fuse it with the geometry at the initial design stage. The object oriented structure of these modellers allow classes of objects within the product hierarchy to be defined and allow inheritance from ancester objects in the structure. This enables models to be defined at the feature level of abstraction in line with many current research methodologies. Having established a total model it is then possible to convert this model and associated logic, through translators, into a form suitable for digestion by downstream CIM sub-systems (geometric modellers, production control systems, computer aided process planning systems). This type of structure also offers the opportunity of generating multiple feature representations for different subsequent uses.

British Aerospace Airbus Ltd have developed several large ICAD models based primarily on functional requirements but experiments with with material rationalisation and tooling generation have also been conducted. The ICAD system is not used exclusively to model all aspects of the product. Instead it has been deployed in areas where it performs best and is used in conjunction with other knowledge based systems and analytical systems as part of a comprehensive hybrid DFM network of systems.

The approach at UWE is somewhat different, Hill, Bleakman, Earl and Miles (1994), Bleakman, Hill and Earl (1992). ICAD has been treated as a modelling environment in which a design by features system has been developed. A features library has been created and a facility for adding user defined features through a production user interface has been provided. Work on another production user interface which allows for menu selection of features as in a conventional CAD arrangement is also being developed. Material selection is automatic in the sense that once the mechanical loading on the components has been specified, a range of materials specifications are identified. The least expensive is taken as the default selection but can be overidden. The cost of components are generated automatically as the features instances are created. Typically, for machined components, these costings are computed by inferring the method of manufacture from the feature description and then estimating the machining time from the material description, the volume of material removed and the tolerance on the features. Design for assembly analysis is conducted interactively within ICAD and is modelled on the LUCAS DFA system, LUCAS Engineering and Systems Ltd (1991). The DFA system assesses the potential for part reduction and then assesses each individual component for feeding and fitting. The feeding characteristics of the component is not particularly relevent for low volume aerospace work but is included for completeness.

One of the key features of the UWE model is the separation of domain specific and generic class definitions. Most manufacturing related classes are generic and can thus be invoked as mixins in the main product structure model. They can therefore be re-used in a object oriented style for different domain product models, thus reducing coding effort and model size.

Research CAD-DFM System

This form/function/manufacture synthesis appears to be the answer to the aerospace designer's problems but has its disadvantages. The effort required to capture this level of knowledge is substantial and results in an inordinate amount of base language code. To overcome this problem a research modeller which has some of the virtues of the ICAD open environment and many of the advantages of a conventional geometric modeller has been developed at UWE by one of the authors, Hill (1994). This feature based modeller (written in CLOS) has a feature generation system (as opposed to a feature library), automatic process selection, automatic material selection, real time cost generation at a sub feature level, semi-automatic DFA analysis and functional kinematic represention. The system also allows the designer to simulate the assembly kinematics. Models are built from combinations of features. Real time costing takes place at the feature generation level and feeding characteristics are generated at a component level. Process selection, inferred from the method of feature generation, component feature maps, material specification and tolerances takes place at feature and component levels. Kinematic relationships relating to mechanical function and assembly kinematics are added when the assembled model is complete. DFA analysis is invoked at this stage. Assembly costs are estimated from the assembly kinematics.

The modeller was not developed solely as a DFM research toolkit but also as a means of investigating feature representation, feature grammars and function-form relationships. Genetic algorithms are currently being deployed for investigations into configuration optimisation and shape generation.

Conclusions

The paper has summarised a range of research and development programmes currently taking place at BAE and UWE. Future work at UWE will involve improvement in local DFM algorithms, development of more sophisticated architectures to support integrated DFM analysis and research into fundamental shape representation issues.

References

Bleakman, M., Hill, T.M. and Earl, C.F. 1992, Architectures for Design to Cost, SERC Workshop on Design to Cost, University of Newcastle.

Forster, R.J. 1994, Design for Cost Effective Manufacturing through Guidance by Knowledge-Based Design, MPhil Thesis, University of the West of England.

Hill, T.M. 1994, Design for Automation through Feature Assessment, Intelligent Manufacturing Competitive Strategy for Tomorow - IA'94, 3rd International Asia-Pacific Industrial Automation Conference, Singapore.

Hill, T.M. 1994, Knowledge Based Strategies in Design for Automation, 4th International Conference on Flexible Automation and Integrated Manufacturing (FAIM 94), Virginia Polytechnic and State University, Blacksburg, Virginia.

Hill, T.M., Bleakman, M., Earl, C.F. and Miles, A. 1994, Knowledge Based Integrated Design/Cost Systems, Factory 2000, University of York.

LUCAS Engineering and Systems Ltd 1991, Design for Assembly, Version 2.0.

Mirza, A.R. 1994, A Knowledge Based Design to Cost System for Sheet Metal Parts, MPhil Thesis, University of West of England.

INTELLIGENT SOLUTION OF CUTTING CONDITIONS USING DATABASES

* M. Pickering, *P.B. Walker and ** T. Newman

Camtek Limited, Great Malvern, Worcestershire WR14 2AJ
University of Wolverhampton, Wolverhampton WV1 1SB

Due to the large growth of computer aided numerical control programming packages there is now the necessity to provide speeds and feeds for individual NC machines for use within CAD/CAM software. This area of technology is described in this paper and is being investigated and implemented by a Teaching Company Scheme of two associates, in conjunction with Camtek Ltd. and the University of Wolverhampton.

Introduction

Some companies and universities have produced individual computer software packages for selection of speeds and feed for use with various materials and cutting tools by using databases. Some cutting tool manufacturers also offer tool cutting parameterised computer software that also includes economics of tool life. Most of this data can be linked by data exchange to Computer Aided Manufacture (CAM) packages for use of preparing Production Planning, Costing and NC Tape files.

The economics of machining concerns the most efficient utilisation of the resources involved for the machining process. Since the two main parts of the machining process are the floor to floor time in relation to the machine hour cost and tooling costs including jigs and fixtures, it becomes important that speeds feeds and depth of cut are optimised. No account of the type of machine is taken into consideration when tool manufacturers give ideal cutting conditions, but machine tools with age have to have a factor built in due to wear. The object of production must be low production costs, high product quality and short production times, and it is the object of this research to achieve some of these priorities with cutting conditions from databases.

---

* M. Pickering and P.B. Walker are Teaching Company Associates at Camtek Limited, in conjunctions with the University of Wolverhampton.
** T. Newman (to whom correspondence should be addressed) is a Senior Lecturer in the School of Construction, Engineering and Technology, University of Wolverhampton.

The developments of CAM systems have seen considerable growth in the last two decades with a vast increase in the rate of development in both CAD and CAM. Advancements in computer technology now means that PC computers have increased processing power which required mini and main frame computers in the past. Also with the reduced cost of PC computers and software the introduction of CAM becomes more viable even for small and medium companies.

PEPS Product

Many Computer Aided Programming Packages for NC machines offer powerful geometrical capabilities and are CAD systems in their own right. They also provide extensive cutter path details with simulation and controlled cutter paths. The selection of tools, speeds and feeds have in the past been left to the user. The programmer uses his experience and cutting data literature to input cutting parameters, but with greater ranges of tools and materials become a time consuming task. By storing cutting data in databases this provides a quick and easy method of interrogating the parameters for machining, for workpiece materials and cutting tools.

The module PEPSCUT for Milling and Turning used for development was taken from the PEPS NC programming system, produced by CAMTEK Ltd. Malvern, England. It is intended to produce an expert system for defined features and as part of the first phase the requirements are for cutting conditions to be optimised for individual modules and machines. An expert system approach is the final objective based on pre-defined rules with the programmer being able to overrule the system and produce his own rules.

One of the main advantages of PEPS is its open architecture allowing customisation by using its own internal language similar to BASIC to develop the full application to include routines to handle set up sheets, planning sheets, machining cycles and parametric geometry. PEPS was one of the first CAD/CAM packages to use the Microsoft Windows environment and is also capable of linking to other packages via Dynamic Data Exchange[1]. The next section describes the structures for the cutting conditions information support for release in the next version of PEPSCUT for the combined databases of milling and turning.

Intelligent Database Support Structure

For any particular machining process, once the workpiece material has been defined and the tool material has been identified, the surface speed may be retrieved from a database containing surface speeds for particular tool materials.

For a machining process, more than one tool material may be applicable to machine the particular workpiece, so the preference for each tool material is also stored. Here, if no suitable tool is found for a tool of the preferred tool material, a search is then conducted on the second preference and so on. Once a particular tool is found by searching through the preferences for the tool or its nearest match, the tool geometry (described later) may be

---

[1]Dynamic Data Exchange is a Windows facility to allow the passing of data from one program to another.

retrieved. This includes fields applicable to the tool. For an end mill, for example, this data will include the tool's actual diameter and actual length.

With these parameters, and also with the knowledge of the workpiece material, tool material and machining process, the horizontal and vertical feed rates and depths of cut may be referenced in the feeds database, which stores the parameters in ranges based on the identifying criteria. The advantage of storing machinability data in libraries is the huge reduction of data and its subsequent management rather than keeping individual speeds, feeds and depths of cut with every tool for each workpiece material.

Figure 1 illustrates the interactions within the database structure.

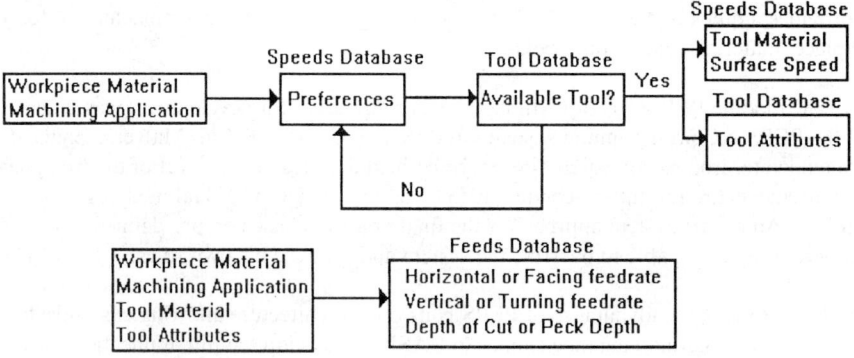

Figure 1. Relationship between libraries and the Tool database.

Tooling Geometry

Every Tool is stored individually in the tooling database. Fields that are applicable to each tool include the prime attribute (diameter, width or length), Tool material, and other attributes such as length, angle (if applicable), number of teeth, tip or corner radius, supplier and any special notes. The database is structured so that any tool, be it a milling or turning tool may be stored, hence a common approach may be used in retrieving data.

Users will not wish to input tools that are purchased in ranges (drills, for example) individually, so intelligent use is made of Windows dialog boxes to simplify this procedure. Users may also wish to access and maintain different data for each machine tool in their shop. This too, is catered for in the system.

Information Supplied at Tool Selection.

The use of Windows dialog boxes allows the user to view available tooling in a concise and effective manner. Figure 2 shows the dialog that is presented to the user at the time of tool selection.

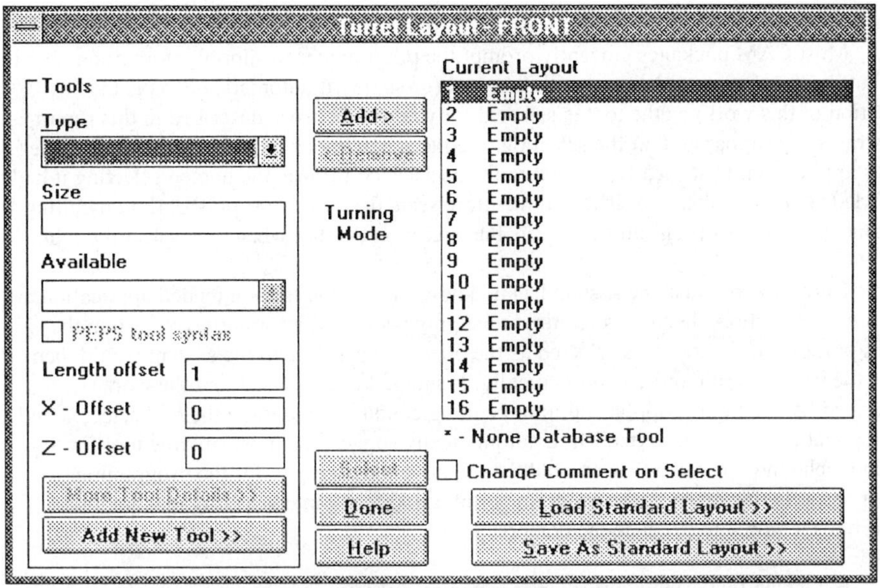

Figure 2. The turret layout dialog.

This dialog allows the user to view the tools currently loaded into the machine's turret. The access to available tools is centred around three key dialog controls. The user first selects the type of tool required (for example drill, end mill, face mill, tap etc.). The user then supplies a size. When the size is given, the system first searches the current turret layout for a tool of that size. If such a tool is not present, the tooling database is searched. If a suitable tool is located, it is placed in the list of available tools. Should more than one suitable tool exist (the same tool in different materials, for example), the most suitable tool is highlighted. If no tools are found, near matches are shown. This would normally be tools with the nearest size smaller than that required. If the highlighted tool is acceptable to the user it is added to the turret layout. Otherwise, another tool from those available is selected and added to the layout.

When the tool is added to the turret layout, the user is prompted to supply information about the speeds and feeds in the form of a Windows dialog. This dialog is pre-loaded with suggestions for speeds and feeds from the databases. The user would normally accept these values but they can be changed if required. Once the tool is added to the layout, it can be selected at any time.

Without this facility, the NC programmer would frequently have to refer to tooling catalogues for this data. The data supplied with these catalogues is usually given in ranges and as such, the exact information used may vary between jobs and NC programmers. The method described here allows the user to supply consistent data for similar operations from component to component. The advantages of having consistent data will be apparent.

Future Work

Most CAM packages currently prompt the programmer to supply information about the cutting parameters to be used with that tool (or supply it automatically as is the intention of this work) as the tool is selected for use. As the work described in this report illustrates, it is apparent that the selection of cutting conditions requires knowledge of the tool's application. Historically, when a tool was selected for use, the person selecting it had knowledge of its application. It would therefore seem desirable to supply optimum cutting parameters when the programmer selects the operation and not when they select the tool.

Currently, because the system has no knowledge of the tool's intended application at the time of selection, the user is required to supply additional information regarding the tool's purpose. If the tool was selected as the operation is selected, the system would then have the information required automatically. It should be recognised that the work described here not only supplies optimum cutting conditions based on the tool selected and its application but can be extended to automatically suggest the most suitable tool for a given machining operation. This would then remove the necessity for the programmer to select tools. In this manner, the programmer moves directly into selecting machining operations required to produce the component.

Extending this principle further, it is common that a given item on a component requires a set of related tooling to produce it. An example of this might be a tapped hole, whereby a hole of the correct diameter must be produced prior to producing the thread. This brings in the use of features to the work. As will be appreciated, this work would readily accommodate such a system, providing a method for defining the required operations to produce a feature can be input into the system.

To complete the automatic selection of tooling, consideration should be given to tool life. The database system used here could be extended to incorporate data related to tool life based on the conditions it is being subjected to. Here, tool turrets could be optimised to include additional tools of the same type, if the life of the tool is insufficient to complete the component with a single tool. Alternatively, the cutting conditions could be automatically altered to increase a tool's life in order that a single tool would be sufficient. A more practical solution might be a combination of these methods.

The work described in this report goes a long way in aiding NC programmers to produce consistent and reliable NC part programs. The suggestions given for future work would greatly assist in the automatic generation of NC code.

A KNOWLEDGE BASED SYSTEM FOR THE SELECTION OF CUTTING TOOLS AND CONDITIONS FOR MILLING

Mr Reza Razfar and Dr Keith Ridgway

University of Sheffield
Sheffield, S1 3JD.

This paper describes the development of a knowledge based system for the selection of cutting tools and the determination of optimised machining parameters for milling operations. Given details of the part geometry the system automatically selects the optimum tool and conditions for each feature using rules elicited from technical catalogues, metal cutting theory and domain experts.

Introduction

Process planning is defined by Chang and Wysk (1985) as the activity which determines the appropriate procedures to transform raw material into a final product. Process planning forms a major portion of the non-machining time in the production of a component thus a great deal of research effort has been applied to the automation of this process. Expert systems are increasingly being used as a tool in this research. The suitability of expert systems for symbolic processing has been greatly utilised and consequently many systems address the problem of machining from a geometric perspective.

Arezoo and Ridgway (1989) suggested that conventional programming was inflexible and inadequate for automating process planning. These systems mix data and logic in one program and hence are hard to understand and extremely difficult to modify and maintain. On the contrary, expert systems maintain the knowledge base separate from the inference engine. Therefore, the knowledge base can be easily modified for different environments by the user. Expert systems are suitable for problems which require a large amount of knowledge, experience and symbolic manipulation for their solution (e.g. tool selection). Chang and Wysk suggest that the major advantage of expert systems versus conventional programming is the elegance of representing data and deducing conclusions from the data. These properties, especially flexibility, provide a great opportunity for development of cutter selection

259

systems where the tool selection criteria need not be permanently embedded in the system.

System Description (Ex-catsmill)

The Ex-catsmill is an expert system which addresses one specific area of process planning, namely the selection of cutting tools and optimum cutting conditions for milling operations. The aim of this research is to remove the routine tasks from tool-selection using an expert system. The system can select the optimum tools and cutting conditions for; face milling, square shoulder face milling, slot milling, pocket milling, edge milling and chamfering operations or for a combination of all. The selection rules have been elicited from technical catalogues, books and domain experts. An algorithm for the determination of optimised cutting conditions (speed, feed, radial depth of cut and axial depth of cut) has been used. Many factors have been considered as constraints such as tolerance of geometric features, surface finish, workpiece material and machine tool characteristics (power, tool holding system, maximum tool diameter and maximum of weight of tool and tool holder). The output from Ex-catsmill includes details of the tool, tool holding system, insert (shape, tolerance, grade) and chip spacing (close pitch, differential coarse pitch and extra close pitch). The optimum machining parameters are calculated using minimum production cost or maximum production rate criteria.

For initial system development work a prototype component has been created which comprises seven features and includes both roughing and finishing operations. The features include face milling, square shoulder face milling, slot milling, pocket milling, chamfered-circle milling, long-edge milling and chamfered-angle milling. Each feature has been specified by eight characters representing the material, set-up number and co-ordinates of the start (X1,Y1,Z1) and end point (X2,Y2,Z2). In addition the tolerance and surface finish criteria (Ra) of each feature has been determined. The part description is entered interactively and translated by the system into a format which matches the tool description.

A number of factors are considered in the selection of the cutting tool.

i)   Component material: more than twenty different materials have been considered and included in the system.
ii)   Type of cutter: The system matches the workpiece feature to the appropriate tool (Razfar, 1985).
iii) Rake angle : the radial rake and axial rake angles can be positive or negative depending upon the type of material, machine power, operation (finishing or roughing) and chip flow.
iv)   Clearance angle: For harder materials the clearance angle reduces.
v)   Cutter diameter: The diameter of the cutter is determined by the width of the feature and type of operation. The cutter diameter should be 1.2 times the width of the feature (e.g. face milling) and less than (as nearest as) the width of the other feature (e.g. slot milling).
vi)   Number of insert: The minimum of number of inserts is selected.
vii) Insert tolerance: For more accurate features more accurate inserts are selected.
viii) Insert length: This is determined after considering the machining duty (heavy or light).

ix) Insert shape: for heavy duty operations and harder materials a stronger insert shape is selected (e.g. square or triangle insert).

x) Chip space: long chipping materials need more space for the chip thus coarse pitch is selected for long chipping materials (e.g. low carbon steel) and close or extra close pitch is selected for harder materials (e.g. cast iron and titanium). If vibration is a problem differential coarse pitch is selected.

In the first step in the selection procedure the system selects tools which can mill the specified material. The system uses geometry based rules to identify the tools suitable for every feature. The remaining tools are eliminated for the next process. The best tool for each feature is selected using priority rules elicited from technical catalogues, metal cutting theory and domain experts.

After selecting all the tools for finishing and roughing operations the appropriate cutter diameter and insert for every feature is selected using rule based constraints i.e.. the maximum tool diameter which the machine tool can accept, the insert tolerance, and the width or diameter of the feature considered. In the final stage of tool selection the system selects the remaining parameters such as: insert grade, tool holding system, number of inserts, and chip space.

Optimisation of cutting conditions

Four factors are considered; radial depth of cut, axial depth of cut, feed per tooth and spindle speed. To obtain the best tool economy, the largest possible axial depth of cut should be selected. The system uses up to 2/3 of the full cutting edge length available (Razfar). If the depth of the feature is greater than this value additional passes are specified. Other constraining factors considered include the power and stability of the machine, the strength of the workpiece and the clamping arrangement. The depth of cut is directly proportional to the power consumption.

To calculate the radial depth of cut ex-catsmill selects only those tools which are wider than the feature for face milling, square shoulder face milling and long edge milling and less than the width for slot and pocket features. The system attempts to mill each feature in one pass if possible.

When selecting the feed several factors are considered including type of cutter, carbide grade, surface finish, workpiece material and machine power. The Ex-catsmill system selects a feed rate as high as possible, within the range specified for each grade, for roughing operations.

If the power consumption becomes more than the machine power available then the feed rate is reduced until the condition is satisfied. For finishing operations, three factors are considered. The first is the surface finish criteria (Ra) for which the arithmetic mean value is calculated using the formula:

$$Ra = (0.0642 / D) . (V_f / n)^2$$

where D is the cutter diameter, V_f is the feed rate and n is the spindle speed in r.p.m. The value of F_t is calculated using the equation:

$$F_t = \sqrt{(Ra . D)} /(0.2533 . z)$$

where z is the number of teeth, Ft is the feed per tooth in mm and Ra is in mm.

The value of F_t must be equal or less than the maximum specified in the knowledge base for finishing operations. The second factor is the length of parallel land. To achieve the best surface finish the feed per revolution must be less than the length of parallel land. Finally, the power required must be less than the machine power available.

The optimum cutting speed for optimum tool life is calculated form the extended Taylor equation:

$$V = C \cdot f_t \cdot f1 \cdot a_p \cdot f2 \cdot t_f \cdot f3 \cdot (a_e / D) \cdot f4$$

where V is cutting speed, f_t is the feed per tooth, a_p is the depth of cut, a_e/D is the ratio of radial depth of cut to tool diameter and f1, f2, f3 and f4 are the exponents which depend upon workpiece material and carbide grade.

Two distinct criteria can be used for choosing the optimum cutting speed for a machining operation; minimum production cost and minimum production time. The tool life for minimum cost (T_c) and minimum production time (T_p) can be calculated.

$$T_c = (1-f_n)/f_n \cdot (T_{ct} + C_t/M)$$
$$T_p = (1-f_n)/f_n \cdot T_{ct}$$

where T_{ct} is the insert changing time, C_t is the cost of each insert, M is the total machine and operator rate and F_n is the exponent of tool life in the tool life equation. Boothroyd (1981) explains that, during milling, the teeth of the cutter are only in contact with the workpiece for a proportion of the machining time. Thus the tool life for minimum cost and minimum time should be modified using a correction factor Q.

$$Q = 1/ \arcsin(a_e /D) \qquad \qquad \text{for face milling}$$

$$Q = 0.25 + 1 / 2 \cdot \pi \arcsin(2 \cdot a_e /D -1) \qquad \text{for end milling}$$

After specifying optimum tool life, feed, axial depth of cut and radial depth of cut the system determines the optimum cutting speed and the power consumption. The power required depends on the workpiece material, metal removal rate, tool geometry and chipsthickness and can be calculated using the following formula:

$$P = a_p \cdot a_e \cdot v_f / (1000 \cdot Q_p \cdot eff)$$

where a_p is the axial depth of cut in mm, a_r is the radial depth of cut in mm, v_f is the table feed in mm/min, eff is the machine efficiency and Q_p is metal removal rate in cm^3/min per kw. Qp is known for most materials for specific rake angle(0' degree) and mean undeformed chip thickness (hm=0.2 mm). Therefore, the above formula must be modified by two factors:

1. Top rake angle: The power consumption decreases by 1.3% per degree of positive top rake angle and increases for negative top rake angle. Ex-catsmill compares the top rake angle of each tool to zero degrees and modifies the power consumption by a factor C1.

2. The mean undeformed chip thickness: The chips produced in milling are not of uniform thickness as in turning. For this reason the term (h_m) mean chip thickness has been introduced and can be calculated by following formula:

$$h_m = \sin K_r \cdot (360/\pi) \cdot (F_t / H) \cdot (a_e/D)$$

where K_r is the entering angle in degrees, F_t is the feed per tooth in mm, H is the angle between tool entry and tool exit in degrees (equal to $2 . \sin a_e/D$), a_e is the radial depth of cut and D is the diameter of cutter. After calculating h_m for each operation Ex-catsmill compares h_m for each operation with (0.2 mm) and modifies the power consumption with the constant C_2. The following formula is valid for all top rake angles and chip thickness

$$P = [a_p . a_e . v_f / (1000 . Q_p . eff)] . C_1 . C_2$$
$$v_f = F_t . n . z$$

Siekmann, Gilbert and Boston (1953) explain the effect of feed rate on power consumption. Thinner chips require a higher unit power according to the formula.

$$P_u = C / f_t^x \qquad\qquad x<1$$

Where P_u is the unit horsepower, f_t is the feed per tooth. After calculating all cutting conditions Ex-catsmill determines the tool holding system and the size of each tool holder.

Overview of Ex-catsmill

The system developed selects the appropriate tools and optimum cutting conditions for a range of milling operations. Due to the flexibility of the expert system approach the system can be easily modified and updated. Ex-catsmill optimises the required tools by eliminating the extra tools for the same feature with different dimensions. It selects the larger diameter for face milling, square shoulder milling and edge milling and smaller diameter for slot, pocket, circle chamfer milling. The next stage of development will consider the inclusion of drilling and tapping operations and the creation of an interface to a CAD/CAM system.

Acknowledgements

The authors wish to thank the Billy Ibberson Trust who support research in industrial change at the University of Sheffield.

References

Arezoo B., 1990, The application of Knowledge-Based System for Cutting Tools and Conditions Selection for Turning operation. Ph.D Thesis, University of Sheffield.
Arezoo B. and Ridgway K. 1989, Knowledge Based Systems for the Selection of Cutting Tools and Cutting Conditions. Proc. Fifth Int. Conference on Computer Aided Production Engineering.
Boothroyd G. 1981 *Fundamentals of Metal Machining and Machine Tools*, McGraw Hill.
Chang T.C., and Wysk R. A. 1985, *An Introduction to Automated Process Planning Systems*. Prentice Hall International, New Jersey
M.R Razfar M.R., 1984. The investigation of rake angle, feed, depth of cut and speed on metal cutting for turning operations .Msc.-Thesis, Amir Kabir University, Iran.
Siekmann, W.W G ilbert and O. W.Boston. (1953) Power Required by carbide-tipped Face-milling cutters., Trans. ASME.

GEOMETRIC ELEMENTS FOR TOLERANCE DEFINITION IN FEATURE–BASED PRODUCT MODELS

Dr James Gao†, Dr Keith Case* and Professor Nabil Gindy**

†*School of Industry and Manufacturing Science, Cranfield University, Cranfield,
Bedford, MK43 0AL*
**Department of Manufacturing Engineering, Loughborough University of Technology,
Loughborough, Leicestershire, LE11 3TU*
***Department of Manufacturing Engineering and Operations Management, University
of Nottingham, Nottingham, NG72RD*

Product modelling is an essential part of all computerised design and
manufacturing activities. A precise mathematical model of the geometry of
products is important, but must be supplemented with technological information
such as the material, mechanical properties, functional specifications and
tolerances. Modern CAD systems can model and manipulate components with
complex geometry. However, technological information is represented as text
symbols on the computer screen or drawing, and subsequent application programs
are frequently unable to use this information effectively. This paper discusses this
problem, and establishes the geometric elements required for the representation of
dimensions and tolerances in a feature–based product modelling environment.

Introduction

Engineering drawings are the traditional way to represent product information,
including the nominal geometry, dimensions and tolerances and other technological
specifications, such as the mechanical properties of materials and method of heat
treatment, initial status of stock (e.g. cast, forged or pre–cut), functional and structural
requirements. Drawings are produced by the design team and passed to the production
engineers who interpret the information to find appropriate manufacturing processes. The
tasks involved in such a traditional design and manufacturing procedure are increasingly
being replaced by computerised systems (CAD/CAM).

The modelling and manipulation of the nominal geometry of components by
computers is a successful and mature technology. However, the technological
information, which is provided as annotation of the geometry of components in
traditional drawings, has not been successfully represented together with the geometry of
components in a single integrated form. In fact, the technological information is
represented as text symbols on the computer screen or on the drawing output from the
computer. Subsequent application programs are unable to use this information effectively
(Juster, 1992). This problem, especially the representation of tolerances in the geometric
model, becomes increasingly important as the product data model is recognised as
playing a vital role in a fully integrated and concurrent computer aided engineering
context.

There are basically three types of 3 dimensional geometric models of components, i.e. wireframe, surface and solid models. Solid modelling is said to be the most "complete" representation of component geometry, in the sense that all required geometric information can be obtained from the model either directly or by simple calculation. There are two types of solid modelling, i.e. Constructive Solid Geometry (CSG) or set–theoretic and Boundary Representation (Brep) (Mantyla, 1988). A Brep model is based on a face–edge–vertex adjacency graph which contains more detailed and more explicit information than any other geometric representation. Even so, a Brep model is still not suitable for representing tolerances and other important technological information. This problem is discussed in the following sections.

Geometric Elements Required for Tolerance Definition

Dimensions and tolerances

There are three types of dimensions, i.e. functional, non–functional and auxiliary. Functional dimensions are essential to the function of an object and are potential candidates for tolerances. Non–functional and auxiliary dimensions would not normally be toleranced, as they are not essential to the function of the object. Tolerances which are applied to dimensions are called dimensional tolerances, and may be linear or angular.

Geometrical tolerances are used to control the variation in the form of features. A geometrical tolerance defines the size and shape of a tolerance zone within which the feature is to lie. There are four types of geometrical tolerances, i.e. tolerances of form (straightness, flatness, roundness, cylindricity, profile of a line, profile of a surface), tolerances of attitude (parallelism, squareness, angularity), tolerances of location (position, concentricity, symmetry) and composite tolerance (run–out).

Features for tolerancing

The basic geometric elements required for tolerance definition are features. Examples of features are surfaces (planar, cylindrical, spherical, conic and free form surfaces), lines (linear and curved) and points. These features are also known as single features as they are the most elementary building blocks of components. Note that surface features may be median planes of other features, line features may be axes of other features, and point features may be central points or end points, etc.

Single features may be combined to form higher level features such as slots, pockets, steps and notches (see figure 1). These combinations are commonly seen in practice and which may have tolerances of position or symmetry defined with respect to their median planes, axes or central lines.

Datums and datum features

Dimensions and tolerances are normally applied with respect to feature datums (except for tolerances of form). A datum is a theoretically exact geometric reference such as a centre point, an axis, a centre line, a plane face or a median plane. Since theoretically perfect surfaces cannot be produced, surface plates, machine tables, axes or median planes of gauges and other equipment used in manufacturing and inspection are usually sufficiently accurate that they may be considered theoretical planes and axes and thus used as datums. A feature of a component in contact with a datum is called the datum feature. Datum features should be chosen by considering the function of the part and with high accuracy. Therefore, form tolerances such as straightness, flatness, cylindricity and roundness are normally required to ensure the accuracy of datum features.

Datum systems

For manufacturing and inspection purposes, a three–plane datum system is required for most non–rotational parts, so that the orientation may be defined and the parts can be located with respect to the datum systems of machining or measuring machines. The three datum planes are mutually perpendicular and are known as the primary, secondary and tertiary datum surface respectively. The primary datum plane is defined by three points contacting the face of the part, and these are known as datum targets. The secondary datum plane is defined by two contacting points and the tertiary datum plane is defined by one point.

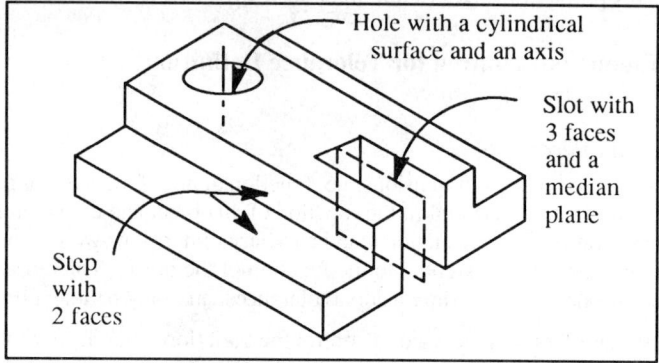

Figure 1. Features used for Tolerancing

Current Geometric Modellers and Tolerancing

As discussed above, the specification of dimensions and tolerances requires a feature–based component representation. However, current geometric modellers are not feature–based. Although Brep models contain detailed information about edges, faces and vertices of components which are essential for tolerance definition, such elements are not distinctively represented as features. For example, during interactive design and editing, the names of the geometric elements of a component will change arbitrarily, especially when Boolean operations are performed.

The latest parametric systems such as Pro–engineer (James, 1991) and CADDS 5 (Computervision, 1991) are able to represent a number of commonly used features such as slots, pockets, holes, steps, etc. Dimensions are represented and manipulated as parameters. However, lower–level features such as points, lines and surfaces are still not represented as distinctive features. In fact, the parametric feature data is stored separate from a Brep model of the same component (although the two data models are associated with each other to certain extent). The former doesn't contain sufficiently detailed information for tolerancing, whilst, the latter suffers from the same problem found with all Brep models.

Feature–based Component Representation

Since the current geometric modellers (even the latest parametric systems) are unable to support feature–based product data models for dimension and tolerance definition, an alternative representation has been proposed and implemented in a prototype feature–based design system. Three levels of features are defined, which are:

ATOMIC FEATURES, which are the lowest level features, i.e. points, lines and surfaces. Point features may be central points or vertices of lines or other features. Line features may be centre lines, axes or edges. Surface features may be real faces, median planes, etc. Atomic features are essential for all types of tolerances. Atomic features are usually treated as the basic constituent elements of higher level features.

PRIMITIVE FEATURES, which are at the second level, are groups of atomic features which form recgnisable geometric entities such as holes, pockets, slots, notches and step. These are commonly described as functional or manufacturing features. A primitive feature can be toleranced by positional and symmetrical tolerances with respect to its axis or median plane(s). Its constituent atomic features can be toleranced or used as datums to tolerance other features.

COMPOUND FEATURES are the top level of features, e.g. patterns of holes, counter–bores, multi–steps, crossed slots. A compound feature is a combination of primitive features and/or atomic features which together perform a single function or may be manufactured by a set of operations. Usually the constituent features, rather than the compound feature, are used for tolerancing.

Using the above feature definition, a component, and hence a product can be modelled and functional dimensions and tolerances can be specified with respect to atomic features and primitive features. The primary advantage of this representation is that the three levels of features are inter–related in a hierarchical fashion, whilst the distinctiveness and consistency of individual features are maintained.

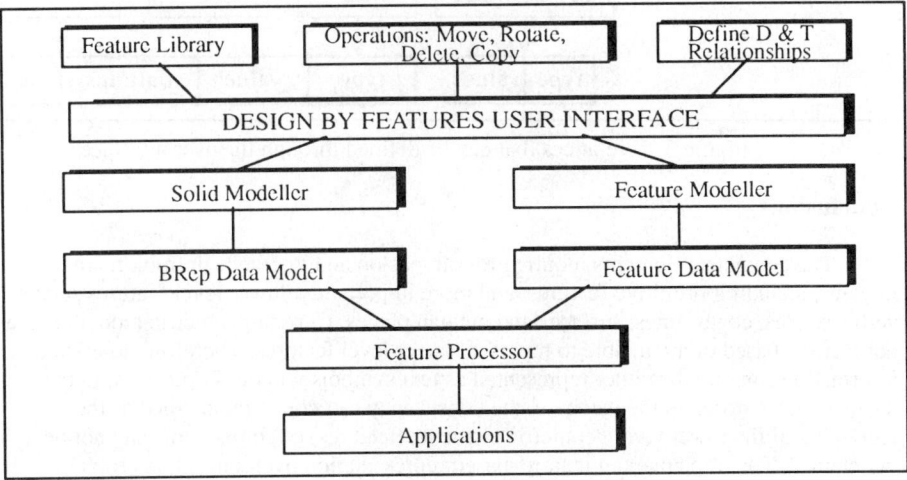

Figure 2. The Feature–based Design System

Implementation

The feature–based component representation has been implemented in LUT–FBDS (Loughborough University of Technology Feature–Based Design System) (Case et al, 1994). The system consists of a user interface, a Brep solid modeller, a feature modeller and a feature processor (figure 2). The primitive features available in the feature library are boss, pocket, hole, through slot, non–through slot, notch, step and surface. Atomic features are not pre–defined in the feature library. However, they are represented as constituent entities of primitive features and their consistency and independence are maintained in the feature data model. For example, A through hole (which is an instance of the hole primitive) is composed of two atomic features, a cylindrical face and an axis. The axis is normally used for tolerancing. Note that for process planning we also define two imaginary faces at the two ends of the hole, but they are not treated as atomic features.

A number of compound features are also pre–defined in the feature library, including pattern holes, cross–slot, counter bore and multi–step. More compound features can be defined by combining primitive features through the user interface. In the database, compound features are treated as relationships between primitive features. Dimensions and tolerances can be defined through the menu–based user interface, which comply with the British Standard BS 308 (BSI, 1988) as shown in figure 3. As the consistency of lower level features are maintained in the database, the validity of the specification of dimensions and tolerances are assured for the whole life cycle of the product.

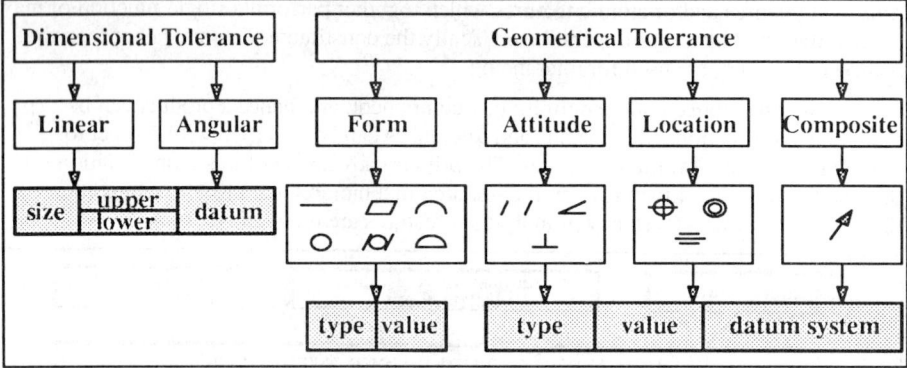

Figure 3. Tolerances that can be defined through the user interface

Conclusions

The geometric elements required for dimension and tolerance definition are features, including primitive features, and more importantly, lower level features such as vertices, axes, edges, lines, surfaces and median planes. Current geometric modellers are not feature–based or are unable to represent lower level features. Therefore, tolerance information can only be either represented as text symbols without implication in the database, or if stored in the database, the consistency can not be maintained as the identifiers of the lower level geometric elements (features) of components can not be maintained. The implemented feature–based representation overcomes this problem to a significant extent.

References

BSI, 1988, Manual of British Standards in Engineering Drawing and Design, 2nd edn (British Standards Institution, Maurice Parker ed).

Case, K., Gao, J. X. and Gindy, N. N. Z. 1994, The implementation of a feature–based component representation for CAD/CAM integration, the Proceedings of IMechE, Part2, Journal of Engineering Manufacture, 208, 71–80.

Computervision, 1991, Introduction to CADDS 5 Parametric Design (Tutorial), (Prime Computervision, DOC38800–1LA).

James, M. July 1991, Prototyping Cad, cover story of CADCAM.

Juster, N. P. 1992, Modelling and representation of dimensions and tolerances: a survey, Computer–Aided Design, 24 (1), 3–17.

Mantyla, M. 1988, An introduction to solid modelling, (Computer Science Press, Maryland, USA).

INTERACTIONS OF CLAMPING AND SEQUENCING FOR 2.5D MILLED PARTS

Yong Yue and J.L. Murray

Institute of Technology Management
Department of Mechanical Engineering
Heriot-Watt University, Edinburgh EH14 4AS, Scotland

Clamping and sequencing are important tasks in process planning. Many aspects need to be considered to create a reliable process plan particularly when small batches of complex parts are concerned. One such aspect is the interactions between clamping and sequencing. This paper deals with the automation of this process considering the interactions and the part stability.

Introduction

Features have been playing an important role in representing the design and transmitting necessary data to various applications. Because of the nature of the process planning tasks, different types of features are required. Two approaches have been used, namely design by features and feature recognition. In the design by features approach, the designer defines the initial CAD model in terms of various form features which are directly relevant to the manufacture whilst by feature recognition, a computer program extracts necessary information from a CAD model and converts it into features. Design by features has been used by many researchers. This method, however, can demand extra work for the designer to tag features. Feature recognition has not reached its maturity although it seems an effective route. More detailed reviews are provided by Bronsvoort and Jansen (1993), Case and Gao (1993), Shah (1991) and Woodwark (1988).

Two major types of parts (i.e. rotational and prismatic) have been considered in the process planning literature. The applications are mainly concerned with machine and tool selection, sequencing of operations, tool approach directions, workpiece determination, and fixture design and setup planning. Although there are often interactions between these tasks, much of the work has concentrated on single activities. Some general surveys can be found in ElMaraghy (1993), Alting and Zhang (1989) and Requicha and Vandenbrande (1988).

An approach for complex 2.5D machined parts has been used to demonstrate a methodology of fully automating the process planning (Yue 1994). Such parts are common in the avionics, electro-optics and precision mechanical engineering sectors. The generation of a process plan involves geometric validation (Murray and Yue 1993), machinability evaluation, workpiece determination, clamping (Yue and Murray 1994) and machining strategies. Due to the interactions between these tasks, it is necessary to consider them in parallel, or to re-evaluate each task when proposing a process plan. This paper deals with the generation of clamping and machining strategies and their interactions.

Clamping Strategy

The clamping methods considered include (a) vice clamping incorporating parallel spacers, (b) machining clamping using a standard platen, and (c) frame bolting using a standard platen. The first two methods use modular fixtures whilst the last is a dedicated solution which was used by the AMP system (Murray and Linton 1986).

Vice clamping is simple but needs opposing parallel faces of sufficient area on the part profile. Machine clamping is more versatile but requires more sophisticated clamp location rules and increased handling time. Frame bolting can deal with the largest range of part geometry although it takes much more material and leads to increased machining and handling times. Vice clamping and machine clamping can be considered as a fixed clamping method (i.e. one setup for all machining operations in the same tool approach direction) or moveable clamping which has been limited to two setup operations at present.

Because of the advantages and shortcomings of these clamping methods, fixed vice clamping and fixed machine clamping are preferred. Moveable vice clamping and moveable machine clamping would be selected next. Frame bolting is the last choice, but it has advantages where parts have complex outline geometry or are inherently unstable. No combined use of vice and machine clamping is assumed currently.

The rules used to propose a clamping strategy are complex and numerous. Vice clamping is taken here as an example to illustrate how a clamping strategy can be defined. All the external vertical faces of the part are examined and parallel vertical faces paired. For a face pair to provide a viable clamping opportunity, the following criteria must be met.

(a) The area of the faces must be sufficient to constrain the part against the machining forces.
(b) The faces must be located to ensure the part stability for clamping and machining.
(c) The faces must be stiff enough to ensure that the part is not deformed beyond an allowable limit.
(d) Access with sufficient safe clearances for features to be machined must not be restricted owing to the vice jaws.
(e) There must be suitable features to ensure precise location of the part or to allow accurate setup by probing.

There are additional requirements which must be met before fixed or moveable vice clamping can be applied besides the general criteria. For instance, the faces in a pair must be on the convex hull of the part for fixed vice clamping or the first setup of moveable vice clamping. A clamping strategy of two setups using fixed vice clamping could be proposed for a 2.5D double-sided part with detailed rules derived from these criteria (Figure 1).

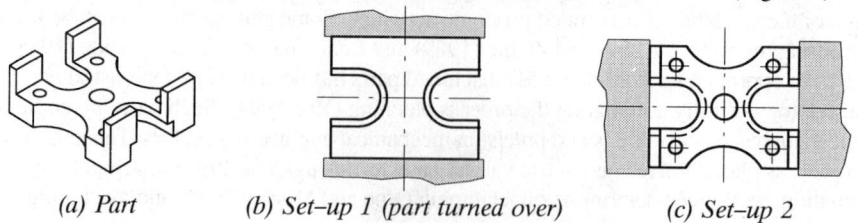

(a) Part   (b) Set–up 1 (part turned over)   (c) Set–up 2

Figure 1 Fixed vice clamping for a double–sided part

Machining Strategy

The machining strategy involves the identification and rearrangement of machined features, sequencing, identification of cutter access, selection of machining operations and tools, and fixture protection for machining through-features.

Machined features can be identified from horizontal faces each of which represents a volume to be removed (Figure 2a, b). Through-features are obtainable by subtracting the part silhouette from the rectangular workpiece (Figure 2c). It is usually necessary to rearrange the features based on sequencing requirements. After sequencing, the tasks of cutter access identification, selection of machining operations and tools, and the protection of fixture for machining through-features can be undertaken.

There are two levels of sequence, the sequence of setups and the sequence of features to be machined in each setup. Two types of criteria must be satisfied of which one is the general requirements and the other the clamping requirements. The general requirements are concerned with vertical and transverse access directions for the cutter for rough and finish machining, and the z levels or depths of features to be machined. They are mainly applied to sequencing features to be machined in each setup. The clamping requirements (described in the next section) dominate the sequence of setups and features to be machined in each setup.

Because there are often interactions between the machined features identified, it is necessary to optimise the machining sequence of these features or rearrange them into suitable groups for machining. For instance, it would not be practical to machine the top feature shown in Figure 2a before the through-features (Figure 2c) as the remaining volumes (Figure 3a) could cause problems. Typically, these problems could include unmachinable sharp internal corners, excessively complex toolpaths and flexible details. There are two ways to ensure the avoidance of this situation:

(a) To machine the top features in the order of z level from bottom to top (the reverse being applicable for bottom features);

(b) To machine the features identified in the reversed order combining volumes which lie within the feature profile and which have to be removed.

The first way may require unnecessary deep machining (e.g. larger depths of through-features in Figure 2c than Figure 3d) which, in turn, could cause unmachinability,

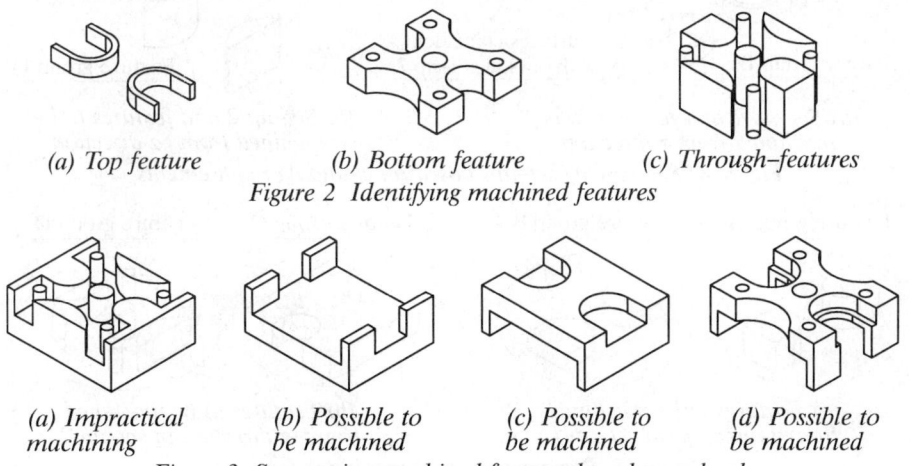

(a) Top feature　　　　*(b) Bottom feature*　　　　*(c) Through–features*

Figure 2 Identifying machined features

(a) Impractical machining　　*(b) Possible to be machined*　　*(c) Possible to be machined*　　*(d) Possible to be machined*

Figure 3 Sequencing machined features based on z–levels

higher cost or lower accuracy. It is therefore expected to machine features in the second way (Figure 3b-d) which, however, requires the features be rearranged.

Interactions of Clamping and Sequencing

Clamping Requirements for Sequencing

The clamping strategy proposed considers mainly the cutter access, safe clearance and part stability. It is obvious that any machined feature which has surface(s) coincident with that of vice jaws, cannot be generated during that particular setup. However, a single feature may have to be machined in two setups. Priority is given to considerations of machining any feature in one single setup.

For the setups proposed in Figure 1, feature groups A and B must be machined in setup 1 (Figure 4a), and C and D in setup 2 (Figure 4b). Because feature groups A and C must be machined in different setups based on clamping requirements, it is impossible to machine feature group C first together with the interactive volumes between the two feature groups (i.e. feature groups A and C). It is also impractical to machine feature group C alone before feature group A as discussed previously in the example shown in Figure 3a. Hence, feature group A must be machined first, i.e. setup 1 shown in Figure 1a must be taken as the first setup as illustrated in Figure 4a. Interactions between feature groups A and D, and likewise between feature group B and feature groups C and D, are evaluated in the same manner to decide the order of setups. If any contradictions are found, different setups will be required or a single feature has to be machined in different setups.

Rearrangement of Machined Features

As discussed previously, multiple setups resulting from clamping requirements may have a strong influence on the machining sequence. When the sequence of setups is made,

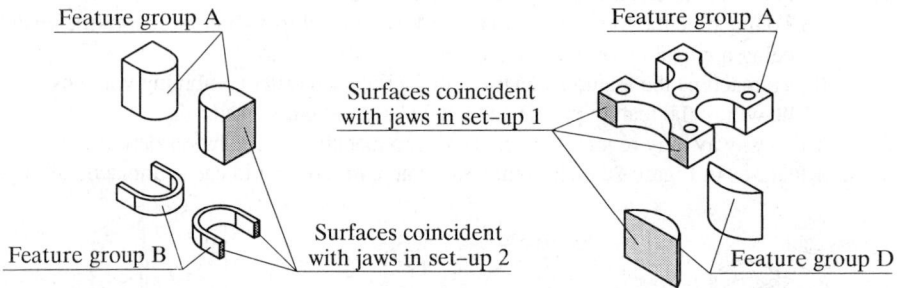

(a) *Set–up 1 and features to be machined from z direction*

(b) *Set–up 2 and features to be machined from –z direction*

Figure 4 Sequencing set–ups based on clamping requirements

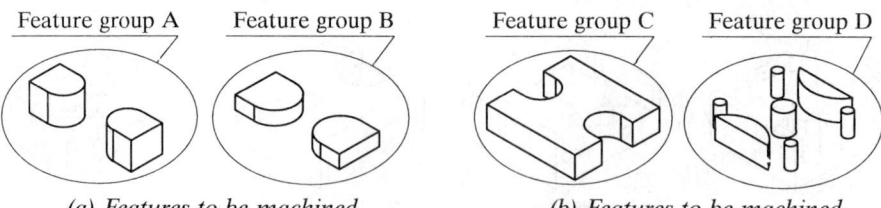

(a) *Features to be machined from z direction in set–up 1*

(b) *Features to be machined from –z direction in set–up 2*

Figure 5 Rearrangement of machined features

the machining sequence in each setup can be determined in the same way as illustrated in Figure 3. Through-features that have no coincident surface with vice jaws in both setups can therefore be machined in either setup. They are considered to be machined last in the second setup to satisfy the general requirements for sequencing (i.e. minimising the cutting depth). Figure 5 illustrates the features rearranged for the setups proposed in Figure 1.

Conclusions

This research is a part of an automatic process planning system for 2.5D machined parts being developed using the ACIS modeller (Spatial Technology 1992) in C++ on an Apollo 425t workstation. The methodology adopted has shown initial success in realising the ultimate goal of full automation without tagging features at the design stage. Low-level features extracted from the model have been very useful in dealing with various process planning tasks. The clamping and machining strategies require complex evaluations and the interactions between them must be considered carefully in generating a viable process plan. It is certain that more needs to be done to progress towards industrial applications.

Acknowledgements

Yong Yue, from Shenyang Aluminium and Magnesium Engineering and Research Institute, gratefully acknowledges the financial supports from the Chinese State Education Commission and the British Council through a TCT award.

References

Alting, L. and Zhang, H.C. 1989, Computer-aided process planning the state-of-the-arc survey, *International Journal of Production Research*, 27(4), 553-585.

Bronsvoort, W.F. and Jansen, F.W. 1993, Feature modelling and conversion - key concept to concurrent engineering, *Computers in Industry*, 21(1), 61-68

Case, K. and Gao, J. 1993, Feature technology - an overview, *International Journal of Computer Integrated Manufacturing*, 6(1&2), 1-10.

ElMaraghy, H. 1993, Evaluation and future perspectives of CAPP, *Annals of the CIRP*, 42(2), 1-13.

Murray, J.L. and Linton, H. 1986, The development of an automatic system for the generation of planning and control data for milled components, *In Proc. Intern. Workshop on Expert Systems in Production Eng.*, Spa, Belgium, (Springer Verlag), 231-245.

Murray, J.L. and Yue, Y. 1993, Automatic machining of 2.5D components with the ACIS modeller, *Intern. Journal of Computer Integrated Manufacturing*, 6(1&2), 94-104.

Requicha. A.A.G. and Vandenbrande, J. 1988, Automated systems for process planning and part programming, In A. Kusiak (ed.), *Artificial Intelligence Implication for CIM (AI in Industry)*, (IFC Publications Ltd, UK and Springer-Verlag), 301-326.

Shah, J.J. 1991, Assessment of features technology, *CAD*, 23(5), 331-343.

Spatial Technology, Inc. Applied Geometry, Inc. and Three Space, Ltd. 1992, *ACIS Modeller Interface Guide*.

Woodwark, J. 1988, Some speculations on feature recognition, *CAD*, 20(4), 189-196

Yue, Y. and Murray, J.L. 1994, Validation, workpiece selection and clamping of complex 2.5D components, In J.J. Shah, M. Mantyla and D.S. Nau (ed.), *Advances in Feature-Based Manufacturing*, (Elsevier Science B.V.) 185-213

Yue, Y. 1994, Automatic Process Planning for the Machining of 2.5D Components, *PhD Thesis*, Heriot-Watt University.

Resolving Feature Interactions in Design for Injection Moulding

R.J.V.Lee, A.H.S. Al–Ashaab, R.I.M.Young.

Manufacturing Engineering Department, Loughborough University of Technology, Loughborough, Leicestershire, LE11 3TU, U.K.

Features have been used to associate functional constraints with the topology and geometry of a designed product to capture engineering meaning. However features tend to address only one viewpoint (e.g functional, manufacturing), and this has been a major limitation on the application of features technology. The work described is in the area of injection moulding, and examines interactions between functional and mouldability features to provide mouldability feedback to the designer. This paper argues that the relationships between functional and manufacturing features can be represented and used to provide information to support Design for Manufacturing applications. Functional features are examined for specific product ranges and their interactions with a general set of mouldability features discussed.

Intelligent design for manufacture CAE systems must reason about the topology and geometry of designed artifacts. Such reasoning is typically in terms of features, and therefore the representation should be in terms of features, Gadh and Prinz (1992). Features derived for one design activity, eg functionality, will not be the same as those for another eg manufacturability, Cunningham and Dixon (1988). Different types of features stem from the differing viewpoint Wierda (1991). The work described in this paper is in the area of injection moulding, and examines the interactions between the viewpoints of functionality and mouldability in order to provide mouldability feedback to the designer.

1.0. Designing with features – Viewpoints in design.

1.1. Functional features.

Achieving a functional product is central to the design process, and therefore using feature based design the designer creates the product using a range of functional features. The definition of a universal set of functional features is unlikely. Each product has its

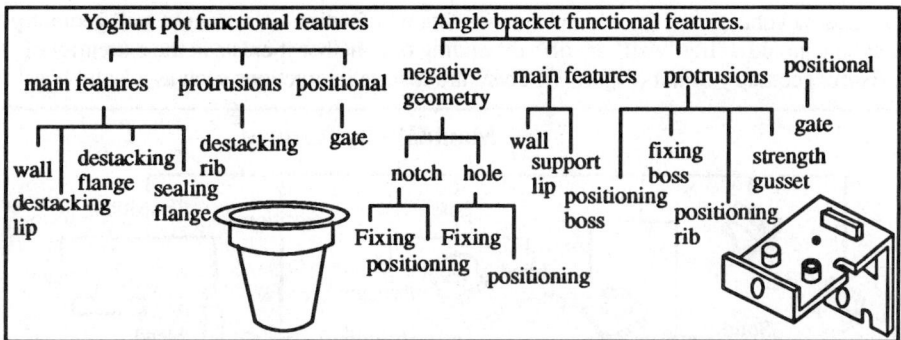

FIGURE 1. Functional features vary between different product types.

own set of objectives that can be translated into functional requirements, as shown by the example in Figure 1. This gives each product its own set of relationships between form and function. Different forms may achieve the same function in the context of different product types and the same forms may achieve different functions. Some forms may have multiple functions. The set of form/function relationships for one product type has no obvious relationship with that for another product type. This problem provides too large a number of permutations to define a universal set of functional feature types.

The functional meaning of specific form features is also changed by its use on a particular product range, as illustrated in Figure 2. For example a hollow boss geometry may be a flange, a rib or a base feature on a 'pot' type product (Figure 2), but can only be a boss type feature on the 'angle bracket' type product shown in Figure 1. Therefore the geometry has to be captured within the product range (functionality) context.

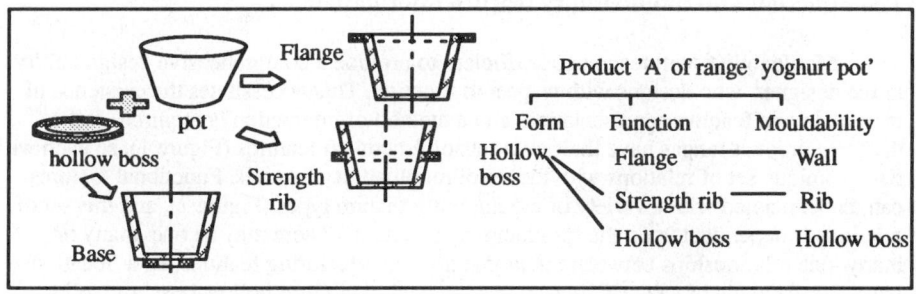

FIGURE 2. Features are not only geometry.

1.2. Mouldability features.

To reason about manufacturability of a set of manufacturing features has to exist for each viewpoint considered. For mouldability a group of features can be identified for which rules are well known (Figure 3). However mouldability reasoning requires a knowledge not only of the new features being created, but also of surrounding features, and this can change the (mouldability) meaning of a feature, eg a straight wall geometry can be a mouldability 'wall' or 'rib' depending on whether it exists at the extremity of existing geometry or not (Figure 3). There are feature interactions such as a 'solid boss'

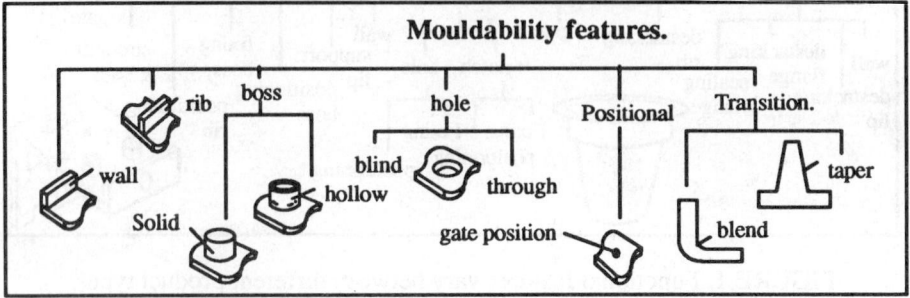

FIGURE 3. Mouldability features taxonomy.

and a 'hole' producing a single 'hollow boss', which has different mouldability constraints. There are three basic types of mouldability features, i) stand alone and 'extending' features, ie those that extend the extremities of the main geometry, these are walls, ii) 'attachment' features that are added to main geometry, these are holes, bosses, ribs and gate position, iii) transitional features that do not correspond to existing forms but exist as a result of their relative position or encroach on the faces of more than one existing form, these are blends and tapers. The type of blend and taper is dependant upon the combination of forms to which they are attached. Orientation data is required for taper application. These feature types provide a basic representation of mouldability which is independant of the product and can be applied to any product in a product range. The problem is how to associate such features with the product.

1.3. Function and mouldability feature interactions.

Mouldability features are not sufficient to provide a comprehensive design ability to the designer, who designs with respect to function. This necessitates the existence of more than one features representation, and a method of interaction is required. Just as different product ranges have their own sets of functional features (Figure 1), so each will have a unique set of relations with the set of mouldability features. Functional features can be associated with a variety of mouldability feature types (Figure 2), and this set of relations is dependant upon the 'product range context'. There may be one–many or many–one relationships between functional and manufacturing features, eg a 'location boss' for the angle bracket (Figure 1) may be a mouldability 'hollow boss' or 'solid boss'. Alternatively both a fixing boss and positioning boss functional feature can be a mouldability 'hollow boss'. For each product range the set of relations is different with no obvious correlation between them. Function can change the mouldability meaning of a feature, eg if a rotational wall on a yoghurt pot has the function 'destack' attached, this will be a (mouldability) hollow boss feature on the base of the pot, whereas with the function 'seal' attached the rotational wall will be a (mouldability) 'wall' feature at the top of the pot. Also form linked to function cannot be simply translated into equivalent mouldability forms because some mouldability considerations require additional forms such as tapers for removal from the mould and radii for constant section thickness (no

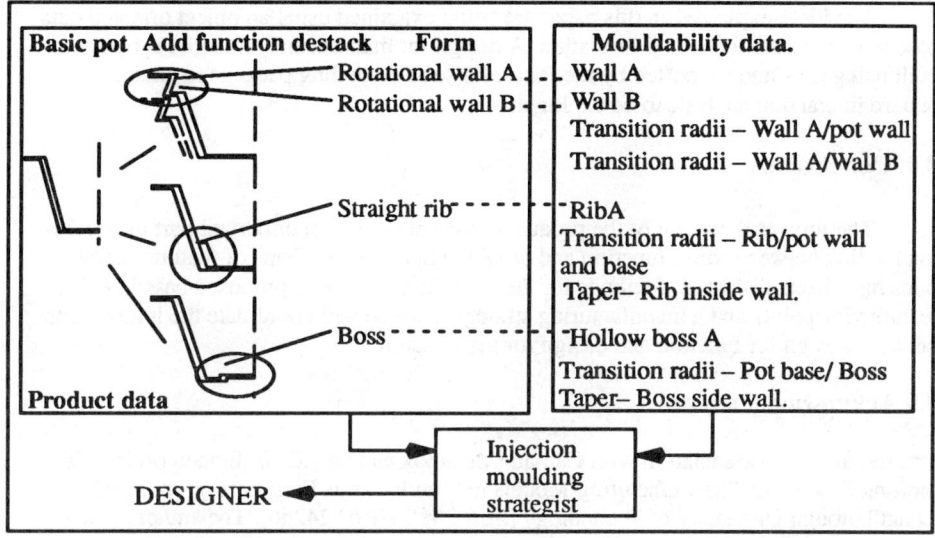

FIGURE 4 – Function and mouldability feature interactions.

shrinkage). Whether these forms are required depends on the new (form) feature and its adjacent geometry and orientation. These relations must be represented in order to provide the designer with manufacturability feedback.

2.0 Managing feature interactions–form, function, mouldability.

The approach taken to functional features has been to define these within product range data. Each product type is associated with a range of customer requirements which can be translated into a series of functional constraints. This provides a finite set of functional features. The designer is then provided with a choice of the means of achieving each product function, for example as shown in Figure 4, the product function 'destack' can be achieved in three different ways from which the designer must choose. Each option involves different forms. This provides the link between product function and form.

The remaining interaction problem is to provide the link of form with mouldability and other manufacturing representations. In this research we argue that this can be achieved by assessing the functional forms through an injection moulding strategist and comparing these with appropriate mouldability representations, as shown in Figure 4. The way in which forms are combined influences their meaning with respect to mouldability. The strategist has to identify and attach that meaning to the features. For example features which extend the form representation are considered to be 'wall' features from the mouldability viewpoint. All other modifications to form will result in mouldability features, which are identifiable by the form eg holes, ribs etc. Having identified the feature, the role of the strategist is to provide feedback to the designer with respect to the mouldability consequences of the chosen functional feature.

The ideas explained in this paper are being examined using an object oriented data base to capture the features information. A design for manufacture strategist is being built using C++ and supported by the Parasolid solid modelling package to enable the feature interaction analysis to be explored.

3.0. Conclusions.

The form and content of the product representation is an important part of providing the link between form, function and mouldability. The problems of feature interactions have been discussed. Methods for their solution have been proposed based on three feature viewpoints and a manufacturing strategist, which will coordinate the interactions between design for function and design for manufacture.

4.0 Acknowledgements.

The research is in association with CarnaudMetalbox and an ACME funded project, 'Exploiting Product and Manufacturing Models in Simultaneous Engineering', pursued at Loughborough University of Technology [SERC ref. GR/H 24266]. The authors would like to acknowledge all those who have provided assistance in the formulation of this work, especially CarnaudMetalbox and ACME.

5.0. References.

Cunningham,J.J.,Dixon,J.R. 1988, *"Designing with features: The origin of features."*, Proceedings ASMS. International Conference Computers in Engineering, 1988, v1, pp237–243.
Gadh,R. Prinz,F.B. 1992, *"Recognition of geometric forms using a differential depth filter."*, Computer Aided design,v24,No11,November 1992,pp583–598.
Weirda,L.S. 1991, *"Linking design, process planning and cost information by feature based modelling."*, Journal of Engineering Design,v2,No1,1991,pp3–9.

THE AUTOMATIC PROGRAMMING OF PRODUCTION MACHINERY FOR DE-FLASHING PLASTIC PARTS

Giles Tewkesbury and Dr David Sanders

University of Portsmouth
Portsmouth, Hants, PO1 3DJ

A new restricted programming language and a structured programming environment are presented. The generic programming language was created in order to automatically program any machinery for defined tasks. This avoids the need to individually program general purpose machines for dedicated actions. As an example the new method is applied to the programming of an intelligent conveyor system and a Fanuc A-600 industrial robot. The programming of these machines to de-flash plastic parts is described.

Introduction

This paper presents a new method for customising virtual task machines by mapping general purpose machinery programming languages to a new restricted programming language. A structured programming environment is used to verify task rules with a task expert, and to control the machines using the restricted language. This new technique has been used to create a virtual task de-flashing machine. The concept of task machines was first presented by Strickland and Hollis (1992) and then by Strickland (1992). Task machines are machines which are constrained for a task, but are not product dependent. A task machine would therefore be specified in terms of task rather than functional abilities, for example, a robot which can 'spray panels' rather than a robot which has 'six degrees of freedom'.

Research by the author has identified that for a task machine to interface to a design level, it not only needs to accept commands and be programmed in terms of a task, but also needs to give interactive advice on prospective designs. This advice may be, 'yes, I can manufacture the part, - it would take xxx seconds', or the machine could offer solutions to enable a production task to be performed, for example, new or alternative orientation information.

Methods of constraining existing general purpose machinery have been

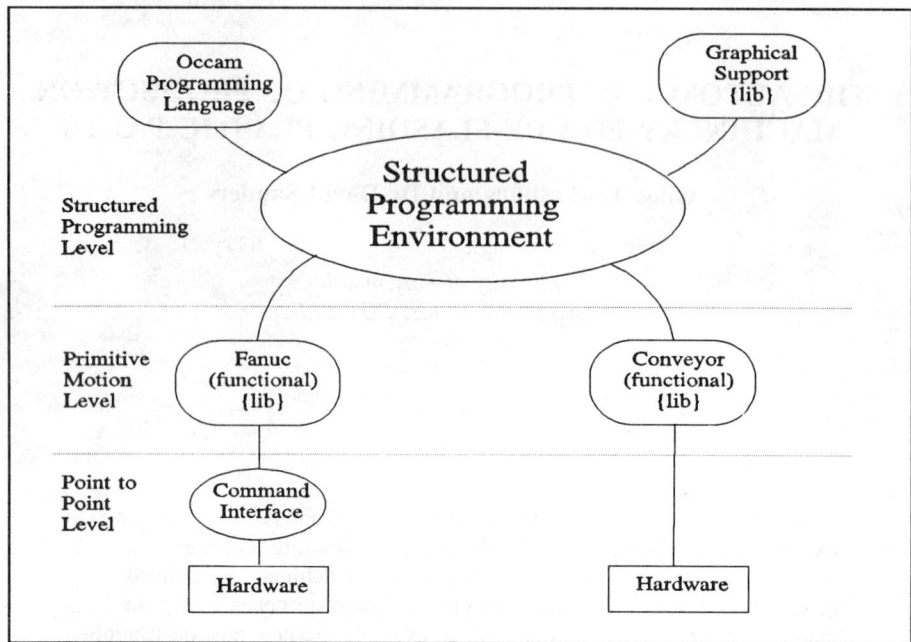

Figure 1. The prototype system

developed. This involved the creation of virtual task machines. Virtual task machines use the existing machinery and controller, but constrain them for a task. A virtual task machine is not a true task machine in that the hardware is not built from modular robotic elements. This means that the structure may not always be the most suited for the task, however, the interface to both types of machine is the same.

Tewkesbury, Sanders, Strickland and Hollis (1993) presented a new programming approach for task machinery. The approach taken was to use the inherently hierarchical structure of a manufacturing system and to match the programming of the different levels to the structure of the system. For example, the de-flashing machinery received information about the mould, and was then able to deduce where flashing would occur. The approach relied on the ability to configure and reconfigure software for a task machine. As task rules were captured and verified with a task expert, so the interface between the machine and the expert would change. Bonner(1983) and Deisenroth(1985), in their surveys, identified many robot programming languages. Since then more have been developed. At the time of writing, a programmer faced with the task of programming many different machines has to cope with various levels of complexity, functionality, and structure in each of the languages encountered. This variance reduces the possibility of modular software design, and increases development time and cost.

To overcome this problem a structured programming environment was created. This can be seen in Figure 1. The occam programming language was used as a base language and graphical windows commands were available in the form of a library, written specifically for this purpose by the author. Libraries for controlling a Fanuc

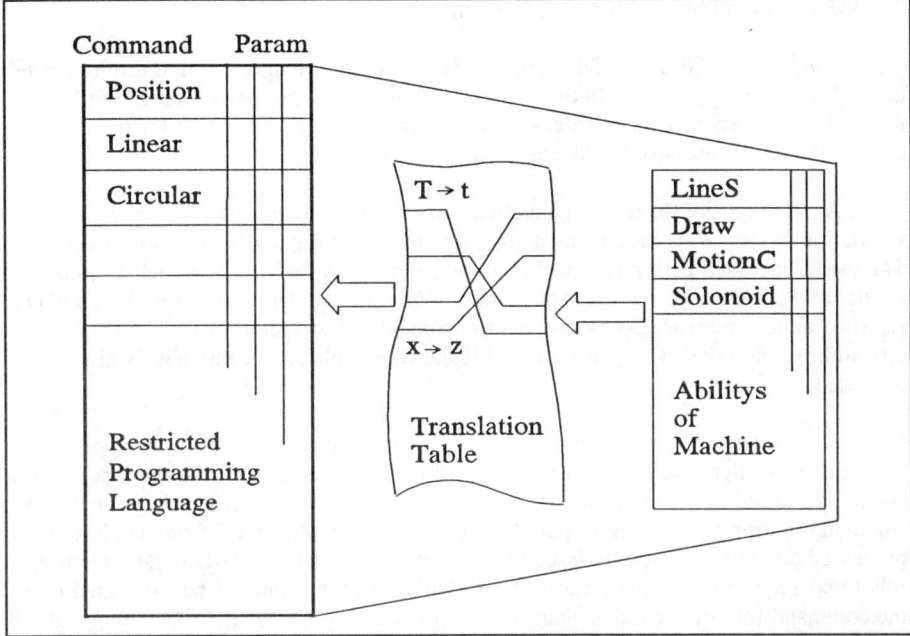

Figure 2. Mapping of functional commands to a new restricted programming language.

robot were written using a new restricted robot programming language, and a library for controlling materials handling task machines was created.

The structured programming environment enabled a programmer to develop a modular approach to the customisation of the virtual task machine. The programmer was able to use the graphical resources to build high quality user interfaces, designed to focus a task expert onto the task. This was necessary to verify task rules and to enable the creation of the virtual task machines.

Mapping the machinery language to a new restricted language

A survey was conducted into robot programming methods and languages, and is documented in Tewkesbury et al (1993). From the survey a new restricted programming language was defined. The language consisted of the primitive motion commands used in the other languages which were examined. Any commands which could be built from other primitive motion commands were not included. Structure commands were also excluded, as program structure came from the structured programming environment.

General purpose machinery was constrained by mapping the controller commands to the restricted programming language, using a predefined format and protocol. The modules not only contained the means to control the machinery, but also a model of the machinery, which allowed simulation. The modules were placed in a library, and were used in the structured programming environment to create a virtual task machine. Figure 2 shows how the abilities of a robot controller were mapped to a broad base of functional commands.

A Fanuc A-600 industrial robot was selected. The Fanuc's native programming method used NC type instructions. This programming language was successfully mapped to the new restricted programming language. The parameters for the commands were translated to the standard for the restricted language.

A pre-requisite for the language was that all commands had to work in a simulation mode. This meant that a complete model of the machine was necessary. The model included both physical limitations, and timing information such as joint acceleration, and data communication time. When in simulation mode each command returned either an error flag to say that the command had failed, or the time for the operation would take. Commands failed if the robots physical constraints were exceeded.

The restricted programming language was more than just a simple mapping from the controller, and a simulation of the robot. The language used the results from simulation to aid in the execution of the commands. This was achieved by having the command interpreter running in parallel with the user application. Commands were processed and stored, but only issued to the robot when an acknowledgment from the robot was required. This allowed the configuration of the robot to be calculated by the command interpreter rather than the programmer. For example, if a command was issued which was impossible because a left elbow convention had been adopted at some previous time, then the language interpreter would look back over the commands which had not yet been executed, and change the configuration to a right elbow at a suitable point.

The creation of a De-Flashing virtual task machine

The restricted programming language was used to create a machine dedicated to the task of de-flashing plastic parts. The virtual task machine was given task data which would have originated from a design level. Figure 3 shows an example of the data for a tool to produce a simple plastic box. (a) and (b) show the two tooling sections generated at a design level, and (c) shows the contact data where the two sections of the tooling touch. This information was used by the virtual task machine to provide manufacturing advice to a design level, and to guide the physical movements of the machinery.

Various design data were tested, some of manufacturable designs, and some of designs which were impossible to manufacture. The de-flashing virtual task machine consisted of the Fanuc A-600 robot with a laser mounted on the end effector. The machine successfully guided the laser to de-flash the plastic parts.

Discussion and Conclusions.

A new method for programming automated machinery has been presented. This was successfully applied to an industrial robot. The virtual task machine created from this robot was programmed in terms of the task to be performed, rather than with functional motions for the task. The machine successfully implemented the task.

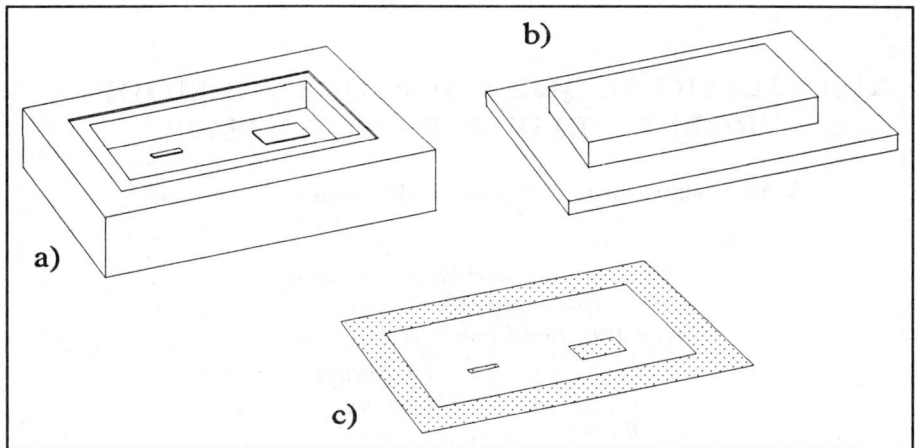

Figure 3. a) & b) two tool sections, c) the contact area for the two sections

The structured programming environment, and the restricted programming language provided a potentially generic method for programming advanced production machinery. This method allows a programmer to concentrate on the job of programming a task rather than on the specifics of a robot language and the structure of the machinery.

The restricted language provided an intelligent interface to the production machinery. This interface not only provided continuity when changing from the programming of one machine to another but also provided a means to simulate the execution of the commands. Without this continuity it would not be cost effective to develop virtual task machinery. The provision of this continuity may allow a programmer to develop virtual task machines which can then be linked to a higher design level. This would allow the production environment to become fully automated while the experience and the knowledge originally obtained from an expert in the task, is held within the system.

References.

Bonner, S.A. 1983, *Comparative Study of Robot Languages*, MS Thesis, Rensselaer Polytechnic Institute, Troy.

Deisenroth, M.P. 1985, A Survey of Robot Programming Languages, *Proceedings of The 1985 Annual International Industrial Engineering Conference*, IIE, 191-194.

Strickland, P. and Hollis, J.E.L. 1992, Task Oriented Robotics, *Proceedings of the SICICI conference*, Singapore, **2**, 835-840.

Strickland, P. 1992, *Task Orientated Robotics*, PhD Thesis, Portsmouth Polytechnic.

Tewkesbury, G.E., Strickland, P., Sanders, D.A. and Hollis, J.E.L. 1992, Product Orientated Manufacturing, *Proceedings of the 25th Dedicated Conference on Mechatronics*, ISATA 92, Florence, Italy, 505-512.

Tewkesbury, G.E., Sanders, D.A., Strickland, P. and Hollis, J.E.L. 1993, Task Orientated Programming of Advanced Production Machinery", *Proceedings of the 26th Dedicated Conference on Mechatronics*, ISATA 93, Aachen, Germany, 623-630.

AUTOMATIC ACQUISITION OF CAD MODELS FROM MULTIPLE RANGE VIEWS

A W Fitzgibbon, E Bispo, R B Fisher, E Trucco‡

Department of Artificial Intelligence
University of Edinburgh
Forrest Hill, Edinburgh EH1 2QL, Scotland
‡Department of Computing and Electrical Engineering
Heriot-Watt University
Riccarton EH14 4AS, Scotland

The project reported in this paper focuses on the automatic acquisition of CAD-like models of small industrial components from multiple range images. The goal of the target system is to generate automatically (or with minimum human intervention) object models suitable for vision tasks of industrial interest, e.g. inspection, quality control, classification, position estimation, recognition and reverse engineering.

Issues in automatic model acquisition

By an *automatic model acquisition system (AMAS)* we refer to a system capable of generating automatically a CAD-like model of an object from a number of images acquired from different viewpoints in space. There is an increasing interest for this technology, and many benefits are expected from the development of reliable AMAS, particularly for manufacturing applications: reverse engineering, product styling, NC machining, classification, recognition and inspection are only some examples.

Our work aims at developing an AMAS based on *range images*, which provide accurate, dense surface measures very efficiently. In order to acquire our own data, we have developed a low-cost, high-accuracy laser scanner capable of acquiring range images within a workspace of about $15cm^3$ with an accuracy of 0.15mm. A simple and efficient direct calibration technique has been implemented. Special consistency constraints discard most of the false measurements generated by reflective surfaces (Trucco and Fisher 1994).

The two essential issues that any model acquisition system must address are *model representation* and *view registration*. Several representations can be in principle adopted to express the models: splines, surface patches, triangulations, volumetric descriptions and finite element models are examples of possible choices. The choice of a representation is in turn linked intimately to the problem of estimating the transformation registering two successive views of the object.

In our system we use two complementary model representations, each of which implies a different solution to the registration problem: a conventional, symbolic surface patch representation (Fisher 1989, Fan 1990, Trucco and Fisher 1992), illustrated in Figure 1 (left) is combined with a B-spline model (Figure 1 (right)). The symbolic model allows fast indexing from a large database, quick pose estimation (due to the small number of corresponding features), and easy specification of inspection scripts (for example, the system can be instructed to "measure diameter of hole 2 in patch B"). On the other hand, pose estimation is limited in accuracy by the small number of correspondences and errors in the surface fitting, and provides only an incomplete surface coverage: only the most stable surface patches are retained in the model. This lack of complete data is undesirable for reverse engineering tasks, and is cured by the use of spline models. Using these models, initial pose estimates can be optimised (albeit expensively), and complete surface models easily obtained.

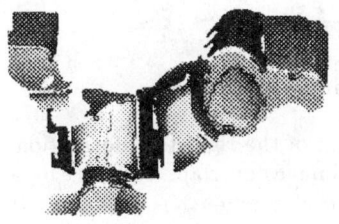 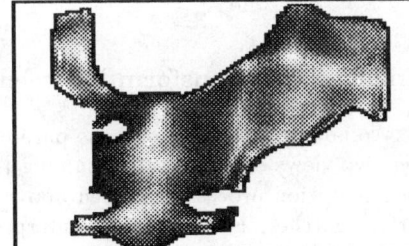

Figure 1: Example models produced by the system. On the left is a surface patch model of the truck part and its environment, on the right a coarse spline model.

In the following sections we give an overview of the computational modules of our system, namely: constructing the surface patch representation and the spline model from a single range image (view); estimating the transform between views by matching the surface patch models associated with two views; refining the estimate using the spline model; merging the views and postprocessing the current object model.

Model construction from a single range view

Building the symbolic model

Objects are initially scanned using a laser triangulation system. Depth and orientation discontinuities are detected from the raw data, and used as *a priori* weights for diffusion smoothing (Trucco and Fisher 1992). Mean and Gaussian curvatures are calculated from the smoothed image, and the data are then divided into homogeneous regions of positive, negative or zero curvature. Each region is then approximated by a viewpoint-invariant biquadratic patch (Fitzgibbon 1993), and finally expanded to include neighboring pixels which are within 3σ ($\sigma 0.15mm$) of the fitted surface. After this segmentation stage, the region boundaries are approximated by polylines and conics, and adjacent regions are intersected to produce more consistent boundaries. The resulting description is converted into a vision-oriented modeling representation, called the Suggestive Modelling System or SMS (Fisher 1987, Fitzgibbon 1992), for visualization and use in our model matching module.

Building the spline model

The spline model is constructed by laying a regular grid of control points on the image and fitting a third-order B-spline to the data. Background and noise points are removed in advance. The density of the grid is currently determined by the required accuracy—a 50 by 50 sampling allows the object in Figure 1 (right) to be approximated to within a maximum error of 0.8mm. An obvious extension is to allocate the knot points more densely at regions of high curvature, as the curvature maps are available from the segmentation process. We plan to implement this in the near future, and expect a significance increase in speed from the reduction in the number of knot points.

Estimating the transform between views

We now wish to estimate the parameters of the rigid transformation which relates two views of an object, assuming the images overlap. We start by applying the segmentation process described above to each image, thus producing two lists of surface patches. From these, an interpretation tree algorithm (Fisher 1994) finds consistent sets of corresponding pairs of surfaces. The pairs allows us to compute the 3-D pose of the object using least-squares techniques. The pose is used as an initial estimate for an iterated extended Kalman filter, which computes the uncertainty of the pose estimate.

The accuracy of view registration is within about 1σ of the noise on the range data if three or more linearly independent planar surfaces can be extracted reliably from the object (See Figure 2). Failing that, biquadratic surfaces estimated about the patch centroids are used to constrain the pose and then translation accuracy falls to about 5mm. If the pose needs to be constrained by using paired centroids, the system is open to error due to occlusion. The rotational accuracy of registration is generally within 1 degree.

Refining the inter-view transform

Given an initial pose estimate from the symbolic model matcher we can now use the spline model to refine the estimate. The pose is optimized using the Iterated Closest Point (ICP) algorithm (Besl and MacKay 1992). We have found in a 2D example that the region of convergence occupies about 25% of the space of all possible poses. In the 3D tests on the object above, a less complete investigation indicates convergence with up to 90 degrees of initial rotation error.

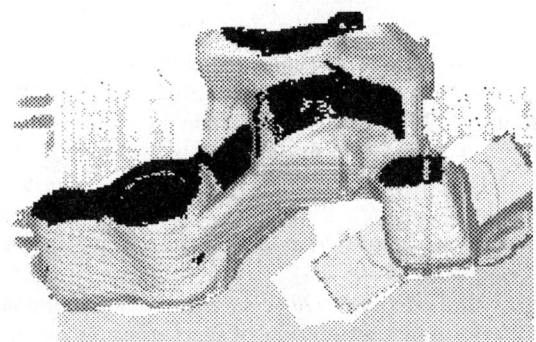

Figure 2: Surface-based matching results. Gray pixels are data, dark grey are model. The image shows a good quality match using independent planar patches and patch centroids.

The disadvantage of this technique is its computational complexity: for each data point, we must locate the nearest point on the model surface, then calculate the registering transform. Locating the closest point is sped up by a combination of multigridding and subsampling the basic gradient descent algorithm. The registration accuracy is "optimal" in the sense that the noise statistics of the residuals are symmetric and white. Non-convergence does occur however, and we are currently investigating ways of further correcting for this.

View registration and model postprocessing

Final processing on the models includes merging the single-view descriptions into a single reference frame. This is done easily thanks to the SMS representation for surface patches, which separates patch descriptions into shape, extent and position. The spline models may be treated similarly, by calculating a new fitting spline for the merged sets of range data.

Discussion

We have described an AMAS capable of creating CAD-like models from range data effectively. The images need not be registered in advance, but should have some overlap between frames. The primary goal of current and future research is to investigate and improve the reliability of the symbolic model matcher, particularly under situations where occlusion and multiple objects pose difficulties. Additional topics being investigated include the extraction of features usable for manufacturing operations, e.g. holes, slots, etc. Further investigation will address the overall reliability of the system, and the improvement of the models in terms of speed and consistency.

References

Besl, P. J. and McKay, N. 1992, A method for registration of 3-D shapes. IEEE Transactions on Pattern Analysis and Machine Intelligence, **14**, 239–256.

Chen, Y., and Medioni, G. 1992, Object modelling by registration of multiple range images. Image and Vision Computing, **10**, 145–154 .

Fan, T.-J. 1990, *Describing and Recognizing 3-D Objects Using Surface Properties*, (Springer-Verlag, Berlin).

Fisher, R. B. 1987, SMS : a suggestive modelling system for object recognition. Image and Vision Computing, **5**, 98–104.

Fisher, R. B. 1989, *From Surfaces to Objects*, (John Wiley, London).

Fisher, R. B. 1994, Performance evaluation of ten variations on the interpretation-tree matching algorithm. In *Proceedings European Conf. on Computer Vision*, Stockholm, September, to appear.

Fitzgibbon, A.W 1992, Suggestive modelling for machine vision. In *Proceedings SPIE-1830 Int. Conf. on Curves and Surfaces in Computer Vision and Graphics III*, (SPIE, Boston).

Fitzgibbon, A.W. and Fisher, R.B. 1993, Invariant fitting of arbitrary single-extremum surfaces. In J. Illingworth (ed.), *Proceedings British Machine Vision Conference*, (Springer-Verlag, Berlin) 569–578.

Trucco, E. and Fisher, R.B. 1994, Acquisition of consistent range data using local calibration. In *Proceeding IEEE Int. Conference on Robotics and Automation*, (IEEE Comp.Soc. Press, Los Alamitos), to appear.

Trucco, E. and Fisher, R.B. 1992, Computing surface-based representations from range images. In *Proceedings IEEE International Symposium on Intelligent Control*, (IEEE Control Systems Soc.), 275–280.

THE INTEGRATION OF CAD AND A KNOWLEDGE BASED TOOL SELECTION SYSTEM

Ms Yue Zhao, Dr Keith Ridgway and Mr Kah Fei Ho

University of Sheffield
Sheffield S1 3JD

This paper describes the development of an integrated system for selecting cutting tools and conditions for turned components. A knowledge base cutting tool selection system developed at the University of Sheffield used a feature based system to describe the geometry of a turned component. The system, written in PROLOG, compared the feature description to the cutting tool capability and selected appropriate cutting tools and conditions. This paper describes a further enhancement to the system that enables the user to describe the component using a commercial CAD system.

Introduction

EX-CATS is an Expert Computer Aided Tool Selection system for turned components developed in pilot form by Arezoo and Ridgway (1989) at the University of Sheffield. The system, which is based on the conventional concepts of the expert system, selects cutting tools, including the tool holders and inserts and determines the optimum cutting conditions for turned components.

The system comprises a user interface, knowledge base (KB), working data base (WDB), inference engine (IE) and explanation facility, which was designed and developed using the Prolog language. The system assumes that the blank type and size and sequence of operations and work-holding methods are already known or found by the user it then selects suitable cutting tools using experience gained from technical guidelines and domain experts. At present the system only considers cutting tools with Tungsten Carbide indexable inserts, which dominate today's modern and semi-modern workshops, are considered.

The cutting tool consists of two main components; the tool holder and the indexable insert. Both tool holders and inserts are internationally coded to indicate the characteristics of each tool. The objective of any tool selection exercise is to determine the following parameters:
a) tool holder (clamping system, type, point angle, hand of cut, size, etc.)
b) insert (shape, size, grade, nose radius, etc.)

c) cutting conditions (feed, speed, depth of cut)
d) type of the coolant (if required)
e) total cost of machining the component.

The turning operation is usually divided into roughing and finishing operations. The majority of the material to be machined is removed during the roughing operation which requires maximum power. During the finishing operation a fine cut takes place which provides the required surface finish and detailed profile. The selection of tool holder, insert, cutting conditions and coolant is dependent upon a number of factors such as the type of operation, work piece material, design, accuracy, finish and power and rigidity of the machine tool. These can be considered as purely technical factors. A further aspect to cutting tool selection is concerned with cutting tool rationalisation. This is often the key to major economic benefits due to a reduction in cutting tool inventories, reduced tool set-up and change over times.

In EX-CATS, the methodology adopted for component representation addresses the problems, and represents the geometry of the component in a form that can be easily understood by the tool expert. All the components are represented manually, by the user, as a set of Prolog facts in an English like syntax. This provides an easy method for planners or machinists who are not familiar with computing to represent and understand the components under consideration.

The main drawback in the use of the above package is the need to describe all components manually. The main objective of the current study is to create a software package that can transfer the CAD database of a turned component to a format compatible with the EX-CATS system.

CADEXCATS Development

The package created, CADEXCATS, is written in the C language and integrates the knowledge based system, EX-CATS, with a commercial CAD package equipped with an IGES pre-processor (AutoCAD). A product model of a turned component developed on a 2D CAD system is first translated into a general purpose IGES (Initial Graphics Exchange Specification) protocol through the IGES pre-processor. Subsequent running of CADEXCATS automatically generates a product model compatible with the EX-CATS process planning system.

As mentioned above, the CADEXCATS software package was developed to integrate EX-CATS with a commercial CAD system. CADEXCATS is designed to read CAD generated data in the IGES format which enables EX-CATS to accept CAD models created on any CAD systems with an IGES pre-processor. Since all the IGES files share the same format which is independent of any CAD systems, IGES generated from AutoCAD Release 12 is used as data sources in this study as an illustration.

CADEXCATS is written in C language and is activated in TURBO C++ Version 3.0. All three packages, i.e. TURBO C++, AutoCAD and EX-CATS are based on an IBM PC.

Work piece representation

An adequate representation of the product model is essential in the development of a satisfactory process planning system. Much of the reasoning needed in process planning is at the symbolic level and involves the manufacturing features of the component. Any generative process planning system must be capable of giving a detailed definition of the geometry of the component. In the EX-CATS package the representation of the turned component takes the format of one single file and follows the following procedural rules:

1) The work piece is represented separately for external and internal operations.
2) The work piece is represented separately for roughing and finishing operations.
3) The roughing geometries are determined by the operation sequence defined by the user.
4) The finishing geometries are the exact profile of the finished component.
5) The external features are numbered in a clock-wise manner.
6) The internal features are numbered in a counter clock-wise manner.
7) When representing the work piece the user assumes that the machining operations are carried out against the chuck.

It should be noted that EX-CATS uses the lower profile of a turned component for geometry description. In the EX-CATS model the origin is always on the intercept of the centre line and the far left hand face of the component representation. The restrictions of the AutoCAD co-ordinate system dictates that an upper profile representation is adopted in CADEXCATS. Consequently points 5 and 6 above must be modified as follows:

5) The external features are numbered in a counter clock-wise manner.
6) The internal features are numbered in a clock-wise manner.

To represent the component, each element is represented by a rule consisting of a Prolog predicate representing the basic primary cutting action required to create the geometry defined, such as longitudinal turning, facing in, in-copying, etc. In addition to the geometric information it is also necessary to represent other associated data such as diametral tolerance, surface finishing and the maximum diameter for internal operation, etc. Each element is represented by specifying the following information in the order shown:
1) type of geometry (external or internal).
2) type of operation (roughing or finishing).
3) identity number.
4) cutting action (longitudinal turning, facing in, in-copying, etc.).
5) set-up number.
6) start point co-ordinates.
7) end point co-ordinates.

Additionally some elements for finishing operations contain surface finishing data Ra, diametral tolerances data (M1 and M2) and information concerning fillets. The fillet information includes the two identity numbers which form the fillet and the radius of the fillet. For internal operations it is necessary to indicate details of the maximum diameters for finishing and roughing operations separately. For rough turning operations the co-ordinates of the beginning and end points of some of the longitudinal and facing features, which do not reside in a recess, are not represented, and only the set-up number is mentioned.

As discussed above, the geometric representation for finishing and roughing operations can be very different for the same turned component, because the roughing geometries are determined by the operation sequence adopted by the user while the finishing geometries are the exact profile of the finished component. Furthermore, the data for finishing operations, compared with that for roughing operation, contains additional non-geometric information. Two separate programs have been created to effectively process data for both finishing and roughing operations.

Example and evaluation

Figure 1 shows a 2D representation (finishing and roughing) of a turned component generated in AutoCAD. It contains finishing and roughing operation profiles for both external and internal operations. Figure 2 presents the automatically generated CADEXCATS's output, "fig1.pro" file. Non-geometric information is also included, such as surface finishing data and diametral tolerance data in the finishing operation. The first line of the fig1.pro data, is:

ext_fin_geom no 1 is_a face_in(1,360,5C,360,80).

This represents geometry identity No.1 for a facing in finishing operation of an external geometry with set-up No.1 and co-ordinates for start and end points (360,50) and (360,80) respectively.

Surface finishing data 1 required by geometric identity No.4 which is for a longitudinal turning operation of an external geometry and the corresponding diametral tolerances data 0.05, -0.05 are expressed respectively as follows:

surf_finish(ext_fin_geom no 4 is 1).
ext_diam_toll(geom no 4,0.05,-0.05).

The description for geometry identity No.9 which is an arc with set-up No.2 in a counter clock-wise manner with start point (70,120), end point (140,120), centre of the arc (105,120) is given as,

ext_fin_geom no 9 is_a arc(2,ccw,70,120,140,120,105,120).

A further example, ext_fin_radii(geom no 5, geom no 6, r is 4) indicates a radius of 4 for the fillet defined by geometry identity Nos. 5 and 6 for the external finishing operation. The file generated is identical with that previously input manually into the EX-CATS system.

Discussion and conclusions

A software package has been successfully developed to achieve full integration between a commercial CAD system and the knowledge based EX-CATS CAPP system which is applicable to the selection of cutting tools and conditions for turned components.

The input to the CADEXCATS system is a 2D model of a turned component in IGES file format. The effectiveness of the interface developed between this package and AutoCAD Release 12 via the IGES format suggests that CADEXCATS can directly access any IGES files for turned components, generated on any CAD system. The CADEXCATS system produces an output file that is compatible, both in file name and data format, with the EX-CATS CAPP system.

CADEXCATS has been tested successfully. The ability to directly access CAD files to obtain geometric information that would have previously been entered manually has resulted in the realisation of a fully automated cutting tools and conditions selection package. It is believed that with minor modification in the output format, the CADEXCATS can be applied to other CAPP systems for turned components.

Acknowledgements

The authors wish to thank the Billy Ibberson Trust who support research in industrial change at the University of Sheffield.

References
Arezoo, B. and Ridgway, K. (1989), An Expert System for Tool Selection for Turning Operations. Fifth International Conference on Computer Aided Production Engineering, Edinburgh, 179-186.

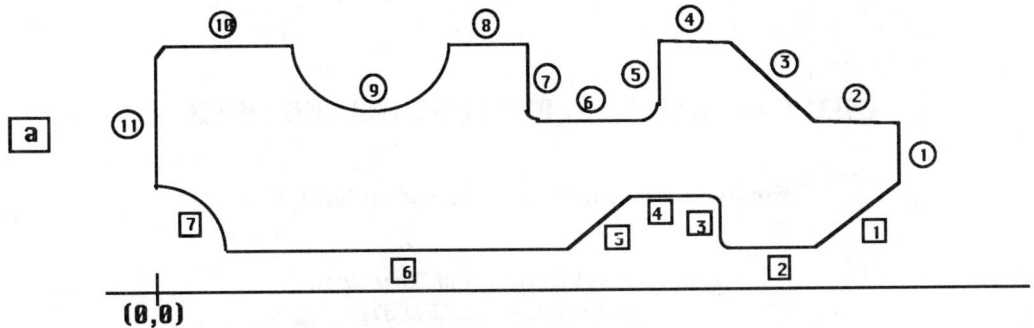

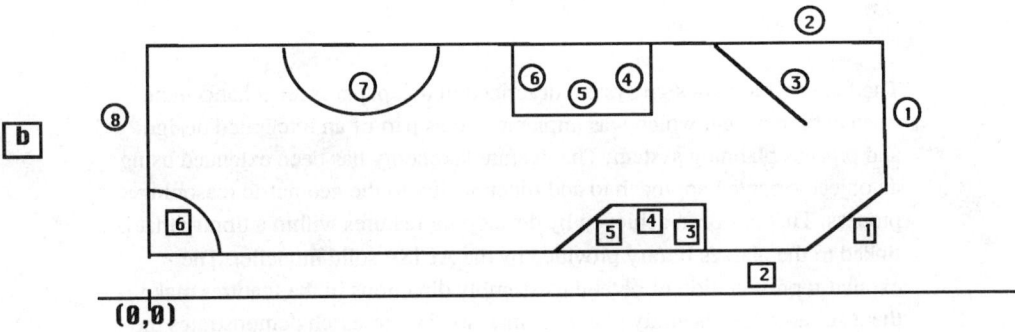

Figure 1: Example of CAD drawings for a turned component, (a) finishing and (b) roughing.

```
comp_name(fig1.pro).
ext_fin_geom no 1 is_a face_in(1,360,50,360,80).
ext_fin_geom no 2 is_a long_turn(1,360,80,320,80).
ext_fin_geom no 3 is_a out_copy(1,320,80,280,120).
ext_fin_geom no 4 is_a long_turn(1,280,120,244,120).
surf_finish(ext_fin_geom no 4 is 1).
diam_toll(ext_fin_geom no 4,0.05,-0.05).
ext_fin_geom no 5 is_a face_in(2,244,120,244,84).
ext_fin_geom no 6 is_a long_turn(2,240,80,184,80).
ext_fin_radii(geom no 5, geom no 6, r is 4).
```
Figure 2: Sample output from CADEXCATS

OBJECT–ORIENTED FEATURE–BASED DESIGN

Wan Abdul Rahman Wan Harun and Dr Keith Case

Loughborough University of Technology,
Loughborough, Leics, LE11 3TU

The feature–based design system described in this paper is an enhancement of an earlier system which was implemented as part of an integrated design and process planning system. The feature taxonomy has been extended using an object–oriented approach to add functionality to the geometric reasoning process. This has been achieved by developing features within a library that is linked to the classes library provided by the ACIS® solid modeller. The explicit representation of potential assembly directions in the features make them suitable for assembly planning analysis. The research demonstrates the possibilities for a single feature representation to support multiple activities within a computer integrated manufacturing environment.

Introduction

The use of features is a prominent trend in research in the application of CAD/CAM systems in recent years. Feature technology is viewed as a key technology for the next generation of computer–aided design and manufacturing systems. Research in this area is aimed at determining alternative component representations which form a suitable basis for a wide ranging set of activities throughout a product's life cycle. Feature technology offers several advantages such as providing a more intuitive way of building up models and capturing more of the designer's intent in the product representation. Comprehensive reviews on feature–based design can be found in Salomons et. al. (1993) and Case and Gao (1993).

This paper discusses the underlying methodology employed in the development of an extended version of a feature–based design system and its application in the assembly modelling environment. Earlier work related to research on process planning and process capability modelling for design and selection of processing equipment (Case 1994). The prototype system is fully integrated with a solid modeller kernel which enhances its modelling capabilities and provides an efficient tool to model a product. A feature–based library has been created and linked to the solid modeller library. The object–oriented approach is employed through the use of the C++ programming language. The explicit representation of potential assembly directions in the current implementation of features means that the features are highly suited to the assembly planning activity. This is obtained through the information provided by the Boundary Representation (B–Rep) model represented by ACIS®.

Feature Definition and Representation

One of the requirements of a feature–based system is a feature representation that is capable of supporting many of the design and manufacturing planning activities. Manufacturing planning systems need to extract feature–based component information from CAD systems both accurately and efficiently. In recent years, object–oriented techniques have been used widely in many computer applications. Object–oriented concepts are seen as fitting well within CAD framework by providing utilities for manipulating data, managing the permanent storage of data and maintaining a user interface (Warman 1990). Using an object–oriented structure provides a general way to model and manipulate features for geometric reasoning (Unger and Ray 1988, Chung et. al. 1988). In this representation, features are conceived as objects that have data and a number of attributes to describe their characteristics. It is readily extensible to include additional data and relationships as appropriate. Examples of the object–oriented feature–based systems can be found in Unger and Ray (1988) and Chen et. al. (1994).

The feature taxonomy scheme used provides the required structure for feature primitives (Case et. al. 1994) and form features are treated as volumes enveloped by a set of real and imaginary faces. The real faces physically exist on the component and are typically surfaces from the original part or the result of manufacturing operations. Imaginary faces can be considered as the surfaces required together with the real faces to form an enclosed volume. Form features are divided into three categories – protrusions, depressions and surfaces. One of the characteristics of this feature taxonomy is the number of orthogonal directions from which the feature volume might be approached. These are known as External Access Directions (EAD's) and all features have between 0 and 6 EADs. Feature geometry is described by defining the EADs, the boundary type (open, closed) and the exit boundary status (through/not through). Compound features and parent–child relationships among features have also been defined.

The above classification of features provide a convenient method for representation using an object oriented approach. Two main base classes are created – features and profiles, based on the feature definition and taxonomy described above. The feature class defines data such as feature identities, feature type and surface roughness. For each feature class, a number of primitive shapes are defined – boss, pocket, hole, non–through slot, through slot, notch, step and real face, as shown in Figure1. Each feature type class has its own data such as the number of EADs, its dimensions, location and orientation with respect to the component. Procedures are also available for feature operators, such as create, edit, move, copy and rotate features. Further procedures determine mating surfaces, define parent–child relationships and the topological data for assembly planning. The modular nature of the object–oriented technique permits feature classes to be extended to allow other procedures to be added in the future.

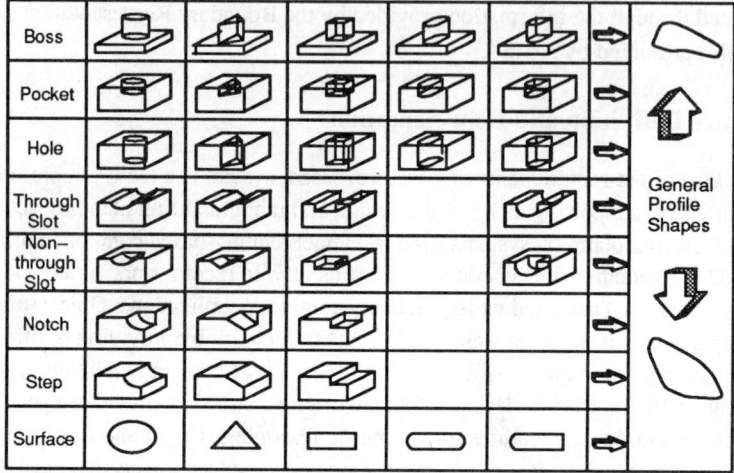

Figure 1. The library of feature primitives

The profile class is a base class for standard profile shapes common to each type of feature, such as circular, rectangular and triangular as well as a user defined profile. For example, a boss may have a circular profile and a pocket a rectangular shape. Other profiles can be defined by the user by entering the appropriate parameters. The primitive shapes of all feature classes form the feature library from which feature instances can be generated by retrieving the feature primitives and specifying the parameters. The two classes are referenced by each other. For instance, a boss would reference a profile object, where this profile object is actually an object of a derived class of profile. The relationship among classes are shown in Figure 2.

The User Interface System

The system utilises the openness of the ACIS kernel modeller, which serves as the main engine of the system. ACIS is an object–oriented geometric modelling toolkit designed for use as a geometry engine within 3–D modelling applications (Spatial 1993). The modular architecture of ACIS provides extensive facilities for the development of a feature–based system. It also provides a rich set of geometric operators for the construction and manipulation of complex models.

The solid model is built within ACIS using class libraries which are accessed through an Application Procedural Interface (API) or direct–object interface to all internal objects, classes and methods. Some of the functions in the feature modelling system such as blending, sweeping and Boolean operations are derived or extended from these classes. The direct interface method allows the topology to be scanned to the

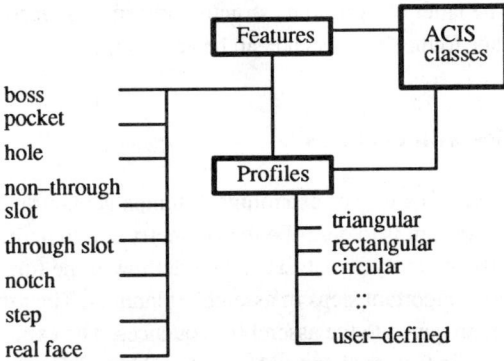

Figure 2. Class relationship

face level while at the same time the geometry is interrogated. ACIS also provides a mechanism to attach classes for assembly modelling attributes to any object in the data structure. Attributes are crucial for extending geometry models to become true product models. This includes complex attributes, which may represent dimensions, constraints on features and references to other objects.

A design by feature user interface to the system has been created to allow designers to generate components using feature primitives and to store attributes in a feature based data structure which is separate from, but associated with the database of the geometric modeller. The design by features interface allows designers to create features by defining the sets of parameters for feature primitives; to perform feature edit operations such as move, rotate, copy and delete; and to define feature relationships such as parent–child relationships and tolerances between features. The user interface to the system is provided by Motif. The system architecture is shown in Figure 3.

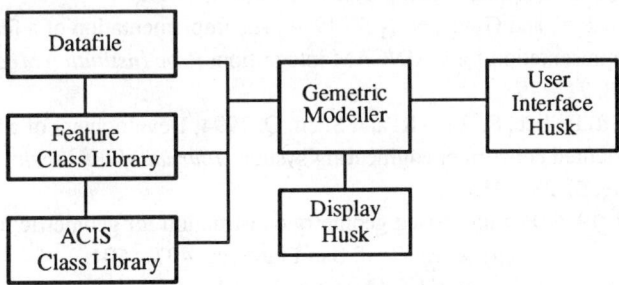

Figure 3. System architecture

A feature primitive is defined as the combination of a set of functions. For example, the create pocket function is the combination of create wire profile, cover the faces and sweep the faces. Once a feature is created, a B–Rep model is generated for the feature. All data on the model and features can be stored in a data file provided by the

ACIS bulletin board. The latter contains data structure information recorded as a result of operational changes made to the features and can be accessed by the user for manipulation.

Modelling Assemblies with Features

Modelling assemblies is a way of examining complex geometric interactions before anything is built. Its aim is to describe the geometry and to define the relations between parts of the final assembly. Analysis and modelling of the final product constitute one of the most important steps in assembly planning. The product model must contain the data required to generate the assembly sequences. The explicit representation of assembly directions in the features is provided by the EADs defined in each feature. The position and the orientation of the feature, which can be accessed through functions provided by ACIS, can determine the location of the mating faces and the interactions of various features in the assembly. The work on this aspect is still under investigation.

Conclusions

The object–oriented approach described above is an effective way to implement geometric modelling in a feature–based system. In general, the modular nature of the object–oriented technique allows modelling at a higher level of complexity than is possible in current systems. It also enables easier extension to the modelling system. The application of the system for the assembly modelling is used to verify the generic nature of the approach.

References

Case, K. and Gao, J. 1993, Feature technology – an overview, *Int Journal Computer Integrated Manufacturing*, **6**, Nos 1 & 2, 2 –12.

Case, K. 1994, Using a design by features CAD system for process capability modelling, *Computer Integrated Manufacturing Systems*, **7** (1), 39 – 49.

Case, K., Gao, J. X. and Gindy, N. N. Z. 1994, The implementation of a feature–based component representation for CAD/CAM integration, *Proc Institution Mech Engrs, (IMechE)*, **208**, 71 – 79.

Chen, C., Swift, F., Lee, S., Ege, R. and Shen, Q. 1994, Development of a feature–based and object–oriented concurrent engineering system, *Journal of Intelligent Manufacturing*, **5**, 23 – 31.

Chung, J.C.H. 1988, Feature–based geometry construction for geometric reasoning, *Proc ASME Computers in Engineering Conf*, San Fransisco, 497 – 504.

Salomons, O.W., van Houten, F.J.A.M. and Kals, H.J.J. 1993, Review of research in feature–based design, *Journal of Manufacturing Systems*, **12** (2), 113 – 132.

Spatial Technology Inc. 1993, *ACIS Geometric Modeler, Programmers Manual.*

Unger, M.B. and Ray, S.R. 1988, Feature–based process planning in the AMRF, *Proc ASME Computers in Engineering Conf*, San Fransisco, 563 – 569.

Warman, E. A. 1990, Object–oriented programming and CAD, *Journal of Engineering Design*, **1**, No 1, 37 – 46.

A STRATEGY FOR EXTRACTING MANUFACTURING FEATURES FOR PRISMATIC COMPONENTS FROM CAD

Mr S. Linardakis and Dr A.R. Mileham

*School of Mechanical Engineering, University of Bath
Claverton Down, BATH BA2 7AY*

A novel approach for interfacing CAD and CAPP for prismatic components is discussed. The analysis presented forms part of ongoing research on a CAD interpreter using the industry standard DXF file format to extract and classify process planning information from a CAD engineering drawing. The DXF file of a typical 3 view engineering drawing of a prismatic component is used as the system input. This is then processed in order to identify a certain range of manufacturing features. The methodology for identifying features is presented, along with a typical application example of how this has been implemented within the system.

Introduction

Recently the concept of using component features for design and manufacturing applications has received much attention and research effort. Describing components by the use of features is seen by many as the key to genuine integration of the many aspects of design and planning of manufacture, particularly in a modern computer-controlled environment incorporating Computer Aided Design (CAD) and Computer Aided Process Planning (CAPP) (Brimson and Downey 1986). Features originate in the reasoning processes used in various design, analysis, and manufacturing activities (Cunningham and Dixon 1988) and are frequently strongly associated with particular application domains. Hence there are many different definitions of features. A broad definition in the engineering domain is given by (Chang 1990) as: " a subset of geometry on an engineering part which has a special design or manufacturing characteristic".

Two approaches to using features in manufacturing applications have been used by researchers: Feature Identification and Feature Based Design. Feature Identification (Choi et al. 1984) is the process of extracting manufacturing features from a CAD database. In Feature Based Design (Shah and Rogers 1988) the designer uses manufacturing features to define an engineering part and build the

CAD database. While this provides a natural transition from design to manufacturing, it also imposes limitations on the generality and extensibility of features, by enforcing an interdependence between design and manufacturing capabilities (Chang 1990).

From the design point of view, one of the attractions of geometric modelling within CAD is that a well constructed modeller is capable of representing all the geometric aspects of a component within a chosen domain. If a complete geometric description is available, then it is clearly possible to use computer methods to interrogate this information and transform it into any desired form. Thus collections of surfaces could be recognised as features and transformed into manufacturing features for CAPP purposes. Much research work has been done in this Feature Identification field.

Mortensen and Belnap (1989), describe a methodology employing feature recognition for rotational parts from a 2D CAD system database (although an interactive graphics system was used to generate the test data). However this system is limited as it lacked automatic dimension extraction. An IGES (Initial Graphics Exchange Specification) file format post processor has been reported (Vosniakos and Davies 1990), which can extract the data that defines a 2-1/2D prismatic component, based on a wire-frame IGES file. However, the awkward points of the IGES standard makes it impossible to transfer the full context of a product model, unless a series of restrictions are applied. An identifier able to extract geometric information from a rotational part data file (generated using a 2D wire-frame CAD modeller) has also been developed (Wang and Wysk 1987, 1988). However, information such as tolerance, surface roughness etc is still typed interactively. The STEP protocol is being formulated by the International Standards Organisation to be the future standard in data transfer. However, since it is still being drafted, it is not yet widely accepted by industry, and incompatible with the current generation of commercial CAD systems (Sivayoganathan, Balendran, Czerwinski, Keats, Leibar, and Seiler 1993). Zhang and Mileham (1990) developed an interpreter using the DXF format for 2D rotational parts that can extract tolerance information automatically. Although all these approaches have contributed significantly to this domain, research on the topic is far from being conclusive.

Interpreting CAD Data for Prismatic Components

The work presented in this paper is part of ongoing research on a CAD Interpreter for Prismatic Components being developed at Bath University (Linardakis and Mileham 1992) aimed at minimising human intervention on the interface of CAD and CAPP. The interpreter identifies the geometric and tolerance data that define a prismatic component from a wireframe-based CAD DXF file. The DXF file format is a current industry standard, enabling data exchange between different CAD packages, hence by using it as an input, the interpreter retains compatibility. An algorithm has already been developed to extract and classify the geometric entities (lines, circles, arcs etc) of a drawing.

For a prismatic part, complete manufacturing information can be provided by a three-view drawing in 2D space using orthographic projection. However, the duplication of the same features on different views of the drawing could confuse the interpretation process. This is avoided by placing the different views of the drawing in different layers. The views are layered in a predefined convention. Since the layer that an entity belongs to is also reflected in its entry in the DXF file, the profile recognition and feature identification task becomes easier.

An algorithm has been developed to identify the external profile of a component (Linardakis and Mileham 1993). This enables the interpreter to determine the shape envelope dimensions and hence deduce the minimum shape envelope of the raw material required for the component to be machined. The objective of the research work is to then convert the CAD geometric entity information into a manufacturing feature based model, which can be used to drive a process planning system directly.

A Methodology for Extracting Manufacturing Features

An algorithm is under development to identify and extract certain external profile manufacturing features from the sorted data file of the CAD Interpreter. In order to locate and classify a feature, the Feature Extractor proceeds in the following manner:

(1) The extracted line entities and their coordinates from the FRONT view are first examined. The edges of the component are formed by the lines with the minimum and maximum X and Y coordinates :
    Line (Xmin, Ymin) to (Xmin, Ymax) forms an edge.
    Line (Xmin, Ymax) to (Xmax, Ymax) forms an edge.
    Line (Xmax, Ymax) to (Xmax, Ymin) forms an edge.
    Line (Xmax, Ymin) to (Xmin, Ymin) forms an edge.
(2) The identified edges are next checked for continuity in a clockwise direction. If there is a continuous line joining two maxima-minima points, then there is no feature on that particular edge. If however, an edge-line is interrupted, then a feature exists on that edge, and the Feature Identification program is called. This consists of several recognition algorithms for different features (currently under development), which are consecutively applied until a particular feature is identified and extracted. The next edge is then examined, and so on, until all the edges have been processed.
(3) The same method is applied to the extracted edges from the PLAN and END views of the component.

A sample component showing the methodology employed can be seen in Figure 1. In this typical application example, a through slot needs to be machined on the part. The through slot recognition algorithm is applicable here, a description of which is given in the next section.

The Through Slot Recognition Algorithm

Consider the component shown in **Figure 1**, on which a plain through slot needs to be machined. This can be seen on the FRONT and PLAN views of the drawing.

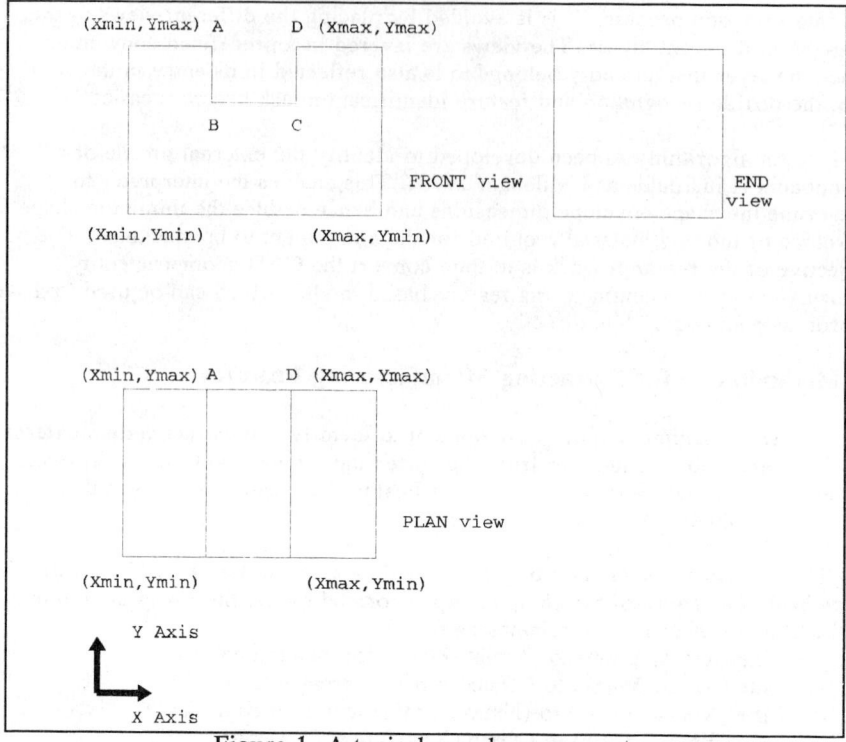

Figure 1. A typical sample component

In order to verify that the feature encountered is indeed a plain through slot, the Feature Extractor proceeds in the following manner:

(1) Point A is located (Xmin < X < Xmax, Ymax) where the discontinuity starts.
(2) Point B is found, and line AB confirmed as a continuous line (no secondary features) vertical to line (Xmin, Ymax)A.
(3) Point C is located, and line BC confirmed as a continuous line vertical to AB. Line BC should also be parallel to edge (Xmin, Ymax)(Xmax, Ymax).
(4) Point D is found, and line CD confirmed as a continuous line vertical to BC and parallel to AB.
(5) Line D(Xmax, Ymax) is found and its continuity confirmed.
(6) The PLAN view is examined. The Extractor checks to see if there are two lines starting from points A and D and finishing at Ymin, parallel to each other and to the two edges. If there are, a through slot is confirmed and extracted.

The above algorithm has been adapted to also detect a blind slot.

Conclusions

A novel approach for extracting manufacturing features for prismatic components from CAD DXF files is being developed. The Feature Extractor presented uses a sorted data file from a CAD DXF Interpreter to locate and classify a range of manufacturing features. This is achieved by consecutively applying various feature recognition algorithms (of which the through slot is discussed here as a typical example) on the component's external profile. The system is being validated over a range of prismatic components and represents a further step towards CAD/CAPP integration.

References

Brimson J.A. and Downey P.J. 1986. Feature Technology: A Key to Manufacturing Integration *CIM Review*, Spring.

Cunningham J.J and Dixon J.R. 1988. Designing with Features: The Origin of Features *Computers in Engineering*, 1.

Chang T.C. 1990. *Expert Process Planning for Manufacturing* (Addison-Wesley Publishing Company).

Choi B., Barash M.M. and Anderson D.C. 1984. Automatic Recognition of Machined Surfaces from a 3D Solid Model *Computer Aided Design*, 16, 81-86.

Linardakis S. and Mileham A.R. 1992. A CAD Interpreter for Prismatic Components *Proceedings of the Eighth National Conference for Manufacturing Research*, 21-25.

Linardakis S. and Mileham A.R. 1993. Manufacturing Feature Identification for Prismatic Components from CAD DXF Files *Proceedings of the Ninth National Conference for Manufacturing Research*, 37-41.

Mortensen K.S and Belnap B.K. 1989. A Rule-Based Approach Employing Feature Recognition for Engineering Graphics Characterisation *Computer Aided Engineering Journal*, Dec.

Pratt M.J 1988. Synthesis of an Optimal Approach to Form Feature Modelling *Computers In Engineering*.

Shah J.J and Rogers M.T. 1988. Functional Requirements and Conceptual Design of the Feature-Based Modelling System *Computer Aided Engineering Journal*, Feb.

Sivayoganathan K., Balendran V., Czerwinski A., Keats J., Leibar A.M. and Seilar A. 1993. CAD/CAM Data Exchange Application, *Proceedings of the Ninth National Conference for Manufacturing Research*, 306-310.

Vosniakos G.C. and Davies B.J 1990. An IGES Post-Processor for Interfacing CAD and CAPP of 2-1/2D Prismatic Parts *International Journal of Advanced Manufacturing Technology*, 5, 135-164.

Wang H.P. and Wysk R.A 1987. Intelligent Reasoning for Process Planning *Computers In Industry*, 8, 293-309.

Wang H.P. and Wysk R.A 1988. AIMSI: A Preclude to a New Generation of Integrated CADCAM Systems *International Journal of Production Research*, 26.

Zhang Y. and Mileham A.R 1990. A CAD Interpreter for Interfacing CAD and CAPP of 2D Rotational Parts *Proceedings of the Sixth International Conference on CAD/CAM, Robotics etc*, 333-336.

INTEGRATED QUALITY PLANNING FOR MACHINED COMPONENTS

Dr J.D.T Tannock, Mr D.R. Cox, Dr H. Lee and Mr J. Sims Williams

Faculty of Engineering
University of Bristol, Bristol BS8 1TR

The identification of, and planning for, the activities needed to control and verify quality is a key requirement in manufacturing. This paper examines the need for quality planning, and its close relationship with the outline process planning activity, then introduces an integrated approach to planning for machined components, making use of CAD to provide a feature-based representation of part geometry. Planning is performed automatically using a theoretical framework consisting of a relational algebra, a problem model for process planning, an operational model of machining, and an automated reasoning approach. The approach is capable of planning operation sequences and associated quality activities for both prismatic and axi-symmetric machined components.

Introduction

The needs of the quality function in relation to the manufacturing process have led to the application of a variety of quality planning techniques, which are used to develop and document the specifications, inspection plans and process control requirements to assure the quality of production activities. Many manufacturing companies do not carry out this quality engineering function effectively, resulting in a range of quality problems which must then be tackled on an ad-hoc basis as manufacturing proceeds.

Some software tools are available to assist with these activities, but they are usually incapable of integration with other computerised systems and fail to integrate quality requirements with process planning, which results in a time-consuming sequence of procedures, sub-optimal planning and a duplication of data and effort. Neither simultaneous engineering nor efforts to achieve quality goals using a multi-functional TQM approach are advanced by these 'islands of automation'. Increasingly, customers require to be satisfied that a supplier is in control of a capable and stable process. Such requirements should be seen as part of an integrated planning activity, and not be implemented as stand-alone activities of the quality engineering function.

Quality Planning

Quality Planning consists of the preparation of a scheme for all quality-related activities involved in the manufacturing process, reflecting quality requirements (such as linear and geometrical tolerances) as well as the nominal specifications. There is a requirement for two types of plan, the stand-alone quality control plan, and the integrated (or combined) quality plan. Both these activities are required by quality systems standards such as BS5750 and the equivalent ISO9000 series.

The Stand-alone Quality Control Plan
   This identifies all quality-related activities in an independent format, and may be used as a framework for:-
   a) Manual or automated quality data acquisition and reporting.
   b) The generation of detailed inspection plans for automated inspection systems.
   c) Process control requirements for critical process or quality characteristics.

The Integrated Quality Plan
   This must be integrated with the process plan, such that all quality activities are shown in the correct position within the detailed process plan, and relate to the features generated at each operation stage. It provides guidance for shop-floor personnel as to the quality activities required to ensure that the product conforms to specifications.

The Process Planning Activity

   While much process planning is in practice a variant activity based on previous product plans, the *ab initio* process planning function can be separated into **Outline Process Planning**, in which outline operation selection is performed and possible sequences are specified by which a component may be manufactured. and **Detailed Operation Development**, where each individual operation must be specified in terms of method, tooling, fixturing and programming required, and a full technical description, including specifications and stage drawings, provided. These two functions - combined with time and cost estimation - represent the traditional bounds of the process planning function. It is widely accepted that the key to effective automation and integration in this area is an appropriate representation of the component features.

Computer-Aided Design (CAD) and Computer-Aided Process Planning (CAPP)
   Many current CAPP approaches concentrate initially on the identification and recognition of component features - entities which have a correspondence with the accepted engineering descriptions of component geometry. Such systems tend to concentrate on detailed operation development rather than outline planning. Generative CAPP systems typically interpret CAD geometry in order to recognise features. An example of a generative CAPP system is GENPLAN (Gindy, Huang. and Ratchev, 1991).

   Several CAD vendors have recently added feature-based functionality to their products. Many CAD products are able to incorporate pre-defined standard features (for example cylindrical holes) into designs, and to group and parameterise sets of geometrical elements (such as faces) which are together identified as 'soft' features.

   An associated area is the exchange of feature-based product data. The latest version of STEP (Standard for the Exchange of Product Data) allows the communication of technical data including geometry, features, tolerances and materials between CAD packages, and to external applications. Implementation of STEP will provide a vehicle for the acquisition of CAD feature data by other applications, and may eventually render redundant the recognition of features as undertaken by current CAPP systems. However, current developments in feature-based CAD and STEP do not adequately address feature connectivity and association - the relationships between features which effect the planning task. Dealing with feature connectivity is a key aspect of CAPP system development, to provide a basis for reasoning about overall part geometry, but current approaches may be insufficient to plan for quality and inspection requirements.   Of particular importance are the problems of geometrical tolerancing, datum reference frame and datum precedence determination.

Computer-Aided Design and Manufacture (CADCAM)
        CADCAM is an established technology which uses CAD component data files to program machine tools.   The main problem with commercially available CADCAM systems is their scope, which is mainly limited to the detailed development of operations for individual machine tools.   They are not capable of performing the overall process planning task where multiple operations on different machine tools are required to complete the generation of the required component geometry.   The enhancement of CADCAM into a full process planning activity is one current direction of development.   One such generative process planning system is BEPPS-NC (Zhang and Mileham, 1991).   CADCAM systems do not tackle quality requirements, although similar approaches, using separate software applications, exist for the detailed programming of Co-ordinate Measuring Machines (CMM) operations (Hassan, Medland, Mullineux, and Sittas, 1992).   A standard for data transfer between CAD and CMMs is available, called the Dimensional Measuring Interface Specification (DMIS).   Software support for programming CMMs via the DMIS is now available from a small number of vendors.

The Acquisition of Product Data for Planning

        Product data must be acquired from feature-based CAD applications and represented in a form which provides sufficient, accessible, information to allow automated planning to proceed. Computervision CADDS5 is used at present, with the aid of the CV-DORS developers resource software.

        Boundary representations (BReps) as generated by many CAD systems, are not suitable as primary models, as they are unable to provide adequate local information concerning individual functional or manufacturing features.   They do, however, provide valuable global information about the overall geometry of a component.   This information is essential for the quality control task, where features will be related together by both linear and geometrical tolerances.   Un-evaluated volumetric representations, such as Constructive Solid Geometry (CSG), provide useful information concerning local volume features, but do not define the topological relationships between features in the context of global part geometry.

        In the absence of methods for the extraction of information about volume features from BReps, a dual approach to feature representation has been developed, consisting of an abstraction of both CSG and BRep primitives.   This approach is implemented using a 'design by feature' front-end to the CAD system, utilising the CAD system BRep engine for evaluation. Shells, which are equivalent to feature boundaries, define the geometric entities that are combined, using operators, into the product.   Both regular (CSG) and irregular (BRep) shells can be used.   The 'stock shell' (which may be an irregular casting or forging as well as a regular shell representing, for example bar stock) is related to the product shell using a range of sorted binary operators for the union or subtraction of volumes.   Product faces are mapped to feature or stock faces.

The Planning Approach

        To develop a quality plan from scratch for a component, it is necessary to undertake a basic level of outline process planning, so that the sequence of machining operations which generate quality features can be identified and represented. The basis for the planning task has been provided by the development of a theoretical framework consisting of the following elements.

The Relational Algebra

Planning relies on successful reasoning with feature representations, taking into account overall part geometry and associated topological constraints. A new relational algebra has been developed, providing the basis for the explicit description and computation of feature relationships and topological constraints, and allowing the development of operation sequences that can be converted to outline process plans. The feature relationships used are adjacency, intersection and ownership.

The Problem Model for Process Planning

The outline process planning approach is based on the proven 'reverse' strategy. Planning starts with the finished part by identifying all possible features to remove. Part geometry is changed by conceptually removing a feature, and identifying the resulting geometry, until the raw material geometry is recovered. A number of potential operation sequences may be produced from the above process. An operation sequence can be successfully developed into a feasible outline process plan only if it can be regularised against a set of topological and practical machining constraints. An operational model of machining has been developed to ensure that feature operation sequences are feasible.

The Operational Model of Machining.

A feasible operation sequence for machining must consist of a consistent series of conceptual removals. Ensuring the correctness of individual conceptual removals is not sufficient to ensure the satisfaction of the overall part geometry, owing to topological constraints between features. In the relational algebra, the necessary feature relationships have been defined to describe topological constraints. These relations are used by the automated reasoning approach to regularise feature operation sequences. This process is known as **topological regularisation**. In addition to topological constraints there are many other practical and machining constraints to which outline plans must conform. Any sequence must also be regularised according to these constraints before it can be developed into a feasible plan. This process is referred to as **machining regularisation.**

The Automated Reasoning Approach

A formal procedure is used for planning, and automated reasoning techniques have been developed to implement this procedure using a generic deduction system. A deduction system for automated planning was previously developed by the authors, (Lee, Sims Williams and Tannock, 1992) based on a synthesis of STRIPS formalism and a production (rule-based) system. This approach had the advantages of rigorous description of actions, easy implementation and efficient problem-solving. However any change of action operators required the time-consuming rewriting of production rules.

In the current system, the deduction of process sequences is performed by representing the topology of the finished part as a logical sentence, and attempting to prove that a sequence of transformations exists by which the topology may be changed to that of the raw material. The proof is obtained by a sequence of logical operations which if attained represent a potential operation sequence. The adoption of generic deduction, based on the Resolution Principle, improves the behaviour of the reasoning model, and enhances and maintainability. As uniform theorem proving techniques, all resolution procedures based on this principle use a single rule of inference, the resolution rule, to perform deduction. It is possible to convert all logical formulae to the conjunctive normal form which involves the smallest possible number of logical connectives (and, or, not), and reduces the number of rules.

Integration

Outline process planning will be performed when the features and geometry of the product are identified.   An outline operation plan is specified, indicating which features are to be generated and inspected at each process stage.   Interim manufacturing and inspection features may be required which are not represented on the product specification.   These features will be defined at this stage, so that suitable graphical representations (stage and inspection drawings) may be generated using the CAD system.

The integration strategy involves the export by the planning system of one or more feasible outline process plans to an existing detailed process planning application.   The user of this system will select the most appropriate outline process plan, and make modifications as appropriate.   The selected operation sequence will be returned to the process and quality planning system, where it will provide a basis for the generation of the detailed quality plans and the definition of intermediate geometry.

Conclusions

The research introduced above provides a practical, integrated approach to quality planning, capable of planning quality activities for a wide range of machined components in the context of the process plan.   The system uses a CAD system to support feature representation. The approach to outline process planning involves a novel method of automated reasoning, which may be applicable to other areas of automated planning

Acknowledgements:

The work described here is funded by the ACME Directorate of the Science and Engineering Research Council.   Support has also been received from Computervision and CIMTEL Ltd.

References

Gindy N.N.Z., Huang X. and Ratchev T.M. 1991, Feature-based Component Model for Computer Aided Process Planning Systems, *Proc. Symposium on Feature-Based Approaches to Design and Process Planning, Loughborough University of Technology.*
Hassan J.H., Medland A.J., Mullineux G, & Sittas E. 1992, *An Intelligent Link between CAD and Automatic Inspection*, Proc. 29th Int. MATADOR Conf.
Lee, H, Sims Williams, J.S. & Tannock, J.D.T., 1992, 'Knowledge-based Inspection Planning', *AI-EDAM* 6,3.
Zhang Y. &Mileham A.R. 1991, BEPPS-NC: An Automated Generator of NC Part Programs for Rotational Parts from CAD Product Model, In D. Spurgeon & O. Apampa (eds.), *Advances in Manufacturing Technology VI)* Hatfield Polytechnic.

SOFTWARE DESIGN
FOR SEQUENCING MACHINING OPERATIONS

Dr. Zhengxu Zhao and Professor Ray W Baines

School of Engineering, University of Derby
Kedleston Road, Derby DE22 1GB

Sequencing in generative computer automated process planning
(CAPP) is a software design problem exasperated because of the
empirical information and knowledge used and the fact that the
optimisation of processes is based on multiple criteria on
interrelated parameters. To cope with these issues, a novel method
of software design has to be considered. In essence, software is
designed to model the changes of features as machining
progresses based on a flexible component concept. A two-
layered optimisation model to facilitate software design in
achieving the optimal sequences of machining operations against
minimum machining time and cost results.

Introduction

Modelling sequencing requires that criteria on which a sequence
depends has to be specified but in practice few common rules exist. In
traditional process planning, sequencing machining operations is usually carried
out to meet particular company-orientated requirements, thus most criteria
become company-orientated. However, in general a sequence of operations can
be considered to depend on criteria as follows.

Firstly, the sequence must produce the required geometry, structure,
surface relationships and technical attributes. We call the sequence at this phase
an absolute sequence, owing to the fact that certain operation sequences or
patterns tend to show up time and time again. For example, the tolerance on a
given hole might require it to be reamed, before reaming can begin, the hole
must be drilled. In this case, a operation sequence of drill-then-ream is
established. Virtually all machining operations tend to follow such standard or
operation patterns, Curtis (1988).

Secondly, the sequence is further refined to meet machining time and
cost criteria. At this stage, the sequence is referred to as a relative sequence
because any single change will effect time and cost.

Finally, the sequence has to be validated according to the overall
production cost and resource available. It is a well-known fact that the decisions
and information provided by process planning determine the flexibility of the

succeeding production activities. To increase flexibility and reduce lead time, a process plan should contain built-in alternative operations ready to respond to any violation or adjustment of the plan when pre-requisites are changed due to, for example, machine failures, Larsen (1993).

There is no sense in considering machining time and cost or production flexibility if a sequence of operations fails to produce the primarily specified shape, size and technical requirements of a component, therefore an absolute sequence should be built first and then relative sequence can be considered. This paper will focus on the establishment of absolute sequences and the optimisation of those sequences against minimum machining time and cost to form a relative sequence.

A flexible component

Components change their geometrical features, feature relations and other technical attributes (such as surface roughness and tolerances) while they undergo a sequence of machining operations. A series of intermediate states are passed through. Linking processes to intermediate states is an essential requirement for representing processes and sequence of operations, Vancza and Markus (1993). If the initial state of the component is raw material (a metal block or bar) and the final state is the finished component, it can be considered as a flexible component changing its shape, size and other technical attributes through a sequence of machining operations as in Figure 1.

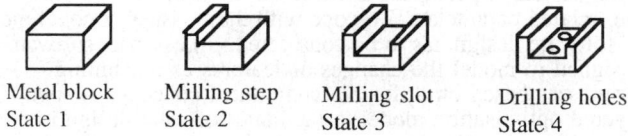

| Metal block | Milling step | Milling slot | Drilling holes |
| State 1 | State 2 | State 3 | State 4 |

Figure 1. Illustration of flexible component

State tree and state paths

Each feature in the flexible component is transformed by an operation from its previous state to the next state until the final state is reached, Zhao (1992). If the process plan is formed by backward chaining, it can be represented by a state tree as shown in Figure 2.

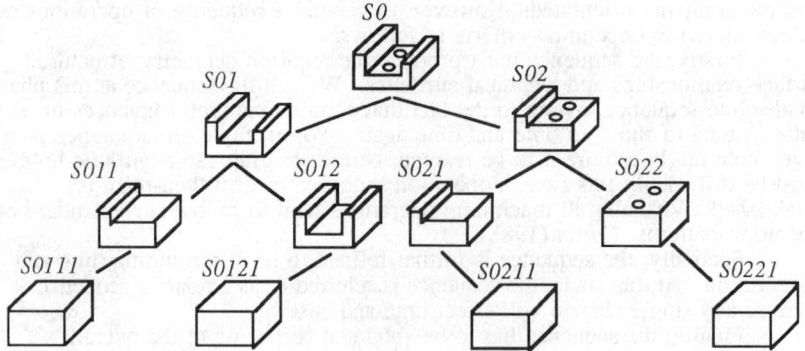

Figure 2. State tree and its representation

The root of a state tree *S0* is the final state (the finished component) of the flexible component and the leaves *S0111, S0121, S0211* and *S0221* are the initial states (here they are the same block). The nodes *S01, S011*, etc. represent the intermediate states and the lines between every two nodes represent a possible machining operation between the two states. A state tree represents the alternative ways of machining the component from the initial state to the final state, each alternative is represented by a path between a leaf and the root in the tree. For example, the path *S0-S01-S011-S0111* shows that the component has evolved from a block to a step, then to a slot and finally to a hole nested in the slot. Such a path between a leaf and the root in a state tree is called a state path. Consequently, if a machining operation for each state in a state path is specified, then a sequence of operations can be automatically achieved. Since there are normally different state paths in a state tree, different sequences of operations can be formed for a single component. These sequences are absolute sequences and can provide a specific sequence domain for process optimisation to select an optimum or relative sequence according to specific criteria such as minimum machining time and cost.

Precedence of machined features

Each absolute sequence of operations corresponds to a state path in a state tree. A state path can be formed based on machining precedence of features. If feature *a* should be machined before feature *b*, then feature *a* has greater precedence than feature *b*.

In a state tree, a feature of greater precedence should always appear in a lower level than those with less precedence (assuming the root state is at the highest level). Therefore, the precedences of the states in a state path are decreasing from the leaf to the root. For example, a thread can only be cut before a cylindrical surface is prepared first. When a state path is to be formed, the cylindrical surface should always be placed at the lower level than that of the thread.

Sequencing model in software design

Once a state tree has been built, absolute sequences of machining operations can be formatted by following the state paths which are sequences of feature states arranged according to their precedence. If machining operations for each feature state is specified, then an absolute sequence of operations result. For each absolute sequence, the actual machining time and cost can be calculated and attached to the sequence as parameters of optimisation. Thus two searching spaces in terms of time and cost are established called sub-domains of sequences that relate to a particular state path. From the sub-domains of all state paths, two major searching spaces in relation to the overall state tree can be formed called the major domains of sequences, see Figure 3.

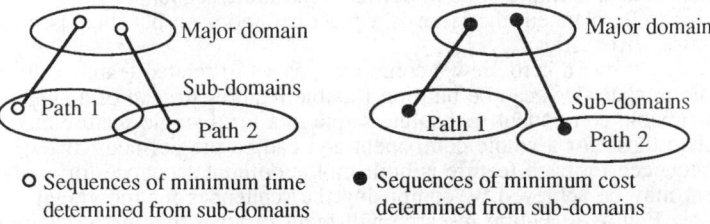

○ Sequences of minimum time   ● Sequences of minimum cost
determined from sub-domains   determined from sub-domains

Figure 3. Domains of sequences for optimisation

Practically, the state tree can be any number of state paths and, for each path, there can be many different sequences of operations because each feature state can be machined by different operations. For example, an endface can be machined by a lathe as well as a milling machine. The number of state paths poses a great problem in terms of memory and computing time.

To help this, a two-layered data model is used, see Figure 3, the lower layer contains absolute sequences related to individual state paths. Optimisation is first carried out within the two sub-domains of each state path to find the sequence with minimum time and cost. When such a sequence is found, it is immediately stored in the upper layer of the model. The upper layer therefore consists of all sequences with minimum time and cost extracted from all individual state paths. At this stage, the two major domains are significantly reduced in size. Virtually, they become two separated sets of sequences with the number of the sequences in each set being equal to the number of state paths in the state tree.

Conclusions and future work

Sequencing of operations is treated in this paper mostly in an optimisation model that simply stores and sorts sequences in individual files (each file is a sub-domain). These files contain absolute sequences with the time and cost calculated and are created during the early stages of process planning as temporary buffers or formed in a process planning data base as permanent files. If these files take the format of a relational data base file, the optimisation at the sequencing stage becomes a procedure of searching for a record with minimum time or cost in a file.

The approach overcomes problems imposed by limitation of computer memory and computation speed, making implementation possible on PC's. The model makes all optimisation occur locally, the lower layer optimisation is carried out purely with independent sub-domains and the upper layer optimisation is performed in two separated sets. This situation allows a 64KB re -usable memory to be used for loading a state tree, a sub-domain or a major domain during the entire optimisation process. Since the size of all domains can be controlled and, more importantly, the optimisation problems in each domain are linear (searching a time or a cost against the index numbers of sequences in a file) , the optimal results can be found quickly.

In this way, a component is considered as being able to evolve from an initial state to the final state by taking a number of alternative potential sequences of operations to form the optimisation domains. Because the component as a whole is viewed as a flexible object changing its shape, size and technical attributes, the absolute sequences have to be determined in advance. From the CAPP system developer's view point, absolute sequences may be manually defined and permanently embedded in the system or may be automatically generated by a program during planning. The former solution is easy to reach but it is impractical to define all absolute sequences for all components to be planned; the latter is a practical approach but increases the programming difficulties.

A way forward is to view a component as set of related features. In this way, different state trees can be built on flexible features instead of a single state tree on a flexible component. Absolute sequences for a single feature are simpler than those for a whole component and can be pre-defined. If the optimal sequence for each feature is built in, the optimal sequence for a whole component may be achieved by composing the sequences for individual features, see Figure 4. Future research will focus on investigating this approach to provide a more effective model for sequencing problems.

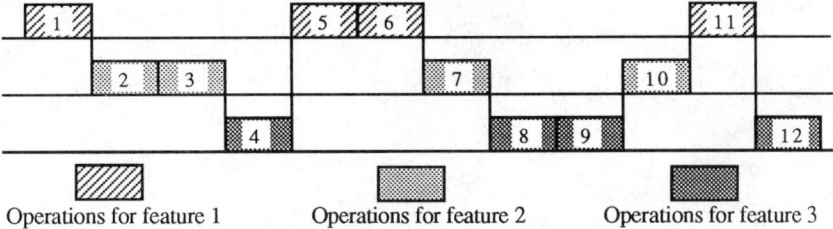

Operations for feature 1   Operations for feature 2   Operations for feature 3

Figure 4. Machining sequences based on features

References

Curtis, M. A. 1988, Process planning, (John Willey & Sons, New York).

Larsen, N. E. 1993, Methods for integration of process planning and production planning, *International Journal of Computer Integrated Manufacturing*, **6**, 152-162.

Vancza, J. and Markus, A. 1993, Features and the principle of locality in process planning, *International Journal of Computer Integrated Manufacturing*, **6**, 126-136.

Zhao, Z. 1992, Generative process planning by conjugative coding design and manufacturing information, *PhD Thesis* (Staffordshire University, Stafford).

Systems integration and modelling

AN INTELLIGENT SCHEMA FOR
MANUFACTURING SYSTEMS INTEGRATION

Allan Hodgson

MSI Research Institute
Loughborough University of Technology
Loughborough, LE11 3TU, UK

Computer integration across the functional areas of manufacturing
enterprises is resulting in the loss of human experience and commonsense
from information-related activities. This paper proposes that an intelligent
layer, within a three-schema distributed information architecture, could
provide the basis for an advanced integrating mechanism incorporating some
aspects of the above human qualities. The paper describes research to-date
towards an intelligent layer. This research utilises the conceptual graph
semantic network formalism, which is briefly described. In addition,
relevant work of other conceptual graph researchers is commented upon.

Introduction

Companies in the high-variety discrete batch production sector have not benefitted
fully from recent information technology advances. Improvements within individual
functional areas have have seldom resulted in company-wide synergy.

Current trends towards increased integration of information systems are resulting in
the removal of personnel and the loss of their commonsense and experience; this is
causing information quality problems. Concurrent engineering approaches, with shorter
time frames and incomplete information, put further pressure on information quality.

An enterprise-wide intelligent schema

General knowledge about engineering science, production capabilities, financial
constraints, and organisational practice could, if available in an appropriate form, provide
integrity and reasonableness checking for decision-support. There is a need for an
enterprise-wide 'commonsense' form of intelligence, as opposed to local expert systems,
to exploit the above knowledge. A logical place for this is at the global conceptual
schema level of a three-schema information architecture, as illustrated in Figure 1.

The three-schema approach has been incorporated into the integration
methodologies of the Loughborough MSI Research Institute and, longer term, it is
intended that the Institute's integration software will be exploited to create a full
intelligent schema.

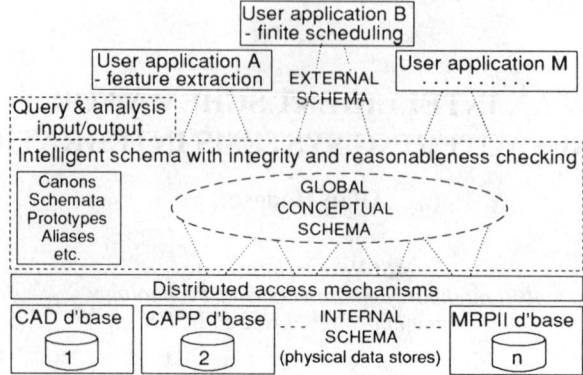

Figure 1: Intelligent conceptual schema

The artificial intelligence formalism selected for the schema

The conceptual graph semantic network formalism, as presented in Sowa (1984), was chosen as the vehicle for the representation of the intelligent schema. A more detailed description than that presented below can be found in Hodgson (1993).

Conceptual graphs consist of concepts linked via relations. A concept can represent a physical or abstract object, state, activity, attribute or even a complete conceptual graph (a subgraph); relations indicate the way in which concepts are connected or related. In graphical form, concepts are represented as rectangular boxes, relations as circles. Figure 2(a) states that the agent of milling is operator J_Smith, the instrument of milling is vertical_mill VM17 and the patient of milling is a part, i.e. 'J_Smith mills a part on vertical mill VM17'. Figure 2(b) presents the linear form of the same graph.

Main elements of conceptual graphs

The *type hierarchy* lists hierarchical relationships between concepts, i.e. supertypes and their subtypes, see Figure 2(c). A concept may have more than one supertype. The *relation hierarchy* is similar to the type hierarchy, but applied to relations, see Figure 2(d). A *type definition* defines a concept in terms of a supertype, as shown in Figure 2(e), which states that an operator is an employee with a machine skill, who is located in a production department. *Schemata* provide optional defaults indicating common associations and situations of concepts, see Figure 2(f); there may be several schemata for a concept. *Prototypes* describe typical instances of concepts, see Figure 2(g).

Conceptual contexts constrain the way in which the concepts stated in the context are related; a negative context can also be used, see Figure 3(a,b). *Relational contexts* define the relations in terms of the concepts to which they relate, see Figure 3(c,d).

The *individual catalog* declares specific instances of concepts, e.g. 'BATCH::#27980' states that an object with the unique identity number 27980 conforms to the concept 'BATCH'. The *composite individual catalog* provides details about individuals identified in the *individual catalog*.

Simple graphs may be added to form a conceptual graph knowledge base.

Conceptual graph processing rules

These rules, or operations, enable the manipulation of the knowledge base to generate new graphs and to ensure that the knowledge base remains internally consistent as further knowledge is added to it. The rules include *copy, restrict, join, simplify, project, expand* and *contract*. When presented with an information item, an idealised conceptual graph processor would use appropriate combinations of the above rules to decide that the information item is known (or can be deduced), is false, is unknown (and can therefore be inserted), or that it exposes inconsistencies in the knowledgebase.

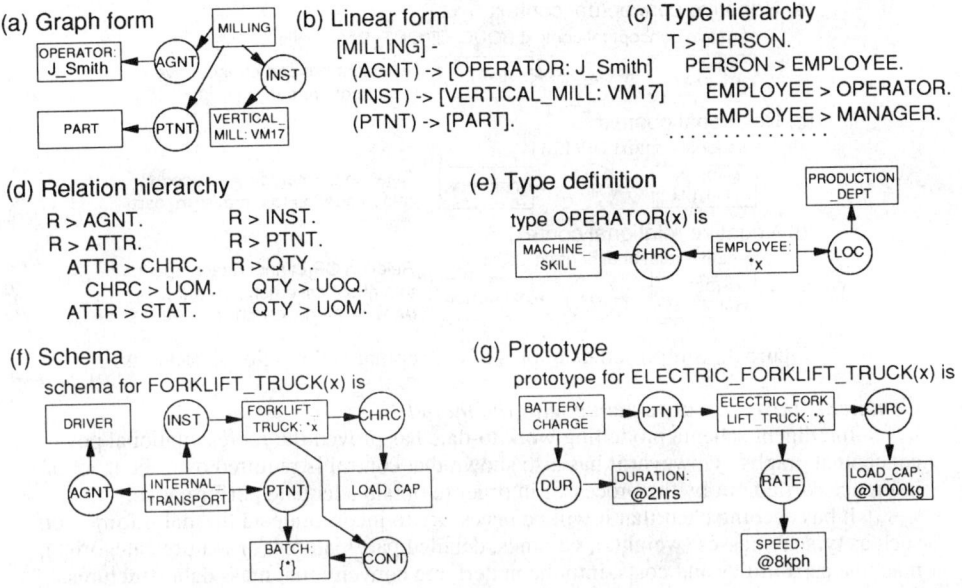

Figure 2: Conceptual graph features

Progress to-date of the intelligent schema research

Development of a prototype conceptual graph processor

The priorities of a manufacturing environment are different to those of the natural language domains representative of many conceptual graph applications. There is less need in the proposed intelligent schema (at least in the short term) for complex modal logic and highly nested contexts; however, features such as relation hierarchies are required to enable the declaration of concise canonical definitions.

The prototype processor incorporates type and relation hierarchies, type definitions, conceptual, relational and negative canons, schemata, prototypes, conformity relations and composite individuals. Processes currently available include limited validation of new graphs and answering simple queries using a form of the projection operator. Input and output is in a linear form similar to the example of Figure 2(b). The processor also features a set of built-in temporal relation operators based on the work of Allen (1983).

Development of a conceptual graph based model

The initial modelling work has concentrated on the production planning and control environment. The model contains details of parts (products and components), machines, workcentres, operators, worksorders, their attributes and relationships with each other.

Some work on the incorporation of terms and definitions of the STEP standard has been carried out, centred around the description of units and dimensions of measure as in draft standard ISO/DIS 10303-41 1993, section 23: measure_schema (ISO 1993).

The built-in temporal relation operators enable some degree of automatic validation of any time-related information, but to-date only very simple temporal components have been incorporated in the model.

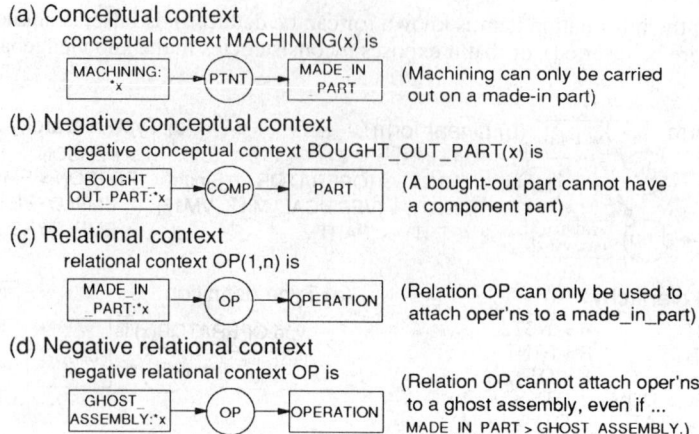

Figure 3: Simplified examples of conceptual and relational contexts

Comments based on experience with the model

Intelligent schema modelling work to-date has proved the representational power of conceptual graphs. However it has also shown that control is required over the level of inference carried out by the processor in order to obtain adequate performance.

It has become clear that it will be necessary to incorporate additional information, such as typical masses (weights), volumes, detailed part features (or feature categories), machine capabilities and costs, into the underlying conventional mass data structures.

Further work on the model

The major medium-term work is concerned firstly with limited development of the conceptual graph processor to enable access to a three-schema information architecture. If an appropriate processor can be obtained elsewhere, the above work may be significantly reduced. Secondly, the product and factory models will be extended.

There is a requirement to incorporate a learning/generalising capability. The conceptual graph 'prototype' provides representative examples of a composite object (such as a part or machine), i.e. it contains typical values for the combination of features (concepts) which make it up. However, as such feature values are seldom completely independent of each other, these prototypes are of little use. It is intended therefore to incorporate some form of pattern recognition capability to utilise these prototypes.

Other conceptual graph developments of relevance to manufacturing

Conceptual graphs offer considerable potential as intermediary representations in data modelling applications. Examples of research in this area include the information schema representation of Creasy and Ellis (1993) and mapping between EXPRESS (the information modelling language of the STEP standard) and conceptual graphs (Wermelinger and Bejan 1993). The American National Standards Institution (ANSI) has proposed conceptual graphs as the basis for the normative language of its conceptual schema for information resource dictionary systems (IRDS), see ANSI (1992).

Conceptual graph theory is being combined with other formalisms to provide it with learning mechanisms, for example the neural network research of Lendaris (1990).

One of the major limitations at the present time is the lack of a general-purpose conceptual graph software toolset. A major international collaborative initiative, the PEIRCE Project, see Ellis (1993), was set up in 1992 to produce a non-proprietary set of robust, portable conceptual graph standards and tools. These include programming and documentation standards, graphical and linear notation editing, conceptual catalogs/-ontologies, inference mechanisms, learning mechanisms, natural language parsers, parallel hardware and vision systems. In addition, the Loughborough University Conceptual Graphs Research Group is developing theoretically rigorous conceptual graph processors, with additional graphical representation/editing tools and efficient storage structures, see Heaton and Kocura (1993) and Bowen and Kocura (1993).

Conclusions

To take advantage of the flexibility that intelligence brings to CIM, it must be re-introduced at the conceptual schema level, rather than as a collection of localised functional elements.

The work outlined in this paper has demonstrated the potential of conceptual graphs for the representation of an intelligent schema, and will therefore be developed further. Conceptual graph implementations are still at a developmental stage, but current international research will result in practical conceptual graph tools and applications during this decade.

References

Allen, J.F. 1983, Maintaining knowledge about temporal intervals, Communications of the ACM 26, no.11 (Nov), 832-843.

ANSI 1992, Task Group X3H4.6/92-091 IRDS Conceptual Schema, American National Standards Institute.

Bowen, B., Kocura, P., 1993, Implementing conceptual graphs in a RDBMS. In G.W. Mineau, B. Moulin, J.F. Sowa (eds.), *Lecture Notes in Artificial Intelligence 699: Conceptual graphs for knowledge representation*, (Springer-Verlag, Berlin) 106-125.

Creasy, P., Ellis, G., 1993, A conceptual graphs approach to conceptual schema integration. In G.W. Mineau, B. Moulin, J.F. Sowa (eds.), *Lecture Notes in Artificial Intelligence 699: Conceptual graphs for knowledge representation*, (Springer-Verlag, Berlin) 126-142.

Ellis, G., 1993, PEIRCE user manual. Available from Gerard Ellis, University of Queensland, Australia, email: ged@cs.uq.oz.au.

Heaton, J.E., Kocura, P., 1993, Presenting a Peirce logic based inference engine and theorem prover for conceptual graphs. In G.W. Mineau, B. Moulin, J.F. Sowa (eds.), *Lecture Notes in Artificial Intelligence 699: Conceptual graphs for knowledge representation*, (Springer-Verlag, Berlin) 381-400.

Hodgson, A. 1993, Conceptual graph based manufacturing systems integration, In M.M. Ahmad, W.G. Sullivan (eds.), *Proc. Third Int. Conf. Flexible Automation and Integrated Manufacturing*, (CRC Press, Boca Raton) 545-556.

ISO 1993, Draft international standard ISO/DIS 10303-41 Industrial automation systems - Product data representation and exchange - Part 41: Fundamentals of product description and support, International Organization for Standardization.

Lendaris, G.G., 1990, Conceptual graphs as a vehicle for improved generalisation in a neural network pattern recognition task. In *Proc. 5th Ann. Workshop on Conceptual Structures*, Boston, USA, 90-91.

Sowa, J.F., 1984, *Conceptual Structures: Information Processing in Mind and Machine*. (Addison-Wesley, Canada).

Wermelinger, M.L., Bejan, A., 1993, Conceptual structures for modelling in CIM. In G.W. Mineau, B. Moulin, J.F. Sowa (eds.), *Lecture Notes in Artificial Intelligence 699: Conceptual graphs for knowledge representation*, (Springer-Verlag, Berlin) 345-360.

A USER-FRIENDLY ENVIRONMENT IN THE CREATION OF PDES/STEP BASED APPLICATIONS

Mr A.S Czerwinski, Mr. J. Drysdale, Dr. K. Sivayoganathan

Manufacturing Automation Research Group,
Department of Manufacturing Engineering,
The Nottingham Trent University,
Burton Street, Nottingham, NG1 4BU.
Email: CD152ASC01@uk.ac.ntu

With the development of the international STEP standard, there now exists a method that will allow information to be transferred from initial concept all the way to customer delivery. Problems arise when straight data transfer is not the goal, and the data is required as the basis of a Concurrent Engineering Environment. In this case, other applications must also be able to query the data. The creation of these programs is a major task, as the data structures developed in the EXPRESS language become very unwieldy, and only someone who is well versed in the part of the STEP standard of interest would be able to fathom it's complex structure. This paper proposes an environment that will allow non-PDES/STEP users to access and query these data files, and so simplify the creation of application programs that will run alongside the CAD packages.

Introduction

The area of application programming within the Engineering field have always had to fight to overcome many obstacles, mainly created by the vendors of Computer Aided Engineering (CAE) packages. The greatest of these barriers was the vendors insistence on using dissimilar information structures to store and transfer their data. This has meant that any application programmer who wanted to use any information created by a CAE package has had to work closely with the vendor to know how the information was stored, and how to access this data.

With the current slant on Concurrent Engineering, a greater emphasis has been placed on the availability of a consistent information source that can be used continually throughout a products life-cycle [Bloor, M.S, Owen, J. (1991)]. This has been achieved by the use of the Standard for The Exchange of Product Data (PDES/STEP), which is a neutral data transfer specification, and independent of any particular computer systems language. It also provides a new methodology for the visualisation and creation of information structures.

This paper concentrates on the development of a simple generic query program that will be used to develop CAE applications. The need for such a package becomes apparent when several models are used in conjunction (such as a Manufacturing Model and a Geometry Model in the case of a Computer Aided Process Planning application), creating very complex information structures. Unless specific knowledge of EXPRESS, and the toolkits which are used to access this information, is available, the programming of applications then become a complex problem in themselves.

The EXPRESS language and it's use in the PDES/STEP standard

The development of the STEP standard also required a method of modelling the information relationships that would facilitate the task of data transfer and storage. The EXPRESS [NIST] information modelling language was therefore developed, and is used throughout the standard.

EXPRESS provides an object-orientated, integrated view of the product data. It supports a schema concept, so that any Part or section of a Part of the PDES/STEP standard can be split into different areas of interest. Each schema contains entities, which in turn contain attributes for data and behaviour.

EXPRESS-G is a graphical sub-set of EXPRESS. It has been specifically developed for the graphical representation of the information models defined in the EXPRESS language, to enable visualisation of the various links between entities. Using EXPRESS-G, an information model is represented by graphic symbols forming a diagram. Basic symbols include: entities (rectangles); optional attribute links (dashed lines); supertype/subtype links, which are similar to parent/child links where all the parents information is also passed to the child, (thick solid lines); all other links (normal lines). The direction of the relationship is shown by a circle (something like an arrow head). All other available symbols are described in more detail in part 11 of the standard, and includes aggregate, select and enumeration types.

In reality, the important information is held in two ASCII files. The first, and most important part of PDES/STEP, is the Application Model. This model is written in EXPRESS and combines the various reference models developed within the PDES/STEP standard. For example, a CAD Application model may require Tolerance information, Geometric information as well as Form Feature data, plus other standard schemas. This information structure is then used to check and create the second ASCII file, the Physical Files. These files hold the data pertaining to a certain part, at a certain instance in time. This is an important point to note. The entire PDES/STEP structure was developed to create snapshots of the part throughout its life span, and is not a dynamic data storage system. This must be remembered when creating applications, especially if Concurrent Engineering is a goal of the user.

Accessing The Information Structures

Complex information structures defined within the EXPRESS environment are very difficult to understand, unless a deep knowledge of the structure has been achieved. Application programming is therefore a specialised task. The question arises, how do you extract, manipulate and use the information found within these physical files [Urban, S.D, Shah, J.J, Rogers, M.T. (1993)]?

One way is to create specific programs to interact with the information, probably using one of the many Express Toolkits available on the market (National Institute of Standards and Technology's fedex-plus, PDES/STEP Tools Inc.'s EXPRESS2C++ convertor). With these tools, the EXPRESS schema for the Application model is converted into an appropriate language which would be used to manipulate the Physical File information.

As EXPRESS is a relatively Object-orientated information methodology, the languages that the schemas are changed into are mainly C++, CLOS, Smalltalk and Modula. Depending on the toolkit the information is either stored interactively within the computer, or uses an object-orientated database to hold the data.

With any of the above methods, the application programmer would need to know which information structures and entities are defined within the EXPRESS schema. This problem is then doubled if a further schema is also required to be accessed within the application. It would therefore be difficult for a non-EXPRESS user to sit down and interrogate the physical files.

Depending on the toolkit employed, the methods that are used to access the information are also fairly complex. A specific knowledge of the toolkit would therefore also be required before any applications could be developed.

It is envisaged that some form of generic tool would be required to query the physical files, allowing non-EXPRESS users to develop applications that utilise the physical file information. This tool would be built upon one of the available EXPRESS toolkits, employing an Object-Orientated programming language, and using an X-based graphics toolkit for extra data input/output, display of the information structures, as well as a means for querying the information within the physical files.

The Application Programming Environment (APE)

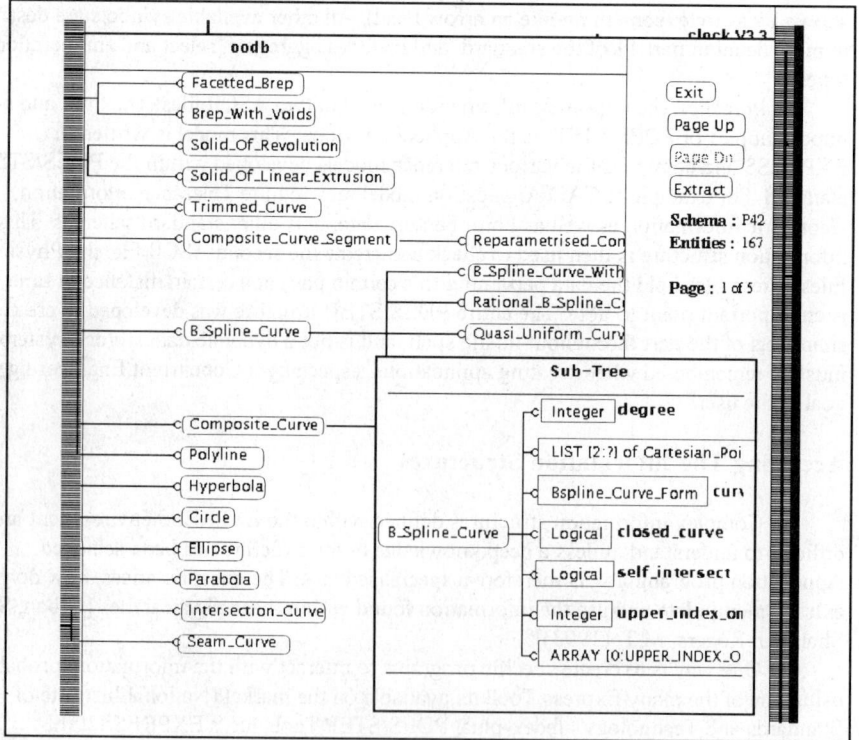

Figure 1 Subtype/Supertype tool, with attributes window

The APE is split into two fundamental areas; the Programming Environment itself, and the EXPRESS data tool. The EXPRESS data tool is a means of displaying the information structure of a certain schema, and is totally independent of the Programming tool. This allows the programmer to access as many schemas as he likes, without the problem of compiling a new APE for each separate application.

This part of the tool is graphically and icon driven, using a pseudo EXPRESS-G structure. The initial display (shown in figure 1) is of the subtype/supertype links from a root entity, that is to say an entity with no supertypes. Using this display, it is possible to access various entities and display their attributes, as well as any attributes passed down the tree structure from it's parents. The tool will also allow the user to extract the structure of the entity's data which would be used in the created application program.

The Programming Environment will allow the programmer to develop the application in a structured manner (see figure 2). A limited number of commands would allow the user to create subroutines (in C++) that would manipulate and use the data within the physical files associated with a certain schema. The commands include looping, if/else/then statements, mathematical manipulation, checking of physical file information to make sure that the required entities for the application are present, development of forms for extra data input, creation of windows for data output, storage of information in new physical files for use in future applications and allow for queries to more than one physical file at a time.

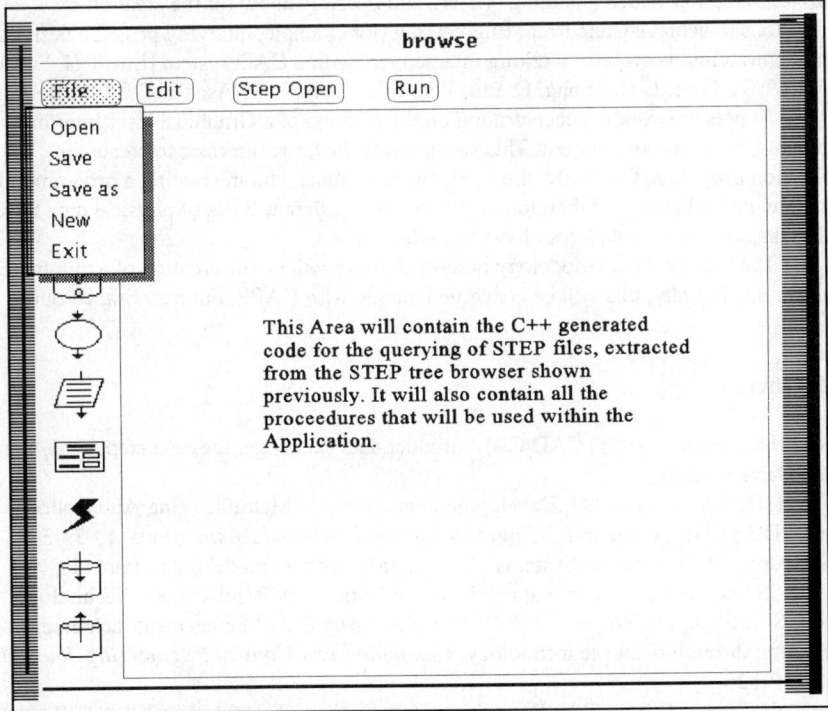

Figure 2 Application Programming Environment

The final application program created will be standalone, that is to say it will be a combination of all the functions created within the APE in C++, any NIST toolkit functions used to access the STEP data, as well as any Query functions developed during this research.

Conclusions

The above technique for the creation of application programs frees the user from the difficulties of :

1. Understanding the EXPRESS language, it's creation and how the information is structured.

2. Understanding how the NIST EXPRESS Toolkit used manipulates, accesses, stores and uses the Physical File data.

It will allow for the creation of any application with the minimum of knowledge of EXPRESS. Possible application areas include; the creation of Coordinate Measuring Machine programs from the data stored within the PDES/STEP Physical Files ie. Computer Aided Inspection (CAI), the creation of Robot Programs for the manipulation of components within an automated environment (using Geometry, Form Features, Robot Manipulator Data) ie. Computer Aided Handling (CAH), and possibly allow for the creation of applications to achieve Concurrent Engineering (for example, querying possible designs for manufacturability, costs, etc) working interactively with a CAD system [Liu, T.H, Fischer, G.W. (1993), Qiao, L-H, Zhang, C, Liu, T-H, Ben Wang, H-P, Fischer, G. (1993)].

At present work is concentrating on the creation of a Graphical User Interface employing an X window Toolkit. This will provide the basic interface to create the application programs. Currently, the work has been successful in creating a program link to more than one schema, and therefore to one or more different types of physical data files, as well as displaying the data trees of various schemas.

The test for the methodology described above will be the creation of a number of applications. Initially, this will be concerned mainly with CAPP, but may branch out to CAI, or CAH.

References

Bloor, M.S, Owen, J. 1991, CAD/CAM product data exchange: the next step, *Computer Aided Design*, **23(4)**.

Liu, T.H, Fischer, G.W. 1993, Developing Feature-Based Manufacturing Applications Using PDES/STEP, *Concurrent Engineering: Research and Applications*, **1**, 39-50.

NIST Industrial Automation Systems - Exchange of product model data - Part 11: EXPRESS language reference manual, National Institute of Standards and Technology.

Urban, S.D, Shah, J.J, Rogers, M.T. 1993, Engineering Data Management: achieving integration through database technology, *Computing and Control Engineering Journal*, **4(3)**, 119-126.

Qiao, L-H, Zhang, C, Liu, T-H, Ben Wang, H-P, Fischer, G. 1993, A PDES/STEP based product preparation procedure for computer aided process planning, *Computers in Industry*, **21**, 11-22.

INTEGRATION AND INTERFACING USING EXPRESS MODELS

Shahin Rahimifard and Dr. Stephen T. Newman

Loughborough University of Technology
Loughborough, Leics, LE11 3TU

This paper reports on the use of the formal data specification language EXPRESS, for the interfacing of scheduling and simulation systems used for modelling machining cells. It considers various options for the transfer of data between different software systems, with the use of STEP standards. This work has been carried out as a part of the Eureka-FORCAST research programme and uses two software packages developed from this programme, namely the SIMAN Tool Issue Strategy Selector (STISS) and a computerised scheduler (PREACTOR). The first package identifies the best cutting tool assignment strategy for manufacturing a defined set of parts in a flexible machining cell, and produces part and cutting tool information to be input into PREACTOR. This paper describes a novel approach for transfer of this information using EXPRESS, a relational database and SQL.

Introduction

Information systems are increasingly playing an important role towards the integration of CAD/CAM software. Information models are constructed to understand the requirements, specification and the functionality of such information systems. Several tools and modelling methodologies have been developed to facilitate the design and generation of information models for complex manufacturing environments. However, none of these tools and methodologies provide a systematic structure that can be used for the development of information models from start to finish(Chadha,1991 & Czernik,1992). Most of these techniques and methodologies have used some kind of graphical representation of a system where data elements, data flows and to some extent the functionality of the system can be illustrated through a set of diagrams. These diagrams are helpful in visualisation and comprehension of the information structure. However, these diagrams cannot easily be coded, compiled and converted in a format usable by computers.

The information generated during the specification, design and implementation of a manufacturing system is used for many purposes. The use may involve many computer systems, including some that may be located in different organisations or even on remote sites. Thus the need to have data models in a computer-interpretable form will become vital in the future. A number of attempts have been made to define standards to be used for information exchange. In recent years, the activity of the STEP committee has been predominate. As a result, drafts of a set of part documents have been produced, known as the ISO 10303-series standard.

This paper reports on the research work that has been carried out in utilisation of such standards to provide integrated software platforms under which a number of CAD/CAM software products can operate. In the next section of the paper, the project background and software packages used for the integration work are outlined, and in section 4 an overview of the ISO 10303-series standard is given. Section 5, describes the use of these standards in the integration/interfacing work.

Project Background / Software Environment

The Eureka(EU358) - FORCAST(Flexible Operational Real-time Control And Scheduling Tools) project began in April 1991 for two years. There were over seventeen industrial and academic collaborators, covering four European countries. FORCAST comprised six sub-projects and the overall project aim was to provide manufacturing and process based companies with software tools, mechanisms and standards for the integration of production planning and control system with simulators, and to introduce an open environment in which control, scheduling and simulation tools may be effectively integrated to improve and support reliable control of manufacturing operations. As a part of this research programme in sub project 4, a need was identified to integrate and link a tool flow simulation model of a machining cell and a finite capacity scheduling software to provide a system capable scheduling cutting tools. The software chosen for this integration work, namely the SIMAN Tool Issue Strategy Selector and PREACTOR are briefly described below :-

SIMAN Tool Issue Strategy Selector

The SIMAN Tool Issue Strategy Selector (STISS), developed at Loughborough in collaboration with The CIMulation Centre (TCC) is a prototype software facility to assist in the selection of a tool issue strategy for a flexible machining cell. It consists of three modules, namely a tool flow simulation module, a manufacturing resources relational database and a decision support comparator module.

PREACTOR Scheduling Software

PREACTOR (TCC,1993) is a commercial simulation based scheduling software environment which has been developed by The CIMulation Centre, as a part of their involvement in the Eureka-FORCAST project. The SIMAN simulation language is used to develop the model required to generate the schedule.

ISO 10303-series standards - An Overview

The STEP initiative is a multi-national effort directed at the creation of a standard to exchange product design, manufacturing and supporting data(Yang,1991). STEP is the name of a set of documents which are known as the ISO 10303-series standards. Additionally, ISO/STEP is the informal name used for the international organisation ISO/TC184/SC4 and its related working groups, the developers of the standard.

The objective of this standard is to provide a mechanism that is capable of describing product data throughout the product life, independent from any particular computer system. The nature of this description enables mechanisms to be defined for physical file exchange, for database implementation, and for direct access to product data by application programs. The product data description is developed using a formal method to define all ideas and concepts uniquely and consistently.

ISO 10303 uses a formal data specification language, called EXPRESS, to specify the representation of product information(Wilson,1992). The utilisation of the product data specification included in ISO 10303 is defined by mapping from the EXPRESS language onto the formal language or formal notation used for the particular application.

Integration and Interfacing Using EXPRESS Models

The subject of integration / interfacing has been of paramount importance within many international and European CIM research and development projects. The need for various computer systems to share their information has resulted in a number of structures, methodologies and tools that aim to provide an environment for computer systems to easily exchange information.

The task of interfacing in the subject of computing, in its most simple from, consists of linking two specific computer systems. In the past, this has been achieved by developing specific file converters / translators (usually referred to as gateways) between the two systems. These file converters were often limited in functionality and inflexible toward changes in files structures and had to be updated as the computer systems were developed further. Additionally, in the applications where a number of computer systems had to be integrated, the task of development and updating these gateways were time consuming and expensive.

More recently, database management systems(DBMS) have been used to interface computer systems. This is done by storing all the inputs and the outputs of the system in a database and linking the software packages not directly, but through the database. However, unfortunately not every software package has the capability of linking to a database, although this capability is becoming more as a demand by majority of the software customers. In addition, there are a number of the commercial DBMS's that have been adopted by the various computer systems, and at times, these DBMS's are not totally compatible with others in their internal structure and in the way they link to other software.

The problem of modelling information, although to some extent similar, is not the same as for designing a database. EXPRESS is not a database design language. However, there is a close relationship between the structure of an EXPRESS data model and that of the corresponding database. This has been recognised by the development committee of the EXPRESS language and those users of the language. Therefore, various sources have developed translators(Rutherford Appleton Laboratory,1992) for EXPRESS that would convert EXPRESS codes to their corresponding code in SQL. The usage of EXPRESS data models for linking / interfacing computer systems are mainly in the two following areas:-

- Generation of STEP physical files using STEP file generators.
- Mapping of EXPRESS models onto a formal language or formal notation such as C, C++ and SQL

The STEP physical file(Altemueller,1988) is an standard file structure that contains both the data formats and the data values. This file structure can be used as means of common file transfer facility between computer systems. To generate STEP physical files, both the EXPRESS data model and data values to be transferred are required, as illustrated in Figure 1. Standard STEP physical file generators have been developed and an increasing number of computer systems are made able to use these type of files. However, in the case of integration between software tools where neither of the tools have the capability to directly use STEP physical files, as was the case considered by the authors, conversion of ASCII files into STEP files would not have helped, but added to the complexity of issues involved. Therefore, a different approach for information transfer was planned using EXPRESS models.

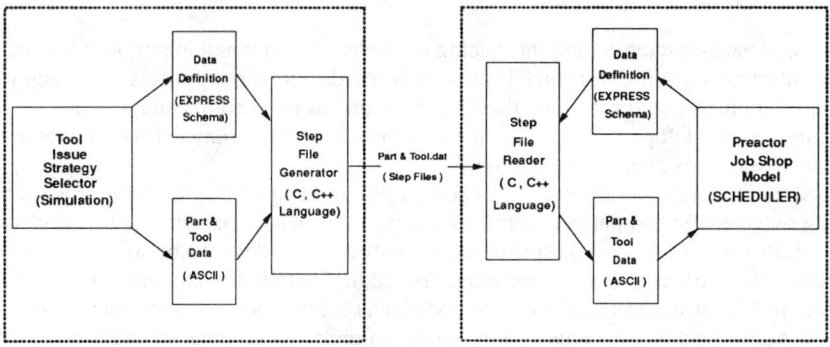

Figure 1 : Possible Information Transfer Between STISS & PREACTOR
via STEP Physical Files

This alternative approach has utilised a Relational Database produced by translating the EXPRESS schema of the data to be transferred into corresponding SQL codes. These SQL codes formed the structure of a database used to store the information to be passed between the software tools, namely SIMAN Tool Issue Strategy Selector(STISS) and the PREACTOR scheduler, as illustrated in Figure 2.

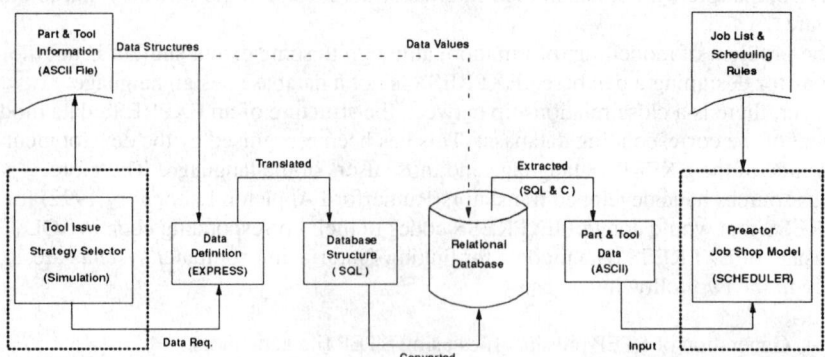

Figure 2 : Information Transfer Between STISS & PREACTOR
via Relational Database

It was decided to test this approach by passing a limited amount of information generated by the simulation models to the scheduler. An EXPRESS schema has been developed for the part information contained in the job list used by the STISS, together with some information on the tooling requirements of the jobs. This EXPRESS schema was translated into SQL. The SQL codes were used to form a database structure in the INGRES database management system. The Part and Tool information produced by STISS was entered into the database and then the appropriate files, in the format required by PREACTOR were generated.

The major advantage of this approach is that in the process of information transfer a database is produced which can store the information as it is being transferred. In addition, an EXPRESS schema of the database structure is available. When standard readers and generators of STEP files become universally available, this EXPRESS schema can be used to pass information via STEP physical files.

Conclusions

The development of a formal data specification language, called EXPRESS under the STEP standard committee, has been a significant advancement towards a computer - interpretable form of a data model. This has provided means of defining data structures and formats of a computer system in a natural format that can be coded and then converted and translated into a more usable format by other computer systems, making EXPRESS an ideal tool for the purpose of integration and interfacing between computer systems.

Using an EXPRESS schema for database design, a number of tool flow simulation models and scheduling software have been integrated. This integrated software environment is to be used for further research into scheduling of cutting tool within flexible machining cells.

References

Altemueller,J. 1988, "The STEP File Structure ", ISO TC184/SC4/WG1 Document N279
Chadha , B. , Jazbutis , G. , Wang , C. , Fulton , R.E. 1991, "An Appraisal of Modelling Tools and Methodologies for Integrated Manufacturing Information Systems", Engineering Databases : Proc. of the ASME Conf., California , U.S.A.
Czernik,S. & Quint, W. 1992, "Selection of Methods, Techniques and Tools for System Analysis and for the Integration of CIM-Elements in Existing Manufacturing Organisations", Int. Journal of Production Planning & Control, Vol 3: No 2, pp 202-209.
Rutherford Appleton Laboratory, 1992, "Code Generation to Produce SQL statements for the direct implementation of a relational database from an EXPRESS model", Oxon, UK
The CIMulation Centre (TCC), 1993, "PREACTOR User Manual", Chippenham, UK
Wilson , P. 1992, "Information Modelling and EXPRESS" , Draft Report, ISOTC184/SC4/ WG5
Yang, Y. 1991, "Three-Schema Implementation in STEP : Standard for the Exchange of Product Model Data", Engineering Databases : Proceedings. of the ASME Conference., California, USA.

INTEGRATED MANUFACTURE FOR SME's.

Dr Charles Wainwright, David Tucker and Prof Ray Leonard

De Montfort University, Leicester, LE1 9BH.
Edbro Limited, Bolton, Lancashire, M69 9BU.
UMIST, Manchester, M60 1QD

Presently, although there exists a substantial body of research relating to the technical nature of CIM, there is a need to exploit the development of ways to make it easier for companies to identify and obtain the available benefits on a low cost basis. Such development is particularly suited to SME's, where the potential benefits of CIM are outweighed by the risk and uncertainty associated with expenditure. The work outlined in this paper proposes that many benefits of CIM may be achieved without full integration, and at low cost, through systems linking. The paper considers GRAI methodology as a process to identify inter-relationships between functions, enabling an implementation sequence to be identified to create a linked system. The paper concludes by illustrating the validity of the process through a case study conducted by the Authors.

Introduction

The rapid pace of manufacturing systems evolution has witnessed *mechanisation* being replaced initially by programmable automation (e.g. NC/CNC) and subsequently by *Islands of Automation* (e.g. CAD/FMS/MRP II). It is considered that the next phase of the evolutionary cycle will centre around the process whereby Islands of Automation are combined homogeneously to enable *Computer Integrated Manufacture*. CIM was first proposed in the early 1970's (Harrington, 1973), but is still under development, as reflected by many authors illustrating there is no universal definition.

Computer Integrated Manufacture

For CIM to succeed requires the cross-functional transfer of data between different software applications in a manner indistinguishable for the user. To achieve this level of operation, in theory requires only a simple solution, however, in practice there are a number of inherent problems (Wainwright *et al*, 1992).

In many companies, computer systems were historically introduced on an individual departmental basis, with little thought given to cross-functional integration. Such systems, which may have been in place for many years, represent vital company assets in terms of stored data. To integrate these systems generates conflict because individual systems are likely to have been developed from different sources, and are therefore likely to use different computing hardware. In addition the systems will have utilised programming languages suitable for specific system operation. For the entire system to operate as a single integrated whole, genuine two-way communication between these different systems must be provided. Current developments are moving in this direction, however, a comprehensive specification is still proving difficult to achieve, and some time may pass before a truly integrated system is operational (Harrison *et al*, 1989). At present CIM development is occurring in three major areas :

- Development of computer standards for transfer of data within a CIM environment (for example, OSI, MAP, STEP etc.).
- Extension of existing systems through additional modules (for example, CAD systems with CAPP capability).
- Individual companies integrating computers for specific purposes (for example development of FMS).

Unfortunately, much of this development is at random (Primrose, 1991), either aimed at a long term perceived need in which benefits are not clearly defined, or as the result of companies attempting to resolve short term problems specific to their own needs. However, any company intending to introduce CIM should evaluate the costs and benefits of currently available technology and compare with possible benefits obtainable with future development. In general, many of the potential benefits of CIM can be achieved with currently available technology through the application of transfer of data through linking between systems. Here data transfer is undertaken from one source to subsequent receivers in a neutral format, usually ASCII code, along a suitable cable. For CAD and CAM, the process of linking is becoming increasingly formalised by the adoption of standards, but problems still exist when linking between other systems. Similarly, within a company, in general the need is not to transfer *data*, but to transfer *information*, which requires neither linking or integration. For example, a designer using a CAD system may indirectly need to use information relating to the stock or WIP of specific components, to enable decision making for direct use in the CAD system. Primrose (1991) quotes typical examples of information use within CIM which include :

- Marketing, which use CAD to produce quotations and require access to MRP systems to determine possible delivery dates.
- Production engineers, when faced with a possible material shortage need to interrogate CAD to determine possible alternatives and MRP to check on stock availability.
- Quality control, when attempting to inspect suspect components need to be able to interrogate MRP systems to find the location of WIP batches and finished stock.

At present, therefore, a comprehensive specification for CIM is proving difficult to achieve, and many years may pass before the true concepts become a viable proposition for the average company. Due to its technical nature there is a significant cost penalty associated with CIM, which may have hampered its introduction in many situations. This is particularly true in the small to medium sized enterprises (SME's) where the potential benefits of CIM are outweighed by the risk and uncertainty associated with expenditure (Leonard, 1988). To minimise this uncertainty, with less risk than direct manipulation, the consequences of particular system development may be revealed through the adoption of suitable modelling techniques. However, to model the dynamic processes involved in a manufacturing system is difficult due to the dominant involvement of people and the number of activities. Thus, any modelling technique must be able to identify functional parts and interfaces; use appropriate descriptive techniques to describe the information and material processing systems; and identify time based elements of decisional activity.

The requirement to identify the functional parts of a manufacturing system, and the interfaces between them is crucial, as functional parts provide the means of incorporating human activity in the operation of the system. The second requirement, the use of appropriate descriptive techniques, is necessary to illustrate the structure of the system and the transformational process. Thus, the model must be both hierarchical and "top-down", to illustrate structure, and functional "bottom-up" in that a sequence of activities must be included to depict transformation processes. The third requirement, to identify the timing of decisions, is necessary to differentiate between strategic, tactical and operational decisions, all of which will be included during the design or modification of a manufacturing system.

GRAI Methodology

There exist a number of structured modelling techniques which have been developed specifically for the design and modelling of CIM systems, for example SADT, IDEF and SSADM. All these techniques, however, focus on the transfer of data, with little consideration given to the informational aspects of decision making. The one modelling technique which does focus on this aspect is GRAI (Doumeingts, 1988). GRAI was developed at the University of Bordeaux 1 in the early 1980's, and involves the development of a conceptual model of the management structure of a manufacturing company. It is pre-dominantly a graphical technique using a decomposition principle comprising a unique combination of "top-down" - "bottom-up" to define decision centres and the information required to make a decision within these centres. The GRAI approach is based upon a conceptual model, developed using tools and representation rules within the framework of an application methodology.

The conceptual model of GRAI was developed from theories on complex systems and organisations. It comprises two parts, to outline the organisation of a manufacturing system and to detail the activities of a decision centre. Using the decision centre as a focal point, the GRAI conceptual model assumes that the structure of a manufacturing system is comprised of the physical sub-system, the decision sub-system, and the information sub-system. The decision making process is such that the decision maker identifies a model representing the desired end state or goal. A decision is then taken which adjusts the system towards this end state. After a certain time interval, the decision maker compares the real situation against that of the model and adjusts his actions accordingly. The structure of the decision centre therefore defines the various activities of a decision centre, the decisions made by the decision centre and the information used by the decision centre. To analyse a system's structure, GRAI uses two graphical tools. The GRAI grid, gives a top-down hierarchical representation of the entire structure of decision centres within the system via functional decomposition. The GRAI nets, enable a bottom-up representation of activities and decisions made at the level of a decision centre, and of links between various centres.

Case Study

To ensure the integrity of application of GRAI to identify suitable linkages within a CIM system in an industrial environment, a model framework was created and tested within a typical multi-product batch manufacturing company. The company selected was a leading UK manufacturer of biscuit making equipment, and was of medium size, employing around 200 people with an annual turnover of £25m. The company, marketing its products on a global basis, was subject to intense competition, heightened during the current recession which saw a trend towards markets in the third world. In such markets the emphasis is placed simply on "a machine to make biscuits", without the strict levels of product quality required in Western markets. In order to respond to increased levels of competition the company management were in the process of introducing a competitive business strategy. The strategy was reduced to a series of functional strategies to include marketing, design and manufacture. The manufacturing strategy expressed a series of goals relating to reduced lead time and cost and increased levels of quality and delivery reliability, and suggested that the goals could be achieved through the introduction of CIM. However, the plant and equipment in the company had been purchased over many years resulting in a series of incompatible systems, including :

- Mini-computer based CAD system operating under PRIMOS.
- Workstation based CAD system operating under AEGIS on a token ring network.

- Workstation based CAM system operating under UNIX with Ethernet DNC.
- PC based CAPM system operating under DOS on an Ethernet network.
- Mini-computer based Accounts system operating under CP/M-80.

Consequently, the Authors conducted an investigation applying GRAI methodology to determine an implementation sequence the company could follow to gain the benefits of CIM without the need for considerable expense in new systems. The initial phase of the process required the creation of two groupings, within the bounds of GRAI methodology, to form the Analysis Group and the Synthesis Group. The Authors formed the Analysis Group. To select the Synthesis Group, personnel were identified within the company management structure, and were used to form the initial basis of the grouping. The subsequent phase in the construction of the GRAI model was to determine the contents of the GRAI grid. This was achieved, following meetings with the Synthesis Group, by an evaluation of functions, decision centres and horizon and review periods. This resulted in the identification of sixteen decision centres. GRAI nets were subsequently constructed for each decision centre.

The initial analysis of the model was based upon examination of the GRAI grid. Observation revealed some inconsistency and, if the grid was examined with the GRAI nets, further differences were exposed. Examination of the grid revealed inconsistencies in terms of transmitted information and relationships between decision centres. In terms of information transmitted, there was generally a lack of transmission between internal sources, observed to a large extent between design and manufacture, but also between design and sales. Similarly within functions there was poor transmission of information within the review hierarchy. Within the manufacturing function, the lack of information from design hindered detailed production planning, and did not allow for jobs to be decomposed into component detail. The main implications of this were that shop floor activities were influenced by unplanned events, leading to reactive actions. Similarly, the purchasing strategy was vague, with purchase orders being initiated without consideration of manufacturing schedules. The capacity planning process was rendered unrealistic through the use of a manual graphical approach which failed to take account of the existing loading. This had a threefold effect :

- The MPS soon became unworkable.
- Detailed information on shop floor capacity was unavailable, resulting in inaccurate decisions to sub-contract or make in-house.
- The decision to tender for orders was taken without any consideration of capacity or current workload.

In terms of the relationships between decision centres, there was evidence of inappropriate relationships. Specifically manufacturing planning was deficient in that only the manufacture of components was planned. When completed, the components "pushed" themselves into assembly, and subsequently from assembly into test. The main problem with this approach is the reactive stance required when unplanned events occur. To achieve the goals of the manufacturing strategy a number of improvement solutions were introduced :

(a)   A new GRAI model was developed to highlight improvements in the transmission of information, and to identify a correct decision making sequence. This resulted in an additional eight decision centres being identified. The new model was then implemented within the company management structure.

(b)   To overcome the difficulties identified with incorrect capacity planning a Rough Cut Capacity Planning (RCCP) computer based modelling facility was introduced. The RCCP model was developed in-house and resided on the CAPM network. However, it was linked to the CAPM system, and also to accounts and the shop floor data collection system (operating on different networks), enabling the transfer of requisite data.

(c)   Direct data transfer links were incorporated between the two CAD systems, and between the workstation based CAD and CAM systems. These links enabled geometric data to be transferred via IGES.

(d)   Indirect information links were identified and implemented throughout the company. One example was the Bill of Material (BOM) link between CAD and CAPM. Here draughtsmen were provided with a link to the CAPM network, which then enabled (on the CAD workstations) a DOS window to be opened, to allow the draughtsmen to enter BOM data directly to the CAPM system. When satisfied the BOM was complete, the file was then transferred to the CAD system database.

At present the advantages of all modifications have yet to be fully quantified. However, clear benefits are being seen throughout the company from the correct sequence of decision making, the identification of effective capacity planning and the quick and easy availability of cross-functional information. This has resulted in increased delivery performance, accurate control of stock and WIP, which has also decreased significantly.

The application of GRAI methodology proved to be the key process in identifying direct and in-direct information links, and immediately highlighted areas of inconsistency. However, it should be noted that GRAI is merely a tool to assist in manufacturing system development and does not itself provide a definitive answer to a specific question, as this falls within the domain of the GRAI analysts. The GRAI analysts do apply the GRAI tools, however, on an iterative basis to simulate improvement to determine an optimal condition.

Conclusions

This paper has briefly discussed the benefits of CIM, which it is assumed are considerable, but unavailable to many companies using incompatible systems, through the lack of developed standards. CIM systems could be implemented on a company specific basis, but the prohibitive cost precludes this option in most SME's. It was suggested that many benefits of CIM are available to most companies simply through the process of linking systems. Such a process may not enable data to be transferred, but allows information necessary for decision making to be obtained quickly and easily. To identify the decision making process in a company, and to outline the transmission of information, GRAI methodology may be applied. The case study illustrated such an application, outlining the difficulties present in the test company and the solutions presented.

References

Doumeingts, G. 1988, "Systems Analysis Techniques", in *"Computer Aided Production Management"*, Ed. Rolstadas, A., Springer - Verlag.
Harrington, J. 1973, *"Computer Integrated Manufacturing"*, Irwin Publishing.
Harrison, D.K., Petty, D.J. and Leonard, R. 1989, "Criteria for Assessing Data Transfer Methods for CIM Systems in Practical Engineering Situations", *Computer Integrated Manufacturing Systems*, **2**, (4), 228-235.
Leonard, R. 1988, "Elements of Cost Effective CIM", *International Journal of Computer Integrated Manufacturing*, **1**, (1), 13-20.
Primrose, P.L. 1991, *"Investment in Manufacturing Technology"*, Chapman and Hall.
Wainwright, C.E.R., Harrison D.K. and Leonard R., 1992, "CAD-CAM and DNC links as a Strategy For Manufacture : Implications for CAPM", *Computing and Control Engineering Journal*, **3**, (2), 82-87.

AN INFORMATION MODELLING METHODOLOGY
APPLIED TO A MANUFACTURING ENVIRONMENT

Shaun Murgatroyd, Paul Clements, Simon Roberts, Christopher Sumpter and Prof. Richard Weston

MSI Research Institute
Loughborough University of Technology
Loughborough Leics. LE11 3TU

This paper describes aspects of work of the MSI Research Institute at Loughborough University of Technology leading to the definition of a methodology for the development of information systems based on a three-schema approach. The approach is centred on the creation of a conceptual model using the EXPRESS data modelling language, and the use of STEP physical file formats to represent information within CIM systems in a manner which is consistent with the conceptual model. In describing the use of this methodology examples are given based on results of an academic-industry collaboration between MSI and D2D. The MSI approach has been shown to yield clear benefits. The paper also outlines the objectives of current MSI research aimed at enabling the approach to be more widely adopted.

Introduction

Information is the life-blood of any enterprise and without timely access to it the enterprise and its component parts cannot function effectively. In the majority of enterprises information is territorially held, so that different departments within an enterprise have severely constrained access and manipulation capabilities with respect to information belonging to other departments, here typical constraints are imposed by the use of non-standard formats and semantics. Hence a key issue of the 1990's has been the need to create enterprise wide information systems which are well structured and can be understood by all necessary parts of the enterprise. Also during the life of any company its information requirements change, hence it is vital that information systems used within the enterprise are capable of facilitating change. Bearing in mind these global requirements, researchers at MSI have identified a number of attributes that the next generation information architectures should possess, these include:

- Underlying model description of the information within the enterprise

- Common interface to all information repositories
- Object oriented view of information
- Ability to handle heterogeneous information stores
- Separation of user application requirements from knowledge of the structure and location of information entities
- Ability to modify the underlying information models

This paper will describe the major components of an information architecture conceived by MSI researchers which possess these attributes. Subsequently it details the use of the architecture in an industrial case study.

The MSI Information Architecture

The underlying meta structure of the MSI information architecture is based on the work of the ANSI/SPARC Database Management Systems Study Group. This proposed a Three-Schema approach to Database Management Systems, Dionysios (1978). The Three-Schema approach embodies three distinct schema commonly implicit in database management, namely;

- Conceptual Schema: This is an abstract definition of the information system. It is used to represent data and relationships between data without considering the physical resources of any system(s) which will be used to achieve storage.
- Internal Schema: This deals with the physical representation of the data and its organisation.
- External schema: This is the user view of enterprise data.

Conceptual Schema

After investigation of various well known techniques for the conceptual design of information systems, (including Yourdon, IDEF1X, JSD etc.). a decision to use the EXPRESS Language, ISO (1993) as the conceptual modelling language for the architecture. The main reasons for this choice were as follows:

- It was about to become an international standard
- It had been produced from the STEP, Smith (1992) standardisation effort which was firmly rooted within manufacturing
- A number of useful models were being produced, these being written in EXPRESS
- It can readily lead to computer readable solutions

An EXPRESS model, consists of entities which themselves are assigned attributes and constraints. Among the useful features of this language are support of supertypes/ subtypes, enumerated types, rules and functions.

Internal Schema

To support this part of the architecture, MSI researchers have created a number of tools which (i) create underlying internal schemas, based on the conceptual schema, and (ii) populate these internal schemas with information. In creating internal schemas, the relational paradigm was chosen, due to its wide and growing use within the industrial community and the acceptance of SQL by the international standards community, as an interface for relational databases. Thus a compiler was written which translates an

EXPRESS model or models into a set of SQL create table statements, thereby defining a set of databases that mirror the structure of the model or models.

For populating the internal schema with enterprise information, MSI researchers chose to use the STEP physical file format, ISO (1994) as this offers a mechanism for describing physical information to be stored within the databases. Thus a parser has been developed by MSI which carries out three primary functions, viz:

- Checks the physical file for syntactic correctness
- Checks that the physical file is semantically correct, with respect to the relevant EXPRESS model(s) and
- Creates the SQL insert statements required to populate the database(s) with information contained within the physical file.

External Schema

A number of necessary features were identified early on in this research programme which are important to the usefulness of the architecture, namely:

- That implementors of application systems should not require detailed knowledge of underlying database organisation
- That creation of the internal schema should be with reference to knowledge of the model created during system design
- That changes in the information system should be allowed

Current emphasis is on providing the user with an object view of the underlying database or databases. Here the automatic mapping of objects in the external schema, onto information entities in the internal schema, is achieved via reference to the conceptual model. This provides the features identified above.

The SPEAR Project

SPEAR is a major D2D development project which focuses on formalising process planning activities of an enterprise related to printed circuit board (PCB) manufacture. This enables the development of information systems which support an integrated design-to-manufacturing approach to product introduction. Reduction in product introduction lead times results form this integrated approach.

Figure 1 illustrates three primary factors which influence how a product or range of products should be manufactured. Lenau and Prinz described the importance of a product model which defines all aspects of the design that will impact on the manufacturing process. Also, all aspects of the manufacturing facilities that determine how the products can be manufactured need to be represented within a factory model. Hence the primary decision making processes are made as a result of considering attributes of the product model and factory model for the range of products to be manufactured in the quantities required (i.e. the orderbook). Essentially therefore this process of deciding 'how' products should be made is a combination of the two traditionally separate functions of process planning and capacity scheduling. For the current state of development of industrial case study system these two functions are independently considered, although a detailed definition of the factory (including standard machine set-ups) is created by considering a range of products and their order profile. Currently in the case study system the product and factory models are implemented on an INGRES relational database which effectively forms the 'core' of the information system for the SPEAR project.

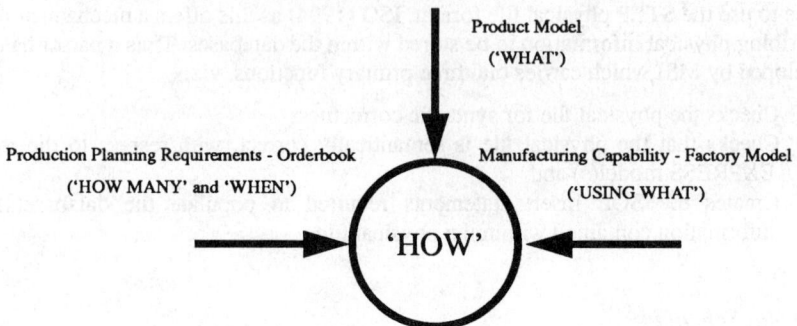

Figure 1 SPEAR Project Philosophy

Case Study Application

The MSI's information modelling approach is in the form of a set of integration tools, was applied to the design definition part of the SPEAR database. This case study application involved:-

- Modelling of the design definition entities, attributes and relationships, thus using EXPRESS, Perkin (1993)
- Creation of an INGRES relational schema from the EXPRESS model.
- Creation of STEP physical files from the design definition section of the SPEAR database, in conformance with the EXPRESS information model.
- Population of the new database schema from the STEP physical files.

The rationale for this work included:-

- Providing means of archiving the SPEAR data. This adds robustness to the SPEAR system by providing an archive which is independent of the database technology.
- Portability of information. SPEAR data can easily be transferred (in file format) and interpreted by external facilities or sites which do not support relational database technology.
- Interfaces to proprietary software. Many proprietary software packages do not directly support a database interface and usually require a proprietary file input. Such a facility can be produced directly from the SPEAR database but file based translators are faster, easier to write (using standard tools e.g. LEX and YACC) and there is the benefit of automatically generating much of the translating code if the format is based on a computer readable information model (such as EXPRESS). In this case study the STEP physical files were translated into GENCAD, Stimson (1993) format files for input into a PC-based board viewing package.

Conclusions drawn from the case study

The D2D case study work has shown that:-

- It is feasible to represent SPEAR Product Models using the STEP physical file format and where global relationships are manufactured through conformance to an EXPRESS model of the SPEAR Product Model.

- The STEP file can be used as an intermediate format for interfacing to proprietary file-based packages.
- The EXPRESS model can be used to structure the rules for STEP file translation (such as in the YACC utility), thereby allowing semi-automatic translator generation.
- EXPRESS provides a more 'natural' way of modelling the information requirements of a particular manufacturing (or other) domain.
- The EXPRESS model can be used to structure relational database schema. Thus STEP files based on the same EXPRESS model can be used to populate these schema.

Ongoing Research

Objectives of ongoing MSI research are (a) to enable mapping of the conceptual schema to existing information resources (b) to enable the methodology to cater for wider scale scenario where both a top-down and bottom-up approach to enterprise information modelling and (c) to provide the presentation mechanisms which allow manufacturing applications to have a model driven application specific view of enterprise information.

Conclusions

An information architecture and its implementation in the form of software tools and utilities which provides a range of novel features which were identified by MSI at the commencement. of a major UK Government supported research project. The architecture has been very successfully applied in a proof of concept form which mirrors and advances product introduction systems in a leading manufacturer of pcbs. The authors wish to fully acknowledge D2D and the ACME directorate of the SERC for their continued support and encouragement.

References

Dionysios, C. 1978, Tsischritzis and Anthony Klug (eds.). The ANSI/X3/SPARC DBMS Framework: Report on the Study Group on Database Management Systems." *Information Systems* 3.

ISO 1993, ISO10303 Part 11 Information Modelling Language EXPRESS.

ISO 1994, ISO DIS-10303 Part 21 STEP Physical File Format.

Lenau, T. 1992, Manufacturing constraints in concurrent product development, Integrated Manufacturing Design: The IT Framework, *Unicom Seminar*, 45-55.

Perkin, J. 1993 SPEAR Data Dictionary: Phase III, D2D Kidsgrove, k007/DD/3, Revision 2.0.

Prinz, F.B. 1992, The product evolution cycle, integrated Manufacturing Design: The IT Framework, *Unicom Seminar*, 6-7.

Smith, B.M., 1992 The STEP Project, *ISO Bulletin*.

Stimson, D.M. 1993 GENCAD Specification, Version 1.2, Mitron Europe Limited.

THE ROLE OF INTEGRATED SYSTEMS IN ACHIEVING FUNCTIONAL ALIGNMENT WITH STRATEGIC OBJECTIVES

David Tucker[1] , Dr .Charles Wainwright[2] and Prof. Raymond Leonard[3]

Edbro Plc. Bolton, [2]DeMontfort University, Leicester, [3]UMIST, Manchester.

This paper presents a case study of a manufacturing company which has recognised the urgent need to address the issue of functional alignment with regard to corporate objectives. Having previously invested in stand-alone elements of Information Technology (IT), the company sought to investigate whether the integration of the various systems could be used to achieve this functional alignment. This paper therefore describes how two separate IT systems, initially installed to enhance the operation of individual departments, are now being integrated specifically to improve the competitive position of the company. It will also show how a relational database (RDB) tool, initially installed on a stand-alone PC in the Marketing department, is now being developed to provide a comprehensive, company-wide, information system which will enhance the strategic decision making process.

Introduction

One consequence of increasing global competition for market share is that manufacturing companies need to ensure the traditional customer requirements of quality, delivery and price, are afforded maximum attention. When a company is organised on a strongly functional basis, the perceived internal goals of different departments are often in conflict, and if this fragmentation is allowed to persist it can obscure the strategic objectives of the enterprise as a whole. This then manifests in sub-optimal competitive performance, and probable loss of customers. To achieve unity of purpose across all departments requires a mechanism which ensures corporate strategy is constantly aligned with the uncertain demands of the market, and that this strategy permeates all boundaries.

When Information Technology is introduced into a functional organisation, the financial appraisal tends to be based on the operational advantages expected to accrue within the department proposing the investment. Whilst this may lead to some localised improvements, maximum benefit will accrue when a holistic viewpoint is adopted, and investments pursued for strategic gain.

Case Study

The company are long established manufacturers of hydraulic equipment located in North West England. In 1993, following a comprehensive reappraisal of the company's market position, the Chief Executive affirmed the most pressing strategic objectives were to respond more quickly and efficiently to initial customer enquiries and to increase the percentage of on time delivery of products to customers.

The company has a history of successful implementation of a variety of advanced manufacturing technologies including CAD, CAM and FMS, which have been implemented to enhance the operations of specific departments. On the basis of improved shop floor scheduling, the Production Control department had, in the late 1970's, installed a mainframe based Material Requirements Planning (MRP I) system. Although noteworthy operational improvements were gained, a major drawback was that feedback from the shop floor was often incomplete and always out of date, sometimes by as much as a week. This procedure was entirely manual, involving machine operators completing progress tickets which were collected periodically by a progress chaser, and input some time later into the MRP system by a clerk. Progress tickets were often lost or overlooked. The Production Control Department were naturally unhappy with the situation and sought to automate the feedback procedure. Consequently they short-listed a number of products that seemed to fulfil the requirements of a feedback system.

At the same time, the Personnel Department were having difficulty recording time and attendance (T&A) details on the shop floor. This too was an entirely manual system using clock cards. Separate bonus cards were used to record pay details. The tasks of accurately recording attendance, and ensuring correct payments to workers, were laborious and error prone, and satisfied neither management nor workers needs. To overcome these problems the Personnel Department proposed the installation of a computer based T&A system.

The company's System Development Committee (SDC), consisting of all Executive Directors and several Senior Departmental Managers, was the forum for evaluating potential IT investments. The usual process was for managers to propose an investment and submit a cost/benefit analysis for discussion. It was felt that whilst this method had led to the establishment of several worthwhile systems, it did not enable a holistic and objective appraisal of a number of potentially beneficial projects, nor did it consider the 'intangible' advantages of integrating existing and future systems. The company wished to investigate the possibility of developing a method which would overcome the limitations of the existing investment appraisal techniques.

Development of a Holistic System Appraisal

The authors were charged with undertaking a comprehensive, company wide, information audit. Initially this involved interviews with Company Directors, Senior and Middle Managers, and Supervisors. Individuals were asked:

* "What are your main job functions?"
* "To function efficiently, what information do you require?"
* "Where does the information come from?"
* "How is it transmitted ?"
* "How is it stored?"
* "What information do you pass on and to where?"

The answers were used to compile a first level data flow diagram (DFD) showing the major information flows and data stores within the company. To develop the company model, a number of alternative modelling techniques were identified from the literature. These were examined to determine their suitability to complete the task. Techniques identified were:

* SADT - (structured analysis and design technique) approach
* IDEF - (integrated computer aided manufacturing definition) methodology
* DFD - (data flow diagram) approach
* SSADM (structured systems analysis and design methodology) approach
* GRAI (groupe de recherche en automatisation integrere) methodology

This list is not exhaustive, but represents the approaches considered by the literature to be comprehensive. All the modelling techniques analysed were suitable for developing structured system specifications. The modularity aspects of all approaches was useful to enable easy refinement and modification of the company model in specific circumstances. GRAI was initially recognised as being the only technique developed specifically for analysis of manufacturing systems. However, its lack of widespread acceptance by industry coupled to the complexity of its operation resulted in its rejection.
SADT was identified as providing very structured models using 'top-down' composition.
This was beneficial in determining the strategic implications of the model, but the approach does not enable inter-action between the implications of decisions and was, therefore, rejected. The limitation of SADT is overcome in IDEF, which was considered a stronger contender for the model development. However, criticisms of IDEF include the difficulty in understanding where data is coming from and where is going ($IDEF_0$), lack of cross-functional integration ($IDEF_1$), and a lack of ability to illustrate control aspects ($IDEF_2$). Consequently the use of IDEF was also rejected.
The SSADM approach is a UK government standard which uses graphical techniques, based on simple notation, to build a system model. The procedural standards of the method are formed from integrating a number of discrete techniques. The concept of SSADM is similar to $IDEF_0$, using functional decomposition to provide greater levels of detail. The main disadvantage of SSADM is that it is deficient in real time systems analysis and design and is expensive to run, making it inappropriate for use on relatively small projects.
One of the techniques applied by SSADM is the use of DFD's to represent a flow of information between the system and the external world. DFD's can be used to model the data flows within an organisation at two levels of abstraction. At a higher level, they may be used to show business processes, and at a lower level to illustrate programs or program modules and the flow of data between these models. Logical DFD diagrams are normally drawn by using only simple notation in graphical format. DFD's have the advantage of ease of application, and are recognised for their excellent potential for real time processing and control. The DFD approach is similar to SADT and $IDEF_0$ although a DFD diagram indicates the source / destination of data as an external entity. As a consequence the application of DFD's was used to develop the company model due to:

. Ease of cross functional integration
. The ability to analyse the implications of 'top-down decisions, reflected in 'bottom-up' functional implementation.
. The ability to express the relationship between different parts of the model.
. The low number of concepts used.
. The ease with which the model may be understood even by non technical personnel.
. A proven methodology with a successful track record.

Having developed the model, subsequent analysis illustrated there were instances when information was being generated and stored in more than one location. This immediately highlighted the probability of inconsistent information being used for decision making purposes. It was also noted different departments sometimes referred to essentially the same information by different names. Further investigation resulted in the compilation of more detailed DFD'S showing the actual interfaces between each department. These were backed up by the compilation of data dictionaries which clarified the terms used in different departments.

Now, for the first time, the company had a visible and usable model which comprehensively and unambiguously documented all information flows throughout the company, and which highlighted anomalies and discrepancies that were not previously apparent. This holistic nature of the diagrams provided a firm basis on which to objectively assess the company's current position with respect to IT investments.

Testing the Holistic Method

The Personnel Department were unhappy with the financial case they had prepared for the justification of the T&A system, yet they were sure major benefits would accrue. The Production Control Department were simultaneously seeking approval to implement a computerised MRP feedback procedure. A DFD was produced which superimposed the T&A system on the existing systems, and showed all resulting information flows. It became apparent the same system could be used to provide the feedback loop into the existing MRP system. Thus not only could computerised T&A be used to reduce the time consuming and error prone manual work in the Personnel Department and increase the accuracy of payments to the workers, it could also be used to more effectively monitor shop floor activity. This would give Supervisors more opportunity to take appropriate remedial action before production problems became serious. The ultimate result would be an increase in the percentage of on-time deliveries to the customer, thus meeting a major strategic company objective.

On this basis, the financial case for the T&A system was greatly enhanced . It was also possible to ensure the hardware and software chosen for the T&A system would be compatible with the existing MRP system. Had this holistic method not been adopted, the likelihood would be the T&A and MRP systems would have been unable to communicate directly. In this situation obtaining the strategic benefits would either incur additional development costs, with the probability it would be prohibitively expensive, or it could have been technically unfeasible to achieve integration. Also, the proposal for a separate MRP feedback system would have probably gained approval, thus resulting in additional and unnecessary expenditure on hardware and software and the establishment of another stand alone IT system.

Achieving Further Strategic Benefit from Systems Integration

Having established the validity of the DFD technique the company now wished to exploit it in other areas. Further detailed analysis of the DFD'S revealed the Sales and Marketing Department has unilaterally installed a relational database (RDB) package on a stand alone PC. This was being used exclusively to record sales history details for individual customers. The RDB was used both to ensure customers were regularly contacted, and to enhance future sales forecasts, and was thus fulfilling a strategic need.

The Sales Department has indicated a problem with responding quickly and efficiently to customers enquires which were received on an equal split between fax and telephone. The DFD revealed the essential part detail information required by the Sales Department was held, in manual from, in the Design Engineering Department. The maintenance of many thousands of records in line with ongoing engineering changes was a huge task, and the Applications Engineers felt they could not guarantee the accuracy of every parts lists. Consequently, each sales enquiry had to be vetted by an Application Engineer before further sales processing took place. This procedure often took several days, by which time the potential customer could have placed an order elsewhere.

Once again using the holistic DFD, it became apparent the RDB package in the Sales Department could be used to computerise the parts lists records held in Design Engineering. All that would be required was to customise the existing RDB format, provide training for the Applications Engineers, transfer the manual data to the database, and to obtain additional software licences. The costs of this exercise, given the potential advantages, was shown to be acceptable. The Application Engineer, freed from the task of vetting enquires, could now maintain the parts database and ensure its accuracy. This was made easy because the RDB had a simple graphical user interface (GUI) enquiry facility which required no knowledge of the structured query language (SQL) usually required for RDB interrogation.. Complex database searches could be undertaken with limited computer knowledge.

To meet the strategic sales requirements, immediate access to the database was now a necessity. This would require the establishment of a computer network. The DFD'S revealed that such a network would also enable the parts lists records to be used to produce and update technical publications for the marketplace. Not only this, a relatively small additional investment would enable a fax gateway to be connected to the network. This would enable Sales Engineers to respond more efficiently to fax enquires from customers. Thus once again, the holistic view of IT investment provided cost effective solutions to strategic needs.

Conclusions

When information technology is implemented on a departmental basis it is likely that some benefit will accrue in the department in which it is installed. However, maximum benefit is only achievable when a holistic perspective is taken and projects selected to meet defined strategic objectives. To achieve this requires a method which comprehensively documents existing systems and accurately defines interfaces between them. The model must be unambiguous so as to be easily interpreted, and thereby provide the basis of consensus decision making throughout the company. Hence the method must be powerful enough to persuade departmental mangers to subjugate disparate departmental objectives in favour of longer term and more lucrative strategic investments.

The enhanced Data Flow Diagrams, as used by the authors, are being developed for this purpose. The progress achieved so far in the case study company has been encouraging. One of the authors has been appointed to the full time task of developing and enhancing the DFD models to include all functional departments as well as incorporating the interfaces between the company and its customers and suppliers. This will be a significant step towards unifying the efforts of the whole company and aligning them with the needs of a volatile market.

One major enhancement will be to use a computer based technique to enhance the clarity of the diagrams. The greater speed of a computerised methodology will enable the company to more accurately asses the potential benefit of new technologies as they emerge, and thereby exploit then to maximum potential. The ability to do this ahead of the competition will be of strategic significance.

THE BUSINESS ENTITY OF SEW-OSA
(SYSTEMS ENGINEERING WORKBENCH FOR CIM-OSA)

Marcos Wilson C. Aguiar, Prof. Richard H. Weston and Ian Coutts

Manufacturing Systems Integration Research Institute (MSI)
Loughborough University of Technology
Loughborough, Leicestershire - LE11 3TU

This paper describes SEW-OSA, an environment which handles the complexity of enterprise modelling when designing and building integrated systems. One of SEW-OSA's main components is the Business Entity: a layer of services executing upon the CIM-BIOSYS integrating infrastructure. The Business Entity plays a major role in transforming a model-based description of a system into interactions among the system components involved. This paper focuses on describing the overall structure of the Business Entity and its key role in using models to 'drive' CIM systems.

Introduction

Based on the findings of a study of currently available Reference Architectures to support design and build activities along the life cycle of an Integrated Manufacturing Enterprise (IME) - Aguiar (1993b), the authors proposed and are consolidating the implementation of SEW-OSA, a workbench that combines CIM-OSA - ESPRIT/AMICE (1993), Petri-Nets, Object-Orientated Design, and the services of the CIM-BIOSYS[1] infrastructure - Weston (1993) to support integration initiatives. SEW-OSA forms a major focus of research effort within "Model-Driven CIM"[2], by seeking to offer the framework of a toolset with a scope which embraces the complete IME life cycle.

Outline of SEW-OSA

Although CIM-OSA offers a comprehensive specification for major aspects of the Enterprise Integration problem, it does not provide solutions for all aspects. This is particularly the case for Function, Resource and Organisation views at the Design Specification and Implementation Description Modelling Levels - ESPRIT/AMICE (1993). Additionally, despite of a number of initiatives be under development - see Katzi (1993), Siemens (1993) and ESPRIT/VOICE (1993), CIM-OSA has yet to provide an organised

---

[1.] CIM Building Integrated Open SYStems: Integrating Infrastructure produced by the Manufacturing Systems Integration Research Institute (MSI) at Loughborough University.

[2.] Model-Driven CIM is a major programme of UK funded research (SERC/ACME research grant) at the MSI.

method implemented in a wide-scope CASE tool which supports graceful migration from requirements definition to a system running upon an Integrating Infrastructure. Thus, SEW-OSA provides the first instance of such a method by providing two classes of capability associated with its design methodology (Fig. 1), namely: **model building capability** and **model enactment capability**.

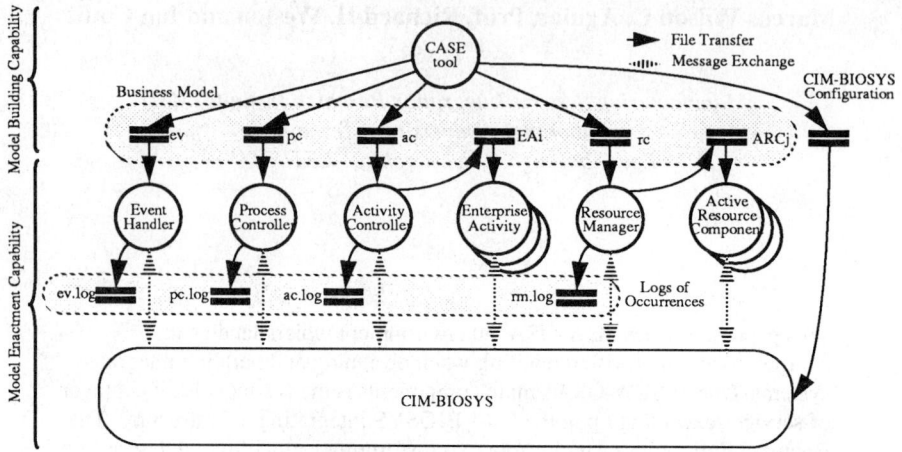

Figure 1 - Model/code flow in SEW-OSA

The Model Enactment Capability

The information formalised in the models (through using the CASE tool) is passed in the form of a number of pieces of interpreted code to the model enactment capability which manipulates it via simulation and rapid prototyping facilities. Figure 1 outlines the flow of models (and code) from the CASE to the major processes that execute those models. Model execution is achieved through:

- Execution of a Petri-Net which represents the behavioural aspects of the business model in simulated time, in order to obtain metrics, such as lead-times and WIP. The methodology adopted for Petri-net execution is discussed in Aguiar (1993b).
- Rapid Prototyping of the system and its (emulated) components: this tests the system structure by executing the business model generated by the CASE tool. Here, model execution is accomplished by the Business Entity components, namely: Event Handler, Process Controller, Activity Controller and Resource Manager, interacting via the CIM-BIOSYS infrastructure (Fig. 1).
- Execution of the physical system by gradually replacing its emulated components by physical ones (i.e. machines, application programs, data storing devices and human beings). In Figure 1, before physical components (i.e. Active Resource Components) are finally deployed, a number of iterations of model building and model enactment may occur. A key contribution of SEW-OSA is that it facilitates the execution of iterations in a consistent manner.

Implementing the model enactment capability involved creating: a link to a simulator for dynamical analysis of the business model; a Business Entity for CIM-BIOSYS for rapid prototyping of a system; and links to other tools and services that address design aspects not covered by SEW-OSA (e.g. information and resource modelling). This paper briefly describes the Business Entity of CIM-BIOSYS. The reader should also refer to Aguiar

(1993b) and Aguiar (1994b) for a discussion of the two remaining elements.

The Business Entity

The Business Entity consists of a model interpreter/debugger capable of driving interactions between system components through interpreting business models[1] generated by the CASE tool. Such an interpreter/debugger comprises four components, as described above. It consists of a layer of services (written in "C" and executing in an X-Windows/Sun Unix work-station environment) which runs on the CIM-BIOSYS integrating infrastructure[2] (Fig. 1). This layer of services achieves a link to the CASE tool and, in so doing, enacts models contained within it to achieve highly flexible but structured integration of IME components.

The Business Entity enables the generation of rapid prototypes of the system structure and system components, the latter being generated in an emulated form[3]. The data flow depicted in Figure 1 illustrates the set of files that are generated by the CASE tool. These hold data which are used to control interactions amongst the resource components, and the code structure for the core functionality of Active Resource Components. The CASE tool generates five sets of files, namely:

- **ev**: describes all events (i.e. asynchronous happenings) in the system.
- **pc**: lists all functions in the form of Domains (i.e. enterprise's major areas), Domain Processes (i.e. objects communicating via exchange of information, material and events), Enterprise Activities (i.e. atomic functions at the requirements level), and Business Process (i.e. an intermediate construct between Domain Processes and Enterprise Activities) and defines procedural rules (i.e. relationships of the type "precedes") inter-relating those functions.
- **ac**: lists all Enterprise Activities and Functional Operations (i.e. methods/messages describing atomic actions at the design level) used by them and the code structure of Enterprise Activities.
- **rm**: lists all Active Resource Components (i.e. people, machines and application programs) and Functional Operations that they are able to execute and the code structure of Active Resource Components.
- **cf**: consists of the information required to configure CIM-BIOSYS in order to recognise the applications associated with the Business Entity as well as those representing Active Resource Components.
- **EAi** and **ARCj** is the code.of Enterprise Activities and Active Resource Components in an interpreted form.

These files are used to configure the four components of the Business Entity, each of which is responsible for certain functions associated with the normal operation of the modelled system. As part of SEW-OSA debugging capability, each component interacts with the designer through user interfaces, which are standard for all four components. These interfaces enable the users:

- to check the list of constructs that are present in the fragment of the model handled by the component (e.g. list of functions in the Process Controller);
- to set the component to automatic mode, so that its functions are executed without

---

[1.] A business model is a description of how the system functionality is organised in terms of content, behaviour and structure.

[2.] An integrating infrastructure functions as a distributed operating system, flexibly mapping application software onto manufacturing and computing resources.

[3.] At the early stages of rapid prototyping, the Active Resource Components and the Enterprise Activities represented in Figure 1 consist of executives of Petri-net models which emulate their internal behaviour.

designer intervention although he can still monitor occurrences;
- to set the component to manual mode where the designer 'controls' the processing of the Business Entity;
- to generate a log of happenings at each component, regarding model execution.

Event Handler

The Event Handler is responsible for starting a tread of business model execution by channelling event occurrences (e.g. order release). A unique number is assigned to an event or a combination of related events. This identification is termed an occurrence and is propagated to the rest of the Business Entity, as a means of individualising it, even if it is competing with a number of other occurrences being executed at the same time.

Process Controller

The Process Controller is responsible for executing occurrences of Domain Processes, this based on the procedural rule sets of its component processes. The Process Controller is able to cope with the execution of several occurrences of the Domain Process at the same time. This is done through spawning local processes representing Domain Processes, which are responsible for the execution of their individual sets of procedural rules. These Domain Processes, in turn, may spawn other processes (this time, representing Business Processes) to execute their particular sets of procedural rules. All those processes are part of the Process Controller and they communicate with each other via Unix sockets.

Activity Controller

The Activity Controller is responsible for coordinating the execution of several Enterprise Activity occurrences, by activating processes which communicate via the CIM-BIOSYS infrastructure which represent such occurrences (see Fig. 1).

Resource Manager

The Resource Manager is responsible for allocating resource components to the execution of occurrences of model treads. In the Resource Manager, the unfolding of a business model is constrained by the resources allocated for model execution. The actual resource scheduling function is not provided by the Resource Manager. In fact, the Resource Manager operates as a bridge amongst the three parties involved, namely: the three other components of the Business Entity, special types of Active Resource Components which provide the scheduling functionality, and the scheduled Activity Resource Components represented as CIM-BIOSYS applications.

In addition to the four main components that can be used for debugging purposes, SEW-OSA provides access to the internal events within the two other classes of components depicted in Figure 1, namely: Enterprise Activities and Active Resource Components.

Enterprise Activity

An Enterprise Activity occurrence at run time is mapped into a CIM-BIOSYS application which executes its internal behaviour (described by means of a Petri-net model - i.e. EAi in Fig. 1). A CIM-BIOSYS application is dynamically started every time an Enterprise Activity is triggered by the procedural rules being executed by the Process Controller. Controlling the creation and interaction with such a process is a function of the Activity Controller. An Enterprise Activity occurrence does not provide the standard debugging functions present in the four components of the Business Entity. However, it does record information associated with the firing of each Petri-net transition and associated actions attached to the transition in the form of Functional Operations. This allows the designer to monitor the evolution of the Petri-net associated with the Enterprise Activity and its repercussion on entities in the business model, this enabling consistency check.

Active Resource Component

During the rapid prototyping stage, the behaviour of active resources is emulated through a Petri-net model. Like the Enterprise Activity, this model is generated by the CASE tool and passed to the Active Resource Component engine as interpreted code (see ARCj in Fig. 1). This code is executed by the Active Resource Component, represented as an independent CIM-BIOSYS application.

Summary

This paper reported on an on-going initiative aimed at bridging the gap between a model description and a running system via the creation of a model executive (i.e. the Business Entity). This represents the first instance of a CIM-OSA model interpreter that transforms business models into service transactions executed by the CIM-BIOSYS Integrating Infrastructure. The debugger associated with the Business Entity plays an essential role in checking the Business Model for any errors not identified at the modelling stage. The Business Entity associated with the CASE tool provides the first instance of an implementation of the Integrated Enterprise Engineering Environment for CIM-OSA - ESPRIT/AMICE (1993), running upon an Integrating Infrastructure. The authors believe that these tools provide a stable basis for incorporating the remaining entities of CIM-OSA, i.e. Information Entity, Presentation Entity, Common Services Entity and System Management Entity - ESPRIT/AMICE (1993), so as to built a complete engineering environment for enterprise integration. Hitherto, models were brought to a stage at which emulated components mimic the behaviour of real systems, in order to test their overall performance. However, integration of actual physical components has also been achieved in this research and has led to proof of concept real world implementations.

References

Aguiar M. W. C. and Weston, R. H. Systems Engineering Workbench for CIM-OSA. *Submitted to the Proceedings of the International Conference on Computer Aided Manufacturing, Robotics and Factories of the Future CARS-FOF/94*. Canada. 1994b.

Aguiar M. W. C. and Weston, R. H. A model-driven approach to enterprise integration. Submitted to the *International Journal of Computer Integrated Manufacturing*, (Taylor &Francis, London) 1993a.

Aguiar M. W. C. and Weston, R. H. CIM-OSA and stochastic time petri nets for behavioural modelling and model handling in CIM systems design and building. *Proc. IMechE Part B*.UK. 1993b.

ESPRIT/AMICE. *CIM-OSA Architecture Description, AD 1.0*. 2. ed 1993.

ESPRIT/VOICE. Validation of CIMOSA (Open Systems Architecture) - a joint ESPRIT projects report. Germany 1993.

Katzy B. R. et al. CIMOSA pilot implementation for technology transfer. Laboratory for Machine Tools and Production Engineering Rheinische-Westfalische Technische Hochschule Aachen - Internal Paper. Germany. 1993.

SIEMENS. CIMOSA/ReMo - Tool for enterprise modelling on requirements level. Germany 1993.

Weston H. R. Steps towards enterprise-wide integration: a definition of need and first generation open solutions. *I. J. Prod. Research*. 1993.

MANUFACTURING CELL CONTROL: A CRITICAL APPRAISAL OF CURRENT AND PROPOSED SYSTEMS

P A Shiner and M Loftus

*School of Manufacturing and Mechanical Engineering,
University of Birmingham, Edgbaston, Birmingham B15 2TT*

Cell control is situated between area and shop floor operations in a factory. The Cell Control Group, at the University of Birmingham, have worked with users and vendors to analyse and specify a generic solution to the cell control problem. This paper examines the important aspects of cell control; namely functionality, architectures, communications and real time operations. The way forward is discussed at each critical point. Current research projects and the state of the market are also examined.

Introduction

It is now widely recognised that the cell control function is an important area for research and development. Sited between the strategic levels of area and workstation, Jones and Saleh (1990) consider it to be a pivotal requirement in the CIM hierarchy. Along with this, cell control has to meet real time requirements of manufacturing devices along with slower responding needs of management information systems.

Franks *et al.* (1993) and Nessa and Mangrum (1991) define cell control functions which range from the complexities of scheduling through to the problems of error handling. Considering the diverse nature of the factory environment, it is not surprising that a standard definition of cell control has not been agreed. What is not in doubt is the need for well managed control systems on the shop floor. Supervisors have always had the unenviable task of keeping the production process in tune.

Functionality

As stated previously, it is not a simple task to define the functions, which should be present in a cell controller, because the functional requirements vary from company to company and industry to industry. A comprehensive generic system would overcome this limitation but the practical difficulties of defining and implementing this solution would be immense. A less grand, but never-the-less powerful, solution based on modular principles could be devised. Franks *et al.* (1990) propose a framework approach to allow the interfacing of the required functions to be facilitated.

From the many functions which have been proposed for cell control systems, dynamic scheduling, monitoring of resources and the coordination of manufacturing equipment, are considered as fundamentals. Pandya (1992) sub-divides these primary functions into three distinct decision making areas: 1) New orders from higher levels; 2) Unforeseen stoppages; 3) Shortages of materials, tools, and accessories. Franks *et al.* (1993) and Nessa and Mangrum (1991) propose that the core functions should be:

- User Interface
- Process Scheduler
- Time and Attendance Monitoring
- Quality Control (SPC and SQC)
- Materials Handling
- Tool Management
- Maintenance
- Error Handling and Recovery.

Control Architectures

Standard architectures have been proposed for the control requirements of cells. Hierarchical and heterarchical architectures are considered to be the main control candidates and this reflects their use in other sectors of control engineering. Hierarchical control systems are based on a pyramid shaped structure with commands filtering down from the top to the bottom. Supervisors are responsible for the control in their respective sub-systems. When difficulties arise in the system, the hierarchical based control structure will keep passing the decision making upwards until a level is reached where the authority exists to give the appropriate response. Sophisticated planning and scheduling functions tend to be executed in the higher levels of the structure, whereas the lower level systems have less control power and, therefore, perform simpler sets of commands. However, the interactive and cumulative nature of these commands can lead to major difficulties. There are concerns that, as the manufacturing systems grow in complexity, the hierarchical control system will become prohibitively complex and unreliable.

In contrast to the centralised philosophy of the hierarchical systems, the heterarchical control architecture employs a distributed approach. This type of architecture, suggested by Veeramani *et al.* (1993), promotes autonomous operation with controllers having no direct line of supervision. Functions, such as planning and scheduling, are negotiated by each machine to achieve mutual agreement on the control requirements. The following three inherent problems have been cited by Jones and Saleh (1990): 1) Deadlock resulting from stalemates in negotiations; 2) Unpredictable shop floor not suited to negotiations; 3) Untried in a dynamic situation involving a stochastic system.

Perhaps the best way forward for control architectures is to adopt a compromise between the best attributes of the hierarchical and heterarchical structures.

Communications and Real Time Control

It is within the area of communications that the need for well defined and accepted standards are essential. The ubiquitous RS232-C serial binary standard has often been quoted as the stock answer to communications within manufacturing devices, but this standard does not address the protocols of information interchange. The Manufacturing Automation Protocol (MAP) initiative, promoted by General Motors, was the first sustained attempt to improve the general requirement for simple interchange rules, procedures and physical communication links in the manufacturing sector. The International Electrotechnical Commission (IEC) Fieldbus standard complements the MAP concept with communications support for field elements and control systems in the process industry.

Unfortunately, the high ideals of the MAP initiative of the 1980's have not been realised and, in some cases, it has been counter productive, because it raised expectations beyond the market capability to supply MAP conformant products. This said, the philosophy behind MAP is sound and complements the European founded initiatives in implementing the ISO/OSI reference model (ISO 7498). The Manufacturing Message Specification (MMS), ISO/IEC 9506, is a valuable by-product from MAP because it provides the vendors and users with an agreed definition of the protocols required for the integration of shop floor devices, such as NC machine tools, robots, programmable logic controllers and the computer systems that control them. MMS is considered to be one of the most important standards to emerge from the international programme of work in computer integrated manufacturing and it will play an increasing role in the future.

Ethernet, DECnet and token bus broadband products are representative of the current range of communication network systems, but a universal standard for manufacturing applications has still to be accepted by industry. According to Franks *et al.*(1993), in their study of companies with cell control aspirations, Ethernet CSMA/CD (Carrier Sense Multiple Access/Collision Detection) is the favoured system because the majority of companies have it already installed for their data network requirements.

It is apparent from all the studies into cell control that there is a fundamental need for real time control provision. The production supervisors have to react dynamically to events, as they occur, but there is often a lack of current, accurate data on which to base an intelligent response. Future cell control products should provide better support in this area.

MAP and its companion MINIMAP, which only uses the first two layers of the seven layer ISO/OSI reference model, are not fast enough to handle real-time response at device level. This is where Fieldbus and ISDN (Integrated Services Digital Network) provide real-time communications between all levels in the CIM hierarchy. Real time communication has to also consider the way in which the network operates, eg token bus and CSMA/CD. Token bus uses a token mechanism to gain access to the network, whereas CSMA/CD employs a priority classification system to assess the relevant importance of the data seeking access to the network. However, token bus and CSMA/CD occasionally encounter missed deadlines which may cause subsequent sequencing problems. Sin (1991) considers that an alternative system, known as polled bus, is more appropriate to real time operations, because it incorporates priority ratings and deadline times. It is interesting to note that the fast reactions required on the shop floor are in contrast to the more sluggish response requirements from higher level systems.

Major Research Projects

Major research initiatives have been funded by national agencies in the USA and Europe.

i) National Institute of Standards and Technology (NIST) - Automated Manufacturing Research Facility (AMRF)

The AMRF programme was established to address the need for standards and investigate the philosophy under pinning "the factory of the future". This initiative included developing shop floor control at the Cell and Workstation levels, as described in the NIST factory model. This model comprises five levels: namely, Facility, Shop, Cell, Workstation and Equipment. The cornerstone of this work is the production scheduler, as outlined by Boulet *et al.* (1991).The scheduler uses selected scheduling rules, which are assessed in dynamic simulations. A degree of comparison and optimisation then takes place before an executable schedule is generated and downloaded to subordinate systems.

ii) Computer Aided Manufacturing International (CAM-i) - MAnufacturing DEcision MAking (MADEMA)

CAM-i prescribe a hierarchical based decision-making architecture for factory control. The structure has four levels: namely, Factory, Job Shop, Workcenter, and Resource level. CAM-i (Intelligent Management Program) developed MADEMA to address the assignment of resources at the shop floor level. MADEMA and the NIST project have a similar approach to scheduling, with both schedulers having reactive properties, which can accommodate changes in resource status. Boulet *et al.* (1991) outlines the functions of MADEMA and the techniques used.

iii) European Strategic Program for Research in Information Technology (ESPRIT) - Project 932

Project 932 (Knowledge-based factory supervision) was initiated in 1986 and currently involves four European countries. There are test sites in Britain and Germany. The factory model, and the hierarchical control structure are founded on the one devised by NIST. The main hub of the project is the "CIM Controller", which is located at levels in the hierarchy where decision-making is performed. The structure of the controller, whether at Cell or Workstation level, remains the same, see Boulet *et al.* (1991).

Market Analysis

The use of computer control technology at the shop floor level is limited when compared to accounting, production planning, warehouse control and CAD applications. With the advancement of computer technology, the integration of information from the well served management levels to the shop floor should become more common. Manufacturing cell controllers, as an integral part of this shift, have been increasing in number, as illustrated by Franks and Telford (1991). This has not been the case with the larger Flexible Manufacturing Systems (FMS) which dominated investment a decade ago. Such systems are now considered to be too complex and expensive to use.

There are now many products on offer that cover cell control. The most prominent ones are Reflex's CIMpics, BAeCAM's CIMITAR and Allen-Bradley's Pyramid Integrator. The Reflex and BAeCAM products are derived from software systems, whereas the Allen-Bradley Pyramid Integrator is founded on the company's programmable logic controller expertise.

A major trend in the discrete manufacturing cell control market has been the move towards generic solutions. Customised solutions are now regarded as being demanding to specify, costly to implement and difficult to maintain. This is in contrast to the process industry sector which still uses customised products. Another trend is towards products which have hardware and software independence. Vendors recognise that users wish to have the freedom of choice when selecting hardware and software systems.

Many cell control products on offer are geared to the higher levels of the CIM hierarchy. Finding a product with complete shop floor, real time, decision making characteristics is difficult. The shop floor environment is volatile and uncertain in comparison with management levels, and vendors find this impossible to accommodate in their products. Vendors and users are also concerned about communication standards, with reliability, stability and capability being of primary importance.

Conclusions

Several key areas of research and enabling technologies are relevant to the future of cell control development. Artificial Intelligence needs to be developed to enable the production of autonomous control systems. Real time control and communications needs to be addressed in greater detail. Their influence on the response times in an error handling and recovery system will be crucial, but the authors believe that the function of error handling and recovery has been neglected. This function is important for the realisation of autonomous operations in discrete manufacturing facilities. The function's operation must be comprehensive in its ability to deal with unforeseen events, such as machine breakdown and tool damage, to enable the goal of unmanned operation to be achieved. The technology is available for autonomous and real time control, but it remains to be seen if it can be utilised.

References

BOULET, B., *et al.* 1991, Cell controllers: Analysis and comparison of three major projects, *Computers in Industry*, **16**, 239-254.

FRANKS, I., LOFTUS, M., and WOOD, N. 1990, Discrete cell control, *International Journal of Production Research*, **28**(9), 1623-1633.

FRANKS, I., and TELFORD, M. 1991, Structured intersecting strategies for vendors and users of discrete cell control systems, *Computer Integrated Manufacturing Systems*, **4**(3), 132-139.

FRANKS,I., LOFTUS, M., and WOOD, N. 1993, Attributes of a cell control system, *Computer Integrated Manufacturing Systems*, **6**(3), 176-184.

JONES, A., and SALEH, A. 1990, A multi-level/multi-layer architecture for intelligent shop floor control, *International Journal of Computer Integrated Manufacturing*, **3**(1), 60-70.

NESSA, E., and MANGRUM, R. 1991, A CIMple claim for automated cell control, *Proc. of The Technical Program of the National Electronic Packaging and Production Conf.*, 1040-1051.

PANDYA, K. 1992, Development of rules for production planning and control at the cell level, *Third International Conf. on Factory 2000 - Competitive Performance through Advanced Technology*, 322-326.

SHIN, .K 1991, Real-time communications in a computer-controlled workcell, *IEEE Transactions on Robotics and Automation*, **7**(1), 105-113.

VEERAMANI, D., BHARGAVA, B., and BARASH, M. 1993, Information system architecture for heterarchical control of large FMSs, *Computer Integrated Manufacturing Systems*, **6**(2), 76-92.

IMDC: AN INTEGRATED ENVIRONMENT FOR THE DESIGN AND CONTROL OF MANUFACTURING MACHINES

Christopher Wright and Dr Robert Harrison

MSI Research Institute
Loughborough University of Technology
Loughborough Leics. LE11 3TU

Current approaches to the design of computer controlled manufacturing machines are characterised by minimal software re-use, poor verification of customer requirements, limited confidence in proposed designs, together with time consuming and costly system maintenance/enhancement. This paper outlines a new approach for lifecycle support aimed at overcoming these problems. An 'Integrated Machine Design and Control' (IMDC) environment is described, populated by a software toolset which operates on global shared data. IMDC enables system builders to concentrate effectively on aspects of potential solutions which add value by providing much needed support for rapid prototyping and 'what-if' analyses. A wide and extendable range of lifecycle activities is supported, including application logic description and analysis, machine modelling, automatic code generation and run-time control.

Introduction

Modern manufacturing strategies demand shorter production cycles, greater adaptability and shorter set-up times. The impact of demand for choice in consumer products with continual updating of specifications requires manufacturing machinery to be revised and reconfigured at increasingly frequent intervals. In many cases, however, current machines are so difficult to modify cost effectively that complete machine rebuild is necessary to accommodate product change. Increasing customer choice leads to a diversity of product variance with a resultant demand for very small batch sizes. For example, in the packaging industry there is demand for high quality cardboard boxes which must be manufactured to order (cut, colour printed and formed) in batches of less than 50, where traditionally only high volumes were considered viable.

Although the use of computer controlled machines is now commonplace in industry there is an urgent need to radically improve the effectiveness of machine and associated control system design/build *and* to enable efficient modification as requirements change. The commercial impact of such improvements would be profound, resulting in more rapid machine design-build-installation-setup, improved product quality, enhanced economics via smaller batches, minimum work in progress, reduced scrap and ultimately reduced overall costs.

The machines themselves consist of suitable combinations of mechatronic (mechanical and control system) elements, with application specific code determining how a particular machine behaves. The inability of current methods to efficiently cater for the visualisation of these mechatronic elements and the design of associated application specific software, particularly as system complexity increases, is seen in spiralling applications costs (Muir (1990)). Using current approaches the application software is difficult to design, maintain and modify, because the designer cannot easily visualise (graphically or otherwise) the effect on machine operation and the potential problems which may result. In limited areas simple graphics based design tools are being created and useful standards are emerging, notably the IEC-1131 standard, which defines a more open approach to the creation of PLC software based around the sequential function chart graphical language (Bekkum (1993), ISaGRAF, APT). In general, however, the widespread utilisation of better design approaches is severely handicapped by a lack of sufficiently open machine control environments with which they can effectively interoperate. This deficiency has prompted a number of research groups to look at more generic forms of industrial machine controllers, for example MOSAIC/Europe, MOSAIC/USA, RCS and UMC (Weston, Harrison, Booth and Moore (1989), Harrison (1991)).

The generally vendor specific and centralised nature of current machine control systems makes efficient modification difficult. Problem decomposition and demands for increased control capabilities require next generation machines to comprise intelligent building elements, physically located at the point where the control functionality is needed. To address this, networks of intelligent controllers are increasingly being designed into high speed machines and usable standards are slowly emerging (Reeve (1993), Heath (1993)). However, in addition to all its advantages, distributed real-time control brings with it yet another new set of design issues and new sets of possible error conditions to cope with (Carpenter (1993)).

IMDC offers a solution to the above problems by providing the means for integrating a suite of diverse machine design and control tools into a single environment. The toolset can be progressively changed, enhanced and extended via a generic set of interfaces. These tools are co-ordinated by a 'Machine Workbench' and operate on a global shared data set. They provide integrated support over the whole machine design and control lifecycle by addressing application logic description and analysis, machine modelling and visualisation, hardware and software topology, information storage, code generation and distributed run-time support. Integration of the tools facilitates rapid prototyping such that potential solutions (and modifications to existing solutions) can be quickly verified. IMDC promotes iterative analysis and revision of these potential solutions - rapid support for 'what-if' analyses enabling fundamental design errors to be quickly identified.

The interchange of data between the tools is effected by data management mechanisms in the IMDC kernel software. The environment is database centred, each tool having a defined view of objects in the database. Tools exchange data with each other indirectly through the database, using the concept of a view provider. Similar principles have been used by Venugopal (1990), Jablonski, Reinwald, Ruf and Wedekind (1990), although the application domains have differed somewhat.

Overview of the IMDC Environment

The overriding aim of IMDC is to provide a highly productive environment for machine life cycle support which is simple and intuitive to use. Conceptually IMDC provides the machine builder with off-line and run-time tools available in the order required, via a single 'Machine Workbench'. Physically, IMDC utilises a network of one or more workstations coupled to an embedded real-time control architecture which resides on the target machine.

The two main user components of the IMDC environment are:
- a software toolset which covers the life cycle of manufacturing machines
- a target distributed control strategy, supported by the software toolset

To integrate and manage the user toolset the environment uses:
- kernel software with a database
- an application interface to the kernel
- an integration platform (distributed services)

The environment is built on top of a set of generic integration services which provide the low-level mechanisms for flexible configuration and management of distributed applications. Applications in this context are any processes we wish to integrate (kernel processes, software toolset processes etc.). These processes fall into two categories:

1. Kernel processes: a set of interacting processes for IMDC system start-up, configuration and management, database access and user access control.
2. User processes: the software toolset used by the machine builder, consisting of off-line tools, run-time tools and the co-ordinating 'Machine Workbench' process. Each of these processes interfaces to the IMDC kernel by means of a function library.

Figure 1 shows how each user process sees the IMDC kernel and figure 2 illustrates the complete IMDC environment. All requests and responses from user processes are routed via a system server process within the kernel which handles permissions, event logging and further routing of information between kernel processes.

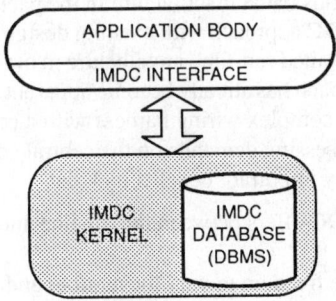

Figure 1: Application Interface To The IMDC Kernel

Implementation

The IMDC environment described in this paper is being implemented as part of a SERC/ACME collaborative research programme based at the MSI Research Institute. The project involves a consortium of seven commercial companies with diverse interests in the automation and control industry. The real-time distributed control methodology for IMDC (embedded in the target machine) is being realised by exploiting and extending the UMC methodology established through previous ACME research (Weston, Harrison et al.) and utilising emerging Fieldbus network technology (Reeve (1993), Anon. (1993)). Integration

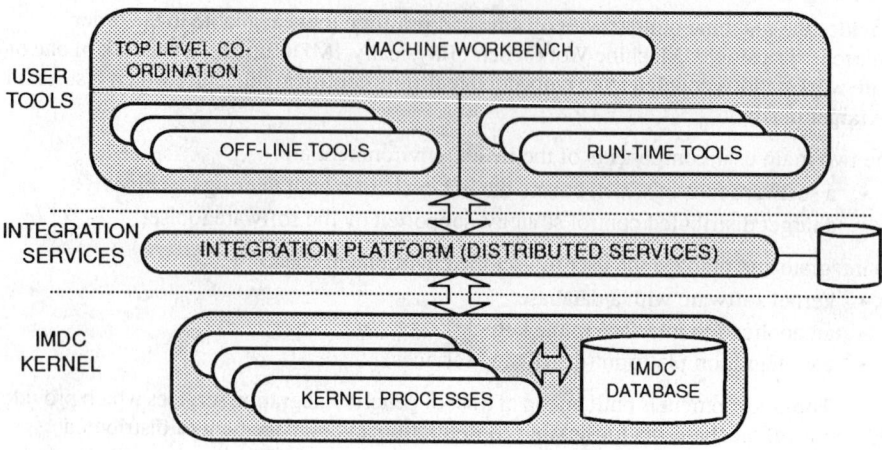

Figure 2: The IMDC Environment

is achieved by means of the CIM-BIOSYS Integrating Infrastructure (Coutts, Weston, Murgatroyd and Gascoigne (1992)), also established through previous research/industrial collaboration. User tools shown in figure 2 are either based on existing commercial tools or are being written according to collaborator requirements.

Case Study

The proof of concept and effectiveness of the IMDC approach is being demonstrated using representative problems from the industrial collaborators. One particular application is an example of semi-continuous batch manufacture in the packaging industry, which is being used to compare the IMDC approach with existing design and control methods. The importance of adopting a distributed run-time architecture in order to effectively realise the full benefits of the IMDC approach has already become apparent from this case study work. The advantages of replacing a complex wiring harness with a control network to reduce weight, assembly cost and processing demands on the central controller are immediately obvious. However, other vitally important benefits include:

- the interface which IMDC offers between the off-line and the run-time parts of the environment
- inherent scaleability: the freedom to vary the number and type of control nodes on a particular machine in order to modify its functionality
- the ability to individually configure and test segments of the machine before they are combined

The case study based evaluation of IMDC is an ongoing process. In particular we aim to assess the efficiency with which the capabilities of the machine can be incrementally enhanced and the effectiveness with which the integrated tools can be utilised throughout the machine life cycle.

Conclusions

This paper has presented an overview of the Integrated Machine Design and Control environment (IMDC) which provides the machine builder with an integrated solution for co-ordinating tools throughout the complete machine life cycle. The commercial benefits of IMDC relate to both its impact through improved manufacturing efficiency for the end user and its potential to provide a stimulus to the fortunes of UK machine and control system vendors by:

- reducing the development cost and time of highly automated applications, and encouraging reuse of software/hardware building blocks
- reducing the cost of eliminating design faults (fast identification) and ease of service and maintenance
- allowing more effective adaptation and alteration of control systems, without resorting to the expensive services of a specific system supplier

The UK regrettably has no major vendors in the machine design and control systems marketplace. However, it does have a large number of innovative small to medium sized enterprises (SMEs) who need to be able to participate collectively in the market for computer controlled machines. IMDC has the potential to enable this by providing an environment for integrating design and control system elements from a diverse range of manufacturers. The approach gives SMEs the opportunity to participate far more effectively in a vast and growing automation market which is still dominated by large vendors of 'closed' systems such as Alan Bradley, Siemens-TI, Gould and Bosch. This is recognised by the industrial partners who are collaborating in the research.

References

Anon. 1993, Fieldbus update: ISP, WorldFIP, SP50, *Control Engineering* **40 No. 9**, 23.

APT, Applications Productivity Tool sales catalogue, Texas Instruments.

Bekkum J. 1993, The Coming of Open Programmable Controller Software, *Control Engineering*, October 1993.

Carpenter G. 1993, Tolerating Communication Failures, *Proceedings of the 1st Euromicro International Workshop on Parallel and Distributed Processing*, IEEE Computer Society Press, January 1993, 394-400.

Coutts I.A., Weston R.H., Murgatroyd I.S. and Gascoigne J.D. 1992, Open Applications within Soft Integrated Manufacturing Systems, *Proceedings of the International Conference on Manufacturing Automation*, Hong Kong, ICMA 92.

Harrison R. 1991, A Generalised Approach to Machine Control, PhD thesis, Loughborough University of Technology.

Heath S. 1993, Echelon - Networking Control, *IEE Review*, October 1993, 363-367.

ISaGRAF Industrial Software Architecture, CJ International, Siege Social, 86, rue de la Liberte, 38180 Seyssins, France.

MOSAIC/Europe, ESPRIT II research project, number 5292.

MOSAIC/USA, Machine Tool Open System Architecture Intelligent Control, contact P.K.Wright, Courant Institute of Mathematical Sciences, New York University.

Muir K. 1990, Stating the CASE for Industrial Automation, *Drives and Controls*, June 1990, 22-24.

RCS, Real-time Control System, Chief Robot systems Division, NIST, MD 20899, USA.

Reeve A. 1993, Plots and Pressure Focus on Fieldbus, *Intech*, **40 No. 7**, 21-23.

Weston R.H., Harrison R., Booth A.H. and Moore P.R. 1989, A New Approach To Machine Control. *Journal of Computer-Aided Engineering*, 27-32.

A 'PROCESS DESCRIPTION' APPROACH TO MANUFACTURING SYSTEM MODELLING.

G. J. Colquhoun* and R. W. Baines**

Liverpool John Moores University, Liverpool, L3 3AF

** University of Derby, Derby, DE22 1GB*

This paper shows how a 'Process Description' approach using the IDEF0 and IDEF3 methods draws upon established structured modelling principles to describe the behaviour of systems in terms of logical, causal and temporal relationships. The approach is demonstrated using a manufacturing case study in a company where a methodology incorporating the two complementary methods provides a basis for identifying problems and implementing the changes required.

Introduction

On established UK manufacturing sites systemic issues such as technology integration, operations streamlining, product quality and cost control, management information handling, product decisions, process technology and workforce stability have to be addressed in the context of diminishing financial margins and increased competition. Re-engineering has become a recognised approach to encourage proactive re-appraisal of the way that a company operates at any, or every level, strategic, tactical or operational. The re-engineering process can be aided by building a model of current practice, the 'AS-IS' state, using IDEF0 and IDEF3 followed by an analysis stage to determine ineffective areas, to compare with best practice, to set targets and to establish measures of performance. This leads to the formation of a 'TO-BE' model that will simplify, remove unnecessary procedures and eliminate waste as a precursor to implementation. The reality of typical manufacturing enterprise systems is that current practices almost always constrain future development and an understanding of those factors that critically effect system efficiency and effectiveness should be established before changes are made.

The case study described in this paper involved investigating high scrap rates in a press tool operation and assessing the implications of the problem for upstream and downstream processes.

The requirement to understand the logical and temporal relationships between processes led to the use of the 'process description' approach the IDEF3 method.

A brief review of the IDEF3 method.

The IDEF3 method (Mayer et al, 1992) has been developed as a result of the US Air Force Information Integration for Concurrent Engineering (IICE) programme sponsored by the U S Air Force Armstrong Laboratory.

IDEF3 is used to describe the behaviour of a system by capturing a representation of what a system does, rather than to predict what a system will do. It uses **Process Flow Network (PFN)** diagrams and **Object State Transition Network (OSTN)** diagrams. The PFN diagram represents processes (where a 'process' could be: a function, an activity, an operation, a decision, an action, or an event) connected by their temporal, causal and logical relationships. An OSTN diagram describes the various states of objects used in processes and relates those states in terms of the processes that cause objects to change state.

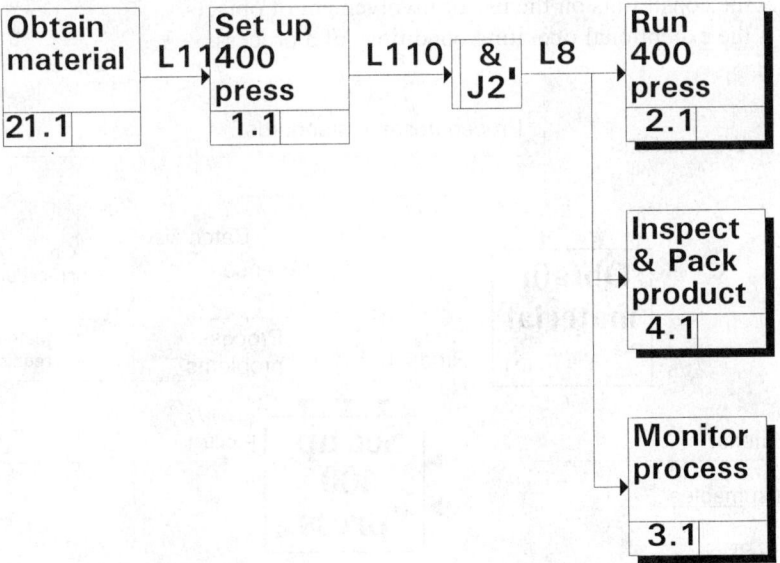

Figure 1. A Process Flow Network - Run 400 Press Line.

Figure 1 is a PFN, the large numbered boxes are **Units of Behaviour (UOB)** that describe processes. Smaller boxes labelled 'J' are **Junctions** used to represent the divergence or convergence of process logic. The box J2 represents an '**asynchronous and**' situation where the concurrent processes 2.1, 4.1 and 3.1 all take place following the sequential processes, 21.1 and 1.1. The principle of decomposition can be applied to 'explode' a UOB into lower levels to give more detail. UOB's that have been decomposed are indicated by shadowed UOB boxes. Junctions and UOB's are flow or a user defined relationship. Each UOB, Junction and Link can have an **'Elaboration Document'** to record more detail in a structured form.

An IDEF3 case study

IDEF3 was used to investigate a production press line prone to producing high scrap levels. Gathering and validating the information took eight hours on a consultancy basis by an analyst with no previous experience of the company. Information gathering focused on the behaviour of the line using an approach loosely based on that of Mayer et al (1993) to identify the following:

the individual processes involved

the conditions to start, maintain and stop each process.

responsibility for the process.

preceding and subsequent processes and the logic of processes.

the nature of the process, i.e. time to complete, type of occurrence :-
continuous, repeating, single instance.

the information, verbal or documented, required to carry out the process.

the objects that take part in the process, tools, material, resources etc.

the constraints on the use or involvement of objects.

the exceptional operating conditions of a process.

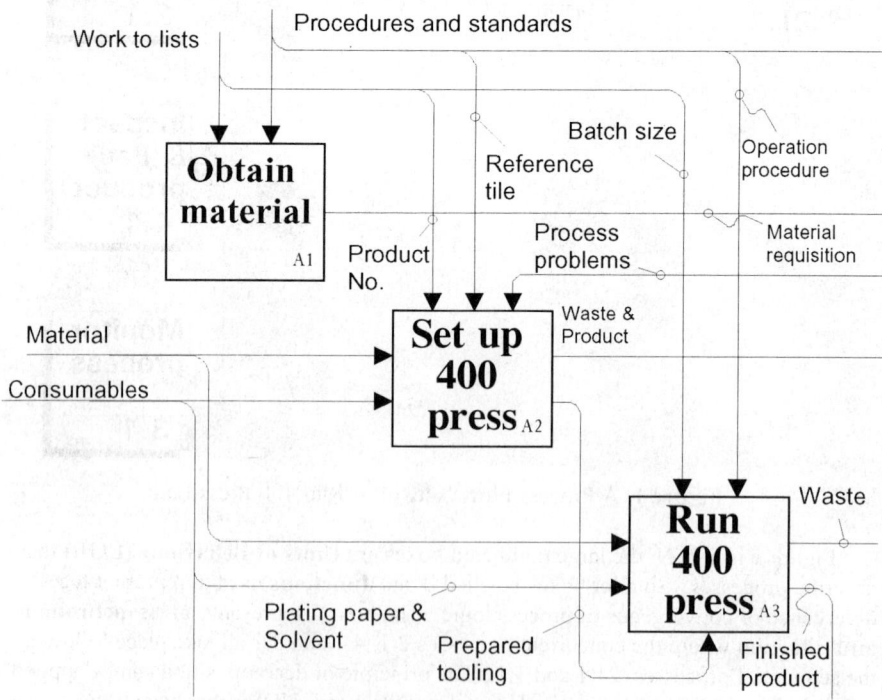

Figure 2. A section of IDEF0 diagram A0 'Run Press Line'

PFN diagrams were developed to two levels which exposed sufficient detail. OSTN diagrams are only produced for important objects and in this case a diagram was only produced for 'Material', the diagram is not shown in this paper. The initial investigation exposed a 'pool' of twenty five related processes. It was evident in the initial phase that a framework for investigation was required and an IDEF0 (US Air Force, 1981, Colquhoun, Baines and Crossley, 1993) model of the press line was used for developing the process description. The key IDEF0 diagram (a section of the diagram is shown in figure 2) provided a process 'viewpoint' of running the line.

Figure 1 shows the top level IDEF3 description of the production press line. The process, 21.1 must be completed before 1.1. The asynchronous junction J2 indicates that the concurrent processes 2.1, 4.1 and 3.1 all take place but do not necessarily start at the same time. An example Elaboration Document for UOB 2.1 related to running the 400 press is shown in Figure 3 .

ELABORATION DOCUMENT		
UOB Label: Run 400 press **UOB No:** 2.1	**Objects:** • Operator 1 • Operator 2 • Roving Inspector • Line supervisor • Plating paper	• Work to list (batch size) • Material • Solvent • Finished Product
Facts: • Throughput fixed at 90 products / min, Predicted efficiency 63%. • Fault free run times variable and not documented.	• Faults types are not documented. • Operators informally exchange tasks every 2 hours.	

Constraints:
• To initiate -Guards Closed
   Proving run accepted
• To continue process -
   Product quality acceptable
   (operator or roving inspector)
   Infeed plating paper available
	Infeed material available
	
	• To stop - Batch complete
	   Shift end
	   Supervisor stop (split batch)
	   Machine fault
	   Unacceptable product	

Figure 3. A UOB Elaboration Document

Discussion and Conclusions

Two aspects of the results can be examined (i) the behaviour of the production press line and (ii) the capability of the analysis method used.

(i) The application of IDEF3 revealed that the line operators were largely monitoring raw material and finished product in order to respond to highly frequent machine operation faults. It exposed that an operator on the output side of the press is required to check seven features of each product and if necessary reject it in a completely unrealistic time window of only 0.7 seconds. It also exposed that the operator on the input side of the press unnecessarily monitors previously inspected in-feed material.

(ii) Rigorous application of the IDEF3 method provided a good basis for a quantitative analysis of system behaviour. It was initially difficult to define the boundaries for the analysis when taking the 'bottom up' IDEF3 approach. This problem was solved by constructing 'top down' IDEF0 model to give an investigative framework the resulting successful methodology in which IDEF3 was used is shown in figure 4.

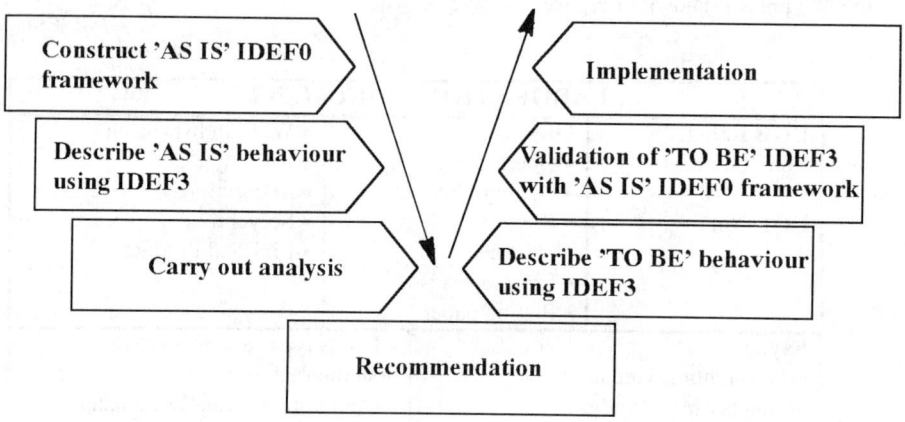

Figure 4. Analysis Methodology

In addition the case study has highlighted two areas of further work, to examine the performance of the process description approach in a higher level manufacturing scenario and to investigate in more detail the OSTN aspect of IDEF3.

References

Colquhoun, G. J. Baines, R. W., and Crossley, R., 1993, A state of the art review of IDEF$_0$, International Journal of CIM, 6,(4), 252 - 264.
Mayer, R.. J. Cullinane, T.P. DeWitte, P.S. Knappenberger, W.B. Perakath, B. and Wells, M.S. 1992, *Information Integration for Concurrent Engineering (IICE) IDEF3 Process description capture method report, AL-TR-1992-0057,* (Wright Patterson Air Force Base, Ohio).
U.S. Air Force, 1981, *Integrated Computer-Aided Manufacturing (ICAM) Architecture Part II, Volume IV- Functional Modelling Manual (IDEF$_0$), AFWAL-tr-81-4023,* (Wright -Patterson Air Force Base, Ohio)

Simulation and scheduling

ANALYSIS OF THE OPERATION AND PERFORMANCE OF FLEXIBLE MANUFACTURING SYSTEMS

Dr Alan Boyle and Dr Andrew Kaldos

Liverpool John Moores University
Liverpool, Merseyside, L3 3AF

The paper examines the extent to which the models, traditionally used to determine manufacturing system performance and operation, can be applied to the special problems generated by flexible manufacturing systems. The particular problems associated with the dynamic performance of the system, the part route scheduling, the workstation loading and the ability of the system to meet specific production requirements need to be examined. In determining the performance of a system the relevance of both mathematical modelling techniques and simulation methods are examined against the production data required to achieve a particular production schedule.

System Performance

A flexible manufacturing system consists of a computer controlled arrangements of independent programmable workstations combined with a materials handling system designed to manufacture a range of different component parts at low to medium volumes. The efficiency of the system is achieved when the workstation set up times are low, the routing of the parts through the system is flexible, pallets are used as holding devices for the material handling system and a range of fixtures are used for part holding and location.

The principle features of flexible manufacturing systems are a number of different workpieces can be machined simultaneously, the machining sequence or route through the system can be different for each component part, the operating sequence of the system is not interrupted by the need for retooling of the workstations, the materials handling and the component flow is not synchronised at a constant preset rate but is geared to the individual machining cycles of the workstations on the route of each component and most importantly the workstations can be accessed on a random basis. The production quantities achieved and the variety of product characteristics make flexible batch manufacturing systems attractive because :-

- reduced set up times can be achieved,
- reduced production throughput times or lead times can be achieved,
- reduced work in progress results,

- high levels of workstation utilisation are possible,
- increased flexibility for the production of component parts is made available,
- increased response to rapid market changes is possible,
- increased flexibility for the introduction of new products is made available,
- the system can be developed progressively.

Flexible manufacturing system operation is influenced by the the need to establish if the system capacity and layout will meet expected production quantities for a specified product mix while satisfying minimum target values and operational levels of performance for workstation utilisation, work in progress, lead times and production rates. The system must be programmed to satisfy a great variety of production schedules based on product characteristics and the immediate market demand. The operational performance is influenced by the component quantities, the route to be followed by each component part through the system, the machining and process cycle time at each workstation on the route, the capacity of the transport units and the physical layout of the system. Further operating factors important to the performance of a system are the scheduling rules for controlling the flow of the individual parts through the system, the nature of the interactions between the component parts in the workstation queues and the workstation tooling capacities which determine the distribution of the operations between the different workstations and hence the workstation cycle times.

The performance indicators for the system can never be totally independent of each other because an increase in the workstation utilisation level is often achieved by establishing at least one queue position on the input to the workstation, while the presence of a queue position on the output side of the workstation provides additional flexibility for the transport unit when moving parts from one workstation to the next. Thus increased workstation utilisation is achieved at the expense of increasing the work in progress, increasing the process lead time through queuing and reducing the production rate.

Production Schedule

Information concerning the system performance is required at the system design stage, is required as part of the production planning function and is necessary when considering alternative manufacturing strategies. Once a production schedule is established information is known about the length of the production run, the combination of parts making up the batch with the quantities required for each different part type, the order of the workstations on each part route and the processing cycle times at each workstation. Several methods are then available to determine the expected operating performance of either an existing or a proposed system.

The flexible manufacturing system given in Figure 1. is required to satisfy the production schedule given in Table 1. in a continuous production run time of 126.67 hours. The system contains six workstations arranged along a bi-directional linear track with a single transport unit. Component blanks and completed parts are loaded and unloaded through the same workstation. The transport unit takes 2.5 minutes to move a part from one workstation to the next and one part is carried on each pallet. Part loading and unloading at each workstation takes a further 0.5 minutes and each workstation contains one queue position at both the input and the output locations. The production schedule is characterised by a large variation in the production quantities, the relatively long workstation cycle times required for milling and inspection compared to the remaining cycle times and the variation in the scheduled route through the system.

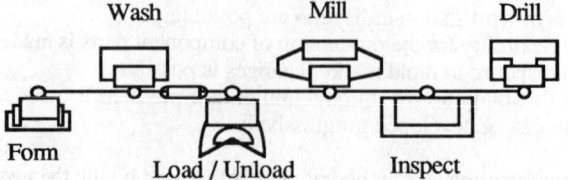

Figure 1.Arrangement of Flexible Manufacturing System

Table 1. Required Production Schedule

Part ID	Production Quantity	Workstation Cycle Time (minutes)						
		Load	Mill	Form	Drill	Wash	Inspect	Unload
P1	62	1.5	25.0	2.0	12.0	15	18.0	1.5
P2	138	1.5	20.0	2.0	-	-	15.0	1.5
P3	230	1.5	21.0	-	5.0	15	12.5	1.5
P4	320	1.5	8.0	4.0	19.0	15	19.0	1.5

Performance Based on System Capacity

The capacity analysis is based on establishing the batch manufacturing time, the production time for each component, the performance of each workstation and the utilisation level of the transport unit from a determination of the workstation cycle times, the quantity produced at each workstation and the part transport time. Once the transport time between workstations, where each part has one less transport move then the number of workstations on the route and the part exchange time between the workstation and the transport system is known, the facilities needed to produce each part can be determined. For each part the total production time is the sum of the individual processing cycle times and the transport time. The proportion of time each part spends at a workstation and in transport is determined as a ratio of the total production time. The operating time for a workstation is computed from the sum of the machining cycle time and the part exchange times with the utilisation value expressed as the ratio of the actual operating time to the available scheduled production time. Once the component transfer time and the production quantities are known the transport utilisation level can be determined from the ratio of the transport working time and the transport available time. If the system is designed to carry one or more parts on each pallet and all parts on a pallet are processed at the workstation before the pallet moves to the next workstation, the expected production quantity per pallet for each type of part can be determined along with the number of pallets necessary to meet the production requirement. If a number of different part types are in the system at the same time then the number of pallets required to produce each different part type can be used as a measure of the average number of parts in progress in the system. The expected performance of the system in attempting to meet the requirements of the production schedule are given in Table 2. and Table 3. Table 2. indicates the production schedule can easily be achieved with the system performance being dominated by the utilisation levels of the milling and the inspection workstations. Significant amounts of idle time are recorded for the remaining workstations. Table 3. shows the balance of the production time for each part is devoted to the actual manufacturing cycle with the transport activity occupying ten to fifteen percent of the production time. The average number of parts in the system, calculated from the number of pallets required, is found to be 4.22.

Table 2. Utilisation Levels of the Workstations and the Transport Unit

Workstation Name	Load	Mill	Form	Drill	Wash	Inspect	Unload	Trans
Processing Time (min)	1500	12075	1940	8280	9486	12516	1500	9985
Idle Time (min)	5000	925	11060	4720	3514	484	5000	3015
Utilisation (%)	23.1	92.9	14.9	63.7	73.0	96.3	23.1	76.8

Table 3. Part Production

Part ID	Total Prod Time (min)	Process Time (min)	Transport Time (min)	Process Time (%)	Transport Time (%)	Production Parts / Pallet
P1	90	75.0	15.0	83.3	16.7	144.4
P2	50	40.0	10.0	80.0	20.0	260.0
P3	69	56.5	12.5	81.9	18.1	188.4
P4	83	68.0	15.0	81.9	18.1	156.6

Performance Based on Queuing Systems

A particular feature of any flexible manufacturing system is the use of buffer storage given by the finite queue facilities at the workstations or by the buffer storage units in the system. The number of parts in the system and the average time a part spends in the system is dependent on the buffer storage, the number of workstations, the system layout and the transport arrangement. The system is represented as a series of independent workstations where workpieces arrive for processing and if the workstation is idle processing starts immediately, otherwise the part joins the end of the input queue. When the workstation finishes processing one part it either commences processing the first part in the input queue or it stands idle until the arrival of the next part. The system thus forms a set of tandem queues in series, as suggested by Hunt (1956). The first workstation in the system is assumed to have either a constant supply of parts arriving at a fixed interval of time or a random supply of parts arriving about a mean arrival time, Brandwajn and Jow (1988), but due to the the finite queue size at the remaining workstations parts moving through the system become blocked when the input queue at the next workstation is full, Berkley (1991). With finite queues between workstations the queuing network becomes very difficult and according to Papadopoulos, Heavey and Browne (1993) as the number of workstations in the system is increased it may be impractical to solve numerically. Using a single workstation model with a mean part arrival time of 15 minutes and an infinite queue at the workstation the operating performance of the manufacturing system given in Figure 1. is given in Table 4.

Table 4. Performance Summary

Part ID	Required Quantity	Mean Process Time (min)	Mean Time in System	Mean Time in Queue	Parts in System	Parts in Queue
P1	62	75.0	1905.0	1830.0	25.0	24.0
P2	138	40.0	1752.5	1712.5	43.6	42.6
P3	230	56.5	4808.0	4751.8	84.9	83.9
P4	320	54.0	6274.5	6220.5	115.9	114.9

Performance Based on Simulation

Flexible manufacturing systems are too complex to have their performance evaluated without some form of computer simulation. Progress in simulation technology, the development of high quality computer graphics for animation and the provision of dedicated and general purpose simulation software has increased the use of simulation in recent years. A simulation of the manufacturing system given in Figure 1., with a constant part arrival time of 17 minutes is shown to produces only 215 of the 750 parts required by the production schedule of 216.67 hours. The results are given in Table 5 and Table 6.

Table 5. Production Summary

Part ID	No Requ	Parts Made	Prod Rate Pts / Hour	Total Prod Time per Part (min)	Proces Time (min)	Queue Time (min)	Transport Time (min)
P1	62	55	0.25	222.2	75.0	102.2	45.0
P2	138	54	0.25	117.5	40.0	52.5	25.0
P3	230	53	0.24	155.5	56.5	69.0	30.0
P4	320	53	0.24	172.8	68.0	52.3	40.0

Table 6. Workstation Performance

Station Name	Proc Time (min)	Proc Time %	Queue Time %	Blocked Time %	Idle Time %	In Queue Time (min)	Out Queue Time (min)
Load/Unload	649.5	5.0	1.7	5.1	88.2	0.50	11.85
Mill	4014.0	30.9	0.8	5.1	63.2	3.88	13.29
Form	430.0	3.3	0.6	0.0	96.1	0.25	16.07
Drill	1937.0	14.9	0.6	0.0	84.5	0.25	13.87
Wash	2430.0	18.7	0.6	0.0	80.7	0.25	6.40
Insp	3471.9	26.7	0.8	0.0	72.5	0.36	8.59

The determination of system performance can vary considerably depending upon the method of evaluation selected. The system capacity approach does not provide sufficient information about the expected system operation and performance and the ability to deliver a specified product mix within a given production schedule. The approach always overestimates the production performance because it is dominated by the static capacity values and makes no attempt to recognise the system dynamics. Queuing networks and simulation attempt to take into consideration the dynamic performance of the system and recognise that dynamically the system always slows down due to the different parts competing for the same workstation resources.

References

Berkley, B.J. 1991, Tandem Queues and kanban controlled lines. *International Journal Production Research*, **29** (10), 2057-2081.
Brandwajn, A. and Jow, Y.L. 1988, An approximation method for tandem queues with blocking. *Operations Research*, **36**, 73-83.
Hunt, C.G. 1956, Sequential arrays of waiting times. *Operations Research*, **4**, 674-683.
Papadopoulos, H.T., Heavey, C. and Browne, J. 1993. *Queueing theory in manufacturing systems analysis and design.*1st edn (Chapman and Hall, London).

INTEGRATED DECISION MAKING PROBLEMS FOR THE OPERATIONAL CONTROL OF FMSs

M.D. Byrne and P. Chutima

The University of Nottingham
University Park, Nottingham, NG7 2RD, U.K.

The purpose of this paper is to address the main factors related to the operational control problems of the FMS including the part entry, machine scheduling, and AGV scheduling decisions. Besides the knowledge of the factors influencing the performance of the FMS, the interactions among these factors are also pronounced. In this study the machining centres are subject to random failure representing a more realistic operating environment. Furthermore, this allows the robustness of the operational control decisions to be tested. The results indicate that the performance of the operational control decisions investigated are significantly different and the interactions amongst these decisions are highlighted.

Introduction

Operational control can be viewed as an integral part of the design of such automated manufacturing systems as FMSs which prescribes to what extent the system components should be managed, integrated, and coordinated to achieve the targeted manufacturing performance. Although an individual system component may be designed to operate optimally in itself, but if it fails to be properly integrated into the overall system, negative effects on the whole system performance may be created. As a result, instead of merely focusing on the design of individual system components, the operational control designer has to consider the function of each component as a part of the whole system, examine its influence on the whole system performance, and analyze how individual components interact with one another, before establishing the operational control strategy of the FMS.

The operational control decisions (OCD) are the answers to the operational control problems consisting of: (i) what part should be loaded next at the load station? (ii) when should the part just loaded be released into the FMS? (iii) if alternative process routes are available, which route the part should progress through? (iv) which

371

operation should be performed first given that no precedence constraints are imposed in the sequences of operations? (v) if the part can be processed by alternative machines, which machine should be selected? (vi) which AGV should be dispatched to move the transport requesting part? (vii) when a machine becomes idle and there are still parts waiting for its turn of processing, which part should be selected for processing next? · We denote the decisions (i) and (ii) as the part entry decision; (iii) and (v) as the part routing decision; (iv) as the part operation sequence decision; (vi) as the AGV scheduling decision; and (vii) as the machine scheduling decision. In fact, there are many more decisions than those mentioned above, however, each of them could be classified as an instance of the main OCD. It is interesting to emphasize that not only the frequency of events that actuates the execution of the related decision but also the inherent characteristics and existing flexibilities of individual FMS's components influence the OCD of the FMS.

It is obvious that considerable research has been focused only on individual components of the whole operational control problem. Very few papers attempt to address integrated solution procedures which consider the relationships and interactions among the control elements (Nof et al. (1979), Hutchinson (1983), Garetti et al. (1990), Ro and Kim (1990), Sabuncuoglu and Hommertzheim (1992), Raju and Chatty (1993), Tang et al. (1993)). Essentially, this consideration can lead toward broader understanding of the total system control.

In this paper, three OCD are simultaneously investigated, namely the part entry, machine scheduling, and AGV scheduling decisions. To be able to test the robustness of the integrated decisions and represent a more realistic operating environment, the machining centres are prone to random failures. In the following sections, the configuration of the FMS, parameters, and OCD rules are described, followed by the results of the experiments, discussion, and conclusions.

Experimental Framework

The FMS consists of eight machining centres operated as closed networks of queues. Eight part types, which follow different routes and require different processing times, are machined on the FMS. Parts are loaded and unloaded onto the pallets at the loading and unloading stations. The preliminary study suggests that allowing 30 palletized parts in the system obtains highest system utilization. Parts are transported from one machining centres to another via two AGVs. Upon completion of a drop-off activity, the AGV checks for a transport requesting part. If no such request is detected, the empty AGV moves to the staging area where it will park and await a future transport request. The incoming and outgoing buffers of each machining centre are capable of accommodating five parts. A direct access part retrieval mechanism is operated which allows any part to be retrieved regardless of its rank. A common buffer area is used for temporary storage of parts which encounter a full incoming buffer and also to alleviate the blocking effects which may lead to the system locking. Also, it is assumed that there always exist production orders, but because of the limited space of the FMS's area, only ten of them are allowed waiting for available pallets and are considered for entering the FMS. As mentioned earlier, three sources of the OCD are the subjects of evaluation. Within each source of the OCD, four heuristics are developed for assigning priorities to parts or AGVs for dispatching. These heuristics include the following.

Part Entry Decision Heuristic

<E1> Select the job with minimum current number of the same part type already loaded into the system, provided that the first machine required is not broken down. If all jobs in the entry queue require failed machines, select the one for which the repair time is smallest. Break ties by using FCFS rule.

<E2> Select the job with minimum total operation time. The rest of the rule is the same as <E1>.

<E3> Select the job with minimum operation time required at the first visited machine, given that the first visited machine is not broken down. In this case, if all jobs in the entry queue require failed machines, none of them is allowed to load. As a result, they have to wait until one of the failed machine is repaired.

<E4> Select the job for which the current average flow time of the same part type (obtained from simulation) plus the current time to repair of the machines needed to be visited are minimum.

Machine Scheduling Heuristic

<M1> Select the job which has the least work remaining and its next operation does not require a failed machine. If all jobs in the incoming buffer have to visit failed machines, select the job for which the next machine is repaired first. Break ties by using FCFS rule.

<M2> Select the job with the smallest ratio of operation time to total processing time. The rest follows <M1>.

<M3> Select the job for which the incoming buffer of the next visited machine contains smallest number of jobs. The rest is the same as <M1>.

<M4> Select the job for which the summation of the operation time at the current machine and the expected waiting time at the next visited machine including the expected time to repair in case the next visited machine is currently broken down. The remainder is the same as <M1>.

AGV Scheduling Heuristic

<A1> Dispatch an AGV to the machine with maximum occupied outgoing buffer spaces and select the job that needs a non-failed machine for the next operation based on FCFS rule. If all parts require failed machines, select the part that requires the machine with minimum time to repair. Ties are broken by using the shortest travelling distance rule.

<A2> Dispatch an AGV to the machine for which the part in the outgoing buffer has spent longest time in the system after arrival. The remaining heuristics follow <A1>.

<A3> Dispatch an AGV to the machine with smallest travelling distance required from the current position of the AGV. The rest of the heuristic is the same as <A1>.

<A4> Dispatch an AGV to the machine for which the part in the outgoing queue requires the smallest expected waiting time at the next machine. The remainder is the same as <A1>.

Obviously, as an FMS comprises a high capital investment of integrated equipments, the sought-after objectives which the FMS's management needs to satisfy are related to maximizing the utilization of resources. In addition, the number of pallets/fixtures residing in an FMS is limited and costly, encouraging the need for achieving a relatively small time to produce jobs. Therefore, in order to appraise the merits of the OCD heuristics, three measures of performance are considered: (i) mean flow time; (ii) mean shop utilization; and (iii) mean number of jobs completed. Another factor considered is the level of uncertainty represented by the mean time before failure (MTBF) and mean time to repair (MTTR). Three level of uncertainly are considered in this study including (i) <960,60>; (ii) <840,90>; and (iii) <720,120>, where the numbers represent the MTBF and MTTR of each machine respectively. Simulation models are developed using SIMAN to compare the relative performance of alternative OCD heuristics. All the tactical decisions of the simulation process, such as identifying warm-up period, minimum run length, etc., are conducted in order to obtain unbiased data that represent steady-state behaviour of the system. Statistical analysis of results are employed to validate that the inferences drawn from the model are accurate and applicable for the system being investigated.

Experimental Results and Discussion

Because of the space restriction, the analysis of results is only presented in brief. Generally, when involving two or more factors, the effects on the system performance can be created by the main effects and interactions between them. The main effect is defined as the change in performance generated by a change in the level of the factor. Whereas, the interaction between the factors can be noticed if the difference in the performance between the levels of one factor is different at all levels of the other factors. To be able to test whether these effects are statistically significant, ANOVA is employed. The null hypothesis (H_o) is that there is an equality in the means among the different levels of each individual factor and also no interactions between the main factors are existed. Whereas, the alternative hypothesis (H_1) is that at least two means of the individual factor are different and there are interactions between the main factors. The results of the ANOVA will be presented in the conference.

As mentioned earlier, three levels of uncertainty are considered in this study. The results indicate a massive decrease in the system performance as the levels of uncertainty increase shown by highest F-ratio. This is due to the decreased production capacity of the shop which results in higher waiting time of jobs. Among the main OCD, the AGV scheduling decision influences the system performance most. Moreover, two and three factors interactions are also significant which emphasizes the need for a proper integrated decision. Another important thing is that the performance of the integrated OCD depends upon levels of uncertainty and criteria being used (see ExMxAxB). However, it is not easy to identify an overall best policy since it depends on several factors. In order to roughly identify which integrated policy can be recognized as a good one that always shows consistent performance, we first rank each policy according to each criteria under different levels of uncertainty, then sum these ranks, and based on these we rank them again to obtain the final results. It is found that the <E1,M3,A3> policy is the most consistent heuristic that provides good performance whereas the <E2,M1,A2> always show poor performance. Interestingly, the heuristic <E1,M3,A3> never ranked first on any criteria and levels of uncertainty. In contrast, the heuristics that rank first in one criteria performed poorly on other criteria. The reason that the heuristic <E1,M3,A3> provides consistent performance may be that it gives the best coordination between decision components.

When ANOVA indicates that there is a difference contributed by main effects or interactions, it is of interest to make comparison between each of the individual effects to discover the specific differences. The least square error method is employed in this regard. The analysis indicates that for the main effects, at 0.05 significant level, the OCD heuristics <E1>, <M2>, and <A3> significantly outperform the others. One may have thought therefore that the integration of <E1>, <M2>, and <A3> should be the best combination. However, as mentioned earlier, the integration of the best individual decisions is not necessarily the best integrated heuristic if their decisions fail to coordinate.

Conclusions

The heuristics for the OCD of the FMS are presented. The characteristics of these rules and their interactions are examined under different levels of uncertainty. The results indicate that good integrated OCD heuristics should be well coordinated and produce synergistic results. Furthermore, stand-alone heuristics that work well in isolation do not necessarily work well with other OCD policies if they fail to make decisions that can be coordinated. Finally, the results highlight that the improvement in the performance of the FMS depends upon the proper selection and implementation of an integrated decision policy.

References

Garetti, M., Pozzetti, A., and Bareggi, A. 1990, On-line loading and dispatching in flexible manufacturing systems, *Int.J.Prod.Res.*, **28**(7), 1271-1292.

Hutchinson, G.K. 1983, The design of an automated material handling system for a job shop, *Computers in Industry*, **4**, 139-145.

Nof, S.Y., Barash, M.M., and Solberg, J.J. 1979, Operational control of item flow in versatile manufacturing systems, *Int.J.Prod.Res.*, **17**(5), 479-489.

Raju, K.R., and Chetty, O.V.K. 1993, Design and evaluation of automated guided vehicle systems for flexible manufacturing systems: an extended timed Petri net-based approach, *Int.J.Prod.Res.*, **31**(5), 1069-1096.

Ro, I.K., and Kim, J.I. 1990, Multi-criteria operational control rules in flexible manufacturing systems (FMSs), *Int.J.Prod.Res.*, **28**, 47-63.

Sabuncuoglu, I., and Hommertzheim, D.L. 1992, Experimental investigation of FMS machine and AGV scheduling rules against the mean flow-time criterion, *Int.J.Prod.Res.*, **30**(7), 1617-1635.

Tang, L.L., Yih, Y, and Liu, C.Y. 1993, A study on decision rules of a scheduling model in an FMS. *Computers in Industry*, **22**, 1-13.

APPROXIMATE MODELLING OF FLEXIBLE MACHINING CELLS

Dr Stephen T Newman* and Dr Robert de Souza**

**Loughborough University of Technology,
Loughborough, Leics, UK*

***School of Mechanical and Production Engineering
Nanyang Technological University, Singapore*

This paper describes the application of an approximate modelling method for analysing the initial performance of flexible machining cells. The model uses Static Capacity ANalysis (SCAN) to compute the gross requirements for stations, transport equipment, manual workers, work-in-process, jobs and tools for planned production. SCAN has been implemented in Lotus 123. This novel implementation illustrates the benefits of utilising spreadsheets for rapid modelling of machining cells.

Introduction

The application of approximate modelling for analysing the initial performance of flexible manufacturing facilities has been reported by a number of authors, for example Wilhelm and Sarin (1983) and Looveren et al (1986). These models though, have had a poor response from industry in their use, due to poor accuracy and limitations when compared to simulation models. Another major constituent to the low usage of these models is that users find the data input requirements a heavy burden, when similar data is required again for more detailed simulation. The need to utilise the strengths of these models, though, is seen as a vital part of the design and assessment process of flexible machining cells.

To this end, SCAN has been built, within a computer spreadsheet package, enabling users to initially balance the performance of the cell specification. On obtaining satisfactory results, the user has the option of running another model, or proceeding to detailed analysis by simulation by also drawing on data from the SCAN spreadsheet.

Static Capacity Analysis Model (SCAN)

Static capacity analysis is a technique which adds together the total amount of work load allocated to each resource, and estimates the performance from these totals or calculates the gross requirements for the resource. The SCAN aggregate model computes the gross requirements of the planned production for flexible machining cells. The model has the following assumptions:-

1) The cell has a number of stations consisting of a storage buffer and the option of a process station. This allows for process stations such as machining centres, turning centres, auxiliary stations, load/unload and temporary storage stations to be modelled.
2) Each station storage buffer is connected to a cell transportation system.
3) Station buffers have a specified time to load/unload to the transport system, and if required, a specified time to load/unload the machine via the buffer.
4) Each part type enters and leaves the cell at the load/unload stations.
5) Part types have a predefined route from station to station with specified operation times at each station together with a defined pallet type.
6) Manual part operations may be defined such as palletising, repalletising, part setup, pallet load/unload and moving the transporter.

The areas of calculation of the model are:-

station requirements;
transport requirements;
manual requirements;
work in process;
job requirements; and
tooling requirements.

Abbreviations

N_p	=	Total number of parts
P_n	=	Part type n
NOP_p	=	Number of operations for part type n
nOP_s	=	Operation time for part type n at station s
BQP_p	=	Batch quantity of part type n
JP_n	=	Number of jobs of part type n
M_s	=	Workstation S
L_s	=	Load/unload time to stations buffer
L_M	=	Load/unload time to machine station from buffer
T_t	=	Average transport time between stations
Ph	=	Planning horizon
L_{SM}	=	Manual load/unload time to station buffer
L_{MM}	=	Manual load/unload time to machine station from buffer
nOP_{sn}	=	Manual operation time for part type n at station S
W	=	Maximum number of workstations

Station Requirements

The station requirement is computed from the cumulation of the total operations assigned to that station according to the part route and taking into account the amount of time required to perform a pallet exchange from transporter to the station and station to the machine if attached, times by the number of each job of the part type to be produced. Other useful performance indicators are the station utilisation for stations and the percentage of time a particular part type is produced on a station. These calculations are shown below.

Number of Stations Required of type S (Ts_S) =   Total Operation Time for Station S

Planning Horizon (Ph)

Total Operation Time for Station S (Top$_S$) = $\sum\limits_{n=1}^{Np} ((nOP_S + L_S + L_M)BQP_n)JP_n$

$Ts_S = \dfrac{TOps}{Ph}$

Utilisation of Station S $= \dfrac{Ops}{Ph} = TS_S$

Percentage of Part Type n = $\dfrac{\text{Total Operating Time on Station S of Part n} \times 100}{\text{Total Operation Time on Station S}}$
Produced on Station S

Transport Requirements

The transport requirement is determined from the cumulation of the total transport time between stations to collect and deliver a pellet via the transporter multiplied by the number of transport journeys required for each part type. The total of all the parts and routes represents the total transporter requirement. These calculations are illustrated below.

Number of Transporters Required = $\dfrac{\text{Total Transport Time}}{\text{Planning Horizon (Ph)}}$

Total Transport Time - Average Transport Time between Stations x Number of Journeys

Average Transport Time between Stations = $T_t + (2x\ L_S)$

No. of Journeys $N_j = \sum\limits_{n=1}^{Np} (NOP_n \times JP_n)$

Number of Transporters = $\dfrac{(Tt + (2 \times Ls)) \times Nj}{Ph}$

Work In Process And Job Requirements

The work in process requirement within a flexible machining cell consists of the time elapsed when a pallet has completed an operation and is waiting to be collected or has been transported and is awaiting the start of its next operation. In this model, the estimated time a part spends in storage is entered as a ratio to the total operation time of the part. Thus a pallet lead time may be determined from the total operation times on a pallet and load/unload times at stations plus with consideration of the ratio of storage time on the pallet. The pallet lead times may be cumulated to determine the total number of pallets required of a particular type over the planning horizon. These calculations are presented as follows:

$$\text{Pallet/Job Lead Time for Part type n } (JLt_n) = \text{Storage Ratio x} \sum_{S=1}^{S=W} (nOP_S * BQP_n) + L_S + L_{SM})$$

$$\text{Number of Pallets for Part type n } (NpL_n) = \frac{JLtn \times JPn}{Ph}$$

Manual Workers And Worker Performance

The manual requirement is determined from the cumulation of the total number of manual operations within the cell. Manual operations considered are the following load/unload, palletising, depalletising, machine setup and transporter moves.

$$\text{Number of Manual Workers} = \frac{\text{Total Manual Operation Time}}{\text{Planning Horizon (Ph)}}$$

$$\text{Total Manual Operation Time for all Parts at Station S } (MOp_S) = \sum_{n=1}^{Np} (nOP_{SN} + L_{sm} + L_{MM})$$

$$\text{Total Manual Operation Time} = \sum_{s=1}^{s=w} MOP_S$$

Tooling Requirements And Tooling Usage

The tooling requirement is determined from the definition of the part routing and from the summation of the tooling times used for the parts routed to that station. This provides valuable static results on the tool usage of each tool, and the number of total tool types required with sister tool prediction for a station type.

Model Implementation - SCAN

SCAN has been validated against a number of modelling platforms and has produced promising results, see Newman (1990). Some typical outputs that can be obtained are shown in Figures 1 to 3. A range of other output formats is available. This model has been implemented using the macro facilities of the IBM PC software spreadsheet Lotus 123. SCAN forms part of an integrated design aid for flexible machining facilities, which includes a number of simulators for multi-level modelling analysis of flexible machining cells, described by Newman (1991). SCAN also forms part of the initial models of this design aid and is integrated with a database management facility for transfer of data between models described by Rahimifard et al(1992).

Acknowledgements

The authors would like to acknowledge the FMS research staff of the Department of Manufacturing Engineering at Loughborough and the industrial collaborators of The

CIMulation Centre, Logica and Rockwell PMC. This work was part of a research programme funded by the ACME Directorate of SERC.

References

Wilhelm, W.E. and Sarin, S.C. 1983, Models for the Design of Flexible Manufacturing Systems, Proc. Ind. Eng. Conf., 564-574.

Looveren, A.J.V. Gelgers,L.F. and Van Wassenhove. 1986, A Review of FMS Planning Models Modelling and Design of Flexible Manufacturing Systems, (Elsevier Pubs.), 3-31.

Newman S.T., 1990, The Design of Flexible Machining Cells, PhD Thesis Loughborough University of Technology, UK.

Newman, S.T. 1991, The Design of Flexible Machining Facilities, Proc. of 5th SERC/ACME Research Conference, Leicester, August, 1991.

Rahimifard, S. Newman, S.T. and Bell, R. 1992, Data requirements for the Design of Flexible Machining Facilities, Proc. of the 8th Int. Conf. on Computer-Aided Production Engineering, (Edinburgh), 436-440.

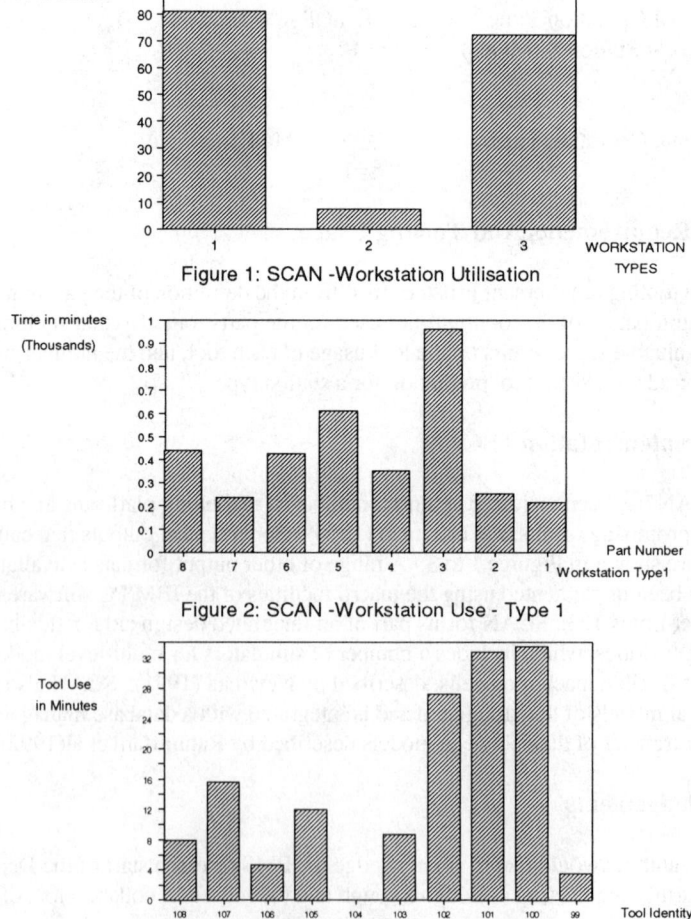

Figure 1: SCAN -Workstation Utilisation

Figure 2: SCAN -Workstation Use - Type 1

Figure 3: SCAN -Individual Tool Usage

BUFFER STORAGE DESIGN IN TRANSFER LINES USING SIMULATED ANNEALING.

J. Navarro Gonzalez and K.J. Aitchison

Cranfield University
Bedford, England, MK43 0AL

This paper presents a new method of solution for the buffer storage design problem to maximise the production rate of transfer lines. The approach combines reduced event simulation and an optimisation strategy based on approximation algorithms. FIST, a Fast Interactive Simulation Tool designed to model transfer lines, is used to simulate different transfer line configurations. The different configurations are generated by following an optimisation strategy based on simulated annealing. A modified simulated annealing algorithm, that includes a local search gradient algorithm, is used. This makes the algorithm flexible enough to move around the space of possible solutions to avoid poor local optima. Performance analysis of the algorithm shows high quality solutions are produced in reasonable time.

Introduction

A transfer line can be seen as a sequence of work stations joined to each other. The pieces to be processed enter at the start of the line, go through each work station in a given sequence and the final products emerge from the end of the line. Normally these lines are highly automated with automated material handling systems transferring the pieces from one station to another. This transfer activity is usually performed by a conveyor mechanism.

Transfer lines, as with many other man made designs, are not perfect. The different operations on the parts are performed with machines that are liable to failure or breakdown. These failures make the line unreliable. By linking the different production stages to form a line, machines become dependent upon on each other. If a machine breaks down, then the next machine in the line stops because it is starved of parts. Similarly, the previous machine becomes blocked because there is not any space available to put its output and also stops.

This blocking and starving of machines continues to propagate up and down the line until the machine that is broken down is repaired. So how can the effects of breakdowns be minimised or eliminated? A solution is to use storage locations or in-process buffers between the machines. Buffers diminish or eliminate the possibility of a stage of the line being forced to stop, by means of their storing and replenishing functions.

Buffer storage design problem

Due to the important role played by buffers in transfer lines, it is extremely important to determine their best size in order to maximise the performance of the line. However, the solution has a constraint, which is usually expressed in terms of the total buffer storage allowed or a maximum budget for buffer installation. In the case of transfer lines, the engineer is usually worried about obtaining the maximum production rate. In this context, the buffer storage problem can be formulated in terms of determining the optimal buffer sizes that maximise the production rate of the line for a given total buffer capacity. The objective of this work is to develop a tool that will help to do this.

Simulation of transfer lines.

Simulation modelling is a useful decision support tool to help analyse complex real life systems. With simulation it is possible to conduct experiments on a representation of the real system. As pointed out by Forrester (1961), "changes in systems behaviour can be traced directly to the causes, because the experimental conditions are fully known, controllable, and reproducible". This feature makes simulation an attractive tool to examine and compare alternative system configurations.

Problems when simulating transfer lines.

Transfer lines present a problem to conventional discrete event simulation - the piece by piece movement, particularly along conveyors, means that the time required to simulate a transfer line can be long. A rough cut modelling tool, based on a reduced event simulation technique, was developed by Cranfield University for the Ford Motor Company Ltd (Aitchison, 1993). This system, known as FIST, improves the speed of transfer line simulation by a factor of 100 whilst maintaining results within 5% of the equivalent discrete event model. This allows a large number of line configurations to be tested in a short time.

Buffer size problem as a combinatorial problem

The buffer design problem can be viewed as a combinatorial optimisation problem. This involves the search for optima of a function of many independent discrete variables. This function is called the "cost function", "objective function", or "score function", and represents a quantitative measure of performance of the system.

Solving a combinatorial optimisation problem consists of finding the optimal solution among a finite number of feasible options. This set of possible solutions or configurations of the system is called the "configuration space", "state set" or "state space". This set is specified by the constraints of the problem. The cost function assigns a real number to each configuration in the set.

In the buffer design problem, the production rate as a function of the buffer sizes assumes the role of the cost function. The set of configurations amounts to all the possible combinations of buffer sizes for a given total buffer capacity to be distributed in the line. This total capacity, together with positive values for buffer sizes are the constraints, which specify the set of feasible configurations.

Simulated Annealing

Annealing is a physical process in which a solid in a heat bath is heated up to a maximum value at which all particles of the solid randomly arrange themselves in the liquid phase followed by cooling through slowly lowering the temperature (Aarts and Van Laarhoven, 1988). As the solid cools, Boltzmann discovered that a system in equilibrium for a given temperature T will assume a configuration corresponding to an energy level E with probability exp(-E/T), known as the Boltzmann factor.

Metropolis et al. (1953), developed a Monte Carlo method to generate the sequence of states that a solid goes through to reach a configuration with minimal energy and low temperature - simulated annealing. It is not difficult to take this idea and apply it to a transfer line. Each of the states of the solid would represent a configuration for the line, and the energy level would represent the cost. Thus as we move through different line configurations (states) we are searching for the configuration with the lowest cost (energy).

Simulated annealing is more than an iterative search, however. The danger with iterative search algorithms is that they can become stuck at local optima. Simulated annealing uses a control parameter that gives a probability of accepting a solution. The probability of accepting a solution that results in equal or higher energy is given by

exp($-\Delta E/(K_b*T)$), where K_b is the Boltzmann constant.

An annealing algorithm based on the ideas above was developed and implemented. However it was tested and found to be slow. The reason for this is that simulated annealing is sensitive to the size of the configuration space. For large transfer lines or large total buffer capacities, a large amount of time is required to find near optimal solutions.

Modified Simulated Annealing

The algorithm that was finally developed in this study tries to remedy this problem by integrating the annealing algorithm with a local gradient search algorithm to speed up the search for the optimum by improving the transition mechanism.

Transition Mechanism

The aim of the proposed transition mechanism is to move more quickly to good configurations than is possible by using the slow transition mechanism of pure simulated annealing. This is done by employing a gradient search technique. The gradient provides local information about the rate of change of the production rate. Transitions to new configurations are made in the direction of the maximum variation at the current point. However, if the new point has a lower production rate, or the new point is not accepted by the acceptance function, a new step size is randomly generated. The interpretation of these variable step sizes is different buffer capacities being moved and redistributed in the buffers. In this way, new paths to different local optima can be started. In turn, each of these paths can originate new branches and so on. The resulting algorithm, based on a descent algorithm developed by Eyler (1979), combined with FIST as the simulator, can be used to quickly test a large number of configurations.

Results.

Two models are presented to compare the performance of the pure and modified annealing algorithms. Both algorithms use FIST as their simulator.

Example 1.

This model consists of a transfer line of 5 machines. The line is sufficiently small to be able to test and compare a large number of configurations, but still has the characteristics of the lines that are to be modelled. In this example a total of 100 buffer spaces can be redistributed to optimise the line. The line configuration is shown in figure 1. Machine parameters are given in table 1.

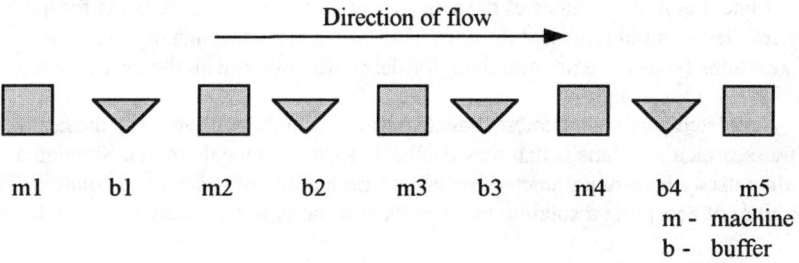

Figure 1. Line configuration.

Table 1. Machine parameters for example 1.

Parameter	m1	m2	m3	m4	m5
Cycle time (minutes)	0.1	0.35	0.32	0.35	0.35
Break down %	0	3	5.7	3	2.6
Mean repair time (minutes)	0	21	13	25	11

The results are shown in table 2. These show that the difference between pure annealing and the modified algorithm is minimal. There is however a large difference in the number of configurations tested.

Table 2 Results.

Parameter	Pure Annealing	Modified Annealing
Production Rate	153.58	153.50
Buffer 1	2	1
Buffer 2	43	39
Buffer 3	34	34
Buffer 4	21	26
Configurations tested	23,920	2,950

Example 2

This is based on a real production line at the Ford engine plant in Bridgend. The line consists of ten machines, and a total of 3104 buffer spaces can be redistributed. The results for the pure and modified annealing algorithms are shown in table 3.

Table 3 Ford line optimisation results.

Parameter	Pure Annealing	Modified Annealing
Start Production Rate (parts per hour)	438.70	438.70
Optimised Production Rate (parts per hour)	453.90	453.50
Time Required	4.33 hours	0.88 hours
Percentage improvement	3.46	3.37

Again these results show that the difference between the solutions from the pure annealing and modified annealing algorithms are minimal, but a substantial saving in the computational time required is achieved from the modified annealing algorithm.

Conclusions.

This paper has presented a new approach to the optimal buffer storage design problem in transfer lines. A modified simulated annealing algorithm that incorporates a local gradient search algorithm is used to generate new configurations to accelerate the search for the optimal line configuration. This has been shown to provide solutions that have minimal differences when compared with pure annealing, but offer substantial savings in the computation time required.

References.

Aarts, E. H. L. and van Laarhoven, P. J. M. 1988, *Simulated Annealing: Theory and Application,* 2nd edn, (D. Reidel Publishing Company, Dordretch, Holland).

Aitchison K.J. 1993, FIST - Fast Interactive Simulation Tool. *IEE Colloquium On Increased Production Through Discrete Event Simulation* (IEE London).

Eyler, M. A. A. 1979, New Approach to the Productivity of Transfer Lines. PhD thesis, Harvard University.

Forrester, J. W. 1961, *Industrial Dynamics*, (The MIT Press, Cambridge MA.)

Metropolis, W. Rosenbluth, A. Rosenbluth, M. Teller, A. Teller, E. 1953, Equation of State Calculations by Fast Computing Machines. *Journal of Chemical Physics*, **21**, 1087 - 1092.

Acknowledgements

The authors would like to thank Mr. J. Ladbrook and Mr. N. Thomas of the Ford Motor Company Ltd., Engine Operations, Dagenham, for their support and guidance.

PC BASED ROBOT SIMULATION APPLIED TO AUTOMOTIVE ASSEMBLY CELLS

I.D. Jenkinson, P.A. Montgomery and B. Mills

*School of Engineering and Technology Management,
Liverpool John Moores University, Liverpool, UK.*

Small robot workcells at Vauxhall Motors Ltd. Ellesmere Port plant are currently being designed and implemented in-house using conventional methods. The recently introduced PC based low cost robot simulation package Workspace 3.2, has initiated an evaluation exercise. This paper shows the development of a robot welding cell using Workspace 3.2. It contains an evaluation of the system as a robot cell design and off-line programming tool and illustrates some of the problems associated with the implementation of such systems to this industry.

Introduction

Robot simulation packages offer the robot workcell planner and programmer many advantages (Montgomery and Jenkinson 1993). They can aid fixture design, help optimise cell layout and reduce programming time. Should design changes to the product require modification of the manufacturing process, changes to the cell layout and reprogramming can be carried out off-line.

There are many robot simulation packages currently available in Europe. Amongst these are IGRIP, Robcad and Grasp (Mahajan and Walter 1986) (Yong and Bennaton 1988). These systems are workstation based and relatively expensive. For many companies the expense is often difficult to justify and such systems are mostly used by subcontractors specialising in the design and installation of large high value automation systems.

Vauxhall Motors Ltd. at Ellesmere Port are currently in the process of automating many small assembly cells. Until now the design, installation and programming of new installations has been subcontracted. In this case the process is being carried out in-house using conventional cell design and robot programming techniques. A simulation system would be a useful tool in the design and programming of these workcells, given that its cost could be justified. The introduction of the low cost PC based robot simulation system Workspace 3.2 prompted an evaluation study. This was undertaken using assembly cells recently developed in collaboration with the University.

Simulation package Workspace

Workspace 3.2 requires an IBM compatible PC, preferably a 486 machine. The package currently offers;

1. a comprehensive 3D solid modelling system,

2. a robot library and range of kinematic solutions including a mechanism for the iterative solution of robots for which the exact kinematics are not available,

3. a file import file facility via DXF or IGES formats,

4. workcell and robot calibration routines,

5. program development in supported robot languages avoiding postprocessing,

6. incorporation of a dynamic modelling facility.

Robot welding assembly cells.

Two robot assembly cells were modelled: the Front Panel Upper Assembly Cell and the Robot Welding Training Cell.

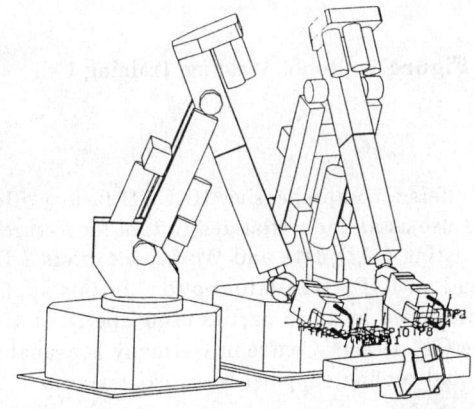

Figure 1: Front Panel Assembly Cell

The Front Panel Upper Assembly cell (Figure 1), manufactures a subassembly of the front end of the Astra car and was recently completed

in-house in collaboration with the University. The cell is designed to spot weld
the six subassemblies that make up the Front Panel Upper Assembly. It
comprises two Cinncinati T3 robot systems equipped with spot welding guns, a
manually loaded fixture mounted on a rotary table and a Gould Modicom PLC
unit. This cell having been recently built, provided an opportunity to assess the
overall suitability of Workspace as an industrial robot workcell design system.

The Robot Welding Training cell (Figure 2) comprises a GMFanuc S-420
robot, weld gun and side body panel located in a simple fixture. The cell is
designed to train operators to use the current range of GMFanuc robots now in
wide spread use throughout the factory. If Workspace were to be successfully
applied as an off-line robot programming system the Karel code produced would
have to be compatible with this range of robots.

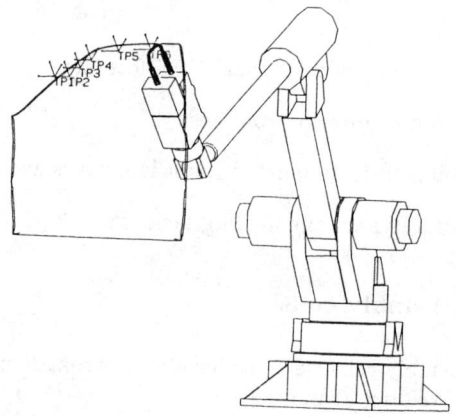

Figure 2: Robot Welding Training Cell

Cell modelling

Workspace 3.2 offers a comprehensive 3D CAD facility. However, it was
found to be of limited use as an industrial design tool for fixtures and tools etc. It
is preferable to use existing CAD data and Workspace offers a DXF file import
facility as standard and an IGES translator option. In this application much of
the model data was produced by third parties using specialist CAD systems. Part
data generated by the GM Design Centre in Germany is available in 3D and
drawings of fixturing and tooling supplied by subcontractors in 2D.

3D part drawings supplied in IGES format were translated into the design
package AutoCAD where they were modified and edited to extract the required
features. These were then converted into DXF format and imported into
Workspace. Only the essential features of fixtures and tools were modelled and
this was achieved using the 2D information supplied.

The results proved satisfactory in this instance where one or two robots and relatively simple parts are involved. The model sizes are relatively small and the PC platform permits good user interaction with the model. Should more complex models require say, a whole car body shell to be used, a PC based system would become unusable.

The robot library is an essential part of a robot simulation package. Workspace in common with other systems offers an extensive robot library. Within Workspace the robot structure is represented in terms of Denavit-Hartenberg notation (Denavit and Hartenberg 1955) and robots can easily be created and existing robots in the library easily modified. In practice it may not be possible to use a specific robot directly from the library without modification. The zero positions of the joints of the library robot may differ considerably from those of the target robot. At Vauxhall motors in-house maintenance of robots results in a repositioning of the joint zero positions. This may be important if a simulation is to accurately model the arm configuration.

Off-line program generation

In Workspace the simulation program is written in the language of the target robot. For the Training Cell using the GMFanuc robot the language used is Karel. Instead of a postprocessor Workspace users must purchase a robot language. The code produced was compatible with that used by the GMFanuc robots at Ellesmere Port. Some aspects of the language are not as yet supported, for example the ability to store a sequence of positions as a path. Thus the following Karel program would not be executable within Workspace:

```
program mike
var
home : position from mainline
topounce : path
begin
$speed =1500
$motype = joint
$termtype = coarse
move to home
move along topounce
move to home
end mike
```

It was not possible to integrate Workspace with the GMFanuc spot welding package Spot Tool widely used throughout the plant. This would be essential if Workspace were to be successfully applied in this instance.

Conclusions

Workspace is easy to use and can produce accurate simulations. In the authors' view its 3D CAD facility was suitable for cell planning and off-line programming purposes. However, it was not as yet sufficiently developed for use as a design tool for parts, tooling and fixtures. At Vauxhall Motors Ellesmere Port, much of the tool and fixture design is subcontracted and part data is passed down from the GM Design Centre in Germany. In this case the ability to make use of existing CAD data using DXF and IGES file import facilities is important.

The Karel language available for evaluation produced code compatible with the GMFanuc robots in use at Vauxhall Motors Ltd. However, it was not possible to integrate the Workspace generated Karel programs with the GMFanuc spot welding application package SpotTool in use at that site and further work remains to be carried out on this problem.

In conclusion Workspace can be usefully applied as teaching aid but has limited application within the complex environment of a car assembly plant.

Acknowledgments

The authors would like to thank Vauxhall Motors Ltd. for their support and assistance.

References

Denavit, J. and Hartenberg, R.S. 1995(June), A Kinematic notation for lower-pair mechanisms based on matrices. *Transactions of ASME-Journal of Applied Mechanics*, **22(2)**, 215-221.

GMF S-420 Descriptions Manual. 1992, GMFanuc Robotics Corporation, Michigan 48057-2090 USA.

Montgomery, P.A. and Jenkinson, I.D. 1993(Nov), The need for Simulation in the design of a robot controlled system. *IEE Colloquium C15 Advances In Practical Robot Controllers.*

Mahajan, R. and Walter, S.E. 1986(Aug), Computer Aided Automation Planning: Workcell Design and Simulation, *Robotics Engineering*,12-15.

Workspace User Manual Version 3.2. Robot Simulations Ltd, Newcastle Upon Tyne, NE4 6UL, England.

Yong,K. and Bennaton, J. 1988, Off-line Programming of Robots Using a 3D Graphical Simulation System, *CAD based Programming for Sensory Robots, Springer-Verlang*, 235-252.

EVALUATING THE PERFORMANCE OF FLEXIBLE MANUFACTURING SYSTEMS

Dr Alan Boyle and Dr Andrew Kaldos

Liverpool John Moores University
Liverpool, Merseyside, L3 3AF

The paper examines the influence system design changes have on the operation and performance of a flexible manufacturing system designed to meet specified production requirements. Decisions relating to the length of workstation queues, the size and location of buffer storage units, the number and type of transporter units and the system layout significantly influences the production rate, the workstation utilisation levels, the work in progress quantities and the lead times achieved. The production rate and the production quantities achieved are used as a measure of performance. Simulation is used to determine the impact of the design changes on the system dynamics and hence on both the operating flexibility and the system flexibility.

Introduction

Traditional manufacturing systems have concentrated on mass manufacturing requirements where the volume of production is the primary organisational and economic factor to be satisfied. This aspect of production is seen in the application of transfer lines using synchronous part transfer methods, automated production lines employing asynchronous part transfer and assembly lines based on continuous part transfer. These systems are characterised by high production rates, high volume of production and a product flow layout with the emphasis placed on system efficiency and the economics of scale. The introduction of Computer Numerically Controlled (CNC) machine tools into batch manufacturing combined with a rapid expansion of computing power, to aid the system organisation, data preparation, equipment control and performance monitoring functions, has promoted the development of flexible manufacturing systems (FMS) aimed at producing low to medium volume capacities combined with a high variety of different parts. High product variety is achieved by employing standard CNC type machining centres, programmable material processing units and automatic measuring, inspection and testing facilities. High volume combined with high variety is achieved by the use of programmable materials handling and part transfer systems. Evaluation of the operation and performance of the system is required for both production scheduling and system design purposes. If the operating life of a flexible manufacturing system is assumed to be greater then the shelf life of the product range used for the original system design, then further demands are placed on establishing the performance statistics for the system.

System Design

At the design stage, system flexibility is required to accommodate the manufacture of future product designs while operational flexibility is required to enable the system to produce a wide variety of different components in varying quantities, as part of the product mix allocated to a particular production schedule. The operation and performance capability and the ability to satisfy a specified production requirement are significantly influenced and critically dependent on a number of design decisions. The workstation selection, the system layout and the number of transport units control the degree of operational flexibility achieved by the system but the performance can not be predicted from an inspection of the system arrangement. Evaluating the system performance for a specified product mix is especially important at the system design stage before the necessary capital resources have been allocated to building the system.

The flexible manufacturing system given in Figure 1. contains a total of fourteen workstations supported by a looped track arrangement with crossover facilities and four transport units. Multiple workstations exist in the form of two loading and unloading units, two CNC facing machines, three buffer storage units and five CNC machining centres. The facilities are completed by a washing unit and a CNC drilling machine which allows short drilling cycles to be carried out on selected components and removes some specialist machining operations from the cycle times at the machining centres.

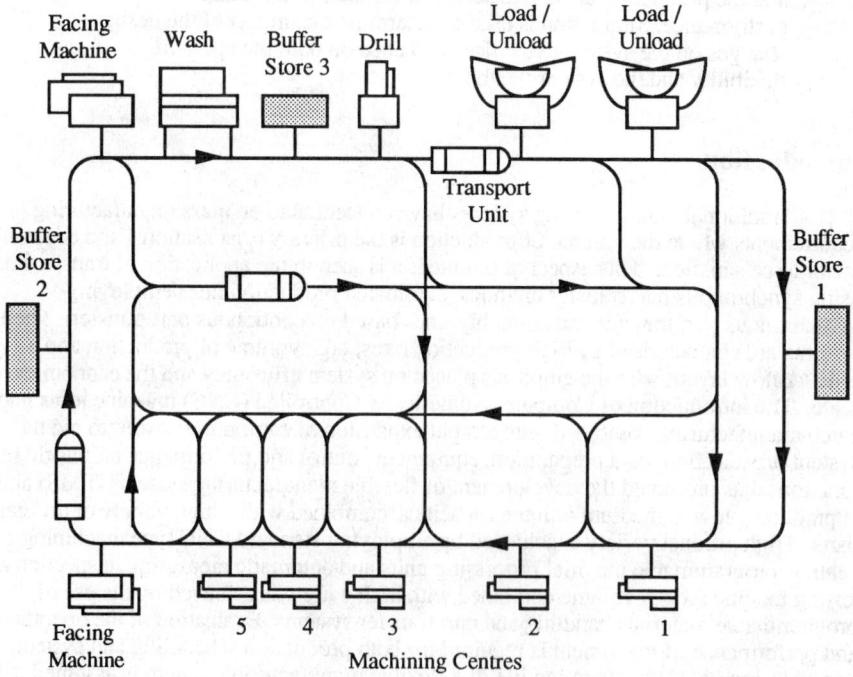

Figure 1. System Layout

Production Schedule

The system is required to produce a batch of 2000 components comprising twelve different part types, within a continuous production time of 80 hours. The production quantities along with the workstation operations, the workstation cycle times and the anticipated buffer storage time between operations are given in Table 1. The part exchange time for loading and unloading a component at a workstation is 0.5 minutes and the transport assignment time for moving a component from one workstation to the next is 3.0 minutes. The multiple workstations have identical tooling sets and tool replacement is carried out automatically during the production run. The total cycle time given in Table 1 for each part is the sum of the processing cycle times and the transport assignment time for moving the part between the workstations.

Table 1. Workstation Cycle Times (minutes)

Part	Quantity	Load	MC	Face	Wash	Drill	Store	Unload	Total
P01	127	1.0	30.0	-	2.0	-	17.5	2.0	64.5
P02	83	1.0	30.0	12.0	2.0	-	22.8	0.5	86.3
P03	103	1.5	26.5	7.0	6.0	15.5	28.5	0.5	112.5
P04	46	0.5	16.0	8.0	-	4.0	-	1.5	45.0
P05	68	2.0	20.0	8.0	4.0	-	-	2.0	51.0
P06	94	2.0	14.0	8.0	4.0	4.0	-	2.0	49.0
P07	87	1.5	22.0	-	4.0	4.0	16.5	1.5	64.5
P08	153	1.0	18.0	8.0	4.0	-	16.0	1.0	63.0
P09	116	1.5	16.0	8.0	-	-	13.5	1.5	55.5
P10	82	2.0	16.0	8.0	4.0	4.0	8.9	1.5	59.4
P11	103	1.0	12.0	-	2.0	3.0	4.6	0.5	38.1
P12	138	1.0	10.0	6.0	-	4.0	5.5	1.0	48.5

It is clear from the processing cycle times for each part type that a conventional batch scheduling approach based upon producing each complete set of parts then retooling for the next part type can not achieve the production quantities demanded in the available production time. A FMS, by producing a mixture of different parts at the same time can achieve significant improvements in the throughput time for the batch. However the information given in the production schedule and in the system layout does not give any indication of how the variations in workstation cycle time will react with each other, of how the system will perform dynamically and if the production schedule can be achieved.

System Performance

Using simulation techniques a performance evaluation of the system is carried out leading to changes in the original system design and operating schedules. The simulation shows a total of 1125 individual components can be produced in the 80 hours allocated to the production, at an average production rate of 14 components per hour In operation, a balance is required at each workstation, between the utilisation level of the workstation, the amount of work in progress in the queue, the lead time required to produce the component part, and the flexibility needed by the transport units to deliver and remove parts, Boyle and Kaldos (1993). The workstation cycle time compared with the amount of time each component spends in the workstation queue is a significant factor in determining the total throughput time of each component and hence of the production rate.

The transport time is a function of the system layout, the transport system selected and the programmed transport velocity. Changes in transport velocity do not have a major impact on the production achieved. The overall performance of the system for the simulated 80 hour period is given in Table 2.

Table 2. System Performance

Part ID	No Req	No Made	Prod Rate (Pts/Hr)	Prod Time (min)	Station (min)	Storage (min)	Trans (min)
P01	127	127	1.59	74.0	35.0	32.6	3.9
P02	83	83	1.02	93.5	45.6	39.1	4.0
P03	103	103	1.29	121.4	57.1	53.0	6.9
P04	46	46	0.57	85.7	30.2	47.7	4.3
P05	68	62	0.77	76.9	36.0	34.6	3.7
P06	94	59	0.74	80.5	34.1	40.9	3.7
P07	87	87	1.09	88.6	33.1	48.2	3.9
P08	153	153	1.91	80.9	32.1	41.2	4.1
P09	116	116	1.45	73.1	27.0	39.4	3.8
P10	82	82	1.02	94.4	35.6	52.1	3.7
P11	103	71	0.89	67.1	18.8	41.8	3.7
P12	138	137	1.71	69.6	22.3	40.7	4.0

Workstation Queues

Increasing the number of queue positions at a workstation increases the utilisation level, the work in progress and the lead time. The presence of a workstation queue at the input to the workstation allows the transport unit to make a delivery as soon as the processing cycle at the previous workstation is complete, a transport unit is available and the track is free. A queue on the output side of the workstation provides a temporary buffer store to absorb the fluctuations in the system caused by the different cycle times necessary to produce the different parts and provides increased flexibility for transport scheduling when selecting the part to be moved next if different parts are competing for the same workstation. The best system performance is achieved when each workstation had one queue position on the input and one queue position on the output. The wash unit, because of the short cycle times, required three queue positions on both the input and the output. The buffer storage units proved to be most useful with five queue positions.

Transport Units

The number of transport units employed need to be optimised for the system capacity and track layout being used. For a simple linear track layout, serving a limited number of workstations, one transport unit is frequently used as this provides the maximum flexibility. It makes the transport scheduling rules relatively simple, it avoids the possibility of the track becoming blocked and allows access to all the workstations in a random order. As the system complexity is increased with more workstations, a greater number and variety of parts and a more complex system layout the number of transport units is increased. As the number of transport units are increased the performance of the system, the production rate, the production quantity and the lead time, are progressively improved until an optimum number of transport units is reached after which the overall performance of the system reduces with an increase in lead time and a reduced production rate. The decrease in the system performance results from a saturation of the transport units as they increasingly block the route to the next workstation.

Multiple Workstations

To accommodate large variations in the processing cycle times of different components, multiple workstations of the same type and capacity are frequently employed. This is particularly important when a relatively long processing cycle time at or near the start of a component route is followed by a number of shorter processing cycle times. Multiple workstations are normally located in close proximity to each other in the system layout as they can then act as machine clusters with attendant advantages for the distribution of operations and tooling sets, Sarin and Chen (1987). This is a particular advantage when the tooling requirement of a particular product mix is greater than the tool holding capacity of any single machine. Multiple machines in close proximity allow all the machines to carry a sub set of tools required for the production of the majority of parts for the long cycle times with the remaining tooling capacity allocated selected tooling sets of the production of shorter cycle times or for a specific type of component.

System Layout

The layout of the system is normally one of three types, Sethi and Suresh (1990). In the case of simple arrangements with relatively few workstations the most common arrangement is a single bi-directional track with the workstations equally spaced along the length. As the number of workstations is increased, usually above five or six, the track layout is developed into a simple loop arrangement with a set of transport units moving in one direction only. This helps to avoid bottlenecks in the system by allowing empty transport units to move ahead of the transport units undertaking an assignment. The advantages of this arrangement are the simple track arrangement, the increased number of transport units in the system, the simple scheduling rules employed and the absence of bottlenecks on the track. The disadvantages are the increased length of the track, the corresponding increase in transport time required to move parts through the system and the reduced flexibility to move components directly to the next workstation on the route. In an attempt to increase the flexibility of a loop layout crossover sections are introduced from one part of the system to another. The crossover sections allow selected workstations to be bypassed should they not be required or if they are not available to accept the next scheduled part. The introduction of the crossover sections provides greater flexibility, reduces the component lead times through the system and increases the production rate but the scheduling rules, the controlling computer system arrangement and the operating software are made correspondingly more complex.

Conclusions

The provision of an input and an output queue at each workstation proved to be a major factor in achieving the production schedule as they provided greater flexibility to the transport system. The provision of buffer storage facilities in the system also made a contribution to the system flexibility, particularly when different components were competing for the next workstation on their route. The multiple workstations require a scheduling rule for loading them based on the component finding the workstation that was idle or had the lowest backlog of work based on machining and queue times.

References

Boyle, A. and Kaldos, A. 1993, Some aspects of FMS design and performance. *Proceedings of the 9th National Conference on Manufacturing Research*, 82-86.
Sarin, S.C. and Chen, C.S. 1987, The machine loading and tool allocation problem in an FMS. *International Journal of Production Research*, **27** (7), 1081-1094.
Sethi, A.K. and Suresh, P.S. 1990, Flexibility in manufacturing:A survey. *International Journal of Flexible Manufacturing Systems*, **2**, 289-328.

SIMULATION IN THE DESIGN AND ECONOMIC JUSTIFICATION OF A PORCELAIN PRODUCTION LINE

S Perez-Castillo, J A Barnes and J M Kay

Cranfield University
School of Industrial & Manufacturing Science
Cranfield Bedford MK43 0AL

The manufacture of porcelain figures has traditionally been craft based with little automation. This study primarily describes the use of simulation to assess the viability of automating the total process. Processes, such as hand painting, which are highly skilled and regarded as essential, may be linked by processes which are capable of automation, such as mould storage and retrieval, glazing and firing. The simulation model developed to represent the proposed line was used to assess its capability, adjust labour and buffer capacities, identify potential bottlenecks and to balance the work flow. The paper will describe the experimental procedures, results, and the comparison with known results for the traditional manufacturing system.

The manufacture of porcelain figures has traditionally been craft based. High levels of skill and artistic ability are necessary to produce the figures to a consistently high quality. The ability to label them as hand-painted adds considerable value in terms of marketability.

The manufacturing company involved in the project described in this paper wished to investigate the possibility of introducing automation where possible but keeping the necessary highly skilled manual tasks to maintain their marketing advantages. The project was to assess, by the use of discrete event simulation, the proposed design. Manning levels, buffer and conveyor capacities, mould retrieval, cleaning and storage equipment, oven and drying area capacities and packing facilities are some of the input parameters that need sizing relative to each other for a range of demand patterns. A spreadsheet model was also developed to compare the economics of the proposed line with the traditional totally manual line. Because of commercial confidentiality, no reference is made to the company or the actual parameters that describe the proposed system. The methods used and lessons learnt are nevertheless of interest to anyone involved in assessing the viability of a proposed manufacturing system.

Description of proposed system with list of processes and variable parameters

The proposed line consists of 5 connected main sections. Each section had a number of processes and buffers. The main sections are casting, assembly, decoration, glaze-firing and shipment. There are 16 processes and 5 buffers. Some of these processes (e.g. dry-to-white) require considerable time and could have been considered as delay buffers which have large capacity. Each pallet in the casting section consists of 9 identical moulds. There is a different pallet for each constituent part. The figures are manufactured in batches of 9. The proposed system has a number of workstations for each manual operation and uses automated guided vehicles (AGVs), robots and conveyors for materials handling. The whole system works a 5-day, 2 shifts per day working pattern with 2 processes, glazing and firing, working 3 shifts per day throughout the 5-day week.

Why Simulation?

There are essentially two choices of how to assess the design of a manufacturing system. It is sometimes possible to undertake direct experimentation on the actual system but this option is not possible before implementation. As this is the case for this proposed design, the other alternative is to create a model of the system. Possible alternatives are scale, mathematical or logical models. Scale models are usually highly specific, difficult to modify and do not show the logical interaction of system components. Their use is limited to the modelling of the physical processes which is not required in this case. Mathematical models are based on the use of algebraic relationships to describe the interactions between the system and its input and outputs. They have advantages of speed of use and limited data requirements but usually require many simplifying assumptions and only produce long term average results. Dynamic variation of demand patterns and the testing of operating logic is difficult if not impossible with mathematical models.

The alternative of simulation has two sub-techniques. Continuous simulation is concerned with the problems of continuous processes usually described by differential equations or rates. To develop a model of a system which is concerned with finite batches, variations in demand and product mix, and a large number of interacting system parameters requires the use of discrete event simulation. These models proceed step-by-step through time as the various activities start and stop. To undertake the simulation experiments in this project, the graphical interactive simulation package WITNESS was selected. (WITNESS Version 307, Release 4.0, (c) AT&T ISTEL Ltd). Unfortunately, the run times for discrete event simulation models, especially if they contain many elements such as conveyors, can be excessively long. For a fuller discussion of simulation see Kay (1984) or Pidd (1992).

Project Objectives

The objectives of the project were to:-

(a)   create a model of the line that would be flexible enough to allow investigation of the interaction of the various system design parameters

(b)   optimise the hardware and manual resources to meet target production rates for an average production mix

(c)   calculate the production capacity for different demand patterns which place excessive loading on certain processes

(d)   develop an economical model to compare the profitability of the automated line with that of the traditional method as functions of production/sales volume.

The Model

In order to create a model of the line the following assumptions and simplifications were made.

(1)   Supplies of materials such as glue and paint were always available.

(2)   All cycle times were set at 5% above the expected value to allow for data logging, information requests etc.

(3)   Labour performance assumed to be 100% and no breaks allowed for.

(4)   No breakdowns or set-ups. (In reality the set-ups are negligible).

(5)   The required labour is always available.

The model was created over 4 windows within WITNESS and used 107 elements. A major task in the model building was the creation of the logic required to assign priorities for the flexible material handling equipment. One AGV, for example, had to choose from 5 separate activities.

The model created produced a number of performance measures. The most important one for the company is the global production rate but the production rates of each sub-section of the line may also be monitored. The modelling system automatically makes available 6 parameters associated with the buffers in the system as well as the utilisation of the various machines and workstations. Throughput times were recorded in an histogram and the work-in-progress (WIP) in a time-series. The model was validated in consultation with representatives of the company.

Experimentation

The first task is to determine the warm-up period i.e. the time for the system to stabilise when starting from an empty state. The weekly average mix was used to investigate the warm-up time. The WIP time-series was monitored for a period of 2 weeks. Because of the mixed shift pattern working, the stabilised system exhibits a saw-tooth WIP pattern. The steady-state appears to be reached after 4 days but for convenience 5 days (1 working week) was selected. All results considered refer to values obtained after the first week.

The experimentation using the model was designed to meet the objectives within a limited time frame. It was therefore necessary to try to minimise the number of runs of the model as each experiment took about 6 hours. The company suggested the 7 main factors which needed to be considered. They were essentially the factors that could be varied easily but which were expensive to provide and were thought to have a major effect on line output. The design values for these 7 parameters were given and an investigative range around them was established. The parameters were 5 buffer capacities and 2 skilled labour capacities.

Full factorial design for this situation would require 128 experiments (or 32 days of experimentation). Multi-level experimental design was used (Barker (1985)). This was available in the form of commercial software DEX (Design of Experiments, Greenfield and Sovas, 1987). This suggested 47 experiments but because there were no statistical variations of process times and no breakdowns, this could be reduced to 23. After running the experiments, a quadratic function of the 7 parameters was created to represent the system production output. The results indicated that the production target could not be met by any combination of the parameters within the selected ranges. The output values in fact varied by only 0.5% around the mean output value. This indicated a previously unsuspected bottleneck in the system. By making use of the comprehensive set of performance measures available, the firing process was identified as being the most probable source of the bottleneck.

A small number of experiments were undertaken to investigate the process in detail. It was essentially an investigation of the effects of conveyor speed and/or kiln capacity. It was quickly realised that the conveyor speed was such that it could not meet the production targets set. As there was no physical reason why the conveyor speed could not be increased this was implemented in the model at a level that allowed the targets to be met. A simple check was also made on the other system elements to ensure that they were capable of meeting the production targets.

The experiments were repeated and a 20 term equation was created. This included quadratic terms and multiplicative terms. It must be emphasised that this equation is predictive but not explanatory.

This equation allows the output of the proposed system to be predicted within the defined ranges of the 7 variables. The DEX software will also produce a list of input parameter combinations that predict output values within a specified range. This is the equivalent to a contour on a map but in this case in 8 dimensional space. It is therefore relatively easy to find which combinations of input variables will just meet the production targert. The "optimum" combination may then be selected. In this case it was selected by minimising the labour for the skilled processes. This configuration was tested by comparing the predicted output with that obtained by the WITNESS model. The results were within 0.2% giving confidence to the process used.

Some production mixes will place disproportionate loadings on certain parts of the line. The equation described above was devised for an average production mix. To test the performance of the system under the different loading conditions, it is therefore necessary to use the WITNESS model. The results were as might be expected. When the excessive loading was placed on a capacity that had been minimised to just obtain the production target, (e.g. one of the skilled manual processes) a bottleneck was created and the output fell. When the loading was placed on a process that was under utilised, there was very little change in output. These results were useful in the development of scheduling rules for the line.

It was also desired to compare the economic performance of the proposed line with that of the traditional manufacturing process. Two spreadsheet models were developed to represent the different costing structures of the 2 lines. Comprehensive data was obtained from the company on the traditional line as a function of production volume. For the proposed automation line, the costs of all equipment, labour, materials and indirect staff were included in the model. The total costs were calculated as a function of the production volume for a range of production mixes. The equipment and labour needed were estimated by using the equation developed by the DEX software and input into the spreadsheet model. Assuming that all production is sold at current prices, financial viability of the proposed line can then be compared with the traditional line in terms of profitability as a function of production/sales.

Conclusions

The use of discrete event simulation can help to develop considerable insight into the behaviour of a proposed manufacturing system. It can, however, be a very slow process if many variables are involved and the run times are long. The use of analytical tools, such as DEX, which help to design experiments which produce a predictory equation can save much time and effort. A salutory lesson in this project was the discovery that variation of the 7 selected parameters produced very little change in output. This is the result of not undertaking the simple check that all elements of the system, in isolation, will be able to meet the required global production rate. The simulation model is used to check the interaction between the elements. Some preliminary checking and experimentation to discover which parameters need to be selected for detailed investigation must be undertaken. The derived equation can be used to product system configurations capable of producing desired target values. These configurations can be used to model the capital and running costs of the system and hence the profitability of the proposed design.

References

Barker, T.B. 1985, *Quality by Experimental Design*, (Marcel Decker Inc.)
Greenfield, A. and Sovas, D. 1987, DEX (Design of Experiments) version 3.0
Kay, J.M. 1986, The Use of Modelling and Simulation Techniques in the Design of Manufacturing Systems, *Proceedings of the International Conference on Development of Flexible Automation Systems*, (IEE)
Pidd, M. 1992, *Computer Simulation in Management Science*, 3rd edn (John Wiley & Sons Ltd)

SIMULATION OF PROCESS SEQUENCE CELL LAYOUTS

David Stockton and Richard Lindley

Department of Mechanical and Manufacturing Engineering
De Montfort University
The Gateway
Leicester LE1 9BH

This paper addresses the methods used to test the validity of Process
Sequence Cell Layouts (PSCL's) in certain manufacturing situations. PSCL's
were initially designed to enable manufacturing companies that manufacture
a high variety of parts in low annual volumes, to implement kanban control
principles on the shopfloor. Essentially a PSCL, is made up of a number of
individual cells each containing a variety of items of processing equipment.
This paper will discuss the types of simulation to be used when evaluating a
PSCL, then describe the benefits to be gained from each type of simulation.
The results of the simulations will give indications of how the constraints of
capacity, lead time and work in progress affect the benefits of PSCL's in
different situations.

Introduction

Process sequence cell layouts were initially designed to enable JIT or lean
manufacturing principles to be implemented within high variety / low volume
manufacturing environments. PSCL's encourage a unidirectional material flow
throughout the shopfloor, and allow a kanban type material control system to be
implemented, (Stockton, Lindley and Bateman 1994).
Simulation of PSCL's will give a greater insight into how they work in practice, with
different workloads, different control methods and breakdowns, etc. The main advantage
in using simulation is in reducing the risk involved in implementing a new system.
Another benefit is that the way proposed system will actually run must carefully
determined and documented.

Process Sequence Cells

Essentially a process sequence cell layout, (PSCL), is made up of a number of individual cells each containing a variety of items of processing equipment. Equipment is allocated to a cell depending on its position within the processing sequence, (i.e. operation route), of components. Initially the process routes, (i.e. equipment codes only), for all components processed by a manufacturing organisation must be entered into a spreadsheet package. Process routes from computerised production control systems, (including MRP and MRP II systems), can usually be exported as an ASCII format data file and could be imported directly into the spreadsheet package overcoming the need for extensive data inputs (Table 1). Ideally each cell should be capable of performing only one operation within the process routes of all the components manufactured on the shop floor. However to improve the efficiency of the system, (i.e. in terms of resource utilisation), cells may be designed to perform more than one operation on a component.

Part Code	Last but Three Operation	Last but Two Operation	Last but One Operation	Last Operation
P1	A	B	C	D
P2			B	D
P3		A	E	B
P4	B	C	D	E
P5			C	E

Table 1 - Process Routing Information

Process sequence cells may provide a way of enabling JIT or lean manufacturing principles to work effectively within HV/LV environments. They have continuous material flow and high WIP visibility attributes inherent in their design. A PSCL will have a general design as described in Figure 1.

To physically keep the batch sizes at the level required a kanban material control system should be implemented. The plant layout method adopted will have a direct influence on the practicalities of using kanban containers and signaling devices to control material flows. In a PSCL environment materials will only move between cells, i.e. along a limited number of specified routes. Material flows will in lean manufacturing terms, therefore, be visible. Hence it should be possible to regulate material movements using kanban signals.

Simulation

Simulation is the use of a model to develop conclusions that provide insight into the behaviour of any real world elements, McHaney (1991). Simulation reduces the risk and uncertainty associated with a given system. It is used to study and predict the performance of new systems or analyse changes to existing systems. It is used to test 'what if' situations and ideas relatively inexpensively, without the need for real life implementation.

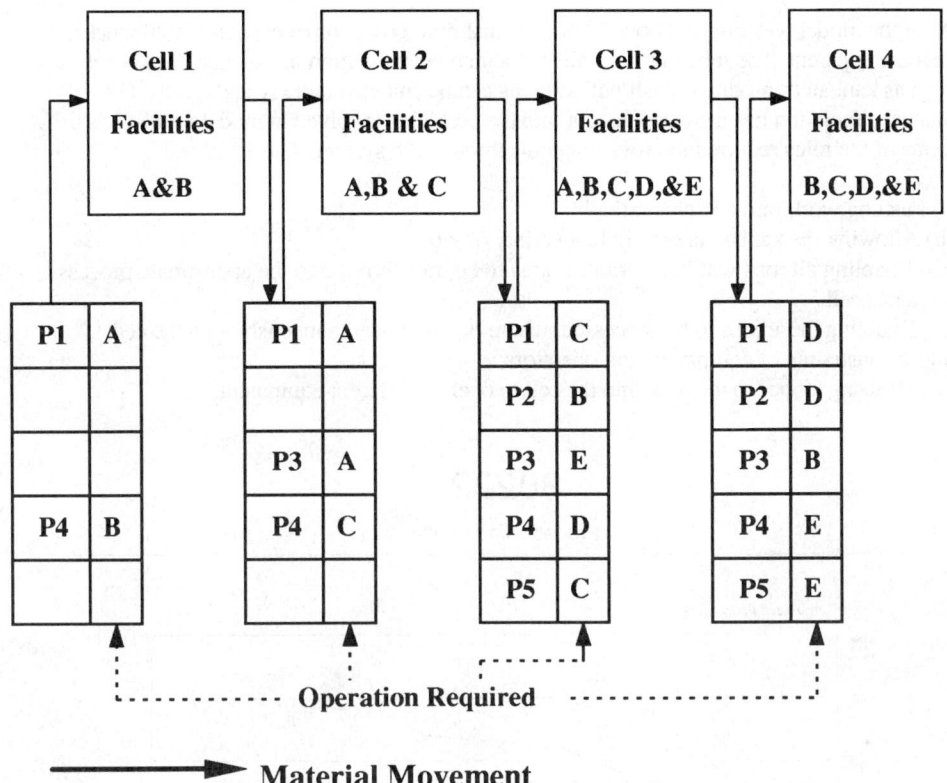

Figure 1 - Process Sequence Cell Layouts

A manual simulation was initially performed using Monte Carlo techniques to gain an understanding of the problems involved in moving materials between process sequence cells. Monte Carlo simulations normally employ random numbers to solve stochastic or deterministic problems where the passage of time plays no role, the actual manual simulation carried out did contain a time element. The use of the manual simulation was in immediately identifying the problems that would occur within PSCL's. The manual simulation carried out used information from a real life situation and a random number table to generate the workload factors.

The manual simulation performed took approximately two days to set-up and run, and used cards to represent the layout of workstations and markers to represent the flow of parts within the system. It was necessary to control the timing of the simulation maintain track of setup times and run times and control the flow of parts and assignments of machines. Cumulative frequency charts of the processing times of every machine were created with which to obtain a processing time from the random numbers (Figure 2).

After the model was run for about 3 hours of real time problems of capacity, bottlenecks, become apparent. The inherent flexibility of a manual simulation means that control rules, such as kanban or modified push/pull systems can be changed quickly and easily. The manual simulation has provided insight into the problems involved in modeling PSCL's in terms of the rules required to move materials through the system. This involves:

(i) Pushing work into a kanban area,
(ii) Allowing the kanban area to hold a variety of jobs,
(iii) Enabling all jobs within the kanban area to be transferred into the appropriate process sequence cell,
(iv) Enabling these jobs to be processed, where possible, simultaneously with the cell under constraints of equipment and operators,
(v) Allowing labour to move within the cell to operate different equipment.

30222

Process time profile

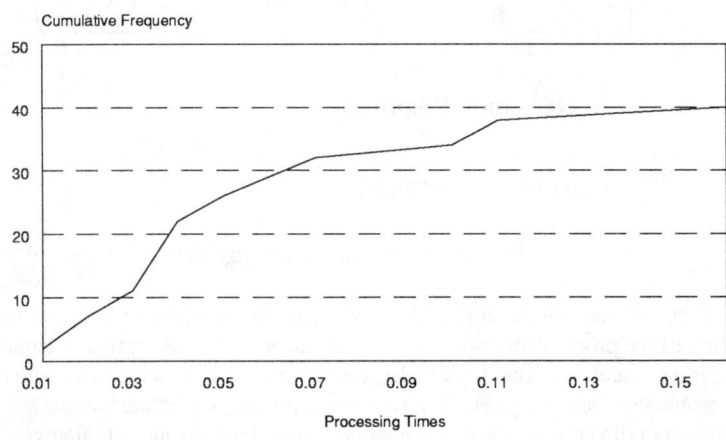

Figure 2 - Cumulative Frequency Chart for Processing Times

The simulation also identified specific short term capacity problems that require more detailed investigation. This will require the use of a computer simulation, in order to enable a complete PSCL system to be modelled.

Describing complex real world systems using only mathematical or analytical models can be difficult or even impossible in some cases. This will necessitate the use of a computer simulation tool to build an effective model of the system. Computer simulation should only be used under certain circumstances, as it can be an expensive and complicated tool to use. Some situations where simulation should be used are when:

1) The real system is theoretical, and it is too costly, time consuming or simply impossible to build a prototype.
2) The real system exists, but experimentation is expensive or seriously disruptive.
3) A forecasting model is needed to analyse long periods of time in a compressed format.

The development of PSCL's has relevance to all of these categories, and is a strong indicator that computer simulation should be carried out at this stage. Computer Simulation is used to obtain accurate statistics on the system being modeled, the varying methods of controlling the material flow through the system assessed by the simulation output. Computer simulation allows a set of conditions to be replicated while varying only the factors which are of interest Flynn and Jacobs (1986).

For the purposes of simulating PSCL's the processing and setup information used in the original formation of the cell design can be downloaded directly into the simulation from the spreadsheet package. The simulation package must therefore be compatible with the data format of the spreadsheet. The capability of the simulation package to handle data from other software packages will save many man hours of data entry, thus accelerating the process of building the model. It is also important that the simulation package has flexible material control rules as they will change as the model is adjusted. The control aspects must complement the layout of the machines.

Conclusion

The manual simulation which was carried out has given an insight into the problems that a PSCL may encounter. The computer simulation will hopefully verify these problems and bring benefits in fine tuning the control rules used in PSCL's. Simulation may also highlight inefficiencies within the layout itself. The model may be useful as an aid to planning, an accurate model can give information on a particular loading schedules.

References

Flynn, Barbara B. and Jacobs, Robert. 1986, *A Simulation Comparison of Group Technology with Traditional Job Shop Manufacturing*, International Journal of Production Research, Vol. 10, No. 5, pp 1171 -1192.

McHaney, Roger. 1991, *Computer Simulation: A Practical Perspective*, 1st edn (Academic Press Limited, London).

Stockton, David. Lindley, Richard. and Bateman, Nicola. 1994, *A Sequence of Cells - Part 1*, Manufacturing Engineer, February. Vol 73, No 1, pp 12-14.

THE APPLICATION OF COMPUTER SIMULATION FOR COST EFFECTIVE RTM PRODUCTION

Bland R.J.[+#], Robinson M.J.[*], Polkinghorne M.N.[+]
and Burns R.S.[+*]

[+]*Plymouth Teaching Company Centre, University of Plymouth, Plymouth, UK*
[#]*Plastech TT, Gunnislake, Cornwall, UK*
[*]*School Manufacturing Materials and Mechanical Engineering, University of Plymouth, Plymouth, UK*

This paper describes the application of computer simulation software (CACI Products SIMFACTORY II.5) for the modelling of the processes involved in RTM production prior to its implementation as an actual production system. The paper examines the necessary requirements for production levels ranging from twenty components to five hundred components per week, in terms of utilisation of materials, equipment, labour and processing. Costs are examined to ensure the cost-effectiveness of the models created. Process optimisation and results are determined by the examination of activity levels of each component in the process chain.

Introduction

Resin Transfer Moulding (RTM) has previously been envisaged as a solution to the mass production of composite material components. However, the characteristics of the process are such that many manufacturers using RTM utilise a large amount of manual labour at various stages. With recent advances in the automation of the technology, increasing levels of production are becoming a reality.

In order to cope with these increased levels of production, analysis of the process is required, as is a method by which that analysis can be carried out. The computer simulation of production processes is a powerful technique for performing extensive analysis of the problems faced by production engineers. Simulation models allow the development of an effective and efficient process design prior to implementation. Once implemented, any process changes can be modelled to determine their effectiveness in terms of productivity and cost. Benefits resulting from the use of this type of simulation include capital, operating cost and lead time reductions, plus faster plant changeovers.

While process simulation techniques have been available for 35 years or more, Hollocks (1992), the amount of computing power currently available has increased the viability of simulation in the area of production management. However the uptake of this production planning method has been extremely slow in the UK with only an estimated 11% of SME's currently using simulation, Simulation Study Group (1991), compared to an estimated 89% in the USA, Christy & Watson (1983).

Resin Transfer Moulding (RTM)

RTM is a rapidly growing process, used for the manufacture of a wide range of high quality Fibre Reinforced Plastic (FRP) components. It is a multiple stage process which involves the procurement of raw materials, fibre reinforcement and a thermosetting liquid resin matrix, followed by the production of a fibre preform. This preform is then placed into a matched mould tool. A sufficient quantity of the resin matrix is injected at low pressure, typically 100 - 400 kPa, into the closed mould cavity thus permeating the fibre pack. Once the thermoset resin matrix has cured, the mould is opened and the product removed, trimmed, inspected and stored.

In order to produce a large number of components simultaneously using this process, it is necessary to optimise the utilisation of each element in the production chain which is possible using traditional production planning techniques. However, the complexity of the process lends itself more readily to computer simulation.

Simulation Process

The objectives of the modelling were to create a simulation which would cover the RTM process from materials delivery to storage of the final component. This model was to be adapted to achieve production rates of 20, 100 or 500 components per week. The additional effect of introducing a new element to the production chain, in this case a fibre pack preforming machine, and it's subsequent production rate implications were determined.

Process Model

For any new process, it is first necessary to create an initial model. To do this, we need to examine the process in its simplest form. A process flow chart for this application was therefore created [Fig 1.]. Once the individual stages of the process had been identified, all relevant data regarding each part of the process was collected. This data included requirements and costs for equipment/raw materials, process times/costs, and operator usage.

The data obtained was utilised to create the initial simple model within the SIMFACTORY software. Once running normally, the scope of the model could be expanded. The first stage of expansion of this model was to introduce interrupts and variances to the process. Interrupts consisted of planned machinery down times, and the workday patterns of the operators e.g. meal and tea breaks. Variances were applied to all stages of the process with the percentage variance depending upon process type and previous experience.

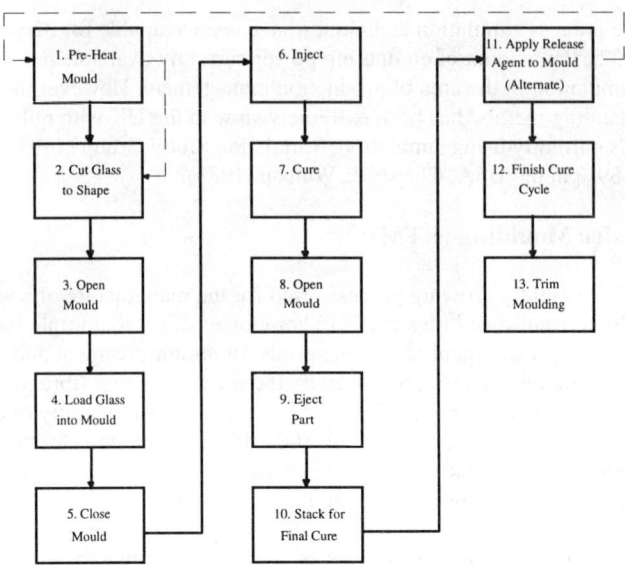

Figure 1. Initial Process Plan for RTM Model

From this stage on, the model was expanded to satisfy higher production rates in accordance with output requirements. During this stage, the following manufacturing issues were addressed:

- *Determination of necessary equipment and personnel*: The effect of increases/reductions in the amount of machinery/personnel can be quickly determined due to the time compression effect of the simulations. The determination of these changes is most effective when production levels are increased, as full utilisation of resources does not equate to process optimisation. The effect of individual elements must be determined as part of the whole process in order to prevent subsequent bottle-necks.

- *Location and size of inventory buffers*: It is necessary to place inventory buffers at selected points in the process where bottle-necks may occur, thus allowing a visual guide to the levels of inventory in the system.

- *Impact of new machinery introduction*: When a new process is added to the production chain, its effect can be determined by the simulation.

- *Impact of random breakdowns*: Random breakdowns can be introduced on specific machinery at variable time frequencies, with fixed repair times. To determine their full effect, these interruptions can be introduced mid process.

- *Throughput and bottle-neck analysis*: This is one of the most important issues addressed by the simulation. Only when bottle-necks are reduced to a minimum, can the maximum throughput be attained.
- *Impact of changing process yields*: The effects of increasing and decreasing output levels on the previous five areas can be determined.

By the iterative utilisation of these issues, the system layout required for the production of 100 components per week, without the use of a fibre preforming machine, may be obtained [Fig 2.]

Process Outputs

Reports were generated by the simulation software at the end of each simulation run. According to the number of replications requested, an average result was produced over a determined number of simulations.

The results produced included the time required to produce a component, the machinery/operator utilisation levels required to achieve the desired output, and the costs of resources, work in progress and final components.

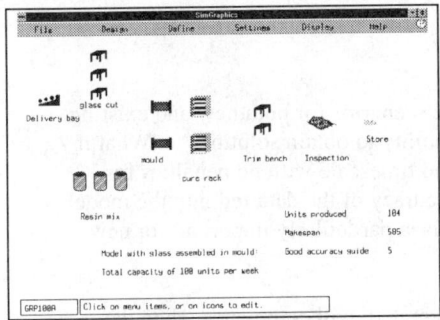

Figure 2. Process model Layout

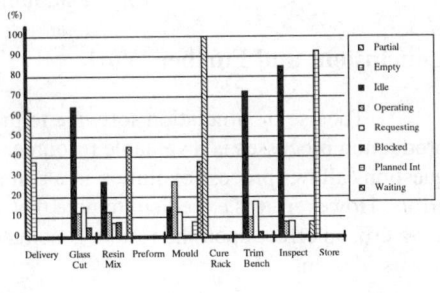

Figure 3. Element activity levels

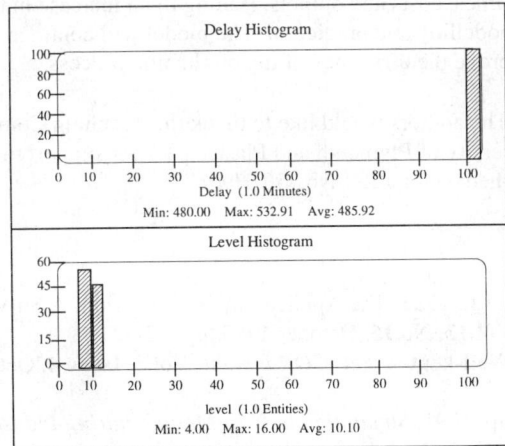

Figure 4. Element Activity Histogram

These results are available in a variety of formats. A process activity output is demonstrated [Fig 3] for each component of the process chain given a production rate of 100 components per week. This type of result can be used to identify over/under utilised elements of the process, leading to an iterative modification of the model. Similarly an activity histogram is demonstrated [Fig 4] for the cure rack element given the production of 100 components per week. This type of output identifies the time period for which an element is occupied, together with its inventory levels, and allows for the iterative optimisation of the inventory buffers.

Results are summarised [Fig 5] for the simulation given the production of 100 components per week.

	Arrival Rate (min.)	Final Parts Produced	Costs per Product (£)	Makespan (min.)	Work in Progress	WIP Costs (£)	Average WIP Cost per Part (£)
100 Components per Week	26	104	16.73	585	25	333.81	13.35

Figure 5. Summary of Results

Conclusions and Further Work

The use of simulation software to produce scenarios for both new and existing production processes is a valuable resource. The ability to obtain solutions to 'What if ?' questions allows process evolution in a compressed timescale with no penalties for errors. However, it is expedient to note that the accuracy of the data fed into the model has a crucial effect upon the results generated. This is particularly important for new process design.

Further work using the simulation model created, will commence when the actual production process is up and running. The two systems performance will be compared in terms of times, costs and outputs, leading to an increase in the experience base of both process modelling and prediction. The model will continue to be used in further attempts to improve the efficiency of the production process.

Acknowledgements: The authors would like to thank the Teaching Company Directorate (SERC/DTI), the University of Plymouth and Plastech TT for support to undertake the Teaching Company Scheme Grant Ref No. GR/H90872.

References

Christy, D.P., Watson H.J., 1983, The Application of Simulation: A Survey of Industry Practices, *Interfaces*, **Vol.13, No.15**, October 1983, pp. 47-52

Hollocks, B. 1992, A Well-kept Secret ?, *OR Insight*, **Vol.5, Issue 4**, October-December 1992, pp.12-17

Simulation Study Group. 1991, *Simulation in U.K. Manufacturing Industry*, Available from The Management Consulting Group, University of Warwick.

INTEGRATED REAL-TIME MANUFACTURING SIMULATION WITH EXPERT ADVICE

Dr Andy Lung and Mr Habtom Mebrahtu

Kingston University, MAP Engineering, Roehampton Vale, Friars Avenue, London SW15 3DW

Manufacturing simulation is a useful management tool which can perform consistent what-if analyses either for an existing system or a proposed system. As the technique has been receiving an immense publicity, most of its development effort has been focused on improving its front end user interface and ease of use. Relatively little, however, has been done to assist the interpretation process of final simulated results. This paper will discuss the feasibility of linking an expert system to a manufacturing simulator at real time in order to achieve continuous improvement with automatic result analyses performed by the expert system.

Introduction

The use of simulation modelling to design new manufacturing systems and to "fine-tune" the performance of existing systems has been gaining publicity in great pace. The increased complexity in modern manufacturing systems, sharp fall in computer hardware costs, improvements in simulation software and the availability of animation have all contributed to its recognition in manufacturing industries.

There are many simulation packages on the market and they adopt various front-end user approach to ease off the processes of problem modelling. Some of these systems require programming skills, whilst others do not. There is however one thing in common: a lot of time and expertise will be required to analyse the simulated results. In addition, there is a missing time-link between the real system and the simulation model. This has greatly hampered the speed of re-simulating the problem for continuous improvement.

The paper will discuss how the amount of decision making required in a simulation run can be minimised with the use of a closed loop 'on-line' data link and an expert system. It suggests that a manufacturing simulation model can be interfaced to a shop floor data collection system for real-time data capturing. The simulator, when it has simulated the user's input data in a chosen scenario, will pass its results to an expert system for analysis. Once the expert system has made recommended actions, the results will be then fed back to the simulator for more 'what-if' reassessments. The process continues to iterate until the simulated results are considered satisfactory. The final processed results and recommendations will be transferred to an integrated database for open access so that data redundancy can be minimised. The expert system will continue to learn so that its knowledge base will grow with the company.

Simulation In Manufacturing

Simulation is a very practical tool for planning a manufacturing system and its operations. With more friendly user-oriented interfaces developed, one can build a mathematical or graphical model much easier than before. The output will be either a report or animated graphics. The engineer is able to evaluate various manufacturing operations without a physical set-up of machines.

There are generally two main areas of concentration in manufacturing simulation software development : to make it easier to create model (front end user interface), and to make it easier to understand the outputs of a model. To achieve the first goal, many software developers have added some front end modelling modules (for example ARENA for SIMAN-IV). Some have used general manufacturing terminologies such as operations, process plans, machines, buffers, operators and so on in the modelling process, instead of using simulation jargons like entities and attributes. To achieve the second goal, many software developers have incorporated animation graphics to make simulation results easier to understand.

Although these efforts have no doubt made a manufacturing problem easier to model, a trained specialist is still required to analyse piles of final reports in order to make some sense out of them. It will be useful if a mechanism, for e.g. an expert system, can be introduced to take care of some common data analyses automatically.

Expert System

Expert systems are computer applications that contain the knowledge, experience and judgement of skilled professionals. Expert system is the embodiment within a computer of a knowledge-based component from an expert skill in such a form that the machine can offer intelligent advice or make an intelligent decision about a processing function (Waterman 1986). A desirable additional characteristic, which many may consider fundamental, is the capability of the system to justify its own line of reasoning in a manner directly intelligible to the enquirer.

The process of building an expert system is often referred to as knowledge engineering. It typically involves a special form of interaction between the expert system builder (knowledge engineer) and human experts in some problem areas. The knowledge engineer extracts from the human experts their procedures, strategies and rules of thumb for problem solving, and builds this knowledge into the expert system. Many of the rules in an expert system are heuristic. The knowledge in an expert system is organised in a way that knowledge about the problem domain will be separated from the system's other knowledge, such as general knowledge about how to solve problems or knowledge about how to interact with the user's terminal or modify lines of data in response to the user's commands. The collection of domain knowledge is the KNOWLEDGE BASE, while the general problem-solving knowledge is the INFERENCE ENGINE.

Expert Systems in Simulation

There are three possible ways in which expert systems can be used to improve a manufacturing simulator. They are (1) Data collection and Model formulation, (2) Control algorithm formulation, and (3) Results analysis. Data collection and model formulation is mainly associated with the front end area. The expert software can provide advice to ensure data entry will be performed consistently and logically. Control Algorithm Formulation - once the data has been provided for a simulation model, the next step is to describe how the manufacturing system operates. This includes decisions of part schedules, station selections, queue priorities, transport selection and operation sequence. The simulation system itself should provide flexible algorithms for the user to define the logic of the manufacturing process. At this point, the effect of an expert system tends to be superseded by the logic builder provided by the simulator itself. Results Analysis - although the on-screen animation graphics help identify any possible blockages, the user still has to perform other analyses with lots of generated files. These analyses include the assessment of machine capacity balance, sizes of queues, resource utilisation and maximum throughput. Here, an expert system can be constructed to read in the output files, group the data according to their functions, and perform some previously defined logical assessments. It will then make final recommendations in accordance with what have been stored in the knowledge base.

The Proposed System

There can be many ways to implement the expert data analyses. To illustrate the concept, the following performance indicators are discussed :

Throughput - The main measure is the quantity of final parts produced by the system;
Resource Utilisation - Working time of a resource is the main measure;
Meeting Due dates - Lead time is the main indicator;

Work-In-Progress Inventory - either queuing time or quantity of parts left in a buffer could be used in the assessment;
Cycle Time - Lead time is the main indicator;
Cost - Production cost is the main criteria.

Depending on the preference of a company, different rule-based criteria could be used in the final expert analysis. To simplify the illustration, "throughput" is taken as the measure of performance as it is more comprehensive than most of the other factors discussed here. In order to improve throughput, "balancing parts flow", "avoiding blockages and bottlenecks" are all considered. The proposed system will try to produce a balanced flow by either improving the capacity of the slower stations or letting them work overtime. It tries to achieve or optimise the required throughput with existing resources. But if the maximum attainable throughput is below what is required, then the expert system will suggest adding more resources at the slowest station. The process continues until optimisation is achieved.

Figure 1 shows the manufacture of universal joints in medium sized batches. The joint consists of five components namely: Fork (2 off), Trunnion (1 off), Bush (2 off), Pin (1 off), and Rivet pin (1 off). SIMFACTORY II.5 was used to model this problem. Here manufacturing of Fork and Trunnion which are two main components of the product were considered. Most of the operations were simplified and some machine centres were grouped together.. Simulation results were assessed manually in accordance with the expert system algorithm indicated in Figure 2. The optimisation process resulted in an increase of throughput from 133 joints/week to 413 joint/week with the slowest station, Mill, working for 23.5 hours a day. To obtain a higher throughput, a resource (machine) should be added to the Mill station. Their working hours would be adjusted proportionally, and optimisation continued. The end result will be compared to similar models in a database. If the result is good it will be included in the list. Otherwise, it will be dumped.

The experiment shows that it is possible to perform data analyses automatically after a simulation run by using set rules. The next stage of the research is to build these rules into an expert system and interface it to the simulation model so that changes will be incorporated prior to the next run. Open access to generated results will also be introduced in the next phase.

REFERENCES:

Carry, A. 1988, Simulation of Manufacturing Systems, (Anchor Brendon Ltd, U.K.)
Law, A.M. 1991, Simulation Modelling and Analysis, (McGraw_Hill, Inc)
Hurrion, R.D. 1986, Simulation, (IFS Ltd, U.K)
McGeough 1989, The 5th International Conference on Computer Aided Production Engineering, Edinburgh, 1989 (Edinburgh University Computing Service)
Pridham, M. and O'Brian, C. 1991, Production Research - Approaching the 21st Century, (Taylor & Francis)
Waterman,D.A. 1986, A Guide to Expert Systems, (Addison Wesley)
Kerr,R. 1991, Knowledge-Based Manufacturing Management, (Addison Wesley)

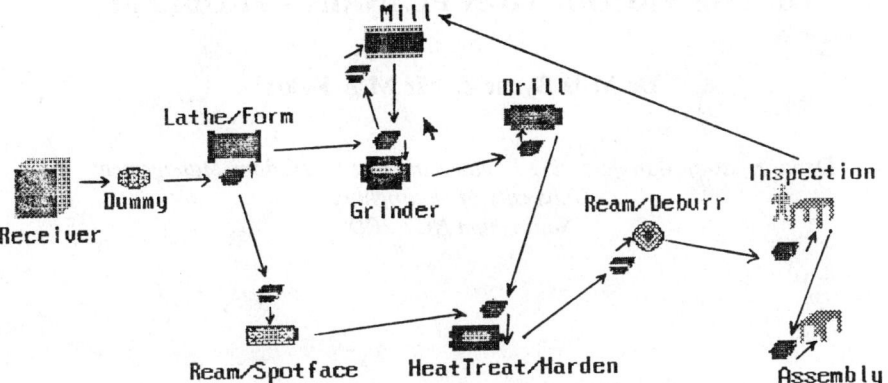

Figure 1   Control algorithm in the expert data analysis system

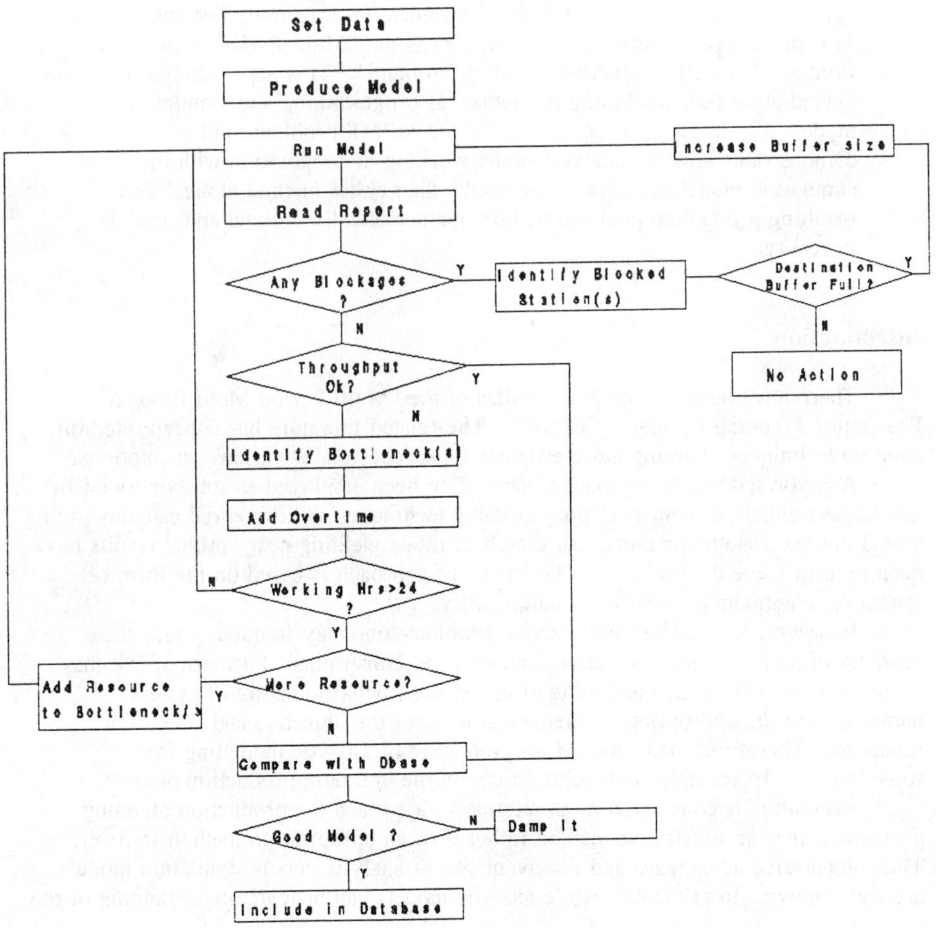

Figure 2   The simulation model used in experiment

HYBRID ANALYTIC / SIMULATION MODELLING
OF THE PRODUCTION PLANNING PROBLEM

Dr M D Byrne & Mr M A Bakir

Department of Manufacturing Engineering and Operations Management
University of Nottingham
Nottingham NG7 2RD

An essential problem of production planning is to match production
activities to fluctuations of demands. In situations where individual
products are not relatively homogeneous, it is important to determine the
optimal variable levels for individual products for upcoming periods.
This procedure is known in the literature as the "Multi Period Multi
Product (MPMP) Production Planning Problem". This paper studies a
hybrid algorithm combining mathematical programming and simulation
models of a manufacturing system for the MPMP problem, and
demonstrates how the analytic model working in cooperation with the
simulation model can give better results than either method alone. The
resulting production plan can be both mathematically optimal and feasible
in reality.

Introduction

There have been a number of studies of the "Multi Period Multi Product
Production Planning Problem" (MPMP). The related literature has concentrated on
solution techniques claiming more efficient approaches under various assumptions.

Analytic modelling approaches have often been employed as solution tools. In
this respect either mathematical programming techniques or differential calculus giving
global optimal results, or numerical search methods yielding near optimal results have
been used to solve the problem. The choice of approach is based on the form of
functions, complexity, linearity or nonlinearity.

However, in practical optimization problems one may frequently face the
problem of more complex objective functions and constraints. Furthermore, it may
sometimes be difficult or impossible to model the stochastic nature of system
parameters in the appropriate mathematical form of the objective and constraint
functions. Therefore, mathematical programming or analytic modelling may
sometimes not be suitable approaches to obtain the optimum production plan.

Simulation is considered as an alternative approach for production planning
problems. It is an illustrative method rather than an optimization method in itself.
The comparative advantages and disadvantages of analytic versus simulation models
are well known. In particular, while analytic models can provide understanding of the

behaviour of a system, and how the system will respond to changes in the resources or parameters of the system, simulation can serve in solving more complex planning models, since it avoids the restrictions imposed by analytic models. Hence, with a simulation model of the system being studied, the problems, such as stochasticity and formulation difficulties, that arise in mathematical programming model can more easily be handled.

In practical studies it has often been assumed that solution of a problem can be obtained only by using either analytic modelling or simulation modelling. Thus, by employing only one of these approaches some of the well known relative advantages of the alternative modelling technique might be lost, although some part of the problem could be solved by using the other approach.

Normally one resorts to simulation only when an efficient implementable analytic solution is not available for the problem at hand. In a pure simulation model one completely simulates the complex system even when a portion of it is simple enough to be analytically solved. However, it is often the case that much insight is obtained using both types of model, with each model giving different perspectives of the system being modelled. A better way, therefore, might involve combining these two modelling approaches.

Such an integrated approach, entitled *"hybrid simulation / analytic modelling"*, has been discussed by Shanthikumar & Sargent (1983). We study in this paper such a hybrid algorithm combining mathematical programming and simulation models of a manufacturing system for the MPMP problem.

The Hybrid Model

The reasons mentioned above make it clear that the best approach may be a marriage between simulation and analytic solution procedures (Hoover and Perry, 1989, p 11).

The focus of this paper is to develop an MPMP model, which minimizes the relevant costs, to meet market demand while keeping inventories as low as possible. We can formulate the problem as a linear programming model which minimises the sum of variable production cost, inventory holding and deficit costs, subject to capacity and inventory balance constraints.

Assigning the machine centre capacities in this analytic model is a difficult task. A significant difference between gross machine centre capacities and the required time to achieve the production plan can be expected because of the fact that the consumed time to realize the production plan is a complex and often non-analytic function of decision variables rather than a well-behaved linear function. Hence, a simple linear formulation of capacity usage may fail to represent the capacity behaviour of system in reality. Different operation policies such as dispatching, sequencing, and order releasing, and system type (flow shop or job shop), can also destroy the well-behaved linearity of capacity constraint formulation. Furthermore, material handling characteristics are difficult to include in the analytic form. Stochasticity in the operation times and / or transportation times with the material handling facilities can also create serious problems in modelling.

It is therefore proposed to use a simulation model to support the analytic model of the manufacturing system. The simulation will provide a possibility to tackle the problems mentioned above. This procedure will reflect the effects of operational characteristics onto the analytic model. Furthermore, the influence of probabilistic characteristics of the system can easily be included in the model. Thus, the procedure can compensate for these characteristics of the real system without sacrificing some of the advantages of the analytic model. This will lead to a production plan which is optimum not only mathematically, but also in reality.

The hybrid modelling logic, which is based on imposing adjusted capacities from the simulation model, is recursive in structure and consists of the following steps:

Step 1.   *Generate optimum production plan configuration by analytic model.*

Step 2.   *Assign optimum production levels of analytic model as input to the manufacturing simulation model.*

Step 3.   *Simulate manufacturing system subject to operation policies.*

Step 4.   *If CTkt ≤ MCkt for all k=1...K t=1...T go to step 7 otherwise go to step 5*

Step 5.   *Calculate adjusted capacities.*

Step 6.   *Go to step 1.*

Step 7.   *Assign analytic model optimum levels of decision variables to the optimum production level.*

Step 8.   *Stop.*

The solution procedure uses independently developed analytic and simulation models together to solve the problem. The simulation model is used as a submodel.

Steps 4-5 provide a link between the two models. Capacity adjustments are made through the following formulations:

$$AF_{rk} = \frac{GC_k}{CT_{rk}} \tag{1}$$

where

$$CT_{rk} = ST_r \cdot NM_k \tag{2}$$

Then the new capacities which will be inserted into the analytic model:

$$ANC_{rk} = ANC_{r-1,k} \cdot AF_{rk} \tag{3}$$

where $k = 1,2,....K$, $r = 1,2,...R$, iteration number, **AF** is adjusting factor, **GC** is gross capacity, **ST** is simulation time, **NM** is number of machines, and **ANC** is adjusted new capacity.

Case Study

This hybrid approach is here applied to a minimization of cost model of a 3 period 3 product MPMP production planning problem under machine centre capacity and inventory balance constraints.

The system comprises of 4 machine centres, each having 1 machine and 1 input buffer. Their capacities are constant at 2400 min / week. The parts are transported between machine centres by two non-accumulating belt conveyors.

The cost coefficients, demand matrix, and the process times are shown in Tables 1-3. All system parameters are deterministic, and known with certainty. Parts 1, 2, and 3 are released to the system with constant time intervals of 2, 3, and 2 min respectively.

The analytic part of the hybrid procedure is modelled as a linear programme. The simulation model of the system is written in the SIMAN simulation language. The optimal production rates through the iterations for each period are given in Table 4.

Cost Coefficient	Unit production cost			Inventory holding cost			Shortage cost		
Periods Product	1	2	3	1	2	3	1	2	3
1	100	100	100	25	25	100	400	400	400
2	150	150	150	30	30	150	450	450	450
3	125	125	125	35	35	200	500	500	500

Table 1: Cost Components

Product Type	Periods		
	1	2	3
1	150	125	160
2	100	150	150
3	125	165	125

Table 2: Demand

Product Type	Machine Centres			
	MC1	MC2	MC3	MC4
1	5	10	4	
2	7		5	7
3	7		10	6

Table 3: Process Times (minutes)

Discussion of Results

The simulation analysis of the system has not validated the original optimal result from the analytic model. This resulted in iteratively running the analytic model subject to the adjusted capacities according to Eq. 1 - 3, which use the simulation model results. It can be seen from the table of results that dramatic changes have occurred in production rates for periods 1 and 3 for products 2 and 3, before agreement is achieved after the 10th iteration. These changes for Period 1 are shown (as an example) in Figure 1.

The simulation model was able to accommodate manufacturing system characteristics, such as queue delays, transportation features, and release time intervals, into the modelling procedure. These features were difficult to include in the analytic model.

This system is completely deterministic. However, the flexibility of simulation modelling could also enable probabilistic operation characteristics also to be considered.

	Iteration Number									
	PERIOD 1									
Product	1	2	3	4	5	6	7	8	9	10
1	150	150	150	150	150	150	150	150	150	150
2	111	90	76	66	60	54	51	50	49	48
3	124	110	101	94	90	86	84	83	83	82
	PERIOD 2									
1	125	125	125	125	125	125	125	125	125	125
2	117	116	116	116	116	116	116	116	116	116
3	120	119	119	119	119	119	119	119	119	119
	PERIOD 3									
1	160	160	160	160	160	160	160	160	160	160
2	105	76	57	48	39	36	36	36	36	36
3	123	104	91	85	80	77	77	77	77	77

Table 4: Results for 10 iterations

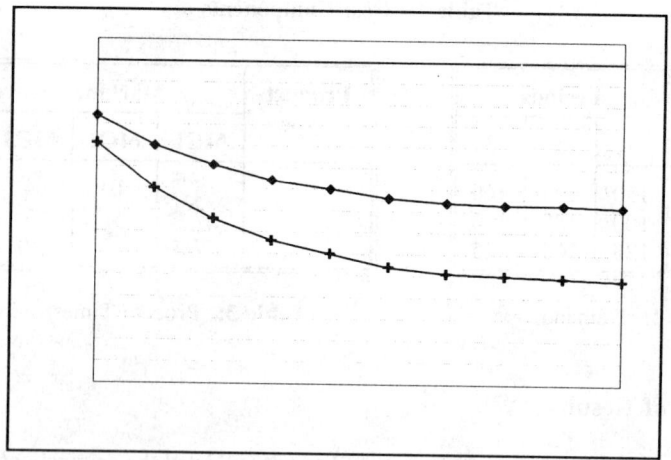

Figure 1: Period 1 production rates

References

Hoover S. V. and Perry R. F. (1989), Simulation: A Problem Solving Approach, Addison Wesley Pub.

Pegden C. D., Shannon R. E., and Sadowski R. P. (1990), Introduction to Simulation Using SIMAN, McGraw-Hill.

Shanthikumar J.G. and Sargent R.G., (1983), " A Unifying View of Hybrid Simulation/Analytic Models and Modelling ", Oper. Res. Vol.31.

DETERMINING NUMBER OF KANBANS AND BATCH SIZE

Liam Quinn and Dr David Stockton

Department of Mechanical and Manufacturing Engineering
De Montfort University
The Gateway, Leicester, LE1 9BH

The importance of batch size and number of kanbans on manufacturing system efficiency and operation has been well documented. Although mathematical equations exist to aid determining these parameters, their ability to consider the dynamic aspects of a manufacturing system are limited. Computer simulation has long been used to model the dynamic properties of manufacturing systems, but they do not possess the capability to improve the system. However in this latter respect Artificial Intelligence is increasingly playing a greater role in the solution of manufacturing problems. This paper investigates the use of one area of AI, Genetic Algorithms (GA's), in conjunction with simulation to determine batch size and number of kanbans within a just-in-time (JIT) environment.

Introduction

Kanban control systems have long been identified as one of the main characteristics of just-in-time (Finch and Cox 1986, O'Grady 1988). O'Grady, Deleersnyder, Hodgson and Muller (1989) identify three major problem areas in implementing and operating a kanban system, ie
a. the identification of flow lines,
b. the flowline loading problem and
c. the operational control problem.

Philipoom , Rees, Taylor and Haung (1987) identified four factors in the production area which influenced the number of kanbans required, the throughput velocity, the variation in processing times, the machine utilisation and correlation of processing times. This paper concerns itself with the area of operational control and the factors identified by Philipoom et al (1987). The main problems in operational control are determining the number of kanbans and the batch sizes to be produced. Monden (1983) used a formula to determine the number of kanbans to be used in a system that used demand, leadtime and container capacity as predictor variables. This formula, however, does not take into consideration the variation or uncertainty which would be present in actual manufacturing systems, such as machine reliability and uncertainty of demand. Few formulae allow these uncertainties to be included in the determination making the implementation of kanban control systems difficult.

Genetic Algorithms

Genetic algorithms (GA's) are powerful search procedures, pioneered by Holland (1975), based on the mechanics of natural selection. The algorithms employ the idea of survival of the fittest similar to that found in the biological world. Genetic Algorithms generate a number of potential solutions to a problem in the form of strings likened to chromosomes in the natural world. These potential solutions are referred to as a population. The solutions are evaluated to determine how good each solution is. The algorithm next swaps parts of chromosomes with those of other chromosomes, a form of mating, using a procedure called crossover with the aim of improving the range of possible solutions and removing weaker solutions from the population, ie survival of the fittest. Occasionally, one element of the string may be randomly changed to allow for new alternatives to be evaluated, this procedure is called mutation. Strings are selected for mating by how efficient their solution is. These three operators, selection, crossover and mutation form the basis of genetic algorithms. It is the crossover procedure that gives the GA its power to search through large search spaces. Figure 1 shows the flow diagram of a genetic algorithm.

The Genetic Algorithm has been successfully used in a number of manufacturing related problems. Cleveland and Smith (1989), Davis (1985) used Genetic Algorithms to schedule flow shop releases and Job shops respectively. Vancza and Markus (1991) used the GA in process planning problems while Minagawa and Kakazu (1992) investigated the line balancing problem with the technique. Stockton and Quinn (1993) investigated the use of the algorithm for identifying economic order quantities. It is the Genetic Algorithm's ability to both search large spaces and also to require no knowledge of the problem domain, (except a function to evaluate how good the current potential solution is), that gives it great potential for solving manufacturing problems.

A computer system has been developed to allow GA experiments to be carried out. This system consists of a Genetic Algorithm which uses information passed from either a simulation package or a mathematical model to determine how good each solution is. The system has the ability to produce the necessary code to construct simulation models. Hence, the system possesses the ability to evaluate numerous alternative solutions without user interaction.

Sample Problem

To allow batch sizes and the number of kanbans to be calculated using the uncertainties that would be present in a real system, a simulation model was built of a sample problem using the simulation package WITNESS (1991). The number of kanbans and batch size were defined as integer variables. The problem described in this paper is relatively simple for ease of description although more complex problems have now been successfully solved.

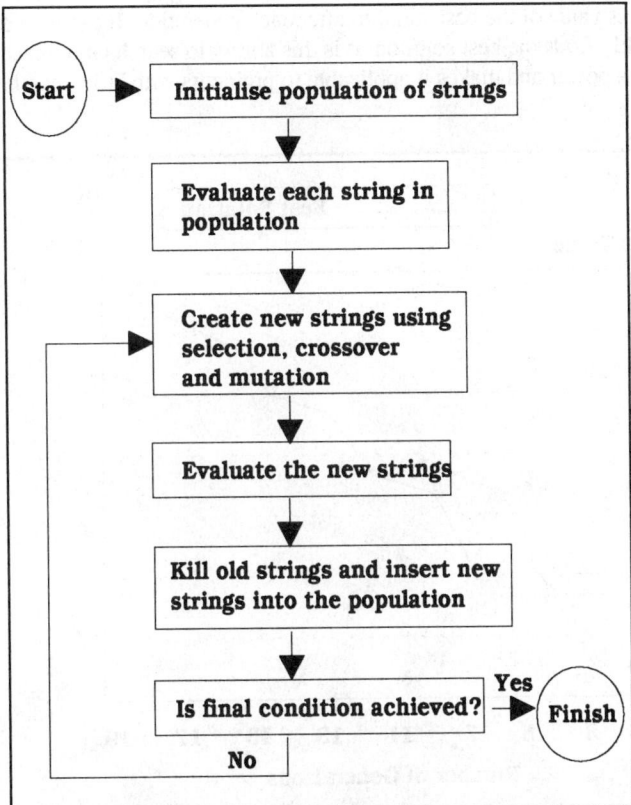

Figure 1 Flow Diagram of a Genetic Algorithm

A part is processed on six work stations, with the flow of material controlled by kanbans. Machine reliability is included in the model. Using the genetic algorithm optimising system described previously, controlled experiments were carried out to determine the best batch size and number of kanbans for the particular problems. The objective was to minimise Work-In-Progress (WIP), minimise the number of late orders and maximise throughput. The model also allowed the user to vary the time between an order being placed and it being dispatched. The problem was also restricted to a maximum of ten kanbans and a batch size of ten, again this is simply for ease of description.

The response curve for the objective function that minimises Work-In-Progress (WIP) and maximises throughput is irregular and complex even with this simple problem which uses only two operational parameters. If the number of parameters is increased then the shape of the resulting response curves will be become more irregular making the determination of the parameters to achieve optimum operation extremely difficult. It is this type of problem that the genetic algorithm is well suited to solve.

The GA algorithm was run until, after each generation, no improvement was being observed, ie the population of possible solutions had converged to one solution. Figure 2 shows the fitness value of the best solution after each generation. It can be seen that the algorithm quickly finds the best solution. It is this ability to search large spaces quickly that gives the GA its power and makes it applicable to problems with large search spaces.

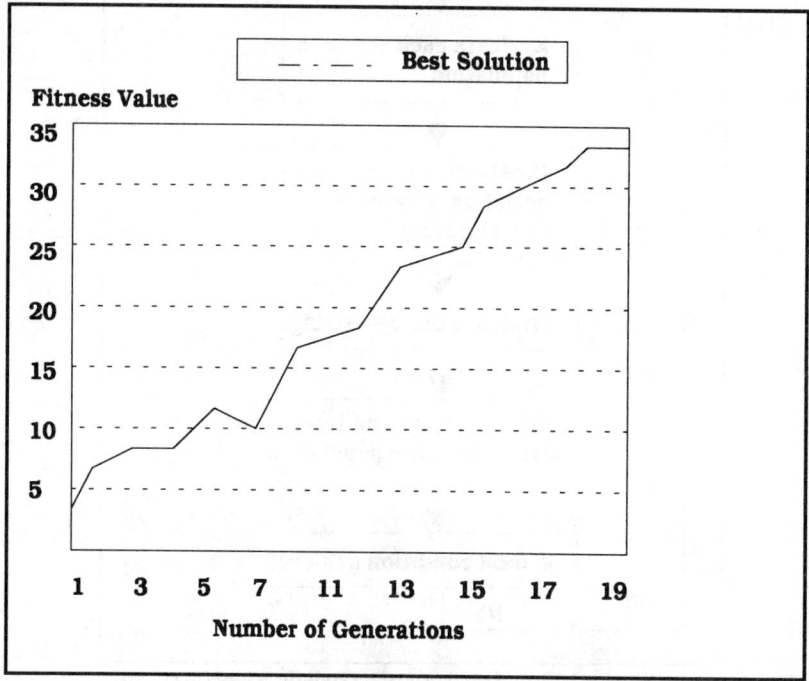

Figure 2 Best Solution found after each Generation

Conclusion

The genetic algorithm used in this problem proved itself to be efficient at finding good solutions to the problem of identifying batch size and number of kanbans within a JIT environment. The algorithm's ability to search large search spaces makes it an ideal tool for planning and designing manufacturing systems where many parameters have to be considered. Using the algorithm in conjunction with a simulation package allows uncertainty to be modelled and used in the design and operation of the manufacturing systems.

References

Cleveland, G.A., Smith S.F. 1989, Using Genetic Algorithms to Schedule Flow Shop Releases, *International Conference on Genetic Algorithms*, 160-169.

Davis, L. 1985, Job Shop Scheduling with Genetic Algorithms, *Proceedings of International Conference on Genetic Algorithms*, 136-140.

Finch, B.J. and Cox, J.F. 1986, An Examination of Just-In-Time Management for the Small Manufacturer: With an Illustration, *International Journal of Production Research*, **24**, 329-342.

Holland, J.H. 1975, *Adaptation in Natural and Artificial System,* University of Michigan Press, Ann Arbor.

Minagawa, M. and Kakazu, Y. 1992, A genetic approach to the line balancing problem, *Human Aspects in Computer Integrated Manufacturing*, 737-743.

Monden, Y. 1983, Toyota Production System: Practical Approach to Production Management, *Norcross, Georgia: Industrial Engineering and Management Press.*

O'Grady, P.J. 1988, Putting the Just-In-Time Philosophy into Practice, *Nichols, New York/Kogan Page, London*

O'Grady, P.J., Deleersnyder, J., Hodgson, T.J. and Muller H. 1989, Kanban Controlled Pull Systems: An Analytic Approach, *Management Science*, **35**, 1079-1091.

Philipoom, P.R., Rees, L.P., Taylor, B.W. and Haung, P.Y. 1987, An Investigation of the Factors Influencing the Number of Kanbans Required in the Implementation of the JIT Technique with Kanbans, *International Journal of Production Research*, **25**, 457-472.

Stockton, D. J. and Quinn, L. 1993, Identifying Economic Order Quantities using Genetic Algorithms, *International Journal of Operations and Production Management*, **35**.

Vancza, J. and Markus, A. 1991, Genetic Algorithms in Process Planning, *Computers in Industry*, **17**, 181-194.

Witness User Manual, 1991, Version 3.1, A T & T Istel Ltd, Redditch, UK.

AN OBJECT-ORIENTED APPROACH TO MODELLING MANUFACTURING PLANNING AND CONTROL SYSTEMS

N.J. Boughton & D.M. Love

Department of Mechanical and Electrical Engineering,
Aston University, Birmingham, B4 7ET.

Modelling, in particular simulation, has long been used to assess operational issues associated with planning and control systems. Although the application of simulation provides obvious benefits, a more fundamental application in the design of planning and control systems is consistently overlooked. This paper discusses the development of a computer-based modelling tool which utilises object-oriented discrete simulation. It is intended that this tool will support the evaluation of both the design and operational issues involved with manufacturing planning and control systems.

Introduction

A manufacturing planning and control system represents a critical part of the infrastructure of any manufacturing organisation. Consequently the design and operation of such systems demands the same close attention to detail applied to other areas of the business. Of the modelling techniques available, simulation is the most suited to the evaluation of complex, dynamic systems (Chaharbaghi 1990); manufacturing planning and control systems clearly belong in this category. Static approaches are to be avoided if the important time-varying inter-relationships between the system and the manufacturing facility and their effect on performance are not to be overlooked.

Much of the production and inventory control research has addressed the relative performance of operational issues, such as lot sizing techniques and scheduling rules, for specific problem situations. This application of simulation does provide obvious benefits. However, the more fundamental application in the design of planning and control systems is rarely considered. Compare the situation to that of manufacturing system design (Shimizu 1991) or control system design for flexible manufacturing systems (FMS) (Bilberg & Alting 1991). In both cases the use of simulation to predict the performance of the proposed system is recognised as an integral part of the design process. There is a requirement for a simulation-based modelling tool which will support the analysis of production planning and control systems from both design and operational perspectives. This becomes particularly relevant in the light of the developing design methodologies, for example Rolstadas (1988) or Bertrand et al. (1993). This paper begins to address the need.

Observed Requirements

The requirements of the modelling tool can be determined from the application domain. From a general perspective, 'standard' planning and control systems will need to be modelled, for example material requirements planning (MRP), kanban or re-order point (ROP) control. However, the characteristics of a company vary and it is not unusual to find that more than one type of control system is in operation; composite systems mean that the modelling tool cannot be restricted to generic solutions. In response to an increasingly competitive environment, manufacturing operations and processes are constantly changing; the development of new and more hybrid systems is inevitable. To support this situation the modelling tool will need both structural flexibility as well as the ability to expand. Furthermore the functionality must appropriately relate to the concepts and terminology found in the real world.

In order to provide essential flexibility the modelling tool is being developed as a set of elementary planning and control building blocks. Once the requirements of the planning and control system have been determined then a model can be constructed using the relevant building blocks. The model may represent an organisation's specific system or, alternatively, a new design altogether. Clearly different control systems will require different building block configurations.

It is essential that the planning and control system is evaluated in terms of the emergent properties between the control system and the manufacturing facility. Therefore, once a system has been configured it will need to be integrated with a model of the manufacturing facility.

Object-Oriented Methods

Object-oriented methods are seen as the most appropriate medium to develop a modelling tool with the attributes which have been outlined. This follows from the recognition that object-oriented techniques offer advantages in an environment where regular design, re-design and update of simulation models occurs (Adiga & Glassey 1991).

The object-oriented paradigm views the world as a collection of discrete 'objects'; each object encapsulates both the data and behaviour associated with the real-world entity it represents. Object-oriented methods not only provide a richness of representation but also a versatility not found in conventional approaches. The properties of object-oriented methods are well documented, as are their applications in manufacturing; examples include Graham (1991) and Rogers (1994) respectively.

The concept of an object (or class) library is particularly relevant to this discussion. An object library is a coherent collection of tested and reusable objects, for example part objects, machine objects, truck objects. A library of 'planning and control' objects will provide the necessary building blocks outlined above; the provision of such a library represents a core theme of this research.

High and Low Level Objects

This application of object-oriented methods to the planning and control domain
has resulted in two types of object: high level and low level. The high level objects
provide less detailed representations of the planning and control functions. Consider,
for example, a high level master production scheduling (MPS) object. This provides
the typical functionality of an MPS; the object records the production programme for
all the end-items in terms of part name, due date and order quantity. This object
interacts with others in the planning and control system; for instance a material
planning object, part object, a stock object and a planned orders object. These
additional objects can be either high or low level.

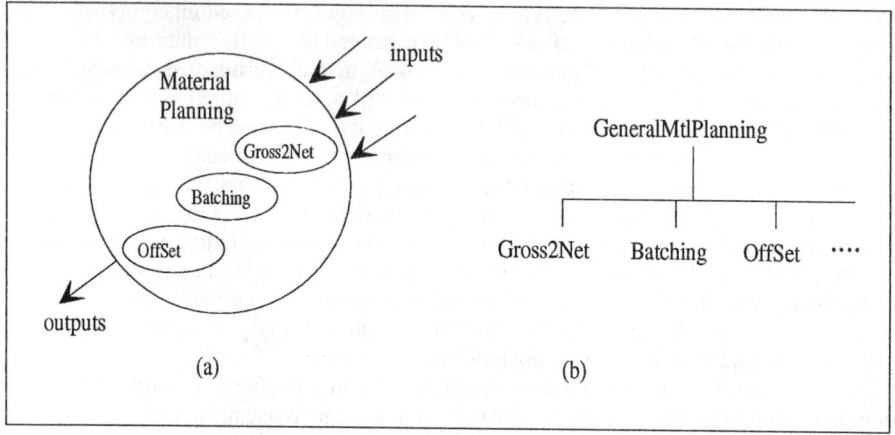

Figure 1 : An example configuration of low level, detailed objects.

It is possible to replace any high level object with low level objects which
provide more detailed and, importantly, more flexible object representations. Consider,
for example, the material planning object illustrated in Figure 1(a); this particular
material planning object represents the core functionality (or mechanisms) of the
material requirements planning (MRP) calculation. The detail of MRP is provided
through the component objects: *Gross2Net*, *Batching* and *OffSet*. The material
planning object is configured from the more elementary low level (component) objects.

This implementation provides real flexibility in object configuration. If, for
example, a company or particular design only required the *Gross2Net* and *OffSet*
mechanisms then only those component objects would be included in the material
planning object. Furthermore if a completely different mechanism were in operation
then this too could be included provided, of course, it is a compatible material planning
object and exists in the object hierarchy (Figure 1(b)). As and when objects are
developed they can be added to the object hierarchy, or similar object hierarchies.

The simulation environment is provided by the Advanced Factory Simulator
(AFS) (Ball & Love 1993). Because all of the objects described here are inherited
from the *Simulation* object found in the AFS architecture, they include all of the
mechanisms to participate and exist in the simulation world. Through the event
mechanism it is possible to provide the natural timing and delays found in a real
system. This ultimately leads to the proper integration of a planning and control
system model and a model of the manufacturing facility.

The planning and control object library will, therefore, contain different types of object: various high level object representations and a host of elementary, component objects. It is the provision of the low level (component) objects which represents a major contribution towards design flexibility. These will cover the range of planning and control functions which exist, from material and capacity planning to shop floor control. In addition, the flexibility of this implementation extends beyond standard solutions such as MRP or kanban; it will be possible to represent the 'actual' company implementation of these systems as well as the company specific mechanisms which support such standard solutions.

An Application Example

The following example (Figure 2) illustrates the scope of the simulation objects available within the domain of production planning and control. For simplicity only key information flows are shown. The planning system model is integrated with a model of a shop-floor facility whose functionality is provided using AFS.

The production planning and control system is made up of the following objects: master production schedule (MPS), bill of materials (BOM), stock status, material planning, planned orders and confirmed orders. All except material planning are high level objects and provide the respective core functionality. The MPS object creates a master production schedule in terms of end-item, due date and order quantity. The basic stock records and transactions are provided in the stock status object. For example booking in/out of stock items and the appropriate update of due-in/allocated parts. All the 'firm' works (or purchase) orders are held in the confirmed orders object. These 'firm' orders are derived from the planned orders object which holds all the planned order releases suggested by the material planning object. The BOM object

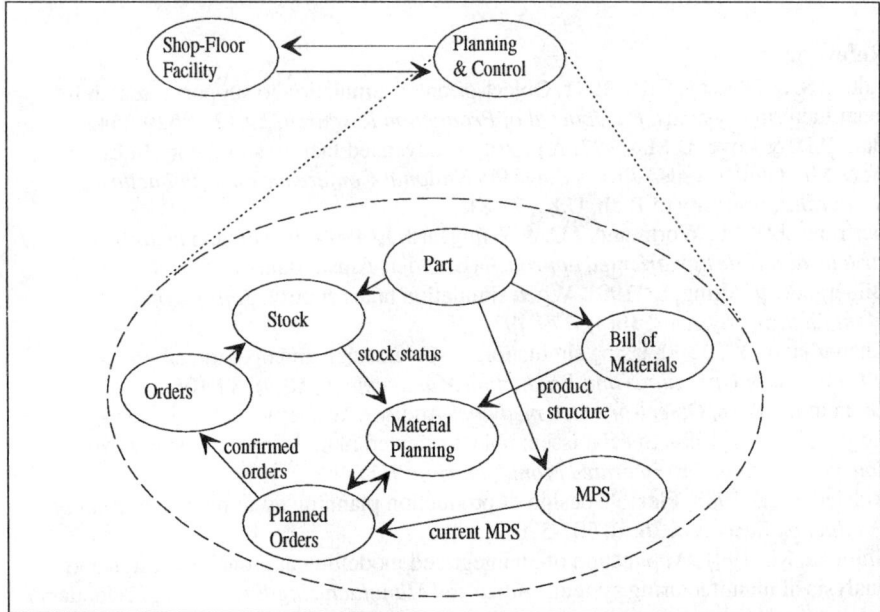

Figure 2 : An application example.

represents a simplified manufacturing bill of materials: the respective parent to each child part and the number required.

The material planning object performs an MRP based calculation and the available options are based on the configuration described above (Figure 1). All, or a selection, of the suggested planned order releases are made 'firm'. This process can be automated or, more likely, left to manual intervention through the user interface. The derived works order schedule is then passed directly to the production department. Actual orders are released on to the various shop-floor departments.

Once an order has been completed in the manufacturing facility, AFS will support appropriate feedback mechanisms. For example, the person responsible for an order completion will advise the foreman, who will in turn advise the production planning and control system and request transportation to move the completed batch on. On arrival of components at the stores the inventory system would be updated.

Summary

This paper has introduced a novel approach to modelling manufacturing planning and control systems, based on object-oriented discrete simulation. It is considered that this approach will offer significant advantages when applied to the design and management of such systems. A library of high level and low level (elementary) objects is being developed; different systems will be configured from these objects. An application example has been provided to illustrate this principle.

Acknowledgements

The authors would like to acknowledge Lucas Engineering & Systems Ltd. for their support.

References

Adiga, S. & Glassey, C.R. 1991, Object-oriented simulation to support research in manufacturing systems. *Int. Journal of Production Research*, **29**(12), 2529-2542.

Ball, P.D. & Love, D.M. 1993, A prototype advanced factory simulator. In Bramley, A. & Mileham, T. (eds.) *Proceedings 9th National Conference on Manufacturing Research*, University of Bath, U.K., 77-81.

Bertrand, J.W.M., Wortmann, J.C. & Wijngaard, J. 1990, *Production control: a structural and design oriented approach* (Elsevier, Amsterdam).

Bilberg, A. & Alting, L. 1991, When simulation takes control. *Journal of Manufacturing Systems*, **10**(3), 179-193.

Chaharbaghi, K. 1990, Using simulation to solve design and operational problems. *Int. Journal of Operations and Production Management* , **10**(9), 89-105.

Graham, I. 1991, *Object-oriented methods* (Addison-Wesley).

Rogers, P. (ed.) 1994, Special issue: object-oriented manufacturing systems. *Int. Journal of Computer Integrated Manufacturing*, **7**(1).

Rolstadas, A. 1988, Flexible design of production planning systems. *Int. Journal of Production Research*, **26**(3), 507-520.

Shimizu, M. 1991, Application of an integrated modelling approach to design and analysis of manufacturing systems. *Advanced Manufacturing Engineering*, **3**(January), 3-17.

A CONTROL STRATEGY
FOR EXPERT SCHEDULING SYSTEMS

B. L. MacCarthy and P. Jou

*Department of Manufacturing Engineering and Operations Management,
University of Nottingham, Nottingham, U.K., NG7 2RD*

This paper addresses the design of the control strategy of scheduling
systems based on expert system approaches. The control strategy is the
core of such systems and may be the key to gain mutual benefits from
AI and OR approaches to scheduling. Scheduling environments are
inherently dynamic, uncertain and complex and consequently are
difficult to deal with. Although many AI techniques may perform
valuable scheduling functions on their own, combining them with OR
techniques in hybrid systems may enhance scheduling systems
considerably. Practical problems in real scheduling environments are
noted and the need for hybrid approaches is emphasized. A generic
control strategy architecture is proposed and described to find a good
match between OR and AI approaches to scheduling problems.

Introduction

The control of manufacturing systems still poses many problems in most
manufacturing environments. There are many reasons for this. A major problem is that
there may be numerous objectives and some of these may be conflicting. Some form
of control system is always necessary.

A control system with one or more feedback paths is called a closed-loop
system The aim of feedback is to reduce the difference between the actual system
output and desired output. One of the advantages of employing feedback is that it can
stabilize an unstable system. In the context of manufacturing systems there is a need
to allow changes not only of the physical elements of the system but also to allow
changes in control rules as the environment changes over time. The priority of the
effect of such internal and external disturbances is difficult to decide. The control
strategy should be the core of the system that handles disturbances to bring the system
back to stability while minimising the disruption and satisfying the constraints of the
moment.

The shop floor control system must be able to react to small changes on the shop floor. Gershwin *et al* (1986) present a view of manufacturing systems from the perspective of control theory. Control theory views scheduling as a dynamic activity, where the scheduling problem is one of understanding how to perform rescheduling.

This paper addresses the design of the control strategy of scheduling systems based on expert system approaches. Scheduling environments are inherently dynamic, uncertain, and complex and consequently are difficult to deal with. Different environments will require different techniques. Although many AI techniques may perform valuable scheduling functions on their own, combining them with OR techniques in hybrid systems may enhance scheduling systems considerably.

Combining O.R. and A.I. in Scheduling

OR employs a systems approach. The emphasis is on algorithms and heuristic methods. The scheduler is required to evaluate and balance constraints in the search for optimum solutions. However, there remains a gap between the systems approach and the real-world problems. Factors not considered in classical scheduling theory include : multiple objectives rather than a single objective; preferential treatment of 'soft constraints'; dynamic environment rather than static; stochastic job processing times rather than deterministic; sequence dependent set up times rather than fixed; the subjective element of human activity. (Khoong (1993) and Farhoodi (1990))

AI started from the approach of trying to model human expertise and capabilities. The earliest contribution from AI to scheduling applications came about in the early eighties. Grant (1986) showed that the key insight of AI scheduling work is the importance of constraints. Fox *et al* (1983) found schedulers spend 80-90% of their time determining the constraints of the system. Only 10-20% of their time is spent on the generation and modification of schedules. Several problem areas arise in using AI approaches in scheduling. These include : domain dependent rather than general purpose systems; partially dynamic rather than really dynamic systems; difficulties in evaluating a dynamic system; difficulties in handling a broad domain, e.g. common sense; systems are limited by lack of contextual knowledge; expertise may not always exist. (White (1990) and Grant (1986))

Grant (1986) concluded that AI and OR need to be appropriately combined in order to alleviate some of the difficulties shown above. Bruno *et al* (1986) provided an example of one such combination, where expert system techniques are used for knowledge representation and problem solving, and queueing network analysis is used for fast performance evaluation. Other specific AI projects reported in the literature are ISA, DEVISER, ISIS, ISIS-2, OPIS, OPAL, FIXER and PLANNET. (White (1990)) This shows that the scope of research is vast. Charalambous and Hindi (1991) reviewed 20 systems, noting that the majority are still prototypes and concluding that hybrid solutions may be the right way forward. White (1990) also advocated a struture based on control theory to define and serve as a problem framework; combined heuristics to generate candidate schedules; simulations to verify the performance of candidate schedules and scheduler to manipulate interactively the simulations.

Control Strategies in Scheduling

The control strategy is the core of such systems and may be the key to gain mutual benefits from AI and OR approaches to scheduling. Ehlers and Rensburg

(1993) indicate that dynamic scheduling solutions and real-time rescheduling are two areas that need more research and proposed an eight rule taxonomy to evaluate dynamic scheduling systems. One of the eight rules is the relaxation strategies that are needed when all the constraints cannot be met (due to conflicting constraints).

Grant and Nof (1991) study the feasibility of automatically scheduling multi-machine complexes on a real-time basis and implementing a unified computer system to provide new scheduling tools in automated systems. They defined 'control' to mean the generation, and subsequent possible adjustment, of achievable schedules and the communication of those achievable schedules to the manufacturing system throughout the production interval. Their framework includes five components defined as follow : scheduler - generates a schedule for orders and objectives; monitor - monitors the current status of the executed schedule; comparator - compares the findings of the monitor with the schedule; resolver - responds to the results of the comparator; adaptor - adapts the current schedule if the resolver decides to replan. The functions of the adaptive/predictive real-time scheduler are : to transform a given cell schedule into a real-time set of work instructions; automatically to develop modified schedules for short periods to implement special requirements; to respond and adapt automatically to certain variations in the real-time schedule execution, following a carefully selected recovery strategy. e.g. leave it as is or shift some tasks or reroute certain operations.

O'Grady and Lee (1988) describe the application of the AI techniques of blackboard and actor based systems (Hewitt (1977)) for intelligent cell control in the framework termed Production Logistics and Timings Organizer (PLATO-Z). The approach is that of breaking the control into a hierarchy where each level of the hierarchy has a more narrow responsibility. Such a hierarchy consists of four levels : factory, shop, cell and equipment levels. The cell control system can be categorized as a part of the planning and scheduling system, but it does require other function such as error handling and monitoring in addition to scheduling. The major functions of the cell control system include the need to schedule jobs, machines and other resources in the cell so as to achieve the commands and goals from the shop level. Four blackboard subsystems are designed : scheduling, operation dispatching, monitoring and error handling blackboards.

The control of the problem solving activities is performed by a control unit called the blackboard controller. The basic control mechanism consists of a three step process. The first step involves checking the status. When the status change matches with the event's data, the monitor generates the blackboard event. The second step is performed by a scheduler which takes the list and selects the highest priority event. The third step is to execute the knowledge source corresponding to the selected event.

Yoshida and Nakasuka (1989) suggest responding to quality-type and quantity-type information to enhance the effect of the hierarchical scheduler. Their algorithm is to formulate 'if-then heuristics'. These consist of inequalities with respect to quantity-type attributes. For example : If $(X+Y-Z < 2.3)$ and $(Y-Z < -11.6)$ and $(Z < 9.7)$ and minimimun makespan is required then use minimum slack time rule where X is tardiness tolerance value, Y is minimum slack time within the buffer and Z is time until the machine completes service.

Proposed Control Strategy for Expert System Scheduling

The struture of the proposed control strategy combines on-line schedule repair and off-line simulation, to provide a decision support tool for shop floor control.

Similar models have been used in several applications. (Farhoodi (1990)) and Efstathiou (1993)). The current research focusses on the important class of scheduling problems which have sequence dependent set up times.

The schedule repair subsystem is a knowledge-based system. After receiving information on changes in the environment, this subsystem functions with an emphasis on rescheduling. The steps include : classify disturbances; define new objectives or constraints according to the classified disturbances; check the feasibility of the existing schedule; relax soft constraints when all the new constraints cannot be met; give priority to some disturbances when all disturbances cannot be resolved; cancel jobs or reduce lot sizes when the relaxation of constraints is invalid and reschedule when it is needed. e.g. shift or resequence jobs in different ranges;

Off-line simulation is carried out to determine the best strategy for the current application among the candidate schedules. Whenever a simulated schedule is obtained the case-base stores data containing problem-solution, configuration and searching techniques. The data in the case base is used to generate meta-rules by machine learning. Afterwords the technique of machine learning is employed to enhance the knowledge acquisition capability. By this technique more powerful and precise heuristics can be obtained. The control module will contain the meta-rule to decide the method of switching between the rules or the decision tree of rules.

Based on the requirements for scheduling, the system approach suggests first six steps to generate a satisfactory solution - choosing the environment, specifying objectives, inputting the required data, generating the schedule, rescheduling, and building case-base by simulation. The use of an object-oriented approach considers modularity and reuseability so that the system can be expanded in a step by step manner. The configuration will include : user interface module, communication module, DB handling module, scheduling module, control module, monitoring module configuration module, event logger module, simulation module and relaxation module.

The proposed system will be suitable for complex multi-machine environments. Users can choose between the different environments listed below : single processor, single stage parallel processors, single stage parallel processors with different capacities, flowshop with a single processor, flowshop with parallel processors, job shop, job shop with parallel processors, job shop with dominant routes. But at first a narrow class has been selected to develop a prototype system.

There are a wide range of performance measures in scheduling. Gupta *et al* (1989) noted that among the most intensely studied criteria were machine/system utilization, throughput/mean flow time and production rate. This system tries to give guidance so that users can specify their own combinations of performance measures. These measures include maximizing machine utilization, minimizing throughput/mean flow time, maximizing production rate, meeting due dates, minimizing mean tardiness, minimizing maximum tardiness, minimizing work-in-process and minimizing average lead time.

Conclusion

This paper addresses the design of the control strategy of scheduling systems based on expert system approaches. The control strategy is the core of such systems and may be the key to gain mutual benefits from AI and OR approaches to scheduling. Although many AI techniques may perform valuable scheduling functions on their own, combining them with OR techniques in hybrid systems may enhance scheduling systems considerably. Such a structure may be based on control theory to define and serve as a problem framework; combined heuristics to generate candidate schedules; simulations to verify the performance of candidate schedules; machine learning to obtain more powerful and precise heuristics; scheduler to manipulate interactively the simulations.

A generic control strategy architecture is proposed and described to find a good match between OR and AI approaches to scheduling problems. Many software issues need to be addressed. Further developments of the system will be reported in the near future.

References

Bruno, G., Elia, A. and Laface, P. 1986, A rule-based system to schedule production, *IEEE Computer*, **19**, 32-40.

Charalambous, O. and Hindi, K.S. 1991, A review of artificial intelligence-based job-shop scheduling systems, *Information and Decision technologies*, **17**, 189-202.

Efstathiou, J 1993, Scheduling the flow of material through an FMS, *Proceedings of the CIM Conference*, Singapore, 587-593.

Ehlers, E. and Rensburg, E.V. 1993, An intelligent object-oriented scheduling system for a dynamic manufacturing environment, *Proc. of the CIM Conf.*, Singapore, 702-709.

Farhoodi, F. 1990, A knowledge-based approach to dynamic job-shop scheduling, *Int. J. Computer Integrated Manufacturing*, **3**, 2, 84-95.

Fox, M.S., Smith, S.F., Allen and B.P., Strohm 1983, Job-shop scheduling: an investigation in constraint-directed reasoning, *Proceedings of AAAI-82*, Carnegie-Mellon University, Pittsburgh, Pa, U.S.A.

Gershwin, S.B., Hildebrant, R.R., Suri, R. and Mitter, S.K. 1986, A control perspective on recent trends in manufacturing systems, *IEEE Control*, **34**, 3-9.

Grant, F.K. and Nof, S.Y. 1991, Automatic adaptive scheduling of multi-processor cells, In M. Pridnam and C. O'Brien (ed), *Production Research Approading the 21st Century*, (Taylor & Francis, London) 165-172.

Grant, T.J. 1986, Lessons for OR from AI, *J. Opl. Res. Soc.*, **37**, 1, 41-57.

Gupta, Y.P., Gupta, M.C. and Bector, C.R. 1989, A review of scheduling rules in flexible manufacturing systems, *Int. J. Comp. Integ. Manuf.*, **2**, 6, 356-377.

Hewitt, C.E. 1977, Control structure as patterns of passing messages. *Artificial Intelligence*, **8**, 323-363.

Khoong, C.M. 1993, The impact of scheduling research on production management. *Proceedings of the CIM Conference*, Singapore, 717-722.

O'Grady, P. and Lee, K.H. 1988, An intelligent cell control system for automated manufacturing, *Int. J. Prod. Res.*, **26**, 5, 845-861.

White, P.K. Jr. 1990, Advances in the theory and practice of production scheduling. *Control and dynamic systems*, **37**, 115-157.

Yoshida, T. and Nakasuka, S. 1989, A dynamic scheduling for flexible manufacturing system: hierarchical control and dispatching by heuristics,*Proceedings of the 20th Conference on Decision and Control, Florida*, 846-852.

SHORT TERM SCHEDULING AND ITS INFLUENCE IN PRODUCTION CONTROL IN THE 1990s

Professor J G Kenworthy, Mr D Little, and Mr P C Jarvis
Liverpool University, Liverpool L69 3BX
Dr J K Porter
Liverpool John Moores University, Liverpool L3 3AF

The role of Finite Scheduling in the modern Production Control system is discussed, from a standpoint that the manual approach is under threat from software led systems, and these may not be necessarily beneficial or cost effective; research to identify "Best Practice" is outlined, across a range of industry sectors and covering a range of scheduling approaches. A Classification System is proposed covering the full spectrum of scheduling methods. Some initial findings from the research are discussed which indicate that in some sectors Finite Schedulers can provide real business benefits but place even greater emphasis on operating disciplines, and calibre of Production Control personnel.

Introduction

It is a fact beyond dispute that any successful commercial enterprise must find the optimum trade off between customer service, manufacturing efficiency and inventory costs. These aspects of a business usually conflict with one another and the point of balance is determined in many different, often irrational ways. With the advancement of Supply Chain Planning as an integrated concept, at least now there is a growing recognition of the importance of such trade offs. The production control function in its various forms and titles, is central to this trade off process; planning and scheduling are at the core of production control, the magnitude of its task depends on a combination of factors both outside and within its' ability to control.

The central importance of planning and scheduling is perhaps implicit when considering the advance of MRP since the early 1970s and its development into MRPII with its systematic consideration of materials and capacity in deriving workable job lists. In the majority of companies a weekly or even monthly MRP run, would give rise to work-to lists which could be used with reasonable confidence to schedule jobs. Such activity was typically manually performed, often at shop floor level, with short term decisions affecting customer service, inventory and manufacturing efficiency being made by quite junior employees.

There is strong evidence [1,2,3] that the MRP approach has given major benefits to firms employing it; however the market pressures in today's business climate make this method of production control increasingly unsatisfactory in many environments. Customers across most sectors of industry are demanding ever shorter lead times and in many cases orders are changing in quantity and required date many times before final delivery. Near-perfect or even perfect delivery service is being increasingly demanded. Product life cycles are reducing, increasingly customised products being demanded, increased production efficiency with associated lower costs are being sought ever more relentlessly and companies are being required to work with ever reducing stocks. These are among the escalating business pressures faced by companies in the 90s. [5] All of this is tending to force companies to seek greater short term control, by more exactly scheduling resources to meet changing requirements yet with the knowledge of the consequences of scheduling decisions in terms of customer delivery performance and cost.

Until comparatively recently short term scheduling was either not done at all (expedite according to latest instructions), performed manually perhaps with decision aids or in a very limited number of cases, use was made of finite scheduling systems either as modules of an MRP system or via a connected scheduler. Decision aids cover traditional style planning boards to computerised versions, sometimes used in conjunction with real time shop floor data collection systems -Leitstands used extensively in Germany. Schedulers as modules of MRP packages are normally of low versatility with their operating algorithms set as single level, often due date based.

Within the last 5-7 years there has been a development of stand alone scheduling packages designed to work independently directly off an order file. In the majority of such systems, the manufacturing processes are developed into a mathematical emulation incorporating work centre capacities, routings, calendar files, etc, which through the application of the Bill of Material files on the sales orders, construct representations of possible outcomes of the manufacturing task, then evaluate these against established objectives, such as customer service criteria, due date performance, etc. Algorithms tend to be complex involving nested or combination rules to determine sequences and thus schedules. The outcomes can be improved via iterative cycling to generate an optimum or best fit against certain preset criteria, and often have quite elaborate graphical displays to show the quality of the schedule that has been generated. Such systems are fundamentally different to MRP in the treatment of the scheduling task, and represent a viable alternative to the generic MRPII models for industries where material availability is not a key issue but where responsiveness to rapidly changing customer requirements is crucial to business success. Computing power available on modern hardware -often a stand alone PC is all that is required- is clearly a key factor in the development of these state of the art schedulers. Applications of such schedulers can be found across a range of industries.[4]

Outline of research

The Liverpool University STS Research [4] group has been engaged for the past two years in a programme of work designed to establish "best practice" in short term scheduling across a range of industrial sectors, and across the whole spectrum of techniques. The research has involved Case Studies, now numbering around 30, at which a series of interviews has been held with a range of personnel involved with production control. This is followed by analysis and scoring against a number of parameters within a model devised by the Group, which then seeks to identify performance against key criteria. The research, like all SERC/ACME sponsored projects seeks to retain industrial relevance at all times, with the aim being to benefit industry through dissemination of findings and sharing of knowledge. The findings are thus based on "real" situations and the conclusions are argued not from an academic standpoint but from pragmatic observation.

The Group believes that a simple production control system classification adequately describes the range of scheduling approaches observable in industry, and this is given in FIG 1.

Broadly the degree of sophistication of the scheduling tools increases from right to left across the spectrum; however any given tool need not necessarily be fixed to a particular point; indeed the Group have seen tools which vendors would claim belong in one category while its' client is clearly using it in a manner consistent with a different category. The distinction between planning led environments and scheduling led environments referred to above, when the concept of stand-alone schedulers was discussed is shown by the division in the classification diagram.

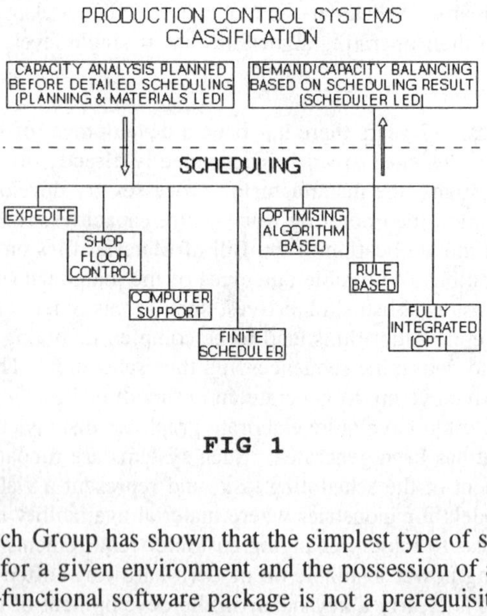

PRODUCTION CONTROL SYSTEMS
CLASSIFICATION

FIG 1

Work by the Research Group has shown that the simplest type of scheduling tool may well be appropriate for a given environment and the possession of a very sophisticated, multi-functional software package is not a prerequisite to best or even good practice in scheduling. Indeed the research has identified surprisingly few examples of good software installations. The choice of scheduling tool needs to be made with regard to the degree of difficulty of the task, and the type of environment in which it must operate. The Groups' findings are at an early stage, yet it is clear that a number of generic conclusions are emerging which can be summarised below:-

Research Findings

*   regardless of scheduling system, good data management and strong disciplines are vitally important ingredients for successful production control activities. The research method has demonstrated this clearly in FIG 2:

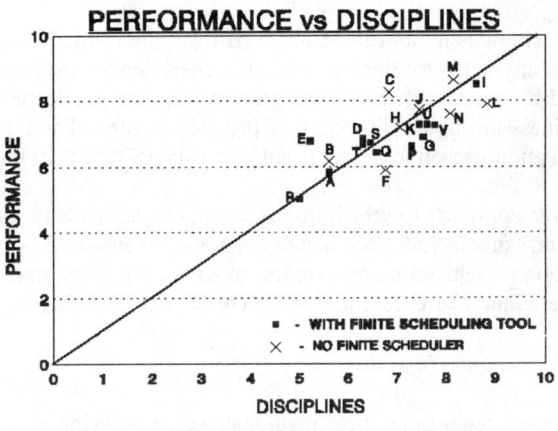

FIG 2

*   there is some evidence to suggest that finite scheduling systems can give improved performance for a given level of effort and at least equivalent performance for less effort (fewer people). In the plot of difficulty (a concept developed in the research in an attempt to relate dissimilar industrial sectors) against production control personnel numbers, this can be seen in FIG 3:

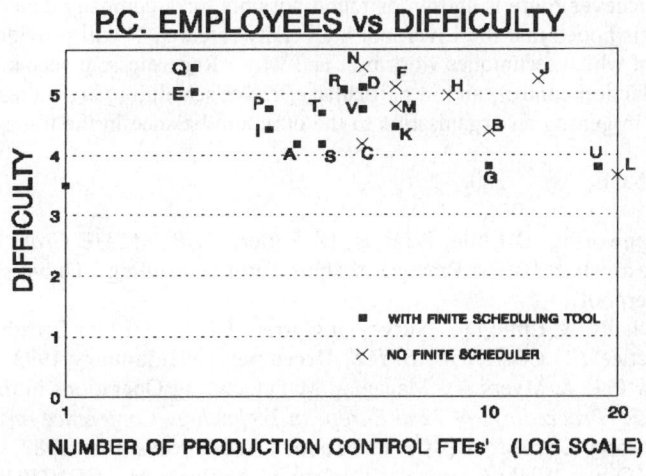

FIG 3

Other findings

*   production control demands high calibre people, appropriately supported and
         recognised for their pivotal importance to commercial operations.

*   major business benefits can be obtained from the critical analysis phase that
         precedes any software implementation; often these benefits can accrue without
         the need to install software. The process of gaining a deeper understanding of
         the business through a dissection of production control activities and subsequent
         simplification can often be as (if not more) beneficial than the software itself.

*   whichever approach to scheduling is taken, strong demand management is most
         important; this includes accountability for sales forecasts, consistent
         application of scheduling criteria (no month end pushes) and acceptance that
         schedule adherence is essential in synchronised scheduling environments.

Conclusions

There would seem to be three main approaches to bring about real
improvements in short term scheduling in a company:-

(a)   Simplify operations (for example by using JIT techniques), so that
         formal planning and scheduling are unnecessary.

(b)   Implement effective high level planning strategies such that material
         and capacity shortages are either eliminated or planned around, so the
         complexity of the scheduling problem is reduced and becomes
         manageable "in the head".

(c)   Develop more sophisticated, possibly computerised scheduling tools.

Whichever route is chosen or found possible for a company depends on many
things; it is hoped that the Liverpool University research [4] will provide some
answers on which techniques work best and why. Regardless, it seems clear that
good production control, and in particular *effective* scheduling are increasingly
important in getting an organisation to the optimum balance in the trade off equation.

REFERENCES

[1]   J Kenworthy, D Little, P Jarvis, JK Porter. *SERC/ACME Grant GR/H/20473;*
       Investigation of Best Practice in Short Term Scheduling. University of
       Liverpool

[2]   Jarvis PC & Little D ; "Survey of Current UK Shop Floor Scheduling
       Practice";   *CONTROL, BPICS,* **December 1992/January 1993,** Pg 24-28.

[3]   New C.C & Myers A.; Managing Manufacturing Operations in the UK: 1975-
       1985. *Proceedings of 22nd European Technology Conference on Production &
       Inventory Control.* **BPICS Blackpool 17-20th November 1987.**

[4]   The Oliver Wight Companies Control of the Business. *CONTROL, BPICS,*
       Coventry, **July 1986,** Pg 31-37.

[5]   PA Consulting Group, "Manufacturing into the late 1990s", *HMSO ISBN 0-11-
       515206-7,* **London, 1989.**

APPLICATION OF AN EXPERT SYSTEM TO SCHEDULING PROBLEMS IN A PRESS SHOP

Kah-Fei Ho and Dr Keith Ridgway

University of Sheffield
Sheffield, S1 3JD

This paper describes the development of an expert system for scheduling the manufacture of laminations in a press shop. The paper analyses the press shop scheduling requirements and compares this to a criterion for the successful introduction of expert systems. The paper concludes with a review of the system developed and the experience gained.

Introduction

In the real production world, the scheduling procedure applied in any particular industry is usually complex, and only understood by the personnel fully involved. The responsibility of the scheduler is to determine when and what resources are required for the jobs to be accomplished. It takes time for a new scheduler to acquire the necessary expertise and skills from an experienced scheduler and understand the working environment. Generally, the experienced scheduler will retain his knowledge and expertise in memory, and produce a workable schedule by relating the status in the production shop floor to this data bank, in which the reasoning process is implicitly logical. Therefore it is possible to foresee the problems that could occur if the experienced scheduler was unavailable.

With the introduction of an expert system, it is possible to alleviate the above scenario. An expert system provides a systematic framework for formulating the explicit logical solution to the ill-structured problem. To demonstrate the implementation of an expert system for production scheduling a case study has been carried out at a UK. Company that manufactures stator and rotor laminations, for electric motors. This paper describes the criteria for the successful development of an expert system. The paper discusses the structure of the scheduling decisions made and describes how the system has been developed to accommodate the informal manual system. The paper concludes with a review of the system developed and describes the experience gained.

Expert system in production planning and scheduling

An expert system can be defined as a computer system that can solve problems using human expertise and knowledge within the specified scope of the environment. A typical expert system has four stages of implementation: knowledge acquisition, knowledge representation, control system, and testing and verification. However, before expert system technology can be applied the problem domain should be examined to determine the applicability from three different perspectives: the characteristics and scope of the problem, availability of domain expertise, and problem solving techniques. In practice, it is advisable to check the feasibility of applying an expert system to the problem rather than to fit the problem to the expert system. Gupta (1990) reviewed various aspects of the successful implementation of expert systems in the manufacturing environment. All these stages will be illustrated clearly in the practical case study.

There are numerous applications of expert system in planning and scheduling of manufacturing systems and they have a key role in enhancing productivity, improving quality, increasing profits and capturing expertise, as reported by Kusiak and Chen (1988). They reviewed more than twenty expert planning and scheduling systems and suggested that the appropriateness to the production areas stems from the scarcity of expertise, the complexity in the decision making process, and the critical significance of accurate and fast decisions.

Problem Definition

The press shop manufactures laminations, from a large coil of sheet metal, through progressive stamping operations in a multi-stage press. Every complete cycle of progressive stamping produces both rotor and stator laminations. These are used in the manufacture of stator and rotor cores respectively of electric motors. The laminations can be classified into standard, losil and unit. Standard laminations are made from normal mild steel. Losil laminations are manufactured from a special heat treated mild steel material which is larger than the standard thickness. Both are used in the normal motor, but the losil lamination is only used in a special design of electric motor. The demand of stator and rotor laminations arises from the downstream electric motor manufacturing facility that receives the motor orders from prospective customers. The number of electric motor ordered in one week is transformed into the quantities of stator and rotor laminations in inches, through the Material Requirement Planning (MRP) system, located on an IBM AS400, using the Bill of Material (BOM). The Company manufactures a total of 123 laminations that comprise 57 stator and 66 rotor laminations and require twelve different sizes (width) of material. The shop floor has eight different multi-stage presses using a total of 35 different dies. A maximum of ten operators work in the press shop at any time and a die setter is assumed to be available at all times.

Everyday the Production Scheduler is faced with the complex task of scheduling the stamping jobs to the shop floor within the manufacturing constraints. Before the Production Scheduler examines the weekly demand for laminations on the computer system, the status of the shop floor needs to be determined in terms. The scheduler needs to determine the availability of labour, material, presses, and dies. In addition it is necessary to know the details of the dies already fitted on each press. The lamination demand is then examined to elicit and prioritise the stamping orders. After getting the list of stamping jobs in order of priority, the Production Scheduler will need to match the requirements to the current manufacturing status. There are many considerations to be handled before the stamping job is released to the shop floor. If a stator lamination is stamped it may be possible to produce more than one type of rotor lamination by changing the die set-up, or vice-versa. Therefore it is necessary to check the demand for the alternative rotor laminations and to confirm which one is to be manufactured. Moreover, each of the laminations may be stamped by more than one press. It is a practice to minimise the die set-up and fully use press labour. It is therefore customary to release non urgent jobs to ensure maximum use of the press labour and press. Thus, the Production Scheduler needs to compromise between throughput rate of stamping orders, customer satisfaction and press labour utilisation within the current manufacturing constraints. Most of the reasoning process for the job scheduling is carried out by the Production Scheduler who has acquired this expertise through years of experience.

Implementation of expert system to press shop scheduling

Many criteria need to be satisfied before the problem defined can be addressed and an adequate schedule achieved. The solution requires a detailed mental manipulation of inputs including the availability of labour, material, presses, and dies and the resources requirement of the stamped lamination. A large amount of inter-related information needs to be gathered before a decision can be made. There will be a number of different solutions depending on the current shop floor circumstances, and the capability and experience of the Production Scheduler. Therefore, the decision made will not be consistent and will vary if a different scheduler is responsible for the scheduling task within the same manufacturing constraints. It takes on average 45 minutes for the experienced Production Scheduler to produce the schedule. Thus the scope and characteristics of the problem are considered to be reasonable for the implementation of expert system. The domain expert was available as a Production Scheduler with experience in press shop scheduling, was accessible for advice as required. The heuristics, at his fingertips, can be explicitly identified, and reduced to a set of rules that could be used in the expert system. Moreover, testing and evaluation of the expert system can be carried out confidently with the expert advice available, providing complete trust and acceptance of the resulting system.

Initially the Production Scheduler was interviewed to determine the method of press shop scheduling used. It was very difficult for him to fully articulate the mental processes that enabled a feasible schedule to be devised and to form these into a set of rules. Flow charting techniques were used to capture his implicitly reasoning procedure through the diversity of facts and constraints imposed by the resources and demand. To assist the development of the rules required scheduling decisions, based on different data, were observed on a daily basis.

After completing the knowledge acquisition phase, it was determined that the knowledge base should comprise rules that can incorporate facts, procedures and control. The facts are the status and availability of labour, material, presses, and dies, and the details of the die-set already in press. These are dynamic facts that are input to the system to represent the actual manufacturing constraints. One example of a particular material factual rule representation is shown below:

IF   Material 101 status is OK
THEN   Material 101 is available

The procedural rule is used to determine the output status after its condition is satisfied. It can be the schedule rule, and any rules requiring numeric manipulation. A schedule rule will generate three different types of schedule depending on the availability of resources as shown:

IF   All resources are available
     Die already in press
THEN   Schedule 1

IF   All resources are available
     Die not in press
THEN   Schedule 2

IF   Some resources are not available
THEN   Schedule 3

The control rule directs the sequence of processing the procedural rules. It can be grouped into lamination, urgent job, and temporary job rules. The lamination rule is used to control the iterations of each lamination in the demand file through the procedural rules. The urgent job rule is used if urgent orders need to be carried out immediately. Each is input manually before the lamination demand file is retrieved so that the urgent order will be scheduled first and preferably selected by the schedule 1 rule. The temporary job rule will be activated if extra labour is available. The schedule 3 list of jobs can be searched to locate suitable jobs that can be selected using the schedule 1 rule and extra labour can be assigned.

Discussion and Conclusion

Since the Production Scheduler's decision making process includes heuristics or rules of thumb that can be expressed in the form of rules, a rule based expert system is desired, as shown in figure 1. The logical reasoning through this set of rules in a particular situation will result in consistent decisions that would have been taken by the Production Scheduler facing the same situation.

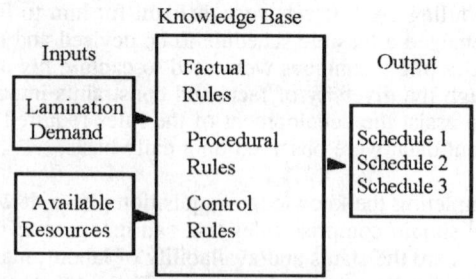

Figure 1: Schema of the expert system developed

Rapid prototyping and simplicity of the expert system were desirable as it was necessary to prove the viability of the project for press shop scheduling. The system was developed using a rule based expert system shell, "Crystal", running on a 386 personal computer (PC) with 4 MByte RAM. The knowledge base was represented using production rules, and commands. The commands were integrated within rules to serve several functions such as assigning values to variables, testing variables, displaying forms, menus and explanatory message, and controlling the rule processing procedure. The inference engine is backward chaining, but it can be forward chaining or a mixture of both. Moreover, it can be interfaced with Dbase and ASCII files.

The backward chaining "Crystal" expert system developed contains approximately 350 rules and 1000 commands and has the following characteristics:

- PC Support to provide a link between the IBM AS400 and PC so that the lamination demand in the IBM MRP system can be downloaded into an ASCII file suitable for use on a PC.
- Database in dBase format containing the standard requirements (i.e. material, press, die, stator or rotor) for each lamination. The priority or preference given to presses, dies or corresponding stamped part for each lamination is sequenced in the record according to shop floor practice.
- Knowledge base constituting the factual, procedural and control rules in an ordered logical fashion. Presented with the standard lamination input the system produces consistently satisfactory schedules similar to those of an expert production scheduler. Three schedules are produced to distinguish whether the jobs can be carried out and giving reasons if the job cannot be started. Schedule 1 is a list of jobs that can be done immediately providing labour is available since all resources are available and dies are already set-up on the presses. In schedule 2 jobs cannot be carried out with all resources available because die set-ups are required on the presses. In schedule 3 jobs cannot be carried out due to lack of resources. All the jobs in the schedules are listed in order of priority. Die set-up time and production times are determined and these help the Production Scheduler to improve the quality of decision making.

The expert system developed is continuously tested and verified with the help of the Production Scheduler. It provides a daily, feasible schedule automatically depending upon the lamination demand in the MRP system and the manufacturing constraints existing. The system has reduced the scheduling time from an average of 45 minutes to just 8 minutes and the expertise is permanently captured and available.

As a result of the active participation of the Production Scheduler, the project is a success. The project demonstrates the importance of ensuring that the problem is appropriate for the implementation of an expert system and that a domain expert is available and willing to transfer his expertise. The developed expert system developed is a good decision aid to the Production Scheduler.

References
Gupta, Y. P. 1990, Various aspects of expert systems: applications in manufacturing, *Technovation*, **10**, No. 7, 487-504.
Kusiak, A. and Chen, M. Y. 1988, Expert systems for planning and scheduling manufacturing systems, *European Journal of Operational Research*, **34**, 113-130.

JOB SHOP SCHEDULING USING MULTIPLE CRITERIA

Simon J. Liang* and Dr. John M. Lewis

Department of MMSE
Napier University
10 Colinton Rd., Edinburgh EH10 5UN
E-mail: s.liang@csu.napier.ac.uk
Tel:++44 (0) 31-447-9242

This paper describes an application of Genetic Algorithms for the generation of job shop schedules. We present results for minimising multiple criteria: mean flow time and mean lateness, with respect to variance of due date assignment. These criteria were used to form a combined fitness function for the genetic algorithm, in which the relative weights of the criteria modifies the schedule generated. In all cases the convergence time for finding a solution is similar to those for a single criterion. In conclusion, we have shown that a Genetic Algorithm can be used to generate near optimal solutions for job shop scheduling with weighted multiple criteria.

Introduction

A common problem for scheduling systems in the production industry is the assignment of customer orders to shop floor processing (i.e. deciding the sequence of orders to be carried out). The majority of research in production scheduling has used time-based performance measures to evaluate various proposed scheduling strategies, such as lateness, tardiness, flow time, percent tardy, and due dates. However, in a practical factory the production performance can not often be measured in isolation, especially when there exists conflict between the measures. Therefore, it is desirable to have a single approach that can consider multiple performance criteria at the same time. Itoh (1993) proposed a method to validate multiple performance criteria, which is based on heuristic dispatch. Sequence scheduling has long been seen as a problem that belongs to the class of NP-complete problem. DeJong and Spears (1989) use genetic algorithms (GA) to investigate such scheduling problems. In GAs, a objective function acts as a survival environment for their candidates, which are normally encoded as genes in a string, or chromosome. A random population pool of chromosomes is created as a first generation. GAs select pairs of individuals from this pool based on their performance in optimising the objective function. The population of low fitness chromosomes decreases over generations, while the population of high fitness ones increases and is selected to bear many offspring from the simulated evolution generation. This procedure is repeated in successive generations resulting in inferior traits in the pool die out due to lack of reproduction, while strong traits tend to combine with other strong traits to produce children.

We apply a GA and use a linear combination of mean flow time and mean lateness as the objective function to generate schedules. In French (1982), the mean flow time is a measure that is often used as a criterion for shop floor resource utilisation, while mean lateness is usually considered as the response speed to the requirement of customer demand. The considered problem is that given n independent jobs $i(i = 1, \ldots, n)$, assume each job has a zero release time and that pre-emption is not permitted. Let t_i = processing time, C_i = completion time, d_i = due date, w_i = waiting time, $L_i = C_i - d_i$ = lateness, $F_i = C_i - w_i$ = flow time. This objective function is, thus, written as

$$\min R = \alpha(\frac{1}{n}\sum_{i=1}^{n} F_i) + (1-\alpha)(\frac{1}{n}\sum_{i=1}^{n} L_i) = \alpha\overline{F} + (1-\alpha)\overline{L} \tag{1}$$

where α is the weighting factor, $0 \le \alpha \le 1$, and $\overline{F} = \frac{1}{n}\sum_{i=1}^{n}(C_i - w_i)$ = mean flow time,

$\overline{L} = \frac{1}{n}\sum_{i=1}^{n}(C_i - d_i)$ = mean lateness. The main objective of equation (1) is to find a minimum, which has best overall performance subject to the prescribed technical constraint.

Experimental Design

Our study considered a set of discrete n jobs with deterministic processing time and due date. The processing times were generated at random with a normal distribution, and the due dates were established and adapted from Rohleder (1993) using

$$d_i = \sum_{i=1}^{n} t_i [0.5k + \beta(1.5k)] \tag{2}$$

where d_i = due date of job i, n = total number of jobs, t_i = processing time, k = allowance multiple, β = beta random variable. In addition, the individual chromosome is a sequence of the set of n jobs, encoded as

$$J = \{j_1 j_2 j_3 \ldots j_n | j, n \in N, j = 1..n\}$$

where j_i is the i-th job in the sequence, and each j represents a job. In equation (1), where $\overline{F}$ and $\overline{L}$ are always not equal in their value scale, the objective function should be converted to a fitness with a fair effectiveness to elude possible dominance in the objective function. Application of the short processing time (SPT) rule has been proved to be able to result in an optimum of mean flow time (denoted F_{SPT}) for sequenced jobs. Using the early due date (EDD) rule a fixed, not necessary an optimum, mean lateness (denoted L_{EDD}) can be derived. In addition, given a set of jobs to be sequenced, both F_{SPT} and L_{EDD} are unique. Therefore, we chose them as the fitness converting factor for mean flow time and mean lateness respectively. The objective function for this study is then rewritten as follows:

$$\min R' = \alpha(\frac{\frac{1}{n}\sum_{i=1}^{n} F_i}{F_{SPT}}) + (1-\alpha)(\frac{\frac{1}{n}\sum_{i=1}^{n} L_i}{L_{EDD}}) \qquad (3)$$

$$= \begin{bmatrix} \alpha \\ 1-\alpha \end{bmatrix} [F_\psi \quad L_\psi] \qquad (4)$$

where

F_ψ = mean flow time converted fitness, and L_ψ = mean lateness converted fitness.

Using equation (4) to validate each possible permutation of sequenced jobs, now depends only on the weight factor α. The chromosome population in one generation can be randomly selected, and our study used sample population of 100 for every generation. In each generation, every chromosome is a possible sequence of n jobs which can be evaluated by the objective function for fitness. If the fitness for that individual is above average for the population, it is selected to mate with another high fitness individual. A partially-mapped-crossover operator (PMX) is then applied to exchange part of chromosome (i.e. genes, which represents a partial job sequence). This PMX operator, Goldberg (1989), is specially designed for generating valid sequence of jobs.

Results and Discussions

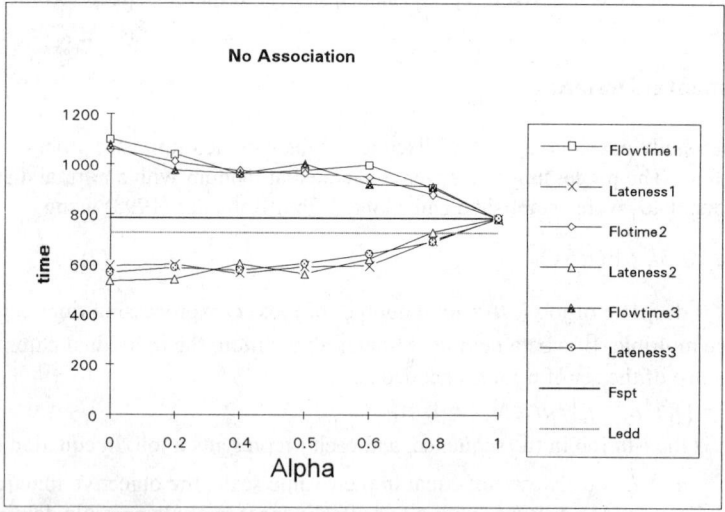

Figure-1 Due date assignment without association

100 random processing times (which have a normal distribution) were generated and two classes of due date assignment were used; random and predictable. For random assignment, processing times (P) and associated due dates (D) take values in the range $\{[P_i, D_i]: 1 \le P_i \le 50, 50 \le D_i \le 200; i = 1, 2, \dots n\}$ where due dates and processing times are assigned independently of each other. These results are shown in Figure-1, in which F_{SPT} and L_{EDD} are also shown to indicate the behaviour of weighting

factor. Setting α to a low value (the mean lateness is more concerned), results in mean flow time being greater than F_{SPT}, and mean lateness smaller than L_{EDD}. As α increase, mean flow time tends to F_{SPT}, and mean lateness increases. Hence, it is observed that mean flow time and mean lateness are of two conflicting criteria for scheduling, and the weighting factor acts as an balance scale between them. The more emphasises on one criteria, the smaller are the results.

For predictable assignment, the established data is of the set

$$\{[P_i, D_i]:1 \le P_i \le 50, D_i = \sum_{j=1}^{n} P_{ij}[(0.5k + \beta(1.5k)]; i = 1,2,..n; j = 1,..(i-1);1 \le \beta \le 10; k = 0.1\}$$

Due dates are dependent upon processing times with this assignment relationship. The results are shown in Figure-2. Since the allowance (k) is set to 0.1, this represents a very tight assignment of due date, and results in a lower L_{EDD}. The behaviour of mean flow time is similar to random assignment because the same set of processing times were used. Moreover, the mean lateness increment is still driven by laying emphasis upon flow time, and vice versa. Comparing to Figure-1, both of results are shown that the α weighting factor dominates the dependency of scheduling parameters (here, flow time and lateness).

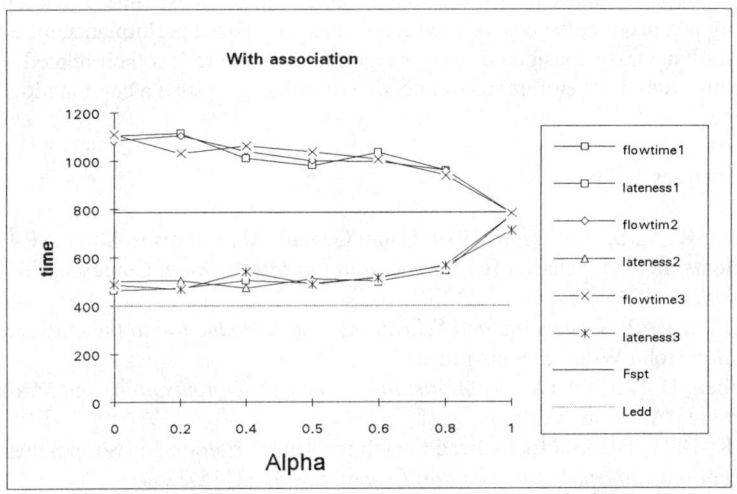

Figure-2 Due date assignment with association

Mean lateness is calculating from the time used by each of the preceding completed jobs. The longer the preceding processing time, the greater the flow time of successive jobs. In addition, processing time for each job can be precisely estimated once the desired manufacturing process has been decided. As a result, processing times for jobs will be member of a set of discrete random variables. Thus, the associated due date assignments are also random variables, which are known to depend most upon shop floor capability and work load, and demands from customers and management. Moreover, the distribution of processing time is different from system to system, and job to job. We observed that the variance of due dates assignment results in a varied level of minimum mean lateness. It is also observed that mean flow time and mean lateness are only dependent upon the weighting factor, which reflect the emphasis of either criteria. However, other factory related parameters such as shop floor capacity and loading will all affect due date assignment. We found that GAs can be used for scheduling using either random or associate assignment, but associate assignment case is mostly required for a real factory.

Conclusions

The use of a genetic algorithm in multiple criteria job scheduling has been presented in this study. The results show that laying emphasis upon criteria can be performed by selecting an appropriate weighting factor (desirable criteria should set a higher weight factor), while combination of criteria can be used to derive a combined emphases. This study has shown that GAs can apply to use multiple criteria for production scheduling. Moreover, the genetic algorithm used in the study validates only the objective function (equation (4)), there is no significant difference between validating single or multiple criteria. The fitness converting factor is an important issue that deserves further investigation to decide its proper value. The factors used in the study are domain dependent which are derived from rule-based results, short processing time and early due date. However, production systems can be evaluated by other important performance measures, such as cost, or jobs release time. From a practical factory perspective, the cost of production and time-based performance measures should be simultaneously considered using weighted factors to reflect their related emphasises. A future study is, therefore, recommended to investigate such a combination of multiple criteria.

References

DeJong K. A., Spears W. M. 1989, Using Genetic Algorithms to Solve NP-Complete Problems, In J. D. Schaffer (ed.), *Proceeding of International Conference of Genetic Algorithms*, 124-132.
French S. 1982, *Sequencing and Scheduling - an Introduction to the mathematics of the job-shop* (John Wiley & Sons Ltd.).
Goldberg D. E. 1989, *Genetic Algorithms in Search, Optimisation, and Machine Learning*, (Addison-Wesley)
Itoh K. 1993, Twofold look-ahead Search for Multi-criterion Job Shop Scheduling, *International Journal of Production Economics*, **31**, 2215-2234.
Rohleder T. R., and Scudder G. D. 1993, Comparing Performance Measures in Dynamic Job Shops: Economic vs. Time, *International Journal of Production Economics*, **32**, 169-183.

*Simon J. T. Liang is a visiting lecturer from National Yunlin Polytechnic Institute, Taiwan, ROC.

Concurrent engineering and design

MODELLING PRODUCT INTRODUCTION: A KEY STEP TO SUCCESSFUL CONCURRENT ENGINEERING APPLICATIONS?

Dr. N.J.Brookes, Dr. C.J.Backhouse and Prof. N.D.Burns

Loughborough University of Technology
Loughborough, Leics, LE11 3TU

This paper examines the need for the modelling of product introduction in the application of concurrent engineering. It presents a review of existing modelling techniques and, from the most appropriate, a variant derived for use with product introduction. It comments on a case-study application of the technique and draws wider conclusions from that experience on modelling product introduction to aid concurrent engineering applications

Modelling Product Introduction for Concurrent Engineering Applications

Concurrent engineering is an approach that is characterised by a fluid definition but it is increasingly regarded at a strategic level as a means of improving product introduction performance, (e.g Sohlenius (1992), Brooks (1992), Harkins (1993), Jacob (1993)). Improved performance is achieved by a loose collection of tools and techniques which fall under the umbrella of concurrent engineering. Voss (1992) summarised these into four generic features :-

    i) <u>Process</u> - parallel and integrated with overlapping problem solving
    ii) <u>People</u> - integration mechanisms
    iii) <u>Tools</u> - CAD,CAM,CAE and analytical methods
    iv) <u>Performance</u> - strong market orientation and leadtime reductions

Voss's summary provides an insight into the multitude of decisions that need to be made in a concurrent engineering application. He identifies the need for a methodology for concurrent engineering applications to overcome this complexity.

Parallels can be drawn between the application of concurrent engineering and the application of CIM (computer integrated manufacturing). Both approaches are trying to achieve improvements in organizational performance by introducing company-wide changes with technological, structural and procedural implications. In developing methodologies to apply CIM, the ability to model the organization was recognised as a key foundation,(Bravaco (1985), Glenney and Mackulak (1985), Chajtman (1988)). It is therefore likely that modelling product introduction for the application of concurrent engineering will be equally beneficial.

The need to model product introduction for successful concurrent engineering is becoming more widely appreciated. Taft (1992) and Zhang (1992) both discuss the use of models in concurrent engineering implementation. However, they are concerned with producing a generic model of product introduction. This paper focuses on the need to generate a model of the specific application.

Identifying the Modelling Technique

In order to identify the particular technique to be used to model product introduction, it is vital to delineate the definitive parameters of product introduction as these are precisely the aspects that the technique will need to capture. In order to determine this for product introduction, it is useful to consider it as, in Checkland's (1981) terms, a *purposeful human activity system*. Checkland argues that this form of system can generally be defined in terms of a transformation process. Furthermore, he indicates that an understanding of a system's behaviour can be achieved through understanding the interrelationship of the transforming process and the structure performing that process. Product introduction can be seen as a transformation process that converts a perceived customer need for a product into all of the information needed to create that product. In order to understand the behaviour of a product introduction system, if one follows Checkland's approach, the following features needed to be identified :-

- The transforming elements within the system and their structure. (In Voss's terms the 'people' and 'tools' aspects of concurrent engineering).

- The interaction of both transforming and transformed elements in individual connecting links to create the overall transforming process. (In Voss's terms the 'process' aspects of concurrent engineering).

These features therefore formed the salient features for understanding the system and hence any chosen modelling technique needed to capture these features. The duality of this approach is very important. Frequently product introduction is considered only in terms of its process and not in the relationship between its process and its structure.

The choice of technique was also determined by its ease of use and how easy it is to understand its results. It by necessity included a mechanism to cope with the system complexity such as hierarchical decomposition.

If product introduction is concerned with the transformation of information, it is possible to use techniques that have traditionally been associated with the automation of information flow to model product introduction. (It is important to remember that only the modelling techniques were compared and not the overall approach of these techniques to systems analysis.) The following existing techniques were reviewed for their suitability :-

i) **Flowcharts** (Yeomans and Chandry(1985))
ii) **Petri-nets** (Passler, Hutchinson, Rudolph and Stanek (1984), Ranky (1988), Petri (1976))
iii) **CORE** (Mullery (1979))
iv) **SSADM** (Cutts (1987))
v) **IDEF** (Hughes and Maull (1988), Le Clair (1982), Baines and Colquhoun (1990))
vi) **IEF** (James Martin and Assoc.(1987)
vii) **Yourdon** (Yourdon and Constantine (1979), Bowles (1990),)
viii) **Jackson** (Craig(1983), Jackson (1983), Mount (1990))

In order to determine the most appropriate of the above techniques to model product introduction, an evaluation matrix was created. The use of an evaluation matrix has been recommended by Blank and Krijger (1983) as an appropriate mechanism for this type of decision. Evaluation was initially carried out in terms of each techniques ability to model the key features of the product introduction systems and the number of separate models that the technique used. (This was taken as a crude measure of the techniques ease of use.) Figure 1 shows the resulting matrix. This indicates that two techniques were capable of modelling all of the required aspects: IDEF and CORE. CORE was rejected as it was judged to be more complex than IDEF.

Although IDEF was considered as the most appropriate technique for modelling a product introduction system from those evaluated, problems were identified in its use. Firstly IDEF, even though it is more accessible than CORE, has a reputation for being difficult to understand and lengthy to use (Wyatt (106), Mackulak (1986), Colquhoun, Baines and Crossley (1993)). Secondly, IDEF2 models detailed time-based aspects of a system which may not be appropriate to the consideration of the product introduction system. Thirdly, although IDEF1 contains the facilities to model the transforming elements structure, it appears only ever to have been used to model the relationships between transformed elements. Because of these problems, a modelling technique for product introduction was devised which, although based on IDEF principles, was altered to overcome the problems outlined above.

Using the Modelling Technique : A Case Study Application

The area used for the application of this modelling technique was the fabrications facility of a large aerospace manufacturer. It employed approximately 400 people. It was responsible for determining how complex fabrications were broken into constituent parts and then assembled together. In that respect it represented the 'tail-end' of product introduction but was considered still worthy of investigation because of the complexity of the components under consideration.

Modelling product introduction helped in implementing concurrent engineering in the following way :-

1) It identified opportunities to further parallelise activities. (These lay in the area of updating 'Bill-of-material' type information.)
2) It identified non-value added activities that could be eliminated
3) It identified the skill types that would need to be accommodated if a team were to be established to perform the product introduction process.
4) It highlighted where the most effective area lay for computer automation.

It is always a difficult in case-study applications to demonstrate that one approach was more successful than any other as precise circumstances can never be replicated in order to carry out such a trial. What can be claimed, in this case, is that by modelling the product introduction system insights were given into how to apply concurrent engineering that were new to the company.

Further Developments of the Modelling Technique

On the basis of the case-study application the following developments were identified :-

1) The modelling technique needs to be more widely tested to prove its efficacy especially in terms of modelling the correct system characteristics.

2) The modelling technique was used at the stage of product introduction where activities tend to be formalised and the flow between them uni-directional. In the initial phases of product introduction, this may not be the case and therefore the modelling technique needs to be amended to reflect this situation.

3) The overall performance measures of the product introduction system need to be related to the model to identify areas of greatest potential improvement.

4) The modelling technique still needs to be placed in the overall context of a methodology for concurrent engineering application. Modelling provides a mechanism for analysing, documenting and communicating concurrent engineering applications. It does not provide a way of creating or evaluating the most appropriate application.

These proposed developments to the modelling technique show that it is premature to make definitive claims for the usefulness of modelling product introduction in concurrent engineering applications. However there is sufficient evidence to make further study worthwhile.

The authors gratefully acknowledge the support of the SERC/ACME Directorate in writing this paper.

REFERENCES

Baines, R.W. and Colqhuhoun, G. 1990, An integration and analysis tool for engineers, *Assembly Automation*, **10**, 141 - 145

Blank, K. and Krijger, M.J. 1983, *Software Engineering : Methods and Techniques*, (Wiley Interscience, Netherlands)

Bowles, A. 1990, Cradle tool for real-time, information systems, *IEEE Software*, 7, 38

Bravaco, R. 1985, Planning a CIM system using systems engineering methods, C.M.Savage (ed) *A Program for CIM Implementation*, (SME) 19 - 35

Brooks, B.M. 1992, Concurrent Engineering - what is it?, *IEE Colloquium on Concurrent Engineering*, (IEE), 1 - 3

Chatjman, S. and Zyzik, M. 1988, The identification of information processes in manufacturing, *Towards the Factory of the Future : 8th Intl. Conf. on Prod. Res.*, (Springer Verlag, Bonn)

Checkland, P. 1981, *Systems Thinking : Systems Practice*, (Wiley, NY)

Colquhoun, G. Baines, R.W. and Crossley R. 1993, The use of IDEF0: A state of the art review, *Int. J. of CIM*, **6**, 252-264

Craig, P. 1983, A structured design methodology, (Rolls-Royce internal report GN 26339)

Cutts, G. 1987, *SSADM : Structured System Analysis and Design Methodology*, (Paradigm, London)

Glenney, N. and Mackulak, G.T. 1985, Modelling and simulation provide the key to CIM implementation philosophy, *Industrial Engineering*, May 1985, 76 - 84

Harkins J. and Dubreil M. (1993), Concurrent engineering in product design/development, *Plastics Engineering*, **49**, 27 - 31

Hughes, D. and Maull, R. 1988, Design of Computer integrated manufacturing systems, *Towards the Factory of the Future : 8th Intl. Conf. on Prod. Res.*, (Springer Verlag, Bonn)

Jackson, M. 1983, *System Development*, (Prentice Hall)

Jacobs, R. 1993, Quality in Concurrent Engineering, *Annual Quality Congress Transactions*, (ASQC, USA), **47**, 902 - 908

Le Clair, S.R. 1982, IDEF: The method, architecture and means to improved manufacturing productivity, (CASA/SME technical paper ref. MS82-902)

Mackulak, G. 1985, An examination of the IDEF0 approach used as a potential industry standard for production control system design, *Automated Manufacturing*, ed. L.B.Gardner, (ASTM, Philadelphia), 131 - 149

James Martin & Assoc. plc 1987, Information Engineering Facility, (course notes)

Mount, R. 1990, Jackson tools build code from specs, *IEEE Spectrum*, 7, 38

Mullery, G. 1979, CORE: A method for controlled requirement expression, *Proc. 4th Intl. Conf. on Software Engineering*, (IEEE NY)

Passler, Huthinson, G. Rudolph, K. and Stanek, W. 1984, Production system Design : A directed graph approach, *J. of Manuf. Sys.*, **2**, 107-116

Petri, C.A. 1976, General Net Theory, *Proc. of joint IBM / University of Newcastle Seminar on Computer System Design*, 7 - 10

Ranky, P.G. 1988, Software Engineering Methodology and Tools for Designing Advanced Manufacturing Systems, *Intl. Conf. on Factory 2000, 31 Aug - 21 Sept. 1988, Churchill College, Cambridge*, (IEE)

Sohlenius, G. 1992, Concurrent Engineering, *CIRP Annals*, **41**, 645 - 656

Taft, J. and Barclay I. 1992, Simultaneous engineering: a management evaluation and establishment model and methodology, *1992 International Engineering Management Conference: Managing in a Global Environment*, (IEEE, NY) 119 - 121

Voss, C.A. Coughlan,P. Hon,B. 1992, Research Priorities in simultaneous engineering in the UK, report from SERC/ACME grants ref. GR/G62400, GR/G63605

Wyatt, T.M. 1987, System methodologies - their application to automated design of manufacturing systems and their usage in the factory of the future, (Internal report, Department of Industrial Studies, University of Liverpool)

Yeomans, R.W. Chandry, A. and van Hagen P. 1985, *Design Rules for a CIM system - Esprit Commision of the European Community*, (North Holland)

Yourdon, E. Constantine, L. 1979, *Structured Design: Fundamentals of a Discipline of Computer Program and System Design*, (Prentice-Hall)

Zhang, H. and Alting, L. 1992, An exploration of simultaneous engineering for manufacturing enterprises, *Int. J. of AMT*, **7**, 101 - 108

MODELLING TECHNIQUES	SYSTEM COMPONENTS REQUIRING MODELLING					MODEL FEATURES	
	PROCESS / ACTIVITY	INPUT/OUTPUT ELEMENTS ROLE IN ACTIVITY	TRANSFORMING ELEMENTS ROLE IN ACTIVITY	TRANSFORMING ELEMENT STRUCTURE		HIERARCHICAL BREAKDOWN	NUMBER OF MODEL TYPES USED
PETRI-NETS	Y	N	N	N		N	1
FLOW CHARTS	Y	N	N	N		Y	1
IDEF	Y	Y	Y	Y		Y	3
CORE	Y	Y	Y	Y		Y	4
SSADM	Y	Y	N	Y		Y	3
IEF	Y	Y	N	Y		Y	5
YOURDON	Y	Y	N	Y		Y	3
JACKSON	N	N	N	Y		Y	2

KEY :-
Y - TECHNIQUE POSSESSES CAPABILITY   N - TECHNIQUE DOES NOT POSSESS CAPABILITY

Figure 1 - The Evaluation Matrix for Existing Modelling Techniques

THE APPLICABILITY OF CONCURRENT ENGINEERING TO DESIGN IN A MANUFACTURING COMPANY

S.J.May[#][*] M.N.Polkinghorne[*] J.P.Bennett[*] and R.S.Burns[*][+]

[#]*Ranco Controls Ltd, Plymouth, PL6 6QT.*
[*]*Plymouth Teaching Company Centre, University of Plymouth, U.K.*
[+]*School of Manufacturing Materials and Mechanical Engineering, University of Plymouth, Plymouth, U.K.*

This paper discusses the application of concurrent engineering to a high volume/low margin manufacturing company. The current design philosophy has been detailed and some modifications have been suggested. IDEF diagrams have been utilised to model existing design strategies with proposals made for an enhanced structure. The benefits to be obtained, and the required company strategy for implementation, have both been discussed.

Introduction

Concurrent engineering is currently being applied to a broad range of industrial manufacturing environments, e.g. a conceptually static product environment, where the manufacturing plant is already in place, or a fully dynamic product environment, where the manufacturing or process system required cannot be specified until the product characteristics emerge, Pugh (1990). Whilst the principles of concurrent engineering have existed for many years, the present drive towards integration of the design and manufacture of related processes ensures that all product elements are given suitable consideration. The entire product life cycle must therefore be closely examined. Traditional 'over the wall' design currently found in many industrial companies greatly restricts the interactive design process causing an increase in the associated design time, overall cost and possibly a reduction in final product quality, Wilson and Greaves (1989). By employing concurrent engineering, interaction between design, manufacture, purchasing and sales during the actual design stage ensures the derivation of a cost effective solution.

Considering a specific application, this paper investigates the utilisation of concurrent engineering to a high volume/low margin manufacturer. Relevant modifications to the existing design philosophy are proposed with a view to accessing the generous performance advantages possible via a concurrent engineering approach. Examples of the required strategy, and of the corresponding benefits, are presented.

After extracting the salient elements, the more general applicational implications are discussed with reference to a low margin/high volume manufacturing company.

The Existing Design Philosophy

The design philosophy within the given company is a common one found in British industry and utilises the traditional approach to design. This traditional method is to break down the company into smaller more understandable functional units, to analyse each one individually and then to optimise the operations of each unit. This type of thinking has resulted in barriers between departments, St. Charles (1990) and is illustrated with the term 'over the wall', which implies teamwork between departments is not fully utilised. The term 'over the wall' has been emphasised, in the given company, by the location of the design centre in Germany, and the manufacturing plant in the United Kingdom. An $IDEF_0$ model of the current process for this high volume/low margin manufacturing company can therefore be demonstrated [Fig. 1], providing illustration of the present problems.

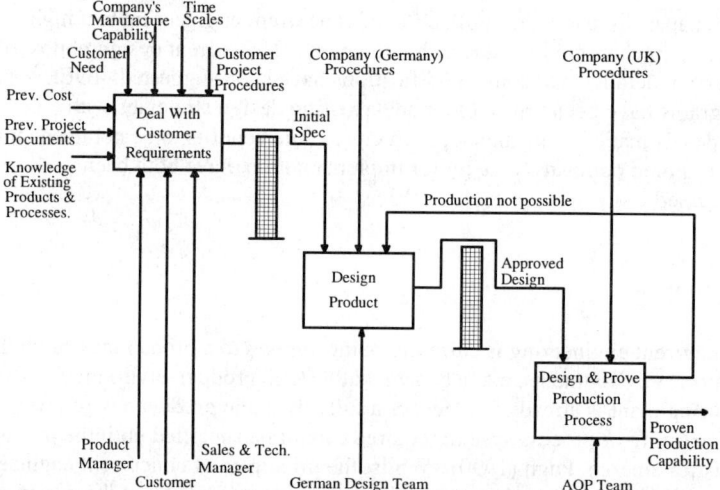

Figure 1. The Current Design Process

The initial requests from customers are assessed in order to develop a requirement, then thrown 'over the wall' to the German design centre, where all design work is carried out, and then thrown back 'over the wall' to manufacturing. The smaller design changes to current products are undertaken within the product engineering department. All of the mentioned design is carried out using conventional empirical methods which are highly laborious and time consuming.

Modifications to the Existing Design Philosophy

It is clear that the current situation within the given company has to change dramatically if concurrent engineering is to be applied. The location of the design centre in Germany causes many problems when looking at the application of concurrent engineering. The process would be significantly easier to introduce if all the design work was undertaken in the United Kingdom. This situation has been modelled using IDEF$_0$ [Fig. 2].

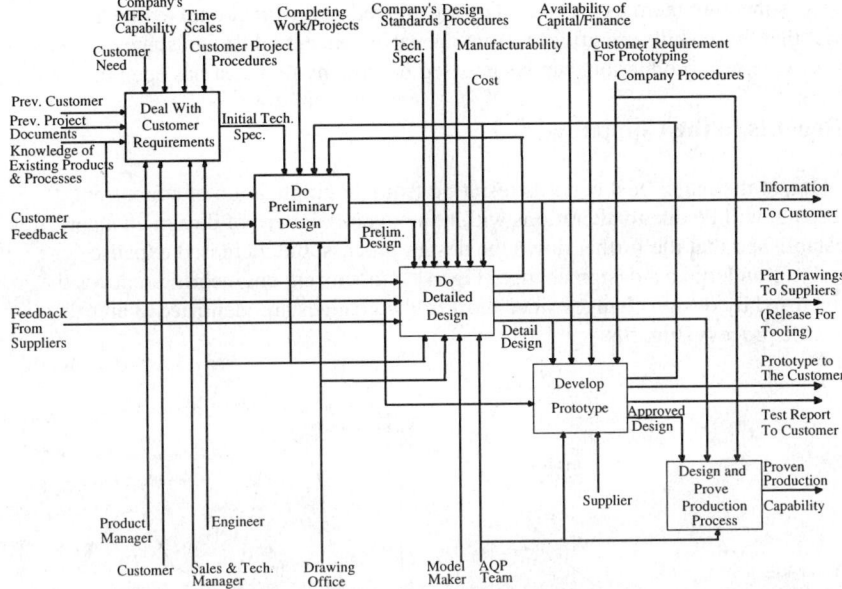

Figure 2. The Revised Design Process (Design Centre in the U.K.)

Parallel to moving the design work to the U.K., it is necessary to implement various concurrent methodologies, Carter and Baker (1992). It is a necessity for top managers to construct product development teams, giving those teams the required amount of power to action decisions. An essential aspect of concurrent engineering is that the development teams have full managerial support to enable attributes such as enthusiasm and commitment to be fully utilised. The teams must be comprised of multi functional disciplines, e.g. members from departments such as design, manufacturing and purchasing. Communication within these teams, and also with management, is of enormous importance in order to realise the potential of the development teams. If this communication does not occur it is likely that relevant ideas get by-passed and possible savings overlooked. In order to achieve the required levels of communication it is essential to minimise the number of available paths [Fig. 3].

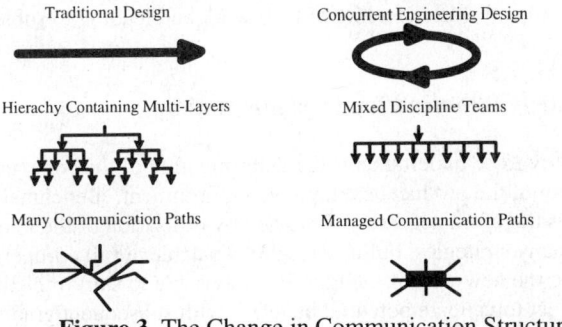

Figure 3. The Change in Communication Structure

Ideally, the customers and suppliers should have an input to the development teams to ensure that the company knows the customer requirements, and that the company is meeting them as required. Having created multi-functional teams, it is essential that these skills are utilised, from the early conceptual design stages, in order that processes, e.g. production, can be assessed before any real cost has been incurred.

The Benefits to the Company

One of the major cost benefits resulting from the implication of concurrent engineering, will be due to alterations within the overall pattern of design changes. It is well established that the further down the design process, the increased expense necessary to undertake a design change [Fig. 4]. Concurrent engineering reduces the costs incurred by design changes since most of the changes are identified at an early stage in the process [Fig. 5].

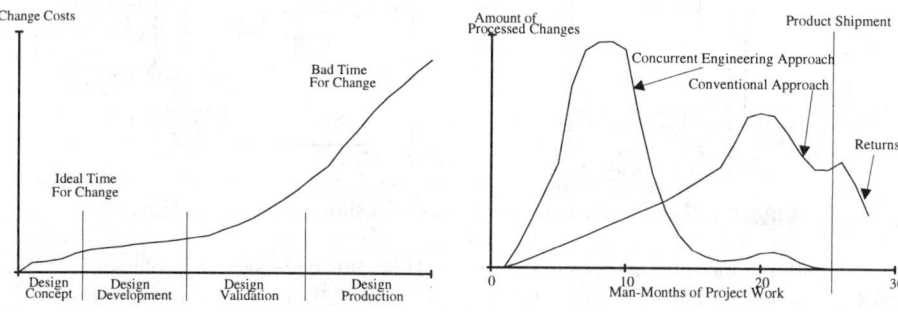

Figure 4. Change Cost .v. Design Life   **Figure 5.** Design Changes : Concurrent .v. Traditional

Due to the fact that most of the design changes have taken place prior to production, and that multi-disciplined teams are utilised, the product going into production will almost certainly be 'manufacturable' first time. Products will be brought to market earlier providing an increased design cost reduction, whilst creating the opportunity for capturing the market/increasing market standing which aids promotion of the company. In addition, there will be higher performance levels with respect to customer needs and quality expectations, St. Charles (1990). Another major benefit from concurrent engineering is the improved communication within the company environment, providing the whole company with a feeling of teamwork and leading to subsequent productivity gains.

The Company Strategy Required for Implementation

The initial activity to be undertaken is the determination of the company's current position, i.e. assessment of the product development environment. Benchmarking against similar organisations is therefore necessary, followed by utilisation of the knowledge gained to update company strategies, Fullmer (1993). To achieve full co-operation, all staff must be trained to the new way of thought. It is advisable to start implementation with a small pilot project to validate potential benefits, with subsequent gradual building on the knowledge gained. To achieve the desired goals from concurrent engineering it is a pre-requisite that targets are set for achievement from the outset, e.g. reducing the lead time by 50 % and reducing the costs by 50 %. Otherwise, the full benefits to be gained from the implementation of concurrent engineering could be missed, Hartley and Mortimer (1990), Wilson and Greaves (1989). When a novel element is introduced to any process, there are many other arising issues which are directly related. Due to the application of concurrent engineering several important side issues become apparent. The engineering and manufacturing capabilities of a company can be cost effectively improved. Similarly, databases and information systems used to maintain design information can be utilised in conjunction with concurrent engineering to enhance product and process design, St. Charles (1990).

Conclusions

It is obvious that the introduction of concurrent engineering to the high volume/low margin manufacturing company in question would generate a range of substantial advantages. These benefits would be gained in several ways, e.g. through cost savings in design and shorter time to market, thus providing the company with an increased competitive advantage. Considering the financial stability of most low margin companies, any cost saving can lead to a substantial improvement in general stability and future investment prospects. Additionally, even small cost savings per unit can equate to large quantities of money when considering the impact on the high volume production required. Increased quality and flexibility of production mean that product viability is increased. The application of concurrent engineering to this sector of manufacturing industry must therefore be considered as an imperative action for managers to undertake.

Acknowledgements : The authors wish to thank the Teaching Company Directorate (SERC/DTI), University of Plymouth and Ranco Controls Limited for their support to undertake the project Grant Ref. Number GR/J55939.

References

Carter D.E. and Baker B.S., 1992, *Concurrent Engineering, The Product Development Environment for the 1990s*, Addison Wesley Publishing Company, U.K., pp 34-64.
Fullmer D., 1993, How to Make Concurrent Engineering Work, *Machine Design*, July 23, pp 21-24.
Hartley J. and Mortimer J., 1990, *Simultaneous Engineering, The Management Guide*, Industrial Newsletters Limited, U.K., pp 107-109.
Pugh S., 1990, *Total Design, Integrated Methods for Successful Product Engineering*, (Addison Wesley Publishing Company, U.K.), pp 171-73.
St. Charles D.P., 1990, Simultaneous Engineering - An Integrated Approach, *Proceedings of CAD/CAM '90 Conference,* NEC Birmingham, pp 59-68.
Wilson P.M. and Greaves J., 1989, Forward Engineering - A Strategic Link Between Design and Profit, *Proceedings of Mechatronics Conference,* Lancaster.

DESIGN RULES FOR RAPID CHANGEOVER

R I McIntosh, S J Culley, G B Gest, Dr A R Mileham

Design and Manufacturing Group, School of Mechanical Engineering
University of Bath, Bath, Avon, BA2 7AY

An ability to rapidly change over manufacturing processes between different products is pivotal in moving ever closer to a goal of a single item batch size. This ideal has been clearly identified as an important competitive feature, both in terms of internal process control, and to greatly enhance a company's flexibility to more precisely serve a full array of customer demands.

Previous work on set-up reduction (SUR) has focused primarily on method improvement. There has been little reference to the contribution that substantive design improvement is able to make. This paper demonstrates the way that method improvement for changeover and improvement by design are related, and indicates the merits of adopting each approach. Generic design rules are given.

Team Based Approach to SUR

SUR is widely approached on the premise that those personnel who operate and have daily contact with their machinery are those who are best placed to understand problems associated with changeover. If properly organised into teams, and with ownership and responsibility for improvement ideas, it is believed that changeover times will be driven down.

This is a simplistic picture. It has been determined that there are a number of factors which have to be taken into account if worthwhile and sustainable improvement is to be generated. Ownership, team awareness of SUR principles, operator belief in SUR, comprehensive factory support for the team (notably in design) are just some areas which are frequently less than optimally achieved. In each case the team's performance will suffer. McIntosh, Gest, Culley and Mileham (1993) have identified that there are further factors which will also serve to drive an SUR initiative off course. There is evidence that SUR is widely approached in an ad-hoc, unquantified and unstructured fashion. The team's efforts are not always targeted effectively and little analysis or documentation occurs.

These structural deficiencies represent major hurdles to SUR implementation. Even when they are overcome, a workplace team can still be limited as experience and expertise within the team are likely to preclude the substantive system design improvements by which ultimate changeover time reductions will be reached. It is these design issues which are now considered.

Eliminating Waste

Once the team has gained a practical grasp of the time wasted during changeover, by careful recording and analysis, it is faced with the task of eliminating it. Time may be saved by organisational improvements and/or by altering the design of the manufacturing system.

Improved organisation is well suited to a team-based approach and Hay (1987) has cited that changeover time improvements of the order of 75% are available by this route. The question arises as to how time savings beyond this occur. How does the team operate when it is required to move from an onus on organisational improvement to one of improvement by design ? Is design improvement appropriate - or can more ambitious targets be achieved by ever-refining the organisation of the changeover ?

The Method-Design Spectrum

Gest, Culley, Mileham and McIntosh (1993) proposed adopting a 'reduction-in' strategy to structure the hierarchy of changeover improvement. This strategy sets out to offer more clearly defined alternatives for directing improvement activity. It extends the previous over-riding internal/external rationale by promoting thought as to *why* activities should be undertaken.

The 'reduction-in' strategy looks to reduce (ultimately, to eliminate) changeover activities in one or more of the following listed categories :
- On-line activity
- Effort
- Adjustment
- Variety

Especially in respect of the latter three categories, this strategy usefully extends the work of Shingo (1985). The strategy allows a far more objective assessment to be made of any specific changeover problem, and thereby to better match a final solution to it.

Each of the four categories listed above may be viewed in respect of organisational (method) or design improvements. Organisational aspects rest particularly well in the category of 'on-line activity', which may be seen to be broadly compatible with the internal/external division proposed by Shingo (1985). Design considerations become more prominent in the latter three categories.

Overall distinctions between method and design are not at all clear cut. The likelihood is that the adopted solution for any specific task will be positioned on a spectrum between pure method improvement and pure improvement by design. There is no clearly defined boundary between the two, as shown in figure 2 below :

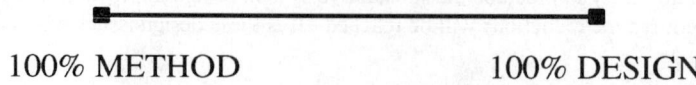

100% METHOD   100% DESIGN

Figure 1. The Method-Design Spectrum

Take a typical simple improvement idea - that of replacing screws securing an adjustable guide rail with quick release devices. The idea is to reduce the total effort associated with this aspect of the changeover. The use of a device moves the idea from being one of pure organisational improvement, yet does not in itself represent a pure design solution because essentially the same manual task of releasing and moving the rail is retained; it is just made easier and quicker to achieve. The solution would therefore lie on some point on the spectrum.

Adopting Design-based Solutions

Design based solutions offer powerful alternatives to relying on method based improvement. By design, a manufacturing system is physically changed and without such a change a changeover is constrained to be reliant on improved procedures. Improved procedures equate to doing the same things better rather than doing better things. It is reasonable to argue that there is a point at which no further changeover improvement becomes possible until design changes are instigated. By this argument a design input becomes essential if the shortest possible changeover times are to be reached.

There are many criteria which must be taken into account before it is decided whether or not an emphasis on design is warranted. In general, design-based solutions will take longer to come to fruition, will require more total effort and will involve greater capital expense. In their favour, design-based solutions offer a greater opportunity to reach ambitious reduction targets and time improvements reached with an emphasis on design should become easier to sustain.

One other important criterion to be identified at the outset is precisely what level of time reduction is sought. As demonstrated with reference to figure 2, the decision is an important one for it is indicative of the level of design input which is likely to be involved. The figure presents a hypothetical case of two alternative target times from a current elapsed changeover time of 10 hours. The two hour target A may be achieved without need to resort to a substantive design input. The same is not likely to be true of the ten minute changeover time of target B. Yet if the company chooses to move to target A and then to target B, much will be lost in terms of time, money and effort because the total design input from either starting point is likely to be similar.

The issue of design expertise also has to be addressed. For some design tasks the workplace improvement team will not be appropriately skilled and the use of external design specialists will need to be considered. The disadvantage of doing this is that improvement ownership is lost to the improvement team. Conversely, by imposing a changed system by design - ultimately to the extent of pressing a button on a console - there is no reason to expect the gains in changeover time to relapse. In such a case complex operator procedures have been replaced by an imposed design alternative. The research team has evidence of high percentage gains lapsing back and being lost when, for numerous reasons, teams have failed to maintain rigorous operational procedures.

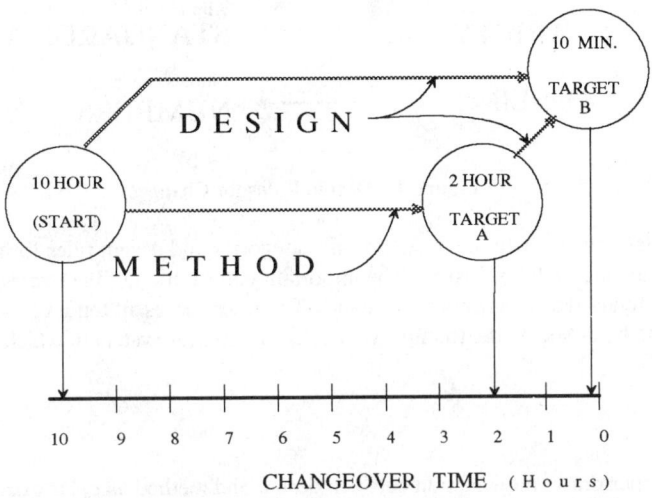

Figure 2. Defining Targets in Advance

Once the assessment of the manufacturing system has been made, at all levels, the design of the product itself should be considered. Lee (1987) demonstrated that minor product changes can be made which greatly assist the system changeover capability.

It is probable that little thought will previously have been given to changeover at the specification stage, either to the product or to the manufacturing system. Any organisation should ensure that this becomes a requirement for the future so that costly and time consuming retrospective improvement is avoided.

Design Rules

Design rules for rapid changeover draw heavily on general good practice in design. The problem for the designer is that of deciding where particular attention should be placed.

The application of the 'reduction-in' strategy assists considerably in this. Reduction in effort, adjustment and variety all contribute to improved changeover performance and each can be approached with a strong design emphasis. Each of the categories in turn may be mapped to particular design rules for changeover.

The on-line activity category in the 'reduction-in' strategy, though more firmly biased to method improvement, may be similarly mapped to a useful design rule. The four design rules are shown in figure 3 :

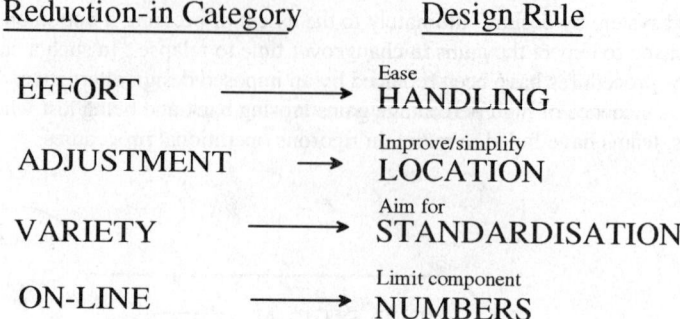

Figure 3. Design Rules for Changeover

The relationships between 'reduction-in' categories and design rules for changeover are by no means mutually exclusive. The important contribution of the strategy is that it enables more focused assessments to be made. The rules represent top level objectives for the designer as he or she works to improve the manufacturing system in which changeover occurs.

Conclusions

This paper draws a distinction between design and method as polar extremes on a spectrum of changeover improvement activities. This greatly assists in determining where SUR effort should be concentrated, where there are arguments in favour of emphasis on both design-biased and method-biased improvement. Any assessment of where the emphasis should be placed on the spectrum should be related to an overall improvement target.

Research by the authors has shown that a typical improvement team will need to be adequately supported as it moves to tackle more complex design issues.

References

Gest G.B., Culley S.J., Mileham A.R., McIntosh R.I., 1993, Classifying and selecting set-up reduction techniques, *procedings of the 9th NCMR conference*, pp 6-10, Bath University, ISBN 1 85790 00 7.
Hay, E.J. 1987, Any machine setup time can be reduced by 75%, Industrial Engineering, Vol. 19, No. 8, p 62-67.
Lee, D.L. 1987, Set-up time reduction; making JIT work, *Proceedings 2nd international conference on JIT manufacturing*, ch. 45, pp 167-176.
McIntosh R., Culley S.J., Gest G., Mileham A.R., 1993, An overall methodology for set-up reduction (SUR) within existing manufacturing systems, *procedings of the 9th NCMR conference*, pp 171-175, Bath University, ISBN 1 85790 00 7.
Sekine K., Arai K., 1992, Kaizen for quick changeover, Productivity Press, ISBN 0-915299-38-0.
Shingo S., 1985, A revolution in manufacturing : the SMED system, Productivity Press, ISBN 0-915299-03-8.

A KNOWLEDGE-BASED PROCESS QUALITY CONTROL PLANNING SYSTEM FOR DESIGN FOR MANUFACTURABILITY

H S Abdalla, C R Goulden and H Pancholi

De Montfort University Leicester
The Gateway LE1 9BH, UK

Quality is imperative for success in today's international market. At one time mass inspection was seen as the way to achieve product quality but it is no longer a realistic option. Rather, the manufacturer must match the capability of production processes to design requirements, and ensure that the processes remain in control during the manufacturing stage. This requires the use of process capability indices and Process Quality Control Plans. The above goal can be achieved through implementing the concept of Design for Manufacturability (DFM), which gives simultaneous consideration to factors influencing manufacturability, reliability and hence cost. Typically, these factors include manufacturing processes, design specifications and materials. The proposed approach employs DFM methodology for estimation of process capability indices and the preparation of quality control plans system.

Introduction

There is a strong body of opinion that quality is now the major determining factor in business corporate success. A number of research papers have confirmed that improved quality leads to reduced costs and prices which result in increased market share and a secure future and improved return on investment. According to Feigenbaum (1991) personal life styles and business practices now mean that quality and not price is the main purchasing criterion. Feigenbaum goes as far as describing quality as "the single most important factor for success in international markets". Business commentators in the popular press increasingly acknowledge the importance of quality in the search for competitive advantage, Roberts (1992).

Quality can be defined as, consistently conforming to customer requirements at the lowest cost", Porter (1990). Such a definition shows that inspection of manufactured goods is no longer a viable option. To inspect out components that do not conform is both time consuming and expensive and also implies that manufacturing capacity is used to produce product that ultimately will be scraped. Clearly inspection and meeting customer requirements at "the lowest cost" are not compatible. Equally such is the reliability of inspection, in identifying and rejecting non-conforming products, that inspection and "consistently" meeting customer requirements are not compatible. However, achieving conformance at the lowest cost demands a knowledge of the variability of manufacturing processes. This knowledge will enable the selection of most cost effective process capable of consistently producing conforming components. The process capability indices C_p and C_{pk} are commonly used to express the relationship between the process performance and the specification limits, Sullivan (1986) and Kane (1986).

$$C_p = (USL - LSL)/6\sigma;$$
$$C_{pk} = min[(USL - \bar{x})/3\sigma,(\bar{x} - LSL)/3\sigma]$$

Where:

USL is the Upper Specification Limit,
LSL is the Lower Specification Limit,
σ is the standard deviation (A measure of the process variability or repeatability), and $\bar{x}$ is the process mean (A measure of the process setting or accuracy within the specification limits).

A knowledge of the numerical values of these capability indices enable the suitability of a process to be assessed. A process with an indices of less than 1.33 would normally be considered unacceptable for the manufacture of automotive components Ford Motor (1990). Whilst a process with a value between 1.33 and 2 may be considered acceptable, special sampling arrangements would be required during manufacture to ensure adequate confidence in the process, Porter and Oakland (1990). This paper describes how the concept of DFM can be implemented to enable the use of capability indices during the design phase to assist in the selection of a manufacturing process. This work enables designers to select processes and specification limits that will ensure consistent conformance to the specification at the lowest cost. Within this work attention is focused on the indices C_p.

The Design for Manufacturabity (DFM) Environment

The DFM concept is a process not a product for converting the traditional sequential time phased product life cycle to a parallel process which concurrently addresses the aspects of design, analysis, and manufacturing. The philosophy is geared towards, reducing manufacturing problems, lowering costs, reducing lead times to market, and producing high quality designs. To achieve this goal, it is necessary to tackle the following three classes of problems: first, traditionally sequential phases in the production cycle must be restructured so that they can be performed simultaneously; second, functional barriers between departments, which have created a strict sequential flow of activity, time wasting and inter-departmental communication, should be removed; and finally, appropriate Information Technology (IT) tools enabling the new approach to be implemented should be developed and adopted. Since, the application of IT can effectively provide support to the proposed DFM approach by integrating the disciplines such as CAD, CAM, CAPP and CAE in which computers already have a well established role.

Why Design for Manufacturability ?

Before we can address the issues of DFM and the approaches that can be adapted, it is important to first identify when and how this design environment can be used and what does the conventional model look like. A sequential model was extended from Molloy's (1993) and Wong et al (1991) models to cover product specification and analysis. In this technique seven important stages can be identified, namely: market needs, product specification, product design, analysis, process planning, manufacturing and sales. This sort of model had suited industries for centuries, but in the days when the sales market was considerably larger then the suppliers could satisfy and this sequential form of departmental communication was acceptable. But in the present day 'first to market' environment, this form of communication is too compartmentalized, assuming the design is the domain of the designer, manufacturing the domain of the manufacturing engineer and so on. The consequence of this can easily justify DFM or Concurrent Engineering (CE), when you consider that the majority of life cycle costs are committed at the conceptual phase. This was the topic of research conducted at the Rensselaer Polytechnic Institute and reported in a paper by O'Flynn and Ahmad (1991), which highlighting the impact that design has on total product life cost. The research also indicated that 75% of a product's cost are committed by decisions made early in the conceptual design phase. Their model also describes the design knowledge build up during the product life cycle. It is this knowledge that DFM is attempting to address at the design stage to implement any down-line changes necessary as early as possible in the cycle.

The Proposed DFM Approach

An integrated Knowledge-based system toolkit (KEE) and a CAD system (Pro/Engineer) was developed by Abdalla and Knight (1994) for establishing a DFM environment. The integration between the solid modeller and the reasoning system was considered as a crucial step towards achieving the objective of this project. KEE itself does not provide an external communication capability but allows complete access to Lucid's Common Lisp language. Common Lisp in turn supports a foreign language interface to communicate with PASCAL, FORTRAN, and C languages. These external languages can then open, read, and write files. On the other hand, Pro/Engineer can communicate to the outside world through the programmatic interface Pro/Develop. In a typical scenario, when a request for a geometric data query is received, KEE will invoke the proper Lisp method which calls a C routine with a command string as an argument. The C routine then puts the command string in a file and goes into a wait and check cycle until complete information comes back from Pro/Engineer. When the C routine receives all the data requested back from Pro/Engineer, another Lisp program is already loaded, and will start immediately to send the data back to the KBS. This system has been enhanced to estimate the capability indices for each process, which assist the user in selecting the most appropriate manufacturing processes.

A Case Study

The design of an automotive brake pedal was considered as a case study, to show the principals of this work. The pedal incorporates an integral pivot in the form of a cylinder. Within the cylinder are two critical dimensions, length, which must be maintained with a tolerance band of 0.5mm, and internal diameter, which must be maintained within a tolerance band of 0.04mm. Nine processes available to the pedal manufacturer are shown in table 1,(all tables are given at the end) together with their natural tolerance band and comparative cost per unit of metal removal (1 being the lowest and 10 the highest).

Cylinder Overall Length
        During the design stage the knowledge based system extracts the required CAD data and identify the shape created as a cylinder. Then it investigates the CAM database for selecting suitable available processes for manufacture the cylinder. For each process the capability indices Cp and comparative costs to a given length is shown in table 2.

Internal Bore
        Following abstraction of the internal bore and investigation of the CAM database processes, five processes suitable for the manufacture of the internal bore can be identified. These processes together with capability indices and comparative costs are shown in table 3.

User Interface Information
        In this example both the features can be satisfactorily manufactured using available processes and hence the designer would be given the information shown in table 4.

Conclusion and Future Work

        This research has discussed a technique for Quality Control Plans for mechanical components within a Design for Manufacturability Environment. The proposed approach ensures that manufacturer must match the capability of production processes to design requirements, and ensure that the processes remain in control during the manufacturing stage, which dictates the implementation of the Process Quality Control Plans and process capability indices. It can be concluded that DFM concept leads to the following benefits reduced lead times, product costs, higher product quality and matching customers requirements.

Further Work
1   Features are considered in isolation we should consider some means of
        optimising process selection allowing for other features on the same part.

2   Process capability feedback into CAM database.

3   Optimising the user interface

4   Advising the designer when significant cost savings are available by a
        reduction in the required tolerance

References

Abdalla, H. and Knight, JAG. 1994, A Features-based Design Environment for Concurrent Engineering, *2nd International Conference on Concurrent Engineering & Electronics Design Automation, CEEDA 94*,(UK).
Feigenbaum, A. 1991, Quality Forum, IQA, London.
Ford Motor Company, 1990, A Guide to Statisical Process Control, Corporate Quality Office.
Roberts, E.,1992, Sunday Times.
Porter, L. 1990, Quality Improvement and Process Capability Indices-A Practical Guide, The European Centre for Total Quality Management,Bradford,England.
Sullivan, L. 1986, Japanese Quality Thinking at Ford, Quality, 32-43.

Kane, V. 1986, Process Capability Indices, *Journal of Quality Technology,* 18, N 1, 41-52.

Porter l. and Oakland, J. 1990, Measuring Process Capability using indices-some new considerations",Quality and Reliability Engineering International, 6, 19-27.

Molloy E, Yang H, Browne J. 1993, Feature Based modelling in design for assembly, *Int Journal Computer Integrated Manufactureing* , 6 Nos. 1&2, 119-125 .

Wong J P, Parsaei H R,Imam I M and Kamrani A K. 1991, An integrated Cost Estimating System for Concurrent Engineering Environment, Computers and Industrial Engineering, 21, 589-594.

O'Flynn M J and Ahmad M M. 1991, The concurrent engineering approach to product development, *Journal of Electronics Manufacturing* , 1, 97-104.

Table 1 Process Available

Process	Natural Tol. Band (mm)	Comp. Cost Per Unit Of Metal Removed	Feat.1 Hole	Feat.2 Slot
Saw	0.8	1		
Drill	0.25	1	Yes	
Plane	0.125	4		
Mill	0.1	2	Yes	Yes
Turn	0.15	2		
Bore	0.05	4	Yes	
Ream	0.01	5	Yes	
Broach	0.01	8	Yes	Yes
Grind	0.02	10	Yes	Yes

Table 2 Process for the Manufacture

Process	Natural Tol Band (mm)	C_p	Comp Cost	Remarks
Grind	0.02	25	10	High cost
Mill	0.1	5	2	Satisfactory
Turn	0.15	3.33	2	Satisfactory
Saw	0.8	0.625	1	Inadequate

Table 3. Summary of processes for the manufacture of the inter. bore

Process	Natural Tol. Band (mm)	C_p	Comp. Cost	Remarks
Broach	0.01	4	6	Satisfactory
Ream	0.01	4	5	Satisfactory
Bore	0.025	1.6	4	Adequate but required N=10
Drill	0.8	0.05	1	Inadequate

Table 4 Summary of designer interface data

Part	Feature	Process	C_p	Comp. cost	Preferred
Brake	Cylin.	Turn	3.33	2	Yes
Pedal	Length	Mill	5	2	No
		Grind	25	10	No
	Inter.	Ream	4	5	Yes
	Bore	Broach	4	6	No

APPLICATIONS OF AI TECHNIQUES TO AID PATHS TOWARDS CONCURRENT ENGINEERING

Brian Parkinson, Steve Hinder, Ulf Ruhr, Alistair Rust and Danny Shwartzberg

University of Hertfordshire
Manufacturing Systems Centre
College Lane,Hatfield,Hertfordshire,AL10 9AB

The traditional design process follows a linear sequence comprising various independent stages from customer through to a finished product. A more desirable approach is that of simultaneous engineering with all relevant departments being involved from the outset. This leads to all concerns being identified early and corrective actions taken, when necessary, at that stage resulting in reduced lead times and hence costs. This paper details the integration of AI techniques into a CAD package, to automatically provide a degree of simultaneousness to the design process. The prime objective was to demonstrate that such techniques can be successfully realised as a feasible PC based system to aid design. This has been achieved using an Artificial Neural Network in conjunction with an expert (knowledge-based) system.

Introduction

Traditionally, product development follows a linear sequence from customer request through to the finished product. Each function performs its own tasks before passing its outcome down the line to the next. Difficulties with this traditional approach can be characterised by a number of factors and may be summarised as follows:

    i) design is goal orientated, often cost based, leading to a depth first design search; alternative designs are not fully explored.

    ii) detail design is carried out too early and at considerable extra cost.

    iii) manufacturability and supportability considerations are left late in the design sequence and often incur additional costs due to the necessity for late modifications.

    iv) design data are fragmented and inconsistent making communication difficult.

    v) cost information is poorly supported.

    vi) design intent is often quickly interpreted as design development.

    vii) design process is iterative and incurs considerable time penalties as a result.

The design phase itself generally accrues minimal actual costs, but can be responsible for very large incurred costs. Inadequate design analysis and synthesis lead to heavy penalties in cost and time later in the product's life cycle. There has, therefore, been considerable interest shown in concurrent or simultaneous engineering methodologies.

Concurrent engineering has a great deal to commend it. However, failure to ensure the necessary effective communications among the product development team can mean considerably extended lead times. This is largely an organisational and managerial challenge: design reviews, information retrieval, the application of value engineering and quality function deployment are all important aspects. With the rapidly developing field of computer aided engineering and design, a new set of tools has presented an opportunity to promote effective concurrency in our engineering product developments. The extraction of existing information to aid current design is an important feature of computer applications to concurrent engineering and may provide a means of reducing the time spent on the design phase. An interrelated system under development in the Manufacturing Systems Centre of the University of Hertfordshire demonstrates how neural network techniques and knowledge based systems may be applied to achieve a concurrent approach to design. The system attempts to emulate the functions of product development team members using its knowledge and experiences to anticipate problems likely to occur in the future life cycle of the product.

Image analysis using neural network techniques

Image analysis has been defined as a transformation of an image into something other than an image ie it produces information relating to a description or decision. The primary task of this system is to identify objects within an image which are unknown and unclassified. Three modules make up the system as follows:

1) Object recognition module: using the PC based CAD system AutoCAD, data extracted is via Data Exchange File (DXF) format. The organising structure of such files lends itself well to file scanning and comparison which was used as the basis within the routine. Co-ordinate information is read from the DXF file and stored in memory as arrays of X and Y ordinates with a count of the number of vertices and the drawing layer information relating to specific geometry.

2) Feature extraction module: having unprocessed co-ordinate information about a shape is inadequate to classify that object directly. The feature extraction module obtains data from the DXF file as represented by the pseudo-code shown below;
* strip information from DXF file
* determine bounded area of geometry
* determine if shape open or closed
* calculate vertex lengths
* sum vertex lengths to determine parameter value
* determine maximum X and Y dimensions
* calculate circularity ratio
* calculate aspect ratio
* calculate ratio of shape to bounding rectangle
* determine area of any cavities
* calculate ratio of cavities to shape

     * calculate ratio of cavities to bounding rectangle
     * encode each angle by Boundary Chain Code
     * determine occurrence of the angles

Information obtained is pared down to be directly used by the classifier module.

3) Classification module: having extracted the features relating to shape, the final stage of recognition is to classify the object based upon these features. Object classification is determined by where the feature vector lies within an n-dimensional space. For a feature vector which has two variables, a point on a 2-D graph may be plotted as illustrated in figure 2. Since each point or sub-space corresponds to a distinct object, the classification effectively identifies the object. The decision theory used is based upon an image which contains one or more objects and each object belongs to one of several exclusive and distinct pre-determined classes; the object exists and can only be labelled under one class.

Classification is achieved with the aid of a Kohonen Self Organising neural network developed at the University of Hertfordshire. Neural networks offer advantages when applied to such a system such as a flexible data structure making for ease of maintenance, inference of subtle or even unknown rules and relationships, generalisation of input data, a non-linear approach enabling accurate solution of complex problems and high processing speeds due to their parallel nature.

Under the control of an algorithm, data in the form of DXF files, are repetitively presented to the net and the neurons modify the inter connections or the weightings between themselves. In this way the network "learns", with patterns of similar classes or categories clustering in particular regions of the map. As no output condition is specified, it is the network which determines where patterns are stored. After the network has learnt, or been trained, on specific data, new data may be presented and classified based upon its internal state.

Having trained a network, the procedure of classifying new and unseen shapes can take place. A new shape is presented to the net and the most sensitive neuron is found to represent the closest matching shape within the database.

Testing the neural network

The system was trained originally on very simple geometric shapes to determine whether the algorithm and data processing software functioned. As defined previously, the data inputs to the system were DXF format files containing shapes of known area. Shapes used for training were an equilateral triangle, a square, a rectangle, a hexagon and a right angled triangle. The effectiveness of classification depends upon the following factors;

   i)The size of the map used; if too small, clusters of classes become indistinguishable, if too large, no clustering readily occurs. The map size is set as a variable within the system code.

   ii)The number of patterns; too many files may cause a spreading of classes such that no individual clusters can be identified. This is determined simply by the number of files used in the training mode.

iii)The number of repetitious training patterns applied; too few loops may result in no satisfactory conclusion. Too many loops will cause inefficiencies if the network settles to a final state reasonably quickly. A control parameter is set in the system code.

iv)The learning algorithm; the algorithm has a number of parameters which determine the rate at which the network learns. Learning too quickly may result in results being unreliable. Time is the only variable if too slow. Again, these parameters are set in the system code.

The corrosion knowledge based system

Arriving at a decision without due consideration of the implications of the outcome is an inherent problem in the traditional design process. It was important, therefore, to demonstrate that a method of analysing the outcome of the neural network with regard to another domain could be achieved. To demonstrate the feasibility of a concurrent approach to design analysis, the corrosion knowledge based system was built. Three methods of system implementation were considered as follows;

i)use of a high level programming language such as C or C++

ii)use of macro statements within a commercially available database application program

iii)use of a purpose built expert system shell and programming language.

Investigation of the first two options indicated two main disadvantages;

a)considerable time would be required to build such a system

b)difficulty was envisaged in making the final system fully compatible with MS Windows which was considered to be desirable.

The third option was therefore chosen and a suitable system was researched and obtained.

Software links

For the corrosion module to function, it was necessary to determine any intersections in the design geometry produced by the CAD system. Intersection implies contact between objects and hence an area of potential corrosion. If no intersection is detected, the module will not be communicated with. Intersections are determined within the system in two stages;

i)derivation of the equation for the line geometry from the point data in the geometry file.
ii)determination of the co-ordinates of intersections by solving the equations simultaneously.

All non-parallel lines intersect; however such intersections may be outside particular line segments under consideration. It was thus necessary for the software to distinguish between intersecting lines and intersecting segments.

Design of the corrosion module

In its simplest form, the module compares the galvanic potential of two materials as indicated by the intersect programme, and informs the user if they differ by more than an acceptable limit. If a problem is highlighted, the user may input property requirement preferences for each object as the basis for a further analysis. Properties offered for selection by the system include;

Galvanic potential	Maximum 0.2% proof UTS
Minimum 0.2% proof UTS	Minimum Youngs Modulus
Minimum temperature limit	Maximum temperature limit
Minimum density	Maximum hardness
Minimum hardness	Maximum fatigue limit
Material cost	Manufacture cost

Alternative materials for each object are then listed that fall within the acceptable corrosion potential tolerance and satisfy the property requirements. Specific selection is entirely the responsibility of the user. Selection results in the material list being updated and further analysis may be carried out for all object pairs.

Summary

The system described was built and a demonstration developed as a means of testing viability. The neural network is able to identify existing parts as alternatives to a new design drawing and associated routines reconstruct the new DXF files. All line intersections between the different objects are correctly identified and this data is passed to the corrosion module. The expert system, having accepted the designer's input relating to various material properties, was able to offer viable alternatives or recommend suitable materials coatings as a means of corrosion prevention.

The expert system used in the development of this project was Knowledge Pro V1.1 by Knowledge Garden using Borland Quattro Pro for Windows to develop the materials database.

The system is undergoing further development to extend the property base and also to allow three dimensional geometry to be read and interpreted.

References

Bedworth.D.D. Henderson.M.R. et al 1991, *Computer-Integrated Design and Manufacture* (McGraw-Hill Inc)
Vernon.D. 1991, *Machine Vision: Automated Visual Inspection and Robot Vision* (Prentice Hall International)
Jackson.T. Beale.R. 1991, *Neural Computing: An Introduction* (Chapman & Hall)
Aleksander.I. Morton.H. 1991, *An Introduction to Neural Computing* (Chapman & Hall)

CONCURRENT ENGINEERING PRACTICE IN UK:
A SURVEY

Dr. Chanan S Syan

Department of Mechanical and Process Engineering
University of Sheffield
Sheffield, S1 4DU, UK

This paper reports on work carried out by the author during the period
from July to September 1993. The work reported is a national survey
undertaken to establish the extent of concurrent engineering (CE) use,
main reasons for adopting CE, actual benefits achieved by the
practitioners, advantages and disadvantages in using CE, and the tools
and techniques used in UK. This paper presents and discusses the
findings of this CE user company survey.

Introduction

The practice of concurrent engineering (CE) allows a company closely to
integrate it's resources in order to achieve bring higher quality, lower-costs products
to the market quicker than the conventional serial approach. Although this assertion
has already been proved and is well documented, the literature available only provides
isolated examples of the CE benefits (Watson, 1990; Syan et al, 1991, Business
Week, 1992). The survey was carried out by mail questionnaires. Despite the low
response rate associated with this approach, the method offered an economically
viable method.

Research Methodology

From an extensive literature review it is clear that CE has been used in four
main industrial sectors i.e. automotive, electrical, electronic and aerospace. The
survey was hence aimed at companies in these four sectors. The FAME (Financial
Analysis Made Easy) CD-ROM database available from The Hallward Library at
Nottingham University was used as it was readily available and accessible. This very
useful system besides giving complete information about the financial and economical
situations of almost all companies in the U.K.

Survey Findings

This survey was carried out during the summer of 1993. From the FAME CD-ROM database, a very large number of companies were identified, after further short listing, 639 manufacturing companies were sent the questionnaire by post.

ITEMS	TOTAL USERS
Total no. of questionnaires sent	639
Completed questionnaires returned	81
Incomplete questionnaires returned	5
Apologies received	23
No response	530
Response rate %	12,8%

Table 1 : Questionnaire Response Breakdown

As shown in the table 1 above, we had a reasonable rate of response from each of the four industrial sectors of manufacturing companies. The average response rate for all the companies was 12.8%, this is well within the expected response from mail surveys. Twenty eight companies which sent apology letters, informing us that they have a policy of not answering any kind of survey. Only seven questionnaires were returned by the post office because the companies were no longer in business.

Company Information

The turnover of 56% of the respondents was between 10 and 100 million pounds per year, and 73.1% having 100 to 500 employees. Less than 10% of these companies can be classified as large (> 300). These companies have their main customers distributed in three basic sectors: automotive (27%); electronics (18.5%); and aerospace (14.8%). Concurrent engineering has been particularly suitable for batch production with fluctuating volumes of high quality and high-technology products. Our findings indicate that the majority (81%) of the respondents are of this type.

Product Management

In terms of the annual rate of new products introduction, we found that a majority of companies introduced 5 or less new products per year. Another significant characteristic identified is that almost 50% of these companies have time-to-market less than 12 months. Therefore, they are required to develop high-quality products quickly.

A large majority of these companies have overrun the schedules for introduction of new products to market. 51% of them reported to have overrun the schedule by over 20%, while only 11.8% claim to have launched their products on time, see figure 1. Therefore, in the majority of cases product development processes require improvement.

A major cause for the delay given is that research and development usually takes longer than expected, see figure 2. Surprisingly, they indicated that specification changes from customers is the second major cause of delay. Many companies also reported difficulties during the introduction of the new product in manufacturing as a cause of delay. This is really a crucial stage. Suppliers were reported to be responsible for causing delay in the product introduction for more than 30% of the companies surveyed.

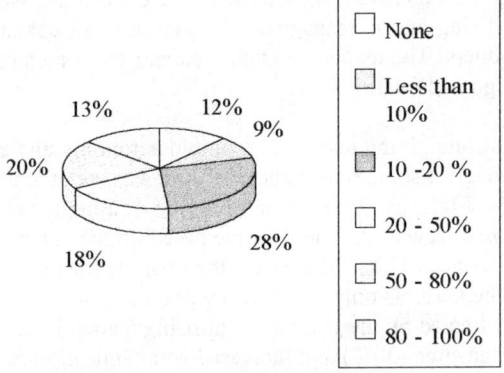

Figure 1: Product Launch Overrun

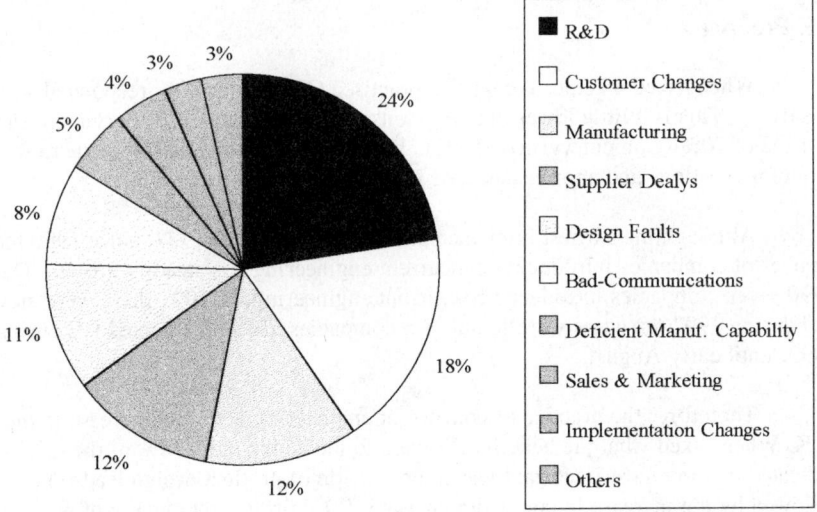

Figure 2: Most Common Causes of Delay

The findings also indicated that the majority of the companies are concerned with their time-to-market. They also seem to understand that it is necessary to develop new products more quickly in order to remain competitive. About 50% of the respondents have pointed out reductions in time-to-market in the last 12 months. The range of improvements reported is large ranging from 5%-100%, but the majority of them improved their time-to-market by 20%, while the average of the time-to-market improvement reported is 35%.

Organisational Structure

Considering the organisational aspects of the companies, we found that almost 50% of them are utilising matrix management organisation (Stephanou, 1992) for the development of products. The teams are mainly composed of members from manufacturing, design and quality areas.

However, despite giving teams the required autonomy and authority, most companies seem to neglect some important structural aspects of teamwork, such as training and rewards. Majority (57%) do not provide training for their team members. The companies reported rewarding their people based on the company performance or on individual performance (39%). Moreover, the rewards for team performance, usually considered the best, is only provided by two companies.   The most important benefits achieved by the companies utilising teamwork are, reduction in the product-to-market lead-time (30%) and increased communication across departments (26%). Increase in the motivation and also improvement in products' quality at 23% was the third reason. Majority claimed to have four or less hierarchical levels of management. Only 10% of them reported to have 5 or more levels of authority.

CE Practice

When asked whether companies practised CE? About 50% responded positively. This is a little lower than previous findings by Computervision survey in 1992 of 70% (Computervision, 1991). This discrepancy may also be due to the different questions and approaches used in the two surveys.

Although the earliest implementation reported date is 1984, since 1990 the number of companies introducing concurrent engineering has steadily grown. During 1990 seven companies introduced concurrent engineering, in 1991 there were eleven, and during 1992 ten cases were found. Six companies had implemented CE during 1993, until early August.

Therefore, the practice of concurrent engineering is on the increase in the U.K. When asked what the benefits CE were to the companies, 78% of the directors indicated that the most important was the reduction of product design lead-time, followed by the increase in competitiveness (66%), improve the quality of designs and products (47%), and also cost reductions (31%).

The main difficulties facing concurrent engineering implementation reported by our respondents are communication problems between departments (100%), difficulties in getting people involved (75%) and project management (53%). The lack of resources is not considered to be a deterrent by the large majority of companies.

The respondents also consider that the practice of concurrent engineering in their companies cause organisational problems (50%).

Majority of responding companies have adopted relatively flat organisational structures. About half (49%) of the companies had three levels of management hierarchy. Only 10% of the companies had more than five levels of management. The mechanisms utilised by these companies to integrate functions, promote communication and information sharing are varied. About 90% utilise the most simple method, just co-locating the people in the same physical area. Other utilise a liaison person (68.7%) who is responsible for integrating them for the development of a new product.

Conclusions

The use of multi-disciplinary teams is becoming a common practice. This seems to be the response to the difficulties posed by fierce global competition. Besides breaking-down the departmental barriers, teamwork is a pre requirement for CE. A large number of companies have been reducing their time-to-market in the last twelve months. However, almost 90% of them still overrun the schedule for introducing new products.

In summary the number of companies implementing concurrent engineering has grown considerably. Although some have achieved significant success , many of them are still striving to overcome structural problems that are hindering full benefits of this practice to be enjoyed.

Acknowledgements

Contribution of Mr. Amilton DeMello, an MSc student, at the University of Nottingham UK, who carried out the work under my direction is acknowledged.

References

Watson, G. F. (Editor), July 1991,"Concurrent Engineering: Competitive Product Development". IEEE Spectrum, pp. 22-37, USA.
Syan, C. S., Rehal, D. S., 1991,"Design-To-Market: An Integrative Approach", Proceedings of the 1st International Conference on Computer Integrated Manufacturing, pp 307-310, October 1991, Singapore.
Business Week - Issue May 8, 1989, USA.
TriMetrix Inc., 1987, USA.
Moser, C. A. and Kalton, G., 1985, "Survey Methods In Social Investigation". Gower Publishing Ltd., Hants - England.
Stephanou, S. E., Spiegl, F., 1992,"The Manufacturing Challenge - From Concept to Production", Van Nostrand Reinhold Publishers, Competitive Manufacturing Series, NY, USA.
Computervision, 1991,"The Manufactuuring Attitudes Survey Report", Undertaken by Benchmark Research, Undertaken by Computervision, UK.

THE PRODUCT DESIGN PROCESS AND CONCURRENT ENGINEERING

Neil Allen and Dr John M. Lewis

*M.M.S.E., Napier University, 10 Colinton Road
Edinburgh EH10 5DT.*

Many authors have proposed models for the design process and the product life cycle. These models have been used for the description of product development in a conventional, linear fashion. With the advent of Concurrent Engineering, the approach to product development has changed, and so we might expect to see a difference in the models of the design process.

Several models of design have been considered for their possible use in representing a concurrent approach. We propose that these models can, when adapted, be applied to both approaches to the design task.

We discuss the modelling of Concurrent Engineering and ask whether it is actually possible to arrive at a generic model for this way of working.

Introduction

In recent years, Western manufacturing companies have recognised that they need to reconsider their approach to manufacturing. This has arisen as a result of stiffer international competition, particularly from the Far East.

In the East, businesses can produce completely new products of a high quality much quicker than has been traditionally possible by their Western counterparts. Such an ability to present products to the market quickly is a major competitive advantage, as studies have shown that the first two firms to enter a market can secure up to 80% of the total market share (Brazier & Leonard, 1990). It is, therefore, very desirable to become one of these market leaders.

The design process has been identified as being of major importance to the competitiveness of a product. It has been estimated that between 70 and 80% of the costs over the life of a product are incurred at the development phase, and it is at this stage, therefore, that considerable benefits can be realised by reducing development time to as great an extent as is possible without affecting product performance.

Traditional approaches to the design process have been sequential in nature. Each department within an enterprise "takes its turn", as it were, at dealing with the design and then passes it on to the next stage, having reached what is considered to be an optimal solution for their aspect of the product.

A common problem with this type of approach is that the parties responsible for design are completely ignorant of the requirements of the manufacturing functions. Designs are generated with little or no consideration of how they will be manufactured, and are then passed over to the manufacturing side of the business. Few of the designs produced in this manner proceed straight through the process. When faced with problems, the people in the manufacturing functions (who, incidentally, are mostly as ignorant of the designers requirements as the designers are of theirs) either make minor modifications to the design themselves, or pass it back for revision.

Iteration such as this is almost inevitable during the development of new products in the traditional, sequential manner as very few designers have the expertise and knowledge to produce a design which will be completely acceptable. As mentioned earlier, a considerable proportion of the cost of a new product can be attributed to the development phase, and this iteration increases the cost unnecessarily. One of the contributing factors to this may be that while each department has its own view of what an optimal solution should be, the various departments' solutions do not necessarily correspond to the requirements of the others. Such a lack of consensus typically leads to confusion and a conflict of interests which can take some time to resolve. We should note, however, that despite its obvious shortcomings, the traditional design process still achieves its aims, but that increased competitive pressure has forced us to find ways to reduce waste at *all* levels.

Concurrent (or Simultaneous) Engineering is a concept which has seen renewed interest lately. As many people have pointed out, Concurrent Engineering (C.E.) is mainly an organisational issue, but it appears that the advent of new tools (such as Q.F.D., and computer applications to facilitate information handling) and competitive pressures have renewed interest in it.

Various definitions have been offered for C.E., but one of the most widely accepted is that of the U.S. Department Of Defence. They state that C.E. is ".... the systematic method of concurrently designing both the product and its downstream production and support processes."(Carter & Baker, 1991). This is achieved by the adoption of an approach to product development which involves teams whose members represent each function which has a contribution to make to the product. Literature in this field typically includes functions such as Marketing, Engineering Design, Manufacturing Engineering, Engineering Analysis, Production Planning, and so on in the product development process.

In a project currently being undertaken at Napier University, we are investigating ways in which we can support the effective sharing of information within multi-disciplinary groups working in a concurrent manner. In order to establish the levels of communication which must be facilitated, we have had to identify a suitable method of modelling the design process which will give us an appreciation of which parties are involved and when.

Modelling The Design Process

Many people, such as Pugh (Pugh, 1991), Roth, Koller (Pahl & Beitz, 1988) and Page (Page, 1992), have proposed models of the design process or product life cycle. In the main, these models have been concerned with the phases involved in the evolution of a product rather than the functions of a business which are involved. Some models are then extended by taking into consideration the steps which are involved at each phase. This can be seen in Roth's model which we have chosen as an example (see figure 1). Many models have been developed, but the consensus among the various authors appears to be that designs progress through a series of phases in a linear and logical manner.

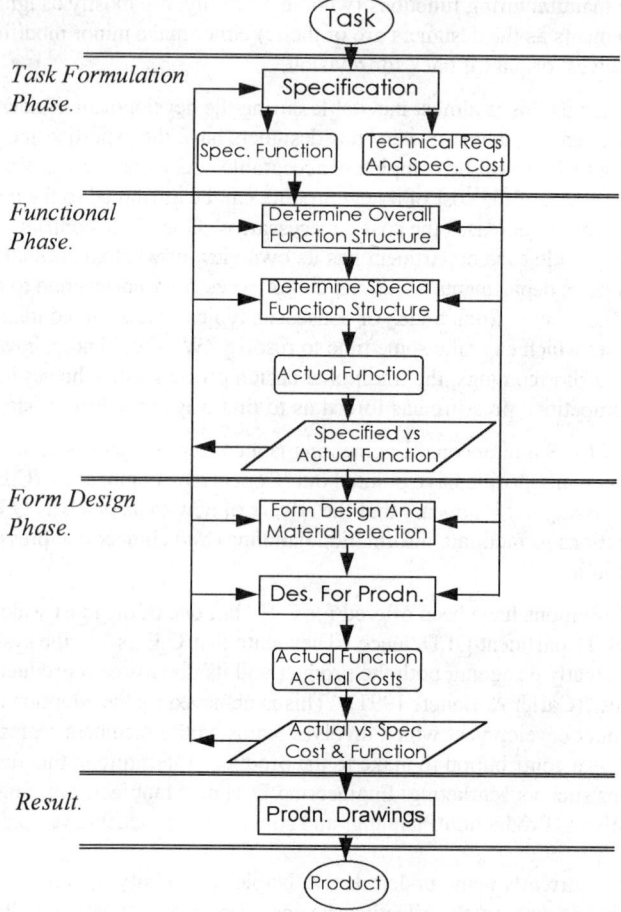

Figure 1 *Roth's Model Of The Product Life Cycle.*

At the beginning of the process, we see a need identified, followed by the generation of a specification which will fulfil that need. Once the specification for the new product is established, work commences on the creation of design concepts which are then evaluated and selected to progress to detailed design. When the design is complete, manufacture of the new product can be planned, and undertaken.

Having examined these models, we propose that they can be used as a basis on which to build new models for the concurrent approach. The boundaries of the phases represent milestones by which all of the information needed for the next phase to be completed should be available. For example, it is widely accepted that specifications need to be defined before designs or concepts are produced (it is possible to create concepts before a specification is determined, but this leads to iteration and wasted resources). If a partial specification exists, some design work can commence, but the fact remains that some form of specification should exist beforehand. A similar series of priorities exists throughout the product life cycle (the design process is a complex set of inter-functional dependencies).

The main thrust of C.E. is the co-operation between disciplines with a view to getting the product "right first time" by having the required knowledge and expertise available to make informed decisions and compromises, thus reducing the need for iteration. It does not demand that the sequence of phases through which the product progresses should fundamentally change, although the boundaries between them may become less clearly defined (as illustrated in other models of C.E. where the phases appear to merge into one development phase incorporating all aspects of the life-cycle).

To allow the model to illustrate Concurrent Engineering it has to be modified to suit our purposes. Instead of considering the process simply in terms of phases, we must extend the model to consider which functions are involved at each stage. Look at figure 2.

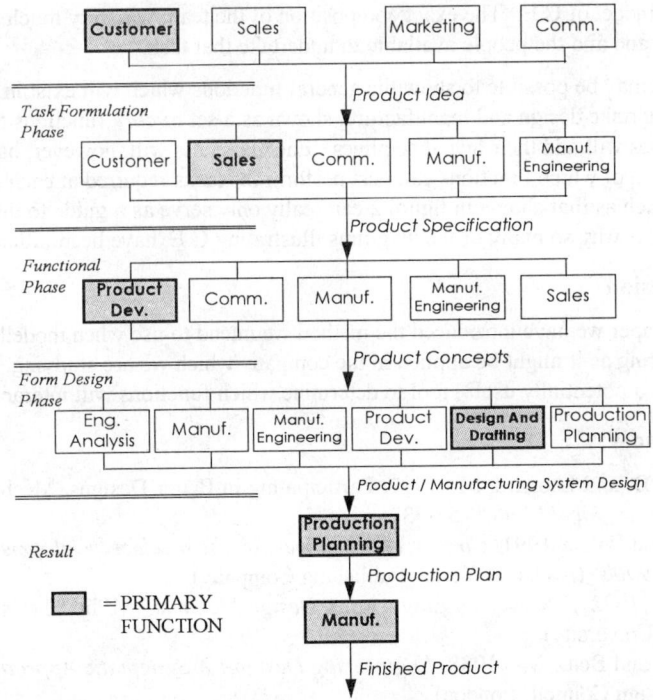

Figure 2 *C.E. Adaptation Of Roth's Model.*

This model is based on the tasks shown in Roth's and the findings of an initial study at a local company. At present, this "functional" model is based on assumptions of which functions become involved at each operation. A detailed study of the company is currently being undertaken in an effort to arrive at a sound model with relevance to an actual case.

Looking at figure 2, it can be seen that when decomposed to a level showing functional involvement, existing product life-cycle models may still be valid tools for the study of C.E. We recognise that as the development of a product progresses some aspects of the design will move through the phases quicker than others, but have chosen to maintain the basic series of phases as the milestones which they represent could be of value in project control. Some functions will have a major role in the process at particular stages, performing the tasks for which they are primarily responsible, then have a reduced, advisory role at other stages.

Additionally, the model illustrates the breadth of knowledge which is required in such an approach, and can give an idea of the scale of the task involved in facilitating the communication which will be needed at each phase.

A Generic Model Of C.E. ?

Given that every company has a unique structure, and that the C.E. project team is a very dynamic entity (with members joining and leaving at unpredictable times according to the need for their skills), we would have reservations about any claim to have established a generic model of C.E. The exact composition of the team will very much depend upon the task in hand and the people available to undertake that task.

It may be possible to show the general functions which will exist in most enterprises who undertake design and manufacture, shown as a set of *core* functions to which most companies will add their own disciplines. Each business will, however, have its own perception of which functions can best perform the tasks required at each stage and so a model such as that shown in figure 2 can really only serve as a guide to the concept. This, perhaps, is why so many of the diagrams illustrating C.E. have been rather vague.

Conclusion

In this paper we have introduced the method we intend to use when modelling Concurrent Engineering as it might be applied in the company which we are studying. We feel that it provides a potentially useful tool to determine which functions will interact.

References

Brazier, D. and Leonard, M. 1990, Participating In Better Designs, Mechanical Engineering, *vol. 112* no.1 52-53

Carter and Baker 1991, *Concurrent Engineering, the product development environment for the 1990's,* (Addison-Wesley Publishing Company)

Page, T. 1992, *A Survey Of Engineering Design Methods*, MPhil Thesis (Napier University)

Pahl, G. and Beitz, W. 1988, *Engineering Design : A Systematic Approach*, Revised ed. (The Design Council, London)

Pugh, S. 1991, *Total Design : Integrated Methods For Successful Product Engineering*, (Addison-Wesley Publishing Company)

EXTENDING DESIGN FOR MANUFACTURE TO EVALUATE THE BUSINESS CONSEQUENCES OF DESIGN DECISIONS

Wells I.E. and Love D.M.

Department of Mechanical and Electrical Engineering, Aston University, Aston Triangle, Birmingham, B7 4ET

The process of 'new product introduction' (NPI) is increasingly being required to be fast and efficient. Design engineering techniques can aid this process, but it is noted that most techniques used to evaluate design decisions do so in isolation from the business. The business implications of concept designs tend to be inferred from abstract cost data.

This paper proposes that Design for Manufacture and Assembly techniques can be extended to generate the data required to model the key business consequences of design decisions made at an early stage. By modelling the actual changes which will be seen in the company's day to day activities the effects of a NPI on the company's performance can be evaluated. A simulator capable of modelling the whole business is currently under development at Aston University.

Introduction

It is recognised that a large proportion of the costs associated with a product throughout its life are fixed early during the design process, (Alic (1993); Whitney (1988); Bogard, Hawisczak & Monrow (1993)). It is therefore important that product evaluation should be seen as an significant aspect of the design process. Product evaluation should be addressed earlier and more thoroughly than is traditionally the case. Later evaluation leaves restricted change options as highlighted by Molley, Yang & Browne (1993).

A number of tools and methodologies are available to aid product evaluation during the design process. This paper will assess some of the more important ones and highlight some areas in which the authors feel they are limited. A tool will then be proposed which would be used address the limitations. The basic information required to use the tool will also be outlined,

Tools and Methodologies Used in the Design Process

Design for Manufacture and Assembly (DFMA)

A number of tools are currently available within the DFMA environment. Probably the best known is Design for Assembly (DFA) (Boothroyd & Dewhurst (1988)). DFA can be summarised as two steps; i) minimisation of the design part count, ii) estimation of handling and assembly cost for each part. By using DFA parts are kept to minimum which can bring substantial savings and costs of assembly can be estimated to compare design iterations. Design for Manufacture (DFM) is generally used in conjunction with DFA to reduce material and manufacturing costs of individual parts. Some of the other tools now available are Design for Service (DFS) and Design for Disassembly (DFD) proposed by Dewhurst (1993), and the Hitachi Assembly Evaluation Method.

Concurrent Engineering (CE)

CE has gained wide acceptance in industry within the last few years. The main philosophy of CE is to carry out the activities within the design process, such as market research, design, production engineering, etc. in parallel instead of sequentially (as traditionally seen) reducing design cycle times and achieving early design evaluation. A major step to achieving this is the use of multi-functional design teams as outlined in Jo, Dong & Parsaei (1992), and Mill (1992), bringing expertise and ownership from various disciplines to the design at an early stage. A number of computer systems are being developed to facilitate the use of CE such as the IT based tools discussed by Stevenson & Chappell (1992) and 'Seamless Design to Manufacture' (Zeidner & Hazony (1992)).

Quality Function Deployment (QFD)

Adopted over fifteen years ago by some Japanese companies, the QFD process uses customer requirements to determine the design quality and functionality of the product throughout the design process. It attempts to ensure that design effort is focused on areas identified as important by the customer. One such tool using this is the 'House of Quality' (Mallick & Kouvelis (1992)) which identifies customer attributes (CAs) and links them to the engineering characteristics (ECs) which can affect them. This is then used to assess all engineering changes throughout the design process.

Limitations of the Current Design Process

DFMA techniques can be said to take a generalised view of manufacturing in so much as very little account is taken of the existing business or any existing or proposed products within the manufacturing process during analysis. Also, once the DFMA cycle has been completed, the results are then taken and used in conventional detailed design with little or no further analysis being done. Likewise CE is seen to look at concept products somewhat in isolation. This may be due to the tools employed in CE being focussed on individual design aspects (Ellis, Bell & Young (1993)) or, more probably because CE tends to be product not business centred. This is evident in software support tools currently available for CE which are based on a central computer-aided design (CAD)/solid model representation of the product, with various analysis tools being applied individually. Little attention is paid to product volumes and business capacities, a problem also found with quality oriented methodologies such as QFD.

Generally no attempt is made to predict the effect design decisions have on the day to day activities of a business. If business implications of concept designs are sought then they tend to be inferred from abstract cost data. The impact of new products on existing manufacturing facilities is not considered beyond the capacity of the machines to produce the items and the affect on existing products not thought of, other than in market forecasts.

Evaluation of the Business Wide Implications of Design Decisions

The Need
New product introduction needs to be evaluated in terms of the business as a whole. New products should be able to fit in with and, if necessary, compliment existing products. They must be capable of being produced by the company. This means that requirements for new manufacturing resources should be identified and the impact on capacity and reliability of existing resources should be assessed. Also the requirements to be placed on the business's suppliers in terms of production and delivery should be clear. Furthermore the activities parallel to manufacturing within a company need to be able to support the new product. For example, is production control capable of managing production of the new product along with existing products, is after sales service in place, can distribution and supplies departments cope with the extra workload and what will be the effect on financial transactions and account balances.

Evaluation needs to be carried out relatively early in the design process for the reasons outlined in the introduction. It should take place in conjunction with, and be seen to complement existing design engineering techniques, such as DFMA, which already bring evaluation to the design early.

Modelling the Effect
Evaluation of the business consequences of design decisions can be achieved by modelling. Modelling the changes in day to day business activities allows the direct effect on the company of new product introduction to be predicted. Direct, resultant and parallel activities within the company can be modelled (Love & Barton (1993)). Various activities can be modelled in different detail dependent on the expected impact placed on them by the new product. For example, manufacturing operations may require highly detailed modelling whereas distribution could be modelled by predicting volumes. Caution is important however, as too little detail in a model may miss some low level effects caused by the new product.

A Tool for Business Evaluation

The characteristics a tool for early evaluation of the business wide implication of design decisions would need can be identified. Some of these are outlined below, but this is not an exhaustive list.

Manufacturing Data
Data for the production of a product is generally produced using process planning, often computer-aided. Basic computer-aided process planning (CAPP) systems give a shell process plan which an experienced process planner would complete. Variational CAPP uses existing standard process plans which are then altered to suit the new product. This is often linked with coding and classification systems used to select process plans (Turnbull and Sheahan (1992)). Generative CAPP is the method of producing process

plans from scratch and is an area in which a lot of research is ongoing, mainly centring on how to obtain geometric data from the design. Examples are CCSPLAN described by Zhao and Baines (1993), feature and expert system based packages. Feature based systems hold a lot of promise, whether using feature based design or feature extraction (Allcock (1993)), but there is still a lot of work to be done. Expert Systems also are being developed, but these currently tend to be in specific types of planning only such as those described by Warnecke and Muthsam (1992).

All CAPP techniques, currently used or being researched, require extensive design detail and process knowledge, whether from the operator or built into the system. This prevents their use early in the design process. It may be possible that a very basic plan with limited detail could be produced at a relatively early stage giving enough information for the evaluation process.

An alternative to process planning is present in the use of DFMA techniques. In carrying out DFA and DFM analysis on designs the user evaluates them in terms of both part reduction and costs which are derived from the type of process, the time taken, etc. Examples of computer packages for this are Boothroyd & Dewhurst (1988) or 'Assembly View' (Adler & Schwager (1992)). Thus once the DFMA cycle is completed information is present which is sufficient to obtain the required manufacturing data for a proposed product. It could be possible to modify such packages to obtain data directly in the format required to be used in the modeller. Manufacturing data would therefore be directly available once the DFMA analysis was completed and is not dependent on detailed design.

Commercial Aspects

Information such as forecast volumes, prices, seasonality, etc. would be obtained from market research which identified the need for the new product. It is envisaged that this information would include the likely effects on the sales of existing products and price changes to them, all of which is important to modelling the whole business. It is likely that the marketing department would consider a number of different scenarios following the introduction of the proposed product and these could be evaluated to predict the actual effects they would have on the business.

To make the model a true representation of the business, the design process itself would also have to be modelled along with any changes in company resources, such as any purchase of new plant.

The Modelling System

A 'Whole Business Simulator' (WBS) (Love & Barton (1993)) is currently being developed at Aston University. A working demonstrator has already proved the principle of whole business simulation (Barton, Bailey & Love (1992)). The object-oriented approach of WBS allows configuration for different and changing companies and tasks coupled with quick and easy expansion for the future. Different functions can be modelled in different detail depending on importance. WBS gives bottom line results for the business in terms of profit but it can also easily give results in schedule adherence, resource utilisation, cash-flow, etc. This would be dependent on company priorities.

It is considered that WBS would be used to model the company with an up to date model being used for various investigations. When a new product introduction is being considered, manufacturing and marketing data as outlined above would be used and the product evaluated using the WBS environment.

Conclusions

This paper has identified that design process tools and methodologies, whilst being important for fast and efficient new product introduction, are limited in their analysis of a product in terms of the business as a whole. A tool has been proposed, and its requirements discussed, which would at a relatively early stage in the design process, be able to evaluate concept products in the context of the whole business. Research into producing the tool is currently underway.

References

Adler, R. & Schwager, F. 1992, Software Makes DFMA Child's Play, *Machine Design*, **64**(7), 65-68.

Alic, J.A. 1993, Computer-Assisted Everything? -Tools and Techniques for Design and Production, *Technical Forecasting and Social Change*, **44**, 359-374.

AllCock, A. 1993, Feature-based Process Planning and NC Code Research, *Machinery and Production Engineering*, **151**(3854), 8.

Barton, J.A. Bailey, K. & Love, D.M. 1992, A Working Demonstrator of a Whole Business Simulator, *Proceedings of the 8th National Conference on Manufacturing Research*, 49-53.

Bogard, T. Hawisczak, B. & Monroe T. 1993, Implementing Design for Manufacture (DFM), *Australian Electronics Engineering*, **26**(2), 60-66.

Boothroyd, G. & Dewhurst, P. 1988, Product Design for Manufacture and Assembly, *Proceedings of the 3rd International Conference on Product Design for Manufacture and Assembly*.

Dewhurst, P. 1993, Product Design for Manufacture: Design for Disassembly, *Industrial Engineering*, **25**(9), 26-28.

Ellis, T.I.A. Bell, R. & Young, R.I.M. 1993, A Design for Manufacturing Software Environment to Support Simultaneous Engineering. In A. Bramley & T. Mileham (eds.), *Advances in Manufacturing Technology VII*, 301-305.

Jo, H.H. Dong, J. & Parsaei, H.R. 1992, Design Frameworks for Concurrent Engineering, *Computers and Industrial Engineering*, **23**(1-4), 11-14.

Love, D.M. & Barton J.A. 1993, Whole Business Simulation: Concepts and Capabilities, *Proceedings of the 1993 Western Simulation Multiconference on Simulation in Engineering Education*, 92-97.

Mallick, D.N. & Kouvelis, P. 1992, Management of Product Design: A Strategic Approach. In A. Kusiak (ed.), *Intelligent Design and Manufacturing*, (John Wiley & Sons) 157-177.

Mill, H. 1992, Total Design and Concept Selection, *IEE Seminar on Team Based Techniques:- Design to Manufacture*, Digest No.**091**.

Molley, E. Yang, H. & Browne, J. 1993, Feature-based Modelling in Design for Assembly, *International Journal of Computer Integrated Manufacturing*, **6**(1), 119-125.

Stevenson, I. & Chappell, C. 1992, IT Based Tools for Concurrent Engineering, *Proceedings of the 8th CIM-Europe International Conference*, 157-166.

Turnbull, B. & Sheahan, C. 1992, Reducing Design to Manufacture Lead Time in High Variety/Small Batch Manufacture, *Proceedings of the 8th CIM-Europe International Conference*, 50-60.

Warnecke, H.J. & Muthsam, H. 1992, Knowledge-based Systems for Process Planning. In A. Kusiak (ed.), *Intelligent Design and Manufacturing*, (John Wiley & Sons) 377-396.

Whitney, D.E. 1988, Manufacturing by Design, *Harvard Business Review*, **88**(4), 83-91.

Zeidner, L. & Hazony, Y. 1992, Seamless Design to Manufacture (SDTM), *Journal of Manufacturing Systems*, **11**(4), 269-284.

Zhao, Z. & Baines, R.W. 1993, CCSPLAN: A Generative Process Planning System, *Proceedings of the MATADOR Conference 1993*, 527-534.

IMPLEMENTATION OF CONCURRENT ENGINEERING USING A PROPRIETARY SOFTWARE SYSTEM

B.Porter and M.Cooke

School of Engineering
Coventry University
Priory Street
Coventry CV1 5FB

The implementation of Concurrent Engineering requires both the enabling technology and a procedural framework. This paper examines the issues involved in achieving the reduction in time from design to market, and follows the implementation experience of a medium size manufacturing company in the UK, which has embarked upon a long term programme using a proprietary Engineering Data Management System. The difficulties encountered are mainly associated with the need to customise a basic generic product to a realistic application environment. Although not yet completed, the implementation provides useful pointers for other similar organisations intending to follow the same route.

Concurrent Engineering-a systems view.

To many people, the most acceptable method of progressing towards Concurrent Engineering (CE) is one which relies heavily upon a procedural basis. This may entail building procedural and managerial bridges between "islands" of Computer Aided Engineering (CAE) systems. In part, at least, this approach is due to the ownership issues surrounding what needs to be a corporate-wide system. Certainly the emphasis given by various functional teams within an organisation will vary, and will also be different between different organisations. Parallels may be drawn here with implementation of other company-wide systems eg. Manufacturing Resource Planning (MRP2) where ownership is often a problem. The difficulties associated with CE and its component disciplines have been highlighted by several authors eg Stark (1992).

These difficulties would seem to occur as an almost inevitable consequence of the nature of the system, requiring the management of Engineering Data (EDM), Engineering Workflow (EWM), and Commercial Workflow (CWM). Crossing the usual functional boundaries inevitably stimulates provocation on such matters as work priorities, utilisation of resources and objective performance measurement. Without clearly defined goals and procedures relating to such matters, the opportunity for data ownership and control difficulties are rife. In an integrated manufacturing environment, a comprehensive set of data is required to define an item master record. This is usually done in a single record which requires input from design, process planning, production planning, purchasing, etc. Creation and maintenance of this data is an organisation problem in itself for many companies.

For most reasonably complex manufacturing organisations the systems development history follows the model shown (Fig. 1):

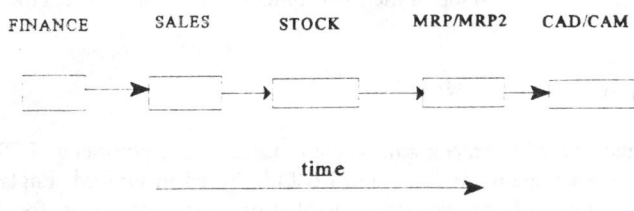

Fig. 1 Systems Development Path.

Where EDM is introduced, it often has its roots in CAD, leaving the organisational problems of linking into other systems conceptually, technically, and procedurally. It is this latter problem which is likely to be the most significant, as technical problems are resolved through open architecture systems and equipment.

If the overall goals for CE are to be expressed as performance measures for the global system, then a corresponding set of sub-system performance measures and goals are required, with inter-system (transfer) performance as an important component. It is thus necessary to generate a typical model for Engineering and Commercial workflow, with cross boundary transfers identified. Objective performance measures can then be specified to a level of detail commensurate with the timescale for implementation and the experience profile of the organisation. By considering the data flow, it is possible to introduce performance controls at the most vulnerable points in the flow eg. where functional boundaries are crossed.

The need to provide concurrency means that the usual sequential stream of operations must be transformed into an iterative process, probably involving the same sequences, but with several, faster passes. The degree to which the software can support this function is in large part a measure of its usefulness in developing beyond an Engineering change recording system to a true CE support system.

For EDM to contribute to the achievement of a reduced cycle time for such a process, a number of requirements become apparent:

access to all relevant data for decision-making in each functional area

status information visible to all

prioritisation of tasks

audit of action taken, and by whom

comprehensive electronic communications.

It is this set of requirements, on top of the EDM functionality, which forms the basis of a CE system.

Implementation.

The authors have been monitoring an implementation of a proprietary EDM system at a manufacturing company, W.Lucy and Co Ltd., based in Oxford, England. They manufacture medium voltage electrical distribution and switchgear for industrial applications. Having already introduced CAD and CAM facilities some years ago, they wished to provide an integration platform as a logical development of their CIM strategy, and to gain improved control of their engineering data. They had dual objectives of improving the quality/integrity of engineering design, and improving the time to production for new/revised products.

The package used is INFOmanager(v 1.6), from EDS Ltd., developed as an integrated database system to provide EDM functionality and links to other factory management software (eg. MRP, CAPP, etc). It is essentially a framework for data integration and applications control. It is intended to be a shell through which other applications are accessed. The benefits of this approach are seen as:

avoidance of data duplication/redundancy

full audit facilities

improved integration through the use of a common interface.

The history of systems development at the company spans about six years, and follows a typical path of separate developments in commercial and engineering areas. For a variety of reasons the company has determined its priorities as the CADCAM area, to be followed by integration of data across the business. It does not have MRP, and scheduling/routing is done manually. With this background, the emphasis in implementing INFOmanager has been in the EDM aspects (engineering change recording, bill of materials, etc). The project is now in its third year, and the system has been run as a pilot on a small number of product groups.

For the initial product, a new assembly was chosen,(with approx 1000 items) to adequately test the software (and the organisation's) abilities. In addition to the main application, two new application modules were required for tooling, and costing. Initial experience of following the original implementation plan was not satisfactory, as technical specifications, and organisational procedures were found to be flawed. A revised approach took a more relaxed attitude to the timescale, and concentrated effort in those departments of the organisation closely involved with the system and its success.

Data entry for the pilot was handled by a dedicated member of the team (ie not part of the product design staff), in order to speed up the process. However, due to a decision to introduce a new numbering system, the data and structures created, although using many standard items, proved to be of little early use, and further work was required to hold "old" numbers under the new item records. Much of the standard reporting from the system is via an SQL report generator, which has proven to be usable by most staff with relatively little training, and which also provides the opportunity to generate specific customised reports. This feature has proven to be important in providing the data for decision-making in different areas of the business.

The general reaction from the Design staff has been positive, with Bill of Materials and "where-used" features being widely used. It was considered that there would be benefits from reduction of number of parts through easier access to previous designs. For the Production Engineering staff, it was much the same, with suggestions to link to Purchasing featuring high on their list of requirements.

Currently, the company is continuing its drive towards a CE implementation, using INFO manager as its basis. There has been considerable need for customising of the software, at a quite low level of code, which has consumed many man-hours of effort. This has also entailed considerable management time and effort, and is proving to be a bigger commitment than originally envisaged. However, they have become increasingly convinced of the long-term benefits to be achieved from the programme, and consider that the investment will pay dividends.

Conclusions.

The implementation of an EDM package into a medium size manufacturing company was intended to form the basis of a move towards Concurrent Engineering. The functionality of the chosen software has provided a framework which would enable such a development.(It should be noted here that version 1.6 on which this work was based is scheduled to be replaced by version 2.3 which provides significant changes to its functionality). The implementation plan for the project was generally well-considered, and such difficulties as have arisen have been largely unforeseeable. The timescale chosen for the project must be considered generous, although in the light of experience, it has proven to be largely necessary. It remains to be seen whether the intended goal of data integration, and of the wider benefits of Concurrent Engineering are to be achieved. It is intended by the authors that this progress will be further monitored and reported in future work.

References.

Stark, J. 1992, *Engineering Information Management Systems* (Van Nostrand Reinhold)

AN IMPLEMENTATION LEVEL APPROACH FOR THE IDENTIFICATION OF SET-UP REDUCTION TECHNIQUES

G B Gest, S J Culley, R I McIntosh & Dr A R Mileham

School of Mechanical Engineering, University of Bath, Bath, Avon, BA2 7AY.

As customers are constantly demanding increased product variety at lower costs, small batch manufacture is becoming the norm in many sectors within industry. This requires that processes can be quickly and accurately changed from one product type to another. Hence set-up reduction (SUR) should be regarded as a high profile business objective. This paper presents an implementation level methodology which has been developed in response to the requirement to consider high level issues such as the inter-relationships between set-up operations. This methodology can be used to highlight the potential improvement areas and define their characteristics. This data is then used to retrieve relevant techniques from a catalogue of set-up reduction techniques.

Introduction

In recent years the trend has been towards a customer demand for increased product variety at lower costs. As a result of this a number of techniques have been adopted in order to ensure company productivity and profitability. Rather than run large batches and hold a high level of stock and work in progress (WIP), small batch and Just in time (JIT) manufacture are becoming the norm in many sectors within industry. These require the processes to be quickly and accurately changed from one product type to another. As a result of this, set-up reduction (SUR) should be regarded as a high profile business objective. Not only does SUR affect the batch sizes, inventory and costs, but it has also been seen to have a significant effect upon the maintenance of machines and the quality of the goods produced together with providing new commercial opportunities (Shingo, 1985).

A large number of SUR implementations have been observed to date. However, most of these implementations have not been undertaken in a rigorous manner. Whilst some methodologies have been utilised in this arena for identifying and solving problems the only coherent structure to be presented for this task was proposed by McIntosh et Al (1993). This has resulted in wide variations in the amount of set-up time saved as well as the approach adopted to obtain it.

In addition to the wide variety of performance, many cases have been observed where techniques have been reinvented across a number of sites within the same company. This in itself is not a great problem, however, in many cases this duplication has been the result of long periods of idea generation and testing. There is thus a case for some means of transferring and retrieving information amongst these various sites.

In order to address both of these problems an implementation level method study approach has been developed which will identify problem areas to be attacked to give the most effective results. This approach functions within the overall methodology proposed by McIntosh et al (1993a).

This approach has been developed from, and is an extension to, a method study technique previously presented at the tenth conference of the Irish Manufacturing Committee (Gest et al, 1993a). The technique has been integrated with a set-up reduction information system to give an overall method for the effective identification of SUR techniques. This has been tested in a number of industrial situations and the resulting feedback was used in the development of the new methodology.

The overall analysis procedure

After a number of trials within industrial companies it was found that, while the original procedure provided a very detailed analysis of machine changeover, it was too detailed to be applied to process lines. As a result of this the analysis procedure has been expanded to allow the identification of the line and machine which are to be attacked. The new four stage procedure is discussed below.

The first stage of the analysis procedure is to identify the line which is to be studied. The result of this identification depends on a number of factors such as the focus of the programme, the resource available and the time reduction required. Information is gathered by the use of an audit of the overall manufacturing system within the factory or business unit (McIntosh et al, 1993b). The following information needs to be gathered.

> Number of changes per line
> Average time for changes
> Number of tools available for each line (Dies etc.)
> Number of tools changed for each line (Dies etc.)
> Number of different ancillary tools required for a changeover
> (Spanners etc.)

Once the line to be studied has been identified, it can be examined in more detail to determine the current procedures utilised during the set-up (if any), any machine inter dependencies which exist and which machine(s) act as the bottleneck or critical path for the set-up process. The following details need to be obtained.

> Line setting procedure - Does one exist?
> Machine interdependencies - is the setting of one machine dependant
> on the output from another.
> Identify the set-up critical path

The next stage is to video or observe the set-up of the machine which is to be attacked. This is primarily used for the analysis phase of the process, however, simply observing the line can highlight any problems or constraints with the line. Once an appreciation of the set-up interdependencies has been gained and the changeover recorded, it can be analysed. This is carried out using the methodology previously presented by Gest et al (1993a). A number of additional symbols have been added to this methodology, see 'The set-up process revisited' below. The 'reduction in' strategy

(Gest et al, 1993b) is utilised as part of this methodology in order to reduce each of the four set-up excesses; on-line activity, effort, adjustment and variety. The following activities need to be carried out.

> Video /observe line changeover - note time
> What operations are carried out
> What operations are carried out in parallel or series
> Analysis of changeover using 'method study' techniques
> Application of the reduction in strategy
> Linking the analysis to existing data sources

This analysis, combined with the classifications used for the fast tool changeover catalogue (Gest et al, 1993c) can be used to generate ideas for possible improvements. The final stage is to re-video the changeover. This performs three functions; firstly, it provides the monitoring feedback loop; secondly, the video can then be used to highlight further possible improvement and thirdly it can be used to spread awareness of the techniques that can be used on other machines in the factory. The following activities are undertaken at this final stage.

> Re-video line & machine
> Compare times with those initially gained
> Advertise improvements

The set-up process revisited

Previously, a set-up was defined as having six elements. As shown below:-
- ◆ Movement elements
- ◆ Adjustment or setting elements
- ◆ Securing or releasing elements
- ◆ Problems
- ◆ Inspect or tryout elements
- ◆ Control elements

In order for a set-up to be completed these should all be completed. However, subsequent work primarily in the packaging industry has highlighted two further elements. These are waiting time and cleaning time. Previously these had been subsumed under the control heading. It was found that they were of far more importance in the packaging industry than had been observed in the pressing and metal cutting industries due to the printing that is typically required.

The waiting aspect was found to be particularly relevant in multi-stage processes. In these cases, it was often found that a machine could not be finally set until the parts that had been produced by it's predecessor were within specification. These parts are then often used to finally set the machine. It is considered necessary to highlight these areas so that such a machine interdependency can be eliminated through the use of, for example, stage gauges or progressives.

The cleaning elements have been found to be particularly important in the packaging industry, but this aspect has also been extensively reported in the injection moulding and process industries. A significant period of time is expended in order to ensure that there is no contamination between different process colours or materials.

As a result of this two new symbols have been added to the set of method study symbols. See figure 1. The overall procedure is summarised in figure 2 below.

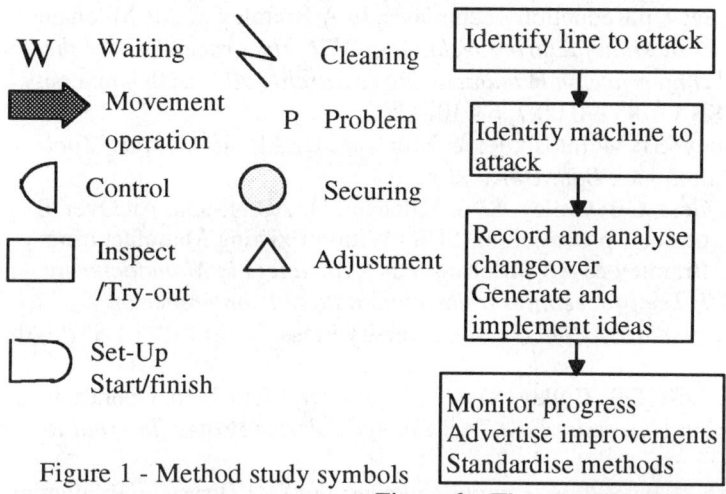

Figure 1 - Method study symbols

Figure 2 - The overall analysis procedure

Conclusions and future work

A modified analysis procedure has been presented for reducing set-up reduction times. This has been developed from initial work, undertaken at the University, into the way that SUR has been implemented within industry. The new approach is part of an overall methodology for set-up reduction (SUR) which has been developed by the Bath University research team.

The analysis procedure has been extended so that lines and machines to study can be identified prior to analysing changeovers to ascertain improvement areas. This analysis can include, for example, videoing, photography and applying the updated method study analysis to highlight possible improvement areas.

The method study analysis has been modified following experience gained within the packaging industry. Critical aspects such as cleaning operations and time spent waiting for previous operations to be finished are now given much greater importance than when they were initially grouped as control operations.

The analysis of the changeovers can be used to develop solutions in conjunction with the catalogue of fast changeover ideas. This catalogue been classified in such a way that relevant ideas can be rapidly identified and targeted through the 'reduction in' system of classification.

The catalogue is being developed at the university in order to produce an advisor system for fast tool changeover ideas. This will include the 'reduction in' classification together with a number of other schema in order to enable full portability of ideas.

References
Gest, GB; Culley, SJ; McIntosh, RI & Mileham, Dr AR. 1993a, Method Study Techniques for SUR. In P. Nolan (ed.), *Proceedings of the tenth conference of*

the Irish manufacturing committee, 1993, Volume 2, (Officina Typografica & Irish Manufacturing Comittee) ISBN 0 90 777563 2, 1001 - 1012

Gest, GB; Culley, SJ; McIntosh, RI & Mileham, Dr AR. 1993b, Classifying and Selecting Set-Up Reduction Techniques, In A Bramley & AR Mileham (eds.), *Advances in Manufacturing technology VIII, The proceedings of the ninth national conference on manufacturing research, 1993* (Bath University Press, Bath) ISBN 1 85790 00 7, 6 - 10

Gest, GB; Culley, SJ; McIntosh, RI & Mileham, Dr AR. 1993c, *Fast Tool Changeover Catalogue, Bath University*

McIntosh, RI; Gest, GB; Culley, SJ & Mileham, Dr AR. 1993a, An Overall Methodology for Set-Up Reduction (SUR) Within Existing Manufacturing Systems, In A Bramley & AR Mileham (eds.), *Advances in Manufacturing technology VIII, The proceedings of the ninth national conference on manufacturing research, 1993* (Bath University Press, Bath) ISBN 1 85790 00 7, 171 - 175

McIntosh, RI; Gest, GB; Culley, SJ & Mileham, Dr AR. 1993b, Operation audit / workbook, *Bath University Fast Tool Change Systems Project Internal Report, RM/DD/145*

Shingo,S. 1985, *A revolution in manufacturing: the SMED system*, Productivity Press, ISBN 0-915299-03-8

Acknowledgements

This research was funded under ACME Grant Number GR/H 21364 with the aid of CarnaudMetalbox. Thanks to Nick Greville of Welton Packaging for his work validating the methodology.

THE APPLICATION OF QFD WITHIN THE DESIGN PROCESS

James Scanlan, Alan Winfield and Gordon Smith

University of the West of England
Frenchay, Bristol, BS16 1QY

This work is part of a larger initiative aimed at providing a dynamic model of the design process by integrating QFD (Quality Function Deployment) matrices with discrete-event simulation.
The objective of this work is to allow the dynamics of the design process to be modelled and understood so that the process can be optimised in terms of resources and timescales.
For validation purposes, a specific design task, namely the structural design of a generic aircraft wing is the focus of this work. The co-operation of BAe Airbus Ltd has enabled realistic data such as target design parameters, timescales and precedence relationships to be used.

Introduction

Many companies now recognise the potential benefits of concurrent engineering but have difficulty in making the necessary organisational and management changes to facilitate concurrent engineering. This is particularly the case in industries with very large or complex design tasks such as the aircraft industry where the number of engineers involved in the development of a new project can run into hundreds and possibly thousands.

The principles of concurrent engineering suggest that, ideally, a dedicated multi-disciplinary team should be established at project inception and be maintained until customer delivery (and possibly beyond).For simple products and designs this is entirely feasible and many classic examples of concurrently engineered products fall into this category.

For large or complex designs, resource limitations rarely allow this. Large organisations often have many new and existing design projects running in parallel and do not have the luxury of being able to focus on one project at a time. Engineers are often over-commited as members of many different project teams resulting in poor levels of personal effectiveness and a lack of project "ownership".

If a company genuinely plans to move from a traditional functionally oriented design organisation towards dedicated, multidisciplinary teams this is likely to require a greater number of engineering man-hours in the short term. However, *in the long term* the benefits of shorter product development times, fewer design changes and easier-to-manufacture products should reduce the number of engineering manhours. This is seen by many senior managers as a an expensive and risky gamble. In simple terms companies are caught in a familiar "resource trap" where they need a major short-term injection of resources to achieve a *long term* improvement in performance or efficiency. Nevertheless, it is likely that the commercial pressures of competition will increasingly cause companies to re-evaluate their engineering organisation. Boeing, for instance, announced with great publicity that they employed concurrent engineering principles in the development of their latest aircraft the 777. They claim that this resulted in significant reductions in design changes, development lead time and a better quality product. "..techniques such as concurrent engineering will become the (aerospace) norm in response to the Boeing-initiated step change challenge" [Parry (1994)].

There are many important organisational and cultural issues associated with making such major changes but these are beyond the scope of this paper. The intention of this paper is to describe some specific management tools that can assist in planning and managing a design project based on the tenets of concurrent engineering.

Concurrent Engineering within large engineering teams

It is useful to try and identify the idealised form of organisational structure consistent with the principles of concurrent engineering for a large engineering team.

Most large engineering organisations currently employ specialists who make design contributions within a very narrow engineering domain. Although many companies recognise the need for engineers with broader product knowledge, specialists will predominate for the foreseeable future. Certainly, the trend within the aerospace industry is currently towards greater specialisation.

It is likely therefore that most specialists will have a fairly well defined "contribution span" within the product development cycle and that it would be wasteful for such engineers to remain dedicated to a project for the full duration of the development cycle. The development of a new product will therefore involve teams with changing populations from a range of disciplines with a range of contribution spans. Within this specialist engineering environment it will simply not be practical for a large core of design engineers to remain dedicated to a project from start to finish as is often assumed by concurrent engineering advocates. The retention of continuity, ownership and accountability therefore becomes a major issue when, in reality, few (if any) team members will remain dedicated to the project for the full life cycle. This is where Quality Function Deployment (QFD) can play a part.

On the surface this proposed organisational model seems to closely resemble traditional design methods. The essential difference is that true concurrent engineering requires that the individual contribution of engineers be carefully scheduled and co-ordinated in order to maximise the degree of concurrency of activity. This concurrency of activity can be identified and measured in terms of "interaction density" [Kusiak and Wang (1993)]. Interaction density refers to the degree of interdependence of design tasks or parameters. In other words, where a collection of design parameters is closely coupled then a dedicated team needs to be assembled to *collectively* make decisions for these design parameters.

Again, this appears to be what currently happens within traditional serial design organisations.However, another important difference is the scope of constraints considered. Traditionally, only a limited set of design parameters or constraints are considered at the design stage. Interactions that concern the so-called "downstream" disciplines of product support, tooling, machining, assembly and testing, to name but a few, are often not considered.

Quality Function Deployment (QFD)

"QFD is a naturally compatible adjunct to concurrent engineering because it replaces the independant department responsiblities with interdependancy and teamwork. The design of new products is transformed from a separated series of steps into a process based on the voice of the customer that is integrated from design to production." [Carter (1992)].

Space does not permit a full description of QFD principles, however, a comprehensive description is given in the Akao (1990) reference.

QFD has gained wide acceptance in Japan and the United States, particularly in the car industry and is now starting to be exploited in the the aerospace industry (notably Hughes, Boeing and McDonnell-Douglas.)

The application of QFD techniques has recently been piloted at Avro International Aerospace (a subsidiary of British Aerospace). Reported benefits are very encouraging: "Each team member has been able to see his own contribution to the whole design. Hence the sense of ownership has flourished. The reasons for making decisions are visible and traceable, thereby nurturing the learning of the team." [Gandhi (1994)].

In summary the benefits of QFD are;

-It clearly identifies the relationships between design parameters. This information can generate precedence relationships so that areas of high interaction density can be identified and used to define the membership of engineering teams.

-It provides a structured way of establishing the relative importance of design parameters so that trade-offs can be arrived at that are consistent with customer priorities.

-Traceability is provided. This will allow future design teams to understand "why" design parameter values were selected for past projects. This is often extremely difficult (if not impossible) to establish from traditional engineering data.

-The QFD framework ensures that all aspects of the product life-cycle are considered at the design stage.

-The QFD charts provide a focus for the design teams and thus help to provide continuity despite the inevitable changing population of team members.

Perhaps the greatest benefit of QFD is that is provides extremely clear and well-defined targets for engineers. "...most design troubles can be tracked down to faulty instructions at the outset" [Constable (1994)].

The application of QFD to the design of an aircraft wing

CUSTOMER REQUIREMENTS	L1 STRUCTURAL MODEL	L2 TARGET FATIGUE LIFE	L2 MANOEUVRE LOAD FACTOR	L2 FUEL VOLUME	L2 GUST LOAD FACTOR	L2 UNDERCARRIAGE LOAD FACTOR	L2 VIBRATION LOAD FACTOR	L2 WEIGHT DATA	L2 PRIMARY MATERIAL PROPERITES	L1 STRUCTURAL DESIGN	L2 WING BOX DESIGN	L3 PARTS CONSTITUTION	L3 PART FEATURES	L4 WING SKINS	L4 STRINGER	L4 SPAR	L4 RIB	L3 CONNECTED PART RELATIONSHIPS
L1 OPERATION	0									0								
L2 MAXIMUM FUEL EFFICIENCY	9			9						0								
L2 LONG LIFE	33	9	3		3	3	3	3	9	120	9	9	3		3	9	9	
L2 MINIMUM PURCHASE COST	3								3	176	9	3	9		3	9	9	
L2 QUICK TURNAROUND	0									24	9			9	3	3		
L3 EASY ACCESS REFUELLING POINTS	0									0								
L1 MAINTENANCE	0									0								
L2 MAXIMUM TBO	22	9	3		3	3	3	1		0								
L2 MINIMUM SCHEDULED MAINT	0									0								
L2 HIGH DAMAGE TOLERANCE	10	1							9	18	9			9				
L2 ACCESS TO ALL CLASS 1 STRUCTURE	0									42	3			9	9	9	9	
L2 EASY TO REPAIR	9								9	25	9			9	3	3	1	
L2 QUICK TO REPAIR	9								9	9	9							
L2 INTERCHANGEABILITY OF PARTS	0									18	9	9						
L2 USE OF SIMPLE HAND TOOLS	0									0								
L2 MINIMUM USE OF DIFFERENT FASTENERS	0									63	9	9		9	9	9	9	9
L1 PERFORMANCE	0									0								
L2 MAXIMUM RANGE	0									0								
L2 MAXIMUM PAYLOAD	28	1	1	3	3	1	1	9	9	0								
L2 MINIMUM FIELD LENGTH	0									0								

(Column group heading above design parameter columns: DESIGN PARAMETERS)

Figure 1. Part of the top-level customer/design parameter QFD matrix

QFD has been applied to the design of a generic aircraft wing. Figure 1 shows a small extract of the matrix which gives examples of customer requirements and design parameters. Although this matrix is still being refined it has raised a number of practical issues:

-The use of a computer for generation of matrices is essential. For this research work a spreadsheet has been used which incorporates such useful features as the ability to enter text and data "sideways" and a facility to manage levels of abstraction. The other major benefit of using a spreadsheet is the ability to output text files easily so that they can be transferred to, for instance, a C program for further manipulation. Macros have also been written that allow the individual parameter ratings to be amalgamated so that a rating summary remains visible when higher levels of abstraction are are being displayed.

Many dedicated QFD packages are available. One of these, QFD/CAPTURE from ITI(International TechneGroup Incorporated), has been evaluated for use within this project. Although these dedicated packages offer many useful tools and facilities that will be highly important for real projects, this research project is currently best served by using a spreadsheet which offers more open data import/export facilities.

-A great deal of training and experience is required to generate elegant and objective matrices that are well-structured and avoid parochial language.

-If QFD is utilised within a real aerospace project there is likely to be a conflict between a need for confidentiality and a desire for wide and open access to the matrices. Much of the data held within a QFD matrix will be commercially sensitive. Any systems used to store, distribute and manipulate these matrices must address this consideration.

-The updating and maintenance of a "master matrix" needs to be managed very carefully especially as QFD would best be implemented as a "distributed" system. This problem is very similar to that of design data storage and issue control. Sophisticated drawing management systems have been developed within aerospace companies to deal with this issue and a similar system would be required for QFD data.

Further work

The next stage is to generate a task precedence network directly from this QFD matrix. A topological sorting algorithm can then be applied to this network in order to identify the optimum task sequence[Kusiak and Wang (1993)]. This task sequence and the relative importance data gleaned directly from the QFD matrix will be used to produce a discrete event simulation of the design task.

References

Akao, Yoji.1990, *Quality Function Deployment: Integrating Customer Requirements Into Product Design.* Cambridge Mass. Productivity Press.
Carter, Donald. 1992, *Customer-Driven Engineering; Quality Function Deployment.* Concurrent engineering: the product development environment for the 1990s. Chapter 7.
Constable, Geoffrey. 1994, *Managing product design; the next industrial revolution.* IEE review march 1994 pages 79-82.
Gandhi, Zafar. 1994, *A house full of innovation: the application of the QFD methodology to the preliminary design of a light aircraft.* Internal report Avro International Aerospace.
Kusiak, Andrew. Wang,Juite. 1993, *Decomposition in Concurrent Design.* Concurrent Engineering; Automation, Tools and Techniques; chapter 19. ISBN 0-471-55492-8.
Parry, Philip. 1994, *Who'll survive in the aerospace supply sector?.* INTERAVIA March 1994 pages 22-24.
Ziemke, Carl. Spann, Mary. 1993, *Concurrent engineering's roots in the World War II era.* Concurrent engineering: contemporary issues and modern approaches; chapter 2 ISBN 0-471-55493-5.

THE USE OF GESTURES IN COMPUTER AIDED DESIGN

Dr Keith Case and Robert Doyle

Department of Manufacturing Engineering, Loughborough University of Technology,, Loughborough, Leicestershire, LE11 3TU

Computer aided design systems are particularly useful in detailing, analysis and documentation but are not well–suited to the very early, conceptual aspects of design. This paper describes investigations of novel methods of interfacing between the designer and his computer system using stereotyped gestures to modify dimensional, positional and orientational parameters for simple three–dimensional geometric models. A prototype implementation using a virtual reality visualisation system enhanced by the provision of a six degree of freedom real–time tracking device is described.

Introduction

Computer aided design (CAD) systems are well–suited to the production of precise and detailed design geometry, can be used for functional analysis and may lead directly into manufacturing planning activities. However, existing CAD systems are not so well–suited to the conceptual aspects of design where the principle objective is to establish general dimensional and spatial relationships in a highly interactive fashion, and where rigidity and precision can be considered to constrain the designer.

The manipulation of 3D shapes within CAD systems to either change their shape and dimensions, or to re–locate and re–orientate them in three–dimensional space is computationally simple, but the way in which the designer specifies his intentions is by no means easy and straightforward. Gestural input replaces (in part) the need for rigid command languages, menu systems, or whatever to define the operation required.

Gestural input is a relatively recent research topic, probably because the necessary supporting hardware has only arrived with virtual reality techniques. Prime (1993) describes 2D work in the context of text editing (e.g. Welbourn and Whitrow, 1988, and Rhyne, 1987) and describes his own work in the use of data gloves for hand–tracking in three–dimensional space. Hauptmann and McAvinney (1993) provide a useful review of current work and also describe empirical studies of the effectiveness of gesture for graphic manipulation which reached the important conclusion that there is commonality between subjects in their use of gestures.

Geometry Manipulation

In the manipulation of 3D shapes two major functions are performed through the user interface: (1) the identification of the object, and (2) the specification of the manipulation. Various techniques are employed to meet the first of these objectives, including the use of a naming scheme, a mouse–driven cursor, etc. The gestural equivalent can be the established methods of lightpens/touch screens or the use of a hand tracking device to determine the location of a 'pointing' finger.

The manipulation usually has one of two objectives, (1) to make some precise change to the model, (e.g. scale by two), or (2) to assist the user in some qualitative way. (e.g. to make something a little larger). Most CAD systems, are geared to the first situation while the second situation requires an imprecise method with a rapid graphical feedback. A means of achieving this control is a form of input which is related to the stereotypes of human movement rather than the existing use of linguistic expression of numerical definitions. These *gestures* must be stereotypes as a system which required the learning of a set of movements is merely a different form of artificial language.

Input Interfaces

Assuming that a useful set of gestural stereotypes can be identified, then some form of transducer is required so that body movements can be translated into commands for the CAD system. The transducer must be capable of sampling at high rates, and needs to be provided with sufficient computational power for real time processing. Similarly the CAD modelling system must be sufficiently responsive so that the user perceives a real time response to commands – characteristics found in virtual reality (VR) systems.

The origins of the work described here lie in some early work (Case & Keeble, 1986) which had the objective of controlling the position, size and orientation of simple objects within a conventional CAD system using human arm movements. This used the UNIMAN posture measuring device (a 'cat–suit' or 'second–skin' garment with in–built strain gauges at the major points of articulation (Paradise, 1980). This work suffered many difficulties including calibration and speed of communication with the host CAD system, but it did demonstrate a potentially useful technique that could be successfully implemented with adequate hardware and software support.

VR systems provide simple 3D modelling and effective communication with the user through a variety of input devices. The work described here is based on the Superscape desktop VR system and a Polhemus 3Space Fastrak device for six degree of freedom sensory input. A small magnetic field sensor is mounted on the hand and obtains positional and orientational information relative to a fixed source. A control unit feeds this information in real time to the VR system where it can processed (eg to filter out extraneous motion) and used to control objects in the virtual world.

Gestures

Gestures are stereotyped and recognisable body postures. For example the 'thumbs–up' sign has almost universal recognition (but beware of cultural differences!). It may be possible to record such gestures in some encoded form such that they can subsequently be 'replayed', and this has been used for example in the Benesh Notation for ballet (Benesh and Benesh, 1969).

The desire to understand the meaning of gestures has a long history and Francois Delsarte (1811–1871) defined laws of expression with defined relationships between

gestures and their meanings. His *Three Great Orders of Movement* associated meaning
with general characteristics of dynamic gestures such that *Oppositions* where any two
parts of the body moving in opposite directions simultaneously express force, strength
and power. Similarly *Parallelisms* where any two parts of the body moving in the same
direction denote weakness, and *Successions* where any movement passing through the
whole body in a fluid wave like motion are the greatest expression of emotion.

Specific interpretations of static gestures by parts of the body can be identified,
and thus the head raised and turned away can mean pride or revulsion and the head
lowered towards a person shows respect. The first of these might be an example of an
involuntary reaction whilst the latter has become a formalised and learned part of
communication in many cultures. Similarly figure 1 illustrates hand postures and
associated meanings.

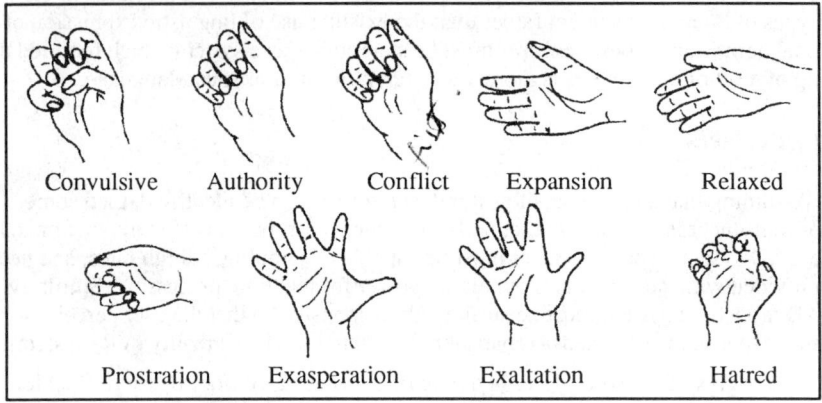

Figure 1. Hand Gestures and Associated Meanings

Geometry Manipulation with Gestural Input

Initial Shape Generation

A substantial aspect of design is the generally tedious creation of initial geometric
shapes. Gestural input at this stage may be feasible by an analogy with physical model
building where shapes may be generated by hand modelling either in a totally virtual
world or through some intermediary 'sensitive' material (where changes to a modelling
material's shape can be sensed and transmitted to the CAD system). However, these
aspects are outside the scope of this work.

Object Selection

Objects are selected by pointing with a stylized model 'hand'. Location of the real
hand is tracked in 3D space and the view of the 3D virtual world tracks the hand to avoid
the need for separate view control. (Figure 2). Dynamic interference detection
capabilities are used for selection by collision of the hand with the object to be selected.

Shape Modification

Shape modification can relate to topology or geometry. Topological changes
present essentially the same challenges as initial shape generation and in our work shape
modification refers simply to dimensional changes. Arm flexion/extension,
abduction/adduction and vertical motion are used to modify the dimensions of the object.

Filtering the tracking device input limits dimensional change to the predominant direction, so that although dimensions can be changed in all three directions within the one mode of working, the user has adequate control over the activity. (See figure 3).

Location

Positional changes to objects are achieved in a similar manner to dimensional changes in that the major hand movements give the required direction.

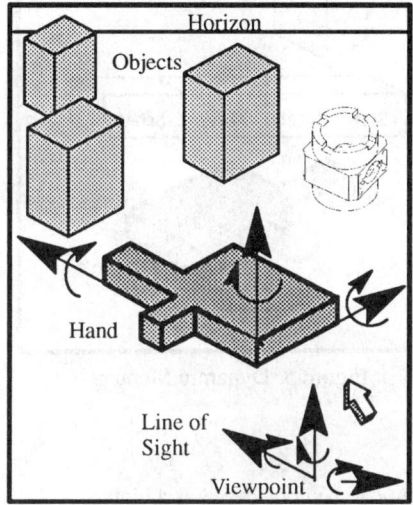

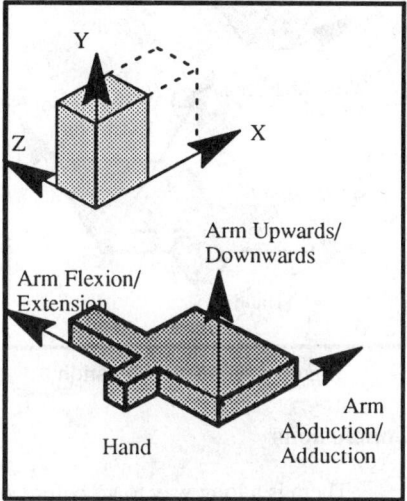

Figure 2. Object Selection   Figure 3. Object Dimensioning

Orientation

Orientation is achieved by tracking wrist abduction/adduction, flexion/extension and pronation/supination. (figure 4). This is an example of the need to amplify the motion of the real hand so as to be able to control orientation of the object throughout the range of motion, as the human wrist does not have full freedom of movement.

Free Move

Combining orientation and location gives full control over the object and effectively the model object will track the real hand position and orientation.

Adjacency

Adjacency refers to the frequently desired ability to position and orientate one object relative to another. A good example would be in kitchen design where a base unit would be butted up to a corner of the room and the floor, and subsequent units would be positioned adjacently. This is achieved by a free move constrained by interference criteria which ensure that the selected object cannot be moved into an already occupied space.

Mode Control

An important issue in the use of gestural input is the need to avoid changes in the mode of control and thus view control is implemented as an integral part of the geometry interaction activity, and the ability to switch control modes (say from positional to orientational control) must also be accommodated in a similar fashion. A dynamic menu which is part of the model space and which tracks the virtual hand position so as to always be available within the 3D space currently in view (figure 5) overcomes this difficulty.

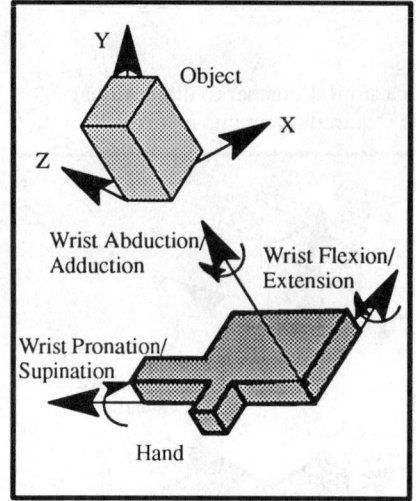

Figure 4. Object Orientation   Figure 5. Dynamic Menu

Conclusions

There is a long way to go before gestural input can be considered as a viable method for interacting with CAD systems, and it should be realised that there are many situations for which the inherent imprecision would be inappropriate. However, this work has demonstrated the feasibility of the approach and further research is required to determine its acceptability to users.

Acknowledgements

This research was funded by theACME Directorate of the Science and Engineering Research Council. (Grant Number GR/J22399, 'Gestural Input for Conceptual Design').

References

Benesh, R. and Benesh, J. 1969, An Introduction to Benesh Movement Notation: Dance, Dance Horizons.
Case, K. and Keeble, J. 1986, Gestural Input to a CAD System, Loughborough University of Technology.
Case, K., 1993, Gestural Input for Conceptual Design, *Proceedings of ACME Research Conference*, Sheffield University.
Hauptmann, A.G. 1993, Gestures with speech for graphic manipulation, International Journal of Man–Machine Studies, **38**, 231–249.
Prime, M. 1993, Hand tracking devices in complex multimedia environments, Workshop on 3D Visualisation in Engineering Research, Rutherford and Appleton Laboratory.
Paradise, M. 1980, Recording Human Posture, PhD Thesis, Department of Production Engineering and Production Management, University of Nottingham.
Rhyne J. 1987, Dialogue management for gestural interfaces, Computer Graphics, **21**.
Shawn, T. 1974, Every Little Movement, Dance Horizons.
Welbourn, K. and Whitrow, R. 1988, A gesture based text editor, *People and Computing IV*, Cambridge University Press, Cambridge.

THE DESIGN OF THE TABLEWARE FACTORY OF THE FUTURE

Ms Nicky Shaw and Dr Keith Ridgway

University of Sheffield
Sheffield, S1 3JD

The paper will examine the requirements of the tableware factory of the future. After describing existing systems and processes the paper will describe the tableware factory of the future. The technological gap between the two factories will be described and the factors inhibiting the introduction of flexible manufacturing systems will be identified.

Introduction

Manufacturing companies are continuously looking for ways to increase profits by adding retail value or reducing manufacturing costs. One approach is to introduce flexible manufacturing systems. The introduction of flexible manufacturing systems can enable the company to reduce the level of work in progress and finished inventory. It can also reduce manufacturing lead time and improve the response to customer demand.

One industry which has been slow to introduce cellular manufacture or FMS is the ceramic tableware industry. This is a divergent industry in which the bill of materials is small in relation to the number of end products. The industry still uses traditional manufacturing processes and Ridgway, Bloor and Ghosh (1989) reported few examples of the successful implementation of advanced manufacturing technology.

The paper will describe the first stage of a project to design the tableware factory of the future. The work presents observations made by the author of sample ceramic manufacturers in Stoke-on-Trent, and describe the process technologies and production methods appropriate to improving the manufacturing flexibility of ceramic tableware.

Manufacturing Flexibility

Manufacturing flexibility can be gauged by a Company's ability to respond swiftly to changes in customer demand. It is made possible by processing small batch quantities, reducing WIP and releasing capital, improving space utilisation and reducing manufacturing costs. An FMS usually comprises new technology and process automation, but can be manual if desired.

Cellular manufacturing (or Group Technology) complements this, operating within a designated area and producing individual groups of parts. The groups are determined by either similarity of machine set-ups or processing requirements, and transportation between work stations is small. Cell operators are responsible for their own scheduling and quality, which has a positive effect on morale. The reluctance of companies to recognise human issues was identified by Ridgway et al (1989) as a key factor inhibiting the introduction of new technology.

Existing Manufacturing Processes

The manufacture of a ceramic product involves a number of discrete steps as shown in figure 1 below:

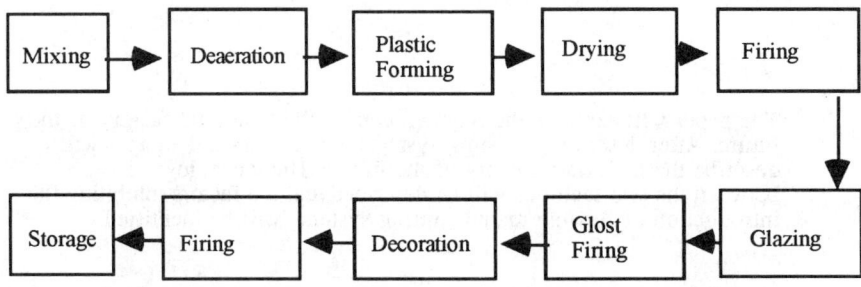

Figure 1: The Tableware manufacturing Cycle

Pottery is an ancient craft, so predictably the current technologies are fully tried and tested and unlikely to change without good reason. The industry is also labour intensive, although technology is not always capable of replicating the processes.

The forming process begins by shaping the clay between a gypsum mould and roller head (flatware) or jollying (cups) machine. The clay is dried on the mould in a hot-air drier, to reduce the moisture content from ~15 to 2%. Hot air is blown onto the formed clay; at the point the air contacts the mould its temperature must be less than 45°C to prevent the gypsum from burning. At ~2% moisture the ware is in its 'green body' state - firm to touch but still very weak, and any seams or edges are removed prior to firing. Conventional drier capacities range from 10-70 dozen items. During a product changeover, all the moulds must be replaced with those of the new product, which for 70 dozen moulds is very time consuming. Each mould yields approximately 150 fills, depending upon the complexity of the pattern being produced. The moulds are brittle and to minimise breakage losses, through excessive transportation, they are used for their entire life in one batch. Thus the inflexibility of the drying phase dictates that the batch size should comprise approximately 126,000 items. Apart from the manufacturing inflexibility, large batch processing at the drying stage dictates the use of large buffer stocks at most subsequent processes to sustain production.

After drying the ware is fired for 8 to 10 hours before glazing and a second stage of firing. The ware is then decorated using a variety of techniques including offset printing and the use of transfers. After decoration the ware is fired for a third time. Tableware factories use a traditional process layout and hold high levels of work in progress. The very nature of a 24-hour kiln necessitates ware being stored for overnight and weekend firing, and high batch quantities exacerbate this problem.

The automatic transfer of ceramics has proved difficult because of the delicacy of the unfired material and number of different shapes produced (Graf and Nicolaisen, 1982). This was confirmed in the study carried out by Ridgway et al (1989) who reported that brick and tile manufacturers have achieved greater success due to the regular shapes of the ware and smaller range of products. Hotel ware comprises up to seventy different shapes per set, ranging from posy vases to divided cookware dishes and saucers to 12" serving platters. Flatware can be handled using suction pads and hollow ware (cups and mugs) can be gripped on the inside rim, although dedicated grippers are required. Cast ware is handled more effectively manually, due to the irregularity of the ware.

The most recent developments in ceramic technology are the fast-fire kiln, and leadless glazes and paints. The former was introduced to shorten the firing cycle, reducing the product lead time and increasing the flexibility of the firing process. Special glazes have been designed so that green-body ceramics (at ~0.2% moisture) can be glazed, then bisquet fired, eliminating the glaze (glost) firing phase. The leadless glazes and paints were developed to satisfy recent legislation limiting the level of lead passing into food and liquid in contact with the pottery.

Potential

Despite the benefits of operating an FMS, the ceramics industry has continued to use existing process technologies, high stock levels and large batch quantities, distancing itself from most automation other than the kilns.

As discussed previously, the drying process is the main source of production inflexibility. However, there are a number of alternative technologies now commercially available that would alleviate this problem. The drying cycle will always be limited by the need to minimise thermal stresses, which results in distortion and cracking, which arise when the ware is dried too quickly or unevenly, :

• *synthetic moulds* can withstand higher temperatures than gypsum moulds, and are more durable and have a longer mould life. They cost ~£12 each as opposed to ~40p for a plaster mould. This cost is offset by a shorter drying cycle and a greater number of fills per mould. Manufacturing flexibility would be improved as the synthetic moulds are less susceptible to breakage and they could be changed more frequently.

• *microwave drying* (dielectric heating) is a more efficient means of drying than conventional hot air, with ~50% electricity being converted into microwaves. The waves permeate the ware, heating throughout rather than via conduction through the ware; obviously this is most beneficial with thicker ware. Microwaves are also attracted to areas of higher moisture content, thus producing a more uniform rate of moisture removal which helps to improve product quality. Microwaves can be used in air or a vacuum depending on the conditions required. Vacuum drying is a slower process because of the loss of convection, but it is desirable for more heat sensitive materials (Metaxas, 1991). Small batches can be processed easily because machine set-ups are short, although it is better to process only one product type at a time to ensure even absorption of the microwaves. The moulds are not very receptive to microwaves and do not become heated, but materials can be 'doped' with non-absorbing materials to prevent unwanted heating if necessary. The higher cost of electricity is offset by increased flexibility and a reduction in drying time (approximately 10%) which leads to a reduction in lead time and inventory.

Microwaves can also be used for firing, again in air or vacuum. The firing time is significantly reduced, with the energy being absorbed solely by the clay. Hybrid kilns comprising both microwave and conventional firing systems can also help to reduce fluorine emissions (subject to more recent legislation) because of the reduced firing time and faster release of chemically-bound moisture (Wroe, 1993).

• *Radio-frequency* (RF) *drying* (also dielectric heating) is very similar to microwave drying. It permeates the ware and is attracted to areas of higher moisture content, giving a uniform heating of the ware. Again it is preferable to dry ware of similar height simultaneously (Riley, 1990).

Both microwave and radio frequency drying are best when used in conjunction with conventional hot air drying The dielectric sources causing moisture to migrate from the centre of the ware, and the hot air evaporates the moisture from the surface. Apart from a shorter drying cycle, a hybrid system also reduces the investment cost as the dielectric heater can be fitted onto the existing dryer system.

• *bi-lateral heat input* uses two heat sources to speed the drying process. Ware is dried (still on the mould) in a hot air drier, each mould being cast with a fully controllable heating element inside. The main investment cost is the moulds, but a faster drying cycle (reduced again to 10%) requires fewer moulds in the round, thus increasing manufacturing flexibility (Boguslavskii, Kryuchkov and Komskii 1989).

A recent development in process technology has rivalled the traditional plastic forming. In dust pressing clay granulate is pressed at high pressure to form a green body. Sheppard (1992) reports that the technology is now so advanced that even thin-walled hollow ware has been successfully pressed by the German press manufacturer, Netzch. Strobel (1991) reports that a demonstration project at Hutschenreuther AG. in Germany, has produced ware by dust pressing, decorating, glazing and firing just once. Although designed for porcelain manufacture there is potential for processing other ceramics on an industrial scale.

The use of clay in granulate form ideally eliminates the need for a drying process prior to bisquet firing, however moisture contents of 2-4% have been detected, which is high enough to create thermal stresses, as Coudamy (1990) suggests. An extra drying process incorporated into the firing cycle would eliminate this problem.

The investment necessary for dust pressing is high because of the cost of the dies. particularly if many iterations are required to achieve the correct fired shape; all pressed ware undergoes a degree of spring back, however CAD modelling would help minimise this problem. The cast of the granulate is also higher than that of slop. The process is suitable for automation, with subsequent sponging and glazing operations easily incorporated into a work cell structure.

For ware transportation, most factories in Stoke on Trent would have difficulty in accommodating large conveyor systems or automated vehicles because of the factory plan. Many companies have excessive amounts of manual palletising to move ware between the processes and buffer stocks, none of which adds value to the ware and increases product lead time. The palletising would be significantly reduced if manufacturing work cells were introduced. Picking and placing the ware would require a robot vision system or other sensor however, and quickly interchangeable robot grippers to handle the numerous shapes and sizes. Flatware could be accommodated more easily using a vacuum head but hollow and cast ware pose a more difficult handling problem.

Comments/Summary

The course of assessing and improving manufacturing flexibility causes inefficiencies to be addressed (or at least acknowledged). This helps to reduce production costs, whilst throughput is free to increase. The reduction of WIP and stocks also releases capital, as well as valuable floor space necessary to house the ware for overnight and weekend firing.

Flexibility is derived from processing small batch sizes, which enables production to respond more quickly to changes in customer demand. Consequently the need for buffer stocks is reduced, which also releases capital. Manufacturing in work cells minimises the ware transportation for firing, items from different cells could either combine for a common firing process, or be fired individually in intermittent kilns.

To summarise, the key process in ceramics for small batch manufacture is drying prior to bisquet firing. Investment in alternative drying methods, realistically microwave or radio frequency plus hot air, means the drying time can be substantially reduced whilst increasing the mould life. Swift product changeovers enable small batch production to operate without increasing the product lead time.

Automatic transfer of holloware occurs on mostly dedicated lines, whereas the transfer of cast ware remains essentially manual. Process automation such as robot vision systems for product inspection, is not able to operate as quickly or accurately as the human operator, so for the time being, certain ceramic processes will remain labour intensive.

Acknowledgements

The authors wish to thank the staff of Churchill Tableware Ltd for their help during the project. The authors also wish to thank the Billy Ibberson Trust who support research in Industrial Change at the University of Sheffield.

References

Boguslavskii V.D., Kryuchkov Y.N., Kulak and G.Z. Komskii G.Z., 1989, Speeding up Plate Drying with Bi-lateral Heat Input, *Glass & Ceramics* **46**, No.11/12, pp484-7.

Coudamy G., 1990. Bisquet Firing of Porcelain Plates Manufactured by Isostatic Pressing, *L'Industrie Ceramique*, 855, No. 12, pp 806-9.

Graf B., and Nicolaisen P., (1982) Industrial robots and their operation in the manufacture of ceramic products, Interceram,**31** (3) 1179.

Metaxas A.C.,1991, Microwave Heating, **5**, No. 5, pp 237-47.

Ridgway K., Bloor I. and Ghosh S. K., 1989, The State of the Art of AMT Implementation in the Ceramics Industries, Chapter 12, pp 101-111, Exploitable UK Research for Manufacturing Industry, IEE Management of Technology Series **9**, G Bryan (ed), Peter Peregrinus Ltd, pp102-110.

Riley B.,1990. Microwave Treatment of Ceramic Materials, *7th Cimtec Part B*, pp 1233-61.

Schulle W., 1991. Trends of Development in Classical Ceramics *Ceram. Forum. Int.* **68**, No. 4, pp 157-65.

Sheppard L.M.,1992. Recent Trends in Tableware Manufacturing, *Am. Ceram. Soc. Bull.* **71**, No. 9, pp 1356-66.

Strobel K., 1991. "Fast Once-firing of Decorated Porcelain" *Interceram.* **40**, No. 6, pp 391-8.

Wroe R., 1993. "Improving Energy Efficiency in the Firing of Ceramics" *Materials World*, pp 446-8.

AN INTEGRATED DATABASE SYSTEM FOR A TENDER TO DESIGN TO MANUFACTURING SYSTEM

Harald Hubel*, Gary Colquhoun**, Peter Hess***

* ABB Turbinen Nürnberg GmbH, Frankenstraße 70-80, 90461 Nuremberg, Germany

** Liverpool John Moores University SETM, Byrom Street, Liverpool L3 3AF, UK

*** GSO-Fachhochschule Nürnberg, FBM, Kesslerplatz 12, 90489 Nuremberg, Germany

The configuration of a steam turbine involves a protracted series of iterative steps to define turbine characteristics and subsequent manufacturing costs. The data necessary for configuration is read from files and in turn written to files and the rapid expansion of programs and files has lead to ever more complex file structures and file handling with the attendant problems of inconsistency and redundancy. An integrated Tender to Design to Manufacturing system with a new approach for the effective storage of engineering data is proposed. This paper uses the configuration cycle for a turbine as an example to describe gradual integration of a relational database at ABB Turbinen Nürnberg GmbH, one of Germanys leading steam turbine manufacturers.

Introduction

With the growing need to compete in an international market quality and design to manufacture time is becoming ever more important. Efficient performance demands an integrated, transparent information flow from tender to design to manufacture and for companies endeavouring to maintain a competitive position the key to control of information is a database (DB) and information management system (Rolland (1990)). Ideally all the data structures and data relationships necessary to run the business must be resident on the system, for most companies however the development of data systems is evolutionary resulting in a variety of hardware and data management systems. In most cases the short term effective substitution of an existing system by a single DB system is not possible. This paper explains the way that ABB Turbinen Nürnberg GmbH (ABB TUR) proposes to solve the problem of data integration and management.

ABB TUR is a company employing approximately eight hundred staff engaged on the design and manufacture of power generation plant, the Nuremberg site is responsible for the development and manufacture of the core unit, the steam turbine. At the present time the tender to design to manufacturing chain of activities for each product is based on a series of several in-house developed programs and data processing steps connected in a unidirectional manner. In the first phase configuration programs carry out calculations to select the most competitive tender for the customer. At this stage the system produces the rotor and blade specifications. Once an order has been placed by a customer the design and manufacturing departments will produce design and detail drawings, manufacturing plans and CNC programs with performance and cost as optimisation criteria. Each stage in the process utilises the data generated in previous stages, for example the thermodynamic turbine data must be processed in the later stage of design. To facilitate direct access to data processed by the chain of programs and to satisfy call backs from previous processes, later revisions, the management's need of information or the automatical generation of a tender and specification paper, an effective secure data handling system is essential.

Problems with the current data processing system

Data used in the technical calculations are stored in directories using ASCII files. At present data is only stored in sequential files to facilitate straightforward programming and to minimise disc space requirements (cf. Amann (1990)). Using this file system data can be processed and transmitted from one stage to the following. In these circumstances the data from one file can be used in several programs and one program can use data from several files. This form of data storage approaches its limits if more and more programs uses more and more single data stored in different files because the program must read and write an entire ASCII file to get the value of the required data. Kalta and Davies (1993) describe the same problem with IGES-files. Their solution with direct access files is inadequate to solve our consistency problem. Consequently an update of one program requires also an update of the files and of all programs accessing these files to guaranty software integrity. For example (cf. Figure 1): Program 2 is used to calculate the diameter X of a 'damping wire' component. There is no possibility using the sequential file storage to transmit X to the 'ASCII input file' automatically. This diameter has to be read from the 'ASCII output file' and in turn written in the 'ASCII input file' by an editor. Using this process data storage is becoming inconsistent as the numerical value X is stored twice, once in the ASCII output file and once in the ASCII input file. In a CAD file the binary value for X appears for the first time when it is taken over by an editor in the ASCII input file and the program 1 is started once more.

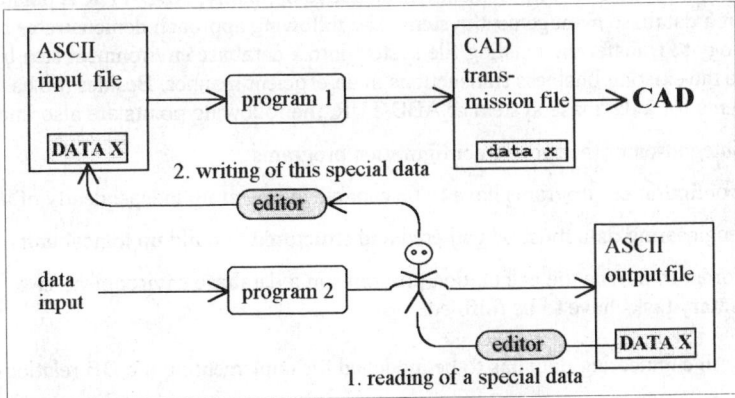

Figure 1. Problems with the file system

Similar problems arise with other components such as the selection of blade foot profile, the optimisation of steam supply nozzles etc. Therefore the easy straightforward programming and the minimal use of disc space of a file system causes additional data transfer via keyboard combined with an increasing error level.

The development of data handling facilities

In a program dependent data storage environment a data file reflects the associated program. Changes in the program therefore also cause changes in file structure, which will consequently influence any other application program accessing the files. The number of programs used in turbine design will continually rise as will the requirement for CAD geometry files. A program independent database is therefore essential in order to develop an integrated system. The following conditions for improved data storage will have to be satisfied:

- data consistency,
  It is acceptable to store a single occurrence of each data item once in the global database, regardless of the number of users or programs utilising with the data item.

- reduced data storage and redundancy,
  No data item should be stored which can be calculated or derived from other data. As a consequence, the amount of data stored will be reduced. E.g. for storing a circle it is sufficient to store the diameter or the radius, but not both.

- easy access to and manipulation of all data,
  It must be possible to have access to each single data item, read it and, where necessary, to update it by the use of an unique key. This key must be defined as briefly as possible and unequivocally, but must be capable of being upgraded and open for future innovations.

- data security,
  The system operator must be able to allow users different privileges for data access.

- independence of data storage and programs,
  Application programs must strictly be separated from data storage. Changes in the programs must not influence the structure of the data pool and changes in the data pool must not influence the structure of the programs.

Database integration

In order to meet the requirements listed above optimally, ABB TUR is planning to implement a database management system. The following approach demonstrates a methodology to transfer the existing file system into a database environment step by step to enlarge the existing business transactions in an efficient manner. Besides the basic requirements for a database system at ABB TUR, the following points are also important.

- The integration of the turbine configuration programs

- The configuration programs have to be capable of operating independently of the DB

- The engineering data must be gathered and structured to build up logical units

In order to pursue the calculation programs in a database environment, two complimentary tasks have to be fulfilled:

Task 1:   All engineering data has to be modelled for implementing the DB relationships.

Task 2:   The configuration programs utilising files have to be linked with the database (cf. Encarnacao and Lockemann (1990)).

To task 1: In order to save the engineering data in logical units it has to be analysed and structured using a data modelling method. By doing so, data relationships can be evaluated using different criteria.

assembly-oriented part list:
```
turbine
    turbine casing
        nozzle segment
        guide blade carrier
            stator cascade
    rotor
        control wheel
        rotor cascade
```

function-oriented scheme:
```
stator cascade  <=>  rotor cascade
nozzle segment  <=>  control wheel
shroud  <=>  sealing strips
```

The aim of data modelling is not to model the geometry data for a turbine, but to model all the data necessary for turbine configuration and to identify and describe the individual parts (cf. Kowal (1992)). The exact geometry data of a turbine assembly part should be stored in the design and detail drawings (CAD-files) which can be accessed by using information of the database. In order to access all engineering data stored in the database, a search key has to be defined. Figure 2 shows an example of a suitable Entity Relationship Data Model.

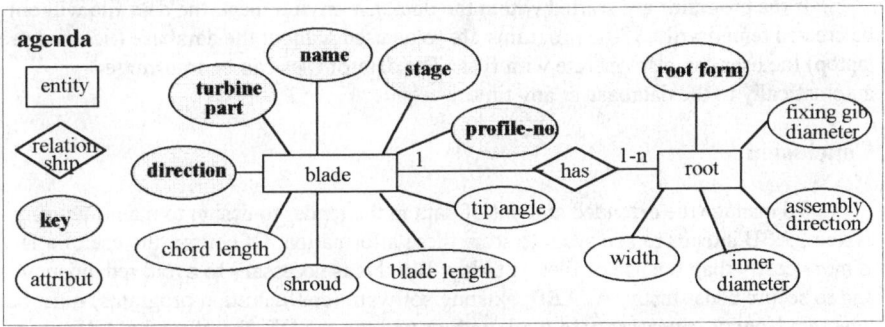

Figure 2. An Entity Relationship Model for turbine blades

To task 2: To be able to access data stored in the database by using turbine configuration programs, it is necessary to embed SQL-statements within these programs. As FORTRAN, C or another 3 GL-programming language has no relationship with SQL, an embedding method has been developed and standardised. This approach enables the programmer of 3 GL programs to intersperse SQL statements combined with a prefix at any line desired of his source code. After finishing the process of coding, the program has to be pre-compiled to translate the SQL statements into the code of the programming language used. The following compilation and linkage of the program-code is carried out in the usual way. A typical embedded SQL of "ORACLE PRO*'3GL' " looks like this:

```
exec sql   select      name, profile-no
           into        :note,  :number
           from        blade
           where       stage = :x
```

In the example the program variables `note` and `number` take the values of `name` and `profile-no` from the database. The colon's function is to separate the program variable from the name of the database column. Proceeding like this is technically perfect, but has the following disadvantages:

• The source code of the calculation programs must be changed

- The calculation programs can only run within the database environment

- Consequently the transposition to a database has to be made very quickly

To circumvent these problems, the interface should not be realised between the calculation programs and the DB, but between the file system and the DB as in Figure 3.

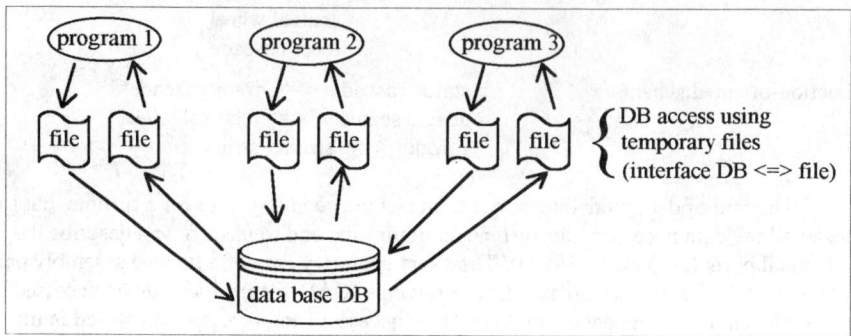

Figure 3. Connection programs - database

If the programs are started within the database environment, the data file will only be created temporarily. If the programs are to be used without the database (licensed use, laptop) the user can also operate with files. The data of files can be transmitted automatically to the database at any time.

Conclusion

To manage the extended amount of data in the tender to design to manufacturing system, ABB intends to use a DB to store these information. Of course, the use of a DB is more costly than the use of files as today. But this is necessary to avoid redundancy and to secure consistency. At ABB, existing software (configuration programs) must be integrated, and it is necessary to work with or without the DB, therefore the ABB concept is designed with temporary files. The adaptation of the first programs to the DB has proceeded well. So it is proposed that the whole system will be reorganised step by step in order to achieve an Integrated Information System using the principles described by Fischer and Biallowons (1993).

References

Amann, J., 1990, *Datenhaltung und Datenaustausch für kommissionsabhängige Daten*, ABB Turbinen Nürnberg GmbH.

Encarnacao, J.L. and Lockemann, P.C., 1990, *Engineering Database. Connecting Islands of Automation Through Databases*, Springer, New York.

Fischer, A. and Biallowons, M., 1993, *Engineering Database als strategische Integrationskomponente*, CIM Management, 5: 4-8, Springer-Verlag, London.

Kalta, M. and Davies, B.J., 1993, *Converting 80-Character ASCII IGES Sequential Files into More Conveniently Accessible Direct-Access Files*, The International Journal of Advanced Manufacturing Technology, 8:129-144, Springer-Verlag, London.

Kowal, J. A., 1992, *Behavior Models. Specifying User's Expectations*, Prentice Hall, Englewood Cliffs, New Jersey.

Rolland, F.D., 1990, *Relational Database Management with ORACLE*, Addison Wesley, New York.

DESIGN FOR RAPID PROTOTYPING: DEVELOPING A METHODOLOGY

R Ian Campbell

University of Nottingham
University Park, Nottingham, NG7 2RD

This paper examines the need for a design for rapid prototyping (DFRP) methodology and describes the work currently being undertaken at Nottingham University to satisfy this need. The reasons that designers should adopt a DFRP methodology are explained by referring to the short-comings of the current design process and the opportunities presented by an extended range of rapid prototyping techniques. An outline of the proposed DFRP methodology is then presented. This is a four stage procedure that runs through the entire design process from specification to optimisation. The benefits of using the methodology are discussed and the paper ends by examining the likely direction of future developments.

Introduction

There have been several attempts to promote the need to take rapid prototyping (RP) into account during the design process. CAD considerations such as the requirement for a fully enclosed surface or solid model have been discussed in detail Smith-Moritz (1992). The impact of using RP on the total design process has also been discussed Meuller (1991), Ulerich (1992). The designer cannot simply produce a CAD model that meets the design specification and then pass this on to the RP service. This "over-the-wall" mentality runs totally contrary to the concurrent engineering philosophy that RP is often used to support. A DFRP methodology that is an integral part of the design process is required. For some time now the need for such a methodology has been recognised Jara-Almonte, Bagchi, Ogale and Dooley (1990), Crawford (1993) but remains unsatisfied. This paper describes the DFRP methodology currently being developed at Nottingham University to remedy this situation.

The Need for a DFRP Methodology

The primary reason that designers should follow a DFRP methodology is to avoid unnecessary redesign at a later stage in the design process. The CAD model used by the RP machine should be right first time. If the RP service is given a CAD model that is unusable then either they will have to send it back for modification or alter it themselves. Both these remedies are undesirable because they add extra lead-time into the product development cycle. Also, the latter alternative may lead to an accidental loss of the designer's intent.

Another reason for designers to adopt a DFRP methodology is the proliferation of commercial rapid prototyping techniques. The way in which a design must be tailored to suit RP will depend on the particular technique being used. If only one technique was available, it would be relatively simple to set out a list of guidelines for the designer to follow. However, there are now many techniques available and each one has its own specific requirements. For example, a prototype that is to be built using stereolithography may require supporting structures to be added to the CAD model because of the liquid material being used. No such structures would be required if a solid material was being used, as is the case with laminated object manufacture. Increasingly, designers will have access to several RP techniques and will need to know how to design for each one.

The final reason for using a DFRP methodology is to be able to take full advantage of the range of RP techniques that are now available. It has already been stated that each technique has its own requirements, the converse of this is that each one offers particular capabilities. An essential part of any DFRP methodology should be the ability to guide the designer in the decision of which technique to use. This choice would depend on several factors including the size and shape of the product, its proposed method of manufacture and the uses to which the rapid prototype model will be put. Therefore the methodology will enable the designer to match the needs of the design with the capabilities of the RP techniques available.

Scope of the DFRP Methodology

The DFRP methodology that is being developed aims to cover every aspect of the design process that is affected by the use of RP. Initially, the format will be a document that will act as a designers guide or handbook. This handbook will be divided into four sections that will follow each other in a logical sequence. Each section will contain a particular stage of the DFRP methodology. The operation of each stage is outlined below:-

Stage 1.   Determining if Rapid Prototyping is Appropriate
The use of RP is not justified in every situation. There may be some designs which can be prototyped more quickly or cheaply by using conventional processes. Likewise, it is possible that RP does not offer the accuracy, surface finish or material properties that are required in the prototype. Designers must know how to assess the feasibility of using RP and how to justify its use to others in the organisation. This stage of the DFRP methodology begins with the designer specifying the desired

attributes of the prototype e.g. size, shape, accuracy, material, number required, function, schedule, etc. These are then compared to the capabilities of both RP and conventional processes such as NC machining and near-net-shape forming. The performance of the various processes against each desired attribute is then quantified enabling selection of the most suitable process.

Stage 2.   Selection of Optimum Rapid Prototyping Technique

Assuming that the outcome from stage 1 is a recommendation to use RP, then the choice of RP technique is the next consideration. To make this choice the designer is provided with a comprehensive comparison of the techniques available. The comparison uses criteria such as cost, speed, accuracy, materials, data format, possible downstream processes and so on. The designer is also given advice on which techniques are suited to particular types of products. By matching this information to the desired attributes that were specified in stage 1, the designer can choose the most suitable technique for each prototype being considered. Different techniques will be chosen for different prototypes. This is in contrast to bench-marking procedures that consider selection to be a one-off decision Kochan (1993). The decision-making processes for stages 1 and 2 are quite similar and may eventually be combined into a single decision support system.

Stage 3.   Level of Information in the CAD Model

The choice of RP technique made at stage 2 will have taken account of the suitability of the data files created by the designer's CAD system. However, using an industry standard such as an STL file is not the only way of transferring CAD information. In fact, a file consisting of a series of triangular approximations to the original geometry is an inadequate method of transferring complex 3d information Guduri, Crawford and Beaman (1993). Other possibilities are so-called "slice files" which contain actual cross-sections of the CAD model, boundary representation, constructive solid geometry and feature-based models. Future developments in RP will require information to be transferred at one of these higher levels. For example, the 3d welding process under development at Nottingham University has benefitted from considering the model as a combination of features (see figure 1). This stage of the DFRP methodology will involve analysing the desired characteristics of the prototype and the selected RP technique to decide on the level of information required. Whichever technique is being used, the designer will be given advice on what information to include in the CAD model. This assumes that future RP systems will be able to accept data in formats other than STL files (e.g. data could be transferred using the STEP standard).

Stage 4.   Generation of the CAD Model

Having decided the level of information that needs to be incorporated, generation of the CAD model can now proceed in the usual manner. However, there are certain extra considerations that must be taken into account because of the decision to use RP:-

- Is the object being designed larger than the working envelope of the RP machine?
- Should the designer be content with a scale model of the design or should it be divided and built in several sections?

- Where should the divisions between sections be made?
- What orientation should the sections have since this could effect the accuracy and surface finish of the rapid prototype model(s)?
- Do some parts of the prototype require increased accuracy?
- What extra geometry is required to provide support structures during the RP process?
- Is there any desirable features that have been made feasible by the decision to use RP?

The designer must approach all these questions in a logical and methodical manner to ensure that the best overall solution is achieved. This stage of the DFRP methodology will consist of sets of rules for each RP technique that the designer should follow while constructing the CAD model. If the design needs to be changed as part of the optimisation process, then any CAD model modifications must still follow these rules. The result will be a design tailored to whichever RP technique has been selected.

Figure 1. 3d welded prototype constructed as a combination of features.

Benefits of Using the DFRP Methodology

The main benefits of using the DFRP methodology will be the faster, more accurate and more consistent production of RP models. This is because the CAD model will be generated in the most appropriate manner for the RP technique being used. Also, there will be a standardisation of the DFRP process which can be considered as an improvement in quality. Finally, there will be a reduction in the number of RP models required since they will be right first time. This will lead to cost savings in terms of time and materials.

Future Developments

The designer's handbook version of the DFRP methodology will contain many questions that the designer must answer and many rules that must be followed. Therefore, it would be logical to incorporate these into an interactive expert system that will lead the designer through the DFRP methodology. This will take the form of a question and answer session for each stage of the methodology. The longer term aim is to integrate the functions of the expert system with other design knowledge modules as part of an "intelligent" CAD system.

Summary

To ensure that designers can take full advantage of RP technology, there needs to be a formalised DFRP methodology. The framework for such a methodology has already been developed and the detailed specification of the methodology is the subject of continuing research at Nottingham University. The anticipated benefits are a reduction in prototyping costs and lead-time and an improvement in quality. Although the methodology will initially be in the form of a textual document, the aim is to develop a computerised version at a later time.

References

Crawford, R.H. 1993, Computer Aspects of Solid Freeform Fabrication, *Proceedings of the 1993 Symposium on Solid Freeform Fabrication*, (University of Texas at Austin) 102-112.

Guduri, S. Crawford, R.H. and Beaman, J.J. 1993, Boundary Evaluation for Solid Freeform Fabrication, *Towards World Class Manufacturing*, (IFIP/SME, New York)

Smith-Moritz, G. 1992, Why is CAD so Bad?, *Rapid Prototyping Report*, December 1992, 4-7.

Jara-Almonte, C.C. Bagchi, A. Ogale, A.A. and Dooley, R.L. 1990, A Design Environment for Rapid Prototyping, *Issues in Design/Manufacture Integration*, DE-Vol 29, (ASME, New York) 31-36.

Kochan, D. 1993, Selection and Evaluation of SFM Systems, *Towards World Class Manufacturing*, (IFIP/SME, New York)

Meuller, T. 1991, How Rapid Prototyping is Changing Product Development, *Rapid Prototyping Report*, December 1991, 5-7.

Ulerich, P.L. 1992, Rapid Prototyping: Cornerstone of the New Design Process, *Proceedings of the 1992 Computers in Engineering Conference - Volume 1*, (ASME, New York) 275-281.

ASKING THE EXPERT IN DESIGN

T R Melling & H L Cather

Department of Mechanical & Manufacturing Engineering
University Of Brighton
Brighton, BN2 4GJ

The link between design and manufacture has traditionally been weak. Recent years have seen the advent of concurrent engineering, whereby the link is strengthened. Designers still need some knowledge of manufacturing techniques and other areas not immediately concerned with design: this is where expert systems come in. With the development of expert system shells it is now relatively cheap and easy to program a computer to act as a decision support system. The incorporation of a data management system allows the use of data held in external files, therefore easing the burden on the programmer. This paper looks at some of the projects undertaken at the University in this field.

Introduction

The classical design process and production cycle is sequential, starting with product specification and finishing with the final assembly of the end product. This traditional model is usually concerned with achieving the basic function, and does not take performance or ease of manufacture into account. The manufacturability and ease of assembly are not considered until late on in the design process: this often necessitates re-design and therefore extra cost.

One way round this problem is to integrate all parts of the design and production cycle so that they are all considered simultaneously: this is known as concurrent or simultaneous engineering. In this system the designer receives comments from other departments as the product is at the conceptual design stage: this saves valuable time and money.

Both these models can be visualised as a flow of information. In the traditional model the information flow is very limited: knowledge is not shared between departments; in the most extreme case the product is handed on and forgotten about. In the concurrent engineering model the information flows around the system allowing any department to have access to the information and knowledge held by any other. The flow of information in this case is facilitated by cross-functional teams, consisting of people from the design, manufacturing, marketing, finance and purchasing departments. The interaction between the departments allows the flow of information and also the education of people in disciplines with which they are not primarily concerned.

There are a number of problems with this set-up: friction between members of the team, the constant asking of questions when problems arise and the inconvenience if one member is missing. This is where expert systems can be utilised. One of the main uses of expert systems is as a consultant, where they are "used by a non-specialist to obtain specialist advice" [1]. They can be used to solve a variety of problems, and have a number of advantages over human experts, including greater accessibility, reliability and consistency.

A simple definition of an expert system is: a computer program that uses knowledge and problem-solving techniques on a skill-level comparable to that of a human expert. A human expert holds a vast amount of knowledge on a particular subject and can use this knowledge to help others. This is the basis of an expert system: it holds knowledge in the form of rules and heuristics, and when given information in the form of answers to pre-defined questions uses this knowledge to give advice.

An expert system consists of three components: knowledge base, inference engine and user interface. The knowledge base contains the rules and knowledge required for the expert system to make decisions in a particular domain. The inference engine controls the operation of the expert system by selecting which rules to use and then accessing and executing these rules. The user interface is the link between the system and the user, providing a framework for the questions the system asks, the replies given and the knowledge acquisition and representation.

Knowledge acquisition is the process by which the knowledge engineer elicits knowledge from an expert, usually by question and answer sessions. Once the information has been gathered, it is coded in a way that the computer can use: this is known as knowledge representation. Production rules (IF...THEN) are normally used as it is possible to represent any rules of thumb the expert may use.

Figure 1 shows how the knowledge base, inference engine and user interface are linked to create an expert system.

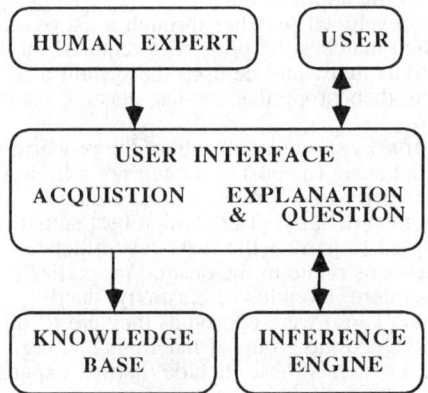

Figure 1: Schematic of an Expert System

It is possible to group the tools for building an expert system into three classes: expert system shells, high-level programming languages and mixed programming environments. All projects discussed in this paper use Crystal, an expert system shell from Intelligent Environments. An expert system shell is a program used to build an expert system; it contains an inference mechanism, an empty knowledge base and facilities allowing the production of user-interfaces. This reduces the time required to develop the system and also means that the building can be undertaken by a non-specialist.

In some cases the data required for the expert system is already available in an existing computer application. It is often preferable for the data to be held in a highly structured environment, like a spreadsheet or database, instead of being part of the expert system program. Having the data contained in an external file allows for easy alteration should any be necessary, and reduces the time taken in building the system. Crystal has links to several standard software products, including Lotus and dBase, which are used in the projects covered by this paper.

The projects discussed in this report can be split into two groups: those to be used during the conceptual design stage, and those used in the detailed design stage. The main difference is the scope: those for concept design have a far wider scope (e.g.. material and process selection) than those used in detailed design (e.g.. selection of fittings and components such as bearings).

Concept Design

Concept design is the first stage of the design procedure. The product specifications will be known and it is the job of the designer to ensure that these are met. Two of the most important parts of concept design are the selections of materials and processes, both of which are covered here.

Material Selection

The choice of material is an important step in the design process as the wrong material could prove disastrous. Additionally, a sub optimal material can be selected which may meet the design loadings but at a heavy cost. There are many factors to be taken into account, and this, coupled with the rarity of human experts, leads logically to the development of an expert system.

There are four stages in the selection process:
(i)   analysis of the product: information needed includes physical attributes, performance requirements and usage conditions
(ii) calculation of the values required for material properties
(iii) selection of candidate materials: once properties have been specified the material search can begin
(iv) evaluation of materials: the selected materials are evaluated to find the one best suited to the application.

The expert system developed searches through a list of materials held in a database to find one which matches the properties required by the user. Because of the vast number of materials that could be used the system places the materials in one of six groups according to their properties: metals, plastics, ceramics, rubbers, glass and woods.

Initially, the system asks six questions about the properties of the required material, plus the preferred cost. The cost, although not a limiting factor, is asked as it can be used to decide between two otherwise similar materials. Using the answers to the questions the system identifies the group which best suits the application. Further questions are then prompted to narrow the field yet further.

The first set of questions relate to the desired properties of the material: tensile strength, operating temperature, modulus of elasticity, ductility, thermal conductivity and electrical conductivity. These were chosen as they are readily available in most material reference books. Once the group of materials has been selected further questions are prompted; for metals these include thermal expansion, yield strength and specific heat.

This system gives a mark to each property of every material identified as a possibility according to how close to the ideal value the answer is. For properties that meet the specification exactly the material would score 100; if the material property is above the necessary value then it would score 10, as it could be used but is not ideal; but if the material property is lower than that specified it would score 1, as it cannot be used.

Assume, for example, that a tensile strength of 550MPa is required. Using Table 1 it can be seen that two groups of materials, ceramics and glass, satisfy this criteria, and so would score 100; metals could be used but are not ideal as they are stronger and would therefore score 10; whereas the other three groups could not be used at all and score 1.

Material	Tensile Strength (MPa)
Metal	600 - 2500
Plastic	30 - 500
Rubber	10 - 40
Ceramic	400 - 850
Glass	400 - 800
Wood	20 - 100

The expert system ranks the materials using the number of ideal and non-ideal properties and, if necessary, the cost. The total score will show how many ideal properties the material has (how many hundreds the score is) and if the material is disqualified (if the final figure is greater than zero). For example, the final scores could be: glass 510; ceramics 420; metals 321; plastics 303; woods 303; rubbers 204. This shows that glass is the group of materials most suited to this application with five ideal properties and one that is better than that specified. Rubbers on the other hand have only two ideal properties and four that do not meet the specifications.

Process Selection
    As for the material selection described above, the selection of the process to be used in manufacturing a product is an important decision for a designer. The choice ultimately decides the cost of the product, and, therefore, how well it fares on the market place.
    This system recommends a suitable production process when supplied with information about the component. The selection procedure is the same as above.
    The consultation begins with an overview of the system, explaining what is needed and how to enter the information required. An option is provided to allow the user to browse through the data held on different types of processes; this is intended to be of some use in the initial stages of design.
    When the system proper is entered the user is asked five questions about the component - material type, production volume, surface finish, workpiece size and workpiece complexity - each of which has a bearing on what process can be used. There are strictly defined limits as to what can be entered; these are displayed onscreen, either as a choice of different answers or a sliding scale of values. A help screen can be accessed at any time if the user is having difficulties.
    Once the input parameters have been entered the system searches through the knowledge held on the different processes. Information on each process is held in a database file; this specifies what values each parameter must take in order for the process to be suitable. The database also holds information describing each production process in more detail. When the search is complete the system will list any process that matches the parameters given. At this point the user will be able to access the detailed information on the processes.

Detailed Design

    Once the concept design has been finished and approved, the detailed design of the product's components can take place. Every component needs to be looked at in detail and the use of expert systems can help enormously.

Selection of Bearings
    Bearings are an integral part of all machinery; their function is to maintain moving and stationary parts in the correct position and transmit loading between them. The choice bearing is crucial, as the wrong could have catastrophic consequences.
    There are many varieties of bearing with each type further divided into different designs. Because of this it is difficult to ensure that the right bearing has been selected without specialised knowledge of bearings. For this reason an expert system has been developed to aid the decision process.
    Due to the large number of bearings, the scope of the system has initially been constrained to two types: deep groove ball bearings and cylindrical roller bearings. These were chosen because of their wide range of applications. There are many different parameters to be taken into consideration in the selection of bearings: for simplicity this system asks only seven questions covering the operating conditions. These are: radial load, axial load, running speed, type of machine, running temperature, shaft size and method of lubrication. The inputs are either entered as figures by the user, or chosen from a selection that the system offers. Help screens are available in case of difficulties.
    The selection procedure is split into four stages:
                    (i)   Data input
                   (ii)   Preliminary selection of bearing size and type
                  (iii)   Calculation of values for bearing life and load ratings

The preliminary selection according to size decreases the selection time as only the load ratings of those bearings chosen are checked. The data for each bearing is held in a Lotus file which is accessed via the Crystal-Lotus interface. Used in this manner, Lotus acts as an information store only, but it is relatively easy for updating without affecting the expert system program.

Once the selection is complete information on any bearing chosen is displayed on the screen. If the user has made a mistake the system allows the data to be changed and the selection procedure will be re-run. Important information such as bearing life (which the system calculates from the running speed that was entered) will also be displayed. The user will have the option to print any relevant data.

Pump Selection

The selection of pumps is often a difficult task due to the wide variety available. Therefore, expertise in the field is needed during the design process.

The selection procedure used for this expert system is similar to that used in the preceding ones. Seven questions are asked: pump type, fluid type, flow rate, pressure, temperature, application and efficiency (see Figure 2). When all the questions have been answered the system searches through the files held in the database and selects the most appropriate pump and its supplier.

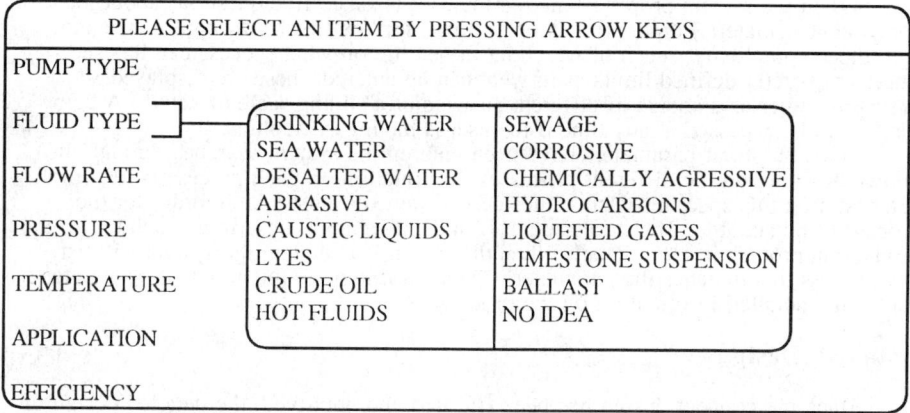

Figure 2: Representation of a Typical Input Screen

Conclusion

Expert systems have an important part to play in the relationship between design and manufacture as an aid to the decision process. Whereas human experts may be scarce or have no time to instruct trainees, computerized expert systems do not have these problems. Companies may need expertise in more than one area: this would prove expensive if human experts were to be employed, but expert systems could be installed for a fraction of the price. It is unlikely that human experts will be totally replaced by computers in the near future due to their ability to incorporate information that at first does not seem necessary, but a computer's enormous benefits will always be attractive.

References

Hasdan, A. On the Application of Expert Systems. In M. J. Coombs (ed.) *Developments in Expert Systems* (Academic Press)
Jackson, M. 1992*Understanding Expert Systems Using Crystal* (Chicester, Wiley)
Corbett, J. Dooner, M. Meleka, J. and Pym, C. 1991 *Design for Manufacture: Strategies, Principles and Techniques* (Addison-Wesley)

KNOWLEDGE BASED SYSTEM FOR CONSTRUCTION DESIGN PROCESSES

C.Kant, J.McPhillimy, and G.Capper

Building Performance Centre,
Napier University, Edinburgh EH10 5AE

This paper presents an overview of the current decision making process employed by the UK flat roof construction industry and indicates possible improvements through the application of knowledge based systems software. This approach will enable the design process to be integrated and lead to overall improvement in design productivity. The decision making process is examined using the flat roofing industry to demonstrate the potential optimisation of design processes through employment of artificial intelligence techniques.

Introduction

Napier University's Building Performance Centre have recently completed a major research project, 'The Assessment of Lifespan Characteristics and As-Built Performance of Flat Roofing Systems'(McPhillimy *et al,* 1993). The project took three years to complete and included physical surveys of 680 large, flat roofed buildings (total surveyed area of 500,000 m²). The sample roofs were drawn from the building stocks of a number of Local and Regional Authorities in Scotland and England. The physical surveys were complemented by analyses of the available maintenance, cost records, specification data, and design information for the roofs and this enabled the features of different roof system types to be evaluated.

The project highlighted the possible application of artificial intelligence techniques to flat roof design methods. In particular, the research has shown that the current flat roof design methods are generally heuristic. The heuristic approach has worked well for a number of years but rapid changes in technology and working

practices are demanding a more structured approach to design in which the
application of artificial intelligence techniques offers certain advantages.

Construction Industry Design Processes

Construction projects are generally considered to be unique in nature with great
emphasis put on the 'old chestnut' triad of time, quality, and cost. The designer may
carry out the decision stage of the design process with little input from the building
client; the client being - in many cases - unclear as to their specific requirements for
individual building elements, or construction in general. Guidance as to alternatives
must be offered therefore from the outset. In addition, the construction industry has
a reputation for legal wrangling, aggression and conflict, in part due to it's
fragmented nature. A large portion of the design management process therefore, is
comprised of managing the 'people' associated with a construction contract. This
further reduces the time available to the designer for the design process itself.

Time constraints may lead to further problems placed on a designer, the
design process being essentially instinctive, the large number of parameters relating to
system components and the overall system itself may be neglected. It is therefore not
generally feasible for a designer to carry out an exhaustive iterative process,
examining each stage of the design, prior to construction of the element, in this case a
roof system.

Knowledge Acquisition in Design Processes

There is currently no single, generally accepted, technique for knowledge
acquisition within the construction industry. It is therefore at the discretion of a
system developer how he/she elicits the required information from the range of
sources available. In the case of the research project into flat roofing, three sources
from the knowledge data path are listed below; each will be described in turn.

- survey data,

- 'expert' interviews,

- literature based material.

Survey Data

Prior to the research project there was no major source of data available on the
in-situ performance of roofing systems in the UK. A study had been carried out

which gave some information on the performance of roofs in the USA (Cullen, 1989). A major component of the project therefore was the physical surveying of roofs. To ease accessibility and subsequent analyses the survey information was recorded on a computerised database which holds all relevant flat roof data.

Observations made during the physical roof surveys, and upon subsequent analysis of the recorded data, led to a number of conclusions with respect to the designs that are currently being utilised within the flat roofing industry. In some cases, design features appeared to be poorly thought out with respect to the overall design and lead to a potential reduction in the roof performance.

Expert Interviews

The primary source for knowledge elicitation must be the 'experts' themselves. The information may be obtained through direct interaction, with a series of in-depth, methodical interviews, once the experts have been identified. A number of previous studies have involved detailed questionnaires designed specifically for this purpose. This approach, however, has had little success.

The presentation of 'typical' design problems to designers is one method of eliciting the 'correct' design decision process but this technique is limited in that this process is language based and many design decisions are not easily described in these terms. Decisions are made quickly eliminating the tiresome process of reexamination of each procedural step to ensure optimisation during the design process. The problem of identifying 'experts' and 'designers' is further complicated by a move in construction towards design-and-build solutions. Whereas, traditionally a designer would be an architect, the introduction of technically more complex systems in all areas of construction has resulted in contractor led design processes. For this reason, in the research project, additional surveys were carried out on roofs in the presence of the contractors who constructed them, and end-users. They were able to provide additional information on maintenance, specification and design data.

Literature Based Material

This information may be readily accessible but often designers will be prevented from making full use of theses facilities due to time constraints.

There are a limited number of technical guides available, such as the PSA Technical Guide to Flat Roofing (Property Services Agency, 1987) and the Flat Roofing - Design and Good Practice guide (Arup Research and Development, 1993). These guides may be infrequently updated to take into account new materials and practices.

There are also substantial quantities of material in the form of manufacturers' technical manuals or, more precisely, product marketing information.   There are however litigation implications from following manufacturers' information without sufficient substantiation and/or examining systems already installed. The unique nature of construction may cause problems in this respect and this is exacerbated by the fact that product information changes rapidly and 'designers' cannot keep-up with the advances.

The above problems were all highlighted during the course of the project.

Expert System Shell for Design

A number of shells were investigated as to their suitability: the Leonardo Expert System shell was chosen as it offers an interactive user environment for the creation and consultation of knowledge bases. Amongst a wide range of facilities available the Leonardo shell offers the familiar; rule-based knowledge representation; forwards and backwards chaining inference mechanism; interfacing to external databases; and simple editing facilities.

For each roofing system the basic components are deck type, insulation, vapour control layer, waterproofing layers and surface protection. With each there are associated performance criteria. The incorporation of these performance criteria within a knowledge based system allows the designer to consider materials data that may otherwise be overlooked or be completely unknown when evaluating design alternatives.

The developed expert system is now at pilot stage, with the individual designers who were involved in the research project testing the prototype system using typical design examples. The designers are providing feedback on the development of the application and an evaluation of its suitability and usefulness in practice, but are continuing to follow traditional design practice with the system being used in parallel.

Further Work

The system is being developed further to include additional product information, revised technical guidelines and legislative requirements, particularly Building Regulations. Consideration is also being given to the effects of European Standards and practices.

References

Arup Research and Development 1993, *Flat Roofing - Design and Good Practice Guide,* (CIRIA/BFRC)

Cullen W. C. 1989, *Project Pinpoint,* National Roofing Contractors Association of America (NRCA)

McPhillimy J. , Allwinkle S. J. , Capper G. , and Kant C. 1993, *'The Assessment of Lifespan Characteristics and As-Built Performance of Flat Roofing Systems',* (Napier University, Edinburgh)

Property Services Agency 1987, *PSA Technical Guide to Flat Roofing,* (Property Services Agency, Department of the Environment, London)

Manufacturing processes and process control

AN APPROACH TO MANUFACTURABILITY ASSESSMENT FOR PRINTED CIRCUIT BOARD ASSEMBLY

A J Clark , P G Leaney and B G Samuel*

Loughborough University of Technology, Loughborough, Leics, LE11 3TU
*D2D Ltd. (an ICL company), Kidsgrove, Stoke-on-Trent, Staffs, ST7 1TL

This paper outlines the approach to the manufacturability assessment of printed circuit boards that is being developed at D2D (an ICL subsidiary and leading UK contract manufacturer) in conjunction with Loughborough University. As well as outlining the system the paper will discuss the application of manufacturability in the contract manufacturing environment and the issues involved. The final system has been defined and its individual modules are currently being developed with a view to implementation by the end of 1994.

Introduction

The constant consumer demand for higher performance electronic products has led to a market hungry for the latest and newest offerings. The companies which are capable of satisfying this need, with quality and timely products, will win market share, Reinertsen (1991). Successful companies will earn the revenue to re-invest in new technologies and stay in the race. The competition is no more fierce than in the global contract manufacturing market for printed circuit assembly (PCA). D2D, a leading UK player in this market, regard best in class quality and service as a key to success. In order to achieve this a manufacturability and testability tool set is being developed in conjunction with Loughborough University. This paper will deal mainly with the manufacturability tools being developed. The aims of the system are to reduce the cost of manufacture of products at D2D whilst improving the quality of these products and the perception of D2D as a best in class manufacturer in the world market. The system aims to be an expedient and demonstrable utility which addresses real needs in an industrial context. Further advantages are seen to address speed of response to a request for quote and in providing a manufacturability assessment service to customers.

Context

ICL and D2D

The approach being adopted for manufacturability assessment is significantly influenced by the industrial context in which the work is being pursued. Design to Distribution Ltd. (D2D) is a wholly owned subsidiary of ICL which, prior to 94, traded as

ICL Manufacturing Operations. ICL became a successful mainframe computer manufacturer dealing with relatively low volume high margin products. However the declining mainframe market of the late eighties led ICL to embark on a new business plan involving the individual ICL divisions becoming autonomous business groups targeting particular markets and showing individual profits, D'Arcy (1994). This meant that ICL Manufacturing Operations had to enter the contract manufacturing arena where product volumes are higher and margins lower. This market was already highly competitive and suffering from apparent over capacity and competition from the Far East, Williams, Conway and Whalley (1993). D2D has become successful in this market turning over $400 million in 1993. The company won the European Quality Prize for 1993.

Mechanical Manufacturability and PCA

Attempts to directly apply mechanical design for assembly (DFA) techniques, Leaney and Wittenberg (1992), are limited as they do not address the peculiarities of PCA. For instance a common mechanical DFA rule stresses the ease of vertical straight down insertions but insertions outside this category in PCA are so rare as to be negligible. The system being developed will make use of a mix of old mechanical and new PCA manufacturability rules. It is worth noting that most ECAD based manufacturability evaluations only address component proximity checks.

Manufacturability and Contract Manufacture

It can be argued that D2D, as a contract manufacturer with no direct product design responsibility, will have little impact upon designs and reap little benefit from design for manufacture (DFM). However there are advantages for the manufacturer in generating accurate quotes for new products and in improving customer perceptions of the overall service available. When tendering for a contract it is important to be fully aware of the manufacturing difficulties that may occur in production so that mutually beneficial quotes can be made. This will involve the designer getting a high quality product at the right price and the manufacturer making a profit on the contract. A good way of identifying these potential problems is by using manufacturability assessment tools. D2D has taken the view that supplying a complete best in class turn key service to PCB designers will give it a competitive edge in the market place. The DFM service will supply the information to customers necessary for them to reduce their costs and hence increase their perception of D2D. The DFM service is viewed as a competitive differentiator.

The Manufacturability System

The manufacturability system has been designed to make a first pass assessment based on a bill of materials (BOM), however a full analysis requires a BOM and part location information usually in the form of an IDF (Interface Definition File). The BOM simply contains a list of part numbers with quantities and ideally component packaging information. Firstly the data is converted to electronic form, if necessary, and entered onto an engineering data base for tight control of the tendering, product introduction and engineering change processes; see Obank, Leaney and Roberts (1993) as well as Standish, Jones and Sumpter (1993). This data base is used to recognise known parts. Each customer will use their own part numbering system and this will frequently reference the same supplier's part numbers so this cross-reference is held on the data base. Once secured on the database the data is passed to the manufacturability assessment system which consists of three distinct sections, the manufacturing route generation system, the quality modelling system (QMS) and the rule checker, each of which will be discussed below. Additionally to these systems a set of paper design for manufacture guidelines are available.

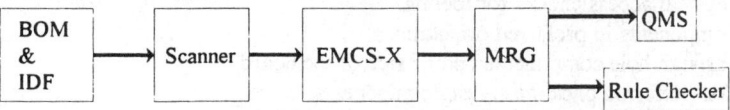

Figure 1 Manufacturability assessment system flow diagram.

The Manufacturing Route Generation System.

This is the first of the systems to run and its output is used to drive the rule checker and quality modelling system. As the complexity of PCB's has increased it has become less obvious how to manufacture them, for instance when a board has components on both sides which do you assembly first, or do you alternate sides. This problem is compounded by the increasing number and types of assembly and soldering machines available on the market and indeed on same manufacturing site such as at D2D. It can be argued that generic optimisation will optimise a board for production anywhere. However this is rarely the case as each process is tailored to deal with particular component packages. With the same electrical functionality being available in differing packages, package selection is therefore dependent upon the manufacturing processes to be used. The manufacturing route generation system has knowledge of all the available processes on site and can select the most appropriate route through them for any given PCB. D2D is currently developing a design to manufacture tool set, which include a process planning tool, see Roberts, Jones and Hunt (1993). The process planning tool will provide the support for the detailed downstream manufacturing engineering activities. However the manufacturing route generation system will be an effective subset that will use the existing factory model and provide the basis for manufacturability assessment early on in the product introduction process. Prior to the development of the full system the manufacturing route has been entered manually.

The Quality Model System (QMS)

The QMS estimates the assembly process yields. The data within the model consists of yield data for particular components assembled using a variety of processes. This data is populated primarily from historic assembly data for known components. For new components and processes the existing data can be extrapolated or new data gained from the manufacturer. The QMS·takes the BOM and the manufacturing route data for its inputs. It then outputs a parts list with each part's estimate yield shown (see table below). From this the low yield, high risk components can be identified and removed from the design if possible. The QMS model is currently being used in the form of a Microsoft EXCEL spread sheet.

Table 1. Example QMS output

Part Number	Quantity	Assembly Defects ppm	Risk
123	10	4000	High
456	1	84	Low
789	7	170	Medium
012	3	0	Low

The Rule Checker

The rule checker automates the checking of the design for compliance with some manufacturability guidelines particularly those requiring a great deal of manual effort. The

checks include the basic mechanical requirements for PCB manufacture, preferred
component types and preferred component locations/orientations e.g.

Physical access checks for tooling.
Components in preferred orientations.
Through hole components on one side of the board.
Non preferred package styles / components.
Components fitted on or over other components

It also calculates the surface area required by the top side and bottom side components to
ensure that components need to be fitted to both sides of the board. The details of the checks
applied are dependent on the manufacturing route selected e.g. the preferred orientations
check is only carried out for those processes where orientation effects the manufacture of a
board. The rule checker has two outputs, a count of the violations of each rule, and a
detailed list of components violating rules and the rules they violate which is used to target
components for redesign. The rule checker software specification has been written and
coding is due to start in quarter two 1994.

The Design Guidelines

These have been in existence for several years in the form of manufacturing standards.
The main difficulty with these standards is that they are process oriented i.e. they are
solderability or surface mount assembly standards for example. This means that at each
design stage the designer has to look through several standards to be certain that board he is
designing is manufacturable. A secondary difficulty with these standards is that they tend to
address the process limitations rather than best design practice. The new guidelines will be
oriented to the design process (i.e. they will contain sections such as board layout rules,
component selection rules) and address best design practice as well as process limitations.
The design guidelines are currently being collated and updated.

Progress

The design of the final system has been completed and documented for internal use
within D2D. Since the completion of the design a survey of software packages on the
market has been carried out to see if any of them matched the requirements for the system.
The closest match was package produced by Boothroyd Dewhurst Inc., however as no
further development was planned on this it was decided to develop in house tools.

As part of the system design exercise an evaluation of the possible cost savings from
the system was carried out. This took the form a board redesign exercise, the redesign
suggested fourteen design changes which together achieved the estimated benefits below.

Table 2. Example savings on a recent board redesign exercise

Estimated Benefit	Saving
Assembly cost reduction	33%
Assembly time reduction	35%
Assembly defects reduction	33%

The design changes included changing certain components to eliminate the need for jigs,
combining the functionality of multiple components into single packages, and reducing the
number of components requiring manual insertion. This information was passed back to the
designers. Subsequent designs from the customer have shown some of the suggested
changes.

The major issues around the implementation are concerned with the capture or availability of design data. Acquiring data in an electronic form that is understandable to our system has proved difficult, and commercial sensitivity of some of the data frequently makes the customer reluctant to part with it until a longer term relationship has been established. Hence the system has been designed to operate on a simple BOM and IDF. Even dealing with the BOM fully automatically when it is available electronically has proved difficult as different naming conventions are frequently used for the same component packages. In effect only where a component supplier cross reference has been built up can a BOM be dealt with automatically. When new customers are encountered the electronic data links usually take time to set up and hence much of the documentation is on paper, in order to deal with this all the tools have been designed with manual interfaces. However work is being carried out elsewhere within D2D to develop scanning and data manipulation tools to ease this task.

Conclusion

A computer based manufacturability assessment tool set is being developed for use by a contract manufacturer and to provide a service to customers. The tool set covers manufacturing route generation (through the production facilities of the contract manufacturer), quality assessment (the QMS model predicts yield) and a rule checker. The tool set compliments the generic DFM functions in ECAD systems by targeting the assessment to the current production facilities and the choice of manufacturing route through them. The possible cost reductions have been demonstrated to design centres and evidence is emerging that designs have changed accordingly. Improvement in quality is sought through the application of the QMS and through its further development. DFM (and design for testability) rules have been defined and their implementation is on going work.

References

D'Arcy, D. 1994, ICL hangs onto a precious commodity - profitability, *Electronics Weekly*, March 16, 1994, 12-13.

Leaney, P.G. and Wittenberg, G. 1992, Design for assembling, *Assy. Auto.*, **12**, 2, 8-17.

Obank, A.K., Leaney, P.G. and Roberts, S.M. 1993, Data management within a manufacturing organisation, Proceedings of the Conference on Managing Integrated Manufacturing Organisation, Strategy & Technology, September 22-24, 1993, 415-426.

Reinertsen, D. 1991, Outrunning the pack in faster product development, *Electronic Design*, January 10, 1991.

Roberts, S.M., Jones, R. and Hunt, I. 1993, The integration of design and manufacture - a quantum leap, *Integrated Manufacturing Systems*, **4**, 2, 1993.

Standish, R.K., Jones, R. and Sumpter, C.S. 1993, Shortening the new product introduction cycle through electronic data transfer, Proceedings of the Conference on Managing Integrated Manufacturing Organisation, Strategy & Technology, September 22-24, 1993, 305-318

Williams, D.J., Conway, P.P. and Whalley, D.C. 1993, Making circuits more than once: the manufacturing challenges of electronics intensive products, *Proceedings of the Institution of Mechanical Engineers - Part B*, **207**, 83-90.

OPTIMAL OBJECTIVE FUNCTIONS FOR AUTOMATED MANUFACTURING PROCESSES

Dr Roland Burns

Plymouth Teaching Company Centre
School of Manufacturing, Materials & Mechanical Engineering
University of Plymouth, Plymouth, UK.

Objective functions are used in a variety of automated manufacturing processes to obtain optimal performance characteristics. The paper reviews some areas where these are employed and gives an example of their use for geometric optimal control. It is concluded that linear control theory can provide a valuable insight into the performance of actual systems operating under an optimal policy.

Introduction

The concept of numerical control (NC) originated in the United States in the period 1947/48. The ability of machines to understand and process directly a numerical language, without human intervention, was made possible by the development of electronic pulse circuits and digital computers, together with servo mechanism design based on feedback principles. In 1953, a government contract was placed with the Massachusetts Institute of Technology (MIT) for a numerically controlled machine tool, and commercial versions were exhibited in Chicago during 1955. Computer numerical control (CNC) was first introduced by Alan Bradley in 1969 as reported by Barnett and Davies (1971).

The fundamental strategy for controlling axes of CNC machine tools has generally remained unchanged over the years. Digital position loops with analogue velocity feedback to provide the necessary damping and stability are still the main components of a CNC control loop.

The optimal control problem is concerned with finding a set of control signals that will cause the plant state variables to vary in such a manner as to minimise or maximise a selected objective or cost function McDonald (1950) first applied the concept to a control system. Deterministic optimal control theory was further advanced by Pontryagin (1959), Bellman (1959), and Kalman (1960).

Since this early work there has been considerable advances in the areas of adaptive, self-tuning and intelligent control, Harris and Billings (1985), Harris, Moore and Brown (1993), Zarrop (1994). A common feature of all these strategies is the definition of an objective function which the control system will maximise or minimise. The block diagram of such a system is shown in Figure 1.

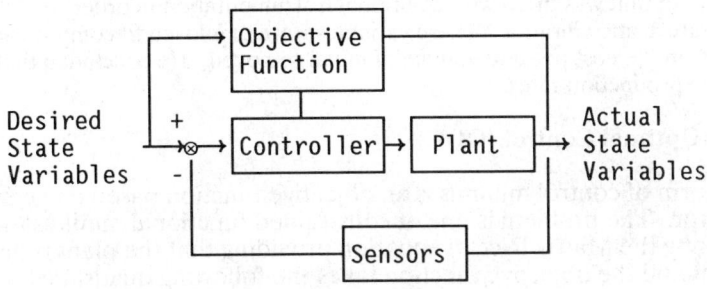

Figure 1. Optimal Control System

The Adaptive Control Problem

In general this is concerned with obtaining an optimal performance when there exists the added complication of varying plant parameters or environmental disturbances. Adaptive control systems have the following categories:

Model Reference Systems
In this case the responses of the actual plant and an ideal mathematical model are compared directly, when both are subjected to the same input as illustrated in Figure 2. An objective function is generated in terms of the difference between the responses which is then minimised so that the controller acts in such a manner as to drive the dynamics of the actual plant to be equal to the reference model.

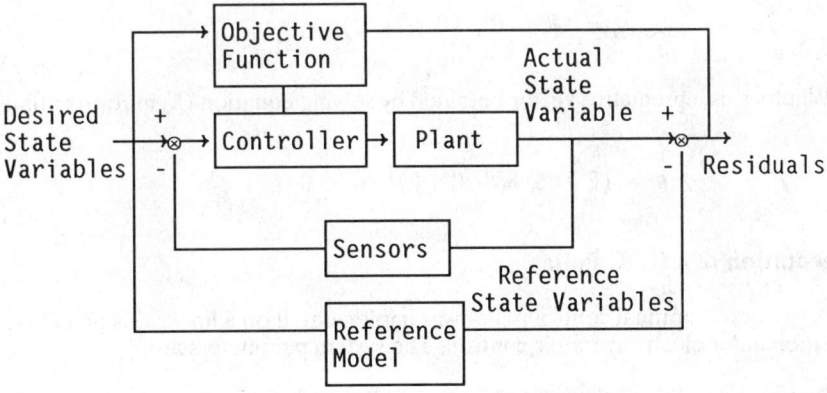

Figure 2. A Model Reference Control System

Self-Tuning Systems
These will normally include an identification facility to update, at frequent intervals the estimated process parameters of the mathematical model of the plant. An optimal control policy is then computed based on the most recent estimate.

Adaptive Control Optimisation (ACO)

This is a technique which minimises an objective function based on the production costs for each workpiece. Centner (1964) describes an early example of AOC. A standard Bendix Dyna-Path unit was fitted with additional instrumentation in order to sense cutting forces, temperature and vibration. These variables were employed to compute an objective function based on the cost per unit volume of metal removed. He concluded that the use of ACO increased production rates.

Geometrical Optimal Control (GOC)

This form of control minimises an objective function based on geometrical kinematic error. The problem is one of constrained functional minimisation which can be solved by the Matrix Riccati equation providing that the plant is linear and time invariant and the objective function takes the following quadratic form:

$$J = \int_{t_0}^{t_1} \left[(x - r)^T \; Q \; (x - r) + u^T \; R \; u \right] dt$$

(1)

The state equations for a linear time invariant plant are:

$$\dot{x} = F \; x + G \; u$$

(2)

Given that the steady-state Riccati equations are:

$$P \; F + F^T \; P + Q - P \; G \; R^{-1} \; G^T \; P = 0$$

(3)

The optimal solution becomes:

$$U_{opt} = -R^{-1} \; G^T \; (P \; x + s)$$

(4)

Where s is a predictive vector obtained by solving equation (5) in reverse time.

$$\dot{s} = (F - G \; R^{-1} \; G^T \; P)^T \; s - Q \; r$$

(5)

Implementation of a GOC Policy

A geometric optimal control policy was implemented on a linear axis of a CNC machine tool under electrohydraulic control. The system parameters are:

Summing amplifier gain = 0.547 dimensionless.
Forward-path attenuator = 0.231 dimensionless.
Drive amplifier gain = 101 mA per V.
Servo-valve gain = 43.8×10^{-6} m^3 per s mA.
Linear actuator effective area = 0.323×10^{-3} m^2.
Position transducer sensitivity = 42.1 V per m
Position attenuator = 0.4 dimensionless.

The open loop frequency response characteristics are shown in Figure 3.

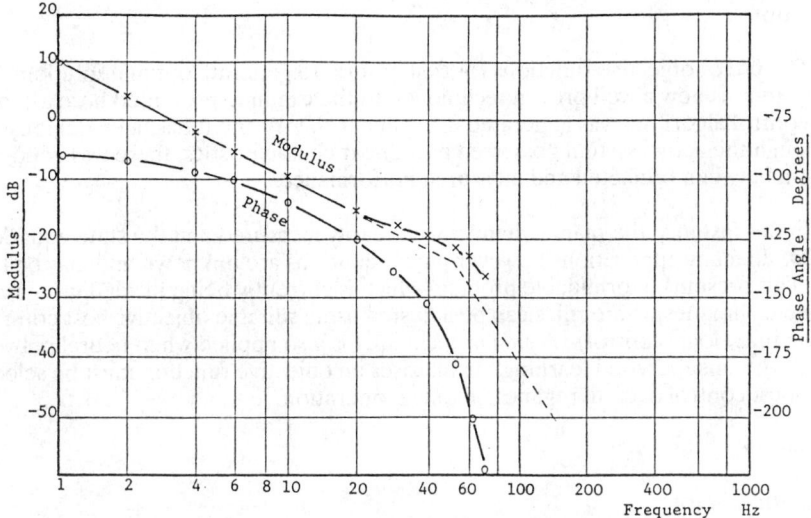

Figure 3. Open Loop Frequency Response Characteristics.

From Figure 3 the plant state equations were obtained.

$$
\begin{bmatrix} \dot{x}_1 \\ \dot{x}_2 \\ \dot{x}_3 \end{bmatrix} = \begin{bmatrix} 0 & 1 & 0 \\ 0 & 0 & 1 \\ -1105000 & -69300 & -372.5 \end{bmatrix} \begin{bmatrix} x_1 \\ x_2 \\ x_3 \end{bmatrix} + \begin{bmatrix} 0 \\ 0 \\ 655000 \end{bmatrix} u
$$

$$(6)$$

The weightings for Q and R in the objective function were adjusted and tested in simulation to give a satisfactory compromise between state error and controller action.

The simulated and actual response of the table are shown in Figure 4.

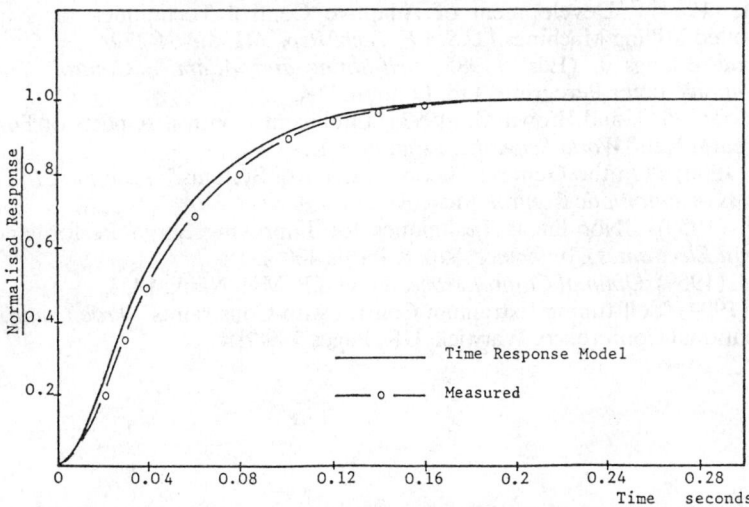

Conclusions

The use of objective functions for cost criteria to generate optimal solutions for control systems is now a well proven technique. In the example presented here a geometrical optimal control algorithm was generated using linear theory and it has been demonstrated that although the actual system possessed non-linear characteristics, there was good correlation between predicted and measured performance.

In this instance the plant dynamics were easily measured and the state equations computed. In many applications however plant equations are unknown and may be time variant. This presents a formidable problem which is currently being tackled using rule-based fuzzy logic techniques where rules can be adjusted using suitable objective cost criteria (self-organising fuzzy logic control). A similar technique is also applied when neural networks are employed with unsupervised learning. In all cases an objective function must be selected to ensure robust control over all regimes of system operation.

Nomenclature

Matrices and Vectors

F   Continuous time system matrix
G   Continuous time forcing matrix
P   Riccati matrix
Q   State error weighting matrix
R   Control weighting matrix
s   Reverse time predictive vector
r   Desired state vector
u   Control vector
x   State vector

Scalar Symbols

x_1 Table position
x_2 Table velocity
x_3 Table acceleration

References

Barnett C and Davis B (1971) "Computer Control of Machine Tools." *Paper 1, Symposium - Birnihill Inst. N.E.L.*

Bellman R.E. (1959) *Dynamic Programming* Princeton University Press, Princeton, N.J.

Centner R.M. (1964) "Development of Adaptive Control Techniques for Numerically Controlled Milling Machines." *U.S.A.F. Tech. Rep., ML-tdr-64-279.*

Harris C.J. and Billings J. (Eds) (1985) *Self-tuning and Adaptive Control: Theory and Applications* Peter Peregrinus Ltd, London, UK.

Harris C.J., Moore C.G. and Brown M. (1993) "Intelligent Control: Aspects of Fuzzy Logic and Neural Nets" *World Scientific,* London, UK.

Kalman R.E. (1960) "On the General Theory of Control Systems." *Proc First International Congress of Automatic Control,* Moscow.

McDonald D. (1950) "Non-linear Techniques for Improving Servo Performance." *Proc National Electronics Conference,* **Vol. 6,** Pages 400-421.

Pontryagin L.J. (1959) *Optimal Control Processes* U.S.P. Mat. Nauk. 14, 3.

Zarrop M.B. (1994) "Self-tuning Extremum Control with Constraints" *Proc Control 94* IEE International Conference, Warwick, UK. Pages 789-794.

ADAPTIVE LOGIC NETWORKS APPLICATION ON ASSEMBLY CELL CONTROL : AN INVESTIGATIVE STUDY

A.Bikos, Dr. R.F.O' Connor and Dr. G.B.Williams

*School of Manufacturing & Mechanical Engineering,
The University of Birmingham, B15 2TT, UK.*

This paper investigates the forecasting potential of high-speed, feed-forward Adaptive Logic Networks (ALNs) when applied to the performance of an assembly cell. Training data were produced by relating varying cell operational factors with key cell performance measures through the use of simulation modelling. The sensitivity of cell-performance forecasts using ALNs with respect to control strategies was tested on a synthetic part assembly using both a pull and a push system control strategy. Comparison of the efficiency of ALN forecasts with respect to cell size was also attempted by increasing the number of internal assembly processes. The findings reported in this paper are deemed to be of key importance towards the development of high-speed, reliable decision-making tools on assembly cell control.

Introduction

A very promising variation of Neural Networks (NNs) is that of Adaptive Logic Networks (ALNs). A thorough discussion on the fundamental principles of their operation is found in Armstrong and Gescei (1979). ALNs are based on binary trees and utilise four boolean operations (AND, OR, LEFT and RIGHT) in order to perform pattern recognition. They are extremely fast, compared to other types of NNs (Pao (1989), Beale and Jackson (1990)), thanks to a host of particularities, the most important of which are :
1) absence of real number mathematics
2) feed-forward processing
3) boolean function short-circuit evaluation properties
4) increasing node functions (tree monotonicity)

Output is obtained as a sequence of bits. Each bit is evaluated by a unique binary tree, the nodes of which implement primarily AND or OR boolean functions (LEFT and RIGHT are discarded once training has been completed). Input variables are transformed into sets of bits and the latter serve as the leaves of the tree. Input variables constitute the domain of the ALN mapping while output variables constitute the codomain. The version of ALNs used in this publication implements random walks in Hamming space in order to represent numerical inputs (and outputs) with sets of bits. The size of each set varies with the quantization levels of the variable it represents. Minimum bit set size is imposed by $\log_2$(number of quantization levels).

In many instances though, the minimum is far from optimum. This depends on the nature of the data to train the ALN. The more the bits assigned to codomain variables, the higher the chance of more refined output (at the expense of training time). Armstrong (1992) promises to eliminate random walks in Hamming space in a commercial version of his ALN software to be released. He also claims that this algorithm improvement will allow the application of ALNs in safety-critical application.

This study applied the current shareware version to performance measure pattern recognition in assembly and found the behaviour of the ALNs to be inconsistent as the size of the data examined increased. Quantization levels and random walk step sizes affected recognition performance. Training time varied considerably with quantization levels, coding width and step size.

Model Details

Two synthetic final assemblies, ASSY2 & ASSY3, were investigated in this paper. Their structure is presented in fig.1. The synthetic assembly cell involved consisted of three workstations, OP1, OP2, and OP3 producing assemblies ASSY1, ASSY2 and ASSY3 respectively. Workstations were connected via a conveyor capable of varying speed. Operation times were assumed equal.

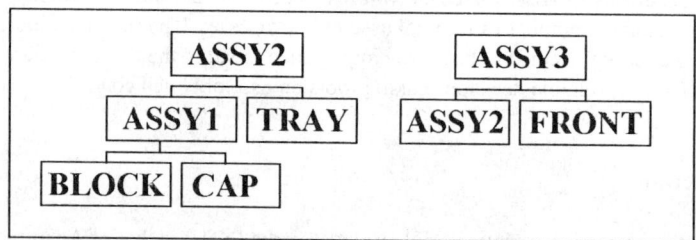

Figure 1. Assemblies ASSY2 and ASSY3

Four models were developed to accommodate capacity-constrained push assembly and two-bin pull assembly of ASSY2 and ASSY3 final assemblies. The models were prepared in ProModel-PC, a proprietary simulation software package, using EPL, an extension to ProModel-PC developed by Bikos (1993). These models were developed so as to allow monitoring of Production Volume over a forty-hour working week (PV), Lead Time of the first final assembly of a batch/kanban to exit the cell (LT1), Lead Time of the whole batch to exit the cell (ALT), Work In Process for all the part types involved in each assembly (WIP1 to WIP6) and, finally, the workstation Utilisations (U1 to U3).

For all four models a production rate of approximately 172 assemblies per hour was assumed. In this case, ignoring downtimes and transport times,

ALT = LT1 + (BS-1) P   (i)

where BS is the batch size and P is the production cycle time. For reasons of simplicity, no downtimes were incorporated in the model. As will be shown in later paragraphs ALNs were able to predict LT1 values quite successfully but failed to do so in the case of ALT values when the domain input variables were both batch size (BS) and conveyor speed (CS) and training data size was significantly higher than 100.

Experimental Designs

The ALN domain input variables chosen were BS and CS. In all four models BS varied between 1 and 100 parts and CS between 1 and 20 meters per minute. Consequently, the number of experimental designs used as the sample space for the various subsequent ALN training sessions was 2000 for each model. The codomain variables were the performance measures monitored (see table 1 below).

Table 1. Performance measures which served as ALN codomain
variables in this study

No of Workstations	Codomain Variables
2	PV, LT1, ALT, WIP1..WIP4, U1, U2
3	PV, LT1, ALT, WIP1..WIP6, U1..U3

The four sample spaces were used to create sub-samples in order to test how efficiently ALNs can forecast performance measurement of rather simple assembly cells. A common policy for the generation of sub-samples was followed for all four models. It is summarized in table 2.

Table 2. Experiment types: Domain variables, number of associated ALN files and
sample sizes for each one of the ALN files of each type

Experiment type	Domain variables	Number of ALN files	ALN file sample sizes
a	CS	5	20, 20, 20, 20, 20
b	BS	5	100, 100, 100, 100, 100
c	BS and CS	5	400, 500, 667, 1000, 2000

In all three experiment types, the test sets were identical with the training sets. Sample sizes varied in size and required very elaborate processing to allow proper syntax of both training and test sets in order for them to be incorporated in correctly formatted ALN files. As training time increased dramatically with training set size, and given that ALN forecast performance varies depending on quantization and coding characteristics, it was decided that the number of quantization levels, coding bits and coding step size would be determined as shown in table 3. In the case of the quantization levels of the codomain variables, a minimum of 100 levels was assumed. The first reason for this choice was the need to cater for output accuracy. The second reason was to work with common coding characteristics for the experiments of type **a** and **b**, the sample size of which was below

100. For type **c** experiments, the number of most codomain variable quantization levels was allowed to be equal to the number of different values detected in each sample, provided that there were more than 100 of them. This was done to force ALNs to their precision limits even though one hundred quantization levels would suffice for forecasting purposes in practice.

Table 3. Determination of quantization and coding characteristics of all domain and codomain variables

Variables	Quantization levels (q.l.)	Coding bits (c.b.)	Coding step size
BS	100	$\log 2(q.l.)+1$	*
CS	20	$\log 2(q.l.)+1$	*
PV	max(100,values)	$\log 2(q.l.)+1$	c.b.-1
LT1	max(100,values)	$\log 2(q.l.)+1$	c.b.-1
ALT	max(100,values)	$\log 2(q.l.)+1$	c.b.-1
WIP1..6	max(100,values)	$\log 2(q.l.)+1$	c.b.-1
U1..U3	100	$\log 2(q.l.)+1$	c.b.-1

* coding step size determination here requires taking into consideration various constraints simultaneously. The latter arise from the utilisation of random walks in Hamming space.

ALNs allow improvement of forecasting in a number of ways. An odd number of trees can be used to establish the value of an output bit by majority ("voting"). Armstrong (1992) mentions that the range of values can extend from 1 to 31, but claims that a value of 7 is used frequently with good results. This technique, though, increases processing time significantly as it multiplies the number of trees utilised for output bits by the number of votes. In the experiments carried out this 'voting' factor was set to 1.

Greater coding widths and step sizes for codomain variables offer higher precision, since more bits with equal significance are allowed to determine the quantization level of a codomain variable. Still, every extra coding bit adds one tree more, and this increases training and testing process times as well.

The positive effect of increasing the number of training epoches on forecast precision is lesser when the number of epoches is high, but it can prove decisive in some instances.

ALN performance

There was no significant difference in ALN forecasting accuracy for every experiment type across all models. This showed clearly that their forecasting accuracy is rather unaffected by both control strategies and shopfloor sizes.

Type **a** experiments (CS constant) showed perfect quantization level forecast accuracy for all codomain variables in all the samples tested.

Type **b** experiments (BS constant) achieved perfect quantization level forecast accuracy for all codomain variables in most of the samples tested. In no exception did the correctly forecast values drop below 97% of the test set.

Type **c** experiments (both BS and CS varied) achieved perfect quantization level forecast accuracy only for PV in all the samples tested. For the rest of the codomain variables, forecasting performance was varying. In particular, although the percentage of LT1 correctly-forecast values was always above 99 (a practically perfect response), the same quantity for ALT ranged between 0% and 30% throughout the type c samples trained and tested. Forecasting performance for the rest of the codomain variables ranged from poor to very satisfactory, depending on the number of different values present in the training sets. The more values per variable in the training set the worse the forecasting accuracy proved to be, especially when their number exceeded the order of 100.

Prediction of values was also attempted, using the same coding assumptions, by purposely omitting some data from the training sets but including them in the test sets. ALNs failed completely when training sizes increased while they performed modestly for smaller sizes.

Conclusions

In the cases where BS or CS were the single domain variables, ALNs performed very satisfactorily irrespective of control strategy or number of operations, even though they were operating stripped of practically all their forecast-enhancing features discussed above. Further runs utilising greater coding widths and step sizes produced zero quantization level errors.

In the case where both domain variables varied, though, ALN performance varied widely as training set sizes ranged between 400 and 2000 for all 4 models. The number of different values per codomain variable present in the training set affected outcome significantly. From the components of formula (i), LT1 is correctly forecast, BS is a domain variable and P is constant. Yet ALNs failed to forecast ALT values correctly and their accuracy dropped as training set sizes increased.

The sharp contrast between the performance of ALNs in type a and b experiments and their performance in type c experiments showed that ALNs can indeed perform extremely well when training sets are small in size. Training data monotonicity is important but for small training sets it is not critical (e.g. ALNs have been trained to produce $\sin(x)$ for $x = 0 .. 2\pi$). Thus, it is advisable to break training to smaller ranges for input variable values. When used in this modular manner, ALNs can be trained faster (smaller coding widths, single voting, fewer epoches) and can forecast far more accurately. With these conditions ALNs promise a safe platform for the development of decision making systems on shopfloor control.

References

Armstrong, W. and Gescei, J. 1979, Adaptation Algorithms for Binary Tree Networks, *IEEE Trans. on Systems, Man and Cybernetics*, **9**, 276-285.
Armstrong, W. 1992, Atree 2.7 for Windows shareware software package on-line help.
Beale, R and Jackson, T 1990, *Neural Computing: An Introduction*, (Adam Hilger,New York)
Bikos, A. 1993, Integrating Assembly Logistics and Product Design, M.Phil. qualifying thesis, School of Manufacturing and Mechanical Engineering, The University of Birmingham.
Pao, Y.H. 1989, *Adaptive Pattern Recognition and Neural Networks*. (Addison-Wesley, New York)

A MOTION DESIGN PACKAGE SUPPORTS THE USE OF A PROGRAMMABLE TRANSMISSION SYSTEM

Andrew Carrott and Dr. Phillip Moore

MSI Research Institute
Department of Manufacturing Engineering
Loughborough University of Technology

To improve performance and simplify construction, it is becoming
increasingly commonplace for modern manufacturing machines to be built
with each motion individually provided by its own motor. This trend adds a
new dimension to the control requirements and has been accompanied and
supported by the increasing use of more intelligent forms of motion control.
Rather than being cast in stone, or rather steel, 'software cam profiles' and
'software gearboxes' are programmable. This paper discusses the
requirement for single and multi-axis motion design tools to support the use
of programmable transmission systems. It describes how a proprietary PC
based motion design package, originally designed to support mechanism
design, has been successfully used to generate complex motion mappings
and motion profiles for a multi-axis machine at LUT.

The Programmable Transmission System

The Programmable Transmission System (PTS) from Quin Systems Ltd. first
became available in 1988, at which time it provided an innovation in motion control. A
PTS system can control from one to ten digital servo axes and comprises a G64 bus based
rack with a host CPU running the central command processing software, Mconsole, and
between one and ten DSC-1M single axis digital servo control cards (3U high, Eurocard
format) each having an on-board motion processor.

The Mconsole host software runs under the OS-9 operating system (Microware,
Des Moines, USA) and provides the user with two different interfaces. The first is a
terminal based command editor which, with its simple command language based on two
letter codes, enables a user to set up control parameters and program axes. Simple
commands may be combined to form complex command sequences, which may in turn
be built up to allow large systems to be programmed in a hierarchical fashion. The second
interface replaces the terminal with an area of common memory, accessible to both

Mconsole and a user's control programs. Control programs are written in the 'C' programming language and compiled for OS-9 using a Microware 'C' compiler. OS-9 'C' compilers are available to run under either OS-9, UNIX, or DOS allowing the option of developing application programs on either (SUN) workstations or PCs prior to downloading to the OS-9 system. User programs send commands to Mconsole by writing them to a particular address in the area of common memory. Each command is acknowledged by Mconsole writing a reply code to an adjacent memory location. All communication is event synchronised, forcing a user's application program to acknowledge receipt of a reply from Mconsole before the next command may be sent.

Move Commands

Simple trapezoidal move commands are supported by the Quin PTS, either to an absolute position or to a position relative to the current position. Also supported are continuous moves of constant velocity and unspecified duration, continuing until either a stop command is issued or a limit condition encountered (in either software or hardware). These simple move commands are programmed by specifying acceleration/deceleration, maximum velocity and (if appropriate) final or relative position.

Complex or non-standard move commands require a far greater amount of data for their specification. The PTS provides facility to program two different types of complex motion; motion profiles and motion mappings. Motion profiles specify motion on a single axis in a position against time format. Motion mappings specify synchronised motion on two or more axes by providing the means to program continuously variable gear ratios in the format of slave axis position against master axis position.

Both motion types are programmed by inputting position data at equal increments of either time (motion profiles) or master axis position (motion mappings). By programming against a constantly incrementing reference parameter, which is entered as the first step of programming the motion, only position data for the programmed axis need be entered (see Figure 1a). As an aside it is interesting to contrast this with the alternative programming method, which would be to have a variably increasing reference parameter and specify each position as a (reference parameter, axis position) pair (see Figure 1b). This would be a more efficient technique in situations where the motion is simple to describe, particularly periods of constant velocity in motion profiles or periods of constant gear ratio in motion mappings, which would correspond to straight line portions of the motion graph.

The specification of complex motions with adequate accuracy can require entering large amounts of position data. With each piece of data taking up two bytes of memory and each DSC-1M having approximately 30KBytes of memory available for storage of motion mappings and profiles, there is a clear requirement for additional software tools to assist with the generation, maintenance and input of motion data. Quin Systems have recognised this need and in the past have used a Lotus 1-2-3 spreadsheet package. A proprietary motion design system has now been used at LUT to specify motions and generate data tables for input into the PTS.

The Motion Design System

The motion design system used was "MOTION" supplied by Limacon. This is a PC based package implemented under a windowing environment. It enables motions to

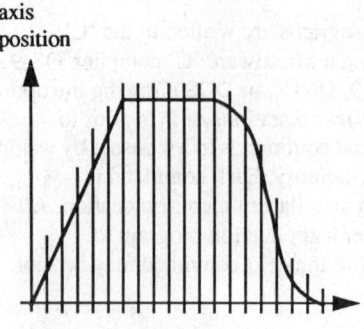

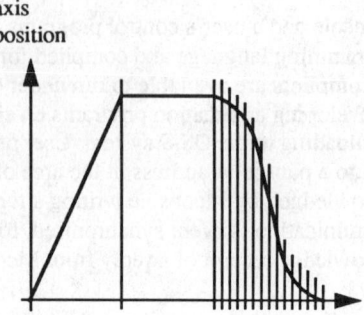

Figure 1a. Programming against a
constantly incrementing reference
parameter

Figure 1b. Programming against a
variably increasing reference
parameter

be constructed from segments of different types, the types available being dwell, throw,
ramp, cycloidal, modified trapezoidal, modified sine, polynomial, triple harmonic,
sinusoidal or user table. The desired motion is incrementally constructed, a segment at a
time, with the package displaying the motion in graphical form. An example of a simple
motion built up from three segments; a ramp, a dwell and a cycloidal is shown in Figure
2.

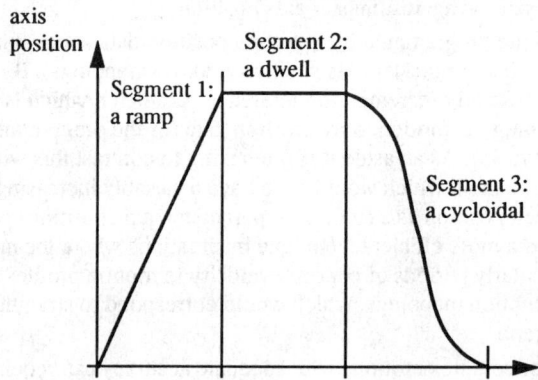

Figure 2. An example of a simple motion built up from three segments

A motion design may be saved as a data file for subsequent retrieval and editing or
may also be saved as a data table for use by other systems. Data tables are in ASCII text
format and may therefore be readily interrogated by other software. They contain simple
lists of motion variables: input and output displacement, velocity and acceleration in four
columns respectively. The quantity of data in a data table is determined by the number of
steps specified in the *Edit/Settings* screen of MOTION. This setting effectively sets the
input displacement interval for which values of output displacement, velocity and

acceleration are provided over the range of the input displacement.

For the purposes of programming the PTS, the only data column of interest is the one containing output displacements. The input displacement data is not required because it is known to begin at zero and increment at fixed interval. The velocity and acceleration data, whilst being of interest during motion design, is not needed to program the PTS. An interface program has been written which reads output displacement values from a data table and enters them into Mconsole, from whence they are downloaded to a specified DSC-1M as either a motion mapping or profile. The interface program enters the data into Mconsole through the area of common memory, as described previously. Note that motion mappings and profiles are downloaded to the DSC-1M motion controllers before run-time to reduce backplane traffic.

The interface program was implemented under the UMC software environment. UMC is a methodology for creating machine control systems which is appropriate to a wide range of different types of manufacturing machines (Weston, Harrison, Booth and Moore, 1989). Software includes utilities, standard function libraries and a run-time software environment: these being based on standardised UMC data structures and standard mechanisms and protocols for interprocess co-ordination and communication.

The LUT Test Rig

A multi-axis test machine was available to test the motion mappings and profiles generated using MOTION. The machine comprised three digital servo motors each turning a marked disk to give an indication of motor position. Each motor was controlled by a Quin DSC-1M motion control card, with Mconsole and the user software running on a Motorola 68020 CPU. The motors were each fitted with an encoder which fed back 1000 counts per revolution.

Programming Considerations

To program a motion mapping it is necessary to specify the required position of the slave motor with respect to the master motor, for all possible positions of the master. From a programming point of view, 1000 encoder counts corresponds to 1000 possible positions and so it was necessary to generate output position data across an input range of 1000. This was readily achieved in MOTION, the input and output changes of each motion segment being simple to enter or modify. Note that when modelling motion mappings the x-axis was treated as having units of encoder counts and when modelling motion profiles the x-axis was treated as having units of time. All output positions were programmed in encoder counts.

The version of MOTION used for the experimentation was limited to producing data tables with a maximum of 360 output positions. If it had been possible to produce data tables containing 1000 output positions then this would have been done. Setting the number of steps specified in the *Edit/Settings* screen to 250 led to a data table being produced which contained 250 output positions at an input interval of four encoder counts. This was acceptable to Mconsole, which allows an input step size other than one to be used when programming motion mappings. The input step size must be an integer value and in this case it was four. When a motion mapping programmed with a step size other than one is downloaded to a DSC-1M, the system interpolates between the values entered to generate all the intermediate motion mapping data values for the slave motor.

Since the data tables contained data saved to seven decimal places and Mconsole is programmed in encoder counts, the user software which read the output positions from any data table into Mconsole had to round each value to the nearest whole number. This would not cause any significant error in a practical system where the resolution, i.e. the motor motion between successive encoder counts, should be an order of magnitude smaller than the required accuracy.

When programming a synchronised motion on more than two axes it was found that only one master axis should be used, with all other axes being mapped to that axis. The alternative scheme would be to have some of the axes acting as both slave and master. Consideration must be given as to whether the motion mapping uses the demand position of the master axis or its actual position. If the motion mapping uses the actual position of the master then using multiple master axes leads to an accumulation of 'electronic backlash'. Electronic backlash is here used to indicate the presence of following error and deadband.

Conclusions

This paper has discussed the requirement for new software tools to assist with the data intensive task of programming software cams and software gearboxes. It has been shown how a proprietary motion design package can be used for this purpose, enabling motions to be constructed from different motion segments conforming to particular mathematical functions. A test machine at LUT, controlled by a Quin Systems PTS, has been successfully programmed and run using this technique.

References

Anon., 1988, Transmission goes digital, *Industrial Technology*, September 1988.

Anon., 1988, Software gearbox replaces mechanical mechanisms, *OEM Design*, July 1988.

Anon., 1988, Software gearbox controls mechanical linkages without physical contact, *Design Engineering*, November 1988.

Dayan, P.S., 1992, *The OS-9 Guru*. Published by Galactic Industrial Ltd., Mountjoy Research Centre, Stockton Road, Durham, DH1 3UR, UK.

Kernighan, B.K. and Ritchie, D.M., 1978, *The C Programming Language*, Prentice-Hall.

Lambe, J., 1992, *Programmable Transmission System Reference Manual*, Rev 9. Supplied by Quin Systems Ltd., Oaklands Business Park, Wokingham, Berkshire, RG11 2FD, UK.

Rooney, G.T., 1992, *Camlinks and Motion*, User Manual supplied by Limacon, Meadow Farm, Horton, Malpas, Cheshire, SY14 7EU, UK.

Weston, R.H., Harrison,R., Booth, A.H., and Moore, P.R., 1989, A new concept in machine control, Computer Integrated Manufacturing Systems, Vol 2 No 2, May 1989. Butterworth & Co. (Publishers) Ltd.

VIBRATIONAL CHARACTERISTICS AS A CONTROL PARAMETER DURING PISTON AND RING INSERTION

Ian McAndrew and Dr James O'Sullivan

University of Hertfordshire
Hatfield, Herts, AL10 9AB

Automating processes does not always improve the quality of
the product unless all the manual methods are controlled by
the replacement equipment. In this paper the problems relating to
the automatic insertion of piston and ring assemblies into a cylinder
block are discussed. The inability to detect missing or breaking rings
was investigated. It can be shown that measuring the vibrational
characteristics of the rings during their insertion process produces
signals sufficiently repeatable to be used as a control parameter for
detecting rejects.

Introduction

Modern electronic technology has enabled machines to become fully
automated and undertake complex tasks which were not previously considered.
This has allowed manufacturing engineers to re-think the processes they use and
decide where automation is now possible. In the quest to automate feasibility has
become a secondary facet; manual processes are viewed as archaic and the sign of
a backward looking engineer.

The process of piston and ring assembly into a cylinder block has traditionally
been manual. In the late 1980s the price of the technology to automate became
realistic and naturally machines were designed and employed based on this point. It
was when these machines were in full production that problems first occurred.

Operators did more than manually push the piston and ring assemblies
through a tapered cone; they indirectly inspected their own work. Missing rings
were detected prior to insertion and when a ring broke or jammed in the cylinder
their dexterity could also insure it was identified for repair. These detection

557

techniques were not incorporated in the machines; the physical action of handling the piston and ring assemblies was only one small part of the operators function. Qualitative methods, FMEAs, may assist in recognising this at an early stage but it is not a guarantee, McAndrew and O'Sullivan (1993). The addition of force transducers only detected extreme interferences in the insertion if the assemblies were jammed. The maximum force setting had to allow for the friction in the tapered cone and bore, which can be close to the forces produced when problems arise.

These failure modes are not prominent, less than 1% will have any problem. However, this low failure rate also makes it difficult for measuring improvements to the process. Long term, the effects have to be tackled and quantitative measurements made to identify all the problems that an operator will detect.

Measurement Problem

The two principal problems can be classified as missing rings and breakage of rings. The severity of both is the same; cylinder pressure will reduce and oil consumption increase. Techniques are commercially available to measure these two types of problems.

Vision systems are sensitive enough to look for the presence of parts; but this is only half of the problem and their cost is prohibitive for the number of parts in an engine set (20 rings on a standard four cylinder engine). Sound detection, in principle, may identify the difference under laboratory conditions. In the middle of a car plant where the machine could be producing a background noise of 85 dB it is unlikely, Whalley & Hemmingway (1989).

A device or system is required that will cover both of these principal areas. It must also be reliable and economical in any complexity that it adds to the controls of the equipment.

Vibrational Experiments

To insert piston and ring assemblies the ring pack has to be compressed to a smaller diameter than the cylinder bore. On entering the bore it is free to expand. The time it takes to naturally expand to the new diameter will be proportional to the stiffness of the ring. Moreover, it will characteristically vibrate; its material properties and geometry produces a type of circular spring.

If the characteristics of the rings are to be a control parameter it is necessary to know their repeatability accuracy. Measuring their natural frequency showed that variations were less than 5% with an average value of 22.5 Hz, and their damping actions were also consistent. Establishing that they have similar natural frequencies indicates that behavioural patterns during assembly will also be reproduced for individual situations. This is as much to do with the reduction in variations of the manufacturing of rings as the design requirements to enable a guaranteed operational performance.

When the rings enter the cylinder bore they will expand in the horizontal plane, see figure 1; therefore the measurement transducer must be placed in this plane to detect the signal required, Bland, Hopkins & Drazan (1989). This constraint limits the possible positions allowed : inside the piston skirt and outside of the cylinder block. There are minor practical difficulties to having the sensor inside the piston during an experiment. For an automated machine this is not a practical solution: the restriction of access for the machine controls and internal components.

Ideally a machined face is the easiest feature for a transducer but these are not always in a close approximation to the cylinder wall. By recording the response with the transducer mounted in different locations to the bore it was apparent that the cast iron block would transmit a clear signal providing it was reasonably close. Although, it would also require being close to the cylinder head joint face where the piston and ring assemblies are inserted.

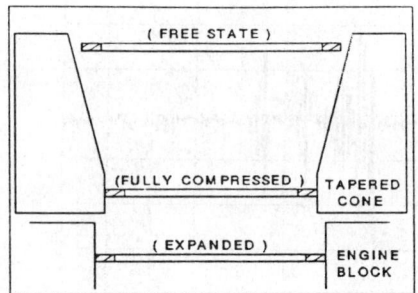

Figure 1. Expansion of rings in a horizontal plane.

Initial trials showed that the speed of the insertion process also affected the recording. A fast insertion speed resulted in all the rings expanding into the bore at a rate which did not allow for the damping actions to subside prior to the next ring, which meant their response signals overlapped. To separate the signal from each other the insertion speed was reduced from 1ms⁻¹ to below 0.5ms⁻¹. Slower insertion speeds separated the overlap and more detail of the response signal could be obtained. Reducing the speed to below 0.1ms⁻¹ allowed the complete damping signal to be shown. However, this speed is too slow to maintain an adequate cycle time for production.

The three piece oil control ring is a combination of two steel scraper rings and a central expander ring to position the scrapers and exert a high force on the cylinder bore, Grove and Davis (1992). This type of ring does not behave in the same manner as the two compression rings and its response signal is heavily influenced by the friction it exerts on the cylinder wall, Davis, Stout & Sullivan (1988). To balance the signal required for this ring and maintain an adequate speed 0.3ms⁻¹ was used for the remainder of the experiments.

Analysis of Results

Using these conditions a typical trace can be seen in Figure 2. The response on the left hand side is that of the oil control ring, the other two represent the lower and upper compression rings. Each of the three clusters has a peak value which damps down in a manner that is consistent to their behavior in a free state. The height of the peak values remains consistent and indicates the ring is present and is not broken; otherwise it would not be possible for the ring to vibrate against the cylinder wall. These peak signals are critical in determining whether a good assembly has been achieved.

Broken rings as a result of the insertion process fail not in the stage when they are free to expand in the cylinder bore, but as a result of over-stressing to insert, see Figure 3. Therefore, the response signal is the same as a missing ring.

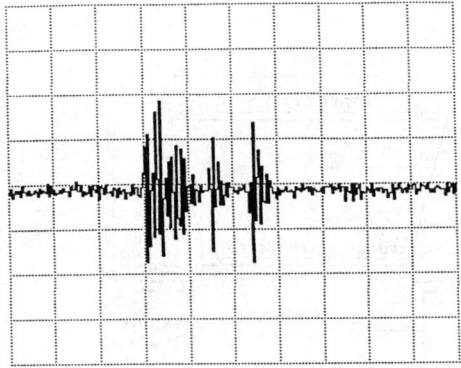

Figure 2. Signal of all three rings during insertion.

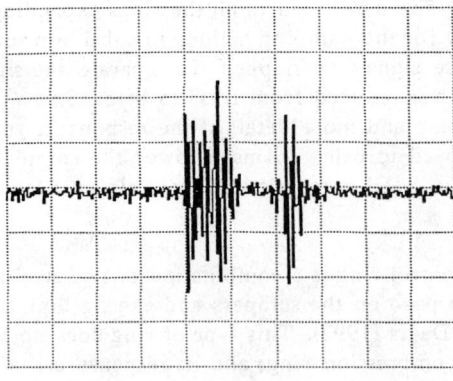

Figure 3. Broken lower compression ring.

Controlling the speed of the insertion and time that the first ring enters the cylinder bore provides sufficient information to identify a successful assembly. The peak signals in defined time bands are the indications of this, if they cross match with each other then it is sufficient to assume the assembly is successful.

It may be that there are certain parameters that will produce signals of a different response to those discussed, this may be for a variety of reasons. The likelihood of other reasons would be low as the manufacturing processes are concurred as having Cpk's of greater than 1.33. In any case it would identify an assembly as reject that would require no action and its interference negligible in terms of cost and time.

Conclusions

Automation is only the answer if all, and possibly more, of the manual tasks can be performed to a high degree of accuracy. For piston and ring assemblies this requires a full analysis of the process. The consistency of the behaviour of rings in their free state and during insertion allows the response to be used as a method for determining if successful assembly has been achieved. Knowing what this should resemble, it is feasible to control the insertion speed and develop windows that all rings must reach to concur a correct assembly.

References

Bland, C. Hopkins, S. & Drazan, P. (Feb 1989), The development of Autonomous devices to aid assembly, *Int Journal of Prod Research*.

Davis, E.J, Stout, K.J. & Sullivan, P.J. 1988. The application of 3-D topography to engine bore surfaces, *Surface Topography*, **1**, p253-267.

Grove, D.M. & Davis, T.P. 1992. Engineering Quality & Experimental Design, (Longman), p267-272.

McAndrew, I.R. & O'Sullivan, J.M. 1993. FMEA A managers handbook, (Technical Communications (Publishing)Ltd.), p 48.

Whalley, G.N. & Hemmingway, N.G. 1989, An application of noise analysis to quality control, *Journal of Applied Acoustics*, **28**, No.2.

IMPROVING SCREEN PRINTING PROCESS CONTROL

M.C.Mitchell, M. Howarth, C.R.Gentle

Manufacturing Automation Research Group
Department of Manufacturing Engineering
Faculty of Engineering & Computing
The Nottingham Trent University
Nottingham, NG1 4BU

Despite a recent increase in research into screen printing, the process is still heavily reliant on skilled operators. This paper examines the need for improved process control within the electronics manufacturing environment, where predictable accuracy during printing is essential. Previous work at The Nottingham Trent University enhanced a Svecia screen printing machine using a PLC. Existing studies of squeegee variables are considered, with an emphasis on the importance of squeegee pressure interaction with respect to other machine controllable parameters. The work experimentally tests a system which can control system performance using pressure. Initial results indicate that variations in pressure significantly affect printed deposits, due to their highly interactive nature with other machine controllable variables.

Introduction

If a manufacturing process is to be successfully employed it must be fully controllable. In the case of screen printing where so many parameters are involved, it is hardly surprising that comprehensive control solutions are non-existent.

The ideal situation for the manufacturer would be to have a control system whereby all the in-process constant parameters, as defined by Fulford, Howarth, Robinson and Lowe (1991), and the desired deposit thickness are entered into a system and then the required machine parameters would be automatically set by the machines's control system. Prior to achieving this, it is necessary to have full computer control and suitable models of the parameter relationships.

Screen Print Process Automation

Only top of the range machines which are commercially available, allow facilities for setting the four main machine parameters automatically. The Svecia screen printer at The Nottingham Trent University (NTU) has also been developed to provide these facilities. The work was begun by Fulford et al (1991) and the process is controlled by a PLC. As described by Fulford et al (1991), the squeegee speed and position are controlled by a proximity encoder, whilst the snap-height is monitored using a linear placement transducer. Currently modifications are being made to the squeegee angle control which includes the use of a stepper motor to enhance the positional accuracy and provide remote control.

The method of control employed for the angle, speed and snap height is adequate for the process requirements. However, the squeegee pressure, controlled using an electro-pneumatic controller (E.P.C) only provides information relating to the pressure applied to the squeegee and does not give an adequate indication as to the pressure created in the ink. It is known that pressure in the ink is a function of the mesh resistance, ink rheology and squeegee deformation as defined by Rangchi, Huner & Ajmera (1986) but thus far has never been measurable.

Parameter relationships

To allow the process to be controlled, the relationship between the parameters must be known.

Several models of the process have been proposed, as summarised by Mannan, Ekere, Lo & Ismail (1993). However, the respective authors point out that the limitations of the models prohibit their practical use, as vital elements in the equations are unknown. For example, Huner (1990) relies on the ink viscosity and the hydrostatic pressure in the ink being known. Ink viscosity is highly susceptible to environmental influences and changes are difficult to monitor (even if they were adequately measured in the first place). Hydrostatic pressure in the ink is also a function of other parameters and to date has not been measurable during the screen printing process. The NTU is currently investigating a method of determining the pressure distribution at the substrate. This information could be utilised in conjunction with existing models, enabling their practical application.

It is necessary to experimentally evaluate the parameters to complement the development of the process models. Work in this area, previously carried out by Brown (1986), Molamphy (1991) and Parikh, Quilty & Gardiner (1991) has concluded that experimental work could greatly assist the engineer to set up a reliable process. However the work is time consuming, does not provide a general process solution and to date extensive results have not been published.

At NTU the significance of small changes in the machine parameter settings has been investigated. Two values for each parameter were chosen based on recommendations for the particular ink viscosity used. The difference between settings is shown in Table 1.

Using ANOVA (Analysis of Variation) the highly interactive nature of the parameters was confirmed. The most significant contributions were the three factor interaction of squeegee speed, squeegee pressure and squeegee angle at 32% and the four factor interaction, including the snap-height, at 21%. The ANOVA response

Table 1. Parameter setting variations

Parameter	Setting variation
Squeegee speed	50mm/s
Squeegee pressure	8kPa
Squeegee angle	10°
Snap-height	1mm

values are shown in Table 2. The highly dependent nature of the variables indicate that significant deposit variations may be expected. This was confirmed by the results as the experiment yielded an average deposit thickness of $34.74\mu m \pm 15\%$.

Table 2. ANOVA response values

Source	Pool	Df	S	V	F	S'	rho%
A	[N]	1	5.17562	5.17562	17.36086	4.87750	3.2
B	[Y]	1	0.95062	0.95062			
AXB	[Y]	1	0.10562	0.10562			
C	[N]	1	10.72563	10.72563	35.97756	10.42751	6.9
AXC	[N]	1	3.70562	3.70562	12.42996	3.40750	2.2
BXC	[N]	1	6.12563	6.12563	20.54753	5.82751	3.8
AXBXC	[Y]	1	0.10562	0.10562			
D	[N]	1	11.05562	11.05562	37.08446	10.75750	7.1
DXA	[N]	1	9.15063	9.15063	30.69445	8.85251	5.8
BXD	[N]	1	1.15563	1.15563	3.87639	0.85751	0.5
AXBXD	[N]	1	48.65062	48.65062	163.19140	48.35250	32.0
CXD	[Y]	1	0.03063	0.03063			
AXCXD	[N]	1	14.63063	14.63063	49.07631	14.33251	9.5
BXCXD	[N]	1	5.88063	5.88063	19.72571	5.58251	3.7
ABCD	[N]	1	33.35062	33.35062	111.86978	33.05250	21.9
e1	[Y]	0	0.00000				
e2	[Y]	0	0.00000				
(e)		4	1.19249	0.29812		4.47181	2.97
Total (Raw)	[-]	15	150.79937	10.05329			

The parameter dependency is a major consideration to the manufacturing engineer. Variations of $\pm 15\%$ are unacceptable in critical conductive applications i.e. printed circuit boards.

The experimental results are in agreement with Frescka (1986) who argues that mesh geometry alone cannot be used to determine the wet deposit thickness.

Small machine parameter adjustments cause significant variations in resulting deposits and, therefore, as the process changes throughout the print cycle, mainly due to ink behaviour, it can be seen that fine adjustment facilities are essential. The machine at NTU has been developed to allow such accurate adjustments, which are controlled remotely. The status of the pressure, snap-height and squeegee speed can be interrogated and modified using a PC. However, to optimise the process of accommodating changes in ink behaviour, further work regarding the affects of the machine parameters on ink behaviour is necessary. In the absence of appropriate models, data collected from experimental work could be used to provide a knowledge based system.

Toward an Intelligent Screen Printing Process

In the short term, the most interesting aspects to be investigated are the squeegee pressure and related changes in the ink viscosity. A major shortcoming of the experimentation described above is that the pressure and angle are only known in terms of the machine values. It has been shown that the effective squeegee attack angle is not the same as the set squeegee angle, Nickel (1993) and it is quite clear that the pressure in the ink is a function of the squeegee pressure, effective attack angle, mesh tension and snap height. Further experimentation will be carried out focusing on these two points, whilst closely monitoring the other machine parameters.

Once reliable models or suitable information systems have been built up the control of the process could be automated. Fulford et al (1991) described the use of an expert system, working in conjunction with a vision inspection system to provide feedback, allowing the screen printing machine to reset itself eliminating the need for the man machine interaction. Another viable option is the intelligent control system proposed by Chandraker, West and Williams (1990) for adhesive dispensing as the processes are so closely related. Alternatively, it may be necessary to look at a system capable of "tweaking" the process just as an operator would, given a specified defective output. A fault classification method would be an essential component. The system should however accommodate expansion to cater for non-machine controllable variables.

Conclusions

The experiment described in this paper confirms previous work highlighting parameter interdependency and the development machine at NTU has been modified to accommodate the small changes necessary. Further experimental work will allow development of an information system which could be incorporated into a useful process monitoring system. However, additional experimentation is required to determine the optimum setting for the process.

It is clear that the screen printing process requires significant improvements in terms of process models and machine development. It is suggested that the way forward is first to examine relationships regarding pressure and angle. Then a control system can be developed, linked to a vision system, which will alter the four machine parameters based on the printed deposit output. The ultimate goal of achieving predictable and repeatable process outcomes would then be attainable.

References

Brown,D.O., 1986. Screen Printing - An Integrated System *International Society for Hybrid Microelectronics*, Reston, Virginia. 582-590

Chandraker,R.,West,A.A., and Williams,D.J.,1990. Intelligent Control of Adhesive Dispensing. *International Journal of Computer Integrated Manufacturing,* 3(1) 24-34

Frescka,T.,1986. Fabric Selection for Screen Imaging. *SITE Magazine* February

Fulford,D.,Howarth,M.,Robinson,M. and Lowe,J.M, 1991. Automating the Screen Print Process, for High Accuracy, High Volume, Printed Circuit Production. *Seventh National Conference on Production Research*, Hatfield 3-5 September

Huner,B.,1990. Effects of In-Plane Permeability in Printing Screens. *The International Journal for Hybrid Microelectronics* 13(2) 35-40

Mannan,S.H.,Ekere,N.N.,Lo,E.K,Ismail,I.,1993. Application of Ink Screening Models to Solder Paste Printing in SMT assembly. *Journal of Electronics Manufacturing* 3, 113-120

Molamphy,T.A.,1991. Application of Experimental Design to the Solder Paste Screen Printing Process. *Surface Mount Int. Conf and Exposition Proc. of the Technical Program* 1991 1, 496-502

Nickel,J.,1993. The Way to Measure the Effective-Squeegee-Attack-Angle and its Significance for the Ink Stream. *Screen Process* February, 25-27

Parikh,M.R.,Quilty,W.F.,Gardiner,K.M.,1991. SPC and Setup Analysis for Screen Printed Thick Films. *IEEE Transactions on Components, Hybrids, and Manufacturing Technology* 14(3) 493-498

THE IMPLEMENTATION OF A DESIGN FOR TESTABILITY RULE CHECKER

*A White, *B Sutton, **C Sumpter and **R Jones

*Design to Distribution Ltd, West Avenue, Kidsgrove.
**Department of Manufacturing Engineering,
Loughborough University of Technology

Over the last decade testing of electronic products has been becoming proportionally more expensive, until in some cases it represents one third of the total manufacturing cost of a product. In order to minimise test times and costs Design for Test rules and methodologies have been devised. These normally have been in the form of "paper rules" that the designers use as checklists during the design process. This paper details the design and implementation of an electronic system for checking a design of a product against these rules.

Introduction

With advances in technology and miniaturisation of electronic components and circuit design, product validation (testing) has become an increasingly difficult part of the manufacturing process. In an ideal world manufacture should strive for a right first time policy with no need for product testing. Unfortunately rapid advances in design and manufacture in electronics means that leading edge technology is consistently in a state of verification and testing is a necessary fact of life. The role of testing has become an important part of the manufacturing process due to miniaturisation and complexity.

These factors have implications on the cost of test, making it proportionally far more expensive to test high technology boards than those using lower or existing technologies. The higher cost comes from use of more expensive test machines that need to operate faster and with more functionality, the need for complex diagnostic aids using more experienced engineers to find the fault and higher scrap rates due to boards that cannot be diagnosed or reworked.

It is not only the cost of test that is affected by these changes, the new product introduction time can be increased which significantly affects profitability. This is due not only to the higher complexity of the test process taking more time to implement but also due to more iterations in the design cycle as designers produces boards that cannot possibly be tested to an acceptable degree of quality.

To overcome these problems more emphasis has to be put on the design of the board which must be optimised to incorporate features that will allow the board to be more effectively and less expensively tested. This process is known as Design For Test (DFT).

Design for Test

For designers to be able to incorporate DFT into their designs they must understand the implications of testing and more specifically what DFT considerations are important to the manufacturing plant where they intend to have the boards manufactured. For DFT to work effectively it needs to be considered from the conception of the design and at every stage thereafter. Trying to apply DFT at the introduction of the product to manufacturing often results in design iterations that are expensive and time delays to market, generally it is not cost effective to change designs at this point or the designer will refuse to change his design as he/she has had problems getting it to work in the first place.

This paper considers the need and development of a design for test strategy for a leading edge electronics manufacturer. Design to Distribution Limited (D2D) is an assembled PCB manufacturer and has been encountering many of the problems mentioned above. They manufacture boards for a wide range of sub contractors as well as their sister companies. The boards they receive from their designers are only assessed for testability at two stages. The first is when the board schematic diagram is finished, the second when the board is released to production for a development build. This system has many shortcomings. The main one being the resistance to change from the designers at late stages but other problems do occur; and the time taken to do these assessments can be substantial. The other main problem is the effectiveness of the manual check. Some aspects of testability cannot be examined using paper copies of the design, without some pre-processing by computer aided design (CAD) systems (i.e. physical access).

Solution

A solution needed to be found that would address all the problems mentioned above. Paper systems were discussed but these didn't reduce the amount of effort required to do an assessment.

An electronic system was chosen because it would be able to make a comprehensive assessment of the board from a set of electronic data. The system would also be able to work from an incomplete data set so that designs could be checked during early stages of design. A further system requirement was a turn round time for assessment in less than twelve hours.

The assessment criteria (the rules) need to be up dateable and expandable, a prime requirement of the system if it is to have any lasting value. This is due to the rapidly changing methods and techniques that are available to the test engineer.

The Advantages of Using an Automatic Rule Checker

The obvious benefits of an automated checker are the speed at which the checks can be made, and the comprehensive nature of the checks. It is a fact that electronically, a more complete set of checks can be made which makes the automated rule checker an attractive proposition. There are also other benefits such as de-skilling, enabling the whole team to become testability experts thus allowing anyone to produce a testability report. A significant benefit is the ease with which the checker will be able to check incomplete designs, by applying the rules that it has information for on its database. The rule checker will also offer a strategic advantage over competitors in the contract manufacturing market by providing a service that the competitors will not be able to offer.

The System

The DFT system considers not only the code which implements the rules but how the use of the rule checker will fit into current procedures and practise. The system is described by figure 1.

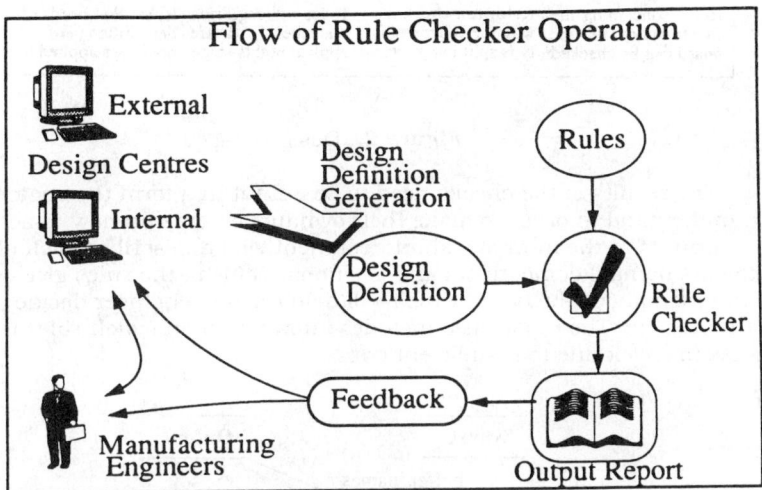

Figure 1. System Overview

Database
The database must have a structure that enables the storage of all ' the information about a board which is to be tested. The main information sets are net lists, placement data, board shape and pads, and component data. The component data contains information such as pin names and pin functions (input, output, bi-directional) as well as physical information. In order for D2D to compete in world markets it has been necessary for them to develop tools and methods which will give reduced and acceptable new product introduction time. The ability to quickly produce manufacturing information that is accurate and conforms to the factories production capability is a prime requirement. An automated process planning system known as SPEAR has been developed which holds factory and product model information within its database. This open systems database will be accessed by the DFT rule checker.

There are further issues surrounding the translation of data into this database for companies dealing with a wide customer base. Information can come in many different formats, with some companies using proprietary systems while others supply paper copies rather than electronic data. Many tools have developed to translate these different data types into the database, such as scanners for optical character recognition, data restructuring and formatting tools and Gerber viewers.

Rule Checking
Once the data has been successfully put onto the database the DFT checks can then be made. The rule checks form several different categories, these categories were chosen in accordance with the information required to complete the rules and fall in line with the design cycle. The categories are: Bill of Materials, Net List (Schematic), Pin Function, and Full Physical. (Manufacturability rules were also added to the rule base as they required the same information sets and could be easily incorporated.)

A White, B Sutton, C Sumpter and R Jones

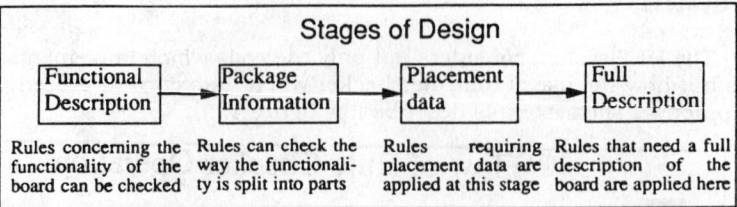

Figure 2. Design Stages

The results of the checks need to be output in a form the engineers can understand in order to make their own judgements on how to act. Thus the output is in the form of a short document which describes each of the problems naming components and locations. Some of the rules give general information about the board which will help the test engineer decide upon test routes etc. The report also includes information on which rules it was unable to check due to insufficient data.

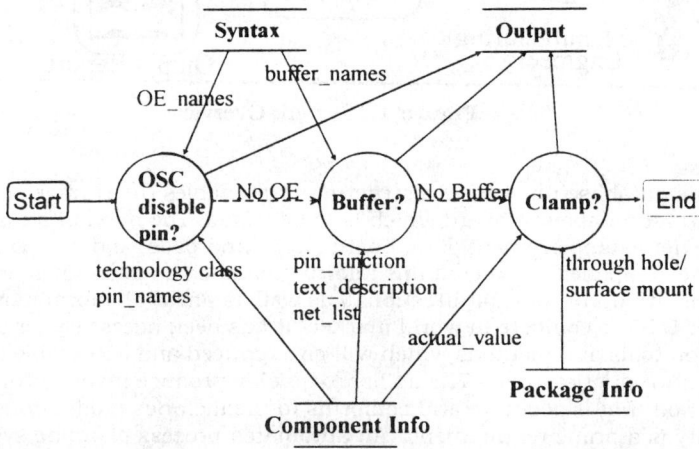

Figure 3. Data Flow Diagram for Clock Control Rule

The rules take on a logical structure, following a flow of decisions made by querying certain attributes of the design. Consider, for example, a testability rule concerning the control of oscillators (as shown in figure 3). This rule first takes each oscillator (detected by its' technology class) and tests to see if the oscillator has an output disable pin. It does this by matching the pin names of the device against a set of syntactic arrangements that describe all occurring names for an output disable pin. If there is a match then the device has an output disable pin. Similarly the other checks are made; the next part examining the device that the oscillator drives to see if it is a device that can be used as a buffer. The last part of the rule checks the frequency of the device to see if it is suitable for clamping and overdriving. There are currently 50 of these rules in the rule base of which about 20 are testability and 30 manufacturability.

Feedback
    With the report generated the engineer can give feedback to the

designers, this report is generated with his comments. Any compromises between design and manufacture can be made at this point thus ensuring manufacturing integrity later in the design cycle.

The Benefits
It is difficult to be specific about the exact cost savings which accrue from improved testability. The main advantages from DFT come from reduced time for assessment which results in reduced time to market. The use of DFT for a sample of 50 board types per year resulted in a manpower saving of 75 days and a reduction in time to market between 2 to 6 weeks. In addition the strategic advantage offered by the rule checker is that of uniqueness, since few other companies can offer this service.

Conclusions

Inability to provide effective testing methods is a costly factor of electronics manufacture. Consideration of test methods by designers during preliminary design will enable significant savings during manufacture. Paper based text rule checkers provide the designer with guide-lines but do not impose a disciplined approach to test. The use of a computer based automated method enables unambiguous access to component and design data and more comprehensive assessments.

Acknowledgement

The authors would like to thank D2D, Loughborough University and the Teaching Company Directorate of SERC for their support for this work.

References

Alunkal, J.M. 1993 Designing an Integrated Corporate Test Strategy, *Electronic Design* Mar. 81-89
Bidjan-Irani, M. 1991 A Rule Based Design-for-Testability Rule Checker, *IEEE Design and Test of Computers*, Mar., 50-57
Isaac, J. 1993 The Design Cycle and Downstream Concerns, *Printed Circuit Design*, Apr., 18-22
Kim, C. and O'Grady, P. et al. 1992 TEST: A Design-for-Testability system for Printed Wiring Boards, *Journal of Electronics Manufacturing* **2**, 61-70
McLachlan, L. 1992 Integrated Manufacturing Test Development, *Printed Circuit Fabrication*, Jan. 22-27
Maunder, C., *The Board Designers Guide to Testable Logic Circuits* (Addison-Wesley)
O'Grady, P. and Kim, C. et al. 1992 Issues in the Testability of Printed Wiring Boards, *Journal of Electronics Manufacturing* **2**, 45-54
Shi,C. and Wilkins, B.R. 1993 Design Guidelines and Testability Assessment *Proc. ICCD IEEE* Oct.
Hunt, I., Roberts, R., and Jones, R. 1993 The Integration of Design and Manufacture - A Quantum Leap, *Integrated Manufacturing Systems*, **4**, No.2

DECISION SUPPORT SYSTEM FOR PCC ASSEMBLY TESTING

L B Newnes, G Dickens and N Weaver

School of Mechanical Engineering, University of Bath, Claverton Down, BATH BA2 7AY

This paper presents a case for the use of a Decision Support System (DSS) within the electronics manufacturing industry. The aim of the system discussed is to operate as an on-line decision tool for the shop floor, focusing on printed circuit card (PCC) fault diagnostics. The focus of the research is described with respect to industrial and academic priorities, and a hybrid framework for the design of such systems is introduced. Finally, a pilot study is undertaken and a prototype decision aid using ExpertRule is validated.

Introduction

Within the manufacturing sector computers are used to aid in ordering, stock control and control of the factory down to shop floor operation. Current research has identified the need to provide decision making at the lowest possible level. This is deemed necessary to meet rising demands in product variability and customer requirements for both make-to-order products and quality expectations. Within the European community some researchers are examining the problems that arise at the bottom of the hierarchy. One such project is the use of Knowledge Based Systems (KBS) as real time CIM-Controllers for distributed factory supervision. A subset of this overall work is a workcell controller for dynamic production, quality and management within electronic applicancies, the tyre and cable industry and finally, Integrated Circuit production [Isen,1990]. Research by Hou, Luxhoj and Shang suggest several areas for DSS implementation at the shop floor level, which include scheduling and control; fault diagnostics, quality and inspection [Hou, Lux, Shang; 1993].

The area of work identified for the research described in this paper is Quality Control. This area was identified after examining current literature and investigating the needs within a few manufacturing companies. The benefits that the use of a DSS could offer to the collaborating company include retaining an experts knowledge, monitoring infrequent batches for quick turn around in board fault diagnostics and, finally, identify from this fault diagnostics any trends in recurring problem areas or equipment.

Design Framework for a DSS

The framework for the definition and implementation of a DSS is described in this section. The approach described utilises the IDEF methodology for structuring the overall design of the DSS and its requirements; Hayes-Roth for identifying the steps to be undertaken in the planning and building of the DSS; and a further technique devised to undertake data acquisition and structuring. IDEF0 [Wu, 1992], has been used to described the overall structure and levels of the system development in terms of input, outputs and the controls and constraints for the system. The overall goal of the research is to design a DSS for the electronics assembly industry. Figure 1 depicts the breakdown of the IDEF parent diagram into the main task boxes. The Hayes-Roth methodology was selected for the planning and building stage of the system development due to its proven research record, [Nar, 1990] where on site validation of a system adopting this approach indicated a 92% success rate. Figure 2 illustrates a summarised version of the steps to be undertaken to enable this definition and implementation. These steps are used to define the task boxes within the planning and building phase of the DSS, depicted in figure 1, task box A02.

A five step knowledge structuring technique was devised to manipulate the rough data into the form of variable and attributes to describe system inputs and outcomes, into a comprehensive set, or truth table, of standard 'examples' to form a knowledge base. This technique is shown in figure 3 and enables progressive data acquisition and structuring as defined by Hayes-Roth. Although it has been designed for the pilot study within the collaborating company, emphasis has been placed on keeping it as generic as possible to provide a structured approach for data manipulation.

Pilot Study

DSS Prototype System

This section describes the use of the above methodology for *Designing a DSS to aid with the manual testing of Printed Circuit Cards (PCC)* within the collaborating company. This system was identified as a key area for DSS application due to a 15% rework rate occurring during final inspection, safety requirements leading to 100% inspection of the product and, customer requests for quality reports. The main benefits anticipated were improving the efficiency of inspection and fault locations, highlighting the causes of inspection failures for rectification and finally, the key area, storing staff expertise. The software package used in building this decision aid is ExpertRule. To ascertain the value of such a system a particular PCC was selected on the basis that it had a recurring history of problems, due to its complexity faults were hard to find, reasonable batch sizes were produced, it was unsuitable for automatic testing due to its bulk and it was seen as being representative of the products under test. Data/knowledge collection involved collating circuit diagrams for the card, manual testing procedure, operator information in terms of data collection of faults and rectification action, and observations and discussions with the operators on all the procedures they adopt; utilising both quantitative and qualitative approaches to knowledge acquisition.

Once the initial information gathering was complete the Hayes-Roth methodology was adopted, the first stage building a final informal description of the problem. Four main sections are covered; Participant Identifications and Roles, Problem Identification, Resource Identification and Goal Identification. The problem was the identification of the appropriate action to take during a fail in the manual testing, the resources being used to overcome this include; personnel, the knowledge information described above, and the ExpertRule software package. The overall goal is to 'Build a prototype model via a pilot study to illustrate the use of the DSS'. Knowledge structuring of the raw data was undertaken utilising the steps depicted in figures 2 and 3. This data refinement resulted in the original 72 records being reduced to 21; 6 Fault Variables, 3

Fault attributes, 5 Cause Attributes, 3 Actions and various Component References. The refined data forms the basis of the truth table for System Building. The knowledge structuring completed during system planning presents data in a suitable format for building the system. The prototype program was developed and fully tested in line with the implementation stage of the Hayes-Roth methodology using the ExpertRule package.

The user is presented with a single screen depicted in figure 4. The details section is for information purposes and monitoring of activities. Fault 'Variable' and 'Attribute' selection lists are located in the top right hand Inputs corner, and allow the operator to classify the PCC fault. Once the fault information has been entered the DSS returns possible cause of failure and suggested actions. For 'Unknowns' the user is prompted to save this as a new fault under user action.

Prototype Testing

This was undertaken in two phases, firstly the testing from the system builders and secondly, testing at the company. The company testing resulted in various feedback comments specific to the particular application. A common theme was the user interface where suggestions focused on minimising the display presented to the operators and making use of function keys. The actions taken for the test information given were approved by manual testing engineers.

Conclusions

A framework for the design of Decision Support Systems is introduced utilising a combination of the IDEF, Hayes-Roth and a Knowledge Structuring methodology. A prototype system in collaboration with an industrial company is built using this approach. The software package ExpertRule is used and the DSS is validated by company experts. Attention focused on the design of this system for use on the shop floor. The overall conclusion was that the use of DSSs for manual inspection failures is recommended. A simple robust user interface was seen as a key feature along with user confidence, to aid in adoption of the system on the shop floor.

Acknowledgements

The authors wish to thank the Nuffield Foundation for the funding of this research and Westinghouse Signals (BTR Group) for their industrial input to the research.

References

Hou, T. et al., 1993, A Neural Based Automated Inspection System with an Application to SMDs, Int. J. Prod. Res., Vol 31, No 5, 1171-1187.
Isenberg, R., 1990, EP2434: CIM-Controller : An Overview, CIM-Europe Workshop, Implementing CIM, 3-4 December 1990.
Luxhoj, J.T. et al., 1993, A Prototype KBS for Selection of Inventory Control Policies, Int. J. Prod. Res., Vol 31, No 7, 1709-1721.
Shang, S. et al., 1993, Output Maximisation of a CIM System, Int. J. Prod. Res., Vol 31, No 1, 215-271.
Wu, B., 1992, Manufacturing Systems Design and Analysis, Chapman and Hall.

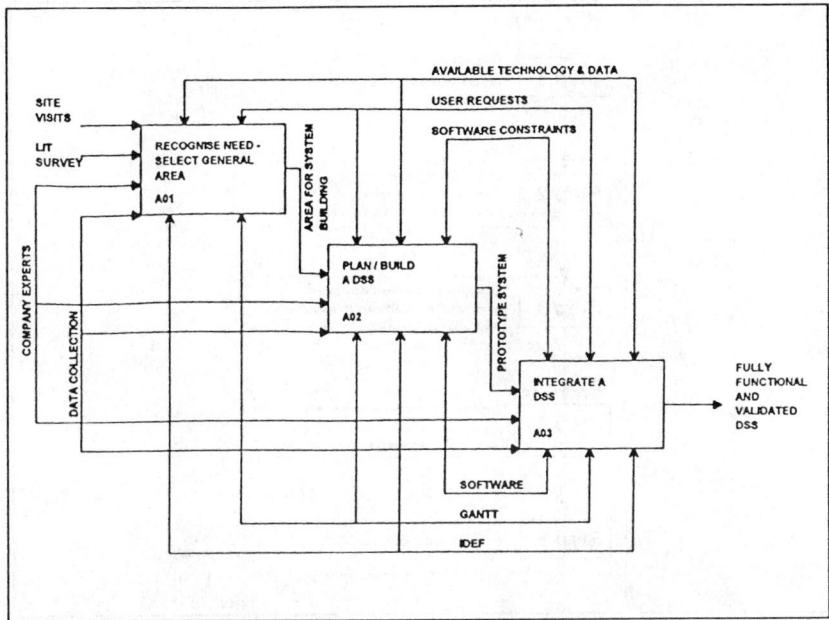

Figure 1   Design a DSS for PCC Inspection

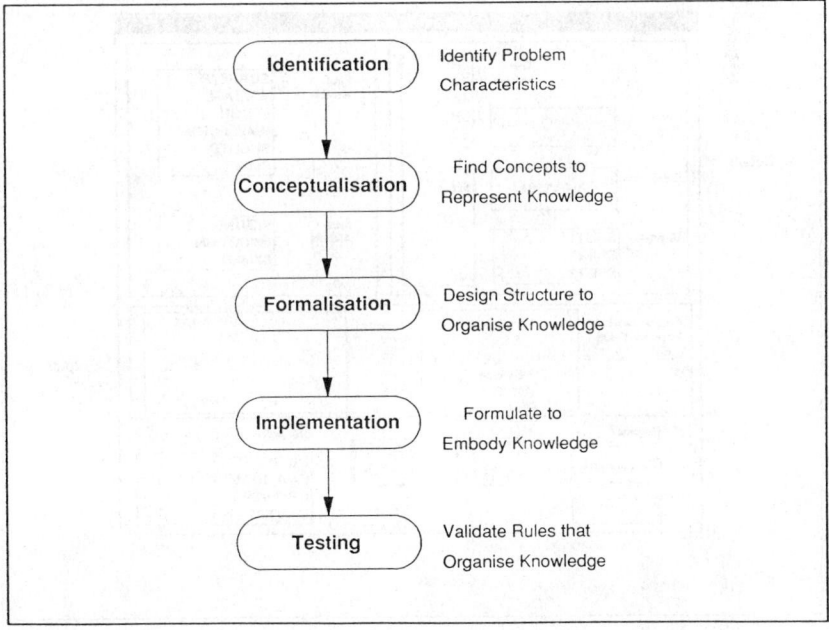

Figure 2   Planning and Building a DSS (Hayes-Roth)

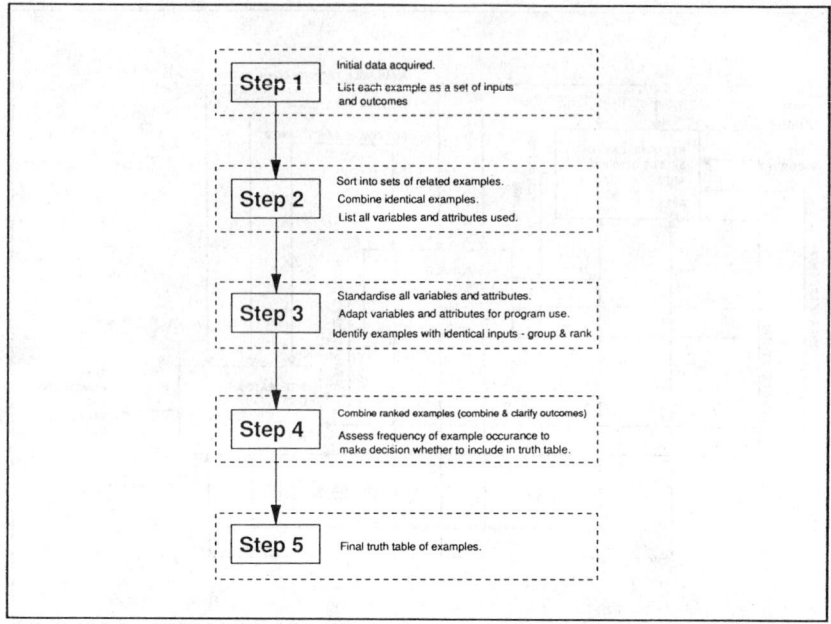

Figure 3   Knowledge Structuring Technique

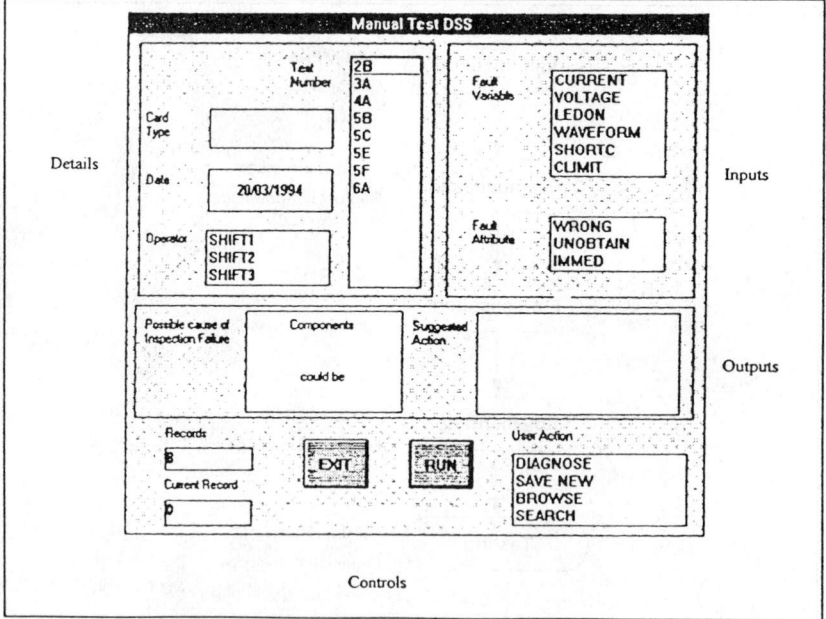

Figure 4   User Interface - Prototype DSS

CASE STUDY OF COMPUTATIONAL FLUID DYNAMICS APPLIED TO DOMESTIC REFRIGERATION DESIGN

Mr. P. Kingston, Dr. N. Woolley, Dr. Y. Tridimas and Dr. S. Douglas.

Liverpool John Moores University,
School of Engineering and Technology Management,
Byrom Street, Liverpool L3 3AF.

There is a growing public demand for healthier foods and foodstuff manufacturers have responded to this by providing preservative and additive free foods. These are inherently more bacterially sensitive. Temperature measurements have shown that in current domestic refrigerators, large temperature differentials occur within refrigerated food compartments. In these conditions traditional and modern foods can become bacterially contaminated. Therefore there is a need to design and manufacture refrigerators that provide closer temperature control. This paper describes how a computational fluid dynamics (CFD) model is used to simulate refrigeration conditions. The aim of this approach is to supplement and improve the general design and manufacture strategies applied to domestic refrigerator production.

Introduction

Recent work carried out by the Ministry of Agriculture, Fisheries and Food (MAFF) into the state of health of the nation [1,2,3,4,8] has focused attention on the domestic handling of frozen and chilled foodstuffs. At present there is no Health and Safety legislation on the storage of domestic refrigerated foodstuffs, although it's acknowledged that much government discussion has arisen on the subject particularly regarding storage temperatures. This has significant implications upon domestic fridge manufacturers who are reluctant to release information on temperature distributions within their refrigerators. Results from experimental work and the MAFF survey indicate an average refrigerator temperature of 6° C with temperatures varying from -0.9° C to a potentially dangerous maximum of 11.4°C. At this maximum temperature cross contamination between foods is potentially dangerous.

The modern domestic refrigerator has evolved over the years using a heuristic design and manufacturing strategy [5,6,7]. If, as is likely, future legislation requires manufacturers to guarantee the design of the refrigerator to provide more uniform temperatures throughout the refrigerated cabinet, then this current design approach will have to be supplemented.

Boundary conditions

The geometrical information used in the analysis derives from a low cost refrigerator currently available on the market. A necessary requirement for the solution of the equations is the specification of known constraints applied in the form of boundary and initial conditions. These include

      (i)   Zero velocities on all solid surfaces.

      (ii)   Specified temperature distribution with time for the cold plate.

      (iii)   Overall heat transfer coefficients for the walls of the refrigerator.

Conditions (ii) and (iii) have been established from experiment. Initial conditions for the transient analysis were all set to 26° C.

Numerical solution

No analytical solution of the governing equations is possible. A numerical method was used requiring the specification of a mesh of grid points. Grading of the mesh was used (see fig 1) to improve predictions in those regions of the flow where changes are occurring most rapidly.

The equations were solved on the specified grid using the commercial CFD code, FIDAP [9] which uses the finite element technique.

Discussion.

Initial experimental data was collected to validate the CFD model. This required the calibration of the refrigeration unit using the ISO standard [6]. Three thermocouples embedded in brass cylinders were positioned as defined by the standard, with two additional sensors at the top and bottom of the refrigeration unit as shown in figure 2. The refrigerator ran for approximately seventy minutes prior to data collection, enabling stabilisation of the system. The operating conditions of the refrigerator were set by ensuring that the average of the three temperatures measured at the ISO points over a period of time was 5°C (figure 2). The two additional temperature sensors illustrate the problem of temperature differentials. It is clear from this graph that a temperature differential of up to 10°C exists.

In order to simulate the unsteady nature of the flow and heat transfer in a refrigerator, it is necessary to start from known conditions. The known conditions in this case were chosen so that the temperature in the compartment was the same as ambient. The predictions and measurements presented in figure 3 show the thermal behaviour of the air in the succeeding 6 minutes. The graph illustrates the temperature variation, plotted nondimensionally, along the vertical line through the middle of the refrigerator. The temperature scale has shifted origin for each timestep. Dimensionless zero temperature corresponds to the original ambient condition. The height axis is normalised. General agreement between computed and experimental results is good . The model is predicting temperature variations between top and bottom of the refrigerator compartment, even at this relatively early stage in the flow development. At two minutes, the discrepancy is due to the experimental difficulty of maintaining a constant initial air temperature. The theoretical model is tending to overpredict the rate of removal of heat after six minutes because of the approximate initial efforts in describing the cold plate temperature conditions.

Physical Problem.

The domestic larder refrigerator, as regards the current analysis, consists of an insulated cabinet and a cold plate through which the heat is extracted. Natural convection of the air will take place because of the temperature differentials established in the refrigerator. This leads to unwanted temperature conditions in the cabinet. The inherent difficulty of the physics is compounded by the transient nature of the airflow. The conditions in the refrigerator investigated were such that the air flow was laminar.

The objective of the CFD modelling presented in this paper is to solve the governing equations for flow and heat transfer within the refrigerator compartment, the output from the analysis being air temperature and velocity distributions. This provides insight into the complex flows produced inside the refrigerator, essential information for the design and manufacturing engineer. In previously undertaken work, the evolutionary design and manufacture approach to a domestic refrigerator has been experimentally investigated [3]. Alternative refrigeration conditions were investigated in a search towards optimal design.

An alternative strategy advocated here, is to evaluate refrigerator design by simulating the performance of the refrigerator using analytical design tools. CFD is such a tool which can enable sophisticated models of the refrigerator to be developed.

Theory

The governing equations for the flow and heat transfer are derived from conservation of momentum, energy and mass and can be stated as follows:

$$\frac{\partial u}{\partial t}+u\frac{\partial u}{\partial x}+v\frac{\partial u}{\partial y}=-\frac{1}{\rho_o}\frac{\partial P}{\partial x}+v\left(\frac{\partial^2 u}{\partial x^2}+\frac{\partial^2 u}{\partial y^2}\right) \quad \text{(1)} \quad \text{-X component of momentum}$$

$$\frac{\partial u}{\partial t}+u\frac{\partial u}{\partial x}+v\frac{\partial u}{\partial y}=-g\beta(T-T_o)-\frac{1}{\rho_o}\frac{\partial P}{\partial y}+v\left(\frac{\partial^2 u}{\partial x^2}+\frac{\partial^2 u}{\partial y^2}\right) \quad \text{(2)} \quad \text{-Y component of momentum}$$

$$\rho_o C_p\left(\frac{\partial T}{\partial t}+u\frac{\partial T}{\partial x}+v\frac{\partial T}{\partial y}\right)=k\left(\frac{\partial^2 T}{dx^2}+\frac{\partial^2 T}{\partial y^2}\right) \quad \text{(3) - Energy}$$

$$\frac{\partial u}{dx}+\frac{\partial v}{dy}=0 \quad \text{(4) - Continuity}$$

Assumptions implicit in the above equations are that the flow
(i)   is two - dimensional, laminar and unsteady.
(ii)   has constant physical properties including density. Density differences enter the equations in the buoyancy force term.

Figures 4,5 and 6 show predicted vector plots of the air flow at times of 2, 4 and 6 minutes respectively from the start. This visualization is important from the design point of view because of the strong linkage between the flow field and temperature distribution.The most noticeable feature of the flows is their unsteady behaviour. Quite different flow patterns are predicted at different times. The general flow comes downward from the cold plate, deflected across the top of the glass to the opposite wall (door) and then up the wall to a greater or lesser extent. Vortices in the strongly shearing flows are continually being generated and destroyed. In general, the main vortex in the lower half of the refrigerator compartment is causing the temperature stratification effect.

Conclusion.

It has been shown that CFD can play an important role in the design and evaluation process. Excellent insight is given to into the nature of the flow patterns and temperature distributions established in the refrigerator compartment. CFD has an important role as an optimization tool in refrigeration design.

References

[1]   Gac, A. 1993, Les industries agro-alimentaires et la chaine du froid - *Institut International du Froid,* International Journal of Refrigeration, 6-9.

[2]   Gibson, G. M. 1991, Microbiology in chilled foods- Is refrigeration enough, James Douglas Memorial Lecture, *Proceedings of Institute of Refrigeration*,4,1-6.

[3]   Gigiel, A. J. Douglas. A, et al, 1994, Improving the Domestic Refrigerator, *Procedings of the Institute of Refrigeration*, 11-17.

[4]   Consumer handling of chilled foods, 1991, A survey of time and temperature conditions, Ministry of Agrigulture, Fisheries and Food.

[5]   Household frozen food storage cabinets and food freezers, 1983, -essential characteristics and test methods. ISO 5155.

[6]   Household refrigerating appliances, 1985, - refrigerator - freezers - characteristics and test methods. EN 28187.

[7]   Performance of household refrigerating appliances, 1985, - refrigerators with or without low temperature compartments. ISO 7371.

[8]   Utilization services report, 1992 - refrigerator control, E.A. Technology.

[9]   Fluid Dynamics International, 1992, FIDAP 7.04 code and Theoretical Manual.

Nomenclature

x,y	-	directions.	t	-	time.
u	-	x velocity component.	v	-	y velocity component.
T	-	Temperature.	T_o	-	Temperature.
k	-	Thermal conductivity.	P	-	Pressure.
ρ	-	Density of fluid.	ρ_o	-	Reference density.
Cp	-	Specific Heat Capacity.	υ	-	Kinematic viscosity.
β	-	Coefficient of volumetric expansion			

Acknowledgements

The authors wish to acknowledge the support of Mr. F. Stephen of E.A.Technology, Capenhurst, Cheshire on this project.

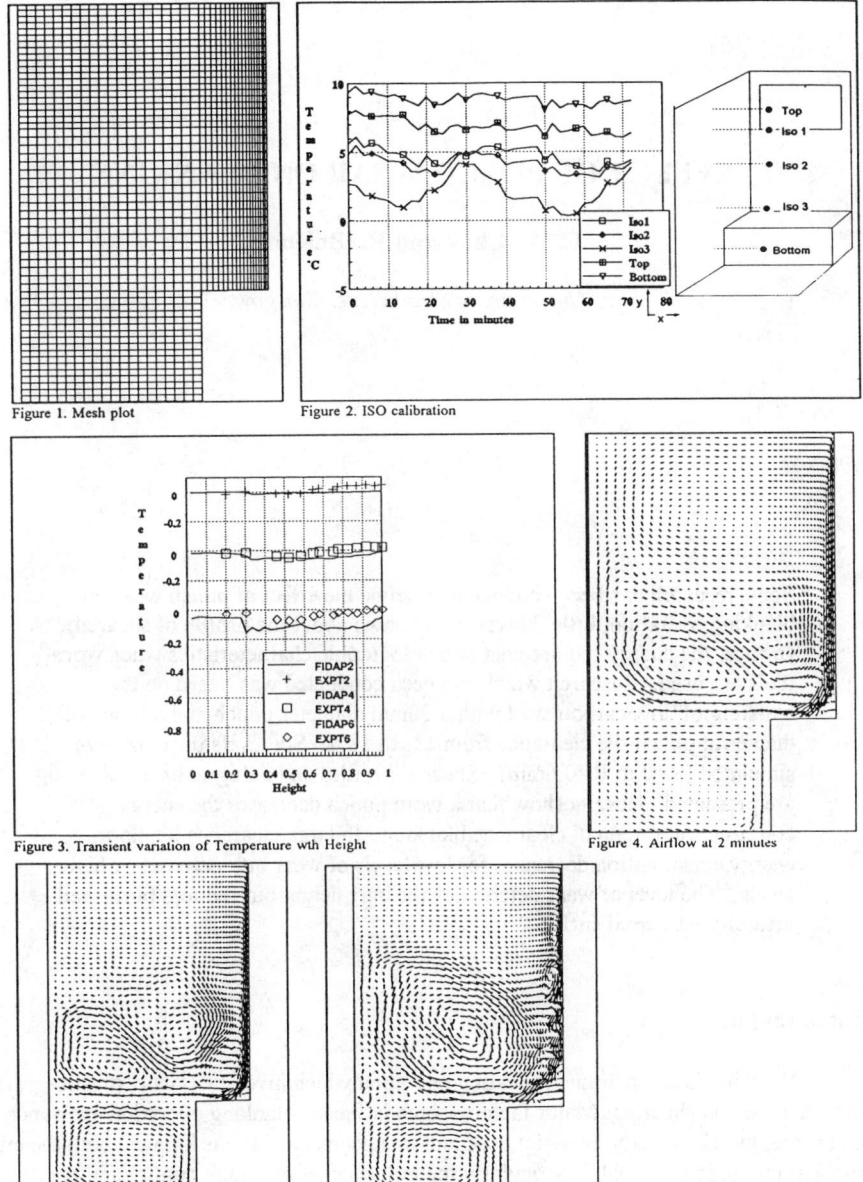

Figure 1. Mesh plot

Figure 2. ISO calibration

Figure 3. Transient variation of Temperature wth Height

Figure 4. Airflow at 2 minutes

Figure 5. Airflow at 4 minutes

Figure 6. Airflow at 6 minutes

EFFECT OF PUNCH WEAR ON BLANKING

C.M. Choy and R. Balendra

University of Strathclyde, Glasgow

Some research has been conducted to define the effect of punch wear on blanking operation; little, however, is known about the profile of sheared surface, the energy requirement and autographic characteristics when worn tools are used. Research which has been conducted was based on the blanking of low carbon steel with a 20mm diameter punch and a range of diametral punch-die clearance from 1% to 18%. Six levels of wear were simulated (R=0 to R=0.6mm). Sheared profiles and autographs of blanking are presented. Results show that a worn punch decreases the energy consumption for small clearance blanking. In large clearance blanking, energy consumption decreases for low levels of wear but increases at higher levels. The level of wear determines the burr height but reduces burnished area of the sheared surface.

Introduction

Blanking is a constrained shearing operation which involves elastic, plastic deformation and shearing. Major factors which influence blanking operation are punch-die clearance, punch velocity, material used and tool geometry, little is known about the effect of wear on the cutting tools, particularly the influence on the blank quality and the operation. Sondershaus, Buchmann and Timmerbeil researched the effect of different levels of wear on the burr height on blanks. Timmerbeil also showed that a worn punch reduces the burnished area at the sheared surface but increases punch penetration, blanking force and energy requirement. However, some aspects of the blanking with worn tools which do not appear to have received serious considerations, are the blanking force and energy requirement at different punch penetration as well as the effect on the sheared surface of blanks; this forms the basis of the presently reported investigation.

Equipment, Material and Procedure

Equipment

A schematic drawing of blanking tool is shown in Fig.1. The blanking die was ground to a diameter of 20.26mm. A total of eight punches were used to provide the diametral punch-die clearances from 1% to 18%. Strain gauges were mounted on the punch holder to enable force measurement. A universal servo-controlled hydraulic press was used at a ram speed of 10mm/min for all test.

Material

Specimens for blanking were produced from 50mm wide strips of 2.95mm thickness flat ground bright mild steel, having the following percentage composition: C=0.160%, S=0.260%, Mn=0.700%, P=0.026%, S=0.015%. Tensile and hardness tests were conducted to identify the property. Nominal Tensile strength and hardness of material were 750 N/mm2 and 93 HRB respectively. Specimens were subsequently subjected to micrographic analysis.

Procedure

A full factorial experiment with eight punch-die clearances and seven punch cutting edge conditions was conducted. Autographic information for each operation was monitored on a data acquisition software and were stored in ASCII format. Five samples were blanked for each test condition. Total of 280 tests were conducted. A blank from each sampling group was subjected to metallurgical examination.

Experimental Results and Discussions

Wear takes place on the punch face and on its free surface (Fig.2). This causes the shearing edges to be rounded. In the experiments, simulation of punch wear did not take both the wear on punch free surface and punch face into consideration. The polished punch radius on punch shearing edge was examined on a profile projector with radius template. The six sets of autographs for specific punch-die clearance are shown in Fig.4. Each set of autographs consists of seven superimposed curves pertaining to respective levels of punch radius. Each curve was obtained by averaging the autograph for the respective sampling group. The universal servo-controled hydraulic press employed for blanking was of high stiffness; metallurgical separation for stock and blank was always accompanied by blanking with an immediate loss of load. This may affect the accuracy of the curve and introduce errors in the value of energy expected. However, it is evident that the difference is only slight.

Energy Requirement

The experimental results(Fig.5a) show that energy consumption for blanking, regardless of punch-die clearance, is reduced considerably when the punch radius increases. These results contradict previous publications, particularly that by Timmerbeil. Work done by blanking may be divided into two parts. The first part(W1) is the energy require to effect elastic and plastic deformation as well as crack initiation and propagation. The second part(W2) is the energy require to overcome mechanical locking interference between stock and blank during mechanical separation. The autographs

(Fig.4) clearly shows that, as punch wear increases, the energy(W2) to overcome mechanical locking interference between stock and blank reduces substantially, in spite of the need for greater punch penetration and energy(W1). By referring to previous research work, published by *Production Engineering Research Association*, crack initiation and propagation can be suppressed when the punch shearing edge is rounded; material is encouraged to flow upward and fill up the space around the punch wear surface. The delay in crack initiation at the punch allows the crack from the die edge to propagate further upwards(Fig.3). With further punch penetration a crack originates from the free surface of the punch. The sequence of this change in the punch crack propagation, besides leading to burr formation, reduces the amount of mechanical locking interference between the stock and the blank. Therefore less energy is required to overcome the smaller amount of mechanical locking interference during mechanical separation. Fig.5a shows that the energy required for blanking with 14%,16% and 18% clearance increases slightly when the punch radius R=0.6mm. Extra energy is required for deeper punch penetration when punches with smaller diameters are badly worn. It is anticipated that further increases in the punch radius will increase the energy requirement.

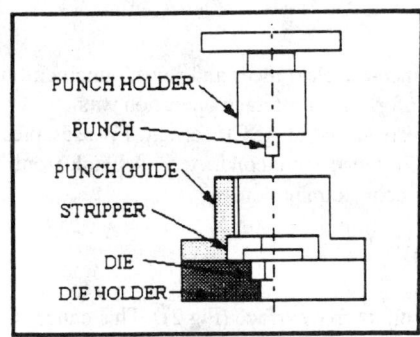

Figure 1. Blanking Tool   **Figure 2.** Punch Wear Profile

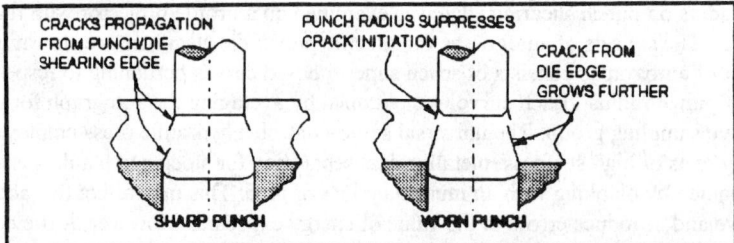

Figure 3. Cracks Propagation from Punch/Die Shearing Edge

Maximum Blanking Force

The results in Fig.5b show that maximum blanking force does not bear an exact relationship to punch radius. However it was observed that there is a general downward trend for maximum blanking force with respect to the level of punch radius. The maximum blanking force is generally decreased by about 5kN for a punch radius of R=0.6mm. On the other hand, for 2% and 16% clearance blanking, the maximum blanking force increases by about 2KN when punch radius is R=0.6mm. These results contradict Timmerbeil's results but are similar to Buchmann's results.

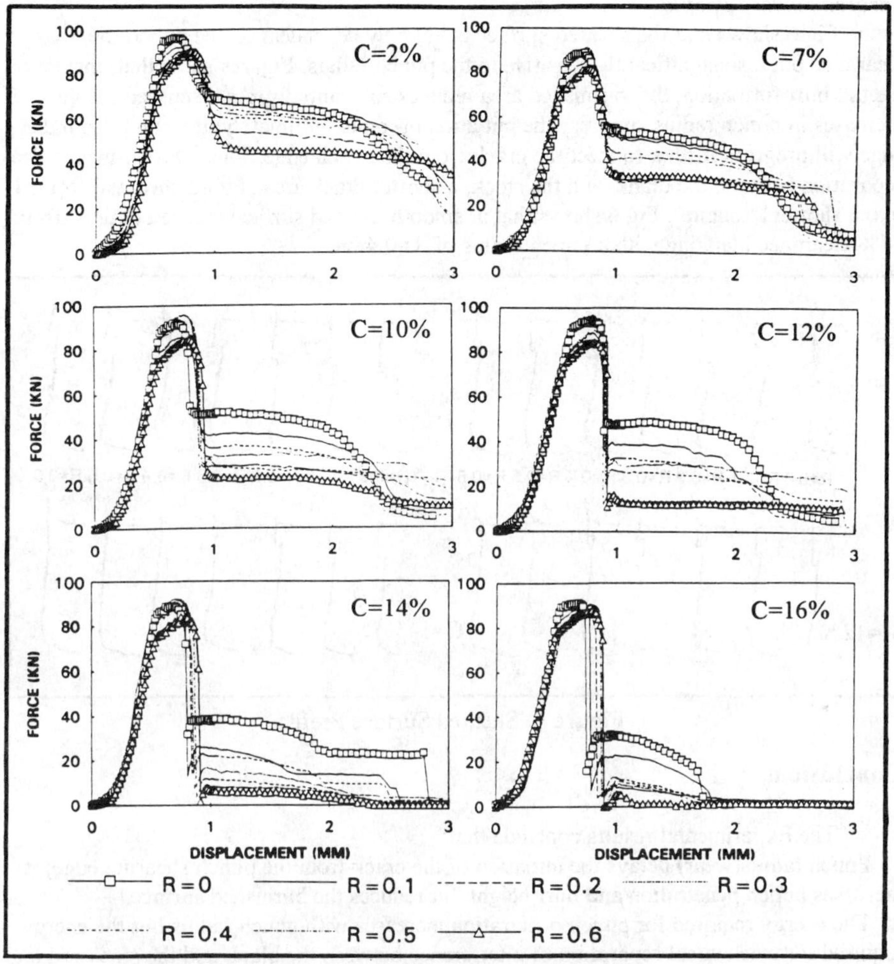

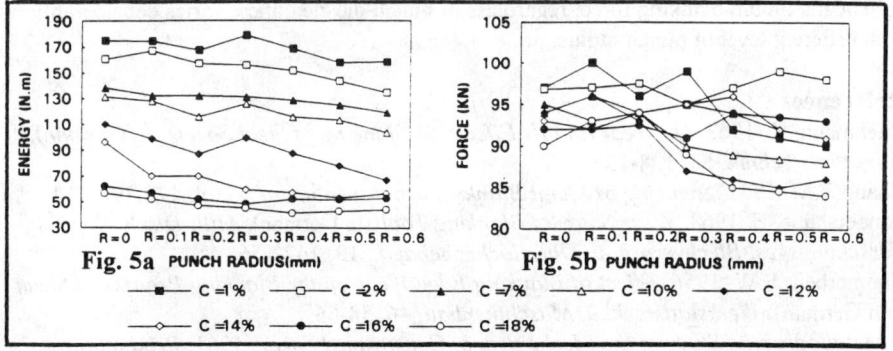

Figure 4. Autographs for Specific Punch/Die Clearance

Figure 5. Blanking Energy and Max. Blanking Force Requirement

Sheared Surface Profile

Fig.6 shows that the sheared surface is not only dependent on the punch-die clearance but also significantly sensitive to the punch radius. Figures show that, inspite of the tall burr formation, the burnished area reduces substantially. As discussed above, increases in punch radius suppress the initiation of crack; ultimately, the crack from die edge will progress further to meet the crack from the punch edge. Thus, during mechanical separation between the blank and the stock, a shorter crack area is burnished and formed into a shorter "tongue". Fig.6 shows that a smooth sheared surface can be obtained from 14% clearance blanking with a punch radius of R=0.4mm.

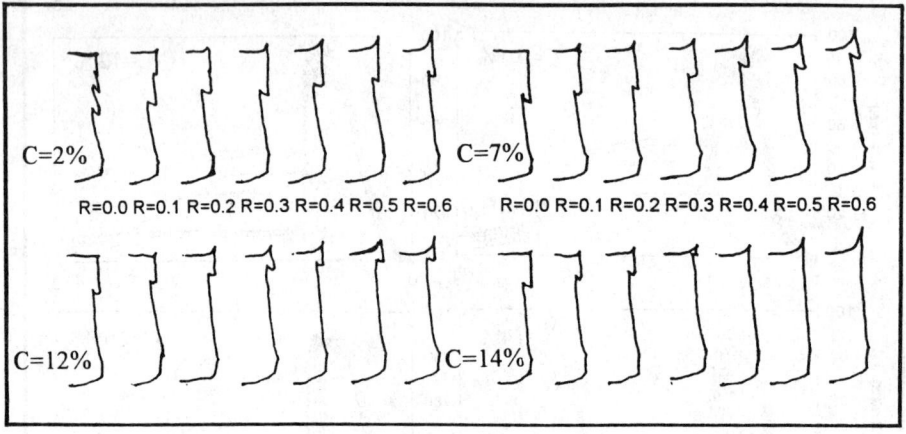

Figure 6. Sheared Surface Profile

Conclusions

The Experimental results conclude that:
1. Punch radius(wear) delays the initiation of the crack from the punch shearing edge; it increases punch penetration and burr height but reduces the burnished surface,
2. The energy required for punch penetration increases with punch radius but the energy required for mechanical separation of interference between the blank and the stock reduces substantially, hence total energy requirement is reduced with the respective level of punch radius.
3. The maximum blanking force, regardless of punch-die clearance, varies considerably with different level of punch radius.

References

Buchmann, K. 1963, *On Wear and Its Effects on Blanking of Steel Sheets (in German)*, *Werkstatttstechnik*, **53**, 128-134.
Chang,T.M. 1951, *Shearing of Metal Blank, Journal Institute of Meta*l, **128**,393-414.
Sondershaus, E. 1968, *Wear Study of Blanking Tools*(in German), *Mitt. Dtsch. Forschungsges. Blechverarb. u. Oberflachenbehand.*, **19**, 142-156.
Timmerbeil, F.W. 1956, *Effect of Blanking Edge Wear on the Blanking Process of Sheet* (inn German), *Werkstattstech. u. Maschinenbau*, **46**, 58-66.
A Metallographic Study of the Mechanism of Constrained Shear, 1961, *Production Engineering Research Association*, **93**.

DEVELOPMENT OF FLAWS DURING
INJECTION FORGING

Yi Qin and Raj Balendra

Manufacturing and Engineering Management
University of Strathclyde   Glasgow G1 1XJ

Several types of failures occur during the Injection Forging of components;
among these, a prominent form results from the unstability of the free length
of the billet. Both FE simulation and experiments show that billets were
unstable at a higher aspect ratio (T>1.65); instability would also occur at
lower aspect ratios when the friction coefficient was less than 0.03 or when a
large exit-radius was incorporated in the injection chamber. Metallurgical
inhomogenity in the material would also result in instability and asymmetrical
die-filling at low aspect ratios.

Introduction

Folding and sliding are the main forms of failure during Injection Forging; this has
been shown to occur when the aspect ratio of the primary deformation zone ($T=t/d_0$, refer
to Fig. 1(a)) is beyond a particular value. Folding occurs as a result of the instability of
the "strut" between the exit from the injection chamber and the anvil; this is similar to the
buckling of cylinders in simple compression, which occurs when the initial aspect ratio of
the cylinder is in excess of 3 (Gunasekera, Chitty and Kiridena (1989)). However, the
critical aspect ratio for injection upsetting has been clearly demonstrated to be 1.65
(Henry(1971), Balendra(1985)). Buckling limits for simple compression were simulated
and verified experimentally, while the evidence for injection upsetting was largely
experimental. The research reported herein, relies on FE simulation and experimental
comparison to analyse the influence of several parameters on the stability of the billet
during Injection Forging.

Equipment, Material

Equipment comprised a compounded injection-chamber, an anvil and spacers to
effect different aspect ratios of the primary deformation zone (T) (refer to Fig. 1(b)). The

test material used throughout was E1CM Aluminium; the flow characteristics were determined by compression tests: flow stress $\sigma = 159 (0.17 + \varepsilon)^{0.29}$ MPa (ε is equivalent strain). All specimens were machined from 12.5 mm diameter bar to -0.08 mm of the current diameter of the injection chamber. The Fig. 2(a) shows three specimens which were deformed at aspect ratio 1.85, 1.75 and 1.64; the specimens in Fig. 2(b) show the examples of lateral sliding of material under low friction condition at the billet/anvil interface.

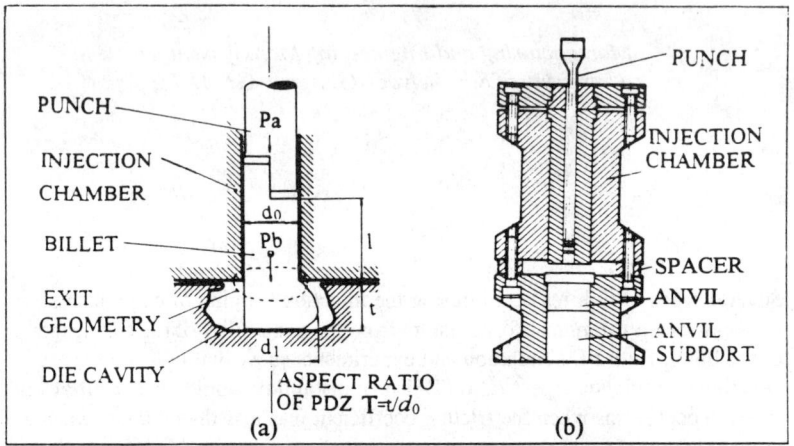

Figure 1. Process model of Injection Forging and research equipment for Injection Upsetting

FE Modelling of Upsetting Injection

Composition of the material is, invariably, assumed to be perfect for general FE simulation - the material is of a uniform characteristic and without geometrical defects. Post-buckling behaviour cannot be simulated using FE methods if a model of perfect symmetry is assumed. Asymmetry in the billet geometry and inhomogenity in the metallurgical structure will always prevail in engineering materials though measures of asymmetry may be difficult to define. Therefore, in the reported study, the geometric eccentricity of the billet in the die-cavity is considered; material inhomogeneity is defined by prescribing different flow strength values to strands of elements in the FE Model.

ABAQUS FE code (Hibbitt, Karlsson & Sorensen Inc.(1992)) was used for simulation. Since the 3-D plastic analysis of material deformation is time-consuming; the plane-strain model was used to enable an approximate simulation of the initiation of flaws. Prior to the extensive use of 2-D simulation for this study, several benchmark tests were performed. The deformation and post-buckling behaviour of the billets were sufficiently similar in both 2-D and 3-D simulation and compared well with the failure forms observed in the experimental programme to justify the use of 2-D FE simulation. Examples of the comparison are shown in Fig.3; figures (a) and (c) are for 2-D simulation while (b) and (d) are 3-D simulation for identical conditions.

Results and Discussion

Examples of both, experimental results and FE simulation are presented in Fig. 2-4 to support the conclusions.

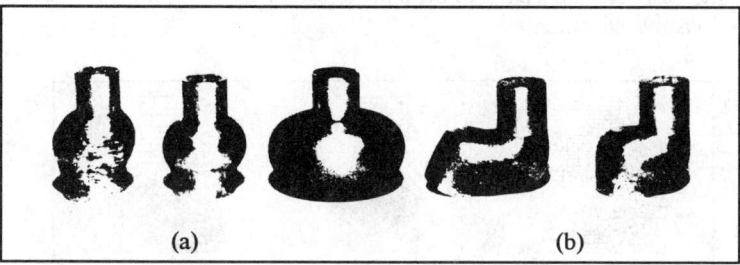

(a)   (b)

Figure 2. The examples of the deformed specimens

Limiting value of pdz ratio

Fig. 4(a-1)-(a-3) shows the deformation of the billet for T=2.5, 1.85 and 1.64 respectively. The billet which was injected with an aspect ratio of 2.5 would always deform asymmetrically; that for T=1.85 shows a small degree of asymmetry. This initial asymmetry does not make the billet unstable; the flow patterns in a fully formed component would, however, not be symmetrical. By comparison with the simulation for T=1.85, that for T=1.64 shows sufficient symmetry as shown in Fig. 2(a).

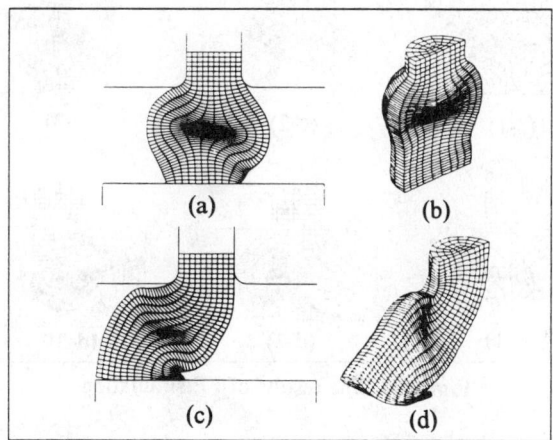

(a)   (b)

(c)   (d)

Figure 3. Comparison of 2D and 3D FE simulation

Friction

Using coefficients of friction of 0.03, 0.02 and 0.01, several simulations were conducted; results are shown in Fig. 4(b-1)-(b-3). At μ=0.03, the billet deforms symmetrically and continues to achieve die-filling. The stability of the billet is seriously affected when the coefficient of friction is reduced by a marginal amount; under a friction condition of μ=0.02 the base of the billet begins to slide laterally while retaining contact over its entire diameter. This will promote the development of a fold at the lower extremities of the billet as is visible in Fig. 4(b-2). When the value of friction is 0.01, the simulation shows that the lateral sliding of the base of the billet is severe - refer to Fig. 4(b-3); this causes part of the base of the billet to lift off the anvil; the specimen shown in Fig. 2(b) shows this condition.

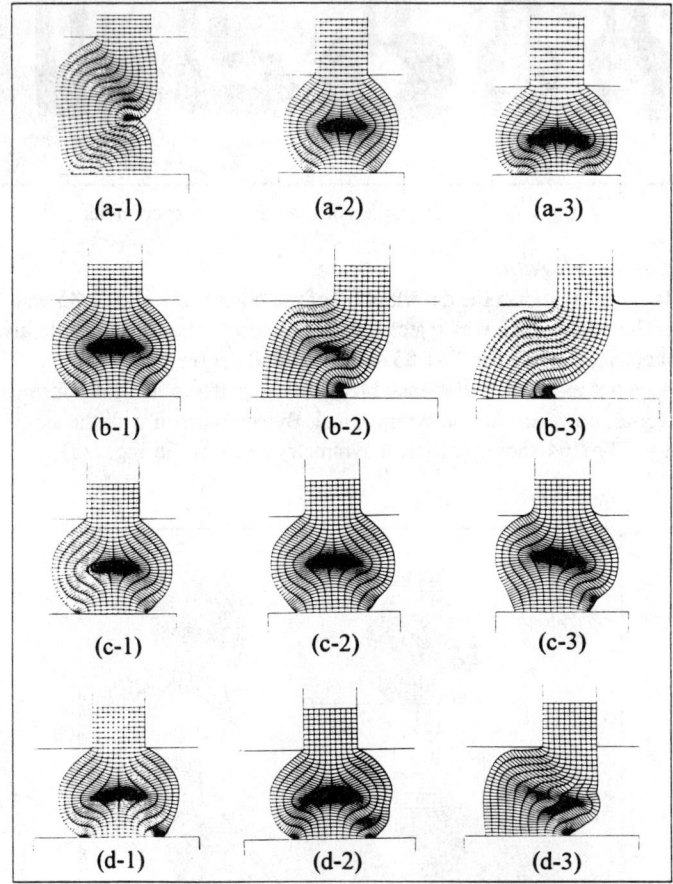

Figure 4. The results of FE simulation

Exit-geometry

The influence of the exit-geometry was evaluated by examining three radii - R (r/d_0, where r - the radii of transition at the exit from injection-chamber) =0.00, 0.126 and 0.24 (Fig. 4(c-1)-(c-3)). The exit-geometry cannot be considered in isolation as increase in exit-geometry results in an effective aspect ratio of the pdz which is higher than that specified. Evidence has already been presented (Balendra(1988)) that using a larger exit-radii would improve the flow of material at the injection-chamber-exit and reduce the punch pressure. The simulation results further suggest that the incorporation of exit-radii makes the billet less stable and hence more likely to deform in a manner which will result in flaws in the component (Fig. 4(c-3)).

Material inhomogenity

There is ample experimental evidence that some batches of E1CM contain material which was not metallurgically homogenous; the material property was not uniform across the billet. Examination of the macro-etched ends of the billets revealed that a portion of it was of a fibrous-grain structure while the remainder was of a coarse-grain form. Such batches were not used for general experimental research. The difference in hardness between these metallurgical forms suggests that the ratio between the flow strengths for such structures was 2. In the FE simulation, metallurgical asymmetry was considered by prescribing different flow strengths for increasing numbers of strands of vertical elements. The simulation indicates that variation in metallurgical symmetry will result in asymmetric deformation - refer to Fig. 4(d-1) and (d-2) - and will in extreme cases of inhomogenity - refer to Fig. 4(d-3) - result in the sliding of the billet which will lead to the development of flaws in the component.

Conclusion

Conditions which are geometrical, interfacial and metallurgical can individually or in combination influence the stability of the billet at the initiation of injection; FE simulation indicates that the deformation is sufficiently stable for pdz aspect ratio $T<1.64$; injection upsetting on smooth platens (friction coefficient $\mu<0.03$) will be initiated with the lateral sliding of the base of the billet which will result in flaws in the component. Increases in exit-geometry ($R=r/d_0 >0.126$) will promote instability even at low pdz aspect ratios; improved flow at the exit has the adverse consequence of generating billet instability. Metallurgical inhomogeneity in the specimen will result in billet instability and asymmetrical deformation of the specimen during injection upsetting.

References

Balendra, R. 1985, Process mechanics of injection upsetting, *Int. J. Mach. Tool Des. Res.*, **25/1**, 63-73.

Balendra, R. 1988, Exit-geometry for injection forming, Proceedings of 4th Int Conf on Manf Eng, Brisbane, Australia, 11-17.

Gunasekera, J. S., Chitty, E. J. and Kiridena, V. S. 1989, Analytical and physical modelling of the buckling behaviour of high aspect ratio billets, *Annals of the CIRP*, **38/1**, 249-252.

Henry, J.C. 1971, An Investigation of the injection upsetting of six steels, *NEL Report 494* (UK Dept. of Industry).

Hibbitt, Karlsson & Sorensen Inc. 1992, *ABAQUS User Manual*.

THE AUTOMATED RECOVERY OF AEROFOIL PROFILES ON SUPER NICKEL ALLOY TURBINE BLADES

J Corbett, D J Burdekin, A M King, V C Marrot and F L Martin

Cranfield University
School of Industrial & Manufacturing Science
Cranfield Bedford MK43 0AL

In accordance with a continual requirement to improve quality and competitiveness, a major aero-engine manufacturer has identified the need to develop new methods of manufacture which accurately recover the aerofoil profile of a range of turbine blades. Currently these 'lost wax' precision cast blades utilise 'pins', to hold a central core, which remain as protrusions on the aerofoil profile and have to be removed. The existing labour intensive hand finishing process to remove these pins also removes material from the blade surface, thus reducing the polytropic efficiency of the turbine. This paper discusses a successful investigation to find an alternative manufacturing method to remove the pins without contacting the blade profile. In addition the proposed method demonstrates a significant cost saving over the current process.

Introduction

Super nickel alloys have been developed for turbine blades which are able to withstand the high stresses and temperatures required for improved performance. A range of blades are manufactured using the 'lost wax' casting process to obtain the precise external aerofoil shape. Complex internal blade cooling passages are produced by the use of ceramic cores which are subsequently removed. These cores are supported during casting by platinum pins which are left protruding from the aerofoil surface as shown in figure 1. These need to be removed without unduly affecting the aerofoil profile which directly influences the blades efficiency. At present the pins are removed using a manual linishing process which also removes up to 100 µm from the blade surface, thus changing the precision cast form and reducing the polytropic efficiency of the turbine. In addition the process is expensive which has led to a programme aimed at identifying a new cost effective manufacturing technique to recover the aerofoil profile without contacting the precision cast aerofoil surface. A maximum protrusion of 25 µm on the finished blade was specified as being acceptable.

Figure 1. A Cast Blade

Removing the Pins

A worst case scenario where a blade may contain up to 30 pins was presented as the objective and used for all subsequent investigation and analysis. In order to be competitive with manual methods, a target time of 3 minutes was allocated for the complete removal of these pins. A wide number of techniques were considered before selecting grinding as being the most appropriate. King (1993) found that conventional aluminium oxide, silicon carbide or boron carbide grinding wheels would require frequent re-dressing and would have a relative short life for the proposed application. Further, an initial study by Burdekin (1993), using CAD solid modelling techniques, indicated that a small ball nosed cutter of only 16 mm diameter was required in order to gain access to remove all pins on the full range of blades.

CBN and diamond grit grinding wheels were identified as being the most promising due to their high hardness and long life potential. For the initial tests CBN was selected for its lower cost and because previous work by Westkamper and Tonshoff (1993) indicated that a vitrified bonded or electroplated carbon boron nitride (CBN) wheel permitted high cutting speeds, a high shape accuracy and high surface quality with comparatively low wear. Electroplated CBN wheels were selected for the trials as the need for truing and dressing would be eliminated.

Machining Experiments

A total of 20 ball nosed grinding wheels were initially produced with grit sizes of 0.125 mm, 0.212 mm and 0.30 mm as advised by grinding specialists, De Beers Industrial Diamonds. A special purpose CNC grinding machine, with a variable spindle speed up to 60,000 rpm, was made available to conduct the following experimental programme in order to optimise - (i) the grit size, (ii) the speed of cutter rotation, and (iii) the feed rate, with regard to wheel life, machining accuracy and cost.

In order to reduce the number of tests required to obtain meaningful results it was decided to design the experiments based upon methods proposed by Box (1978). This led to the use of a 'face centred cube' model which was appropriate for establishing the effects of the three input variables over a range advised by De Beers. The three input variables were set at the following values.

1. Grit sizes (μm)   125   212   300
2. Grinding wheel speed (rpm)   20,000   40,000   60,000
3. Feed rate (mm/s)   1.0   1.5   2.0

These parameter values were keyed into the 'experimental design' menu within Statgraphics (a software analysis package) to create a random programme for the experimental tests. The interaction of the input variables was shown via a number of surface plots, similar to figure 2. This shows the predicted effect of the variables of grit size and spindle speed on the corresponding wear of the cutter. The 'wear' was measured using the optical 'shadowgraph' technique.

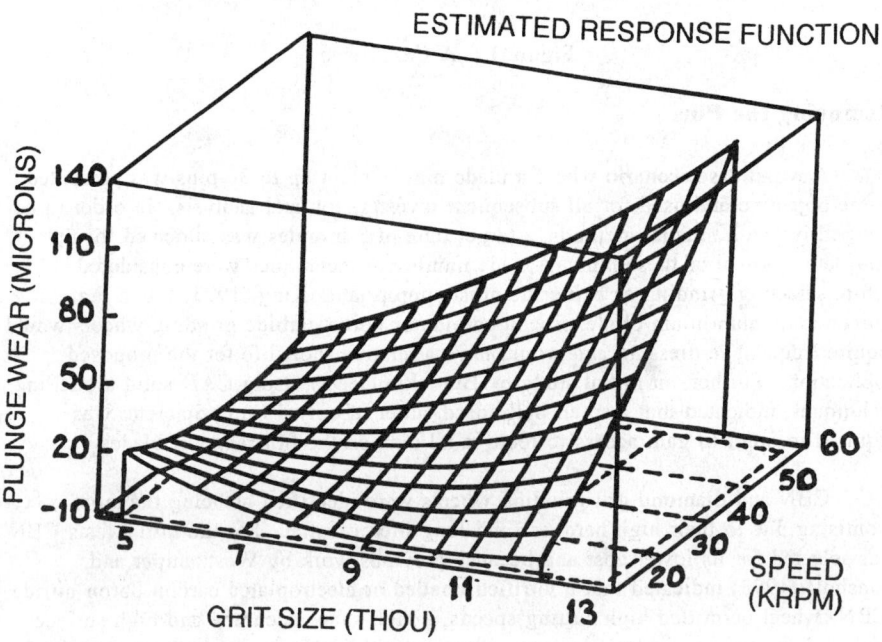

Figure 2. Statgraphics' Response Plot

Summary of Results
   1)   Grit size - The largest grit size demonstrated the lowest wear and was capable of meeting the specified 0.8 μm surface finish requirement.
   2)   Grinding wheel speed - The highest ball speed, 60,000 rpm, resulted in dramatic wear of the ball nose tool. This contradicted results of earlier CBN trials on different materials undertaken at Bristol University (1992) which indicated that higher speeds gave better results. However on closer examination of the process it was

apparent that the cutting fluid was not reaching the point of cutting due to a turbulent air boundary layer around the tool. For the final machine a high pressure pump was proposed to ensure coolant penetration of the boundary layer of air.

    3) Feed rate - This was found to have little effect on the response values, although it was noted that a marginal improvement on wear occurred at the lowest feed rate.

    4) Wear rate - A conservative estimate was that twenty turbine blades could be machined before replacing the ball nosed grinding wheel.

    5) Force measurements - During machining a maximum force of 3N was exerted on to the grinding wheel.

    6) Power requirements - The maximum power requirement for the grinding spindle rotating at 60,000 rpm was estimated to be 0.5 kW. This was determined by using a 3-phase meter and measuring the voltage across each phase, together with the current through one of the phases.

Locating the Pins

    Due to the casting tolerance for the blade geometry and pin positions, coupled with the blade location and clamping errors, each pin had to be 'sensed' independently. Further, it was required that all pins and the 'local' surfaces of each blade should be 'mapped' within a timescale of one minute.

    An error budgeting technique described by Treib (1987) was used to consider and allocate suitable values for each significant error source, ie machine, clamping and sensing errors. This resulted in the need for a sensing system with a 10 μm accuracy capability in measuring the height of each pin relative to the blade surface. The only sensing technique found to meet the specified accuracy coupled with the speed requirement was the structured light method.

    The structured light method utilises a CCD camera similar to traditional object recognition systems. However with the structured light system a laser beam is set at a precise angle relative to the camera. The laser emits a light 'stripe' on to the object and the camera captures the data from the illuminated object. In order to 'see' the laser line on the object the camera is equipped with a filter. The diffracted light is filtered by a narrow band optical filter and recorded by the CCD array sensor. The profile of the scanned object is generated and represented by points in space. Such a system achieves a higher accuracy than the traditional object recognition system because it is possible to focus the camera precisely on the area illuminated by the last stripe.

    The blade profile, pins and grinding wheel will be scanned on the machine and the coordinate measurements processed by special software written to store the location of the blade surface and pin coordinates. This will in turn be used to control the machine for the subsequent machining of pins.

Proving Trials

    A test programme was undertaken in order to 'close the loop' between the sensing and subsequent machining of the pins. This was done by fitting the 3D scanning system to a CNC machining centre which was fitted with a specially designed rotating fixture to clamp the blade during the machining operation. The tests demonstrated that the scanning system had an accuracy capability of 6 µm compared with the specified requirement of 10 µm. All the pins machined away were within 20 µm of the surface compared with a specified tolerance of 25 µm.

Financial Feasibility

A 4 axis CNC grinding machine was proposed and a machine concept developed, complete with layout drawings and specifications for the drive and measurement system for each axis. This was used to obtain a cost estimate for the development and subsequent manufacture of machines.

The cost effectiveness of developing and applying the method was based on a nominal 150,000 blades per year, with the worst case of 30 pins. In order to produce this number of blades the company would require 7 machines. A cash flow analysis, using the company's accounting principles, assumed that the first machine would be developed within 2 years with the remaining 6 machines being built in year three. Payback for the machines would occur during the following year. The internal rate of return (IROR) for the company was estimated to be between 22 and 30 per cent, over a 10 year period, depending upon the quality control procedures adopted. A project with an IROR in excess of 12 per cent is currently considered to be economically attractive.

Conclusions

The research programme clearly demonstrated the technical feasibility of automating the pin removal process. Further with the proposed special purpose grinding machine, and sensing system, a significant accuracy improvement should be obtained, leading to an improvement in the blade's operational efficiency. The main machining parameters were identified and it is felt that further work would lead to further optimisation of grinding wheel wear. As well as being technically feasible the project demonstrated the potential to make substantial cost savings over the methods currently used.

References

Box, G. 1978, *Statistics for Experimenters : An Introduction to Design, Data Analysis and Model Building.* (John Wiley, New York).
Burdekin, D. 1993, *Development of an Automated Machine to Recover the Aerofoil Profile of Turbine Blades - Design*, MSc Thesis (Cranfield University).
King, A. 1993, *Development of an Automated Machine to Recover the Aerofoil Profile of Turbine Blades - Machining*, MSc Thesis (Cranfield University).
Pearce, T. 1992. *Private Communication.* Institute of Grinding, Bristol University.
Treib, T W. 1987, Error Budgeting - Applied to the Calculation and Optimisation of the Volumetric Error Field of Multiaxis System. *Annuals of CIRP*, Vol. **36**, No. 1 (CIRP, Paris) 365-368.
Westkamper, E and Tonshoff, H K. 1993, CBN or CD Grinding of Profiles. *Annuals of CIRP*, Vol. **42**, No. 1. (CIRP, Paris) 371-374.

GRINDING WHEEL SELECTION USING A NEURAL NETWORK

Y. Li, B. Mills, J. L. Moruzzi and W. Brian Rowe

Liverpool John Moores University
Liverpool, L3 3AF

This paper describes the development of a neural network system for grinding wheel selection. The system employs a back-propagation network with one hidden layer and was trained using a range of data from manufacturers catalogues. The user inputs the type of machine, workpiece material, hardness, surface roughness required and the severity of the grinding operation. The system then selects a suitable grinding wheel, specification for the abrasive type, grade, grit size and bond. The system was developed on a PC using the language C and has a user-friendly windows interface. For further training the system also includes a back-propagation neural network prototype which has 1 to 3 hidden layers.

Introduction

It is difficult to make the optimum choice of a grinding wheel since many factors affect grinding wheel performance. The main factors affecting grinding wheel performance are the material to be ground and its hardness, the stock to be removed and the finish required, the nature of the coolant (if any), wheelspeed, the area of grinding contact and equivalent diameter, the severity of the operation, and the power available for grinding, King and Hahn (1986). Each of these factors affect the selection of the optimal grinding wheel. Considerable knowledge and experience is required to select a suitable grinding wheel for a given grinding operation. Artificial intelligence is therefore appropriate for an optimal approach. Midha, Zhu and Trmal (1990), Zhu, Midha and Trmal (1992) developed a rule-based system for wheel selection. The system works well but a very large number of rules are required. The rules represent the domain knowledge of one or more experts. Sometimes these rules are difficult to express explicitly and their development requires much time and effort. Often, reasoning is weak and slow in a rule-based system.

Neural networks have been successfully applied in many fields. Unlike traditional expert systems, neural networks generate their own rules by learning from a set of training examples. Learning is achieved through a learning rule which adapts or changes the connection weights of the network in response to the training inputs and the desired outputs of those inputs. In addition, neural networks employ distributive memory, which decreases the size of memory required. The development of the system for grinding wheel selection using a neural network was relatively fast and the system is flexible, allowing for further training as new data become available from manufacturers or other experts.

The Structure of the Wheel Selection System

The wheel selection system was developed on a PC using the language C. The functional modules are illustrated in Figure 1. To simplify the description, selection of conventional wheels for cylindrical grinding is given as an example. Selection of grinding wheels for other operations follows a similar process.

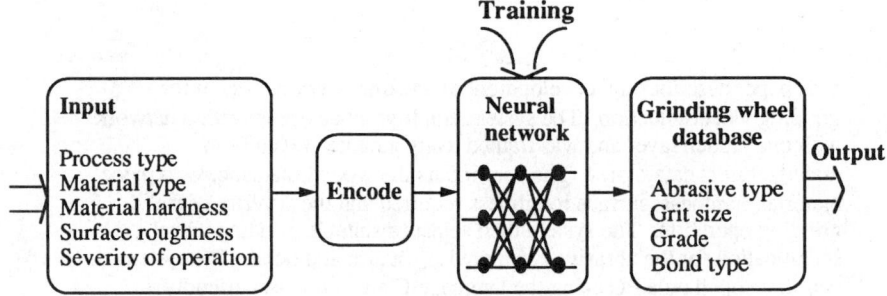

Figure 1 Functional modules of the wheel selection system

Input

The user inputs the grinding conditions. The system employs graphic window menus so that it is convenient for the user to input data. Each input group includes a menu of items for the user to select from. For example, severity of operation has 6 items: rough cast or forged condition, interrupted cut, large diameter workpiece, small diameter workpiece, wide wheel and light pressure, and narrow wheel and high pressure.

Encoding

The input information is then encoded into a form that the neural network can recognise. The encoding methods are important for the effectiveness of the neural network. Suitable methods ease the training process and allow higher accuracy. The same encoding rules are employed during training and use of the neural network. The codes are composed of decimal digits from 0 to 1. The encoding rule is one-to-one, namely, each input item takes a value from either the workpiece material type, the material hardness or the surface roughness. For the grinding method and severity of operation groups, five items are chosen simultaneously, namely, traverse grinding or plunge grinding, wide wheel and light pressure or narrow wheel and high pressure, large diameter workpiece or small diameter workpiece, rough cast or forged and interrupted cut.

The use of one-to-one encoding is possible, but more input neurons would be required, so that the system becomes more complicated. Furthermore, the severity of operation simultaneously affects grade and grit size and the effects on each of them are different. Larger sets of training patterns would therefore be required. For simplicity, a one-to-two encoding rule was employed, namely, one item is given two codes. One code corresponds to the effect on grade while the other code corresponds to grit size. A simple encoding rule is used to combine the initial codes into two resulting codes which are used as input to the neural network. The rule is more complicated than a one-to-one encoding rule, but the neural network is greatly simplified and the sets of training patterns required are decreased. Table 1 shows the input codes for cylindrical grinding. Test results show that the encoding methods work well.

Table 1 The codes for input to the neural network for cylindrical grinding wheel selection

Workpiece Material	Codes	50Rc - 57Rc	0.4	Grinding Methods	Codes
General steels	0.1	58Rc - 63Rc	0.6	Traverse grinding	0.1, 0.1
High alloy steels	0.2	Surface Roughness	Codes	Plunge Grinding	0, 0.2
High speed steels	0.3	1.4 - 1.6	0.1	Severity of Operation	Codes
Martensitic stainless	0.3	1.1 - 1.4	0.2	Wide wheel, light pressure	0, 0.1
Cast iron	0.4	0.7 - 1.1	0.3	Narrow wheel, high pressure	0.2, 0.1
Austenitic stainless	0.4	0.4 - 0.7	0.4	Large diameter workpiece	0, 0
Non-ferrous	0.4	0.2 - 0.4	0.5	Small diameter workpiece	0.2, 0.2
Material Hardness	Codes	0.14 -0.20	0.6	Rough cast or forged	0.2, 0
<50Rc	0.2	0.10 - 0.14	0.7	Interrupted cut	0.3, 0.2

The C-based Neural Network

Although commercial neural network software tools are available, there can be size and cost problems in integrating a commercial package into an industrial application. It is possible that the training results obtained using a commercial package can be employed in an application system. However, retraining is then difficult and inconvenient. A multi-layered feedforward neural network prototype which has 1 to 3 hidden layers was therefore developed using the language C. The training algorithm was based on back-propogation. The multi-layered feedforward neural network is the most widely used structure. It has proved to be capable of approximating any non-linear function with arbitrary accuracy, Hornik, Stinchcombe and White (1989). The neural network prototype is intended to be used as a subsystem in a system for intelligent selection of grinding parameters. However, the prototype can also be used for other purposes.

A three layer network was employed for grinding wheel selection. The function of the network is to map input vectors to output vectors. The network is illustrated in Fig. 2. The neural network was composed of interconnected neural processing units. Each neuron is represented by a circle and each interconnection, with its associated weight, by an arrow. It has an input layer, an output layer and 1 hidden layer. In practical application, a user can use the menu to select the numbers of input nodes, output nodes, hidden layers and nodes in each hidden layer. In the example, workpiece material, material hardness, surface roughness and severity of operation are the elements of the input vector. The output vector consists of abrasive type, grit size, grade and bond. However, there is no good rule for the selection of the optimal number of neurons in the hidden layer. A straightforward way to deal with the problem is to start from a small number of hidden neurons, then , gradually increase the neurons in the hidden layer until an acceptable accuracy is obtained. Six neurons were required in the hidden layer.

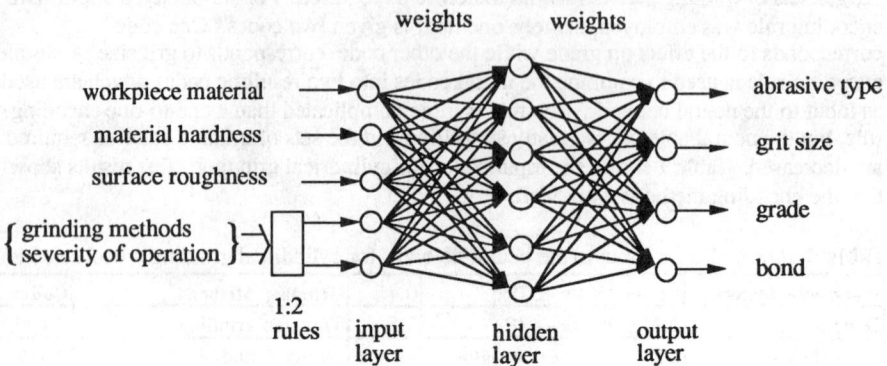

Figure 2. Neural network for wheel selection

As a feedforward neural network, the network uses the following relationships to map inputs to outputs.

$$out_j = f(\, net_j\,), \quad f(\, net_j\,) = \frac{1}{1 + e^{-net_j}}, \quad net_j = \sum_i w_{ij}\, out_i - \theta_j$$

where:
- out_j output of the jth node
- net_j unshaped output of the jth node
- $f(net_j)$ shaping function of the jth node
- w_{ij} the weight of a connection between the ith node in one layer and jth node in the next layer
- θ_j threshold of the jth node

Parameters w and θ are decided through training. Training is accomplished using the back-propogation technique with sets of inputs and desired outputs, Rumelhart and McClelland(1986).

In order to train the neural network, sets of data containing sufficient information were required. Sets of data were obtained from manufacturers handbooks. In the cylindrical grinding example, 25 sets of data were used for training. Another 10 sets of data were used to test the network. After training was completed, the neural network model was ready to be used for wheel selection. If required, further training is possible and may be conveniently carried out as new sets of data become available.

Grinding Wheel Database and Output

The grinding wheel database includes abrasive types, grit sizes, grades and bond types. The output module uses the output of the neural network to identify a suitable abrasive type, grit size, grade and bond type in the grinding wheel database. The outputs are presented by the conventional symbols and displayed on the screen.

An Example

A typical input to the graphical input menu may be as follows

Process type:	Cylindrical grinding
Workpiece material:	General steel
Material hardness:	50 - 57 Rc
Surface roughness:	0.4 Ra (μm)
Grinding method:	Traverse grinding
Severity of operation:	Large diameter workpiece and interrupted cut
Wheel type:	Conventional
Manufacturer:	Universal

The result displayed on the screen is

The recommendation is
48A60LV

Conclusions

A neural network approach to modelling the grinding wheel selection process has been investigated and a system developed. The neural network approach has the following advantages:
(i) A neural network approach is more easily developed than a rule-based system.
(ii) Knowledge is obtained by learning from examples. As a result the development of a wheel selection system using a neural network is relatively fast.
(iii) The system is flexible, allowing for further learning from new data to enlarge and improve the knowledge.
(iv) Knowledge is distributively stored in the whole neural network. A large amount of knowledge requires only a small amount of memory. For the network in Figure 2, only 54 weights and 10 thresholds are needed.

References

Hornik, K.,Stinchcombe, M. and White, H. 1989, Multistage Feedforward Networks are Universal Approximators, *Neural Networks,* Vol. 2, 359-366
King, R.I. and Hahn, R. S. 1986, *Handbook of Modern Grinding Technology*, (Chapman and Hall, New York)
Midha, P.S., Zhu, C.B. and Trmal, G.J. 1990, An Expert System for Wheel Selection for Cylindrical Grinding Operations, *Proceedings of the 6th International Congress on Computer Aided Production Engineering*(London), 445-450
Rumelhart, D. and McClelland, J. 1986, *Parallel distributed Processing,* Vol 1 (MIT, Cambridge, MA)
Zhu,C.B.,Midha, P.S. and Trmal, G.J. 1992, Development of a Knowledge Based Abrasive Wheel Selection System Using a Proprietary Expert System Shell, *Proceedings of 4th International Congress on Condition Monitoring and Diagnostic Engineering Manufacturing,*(Selis), 318-325

TOOL DESIGN FOR ELECTROCHEMICAL MACHINING

Dr Peter Lawrence

De Montfort University
The Gateway, Leicester, LE1 9BH

One aspect of designing tools for Electrochemical Machining is
considered, namely the specification of the precise shape of the tool
surface. The difficulty of finding solutions to the mathematical model of
the process by pure analysis has lead to the use of numerical methods.
This paper reports on a revised version of a novel method for modelling
this type of problem. Starting from a defined workpiece profile, a model
of an electric field can be constructed in which each new equipotential
surface is a possible tool electrode surface. Results obtained with the
software are compared with an analytical solution confirming the veracity
of the method.

Introduction

Electrochemical Machining(ECM) employs the phenomenon of electrolysis to
remove material from metal workpieces in a controlled manner. This is achieved by
passing a heavy electrical current through a shaped tool electrode, which is traversed
towards the workpiece while an electrolyte solution is pumped between the two.
There are several major problems to be resolved when designing practical tools for
ECM. One of them is to define the precise shape of the profile or surface of the tool,
since it is not simply a matter of making a constant offset or clearance from the
required workpiece.

The case to be considered is that when the shape of the workpiece is constant in
time, since this is the condition most commonly used in practice. The so called
equilibrium machining condition is usually achieved by holding all the machining
parameters , such as the tool feed rate and the applied voltage, constant. Alternatively,
in some cases, a control system is employed to alter one or more parameters in
proportion to changes in others.

The operating conditions can be chosen arbitrarily within a range of suitable
values. Considering only the production of the correct shape in the workpiece, any

combination of applied voltage and tool feed rate etc. which results in a specified equilibrium gap between the electrodes will suffice. Since the magnitude of this gap varies over the machining area, it is specified as an operating condition at a specific point. Therefore the equilibrium end gap at which the tool should be operated is usually chosen, this being the distance between the electrodes, measured in the direction of the tool feed, at a point on the tool surface which is perpendicular to the direction of the tool feed. This is usually, but not necessarily, the shortest distance between the electrodes. Practically the equilibrium end gap will usually lie in the range 0.05 to 2.0 mm.

The actual combination of machining parameters used to achieve the specified end gap will be influenced or determined by considerations such as the output voltage of the power supply; the current capacity of the power supply and the area of the cut; the surface finish to be achieved and the conductivity of the electrolyte solution to be used etc..

Analysis

In mathematical terms, the problem can be stated as finding solutions to the partial differential equations which model the equipotential and current flux distribution in the electrolyte solution between the tool and workpiece. Laplace's equation governs the potential and flux distribution within the region bounded by the electrodes at which mixed boundary conditions have been shown to apply by Lawrence (1977). Both electrodes are considered to be equipotential surfaces and additionally, the potential gradient normal to the work surface is proportional to the cosine of its angle of slope.

An analytical solution published by Lawrence for a semi-cylindrical workpiece of radius unity, shows that the family of equipotential surfaces with potential V, is described in polar co-ordinates by:

$$V = \tfrac{1}{2}(r - r^{-1})\cos\Theta$$

However, solutions to the problem for other general workpiece shapes seem to be intractable by analysis and for this reason attention has focused on numerical methods.

Continuity Method

A novel approach to the tool design problem was published by Lawrence (1981) which rests on two properties of an electric field:

    1. the equipotential and flux lines are a set of orthogonal curves.
    2. the current flowing between a pair of flux lines is constant throughout the field.

The second property is a statement of the condition of continuity and, because it is the property which enables the construction of the field model, the procedure is called the Continuity Method.

The workpiece surface is represented by a series of straight line elements whose nodes lie on a smooth continuous curve which can be described by a single function. Practical work shapes which do not meet this condition cannot me machined by a tool consisting of a continuous equipotential surface. Practical work profiles which are smooth but not continuous, can usually be very closely approximated with polynomials using available curve fitting software. It is not required that all the elements are of the same length, but it is usually convenient to make them approximately so.

A new equipotential tool surface is constructed from the existing equipotential (initially the work surface) by erecting normals at the nodes as shown in Fig.1. The lengths of the normals, which model flux lines, are inversely proportional to the normal potential gradient at the nodes. The projecting ends of the normals are the element nodes of the new equipotential. However, to comply with the first field property, the inclination of the normals must be adjusted to meet both the original and the new equipotential at the same angle. This in turn makes a small alteration to the co-ordinates of the new nodes.

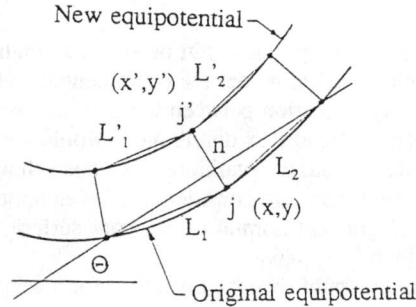

Figure 1. Construction of equipotentials.

Initially, at the work surface, the normal potential gradient is proportional to $\cos\Theta$, this being the boundary condition for an equilibrium shape as previously stated. To determine the normal potential gradient at the new nodes, it is necessary to make use of the second field property, noting that current(flux) density, j, is proportional to normal potential gradient. Thus with reference to Fig.1., the current density, j', at the corresponding new node can be calculated using the appropriate element lengths, L, from:

$$j' = j(L_1 + L_2)/(L'_1 + L'_2)$$

The process of constructing a further new equipotential can now be repeated as often as required.

Software originally written in ALGOL to carry out this procedure, has been re-written in GWBASIC to operate on an IBMPC and has been refined in the process. Apart from very much faster operation, it is now possible to see on the VDU the required work surface and the equipotential tool surfaces in their correct relationship as they are constructed. This enables many alternatives to be quickly explored and assessed before plotting and printing co-ordinates of the chosen solution. The flux lines can also be displayed since these may be required as an insulated portion of a tool surface. The ability to see the flux lines serves another purpose in that they give an immediate visual indication of the stability of the field model. Moreover, the magnification of the display and plots can be varied at will and selected areas of the field can be displayed for particular scrutiny.

Comparison of Analytical and Numerical Results

A semi-cylindrical work profile, represented by 200 elements, has been processed to illustrate the output from the software and to establish the veracity of the field model.

Fig.2. shows the plot of the workpiece and computed tool shapes to operate with end gaps of 1, 2 ,3 and 4 percent of work radius. The flux lines have been suppressed for clarity at this scale.

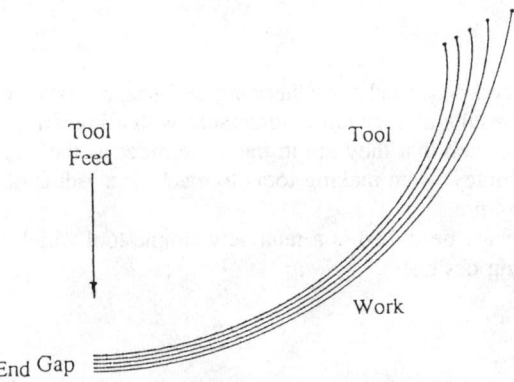

Figure 2. Circular work and tool surfaces.

An expanded plot of the upper portion of the field is shown in Fig.3. where the individual elements and flux lines can be seen. Polar co-ordinates for two arbitrary points, A and B, on each of the tool equipotentials were compared with the radius derived from the analytical solution using the same angle. The results are shown in Table 1.

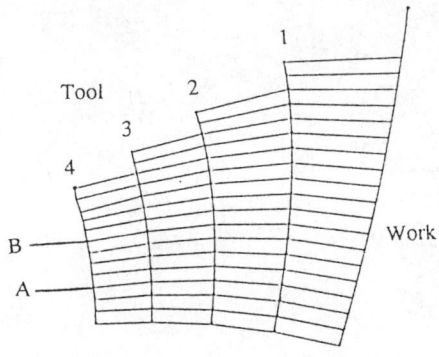

Figure 3. Expanded field plot.

Table 1.Differences as % of Work Radius

End gap(%tool rad.)	point A	point B
4	0.00117	0.00171
3	0.00133	0.00202
2	0.00110	0.00170
1	0.00057	0.00081

Conclusions

The method described for generating tool shapes corresponding to given work profiles gives results in very close agreement with a known analytical solution. Differences are such that they are in the same order as the attainable practical machining accuracy when making tools to machine a radius of 100mm operating with an end gap of 4mm.

The software described is a relatively simple tool which it is hoped will prove of use to practising designers.

References

Lawrence, P. 1977, Prediction of Tool and Workpiece Shapes.*Proceedings of the International Symposium for Electromachining-ISEM 5,1977,Switzerland,* 101-104.
Lawrence, P. 1981, Computer-Aided Design for ECM Electrodes.*Proceedings of Twenty-second International MTDR Conference, UMIST,1981,*379-384.

A COMPARISON BETWEEN BENCHMARKING PROCESS MODELS

Abby Ghobadian, Hong Woo and Jonathan Liu

Middlesex University Business School
Centre for Interdisciplinary Strategic Research
The Burroughs, Hendon, London NW4 4BT.

Structured business benchmarking is a relatively new organisational improvement tool. The idea is simple and potentially attractive. In general term, it is an attempt to encourage organisations to move away from the syndrome of "not invented here" towards learning from one another and implementing the best practices of the leading companies. A simple concept, but not so easy to plan and effectively implement. To realize the aims of benchmarking, its industrial exponent such as: Alcoa, AT&T, IBM, Rover Group, and Xerox have designed and implemented formal processes. This paper compares these processes using a structured framework. The deductive research shows that it is possible to design a versatile and simple benchmarking process.

Introduction

Benchmarking is a continuous process of measuring products, services, and process against the company's toughest competitors or those companies renowned as world class or industry leaders (Camp 1989). Most organisations have their own definition. Alcoa defines benchmarking as " 'managing for quality' process that uses the talents of those responsible for the process, service, and/or product to assess current activities and set up future priorities where competitive advantage can be gained" (Alcoa 1990). 3M defines benchmarking as "a tool used to search for enablers that allow a company to perform as a best-in-class level (benchmark) in a given business process" (Willhite 1991). Xerox pioneered the use of benchmarking in 1979. It helped Xerox to revive its fortune. Benchmarking is employed by more and more companies (Geber 1990). A recent survey of Fortune 1000 companies revealed that 65 percent of respondents used benchmarking (Foster 1992). In many companies benchmarking is a key component of the TQM process (Whitting 1991). The importance of benchmarking as an improvement tool is recognised by leading quality awards. For example, in the case of Malcolm Baldrige Award, a quarter of the total points are allocated to benchmarking.

Despite the increasing attention paid to benchmarking, many organisations do not know how to conduct benchmarking in practice or what it entails. A common framework or model is necessary because it gives consistency throughout the company. It provides a structure and a common language to the organisation (Spendolini 1992). Spendolini argues that within the framework of a model, all types of variations are possible, and can be tailored to fit the specific requirements of the individual, groups, and organisations that use it. A model enables practitioners from across the business to discuss issues of benchmarking. For example, employees from different disciplines or departments can more easily discuss benchmarking, if they both use the same model, although what they may be benchmarking is unknown to each other.

Ghobadian and Woo (1994) developed a model to conduct a number of benchmarking exercises. For convenience this process is referred to as GW model. This process was used as the framework to examine the degree of fit between steps identified in the most widely known benchmarking models and the proposed model.

Key Elements of the GW Model

Figure 1 depicts the model developed by Ghobadian and Woo (1994). It consists of five key stages: Plan; Identify; Collect Data; Analyze; and Implement Improvement Programme. The planning stage forms the core and the periphery of the model. This is because development of an overall plan and a plan for each stage is the key activity. At the planning stage the key objective(s) should be defined; quantified; and the time span for the attainment of objectives specified. These five stages apply to any benchmarking activity irrespective of contingencies. The contingencies influence and determine the necessary step at each stage. The stages of the GW model is described below.

The aim of the "identify" stage is to select the process or product to be benchmarked. Benchmarking models generally have planning as the first stage. The authors believe that "identification" should proceed "planning" because benchmarking frequently does not form a part of the organisations' corporate plans or strategies. This may inhibit employees to propose benchmarking. In addition, the choice of the process/product to be benchmarked influences the content of the benchmarking plans. "Planning" is a very important stage in a benchmarking process. Mustafa Pulat, a benchmarking coordinator at AT&T Oklahoma is on the record as saying that "planning accounts for about 50 percent of the overall time attributed to a typical benchmarking project (Ghobadian & Woo 1994). It is a very important aspect of benchmarking and should neither be neglected nor under estimated". "Data Collection" is concerned with the identification and collection of comprehensive data relevant to the particular benchmarking study. It includes both data from within and outside the organisation. The penultimate stage is to "analyze" the information and data. This analysis forms the basis of the action and implementation plan. It must be remembered that for benchmarking to be truly effective, the intention of the benchmarking exercise should be the implementation of the findings. The final stage is to "implement improvement programme".

Comparison of Benchmarking Models

The following is a comparison between the six benchmarking models using the GW Model as the framework. Where applicable, the step number of the individual models are shown in brackets. This will provide an idea of how different processes assign different order to similar tasks.

Identify
ALCOA: Deciding what to benchmark (1);
AT&T: Project conception (1); Best in class selection (4)
IBM: Clarify customer and outputs (1); Define appropriate measurements (2); Prates/select what is to be benchmarked (4); Choose benchmarking partner (5);
Rover Group
Xerox: Identify process (1); Identify comparatives (2);

Plan
ALCOA: Planning the benchmarking project (2)
AT&T: Planning (2); Implementation planning (7)
IBM: Review and refine the process (3); Set a level of data collection (6).
Rover Group: Plan Investigation (1); Plan and Implement Actions (4)
Xerox: Develop action plans (8).

Collect Data
ALCOA: Understanding your own performance (3); Studying others (4)
AT&T: Preliminary data collection (3); Best-in-class data collection (5);
IBM: Collect and organise data (7).
Rover Group: Measure and Analyze (2)
Xerox: Data collection (3)

Analyze
ALCOA: Learning from the data (5)
AT&T: Assessment (6)
IBM: Calculate gaps (8); Estimate future achievement (9); Present benchmarking results (10); Set goals and action plans (11)
Rover Group: Measure and Analyze (2); Review and Recalibrate (5)
Xerox: Determine gap (4); Project future performance (5); Establish functional goals (7);

Implement Improvement Programme
ALCOA: Using the findings (6)
AT&T: Implementation (8); Recalibration (9)
IBM: Implement actions and assure success (12)
Rover Group: Plan and Implement Actions (4)
Xerox: Implement & monitor (9); Recalibrate (10).

The comparison showed that steps identified in other models fell within one of the stages of the proposed GW Model. The advantage of the GW Model is twofold.

First, it consists of only five key stages and thus does not appear as daunting as some of the other models. Second, it is not prescriptive in terms of prescribing the steps necessary to carry out a benchmarking exercise. Furthermore, it recognises that detail steps and plans are significantly influenced by contingency factors. Thus, it is a more flexible process.

Benchmarking with the GW Model

The comparison carried out in the previous section shows that the GW Model contains the salient steps of the benchmarking processes reviewed. The key stages identified in the GW Model can be subdivided into a number of steps. The proposed steps within each stage are assigned with a number running consecutively from step 1 in the "Identify" stage to step 18 in the "Implement Improvement Programme" stage.

Identify Step 1, "project conception". This step is an appropriate first step because it enables benchmarking to be initiated at both corporate and process levels. Step 2, "clarify customer and outputs" can be used later to help prioritise what is to be benchmarked. Step 3 is to "identify critical performance measures". This step forces the benchmarker to identify the elements that are essential, and to determine measures appropriate for each operation. Furthermore, this step can also help prioritise what is to be benchmarked. Step 4 is concerned with "deciding what to benchmark". Step 5 is "select benchmarking partner".

Plan Planning takes place in this stage. The GW Model is sufficiently flexible to enable the benchmarker to plan for the study before identifying the subject for benchmarking, or to firstly identify what to benchmark and then plan for the remaining steps and stages. Step 6, "build leadership team commitment", brings in leadership commitment and involvement. Step 7, "commission the benchmarking team" involves identifying team members and providing the necessary training. Step 8, "determine data collection method".

Collect Data There are three steps in this stage. Step 9 "preliminary data collection" is used to provide an idea of the process for benchmarking. This is linked to the step where the process to be benchmarked and the selection of the benchmarking partner is decided. Step 10 is to "collect internal data" and understand the process to be benchmarked. Step 11 is to "collect best-in-class data" and organise the data into usable form for analysis.

Analyze This stage is concerned with the analysis of data, both internally and externally collected. This stage contains 3 steps. Step 12 "analyze and compare data and determine gaps" that exist; Step 13, "estimate and project future achievement"; and Step 14 "establish functional goals and develop action plans".

Implement Improvement Programme Four steps are involved in this stage: Step 15 "communicate findings and recommendations"; Step 16, "implement plans"; Step 17, "monitor progress"; and Step 18, "review and recalibrate".

Conclusion

The GW benchmarking Model has 18 steps. This, compared to IBM's 15 step process and Xerox's 10 step process, appears to be complicated. However, the 18 steps are only suggestions rather than a prescription. Organisations can pick and choose their own steps within the wider framework. Benchmarking is not an easy and straight forward procedure to carry out. The findings suggest that benchmarking does not necessarily follow a predetermined path. It is possible, when necessary, to take a loop back to a previous stage. Planning is central to a benchmarking study.

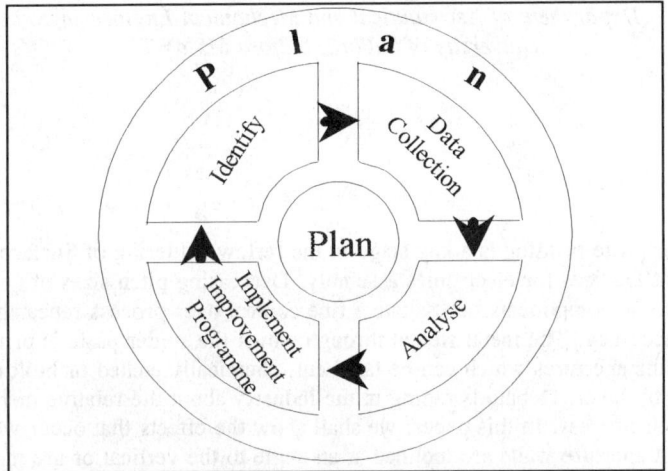

Figure 1 The GW Benchmarking Model

References

ALCOA 1990, *Benchmarking: An Overview of Alcoa's Benchmarking Process*, pamphlet no. A15-15193, Aluminium Company of America.
Camp., R. C. 1989, *Benchmarking: The Search for Industry Best Practices That Lead to Superior Performance*, 1st edn (ASQC Quality Press, Milwaulkee, WI).
Foster, T.A. 1992, Logistics Benchmarking: Searching for the Best, *Distribution*, March 1992, pp 31-36.
Geber, B. 1990, Benchmarking: Measuring Yourself Against the Best, *Training*.
Ghobadian, A. and Woo, H. 1994 Cases in Benchmarking, Unpublished Research, Centre for Interdisciplinary Strategic Research, Middlesex University 1994.
Spendolini, M. 1990, *The Benchmarking Book*
Whitting, R. 1991, Benchmarking: Lessons from the Best-in-Class, *Electronic Business*.
Willhite, C. 1991, The Migration from Business as Usual to World Class (and on to World's Best), 3M, Weatherford, OK, *Proceedings of the Fourth Annual Oklahomans for Quality Conference, Oklahoma City, OK.*

EFFECT OF STENCIL APERTURE WALL FINISH AND SHAPE ON SOLDER PASTE PRINTING IN SMT

Dr Ndy N Ekere, Dr Samjid H Mannan, Miss Ismarani Ismail and Mr Mark A Currie

Department of Aeronautical and Mechanical Engineering
University of Salford, Salford M5 4WT

Solder paste printing is a key stage in the reflow soldering of Surface Mount Devices for electronics assembly. Decreasing pitch sizes of electronic components necessitate a fine control over process repeatability and accuracy. The metal stencil through which the solder paste is printed contains apertures which can be laser cut, chemically etched or built up, layer by layer. Debate is raging in the industry about the relative merits of each process. In this paper, we shall show the effects that occur when stencil aperture walls are inclined at an angle to the vertical or are more irregularly shaped. Similarly the effects of wall surface roughness are examined to see what type of roughness will cause solder paste to adhere to the walls instead of passing through.

Introduction

The printing of solder paste onto a printed circuit board at an acceptable quality standard is essential to successful reflow soldering of surface mounted components. Stencil printing is currently the most popular technique for depositing solder paste. The stencil defines where and how much solder paste is deposited onto the PCB pad. Stencil masks are made of solid metal foil traditionally chemically etched to provide a defined pattern of cavity openings corresponding to the component lead pads. Problems that may occur with the printed deposit include skipping; paste is not fully transferred through the stencil and also bridging; neighbouring deposits coalesce.

Stencil Material

An important consideration in stencil material selection is its mechanical and chemical properties such as strength, durability and the ease of fabrication. The most widely used materials for fabricating stencils are: brass, stainless steel, electroformed nickel and molybdenum. With brass, nickel plating helps to reduce the chamfers on aperture walls left by chemical etching. Brass is however easy to etch and readily available in all sizes. Stainless steel with its ability to resist harsh chemicals is one of the preferred materials, as companies move away from CFC cleaning. Stainless steel is also more resistant to coining, does not oxidise, and tolerates handling better than brass. Stainless steel is more suitable due to its higher strength and durability.

Stencil Fabrication

Three major stencil fabrication techniques are currently used in the industry: Chemical etching, Laser cutting, and Electroforming (not covered by our experiments; see eg. Coleman (1993)).

Despite printing trials with several chemically etched and laser cut stencils no consistent trend has been found by the authors in either bridging or skipping. Chemically etched apertures are generally wider at the ends than the middle (by ≈ 13%), which may cause problems with skipping or bridging at fine pitch, but they are smoother than laser cut stencils; see figure 1. Figure 1 shows surface irregularities of 1μm, with larger features (>2.5μm) at a density of 1/2500μm², and a central ridge which can be markedly reduced for fine pitch apertures.

Figure 1. Chemically etched aperture (x 400)

Laser cutting involves the use of a finely focused laser beam for sequentially cutting out apertures. The aperture walls are relatively straight, although the walls slope outwards. Laser cutting is very flexible (both aperture size and location can be changed easily, and consistency in aperture width is high; ≈3% deviation was found). The sequential nature of the process makes it rather slow and expensive, and other areas of concern relate mostly to the aperture wall smoothness; fig.2 shows a laser cut aperture; surface undulations, 30μm in width occur.

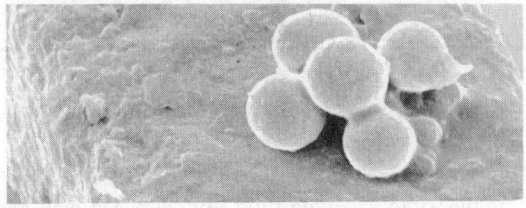

Figure 2. Laser etched aperture (with solder paste particles attached to walls; x 400)

Analytical Study of particle - wall drag

Particle Packing Near Boundary

Solder paste consists of solder spheres (20-38μm in diameter) suspended in a liquid. By printing onto a glass slide and viewing through the glass, the distances from the particles to the boundary and the average number of neighbours of each particle at the boundary were determined, and compared to the statistics produced by a computer which generates randomly distributed spheres with the same size distribution.

Figs.3 and 4 show that while the solder paste particles are almost randomly distributed, spheres tend to group in a hexagonal pattern (6 neighbours), reminiscent of close packing. 70% of particles are within 5 microns of the wall, so that stencil irregularities larger than 2-3 microns may have a significant effect on particle motion.

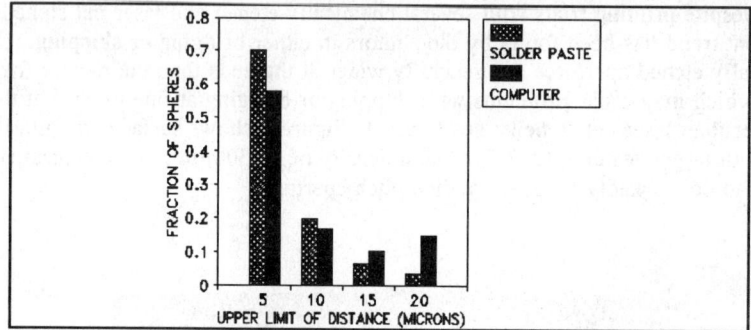

Figure 3. Distance from bottom of sphere to boundary

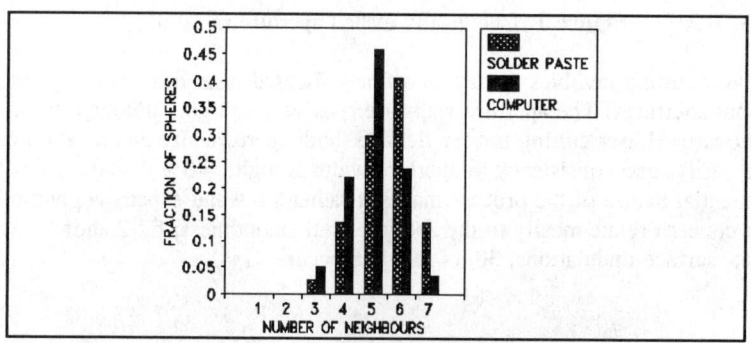

Figure 4. Average number of neighbours surrounding sphere in 2-D

Effect of surface roughness

Using lubrication theory to calculate fluid flow between wall and particle Goldman, Cox and Brenner (1967), the drag force on a single particle can be estimated. For a single particle the drag due to movement parallel to wall, F_D, is;

$$F_D = 6\pi\mu ua\frac{1}{2}\ln\left(\frac{h}{a}\right) \tag{1}$$

where μ is the liquid viscosity, $\mathbf{u}$ is the particle velocity parallel to the wall,

a is the particle radius, and **h** is the gap between particle and wall. This equation represents the leading contribution to the drag as **h** tends to zero, allowing the addition of forces acting on each particle separately. Eq(1) must be summed and averaged over all the particles at the wall to give the average force acting on a particle.
Using the results of the previous section we estimate 70% of particles with **h** = 2.5μm microns and 20% of particles with h = 7.5μm (particles further away contribute negligibly). Hence we obtain an average force of F_{aveD};

$$F_{aveD} = 6\pi\mu ua \times 0.67 \tag{2}$$

Similarly the adhesive pull of the substrate, F_P is given by;

$$F_P = 6\pi\mu ua\frac{a}{h} \tag{3}$$

Averaging over the same values of **h** as before we obtain an average force, F_{aveP}, of;

$$F_{aveP} = 6\pi\mu ua \times 3.92 \tag{4}$$

The ratio of drag to pull forces on the paste, **R**, is hence given by;

$$R = \frac{A_{wall} \times F_{aveD}}{A_{pad} \times F_{aveP}} \tag{5}$$

where A_{wall}, A_{pad}, are the aperture wall and pad (substrate) areas respectively. For a fine pitch aperture (eg. 0.15mm wide, 0.15mm deep, 1.5mm long) aperture, A_{wall} is 0.495mm², A_{pad} = 0.225mm², hence **R** = 0.37. Thus it seems unlikely that skipping occurs because adhesive pull is insufficient. In fact the tensile stress in the paste inside the aperture becomes too high; this is supported by the fact that when skipping occurs, a thin layer almost always adheres to the pad, the rest clogging up the apertures.

The fraction of area of a plane taken up by a particle in a close packed situation is 0.74 whereas in the computer generated random distribution of the previous section, the fraction is 0.60. Since the observed distribution is intermediate between these two, we average the above to obtain the number of particles at a wall in area A to be 0.67A/(πa²). The tensile stress in a paste layer adjacent to the pad, **T**, is given by;

$$T = 6\pi\mu ua \times 0.67 \times \frac{0.67}{\pi a^2} \times \frac{A_{wall}}{A_{pad}} \tag{6}$$

and when this expression exceeds the tensile strength (experimentally determined to be $\approx 10^4$Kg/m²) the paste breaks off in the aperture. The situation is complicated by the fact that the solder paste layer nearest the walls is slowed down by the wall drag, and may lose contact with the pad. This leads to **T** exceeding the value in eq.(6) by up to 74% in the case of the fine pitch aperture with dimensions quoted above (caused by the decrease in effective pad area when a 30μm wide paste layer around the edge of the pad is subtracted from the pad area). A computer simulation which takes into account only the leading order lubrication forces amongst the particles, is at present under development, but preliminary results do indicate that the wall layer becomes disconnected from the pads.

Increasing surface roughness has two major effects. Firstly the particles near the walls will tend to roll more readily. This can easily lead to increased stresses in the paste. Secondly roughness enhances skipping due to reduced average values of **h** (eq.(1)) for the first prints. On subsequent prints a layer of liquid will be trapped by the surface features, possibly increasing the average value of **h**, depending on whether the irregularities are dense enough and uniform enough to present an effective plane wall to the particles as they approach the wall, thus keeping the particles at a distance.

Effect of inclination

Aperture walls can be inclined to form a ridge (fig.1), or a trapezoidal shape with the top smaller than the bottom. The ridge is undesirable; as the gap through which the particles must pass narrows, jamming can occur. To see when this might occur, assume the normal aperture width to be X_{max} and the minimum (at the ridge) to be X_{min}. If the volume fraction of solid at the ridge (Φ) is greater than 58%, it is known that the paste ceases to deform as a fluid. Particles will be pushed into closer proximity at the ridge, and if we assume that Φ is inversely proportional to the area enclosed by the walls (ie. assuming minimal rearrangement of particles) we obtain;

$$\frac{X_{max}}{X_{min}} < \frac{0.58}{0.51} \tag{7}$$

where 0.51 is the normal value of Φ and eq.(7) expresses the condition that jamming does not occur. If $X_{max} = 150\mu m$, then X_{min} must be larger than $132\mu m$ for jamming not to occur; the ridge height must be below $9\mu m$; achievable by today's technology.

For a trapezoidal aperture, lateral paste motion is expected. This allows easier sliding of layers over each other, so easing skipping. If the paste is forced to leave the wall vertically the force preventing this, F_V, is (using eq.(1) and eq.(3));

$$F_V = 6\pi\mu ua\left(\sin^2\theta\frac{a}{h} + \cos^2\theta\frac{1}{2}\ln\left(\frac{a}{h}\right)\right) \tag{8}$$

where Θ is the angle of inclination and eq.(8) exceeds eq.(1) by 36% at $\Theta = 15°$. Thus paste will be dragged laterally to the edges of the deposit causing a build up of paste there, and possibly leading to the bridging reported by Xiao, Lawless and Lee (1993).

In summary we have examined how different aperture characteristics affect paste withdrawal, concentrating on skipping. Observations of particle packing at the boundary were fed into equations for stress which can be used for predicting skipping.

References

Coleman, W.E. 1993, Photochemically Etched stencils, *Surface Mount Technology*, June, 18-24.

Goldman, A.J., Cox, R.G. and Brenner, H. 1967, Slow Viscous Motion of a Sphere Parallel to a Plane Wall, *Chem. Eng. Sci.*, **22**, 637-651

Xiao, M., Lawless, K.J. and Lee, N.C. 1993, Prospects of Solder Paste in the Ultra-fine Pitch Era, *Soldering and Surface Mount Technology*, **15**, 5-14.

GAS SHIELDED METAL ARC WELDING, IMPROVING QUALITY BY MINIMISATION OF DEFECTS

C S Kirk, A R Mileham*

Norton Radstock College, South Hill Park, Radstock, Bath, BA3 3RW.

* School of Mechanical Engineering, University of Bath, Claverton Down, Bath, BA2 7AY.

There is a need to increase quality in weldments effected by the Gas Shielded Metal Arc Welding Process, (GSMAW). This need exists because inherent defects in the process make production quality unreliable on a consistent basis. This is evident in all sectors; from the Onshore/Offshore Petro-Chemical Industry, through structural building projects, to manufacturing in the wider sense. Problems are exacerbated by the fact that new European Standards, which encompass welder and procedure approval, demand stringent quality levels in all cases.

This paper reassesses the GSMAW process in the light of the new Standards, and shows, with new techniques, retraining of operators, and a detailed assessment of flux cored wires and shielding gases, that the process could be capable of achieving requirements on a consistent basis and moreover, that the quality levels can, indeed, be increased.

Introduction

As industry comes out of recession, and the consequential increases in production are seen, process improvement could be an effective route to competitive advantage. This is particularly important in the pan-disciplinary engineering sector of joining by welding.

The worldwide desire to raise quality standards promoted by the ISO 9000 series (in Britain and Europe, BS 5750, and EN 29000), has led to an increase in the use of full quality systems. As these are principally organisational, they place greater reliance on procedures and instruction sets, and in welding these are now evident mainly in the form of the ASME IX and BS EN 288 standards. The European Standard BS EN 288 substantially replaces BS 4870 and has had full effect since 1992. Whilst it is possible, by reference to BS EN 288 Part 1 (5.2), to approve a current Welding Procedure Specification by reference to satisfactory authenticated welding experience (such as procedures approved under BS 4870, BS 2633, BS 2971, or BS 5135), this could well lead to the use of procedures which do not comply with the stringent requirements of BS EN 288 Part 3, and BS EN 25817. This is particularly likely if the test piece is of thin walled section. In the main therefore, and in any event all new procedures, will need to be re-assessed and re-tested before approval may be certificated.

Generally, the tests follow the accepted global restrictions on fusion, penetration, micro-fissures/cracks, and porosity, but the author is of the opinion that the tightened restrictions on cap and root bead height exacerbates the problems encountered with inherent defects common to GSMAW process.

Guidance on quality levels for imperfections is expounded in BS EN 25817:1992, and is used to establish the acceptability of welds when seeking compliance with BS EN 287 or BS EN 288. Table 1 summarises common quality levels across different standards.

Table 1: Comparison of Quality Guidelines in Welding Guide for Butt Welds 3mm to 10mm in thickness

	BS 4870 BS 4871	ASME IX	BS EN 287/8 BS EN 25817
Volumetric NDT	100%	Optional	100%
Indiv. Solid Inclusions	l=0.50t 6mm max w=1.5mm max	3mm max	0.30t 3mm max
Lin. Solid Inclusions	8% of 12t by length	1t of 12t by length	N/A
Individual Porosity	0.25t 3mm max	0.20t 3mm max	0.30t 3mm max
Group Porosity	2% by area	7% by area approx	4% by area
Elongated Indications	l=6mm max w=1.5mm max	0.20t 3mm max	0.30t 2mm max per 100mm
Cap Height	Smooth, no max	N/A	1mm + 0.1 cap dim. max 5mm
Excess Penetration	3mm max, some excesses allowed	N/A	1mm + 0.3 bead dim. max 3mm

Defects - Porosity and Elongated Indications

Having observed that many new restrictions are now in place, it is appropriate to re-examine the current state of GSMAW to establish the feasibility of its use as a process capable of high quality output, and in particular to provide direct practical suggestions to overcome problems where they are most evident.

There exists a substantial interest in this area, as designers in all sectors seek to embody speedier more cost effective joining technology whilst maintaining objective, critical, requirements for high quality. This requirement extends to the Petro-Chemical Industry, including Onshore and Offshore work, together with structural building projects and manufacturing in the wider sense.

Inherent defects exist with the GSMAW process, (AWS Welding Handbook, 1991). These are principally porosity, elongated indications (wormholes), and lack of fusion, (BS EN 26520 terminology).

The use of flux cored electrodes markedly assists in the reduction of porosity, as the chemical composition of the gas in solute which would give rise to the cavity, has been reduced by the interaction of flux agents, which have deoxidised or denitrified both the liquid metal from the electrode, and also in the weld pool prior to solidification, (AWS Welding Handbook, 1991). This is also a particularly efficient reaction, assisting the plasma column equilibrium by causing multiple ionisation, since the first, and sometimes the second stage ionisation potential of the most common core elements within the electrode are all below that of the first stage ionisation potential of the argon shielding gas, (Anon, The Physics of Welding , 1986), (AWS Welding Handbook, 1991).

It has been reported, (Lancaster, 1987), and (Raja A et al, 1989), that the outer metal sheath of the cored electrode, melts at a faster rate than the flux core, thereby acting in a manner which is the reverse of a manual metal-arc electrode. Further, by photography, it has been shown that the central flux tapers, and is surrounded by the plasma column, as the arc root is at the metal sheath, not the flux core. This suggests that any deterioration in the cover provided by the shielding gas, would permit instantaneous reaction of the liquid portion of the electrode with the atmosphere. It is therefore reasonable to assume that the gas, (oxygen and nitrogen), in solution, are transferred into the weld pool directly, not by poor reaction with the flux agents in the plasma column, but by inconsistent protective gas shielding. Even with adequate shielding, (as exists most of the time), care must be exercised with observations that ideal reactions are wholly beneficial, since an oxidised metallic element is non-metallic and will affect the mechanical properties of the weld, notwithstanding that this assists compliance with the standards by minimisation of porosity.

The delivery of the shielding gas to the area surrounding the arc and the weld pool has remained substantially unaltered since the process was introduced in the 1950's. It is considered that this could have a major impact on weld quality and is therefore being actively researched by the author. The delivery of gas at reasonable flow rates through a tubular shroud is effective in reducing defects, but is extremely susceptible to side draughts. Additionally, chemical composition of the gas is extremely important in the effective maintenance of the arc column and internal plasma jet stream. The action of the arc is extremely violent in comparison to it's size, (Lancaster J F, Chapter 7, The Physics of Welding, 1986), even in relatively stable working conditions such as those experienced in gas shielded tungsten arc welding, and even more so when the electrode composition is constantly being changed from solid through liquid to vapour and undergoing severe magnetohydrodynamic forces. The result of this violence is that the surrounding gas shield is subject to turbulence at precisely the time when it should be at it's most stable if it is to contribute effectively to the maintenance of virtually total thermal equilibrium by ensuring optimisation of electron collision through the ionisation of the gas.

This whole problem is compounded by the fact that the shielding gas shroud and contact tip is being moved in relation to the work piece as welding progresses. It has been reported by (Okada T et al, 1979), using a He-Ne laser as a light source with a band pass filter, that eddy turbulence occurs behind the gas shroud, particularly at low flow rates. This reporting further confirms that it is reasonable to assume that oxygen and nitrogen from the atmosphere can combine with the molten electrode moments before it strikes the weld pool, and furthermore in sufficient quantities to induce both porosity and cavity creation, notwithstanding the fact that the introduction of diatomic gases to argon may alter the rate of absorption, (Lancaster J F, 1987). Porosity will occur, (Lancaster J F, 1987), when nuclei for the heterogenous nucleation of bubbles are present in the supersaturated weld metal, and the rate of escape of the bubbles is less than the rate of solidification. Under marginal conditions, this will form at the weld boundary, whereas elongated indications will form at the solidus/liquidus boundary, and progress as solidification occurs. Since the material transferred across the arc may be supersaturated with atmospheric gases if the shield has been disturbed, the efficiency of the chemical reactions provided by the flux core are reliant upon the slag-metal interaction in the weld pool. Should the supply of reagent be insufficient, porosity will occur to some degree. Progress has been made in the development of flux compositions, but it is important to give further attention to ensuring the adequacy of shielding gas as welding takes place.

In this regard therefore, further experiments are in hand for the re-design of the contact tip/ gas shroud assembly, necessarily including the consideration of a second supply of gas in line with the wire feed, whilst having due regard to the positional aspects of the torch, which, in site welding give rise to increased incidence of clogging by spatter. It is believed that this will enhance protection of the electrode precisely at it's most vulnerable point whilst at the same time assisting with the maintenance of optimum electron collision in the plasma column. Theoretically the result will be a more efficient superheated arc, which will further enhance penetrative aspects without promoting burn-through caused by additional arc force.

Defects - Lack of Fusion

Lack of fusion at the root, side wall, or inter-run, causes inconsistencies in the quality of the process as a whole. Despite recent advances in the creation of synergic pulsed MIG and semi-robotic machinery, one must make a basic observation that the size of the arc emanating from say a 4mm manual metal arc electrode to that of say a 1.2mm solid or cored wire is very different, and they have as their base a different volt-ampere output characteristic. This leads to the supposition that fusion problems are being encountered through the inconsistencies in overall size and the directionability of the arc, and this markedly affects the ability to maintain optimum temperatures at the faster travel speeds in GSMAW welding. The advancements in a number of cored electrodes enhance deposition, but these are generally unsuitable for thin walled sections.

A welding power source has individual volt-ampere output characteristics, and in GSMAW these are principally straight line. Subject to the arc length being self adjusting, variations in voltage affect the spread of the arc, and the area of penetration appears to be determined by the heat of the molten metal transferred as droplets or vapour, and the depth of penetration by their impact in the molten weld pool, (Essers W G and Walter R, 1979). The mode of metal transfer changes with current output, and progresses through transitional currents which are reduced as electrode extension increases, since the ohmic heating effect weakens the boundary between the liquid drop and the solid tip, resulting in an increase in the drop frequency, (Rhee S and Kannatey-Asibu Jr E, 1991).

With the increased use of cored wires, further research needs to be conducted, particularly in regard to the practical aspects of welder control over optimum electrode extension, and the author is currently conducting further experiments in this field. Such experimental work naturally differs from previous work done in connection with solid electrodes due to the presence of the inner non-conducting core. This results in substantial differences in the pre-heating and transfer characteristics, and as many manufacturers of cored wires give optimum electrode extension dimensions these must be maintained consistently if high quality deposition is to be maximised. This has led to examination of the contact tip itself with it's potential for re-design, incorporating the use of a non-conducting end piece through which the pre-heated wire will pass. Additional experimentation will permit further controlled practical study of the ohmic heating effect in cored electrodes whilst providing valuable assistance to those engaged in site work using this process and will provide data which will be of interest to those engaged in the welding of thin walled sections. It is hoped that this will contribute to alleviating the problems of inadequate fusion, particularly at low amperages.

Conclusion

Much of the foregoing re-assessment of the process has been necessitated as a result of observations regarding the difficulty with satisfying the stringent criteria embodied in the European Standards. It is hoped, however, that benefits from the author's research will traverse the whole welding field and not simply be concentrated in the specialist welding of thin walled sections. This is vital today, because, amongst other things, designers continue to select structural hollow sections to match strength and suitability in service with cost and ease of construction. Welding methods must be available to achieve a balance of speed and cost effectiveness, and, above all, quality, to match the choice of this material. If the inherent defects in GSMAW can be minimised on a consistent basis to enable quality to be raised to at least the stringent levels of BS EN 25817, then the whole process will experience an elevation as a qualitative process providing the answer to requirements of design engineers desperate for more effective control of weld time and costs.

References

American Welding Society, 1991, *Welding Handbook,* 8th Edition, Section 2, Miami,AWS
Anon, 1986, *The Physics of Welding,* 2nd Edition, International Institute of Welding, Study Group 212, Oxford, Pergamon Press.
Essers W G and Walter R, 1979, *Some aspects of the penetration mechanisms in MIG,* Proc. of the International Conference on Arc Physics and Weld Pool Behaviour, Cambridge, The Welding Institute.
Lancaster J F, 1986, *The Physics of Welding,* 2nd Edition, International Institute of Welding, Study Group 212, Oxford, Pergamon Press.
Lancaster J F, 1987, *Metallurgy of Welding,* 4th Edition, London, Allen and Unwin.
Okada T, Yamamoto H, and Harada S, 1979, *Observations of the shielding gas flow pattern during arcing by the use of a laser light source,* Proc. of the International Conference on Arc Physics and Weld Pool Behaviour, Cambridge, The Welding Institute.
Raja A, Rohira K L, Srinivasmurthy K S, Ramura K S, 1989, *Some aspects of the penetration mechanisms in Flux Cored Arc Welding,* Tiruchirapalli, India, Welding Research Institute.
Rhee S and Kannatey-Asibu Jr E, 1991, *Observation of Metal Transfer During Gas Metal Arc Welding,* Proc. of the Winter Annual Meeting of the ASME, New York, ASME.

A GENERIC COMPUTER MODEL FOR HIGH SPEED MACHINING WITH AN INTEGRATED DATABASE

Mr Israel Dagiloke, Dr Andrew Kaldos, Dr Steve Douglas and Prof Ben Mills

School of Engineering and Technology Management
Liverpool John Moores University, Liverpool, L3   3AF

The selection of optimum cutting parameters for high speed machining (HSM) is absolutely critical and dependent on the adequacy of the cutting model employed. The general equations best describing the high speed machining process form a set of non-linear and partial differential equations which are too complex to solve analytically. A numerical solution is therefore required involving a reliable technique for developing an adequate computer model. In this paper an interactive generic computer model for evaluating the cutting process at high cutting speed complete with an integrated database is developed and discussed.

Introduction

The fundamental reason why there has been significant interest both in the metal cutting industry and research community in recent years in high speed machining technology is the claim that it can dramatically increase metal removal rates resulting in reduced machining times and increased productivity, Flom, Komanduri and Lee (1984). Despite the importance of the high speed machining process, very little has been published on the cutting process prediction at high cutting speed. This can be attributed to the fact that the high speed machining process is, perhaps, one of the most complex of the manufacturing methods employed today. It involves various input and output parameters linked together through a multitude of complex interactions.

It is well documented by King and Vaughn (1984), that as the cutting speed increases above the conventional speed range, new dynamic ranges are encountered, for example, in respect of the cutter and material interface, the basic chip morphology changes as new cutting phenomena are experienced which include the momentum force effects caused by momentum change of the chip from undeformed to deformed chip in the shear zone. The emergence of momentum force at high cutting speed causes an increase in the cutting forces and the cutting power. The result of this is that machine tools must have adequate power, speed control and that the prediction of process parameters is vital in the machine tool design process.

Experimental data and a satisfactory mathematical simulation model are therefore vital in determining the factors which influence productivity at high cutting speed. Due to the complexity of the physical processes involved at cutting speeds above 600 m/min, reliable analytical and experimental data on cutting forces and temperatures are virtually non-existent or limited in the public domain. Information on cutting process parameters at high cutting speed is therefore of both practical and fundamental value.

An in-depth literature review and the study of the manufacture of large aerospace components suggests that efficiencies of HSM have been investigated to some extent by King and McDonald (1976). Although the purpose of HSM has been based on economies in production, most of the previously published work has focused on the difference in cutting mechanism between high speed and conventional low speed machining rather than in improving the economics of production von Turkovich, (1979), Arndt (1973), Okushima, Hotomi, Ito and Narutaki (1965). A neglected aspect of high speed machining technology has been the implementation of the technology in industry based on a systematically integrated total analysis of the high speed machining process.

The purpose of this paper is to describe a generic computer process model with integrated databases that can be used to predict the effects of the input cutting parameters (cutting speed, feed, rake angle and depth of cut) on the cutting process and on the selection of machine tools, cutting tools and workpiece materials in high speed machining in an interactive way.

Numerical modelling techniques of HSM and program design

In developing the cutting process model, several areas of workpiece behaviour and the mechanics of machining were explored. As an essential part of this research work, relevant cutting models related to this study were compiled and in some cases critically modified to suit conditions of the present application. The philosophy used was to create and implement an extensive and comprehensive theoretical model, based mathematically on basic cutting process parameters. Such a model makes it possible to show at a glance what the dependency of the cutting parameters and their effect within the simulated results. Further, by solving the equations, simultaneously as a set, it is possible to show how the user's choices affect the results of the model.

A computer program for modelling high speed machining processes partly based on general equations published by Dagiloke, Kaldos, Douglas and Mills (1994) and some other basic equations of the mechanics of cutting has been developed to obtain the cutting power, forces, temperature and other process parameters. To establish a broad data base for future access, all the data acquired for workpiece and tooling mechanical and thermal properties, cutting conditions, workpiece geometries, tooling geometries, machine tool parameters, generated numerical and graphical results data were stored on a DEC Vax computer system.

Numerical modelling structure

Figure 1. illustrates the structure of the cutting simulation system which consists of three major modules: integrated input database, numerical simulation and output modules.

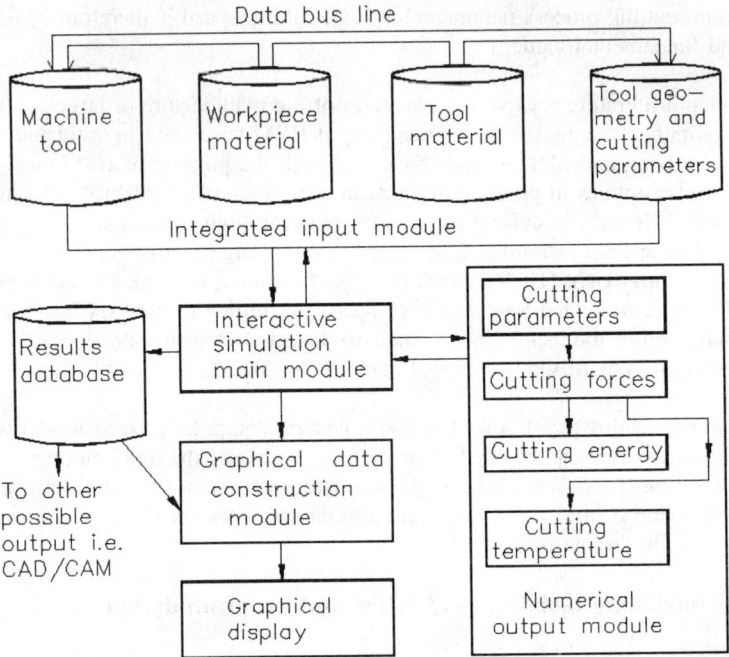

Figure 1. The structure of the cutting simulation model.

The input module has an integrated database on the following modules:
(i) Workpiece and tool materials thermal and mechanical properties, e.g. workpiece density, thermal conductivity, specific heat capacity, initial workpiece temperature and Brinell hardness.
(ii)Tool geometry:Tool diameter, number of teeth or flutes, tool shank diameter, rake angle and flute length.
(iii) Workpiece geometry : Blank size, length, height, width and geometrical features.
(iv) Cutting parameters : Feed, feed rate, depth of cut, width of cut and cutting speed.
(v) Machine tool parameters: Range of operation for feed rate in x,y and z axes, spindle speed and power, tool length/diameter in x,y and z axes, i.e. working envelope.

Having identified the set of input information required, the program was designed to transform the geometric, mechanical and thermal properties information of the workpiece, tools, machine tool and cutting conditions from the databases into a suitable form for simulation. Alternatively a new set of input data can be entered into the model. The workpiece, tool, machine tool and cutting conditions data files in the modules can be edited making it possible to update and expand the databases. The input module can also be edited prior to cutting simulation.

The numerical simulation module has a built-in numerical data controller with four sub-modules which are to be selected in pre-determined order. The output module generates the numerical output data and graphical representation of the cutting process based on the numerical data obtained through the "Results database". Typical examples of graphic outputs obtained during the cutting simulations are shown in Figure 2.

Numerical simulation results

Tool forces play an important role in the economics of the metal cutting process. They control the energy consumed as well as the power requirements of the process. Therefore, the chosen simulation results which demonstrate the capability of this cutting model is the relationship between the principal cutting force, the momentum force and the cutting speed as illustrated in Figure 2. It can be seen that at low cutting speed the principal cutting force remained relatively constant, whilst, the momentum force magnitude is very low. However, as the cutting speed approaches 1200 m/min, momentum magnitude starts to increase, consequently causing the principal cutting force to increase due to the fact that the vector sum of the momentum force has to be added to the values of the main cutting force. From previous work Dagiloke, Kaldos, Douglas and

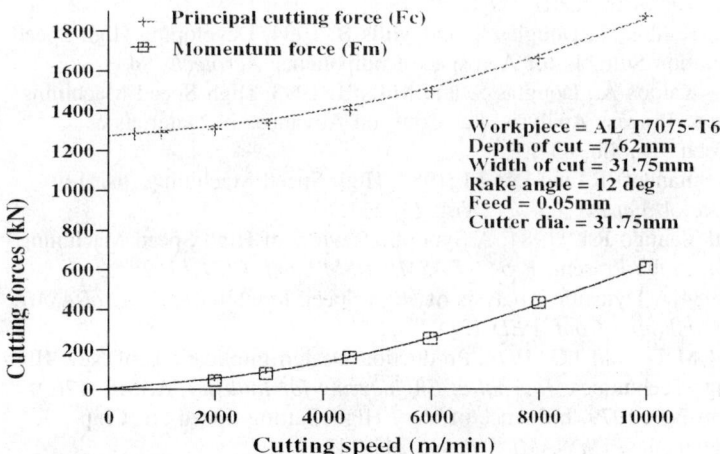

Figure 2. The relationship between the cutting speed and the cutting forces.

Mills (1993) the relationship between the cutting speed and the cutting temperatures was simulated. It was found that the workpiece and the average shear zone temperatures decrease with increasing cutting speed whilst chip and tool rake face temperatures increase. The increase in chip and tool rake temperatures can be attributed to the increase in energy dissipation at the tool face. Dagiloke et al (1994) suggested that as the cutting speed increases the energy required to remove the deformed chips reduces, however, not linearly, whilst the power required to remove the undeformed chips increases due to the effect of the momentum force. The simulated results of this model compared well with experimental data Arndt (1973), King et al (1976), Okushima et al (1965), Ber, Rotberg and Zombach (1988).

Conclusion

A generic computer model with an integrated database suitable for high speed machining has been developed and assessed by numerical simulation. Firstly all the necessary input elements were identified and discussed followed by the formulation of the required equations and programming design. In this study an extensive and comprehensive theoretical model, based mathematically on basic cutting process parameters has been created. This model makes it possible to show at a glance the dependency of the cutting parameters on cutting speed. Further, by solving the equations, simultaneously as a set, it is possible to show how the user's (process engineer) choices affect the results of the model.

References

Arndt G. 1973, Ultra-high Speed Machining : A Review and an Analysis of Cutting Forces, *Proc. of Inst. of Mech. Eng. (London)*, 1973, Vol 187, No 44/73, 625-634.
Ber A. Rotberg J and Zombach S.A. 1988, Method for Cutting Force Evaluation of End Mills, Annals of the CIRP, 37/1, 39.
Dagiloke I.F., Kaldos A., Douglas S. and Mills B. 1994, Developing High Speed Cutting Simulation Suitable for Aerospace Components, *Aerotech '94*.
Dagiloke I.F., Kaldos A., Douglas S. and Mills B. 1993, High Speed Machining: An Approach to Process Analysis, Int. Conf. on Advances in Materials & Processing Tech., Dublin, 63-70.
Flom D.G. Komanduri R. and Lee M. 1984, High Speed Machining (hsm) of Metals, *Ann.Rev.Materials Science,* Vol. 14, 231.
King R.I. and Vaungh R.L. 1984, A Synoptic Review of High Speed Machining from Salomon to the Present, *Proc. of ASME HSM Conf. PED 12*, 2.
Recht R.F. 1984 A Dynamic Analysis of High Speed Machining, *Proc. of ASME High Speed Machining Conf. PED 12*, 83-93.
King R.I. and McDonald J.G. 1976, Production Design Implications of New High-Speed Milling Techniques, *Journal of Engineering for Industry,* ASME, 176.
Turkovich von B.F. 1979, Influence of Very High Cutting Speed on Chip Formation Mechanics, *7th NAMRC Proc.,* 241-247.
Okushima K, Hotomi K., Ito S and Narutaki N., 1965, A Fundamental Study of Super-high Speed Machining, *Bul. Jap. Soc. of Mech. Eng.,* 1965, 8, No 32, 702.

APPLICATION AND DESIGN OF BURR TOOLS

Dr D Ashman*, A Hayward†, Dr W A Draper*, J Kempson*

** - University of Central England in Birmingham,
Birmingham, West Midlands. B42 2SU.*

*† - Morrisflex Ltd., London Road, Braunston,
Northants, NN11 7HX*

The paper describes the role and application of manual techniques for deburring currently used in metal working industries. The development of an automatic test rig is presented which accurately simulates the manual operation of burr tools, together with a discussion of the merits of using an experimental programme based on the Taguchi method. Finally, using the results from the experimental programme, the paper discusses the effects of changing the design parameters on the performance criteria of metal removal rate and surface finish.

Introduction

Hayward (1992) described the process of deburring as the removal of burrs i.e. rough edges left on cast, machined or punched material, in order to avoid dangerous sharp edges or the prevention of assembly interference. There are a wide range of automatic, semi-automatic and manual deburring processes including mechanical, electrochemical, chemical and thermal. Hignett (1984) showed that deburring utilising hand tools was by far the most frequently used method.

The use of hand tools can be separated into two groups namely power assisted, including rotary burrs, rasps and files, mounted abrasive points, cylindrical grinding wheels and hones, flexible abrasive wheels and brushes, and filamentary brushes, and non-power assisted, including files, scrapers and abrasive cloth. Hand-held power tools can be driven by electric motors, via flexible shafts, or by compressed air. Generally, air tools are used at operating speeds between 15,000 and 100,000 rpm with electrically powered devices used for tools requiring slower speeds but greater torque. However, slow speed/high torque air guns may be used where electric power is not available or undesirable, whilst hand tools are available where small electric motors are contained within the gun, thus negating the requirements for a flexible shaft, which can operate at speeds well in excess of 15,000 rpm.

With respect to the commonly used power assisted manual processes, a selection of rotary deburring tools is given in Figure 1. In these tools each tooth can be considered to be equivalent to a single point cutting tool. However whilst the basic principles of the cutting action are well established (Trent 1984), there are still a number of areas controlled by experience rather than by well established rules, and this is particularly true for the burr tool (Hayward 1992).

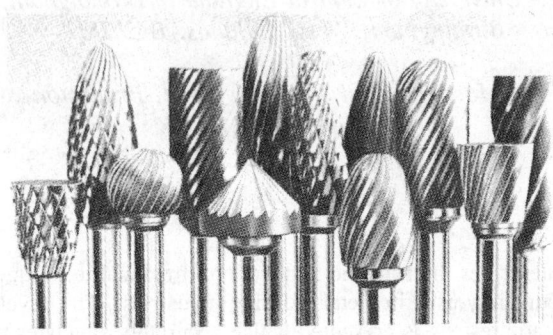

Figure 1. Burr tool types

Further examination of Figure 1 shows that a burr tool can be described as a relatively small multi-toothed cutter. Burr tools can be made of solid tungsten carbide/high speed steel/tool steel, or steel shanks tipped with tungsten carbide. Tungsten carbide burr tools are manufactured using a diamond grinding wheel whereas steel tools are more commonly milled (Dallas 1988). The size and shape of burr tools is wide with selection depending on the type of application. Small 3mm diameter "tree" shaped burr tools can be used to radius turbine blades whereas larger (up to 19mm diameter) "ball-nose cylinders" could be used to remove flashings from castings. Burr tools are commonly used on cast and stainless steels, cast iron, tool steels and other difficult-to-machine materials such as nimonics and titanium alloys. To a lesser extent they are also employed to deburr components manufactured from more easily machined materials such as brass, copper and aluminium. Whilst burr tools are predominantly used by hand, there is an increasing tendency for them to be utilised in CNC machining centres and robots (currently around 10% of the market).

The literature search, summarised in this paper, revealed that whilst on average manufacturers stated that more than 5% of their manufacturing costs were attributed to deburring and mechanical finishing, industry was investing considerably less than 5% of engineering talent to improve the situation; nor was an appropriate proportion of capital being invested in improved finishing methods.

This paper addresses the design of burr tools in order to improve their performance and assesses the relevance of appropriate experimental procedures, namely full-factorial, single factor and Taguchi arrays.

Burr Tool Testing

Traditionally, burr tool testing has been carried out by hand and whilst this is representative of their use in industry, it does not provide repeatable results due to the difficulties involved in applying a constant engaging force and feed rate. For these reasons a dedicated test rig was designed and manufactured consisting of a fixture for holding the air tool or other prime mover, with a reciprocating table on which was located the workpiece.

The air gun fixture pivots on a central support on which was hung a mass of 2kg in order to provide the necessary constant downwards cutting or feed force. The burr tool is lowered onto the workpiece, held in a fixture on the machine table, which in turn is reciprocated prior to loading via a hydraulic cylinder operated on an air-oil system. Smooth movement of the workpiece table is ensured by using air pressure to pressurise oil which then operates the cylinder. The stroke of the table is controlled by limit switches and the rig is provided with a quick stop facility which rapidly lifts the burr tool clear of the workpiece in case of emergencies.

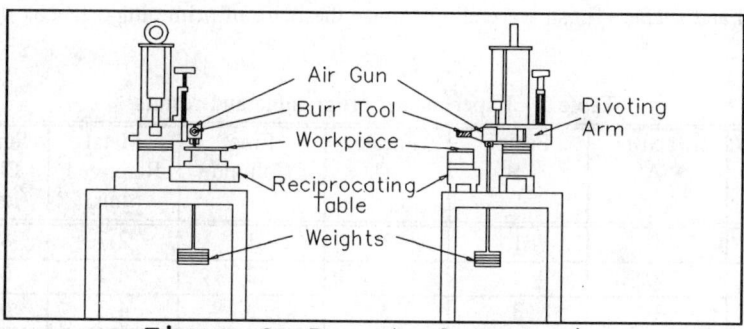

Figure 2. Burr tool test rig

Test Programme

Wu and Moore (1985) published a summary of the work by Taguchi and Wu (1985), and argued that factorial analysis techniques were unsuited to the needs of industrial experimentation and gave details of the Taguchi technique which overcame the following problems identified with conventional experimental methods :-

i) Expense - To conduct a full factorial analysis involves a relatively large number of experiments which may be extremely costly in both time and money.

ii) Errors - In the case of single factor experiments only two results can be used to determine the effect of one factor. Therefore the results would be subject to a large degree of statistical uncertainty.

iii) Analysis - The data from the results of a full factorial analysis may be difficult to analyse.

The Taguchi method analyses the effects of differing factors without having to test every combination. This is achieved by the establishment and solution of a number of simultaneous linear equations by using standard tables detailing which combinations should be tested for a range of differing factor conditions. A full description of the technique is given by Taguchi and Wu (1985). The burr tool and workpiece materials were high-speed steel and mild steel respectively. The choice and level of control factors were as follows : -

Table 1 - Choice and level of factors

Factor	Level 1	Level 2	Level 3
A - No. of teeth	12	24	36
B - Helix angle (°)	0	15	30
C - Rake Angle (°)	-5	0	5

Table 2 shows the standard L_9 Taguchi array used to determine the effects of three factors. The fourth column, labelled "Free Column", was not allocated and so used to determine any possible errors in the tests. The quality criteria of metal removed and surface finish were chosen since these are of prime importance.

Table 2 - Experimental programme and results

Test No.	Factor A	Factor B	Factor C	Free Column	Metal Removed (gms)	Surface Finish (μmRa)
1	1	1	1	1	2.7	8.74
2	1	2	2	2	4.1	6.52
3	1	3	3	3	4.3	6.93
4	2	1	2	3	28.0	4.41
5	2	2	3	1	63.0	3.21
6	2	3	1	2	19.9	1.92
7	3	1	3	2	46.3	2.11
8	3	2	1	3	24.0	1.39
9	3	3	2	1	27.0	1.57

Diameter of cutter	= 16mm	Rotational speed	= 9000rpm
Load	= 20N	Stroke length	= 55mm
Axial length of cut	= 16mm	Table speed	= 3.2m/min
Duration of test	= 10min.		

Results and Analysis

The results in Table 3 show that surface finish improves with an increase in both the number of teeth and helix angle, whilst the metal removed increases with a more positive rake angle. These findings are typical of the results obtained in conventional peripheral milling experiments (Trent 1984).

Examination of the Taguchi analysis results (see Table 3), shows that when the largest values of metal removed are chosen for each factor, the optimum parameter settings are A2, B2, C3, which are supported by the findings of a full factorial experiment. The optimum parameter setting for surface finish (smaller the better), as specified by the Taguchi analysis, was A3, B3, C1. In a full factorial experiment this combination was measured at 1.88µmRa. However, the optimum combination was A3, B2, C1, which produced a surface finish of 1.39µmRa. Examination of the results for the "Free Column" indicates that a factor or combination of factors was not included. This could be addressed at a later stage.

Table 3 - Metal removed, MR (gms) and surface finish, SF (µmRa)

Level	Factor A		Factor B		Factor C		Free Column	
	MR	SF	MR	SF	MR	SF	MR	SF
1	3.7	7.40	25.7	5.29	15.5	4.04	30.9	4.51
2	37.0	3.18	30.4	3.71	19.7	4.17	23.4	3.52
3	32.4	1.69	17.1	3.47	37.9	4.08	18.8	4.24

Conclusions

Whilst burr tools are primarily used in a manual mode, an automatic test rig is required to ensure reliable and repeatable results. The most appropriate method of assessing burr tool performance is by measuring the amount of metal removed and surface finish produced. The application of the Taguchi technique to the testing of burr tools demonstrates that this method can be used very successfully. Whilst it can be seen that the optimum combination for surface finish was not found by the Taguchi programme, it could be argued that the difference between the Taguchi derived combination and the full-factorial is small and not worth the additional amount of testing necessary to produce a similar result. In addition, the extrapolation of the results using the Taguchi method, ensures that the probability of achieving a good result is higher than for a single factor experimental programme.

Acknowledgements

The authors would like to thank the University of Cental England and Morrisflex Ltd. (Northants) for the use of their facilities.

References

Dallas, D.B. 1988, *Tool and engineers handbook*, McGraw-Hill.
Hayward, A.S. 1992, *The cutting performance of burr tools,* MPhil Thesis, University of Central England, Birmingham, U.K.
Hignett, J.B. 1984, *The technology of deburring and surface conditioning in the U.S.A.*, SME Tech. Paper, MR84-947.
Taguchi, G. and Wu. Y. 1985, *Introduction to quality control*, Central Japan Quality Control Association.
Trent, E.M. 1984, *Metal cutting*, Butterworths.
Wu, Y. and Moore, W.H. 1985, *Quality engineering : Product and process design optimisation*. American Supplier Institute Inc.

A NOVEL MACHINABILITY ASSESSOR FOR FERROUS ALLOYS

Mr Ian Carpenter and Dr Paul Maropoulos

School of Engineering, University of Durham
Science Laboratories, South Road, Durham, DH1 3LE

This paper describes a machinability assessor for ferrous alloys. The system uses a combination of rules which are automatically generated for the machinability classification of materials and statistical methods for the generation of appropriate initial cutting data. The rules are used for placing any ferrous alloy within a certain class of steel and each class has cutting data assigned to several sub-classes which are categorised by mechanical properties. Multiple regression is used to derive the relationships between hardness and the main parameters of machining such as cutting velocity, feed and metal removal rate. The operation of the system is based on an initial comprehensive database of over 750 steels with associated mechanical and chemical properties and has been tested with promising results.

Introduction

Within the field of Computer Aided Process Planning (CAPP) much research work has centred upon the tasks of tool selection and calculation of efficient cutting data (Carpenter & Maropoulos 1993). Most systems require some form of applied mathematical model that is used to simulate the milling process (Smith & Tlusty 1991). More recently, attention has been drawn to the possibility of using artificial intelligence techniques to select process parameters in intelligent machining environments (Chryssolouris & Guillot 1990).

Most of these systems are only effective for workpiece materials that have well defined machining characteristics and for which cutting data is stored within the system's database. However, in the modern global manufacturing environment machinists are likely to encounter foreign ferrous alloys that are not well known or fully specified. Some of these new materials may require new or innovative methods of machining (König, Cronjäger, Spur, Tönshoff, Vigneau, & Zdeblick, 1990).

Also the wide variety of national and international tool standards and the emergence of many specialised engineering materials within certain areas such as the aerospace and the medical industry makes it increasingly important to categorise and assess the level of machinability of new materials in order to improve the efficiency of machining processes. Hence, a reliable and quick method for suggesting initial cutting data for a wide range of newly developed or partially defined materials is required.

The research work described herein aims to provide a computer-based assessment of the machinability of engineering ferrous alloys, whilst permitting the user a great deal of flexibility in the data required and the level of efficiency of the suggested cutting data.

Overall structure of the system

This paper describes the structure and operation of a machinability assessment system that produces reliable cutting data for milling operations performed on new or partially specified ferrous alloys. The system is written in FoxPro for Windows and runs on an IBM-compatible PC. In operation the user is presented with a user friendly graphical interface that allows a large amount of flexibility in what data is required to input into the system. The material classification and cutting data calculation process is altered based on how much information the user has specified as shown in Figure 1.

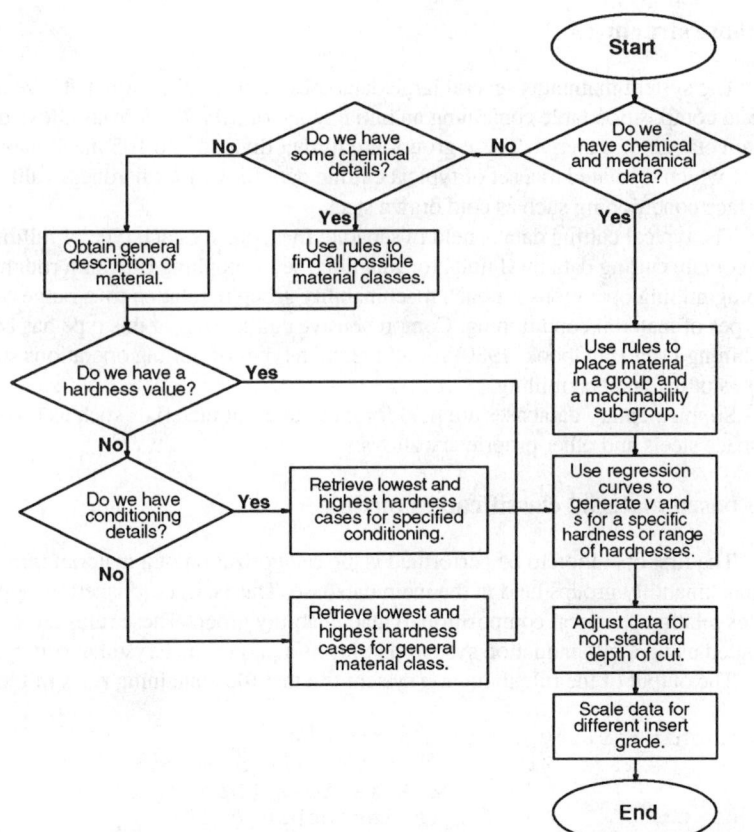

Figure 1. Overall decision making process of the machinability assessor

The user inputs data about the material that may include chemical characteristics such as the percentages of alloying elements in the alloy, and mechanical properties i.e. surface hardness. If the user is unable to give full details of the material then the system allows the user to input whatever information is available and guide the assumptions that the system makes using mainly qualitative data about the material.

For instance, if the chemical composition of the alloy is unknown then the system presents the user with a list of possible categories of alloy, along with the critical properties of each group. Similarly, if the user has input some chemical information but does not possess a figure for hardness then the system suggests a range of hardnesses for the appropriate group of alloys and the specific surface treatment of the given material. Cutting data are calculated for either a specific hardness or for the limits of the hardness range using the regression curves of the specific machinability class.

The assessor generates a relative machinability index that classifies numerically how easy the material is to machine within its own material group. Also an absolute metal removal index is generated that classifies the material amongst all ferrous alloys by considering what metal removal rate is achievable for a standard tool life when using an appropriate insert grade.

Database structures

The system maintains several large databases in dBase file format. These include a material composition table containing an initial range of over 750 ferrous alloys divided into 29 main classes of material. These groups are further divided into 105 machinability groups, each of which is related to a set of typical cutting data for various hardness values and types of surface conditioning such as cold drawn steel.

The typical cutting data is held in separate files, one for each type of milling operation. They contain cutting data in SI units for roughing, semi-roughing (medium roughing) and finishing milling operations for each machinability group in relation to a range of hardnesses and types of material conditioning. Comprehensive cutting data of this type has been collected (Machining Data Handbook, 1980) for all the main types of milling operations such as facing, slotting and end milling.

Supplementary databases are held for more unusual materials such as high alloy aerospace steels and other proprietary alloys.

Rule-based material classification module

The first operation to be performed is the categorization of a material into one of the 105 machinability groups held in the main database. This is accomplished by applying a set of rules relating chemical composition to machinability group. These rules are automatically generated using a rule induction system that forms a part of the Crystal expert system shell.

The output of the rule induction system is a text file containing rules of the form,

```
Machinability_Group_No is 1
        IF    Si is less than 0.275
              AND   NOT P is less than 0.035
              AND   C is less than 0.295
              AND   Fe is less than 98.940
              AND   NOT S is less than 0.225
              AND   NOT C is less than 0.160
              AND   Fe is less than 98.305
```

The input to the rule induction system is a large table of examples, each consisting of eighteen numeric values representing the percentages of alloying elements within the material and one symbolic output field containing the number of the machinability group for that material. The whole materials database is processed by the rule induction system to produce the rules file.

This text based rule file is parsed by a custom filter program to produce a summary data table for FoxPro, as shown in Table 1, that lists the minimum and maximum values for each alloying element within each machinability group.

Table 1. Machinability group rules in table form

Machinability Group	Fe Min (%)	Fe Max (%)	C Min (%)	C Max (%)	Cu Min (%)	etc ....
1	0.000	98.940	0.000	0.295	0.000	....
1	0.000	98.305	0.160	0.295	0.000	....
4	98.515	98.620	0.295	0.455	0.010	....
4	0.000	98.515	0.295	0.575	0.000	....

As can be seen in Table 1, there may be several different rules all relating to the same machinability group. The default minimum and maximum values are 0% and 100% to allow any material to fulfil a rule which does not specify limits for all the alloying elements. The chemical composition given by the user is used to search the rules table for all the machinability groups that the material might lie within. Generally only one rule will be fulfilled but on the rare occasion that several machinability groups are possible the user is prompted with details of each group and asked to select which he believes to be more suitable.

As the rules are generated from the main materials database it is possible to add new materials to this database and then regenerate all the rules to further refine the accuracy of the material categorization process.

Cutting data regression module

After a material has been categorized into a machinability group it is possible, using some details of hardness or conditioning, to suggest initial cutting data for a few depths of cut (typically 1, 4 or 8 mm) and standard tool materials by extracting the information from the cutting data database. However the cutting data database is generally only a conservative guide and some further modification of the cutting data will be required to customize it for specific operation details.

As cutting data is only stored for certain incremental discrete values of hardness it is necessary to use a regression technique on the stored cutting data to generate a mathematical relationship between material hardness and the critical cutting parameters, such as cutting velocity, feed rate and the corresponding metal removal rate. This regression equation is then used to generate a more precise value of the cutting parameter with regard to an exact value of hardness. A similar technique is used to interpolate a value for cutting velocity and feed for non-standard depths of cut.

The cutting data database holds information for only a small range of the most widely available tool insert grades. The user may want to use a particular more efficient grade and in this case the cutting data is converted using a table of scaling factors derived from statistical analysis of manufacturers' data. These scaling factors allow the cutting velocity to be adjusted for the use of any other suitable carbide grade and are shown in Table 2.

Table 2. Cutting velocity scaling factors for finishing facing - Material Group 1

Carbide	T25M			S10M			S25M		
	F	SR	R	F	SR	R	F	SR	R
T25M	1.00	1.00	1.00	0.93	0.93	0.93	0.83	0.84	0.84
S10M	1.08	1.08	1.08	1.00	1.00	1.00	0.91	0.90	0.91
S25M	1.20	1.19	1.19	1.10	1.11	1.10	1.00	1.00	1.00

where F = Finishing, SR = Semi-Roughing, R = Roughing

Discussion and Conclusions

The machinability assessor described herein forms part of a closed loop Intelligent Tool Selection system for milling operations. It enables the main optimisation modules of the system to operate when materials have no machinability categorization.

The computer based system uses a powerful combination of rule-based reasoning and statistical modelling to provide a reliable and robust method for assessing the machinability of a new or partially defined ferrous alloy. The rule base is created automatically by the rule induction system using a materials database which can be constantly updated to refine the material categorization rules. Similarly, the regression methods are used to further optimize the suggested process parameters for a specific depth of cut and carbide grade and they fully take into account any new information entered into the cutting data database from newly performed operations. Consequently, appropriate parameters are recommended for the workpiece material and operation details specified by the user.

The machinability indices generated offer a user friendly way of gauging how easy a material will be to cut. The method can be extended to deal with non-ferrous alloys and also include other operations such as turning.

Acknowledgements

The authors wish to express their gratitude to Seco Tools (UK) Ltd. and SERC for supporting this research. They would also like to thank Mr. M. Kingdom, Technical Manager, Seco Tools for his considerable help during this phase of the work.

References

Carpenter, I. D. and Maropoulos, P. G. 1993, A Decision Support System for Process Planning Milling Operations. In A. Bramley and T. Mileham (ed.), *Proceedings of the Ninth National Conference on Manufacturing Research*, (Bath University Press) 22-26.
Smith, S. and Tlusty, J. 1991, An Overview of Modelling and Simulation of the Milling Process, *ASME Journal of Engineering for Industry*, **113**, 169-175.
Chryssolouris, G. and Guillot, M. 1990, A Comparison of Statistical and AI Approaches to the Selection of Process Parameters in Intelligent Machining, *ASME Journal of Engineering for Industry*, **112**, 122-131.
König, W., Cronjäger, L., Spur, G., Tönshoff, H. K., Vigneau, M. and Zdeblick, W. J. 1990, Machining of New Materials, *Annals CIRP*, **39**(1), 673-681.
Machining Data Handbook 1980, 3rd edn (Vols I and II) (Metcut Research Associates, Cincinnati).

Turning Of Metal Matrix Composite (MMC)

A.Jawaid[1] and A.Abdullah[2]

[1]*Coventry University, Coventry CV1 5FB, U.K*
[2] *International Islamic University, 46700 Petaling Jaya, Selangor, MALAYSIA*

The superiority of the mechanical and physical properties of metal matrix composite (MMC) has indicated that these materials play a significant role in applications for engineering components. The aim of this investigation is to enhance the knowledge of the machinability of 2618 Aluminium alloy reinforced with 18 vol.% SiC in particulate form. Different grades of cemented carbide cutting tools were used in the machining tests. The objective of the investigations was to study the effect of tool geometry, tool life, tool wear and surface roughness.

1.0 Introduction

Metal Matrix Composites (MMC's) have been the subject of considerable attention for over thirty years. These materials have considerable potential for providing lightweight components exhibiting high strength, high stiffness and good wear resistance. MMC has long been recognised as difficult to machine materials compared to standard wrought aluminium alloys. Limited data is available on the machinablity of MMC material. Chadwick and Heath (1990) have illustrated that the tool life of polycrystalline diamond (PCD) tools is at least two orders of magnitude greater than that of cemented carbide tools for machining of the unreinforced Al-20% Si alloy, and the quality of the cut surface produced by PCD tools is significantly better than that produced by cemented carbide tools.

For these experiments, two different types of cutting tool were used, coated and uncoated cemented carbides with positive and negative rake angle geometry's to machine Aluminium-based 2618 MMC with 18% vol. of SiC as the reinforcement. The performance of the cutting tools was observed. The surface finish on newly machined surface were also recorded. The effects of depth-of-cut, feed rate , cutting speed and chip-breaker were also studied and analysed.

2.0 Experimental Procedures

2.1 Workpiece Material

Aluminium 2618 particulate reinforced aluminium Metal Matrix Composites (MMC) manufactured by ALCAN via Co-spray deposition and subsequent extrusion route, has been used throughout the course of this investigation. Table 1 shows the composition (wt.%) of 2618 Aluminium metal matrix composite. Figure 1 shows the microstructure of the 2618 MMC reinforced with vol. 18% SiC.

2.2 Cutting Tools

The cutting tools used in the tests were produced by Kennamatel Inc. The grades of carbides tools used for these tests were K68, KC910 and KC810. Two tool geometries were chosen, one negative and one positive rake angle with and without the chip-breaker. ISO standard inserts SNMA 12-04-08 and SPGN 12-03-08 clamped in a MSDNN-2525M 12K and CSPDN-2525M 12 tool holders were used for the experiments. Basic data for the cutting tools used is given in Table 2.

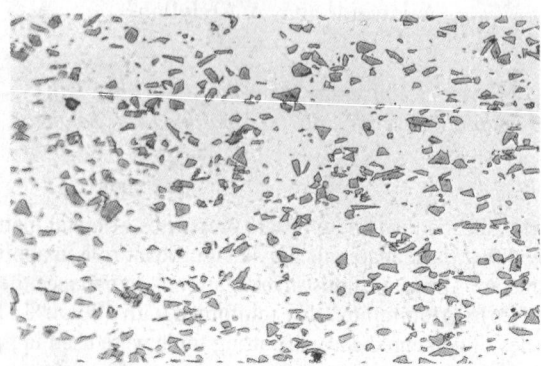

Figure 1: The microstructure of the 2618 Aluminium alloy reinforced with 18 vol% SiCp

Table 1. Chemical composition (wt.%) of 2618 Aluminium Matrix

ALLOY	Si	Fe	Cu	Mn	Mg	Ni	Zn	Ti	V	Al
2618Al MMC (18% SiC)	0.21	1.05	2.67	0.02	1.50	1.06	0.03	0.08	0.01	Rem

Table 2. Properties of cutting materials used in the machining tests.

Cutting tool material	Type of coating	Binder wt. %	Grain size (μm)	Hardness (HRA)	Coating Thickness (μm)
K68	Uncoated K grade	5.7	1-4	92.7	-
KC910	TiC + Al₂O₃	6.0	1-6	91.8	9
KC810	TiC + Ti(C,N) + TiN	8.5	1-8	91.2	10

2.3 Machine Tools

Continuous turning tests were carried out on a CNC Cincinnati Milacron 10CU lathe. The machine is driven with a variable speed, DC motor capable of 30 - 3,000 rpm and controlled with Acramatic 900TC controller. The motor was capable of supplying 22 kW(30hp) over a 30 minute duty cycle.

2.4 Test Procedures

For the turning tests, a bar of 2618 Aluminium Metal Matrix composite reinforced with vol.18% SiC particulate and 120mm in diameter with 540mm length was used. The cast surface of each bar was removed (1.5mm) prior to the tool life testing..

In turning tests no cutting fluids were used, the cutting speed, feed rate and depth of cut were varied to find the optimum machining conditions. Table 3 shows the cutting conditions used in these experiments. The tool life criterion was based on the ISO standards 3685.

Table. 3 The cutting conditions used in the machining tests.

Feed Rate(mm/rev)	Depth of Cut (mm)	Cutting Speed (m/min.)			
		15	20	25	30
0.2	2	X	X	X	X
0.4	2 and 4	⊗	⊗	X	X
0.6	2 and 4	X	X	⊗	⊗

Note: X = Single test, ⊗ = Tests which were replicated

Before any machining, all cutting tools were examined under optical microscope at 10X magnification to check for any manufacturing defects. Machining was stopped and the wear on the tools measured at intervals appropriate to the tools and cutting conditions being used. Both maximum and average flank wear land, and notch at the depth of cut were measured using a 10X magnification microscope. A reading of an average surface roughness values (Ra) on the workpiece was taken at right angle to the feed marks after every cut using a SURTRONIC 3, portable stylus instrument. Scanning Electron Microscopy (SEM) was used extensively to analyse the cutting tools and a workpiece sections. Standard methods were used to prepare the tool and workpiece samples.

3.0 Results and discussion

3.1 Tool Wear And Tool Life

K68, KC810 and KC910 inserts, with positive and negative geometries, with and without chip breakers failed due to the flank face wear. Flank wear was the dominant failure mode under all cutting conditions. Fig. 2 shows the typical flank wear that occurred on all the inserts used for experiments. Slight cratering on the rake face and notching were observed, but they were never the tool life limiting factors. SEM examination revealed that only one wear mechanism operated when machining Aluminium 2618 MMC with uncoated carbides tools i.e., attrition wear. The cross-sections through the tool revealed that the WC particles were lifted off during the process of machining from the flank face and once established such a process enhances the attrition wear (Fig.3).

Fig. 4 shows the rate of flank wear versus cutting time when machining with KC910 cemented carbide tools, the flank wear of 0.4 mm was reached after 10 mins. of cutting at 20 m/min. Whereas a tool life of 26 min was achieved at 30 m/min. Such a small tool life may not be acceptable for the use of the tools in industry. Therefore, the effects of feed rate and depth of cut

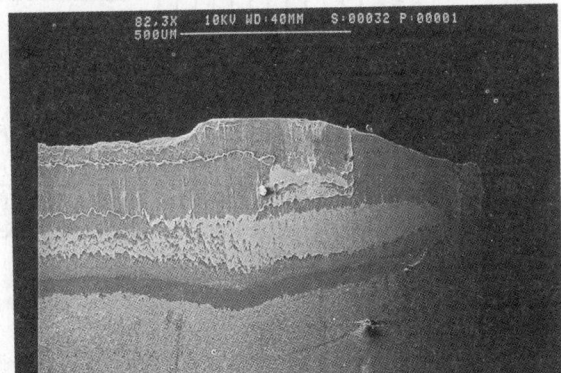

Figure 2. The flank wear pattern on the carbide cutting tools

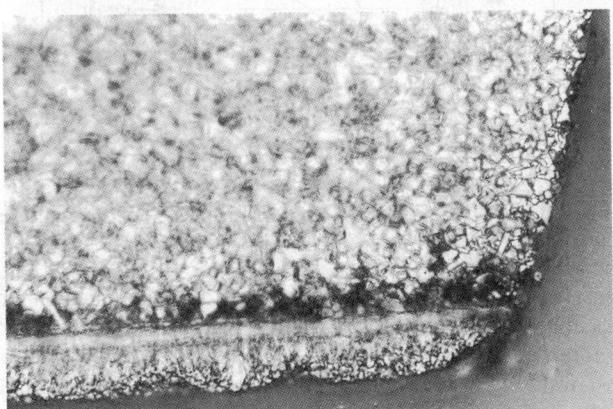

Figure 3. Cross-section through the flank face of K68 insert showing attrition. (1500X)

were investigated. It was found that increasing the depth of cut from 2mm to 4mm did not significantly affect tool life at any cutting speed, suggesting that a higher metal removal rate (MRR) can be achieved in this way without effecting the tool life. Increasing feed rate from 0.2 mm/rev to 0.6 mm/rev also increased the tool life (Fig.5), again suggesting that high MRR can be achieved at higher feed rates. Combination of such factors make use of carbides an alternative preposition for industry for the machining of MMCs.. The same phenomenon were encountered by Chambers and Stephens (1991) when machining Al-5Mg reinforced with 15 vol.% SiC using K10 and coated P15 cemented carbide tools.

Of the three different grades of carbides used in the tests, KC910 gave the longest tool life at most of the cutting conditions. Figure 6 shows the comparison of the tool life for K68, KC910 and KC810 at feed rate of 0.4mm/rev and with 2mm depth of cut for all cutting speeds. Negative rake inserts with a chip-breaker gave slightly longer tool life at lower cutting speed. As cutting speed increased, the flat-top insert gave longer tool life than the ones with the chip breaker.

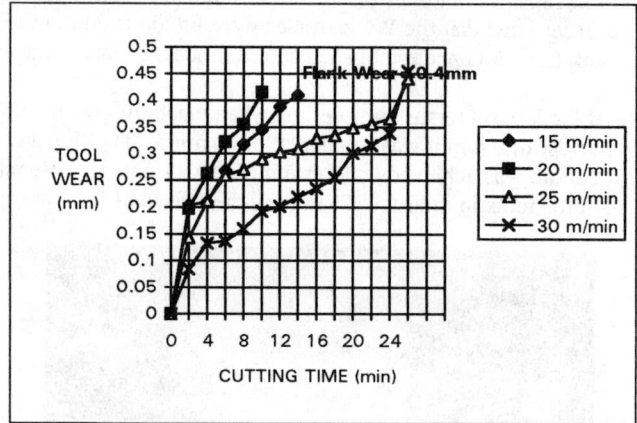

Figure 4. Tool wear versus cutting time for KC910 insert for feed rate, f=0.4mm/rev and DOC=2mm showing how tool lives to 0.4mm flank wear were determined.

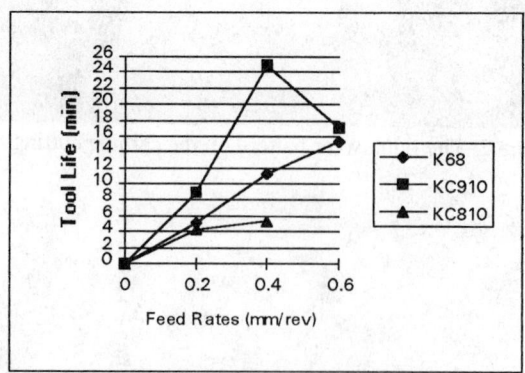

Figure 5: The effect of feed rate upon tool life of cemented carbide tools.

3.2 Built-Up Edge (BUE)

Cronjager and Biermann (1991) has shown that BUE was formed on most of cutting conditions when machining several composites with a matrix of aluminium wrought alloy (AlMgSiCu, AlMg3) reinforced with SiC or B_4C particles. In these tests the formation BUE was observed under most of the cutting conditions when machining Aluminium 2618 MMC with K68, KC810 and KC910. Most BUE observed were of a stable type, at some instances its presence was observed to have increased the tool life. The BUE disappeared as the cutting speed was raised.

Increased tool life was recorded, where the BUE was present, but its presence resulting in a poor surface finish. Tomac and Tonnessen (1992) found that cutting with the presence of BUE under certain conditions produced a good surface finish.

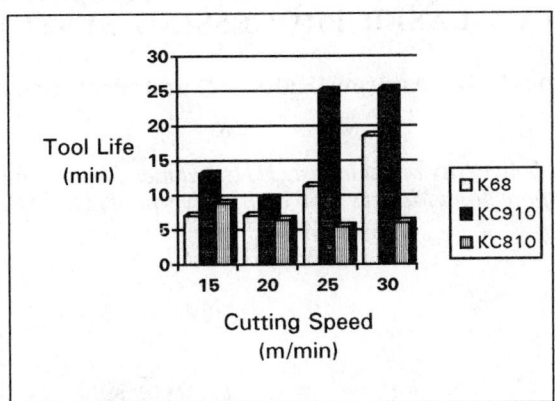

Figure 6. Tool life for K68, KC910 and KC810 at feed =0.4mm/rev,doc= 2mm for all speed

3.3 Surface Roughness

The surface roughness value which were recorded were always above the ISO recommended standard of 1.6 µm. At some instances, the value of surface roughness could not be taken due to very uneven surfaces produced. The 'tearing off surfaces' were observed under most of the cutting conditions. The BUE on the cutting edge perhaps has been responsible for such poor surface finish readings.

4.0 Conclusions

1. KC910 insert with negative rake geometry gave an overall better performance than those given by KC810 and K68 grade inserts under most cutting conditions.
2. Flank wear was the dominant failure mode under all cutting conditions when machining particulate Aluminium 2618 MMC.
3. Attrition wear was observed during the experiments when particulate Aluminium 2618 MMC was machined with cemented carbide tools.
4. Built-up edge (BUE) was observed under most cutting conditions, and it significantly improved the tool life but the surface finish generated during such conditions was poor.
5. Chip breaker did not have a significant effect on the tool performance.

References

1. Chadwick G A and Heath P J, 1990, *Machining of metal matrix composites*, Metals and Materials, 73-76.
2. Chambers A R and Stephens S E,1991, *Machining of Al-5Mg reinforced with 5 vol.% Saffil and 15 vol.% SiC*, Material Science and Engineering, 287-290.
3. Cronjanger L and Biermann D, 1991, *Turning of Metal matrix Composites*, Proc. 2nd. European Conf. on Advanced Material and Processes, Cambridge, Vol.2, 73-80.
4. Tomac N and Tonnessen K, 1992, *Machinability of Particulate Aluminium Matrix Composites*, Annals of CIRP Vol. 41/1/1992, 55-58

DEVELOPMENT OF A MULTI-FUNCTIONAL CAD/CAM LASER PROCESSING SYSTEM

Dr Janos Takacs*, Dr Andrew Kaldos** and Prof Ben Mills**

*Technical University of Budapest, 1111 Budapest, Bertalan u.2
**Liverpool John Moores University, Liverpool, L3 3AF

The industrial applications of laser processing technology have increased significantly in recent years, particularly in areas where laser processing has increased versatility and economic advantages over conventional techniques. The range of applications of laser processing and inherent flexibility give rise to the concept of a multi-functional CAD/CAM laser processing system complete with on-line laser based diagnostics, data capture and processing facility with an integrated database for the optimisation of the laser processing techniques. The concept of such a centre is presented in this paper and Stage I of the concept, which involves a two-dimensional laser cutting facility, is described,which has 5 axes of control. The concept will be developed further in the future with an increased number of axes of control.

Introduction

It is widely accepted that in the discrete product manufacturing industries a high degree of flexibility is required to meet the ever changing market demand, to respond quickly to market conditions and to maximise available resources to the highest possible degree. A high degree of flexibility means that the applied manufacturing system must be able to change various elements, processes, tools and information within the system without changing the physical layout of both production machinery and ancillary equipment. The flexibility of this system allows for the manufacture of many different parts employing different processes.

Analysis shows that components spend approximately 15% of the total time available on the machine tool, the majority of time is spent on loading/unloading, transfer, waiting in queues ie. in non-value added activities. The concept of flexibility and the real need to reduce non-productive time gave rise to the concept of 'single hit' manufacturing whereby the number of set-ups is reduced to the bare minimum ie. to one. This concept is not entirely new, however, it is easy to recognise that a high degree of process concentration is required which may cause extreme difficulties due to the requirements and conditions of these various manufacturing processes. In addition, the component parts must be designed for the 'single hit' manufacturing concept, therefore the design and manufacture must be integrated under the philosophy of all activities and facets of Computer Integrated Manufacture.

The multi-functional/multi-process capability, the competitive performance and flexibility frequently present conflicting requirements for the traditional machining approach, particularly when processes based on entirely different physical principles are to be applied in one single system. The problem of process integration is a very complex task but not impossible to resolve. However, a step by step approach seems to be a reasonable approach to achieve multi-functionality/multi-process capability for which laser technology is a prime candidate, Steen (1991). In recent years laser systems have been developed for cutting, welding, surface treating, micro-alloying and laser systems are also used for quality assurance, parts recognition and measurement. Inherent to laser technology is that mechanical forces are not generated directly on the manufacturing centre which eliminates tool vibration, the problems of tool wear and failure are also . absent. The present paper therefore addresses the problems of the development of a multi-functional CAD/CAM laser processing centre as a first step in the process of developing a single hit multi-purpose manufacturing centre.

Concept Design

The present research is aimed at the development of a multi-functional CAD/CAM system capable of operating with the minimum number of set-ups, ideally in a 'single hit' way. This work has been divided into three stages, the first stage covers 2D laser cutting processing with an overview of all aspects of laser processing and a possible extension to integrate other manufacturing processes into the system, the second stage covers the integration of all laser processing and the third will integrate all other processes felt to be necessary to produce component parts in a flexible and economic way into the system. This CAD/CAM system must be capable of producing both rotational and prismatic components. However, the first stage of development concentrates on the manufacture of prismatic type components with special reference to the laser cutting of sheet metal 2D parts.

Machining System Configurations

The laser machining system set-up is shown in Figure 1. The chosen fast axial flow CO_2 laser complete with oscillatory mirror is capable of cutting, welding and heat treatment operations. The power chosen is a compromise; when the system is complete and proven, a larger capacity laser head can be applied for speeding up operations and enhancing performance. The workholding part of the system provides three axes for the two linear slides and one rotary table, the rotary motion being about either of the two axes or about any of the other horizontal axes by changing heads. All drives are of stepper motor type open loop systems for it was felt that since forces are not generated in laser processing the accuracy and reliability of the open loop control system is sufficient. The system is equipped with a FRANKE SM2000 8 axes control system controlling the three motion axes, the mirror oscillation and the in-line process control data via the laser control system. It is anticipated that the mechanical system will be replaced with a 6 axes Cybaman Universal Machine which enhances not simply the geometrical machining capabilities of the processing centre but also provides a superbly shaped work envelope for intricate works, Kaldos and McLean (1993).

The present laser processing system is capable of performing cutting operations and limited welding and heat treatment operations with the power available.

Structured CAM Processor and Database

Although the multi-functional system described is suitable for 3 axes continuous path machining and two more parameters are controlled simultaneously, the modified version will be able to control 8 axes simultaneously and it is good practice to build and prove such a system in a step by step mode.

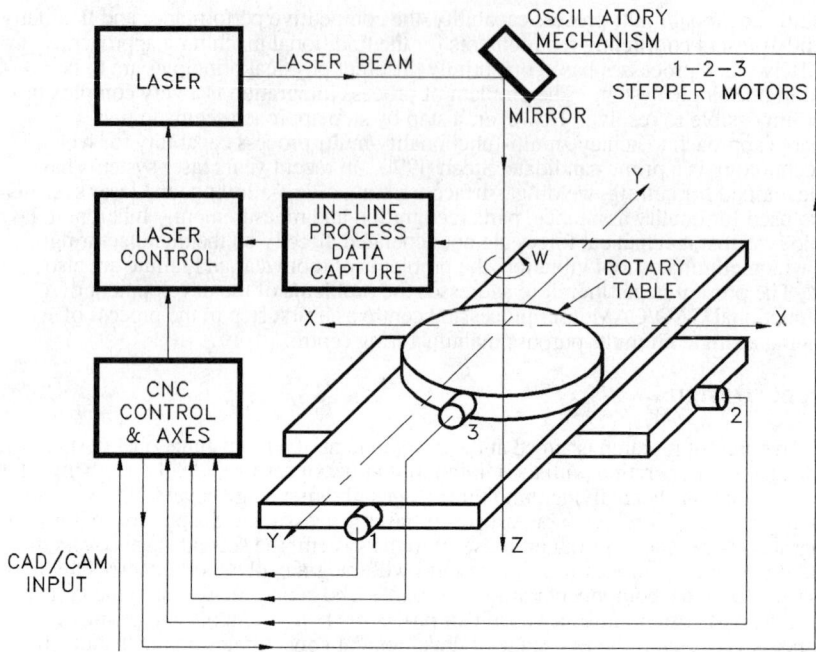

Figure 1. System Set-up.

Therefore 2D machining has been tested and evaluated for which the laser cutting process was used to produce sheet metal parts. This task involves part program generation and nesting for which a CAD system for drawing and nesting the parts and also an automatic CNC part program generating interface were used. There is also a need to incorporate technological aspects into the CNC part program which leads to the development of a sheet cutting CAM system with the following provisions:

(1)   sequencing the laser cutting motion to maintain stability of the remaining sheet frame and to minimise the cutting time;
(2)   analysing the part geometry and add special features to the contour to improve the quality of the cutting process;
(3)   optimising laser cutting parameters for individual components according to the part features and technology database;
(4)   graphically simulating the cutting process to verify the cutting sequence;
(5)   generating CNC part programs directly from the CAD database and transmitting it to the machine tool controller through an RS232 interface.

The structure of the CAM system and integrated database is illustrated in Figure 2. The first step to generate a CNC part program is to invoke the CAD database interface module which extracts the essential parts geometry information directly from the CAD database and converts it so that it is compatible to the CAM processor unit.

Geometry Processor

The geometry processor analyses the part geometrical elements and provides an optimised cutting sequence based on both the stability of the remaining sheet and on the cutting time. The operation of the geometry processor is rule driven according to the following sequence:

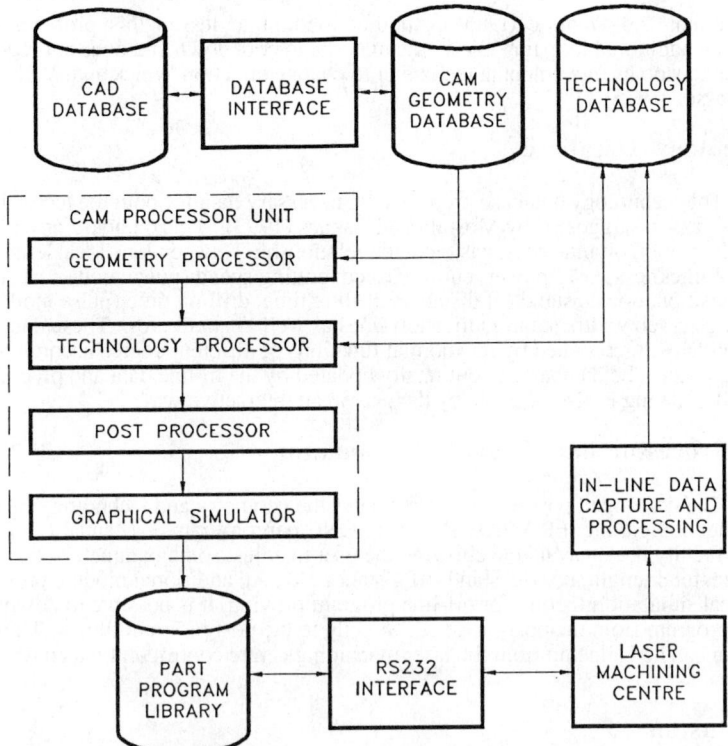

Figure 2. The CAM system structure.

(1)   To select part geometrical elements by removing the overlapped elements and joining together any close vertexes if the distance between them is less than the width.

(2)   To convert the individual geometrical elements to polygons by sorting and joining together any two identical vertexes, then identifying all generated polygons as close or open.

(3)   To generate a vertex search space list that contains start and end vertex co-ordinates of all open as well as all closed polygons.

(4)   To sequence the cutting motion by selecting the polygon from search that has the closest vertex to the previous laser beam position.

(5)   To remove the selected polygon from search space list and repeat step number 4 until the end of all nest parts.

Technology Processor

The subsequent technology processor takes up the processed geometry information and generates some additional features to the part geometry based on technological constraints. The technology processor is an information based module that helps to improve the quality of the cutting process. Difficulty arises when cutting complex contours having acute angles or sharp edges, starting the cutting from inside the shape, cutting narrow webs and cutting small size contours, either circles, lines or arcs. Cutting acute angles causes the material to be burnt off due to the energy concentration, since the actual cutting speed becomes nearly zero at the corner with directional conversion, and yet the laser power is kept constant. This problem can be alleviated by controlling laser power, output mode and cutting speed adaptively Moriyasu, Hiramoro,

Hoshinouchi and Ohmine (l985). Another method to lessen this problem is simply cutting an additional loop that allows the material to cool down and does not require the machine to stop its movement at the corner to change direction Weick and Wollermann-Windgasse, (1985).

Technology Database

The technology database produces the necessary data for both the technology and post-processors suggested by Mostafa and Takacs (1993). This database provides up-to-date information on materials, gas type, the relationship between lens focal length and material thickness, laser power, cutting speed, cutting speed (pulse mode), gas pressure, focus position, nozzle stand off distance, drilling time, drilling time (pulse mode), acute edge angle, web width, small radii, short lines as well as small arcs. These data are automatically interpolated by polynomial functions of different degree according to the user's choice. The database is continually updated by the in-line data and processing unit and updating is also possible by the user in an interactive way.

Post Processor and Graphical Simulator

The developed post processor translates the geometry and technology files to the controller compatible (FRANKE SM2000) CNC part programs . Manual Data Input (MDI) facility is also available allowing the user to adjust some parameters, eg. gas pressure, total length, nozzle stand-off distance etc. An additional module provides graphical simulation facility for off-line program proving. It is possible to download the actual program from the post-processor directly to the post program library. The stored program can be called up from the laser machining centre control unit via an RS232 interface.

Conclusion

The laser CAD/CAM multi-functional laser processing system developed after the first stage offers interactive computer aided programming for CO_2 laser sheet cutting. This system presents an easy and fast way of generating and modifying a CNC part program directly from a CAD database. This system has been successfully used to recognise nesting features and to maintain the relationship between geometry and technology data and to cut mild steel sheet parts under optimum conditions. Based on these achievements the system is being further developed in the next phases to accommodate all conceivable laser processes and therefore an 8 axes Cybaman Universal Machine will be incorporated followed by a further integration of other manufacturing processes to make the system fully multi-functional/multi-process system.

References

Kaldos, A. and McLean, J. G. (1993) Control software for a new generation six axes machine. *Proceedings of the 9th National Conference on Man. Res.* 227-231.
Moriyasu, M., Hiramoro, K., Hoshinouchi, S. and Ohmine, M. (1985) Adaptive control for high-speed and high-quality laser cutting, *Proceedings of 5th International Congress on Application of Laser,* 129-136.
Mostafa, A. and Takacs, J. (1993) Automated programming system for CO_2 laser cutting.*10th Int. Coll. on Adv. Manu. and Repair Technologies.*, Dresden 28-36.
Steen, W. M. (1991) Laser material processing. *Springer-Verlag.*
Weick, J. M. and Wollermann-Windgasse, R. (1985) Trumpf GmbH & Co. West Germany. Laser cutting of sheet metal in modern manufacturing. *Proceedings of Third International Conference on Laser in Manufacturing LIM-3,* 47-56.

Inspection

3-D DATA COLLECTION FOR OBJECT RECOGNITION

J. Keat, V. Balendran, K. Sivayoganathan, and A. Sackfield*

Manufacturing Automation Research Group,
Department of Manufacturing Engineering,
**Department of Mathematics and Operational Research,*
The Nottingham Trent University,
Burton Street,
Nottingham,
NG1 4BU.
email: man3keatjs@uk.ac.ntu

Traditionally the collection of 3-D coordinate information from an object has been done by coordinate measuring machines (CMM's). This process is time consuming when taking discrete points. Recent improvements in laser line scan hardware have made possible rapid data collection, but with a loss of accuracy. This improvement in data collection rate has allowed the use of 3-D surface data in `real-time' manufacturing applications.

This paper describes the development of a laser-line scan 3D data collection system which is to be used as a front end for our work on the matching and recognition of free-form surfaces. The future direction in the use of such data for 3D shape recognition is discussed along with the future aims of this work.

Introduction

The problem of automated fettling was discussed by Balendran(1992). From his work a project has evolved looking at the problem of 3D object recognition and free-form surface matching. The aim of this work is to develop a solution for use in a manufacturing environment, specifically, but not limited to, automated fettling. The initial work has centred on the development of the front-end. The following requirements for the front-end were determined:-

- Real-time data collection rates,
- The capability to measure a wide range of object complexities,
- Must not be reliant on the nature of the objects surface,
- View point independent data collection,
- Small enough to be mounted on a robot arm,
- Robust.

The accuracy of the measurement system is not critical for the recognition process which is only interested in the relative form of the object.

Data Collection Techniques

Data collection techniques can be divided into two categories; Contact and Non-contact. Contact measurement is done by Coordinate Measuring Machines (CMM's) using a touch probe. CMM's generally collect data a point at a time. Such a process is slow, and because it is a contact process, it is limited to measurement and calibration applications where accuracy, rather than speed, is paramount. Although our later object recognition work will be capable of using CMM point data, it is not suitable as front-end for a real-time recognition system.

Non-contact surface inspection can be sub-divided into optical and non-optical methods. For our work we have limited ourselves to optical methods. Non-contact optical surface inspection can be divided into two classes of techniques:-

● Flying spot scan methods. This is done by the projection of a single point of laser light which is viewed at an oblique angle. The offset of the point relative to a datum is measured and if calibrated correctly can be given as a coordinate.

● Pattern projection methods. This is by far the largest category of optical methods. The two major techniques are laser line or grid projection and interferometric fringe techniques. The laser line (or grid) method is similar to the flying spot technique but instead the line (or grid) offset is measured. This allows more points to be measured at a time. Such a technique is shown in Figure 1. and described in the next section.

Interferometric techniques rely on the interference of two beams of light. Figure 2 shows a technique known as pattern projection Moiré contouring. A reference grid is projected onto the surface of an object. The object is then viewed through a grid with a similar pitch frequency and a series of dark and light contours are superimposed on an image of the surface. These lines

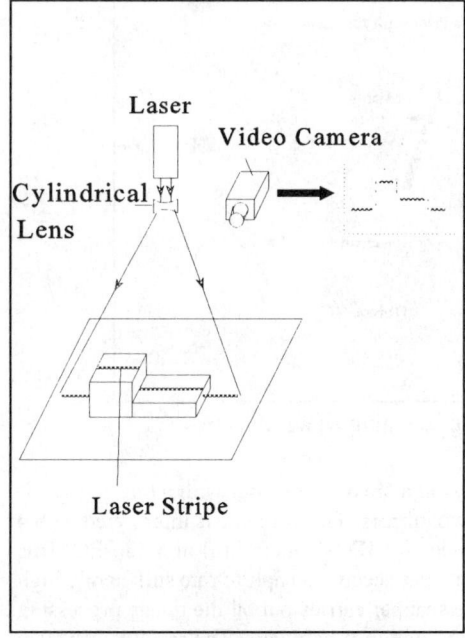

Figure 1 Laser line scanning

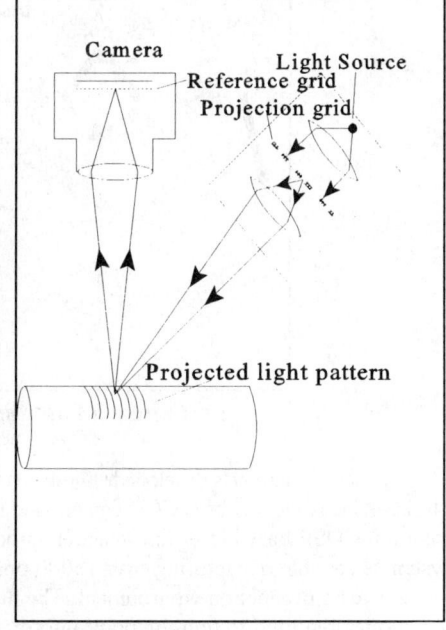

Figure 2 Moiré contouring by the pattern projection method.

correspond to a series of imaginary planes whose normals lie along the viewing axis of the instrument. The departure of a fringe from a straight line implies a change in surface height. Similarly a series of parallel straight fringes implies a flat surface. The resolution of the process is limited by the object depth or contour range divided by the number of fringes resolved by the viewing system. This type of process has the advantage of capturing the surface of the object in one go. The disadvantages are the relatively large amount of processing that is required to convert the image data to point data and the system is not view point independent or able to cope with a wide range of surface types.

Laser Line Scanning

A laser line scanning system has been developed within the Department of Manufacturing Engineering (Balendran(1989)(1994)) for use as a tool for the inspection of cosmetic quality of car body panels. In this system the laser stripe is directed across the surface of the panel at a shallow angle instead of at a normal, and the 3-D information is gained from the interaction of the image of the stripe on the panel and its reflection onto a viewing screen. The use of two images of the stripe gives greater detail but relies on the reflective properties of the subject. This limits the system to the measurement of relatively flat surfaces.

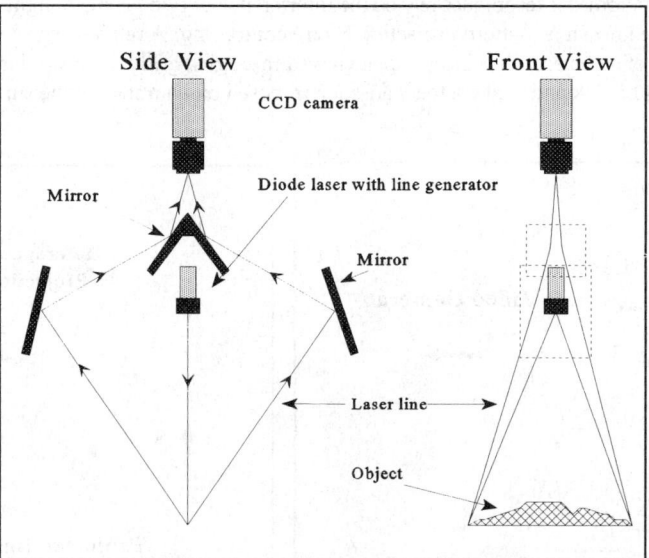

Figure 3 Laser line scanning of an object

A new system was developed (**Figure 3**) using a 5mW diode laser with a line generator. The laser line is viewed by a CCD camera and two mirrors. The laser line is interpreted with a 'real-time' DSP based laser line scanner (produced by 3D Scanners Ltd) in a 486 PC. This system is capable of capturing over 15000 points per second, a capture rate sufficiently high enough for a production environment. The line scanner carries out all the image processing techniques required to find the point information within it hardware. The result is a system much faster than is possible using the interferometric techniques, and which is capable of being used with a wider range of object types.

If the laser line is viewed from one angle valuable surface information can be lost. This problem is overcome to a degree by viewing the projected laser line from two angles using mirrors.

Surface Segmentation

The work described above is only the initial work for use a front end for use in the recognition of free-form surface. As in any object recognition task the input data has to be segmented and put into a form that can be used by the recognition system. The aim of our work is to develop an invariant surface recognition system, therefore the segmentation process has to be view point independent. There are eight view point independent surface types Besl and Jain(1986) and their use in the fettling problem was outlined by Balendran(1991). Local windows, in our case 3x3 points, of the surface are taken and the mean and Gaussian curvatures calculated. By using only the signs of these values it is possible to define eight view point independent surface types as shown in **Table 1**. These surface types have been well used in medical imaging systems for modelling the effects of reconstruction surgery and for body scanning (Coombes et al. (1990)).

Table 1 Elementary surface types definitions

Surface Type	Mean Curvature	Gaussian Curvature
Peak	negative	positive
Ridge	negative	zero
Saddle Ridge	negative	negative
Plane	zero	zero
Minimal	zero	negative
Saddle Valley	positive	negative
Valley	positive	zero
Cupped	positive	positive

Conclusions

We have a system that is able to collect surface data at a rate of 15,000 points per second and segment this data into the eight fundamental surface types. This data collection system is not an alternative to CMM's as our future work (outlined below) does not require the accuracy that a CMM can produce. At present the system is mounted on a stationary gantry with the object being passed under the laser using a stepper motor controlled table. To be used for the fettling problem the system will be mounted on a robot arm. Another project within the Department of Manufacturing Engineering has converted the surface point data produced to an IGES format for input to both AutoCad and Unigraphics, this shows another use for this work as a tool for reverse engineering.

Future Work

The human brain and eye is able to discriminate between like and dislike complex shapes irrespective of pose with relative ease, this cannot be said for the case of computerised shape comparison and matching. Until recently the majority of work carried out on 3-D shape matching has been with the use of constrained geometric shapes. These shapes (such as polygons, quadric surfaces, planar curves etc.) are matched with aspects of the object that is being analyzed to check for correspondence. The lack of work in this area is outlined by Besl (1990) and Stein and Medioni (1993).

Work has been undertaken within the Department of Manufacturing to assess cosmetic quality of car body panels (Balendran(1994)) with the aid of Neural Network Systems (NNS). This has formed a starting point for the work . Initial investigations have looked at work done within the areas of 2D and 3D invariant pattern matching, with traditional and NNS methods. There are two approaches to invariant object recognition using NNS. The first is to use a preprocessor to remove invariance before the information is put to the network. The second approach is to incorporate features within the NNS to take into account the invariance. This work will aim to follow the second approach. This implies that NNS methods for data segmentation, invariant feature extraction, and object recognition will have to be developed. It is also the aim of this work to be of use within a manufacturing environment. This implies working on-line in real (or near real) time. All approaches investigated will have to take this into account.

References

Kyongate T. Bae and Martin D. Altschuler, High resolution fully 3-D mapping of human surfaces by laser array camera and data representations, *SPIE Vol. 1380 Biostereometric Technology and Applications*, 1990.

Balendran, V., Sivayoganathan, K., and Al-Dabass, D., 1989, Detection of flaws on slowly varying surfaces, *Proceedings of the Fifth National Conference on Production Research*, Editor J. Chandler, (London: Kogan Press), pp 82-85.

Balendran, V., Sivayoganathan, K., and Howarth, M., 1992, Sensor Aided Fettling, *Proceedings of the Eighth National Conference on Manufacturing Research*, Birmingham, pp 132-136.

Balendran, V.,1994, Cosmetic Quality: A Computational Approach, *Ph.D. Thesis*, The Nottingham Trent University.

Paul J. Besl and Ramesh C. Jain, Invariant surface characteristics for three-dimensional object recognition in range images, *Computer Vision, Graphics, Image Processing*, January 1986, 33(1), pp. 33-80.

Paul J. Besl, 1990, The Free-Form Surface Matching Problem, *Machine Vision for Three-Dimensional Scenes* pp 25-71, Academic Press.

Anne M. Coombes, Alfred D. Linney, Robin Richards and James P. Moss, 1990, A method for the analysis of the 3D shape of the face and changes in shape brought about by facial surgery, *SPIE Vol. 1380 Biostereometric Technology and Applications*.

F. Stein and G. Medioni, 1992, Structural indexing: efficient 3-D object recognition, *IEEE Transactions on Pattern Analysis and Machine Intelligence*, Vol. 14, No.2, February 1992.

AUTOMATIC SENSOR POSITIONING FOR ACTIVE VISUAL INSPECTION

Dr Emanuele Trucco, Marco Diprima, Dr Andrew M. Wallace

Department of Computing and Electrical Engineering
Heriot-Watt University
Riccarton EH14 4AS, Scotland

The ability of active vision systems to reconfigure themselves in order to improve their performance at visual tasks can make advanced inspection more powerful and flexible. In this paper, we address the problem of computing the sequence of positions in space from which a robot-mounted sensor can perform a given inspection task optimally. We call such sequence an *inspection script*. We present a sensor planning system, GASP, capable of generating inspection scripts for a variety of tasks, objects and sensors, and show some examples of inspection with a mechanical component in a synthetic environment.

Sensor planning and inspection

A key feature of *active vision systems* is the ability to reconfigure themselves in order to improve their performance at given visual tasks. This ability can be exploited in advanced inspection, for instance by directing robot-mounted sensors to positions in space from which an inspection task can be performed optimally, e.g. the features to inspect are maximally visible, or most reliably detected by the vision system, or both. In general, several parts (features) of an object require inspection, and each one is best observed from a different viewpoint. Therefore, a sequence of sensor positioning actions (an *inspection script*) must be planned to perform the task (Trucco, Thirion, Umasuthan and Wallace 1992). This paper is a brief overview of GASP (General Automatic Sensor Positioning), a sensor planning system capable of generating inspection scripts for a variety of tasks, objects and sensors. We sketch GASP's main design features, the inspection tasks currently considered, and show some examples. GASP was started as part of a cooperative research action, LAIRD, investigating the use of range data for advanced 3-D inspection (Buggy, Bowie, Green and Clarke 1992).

Inspection scripts and inspection tasks

Visual inspection strategies and their automatic generation depend on the type and number of both sensors and object features to be inspected: therefore, models of both sensors and objects must be available to the planner. At present, GASP can compute inspection scripts for both *single* or *multiple sensors*. A single sensor can be either an intensity or a range imaging camera. The only multiple sensor currently modelled in GASP is a stereo pair of intensity cameras (or *stereo head*). The inspection scripts that GASP can generate at present are the following.

- *Single-feature, single-sensor scripts:* find the position in space from which a single imaging sensor (intensity or range) can inspect a single object feature optimally.
- *Single-feature, multiple-sensor scripts:* find the position in space from which a stereo head can inspect a single object feature optimally.
- *Multiple-feature, single-sensor scripts:* (a) find the position in space from which a single imaging sensor (intensity or range) can simultaneously inspect a set of features optimally; (b) find the best path in space taking a single imaging sensor (intensity or range) to inspect a set of object features from optimal positions.
- *Multiple-feature, multiple-sensor scripts:* find the best path in space taking a stereo head to inspect a set of object features from optimal positions.

Overview of GASP

We have designed a representation, the FIR (Feature Inspection Representation), which makes inspection-oriented information explicit and easily accessible. In the next sections FIRs are used by GASP to create inspection scripts in a simulated environment, written in C on SPARC/Unix workstations.

A FIR partitions the (discrete) set of all the viewpoints accessible to the sensor (the *visibility space*) into *visibility regions*, each formed by all the viewpoints from which a given feature is visible. Viewpoints are weighted by two coefficients, *visibility* and *reliability*. The former indicates the size of the feature in the image (the larger the image, the larger the coefficient); the latter expresses the expected confidence with which the feature can be detected or measured from a viewpoint by a given image processing module. For instance, estimates of the curvature of cylindrical surfaces from range images become seriously unreliable if the surface appears too small or too foreshortened (Trucco and Fisher 1994).

Visibility and reliability are combined into an *optimality* coefficient, which quantifies the global merit of the viewpoint for inspecting the feature. The relative importance of visibility and reliability can be adjusted in the combination: it is therefore possible to bias GASP scripts according to task requirements. For instance, one might want to emphasize visibility when simply checking that an object feature is present; to ensure that a measurement is taken with maximum confidence, reliability should get a high weight.

In GASP, the visibility space is modelled by a *geodesic dome* centered on the object. The viewpoints are the centers of the dome's facets. The algorithm to compute the FIR (Trucco et al 1992) generates a geodesic dome around a CAD model of the object to inspect, then raytraces the model from each viewpoint on the dome. Sensor models incorporated in the raytracer are used to determine the visibility and reliability for each viewpoint. FIR sizes vary with the number of

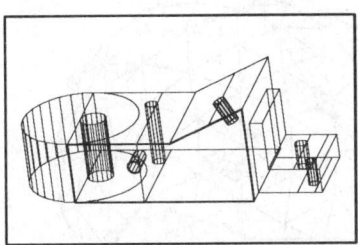 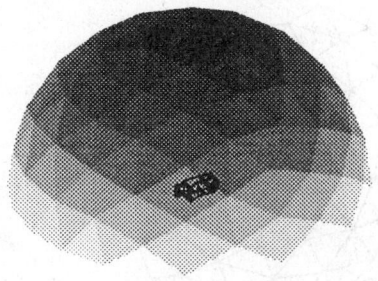

Figure 1: CAD model of a widget (left) and visibility region for (top) highlighted planar patch (right).

object features, the resolution of the raytracer images, the density of the viewpoints on the dome, and the regions reachable by the sensors. A FIR for the widget shown in Figure 1 (left), observed from 320 viewpoints and with image resolution of 64x64 pixels, occupies about 80 kbytes. The widget is about 250 mm in length.

Single-sensor scripts

Single-Feature Inspection

In a FIR, the viewpoints belonging to a visibility region are arranged in a list ordered by optimality; therefore, the best viewpoint for inspecting a given feature is simply the first element of a list. Figure 1 (right) shows the visibility region of the planar patch highlighted in Figure 1 (left) given a single intensity camera of focal length about 50mm and image resolution 64x64 pixels. The widget has been magnified for clarity. The object-camera distance (dome radius) was determined by GASP as the minimum one such that the whole widget is visible from all viewpoints; other choices are possible for the choice of the dome radius. Viewpoints have been shaded according to their optimality (the darker the better), assuming equal weights for visibility and reliability in the optimality computation.

Simultaneous Inspection of Several Features

In this case, a sensor must inspect several features simultaneously from one position in space. The region of the visibility space from which a set of features is simultaneously visible is called *covisibility region*, and is obtained as the intersection of the visibility regions of the individual features. This is done simply and efficiently due to the FIR's structure. Notice that the merit (optimality) of a viewpoint for inspecting the set of features must be defined as a function of the optimality of the viewpoint for inspecting each *individual* feature (Trucco et al 1992).

Sequential Inspection of Several Features

Given a set of features, we now want to find a *path in space* which takes the

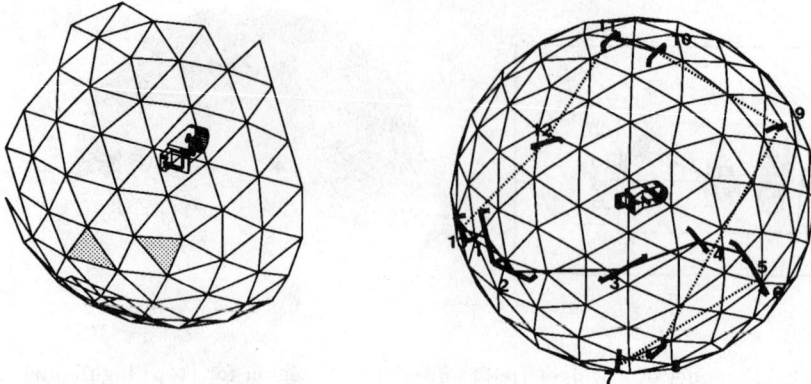

Figure 2: Stereo head inspection (left) and inspection path (right).

sensor through the optimal inspection positions associated to all the features in the set; the path must also be as short as possible since we wish to minimise the number of sensor repositioning actions. Planning the shortest 3-D path through a given set of points is a NP-complete problem, and only approximate solutions can be found. We considered three TSP algorithms: CCAO (Golden and Stewart 1985), simulated annealing, and the elastic net method (Durbin and Willshaw 1987), and chose the best one for our purposes by implementing them and testing their performances with a large number distributions of viewpoints to visit. We found that, with up to 100 viewpoints, CCAO outperformed the other two in terms of path length and distance from the overall optimal solution. CCAO was therefore incorporated in GASP. An example of inspection path is given in the section on multiple-sensor, multiple-feature scripts.

Multiple-sensor, single-feature scripts

At present, the only multiple-sensor configuration considered in GASP is a stereo camera pair, or stereo head. In this case, we want to determine the position in space from which the head can observe at best a given object feature. To do this, both cameras must be placed in good viewpoints for inspection, which is not trivial because good visibility for one camera does not necessarily imply the same for the other. Using a FIR and information about the head geometry, GASP selects efficiently the most promising admissible positions of the head inside the region, and picks the best one. We define the position of the stereo head as that of the midpoint of the line connecting the optical centers of the cameras (the head's *baseline*). Since the two cameras are placed in two distinct viewpoints, the optimality of the head position (*stereo optimality*) is defined as a combination of those of the viewpoints in which the two cameras are placed. Figure 2 (left) shows the optimal placement for a stereo head found by GASP in the large (compared with the head's camera-to-camera distance) visibility region of one of the widget's back planar surfaces.

Multiple-sensor, multiple-feature scripts

The final class of inspection scripts finds the shortest inspection path through the viewpoints from which each feature in a given set can be optimally inspected by a stereo head. Since in GASP the head's position is represented by a single point, it is possible to apply again the path-finding module used for the single-sensor case. An example of shortest 3-D inspection path is shown in Figure 2 (right), which shows only the visible part of the dome. Solid lines connect points on visible dome facets, dotted lines lead to or from occluded viewpoints. The stereo head is represented by the baseline segment. The path was computed simulating two intensity cameras of focal length about 50 mm and image resolution 64x64. The dome radius (object-camera distance) was then fixed by GASP at 2400 mm in order to ensure full visibility of the widget from all views. The head's baseline was 500 mm. In these conditions, 13 surfaces of the widget were visible to the cameras (others were too small or always mostly occluded, e.g. the bottom surfaces of holes), and each surface was associated to an optimal inspection viewpoint for the stereo head. The 3-D path shown is the shortest one found through such viewpoints.

Conclusions

Planning inspection scripts, i.e. the positions in space from which a mobile sensor can perform visual inspection tasks optimally, is an important capability of flexible robotic inspection systems. We have presented briefly the main design issues of GASP, a sensor planning system which generates inspection scripts for various tasks, objects and sensors, and shown a few examples. Current and future work on GASP include extending the sensor and workspace models, as well as running GASP-generated scripts in a real inspection setup.

References

Buggy, T.W., Bowie, J.L., Green, S.M., Clarke. A.R., 1992, Intelligent Control of a Multisensor Inspection System. In *Proceedings IEEE International Symposium on Intelligent Control*, (IEEE Control Systems Soc.), 592–598.

Durbin, R. and Willshaw, D., 1987, An Analogue Approach to the Travelling Salesman Problem. Nature, **326**, April, 689–691.

Fekete, G. and Davis, L.S., 1984, Property spheres: a new representation for 3-D object recognition. In *Proceedings IEEE Workshop on Computer Vision, Representation and Control*, (IEEE, Inc.), 192–201.

Golden, B.L. and Stewart, R., 1985, Empirical analysis and euristics. In Lawler, Lenstra, Rinnoykan, Shmoys (eds.), *The Travelling Salesman Problem*, (Wiley Interscience Pubs).

Trucco, E., Thirion, E., Umasuthan, M. and Wallace, A.M., 1992, Visibility scripts for active feature-based inspection. In *Proceedings British Machine Vision Conference*, (Springer-Verlag, Berlin), 538–547.

Trucco, E. and Wallace, A.M., Using viewer-centered representations in machine vision. In V. Roberto (ed.), *Intelligent Perceptual Systems, Lecture Notes in Artificial Intelligence vol. 745*, (Springer-Verlag, Berlin), 307-321.

Trucco, E., and Fisher, R.B., 1994, Experiments in Curvature-Based Segmentation of Range Data, IEEE Trans. on Pattern Anal. and Machine Intell., to appear.

LIGHT SCATTERING AND SURFACE FINISH

Dr. Brian Griffiths, Russell Middleton and Dr. Bruce Wilkie

Brunel University,
Uxbridge, Middex. UB8 3PH

This paper is concerned with the assessment of surface finish using light scattering techniques. Much work has been published in this area, particularly for smooth surfaces. However, piece-parts manufactured by conventional processes are usually comparatively rough and previous research has illustrated the complexities involved in light scattering. Previous attempts to determine surface finishes have been process specific and do not provide a reasonable separation. This research programme has concentrated on the interpretation and analysis of light scattered in three dimensions rather than two dimensions. This approach has provided more data than has been the case and has allowed a separation of surface finish as well as texture so that different processes and finishes can be separated.

Introduction

The need for industrial inspection for consistency and quality have grown over the years, what was once precision machining is now in the realm of normal machining and this trend will continue, today's precision machining becoming tomorrow's normal machining. With the increase in the precision demanded by normal machining, so do demands in the normal metrology employed by manufacturers. Components such as compact disk optics are required to be mass produced to very high dimensional tolerances. High performance car engines are examples of mass produced components with demanding surface roughness and topography specifications.

During the early 20th century most surfaces were finished according to experience and trial-and-error. This method was adequate until the demands of the emerging aerospace industries from the mid forties onwards. It is not only aerospace industries that surface topography is important, figure 1 shows the relevance of topographic features to various aspects of surface performance. From this table it can be seen that the surface height parameters (R_a, R_q etc.) have considerable significance to a wide range of applications. It is important that industry identifies what performance features it demanded from surfaces and hence identify the correct topographical features to be measured.

FUNCTION	Heights R_a, R_q, R_t	Distribution & Shape R_sk, R_ku, BAC	Slopes & Curvatures Δ_a, Δ_q, Peak Curvature	Lengths & Peak Spacing λ_a, λ_q, Pc Correlation Length	Lay & Lead
Bearings	■	■	□	□	■
Seals	■	■	■	□	■
Friction	■	■	■	■	■
Joint Stiffness	■	■	□	□	□
Slideways	■	■	□	■	■
Elect/Thermal Contacts	■	■	■	■	
Wear	■	■	■	■	■
Galling	■	□	■	□	
Bonding & Adhesion	■	■	□	□	□
Painting & Plating	■	□	□	□	
Forming & Drawing	■	□	□	■	□
Fatigue	■	□	□		■
Stress and Fracture	■	□	□		■
Reflectivity	■	□	■	□	
Hygiene	■	□	□		

Key: ■ = much evidence; □ = some evidence; □ = little or circumstantial evidence

Figure 1. Connection between surface parameters and function

Having identified that it is the surface height deviations that are the most important to most industrial applications, systems need to be designed to measure these surface deviations. There are two main classes of metrological systems, qualitative and quantitative. The former tend to be simple systems, such as the glossometers described by Westberg (1967). In these systems a parameter is generated that can be related to surface roughness after calibration with the samples to be measured, but they are not absolute measures of surface roughness. The latter measurement method produces parameters which are absolute measures of the surface roughness within the constraints of the system resolution, the most common of these is the stylus profilometer. This instrument is suitable for off-line inspection but because physical contact is made with the surface under inspection and that some time is required to make the measurement the system is not suitable for on-line or in-process applications.

Light scattering

There has been much interest over the years in the use of scattered light in the measurement of surface roughness both from a theoretical and an experimental point of view. Traditional approaches have taken three general paths. Firstly, a well understood area of scattering theory is used to constructed an instrument that will give absolute measurements of roughness when a number of conditions are met. Secondly, if a more general theory is taken, a number of assumptions are made that mean that no hard and fast connection with a particular parameter can be made, but that the instrument can be used to measure a wide range of surface finish and roughness. The third approach is application specific, an instrument is designed that solves the problem at hand. The reason for this confusion is that there is no all embracing theory of light scattering, usually limits are placed on the applicability of the theory by the range of roughness values that can be covered. Theories exist for very rough random surfaces, Macaskill (1991) or for very smooth surfaces, Stover (1975). Usually the most easily related surface parameter to light scattering is R_q, see Marx and Vorburger (1990) for more details. Due to the lack of an all embracing light scattering theory the qualitative and ad-hoc approaches have dominated the field. The immediate need of industry is repeatable go/no-go gauges and for this application complex light scattering theories are not required, but a clear understanding of the inspection problem. A review of the application of light scattering to surface finish measurement can be found in Griffiths, Middleton and Wilkie (1994)

Recognition Methodologies

Logical Neural Systems

The rapid generation of results from inspection systems is an obvious industrial need. Real time inspection is easy achievable by using hardware. The WISARD system, described in Aleksander (1983) was developed from the n-tuple technique developed by Bledsoe and Browning (1959). The WISARD architecture has been used for a number of complex classification tasks such as face recognition, Farahati et al (1992) and piece part recognition, Elliot and Griffiths (1990). More recently WISARD has been applied to the recognition of scattered patterns produced by machined metal surfaces as in Middleton, Wilkie and Griffiths (1993a and 1993b) . When classifying between finishing methods that alter the lay of the surface from random to periodic then a considerable change in the shape of the scatter pattern takes place as can be seen in figure 2 which shows the three dimensional shape of the intensity pattern produced by parallel and cross lapped surfaces. The classification of shapes according to roughness is complicated by other surface features such as form errors. Figure 3 shows the effect on the light scattering pattern of a form error along with the three dimensional profiles of these surfaces. This highlights one of the problems encountered when relating stylus based parameters to light scattering results. Parameters such as R_q have been derived from mathematically flattened profiles with the long wavelength components removed by filtering.

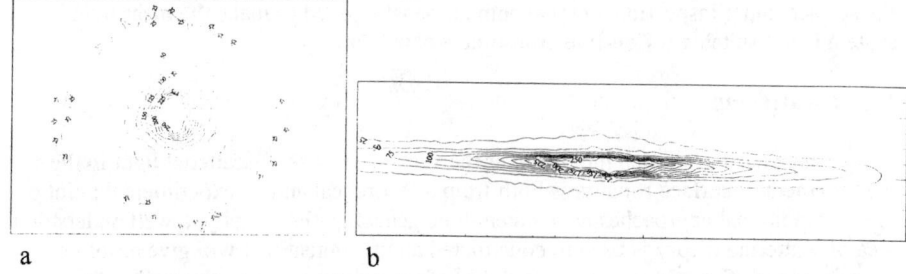

a   b

Figure 2. Difference between scatter patterns generated by (a) random (b) periodic lay.

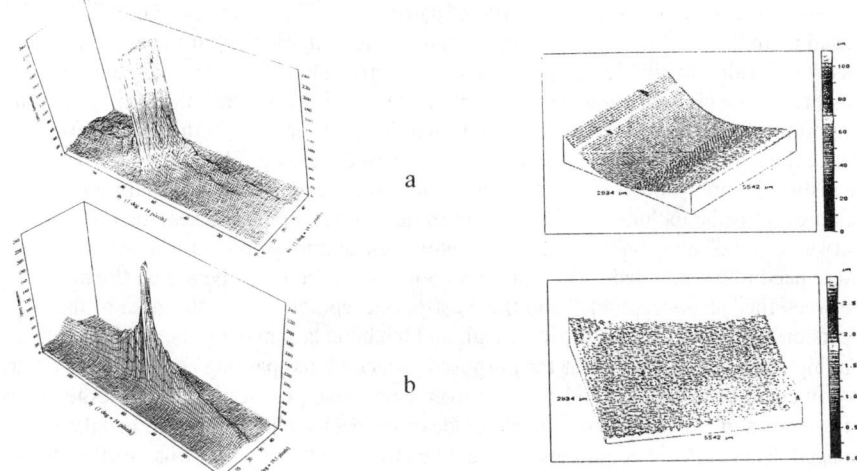

Figure 3. Scatter pattern from surfaces with (a) with form error, (b) no form error.

Light scattering on the other hand contains information about the surface roughness, form, reflectivity, contamination and conductivity. This goes along way to explaining some of the problems encountered when relating light scattering and stylus results.

Colour Content of Images

Observation of a machined surface under diffuse white light shows that the light scattered from the surface is coloured. As this seems to be a general feature of machined surfaces it should be possible to use the coloured content of scattered light for surface inspection. Using software developed for coloured object inspection, see Wang and Griffiths (1991), the colour content of a series of scatter patters from a set of shaped surfaces of different roughness was investigated. The results can be seen in figure 4. It can be seen that the shape of the colour cluster changes as the roughness decreases.

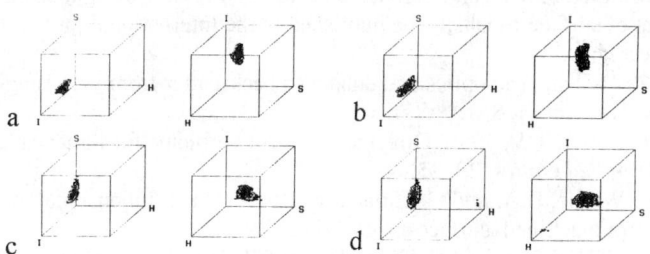

Figure 4. Colour clusters of shaped surfaces. Ra = (a) 0.8µm, (b) 1.6µm, (c) 3.2µm, (d) 6.3µm. H= Hue, S= Saturation, I= Intensity.

Discussion

A number of methods of surface assessment have been discussed. Each illustrates how features of the light scatter pattern produced by surfaces can be used to provide parameters that can serve as a measure of particular surface features. These methods are based upon the n-tuple technique which can be easily implemented in hardware as it is based on readily available computer memory chips. This allows the development of low cost, real-time classification systems. When developing any classification system, it is important to bear in mind that the system will have to function in the "real-world" hence the test and training sets that are used when developing the system must be realistic. Figure 3 shows what an effect form errors can have on the final image. These features must be included in the test and training sets. In most cases, the distinction between "good" and "bad" surfaces is easily measurable, so only a few examples of these need to be included. The task of a go/no-go gauge is to operate on the boundary between the "just-acceptable" and the "just-not-acceptable". It is this area of the decision surface that is the most critical, and the boundary may not be so clear cut. Failing components that are fit for purpose is wasteful and passing components that are not fit could be dangerous and can lead to embarrassing product recalls. A large number of examples of surfaces that fall either side of the decision plane have to be taken and imaged under realistic conditions. It must be clearly understood on what surface features are the pass/fail criteria to be based

Acknowledgements

The support of the SERC-ACME directorate, the Ford Motor Company Ltd. and Integral Vision Ltd. is acknowledged.

References

Aleksander, I. 1983, Emergent intelligent properties of progressively structured pattern recognition nets. *Pattern Recognition Letters*, 1.

Bledsoe, W.W. and Browning, I. 1959, Pattern recognition and reading by machine, *Proc. Eastern Join Computer Conference.*

Elliott, D.J.M. and Griffiths, B.J. 1990, A low cost artificial intelligence vision system for piece part recognition and orientation. *Int. J. Prod. Res*, 28, 6.

Farahati, N. Green, A. Piercy, N. and Robinson, L. 1992, Real-time recognition using novel infrared illumination, *Optical Engineering*, 31, 8.

Griffiths, B.J., Middleton, R.H. and Wilkie, B.A. 1994, A review of light scattering for the measurement of surface finish, To be published in the International journal of Production Research.

Macaskill, C. 1991, Geometric optics and enhanced backscatter from very rough surfaces, *J. Opt. Soc. Am. A*, 8, 1.

Marx, E. and Vorburger, T.V. 1990, Direct and inverse problems for light scattered by rough surfaces, *Applied Optics*, 29, 25.

Middleton, R.H., Wilkie, B.A. and Griffiths, B.J. 1993a, Three dimensional analysis of light scattered from machined surfaces, *COMADEM '93.*

Middleton, R.H., Wilkie, B.A. and Griffiths, B.J. 1993b The classification of machined surfaces using three-dimensional analysis of scattered light. *Advances in Manufacturing Technology VII.*

Stover, J.C., 1975, Roughness characterisation of smooth machined surfaces by light scattering, *Applied Optics*, 14, 8.

Wang Y.S. and Griffiths, B.J. 1991, Position invariant colour information for N-tuple classifier. *Proc. 11th Int. Conf. on Production Research.*

Westberg, J. 1967, Development of objective methods for judging the quality of ground and polished surfaces in production. *Proc. Instn. Mech. Engrs*, 182, 3K.

GEOMETRIC ASSESSMENT OF
SCULPTURED FEATURES

G. Smith, T.M. Hill, S. Rockliffe, J. Harris

Intelligent Manufacturing Systems Group,
Faculty of Engineering,
UWE, Bristol, Avon. BS16 1QY

Sculptured features are finding an increasingly wide range of
applications in consumer products. However, there is still no
universally accepted standard for assessing the geometry of
products with sculptured parts. This paper reviews those
techniques currently used to assesses surface geometry and
presents a summary of the relative advantages of these.

Introduction

This paper presents a review of methods available for the geometric
assessment of manufactured surfaces. The review looks at how these methods are
applied, why they might be selected and what considerations should be made when
selecting a particular method for the geometric assessment of sculptured features.

The need for surface assessment has arisen from the desire by engineers to
create products with sculptured surfaces with specific functional and/or aesthetic
properties. Typical functional requirements being; drag characteristics of motor
vehicles, lift form aerofoils and the hydrodynamic performance of ships hulls. The
use of sculptured surfaces in engineering structures has been long established.
Originally surfaces were mainly aesthetic appendages, or they provided protection
form wind and water. One notably exception was the shipbuilding industry were
surfaces played a crucial role in the functionality of any vessel. Models and
prototypes are used to prove hull shapes, generally a system of complex curves,
which are then scaled up for actual hull constructions. At some point the work of
the designers is transferred to the builders and it this transference that ultimately
leads to the need to check the accuracy with which the design has been followed.

In the early part of this century surfaces became a more crucial part of the engineers' remit with the invention of the aeroplane and the use of aerofoil surfaces to provide lift. The shape of these surfaces was as critical as those employed in ship building. It is interesting to note that both industries employed similar construction methods. The method used was to construct a frame work of runners and stringers, over which a material was stretched, thus creating a surface.

This method has influenced assessment techniques that are still in use today. In particular in the automotive industry were surface assessment is realised through the combination of features assessment (trimlines, attachment lugs, etc) and direct (template) surface assessment. Anthony (1986) described the limitations of direct assessment techniques when tolerances are used. He discussed the use of microscopes to measure visible gaps between templates and manufactured surfaces, and concluded that this process was very dependant upon the skill of the inspector for its accuracy. The problems increase when the manufactured surface exceeds the nominal volume.

Indirect Assessment Techniques

The 1960's saw the beginnings of work that is still continuing to represent curves and surfaces in a computer environment. This heralded the beginnings of the desire to assess surfaces themselves, rather than secondary templates. It also allowed the use of so called in-direct methods to assess surfaces. The problem posed for anyone working in the field of surfaces and their assessment is to choose an appropriate combination of representation methods and assessment methods. Considerations in this choice include speed, accuracy and ease of operation.

Indirect assessment can be further sub-divided in to contact and non-contacting devices. Essentially contacting devices are tactile probes, either scanning or touch trigger. The non- contacting approaches include the use of cameras and lasers. They are also often referred to as Optical methods. Most indirect assessment techniques are associated with the use of Coordinate Measuring Machines (CMM).

Indirect Surface Assessment Techniques

This section considers, in more detail, the following assessment methods:

      i, Laser Scanning
      ii, Optical Scanning
      iii, Scanning Probes
      iv, Touch Trigger Probes

i, Laser Scanning Devices
Laser scanning heads and software have been available since the early 1980's Goh, Phillips and Bell (1986). Their development has been slower than that of techniques involving touch trigger probes due to the relative expense and

bulk of laser/optic devices. However, the combination of advances in laser/optics technologies have reduced the cost and bulk of these devices such that they are now a available for use on CMMs. As mentioned earlier, laser systems are generally used in conjunction with a coordinate measuring machine. Laser probes, and touch trigger probes, rely upon the principle of digitization of number of discrete points on the assessed surface. Laser probes differ in the method for performing this digitization and the speed at which data is collected.

Laser probes make use of the triangulation principle as described by Somerville and Modjarred (1987). However, these probes are not without their drawbacks. For instance their accuracy is approximately an order of magnitude lower than that of touch trigger systems at around 10 microns, Chenggang (1992). Laser probing systems also suffer problems associated with surface finish variation; ie the surface finish can affect the way that light is reflected back to the detector and hence the refraction angle that is detected. The cusping effect produced by some surface machining processes may lead to particularly acute problems with the angle of refracted light beams. Laser probes also suffer problems in identifying the edges of components. Finally, their cost is reported to be higher than conventional touch trigger probes.

There are two distinct advantages with laser probes; one is their non-contacting operation which is attractive to certain industries concerned with the manufacture of fragile components, the other is the speed at which they can collect data and digitize surfaces. Somerville and Modjarred (1987) reported that whilst traditional CMM's and tactile probes would take around 12 minutes to digitise the section of an F-16 aircraft wing, a laser probe could perform the same task in 11 seconds.

ii, Optical Devices
The particular method considered in this review is the system known as ECOSCAN 100, marketed by Hahn & Kolb. This method uses a projection moire, a laser triangulation head and a solid state video camera. The system works by building an image of the measured component in layers, thus it is well suited to rapid phototyping using stereolithography processes. The optics head is mounted on a CMM and the component to be measured is presented in an appropriate manner. The optics head can also be interchanged with a conventional touch trigger probe. This method is particularly useful in the area of reverse engineering as measurement points can be gathered for use as CAD data very quickly. It is also useful to note at this point the method proposed by Ip (1993). Ip proposed that surfaces could be assessed using monochrome image recovery from shading. This method involves the use of the photometric stereo technique; essentially consisting of 2 cameras and a moveable light source.

Both of these approaches have their major advantage in the speed at which they can collect data. In the case of the ECOSCAN 100 it can acquire data at the rate of 100,000 points per second, compared with approximately 1 per second for touch trigger methods. However the quoted accuracy for this system is as low as 100 microns. The photometric stereo technique is also very rapid, but lacks the accuracy essential for many engineering applications.

iii, Scanning Probes

This technique resembles the touch trigger technique in appearance and is similar in its operation. Scanning probes are a fairly recent development of touch trigger technology, and are designed specifically for use with the assessment of sculptured surfaces. Unlike the touch trigger probe, the scanning probe unit is an analogue system that notes the displacement direction of the probe. It is this displacement direction that can be used by the measuring system to compensate the probe and calculate the direction cosines for the collected points. It further differs form the touch trigger probe in that because of its analogue operation it is not required to 'back-off' between points thus greatly enhancing its operating speed. Little academic work appears to have been performed using these probes but possible drawbacks may be their inability to compensate for rapid changes in curvature or vertical edges on components.

iv, Touch Trigger Methods

Touch trigger probes are essentially complex, micro switches that are mounted on CMM's. Attached to the probe is a stylus that has a small industrial ruby at their tip. It is this tip that makes contact with the surface being inspected. Their use in industry for feature inspection is currently widespread and is covered by the British Standard BS7172 (1989). Unfortunately this document does not extend to the assessment of surfaces.

The LK company proposes 2 approaches to the assessment of complex curves. The first is restricted to planar curves, it operates on a single scan path and is thus unable to compensate for surface normals other than those in the scan plane. This technique is simple and fast, but, it is not recommended for the assessment of non-planar curves or surfaces. Sculptured surfaces are in general composed of non-planar, complex, curves thus this technique is not well suited to their assessment. However, it is favoured by some sectors of the motor vehicle industry where specific areas on a surface (often trimlines or arbitrary datums) are scanned in this manner and the coordinates of the measured points are compared with those of the nominal points, obtained from a geometric CAD representation of the product. The assumption being that the surface geometry is acceptable if the arbitrary datums are within tolerance.

The second method is for use with complex curved surfaces. In this method a series of scans are made before compensating for surface contact vectors. The use of multiple scans paths facilitates the determination of surface normal vectors and so the 'direction' that the curve is progressing in can be calculated by compensating for probe radii and contact vectors.

Collection of data with these methods is relatively slow; the first method collecting 2-3 points a second and the second method effectively collecting data at less than half of this rate. The touch triggers advantages, however, lie in its accuracy (1 micron) and its flexibility. The touch trigger probe can access more of a component (through suitably alignment of the probe) without the need for specialist fixtures or realignment of the measured component.

Summary

Various assessment techniques are available and the selection for a particular application is largely dependant upon the nature of the product. Speed can be found in the optical techniques whilst accuracy requires the use of contact probing methods. When considering a complete system for the assessment of components containing both sculptured and conventionally dimensioned features, it may worth considering the use of interchangeable heads giving some of the advantages of each method. Table 1 summarises the features of various CMM based probing systems.

Table 1. CMM Based Methods

Method	Speed (ppm)	Accuracy (micron)	Flexibility	Cost (£)
Touch Trigger	2	1	High	10,000
Laser Probe	50	10	Low	30,000
Ecoscan 100	1.10^5	100	Low	40,000
Scanning Probe	80	100	Med	20,000

References

Anthony, D.M. 1986, *Engineering Metrology* (Oxford Pergamon, London).
Chenggang, C. 1992, Scanning compound surfaces with no existing CAD model by using a laser probe on a coordinate measuring machine (CMM). *SPIE* **1979**, 57-67.
Goh, K.H., Phillips,N., Bell,R. 1986, The applicability of a laser triangulation probe to non-contacting inspection. *International Journal of Production Research.* **124 no. 6**, 1331-1348.
Ip, W. 1993, Enhanced procedures to model machine and measures free-form surfaces using analytical and artificial intelligence techniques. Phd Thesis, University of Birmingham, U.K.
B.S.7172, 1989: *Guidance to Operators and Manufacturers of CMM.*

CHARACTERISTIC VIEW MODELS FOR OBJECT LOCATION

Dr Andrew M. Wallace, Frederic Lavorel and Dr Emanuele Trucco

Department of Computing and Electrical Engineering
Heriot-Watt University
Riccarton EH14 4AS, Scotland

We investigate the use of Characteristic View Models (CVMs) for the recognition and location of objects in depth images, acquired as an initial stage in a system for location and inspection of metallic components using range data. Each Characteristic View (CV) of an object may be either a prototypical view contained within the set, or a composite representation which contains data about the common features of all views within the set. The CVM is derived automatically from a standard Computer Aided Design (CAD) package, and used for pose definition in 3D space.

Derivation of a Characteristic View Model

A CVM represents a balance between the two extremes of single object centred and aspect graph models. It is arguable that a single object model leads to unnecessary complexity in matching model to scene primitives, since not all model primitives can be present in a single view. At the other extreme, aspect graphs encode separate representations for all views separated by distinct visual events (Bowyer and Dyer, 1990). The number of aspect nodes for even moderately complex objects is prohibitive. Furthermore, when visual scene data is segmented, many scene primitives may be missing or poorly defined, and it is not necessary to match all primitives to recognise and locate the object. These factors make aspect graphs inappropriate.

To derive a CVM, an image of the CAD model is raytraced from all viewpoints on a geodesic dome surrounding the object, giving 20, 80, 320 or 1280 approximately evenly spaced views. This information is encoded in a "visibility table" which includes a description of the area of each primitive which is visible from each viewpoint. From the original object model and this visibility table, a number of continuous variables (range 0 to 1) are derived for each primitive and each viewpoint.

Visibility: the ratio of the visible area of the primitive in the view to the maximum visibility of the primitive. This encodes the idea that views are characterised by how much of and which primitives can be seen.

Area ratio: the ratio of the primitive area to the total surface area of the object, known from the object model. If the primitive is not visible this is defined as zero. This encodes the idea that larger primitives have more weight in defining characteristic views.

Rarity: this is defined as $1 - P_c$ where P_c is the probability of occurrence of the primitive. This is the number of views in which the primitive is visible divided by the total number of views. If the primitive is not visible, this is again set to zero. This encodes the idea that rarely seen primitives have more weight in defining characteristic views.

Covisibility: Adjacent primitives are those which are connected in the three dimensional model; in a given view there may be additional primitives which appear adjacent but these are not considered useful. For N primitives, there are $\frac{N(N-1)}{2}$ possible pairs; for a given primitive those which are adjacent are known a-priori from the model. The covisibility is the ratio of those adjacent pairs which occur in the view divided by the total number of adjacent pairs. This encodes the idea that it is the geometric or topological arrangement of groups (or in this restricted case pairs), as much as isolated primitives, that is significant in defining characteristic views.

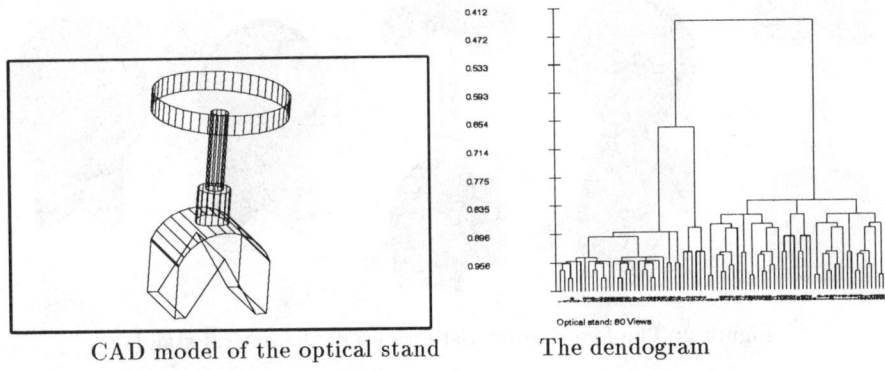

CAD model of the optical stand   The dendogram

Figure 1: Generation of Characteristic Views

Hence, for each view of the object, with associated viewing coordinates, and for each feature defined in the object model, there are 4 measures of the "importance" of that feature in defining the characteristics of the view. This allows the formation of a *similarity matrix*, which has a number of rows equal to the number of ray-traced viewpoints, and a number of columns equal to four times the number of features. Using this matrix of continuous variables in the range 0 to 1 we can form clusters of similar views. Each cluster, or alternatively a representative view within each cluster, is a characteristic view.

The key parameters are the choice of variables, defined above, the measurement of similarity and distance, and the technique for forming clusters. The choice of measure of similarity or dissimilarity can affect the final classification significantly (Everitt, 1977). We use the product moment correlation coefficient to measure similarity between individuals, although it has been criticised on the grounds that it favours similarity of data profiles rather than average levels. In our case, we believe the profiles and levels are closely related. Using agglomerative hierarchial clustering, the single link or nearest neighbour aproach reduces the number of groups progressively by one in each iteration. Following each reduction, a new similarity matrix is formed. The hierarchial process can be represented in the form of a dendogram, or tree diagram, in which the progressive agglomeration of views is plotted against correlation coefficient.

The threshold is the coefficient value at which the clustering process is stopped. This determines how similar views of an object should be to be capable of representation by a single characteristic view. Informal methods for determination are generally based on the idea that a significant change in coefficient level between fusion is indicative of the correct number of groups.

The CAD model of the metallic test object used as an exemplar here is illustrated in Figure 1. The corresponding dendogram shows the progressive fusion of the initial 80 views. The five characteristic views which correspond to a level of 0.840 are shown in Figure 2. Broadly speaking, these 5 views are sufficiently different to avoid unnecessary duplication of the matching process, yet show sufficient simultaneously visible combinations of primitives to allow location of the object in any arbitrary view.

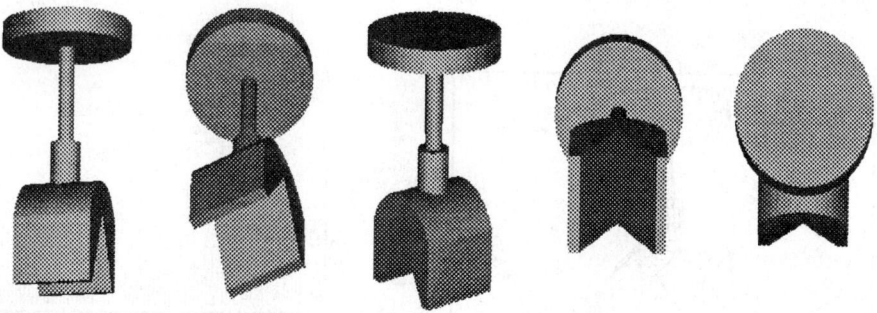

Figure 2: The five characteristic views of the optical stand

Matching of object and characteristic view models

To locate the object in an arbitrary view, there are 3 stages, selection of matching primitives, computation of the pose hypothesis, and verification of the back projected model in comparison with the scene. There are several constraints which may be employed to select matching primitives of similar type, of which the most common and effective are probably the use of pairwise geometric constraints (Wallace, 1987) and concentration on local primitive groups (Bolles and Cain, 1983). In this study, we consider only viewpoint modelling, i.e. constraining the selection to include only model features which are present in the given object or CV model. Each scene or model primitive is represented by a point vector, so that a minimum of two matching primitives is required to solve for the pose. Normal and axis vectors, and central and central axis points are used for planes and cylinders respectively. Pose determination is based on minimisation of the least square error between the transposed model and scene point vectors using singular value decomposition (Haralick, Joo, Lee, Zhuang, Vaidya and Kim, 1990).

To assess the accuracy of pose determination, we have measured the residual positional error in the solution of the pose of the object in an arbitrary view in several hundred trials, using both an object model and a CVM. When random Gaussian errors were added to the source data, there was no significant difference in the mean accuracy of pose determination using the object or the CVM, although the spread of values was slightly higher in the latter case.

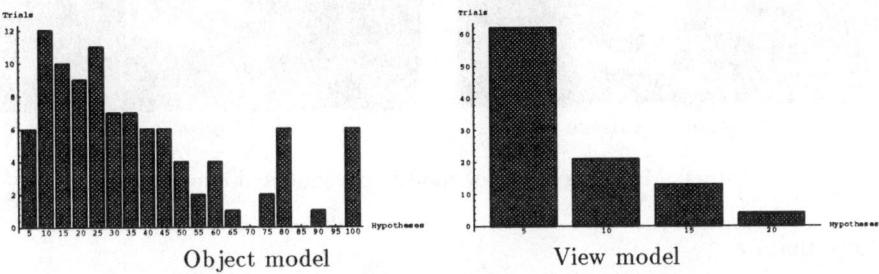

Object model   View model

Figure 3: Complexity of the matching process

To evaluate the reduction in complexity achieved by viewpoint modelling, we have performed several sets of trials using random and ordered subsets of features from the model and scene respectively. Figure 3 illustrates the result of one experiment using ordered subsets of 3 model and 3 scene primitives. Each bar of the histogram represents the number of trials in which n hypotheses were generated before the correct solution was verified. The x-axis defines $n-5$ to n, i.e. a range of values. For example, examing the object data, there were 11 occasions on which between 20 and 25 hypotheses were generated before the correct one was found. The value "100" is used to represent 96 or more hypotheses. The right graph shows that in 62 per cent of the trials, 5 hypotheses or less were sufficient. However, it should be noted that this reduction in complexity is achieved on the assumption that the correct viewpoint is known or invoked by view-specific characteristics.

Finally, we show an example of a matching task of a real image in Figure 4. Segmentation of the depth image on the right produced a typically fragmented scene containing eight planar and five cylindrical patches, obtained by the robust method of Jones, Illingworth and Kittler (1991). The pose was estimated by matching 5 primitives in the scene against 5 primitives in the central view of Figure 2. The CAD model has been superimposed on the right image to show the pose estimate, since the viewing parameters of the camera system are known. This shows some error in the computed pose, due to the inaccurate definition of the scene control points; there are errors in defining the positions of the planar patches and in the radii of curvature, and hence the axes of the cylindrical patches.

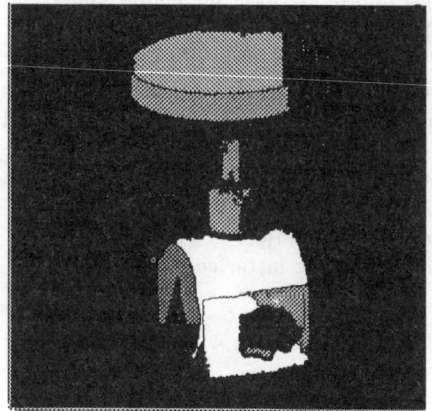 

Segmented surface patches   Computed pose

Figure 4: Matching the view model to segmented depth data

Conclusions

We have defined criteria for Characteristic View Models, and shown how these may be formed automatically from a source CAD object model. The matching complexity of the restricted subset of features included in a single characteristic view may be reduced in comparison with a full object-centred model, without significant reduction of pose accuracy. Although Characteristic View Models have a possible role to play in rapid location of objects, this does require invocation of the correct view, and does not imply that viewpoint constraints are preferable to other methods of reduction of search complexity, for example geometric constraints, nor that the combined effect of viewpoint and geometric constraints is necessarily greater than geometric constraints in isolation.

References

Bolles, R.C. and Cain, R.A. 1983, Recognising and locating partially visible objects: the local feature focus method, in A. Pugh (ed.) *Robot Vision*, (IFS Publications).

Bowyer, K.W. and Dyer, R.C. 1990, Aspect graphs: an introduction and survey of recent results, International Journal of Imaging Systems and Technology, **2**, 315-328.

Everitt, B.S. 1977, *Cluster Analysis*, (Heinemann).

Haralick, R.M., Joo, H., Lee, C-N, Zhuang, X., Vaidya and Kim, M.B. 1990, Pose estimation from corresponding point data, IEEE Trans. Sys. Man. Cybernetics, **9**, 1426-1446.

Jones, G., Illingworth, J. and Kittler, J. 1991, Robust local window processing of range images, Proc. of 7th Scandinavian Conf. on Image Analysis, 419-426.

Wallace, A.M. 1987, Matching segmented scenes to models using pairwise constraints, Image and Vision Computing, **5**, 114-120.

MEASUREMENT OF FEATURES
BY A SAMPLE-GRID APPROACH

A.J. Medland, G. Mullineux and A.H. Rentoul

Centre for Geometric Modelling and Design
Department of Manufacturing and Engineering Systems
Brunel University, Uxbridge, Middlesex, UB8 3PH

Traditional techniques used within inspection have evolved
from a combination of height measuring techniques and the
use of comparative devices for profiles and forms. The
introduction of the coordinate measuring machine has provided
an alternative approach by sampling points on the surface of
the measured part. An inspection procedure is proposed for
use with coordinate measuring machines that allows differing
features and tolerance requirements to be deduced from an
array of probing points spread across the surface of the
measured object.

Background

A recent research programme into the intelligent integration of
manufacture and inspection processes has led to a re-evaluation of the ways
in which tolerances on geometric features are defined and assessed. Within
this programme the aim was to determine means whereby not only the
errors in geometry could be detected, but also the manufacturing processes
giving rise to the errors, Medland, Mullineux and Rentoul (1993).

Within this study it was necessary to investigate the tolerancing
procedures. These were based on the British Standard BS308 (1986) and
derived from traditional manual inspection techniques. These processes
depend upon the recognition of forms and groups of features that can be
measured by comparative instruments, such as gauges and callipers. They
are based on two distinct approaches. Firstly global measurements are
obtained by mounting the object on a reference surface and measuring the
height to selected features. These measurements can then be combined, and
the component orientated, to allow both relative positional and angular

673

values to be obtained. The second measuring approach is based upon the direct gauging of the desired geometric feature. Here 'go' and 'no-go' gauges are created to allow the range of acceptable variation to be determined.

Inspection strategy for coordinate measuring machines

The use of coordinate measuring machines (CMMs) has provided a new way of measuring features. The approach is to sample points and from these infer information about geometric features. Here we propose a procedure for use with CMMs that allows differing features and tolerance requirements to be deduced from an array of probing points. It requires that all tolerance conditions are recast into a sequence of probing points. This allows all measurements to be made without reference to imaginary points or features, such as axes of holes. Measurement points are offset by the radius of the probe allowing tolerances to be assessed by direct comparison between theoretical and actual probe centres.

For simplicity, all deviations are calculated in terms of their absolute values and all tolerance checking made against a specified shperical deviation. Traditional tolerances can be recast into this form, giving a sequence of probing points and their allowable deviations, Medland, Mullineux and Rentoul (1994a). A number of points has to be chosen over each feature to allow the chosen tolerance conditions to be assessed. By using a common grid of points over a selected feature, the same point (with different allowable deviations) may appear in several different tolerance checks. The strategy is to select the correct combination of points and deviations to establish whether the feature meets the specified tolerance condition.

Evaluation system

The sample-grid procedure has been applied within a constraint based approach to inspection, Rentoul, Medland and Mullineux (1994b). Here objects are created as solid models from the process plan. This is achieved by the integration of the ACIS solid modeller (1992) into the RASOR constraint resolver, Medland and Mullineux (1993), and using procedures to read and interpret the plans. The component is created by transforming the tool motions into solid sweeps that are subtracted from the original solid block (Figure 1).

In an investigation of inspection strategies, the sample-grid technique was applied. A network of reference points was applied to specified inspection surfaces of the individual primitives used in the creation of the part (figure 2). The grid points were initially specified at a high density (with separations approaching twice the expected maximum probe diameter) with no points being closer than a probe radius to any boundary. After combination of the primitives, points lying inside or remote from the final solid were eliminated (figure 3). The number of grid points is substantially reduced. This is reduced further by taking into account the final geometric form and the tolerance measuring requirements. Any point lying within the maximum probe radius of an adjacent feature must be moved or eliminated.

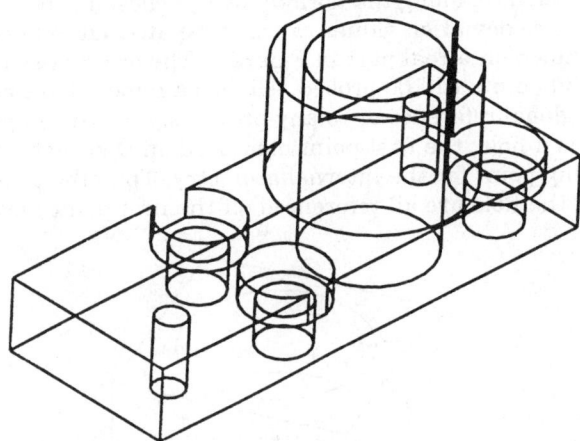

Figure 1. Test block created from machine operations.

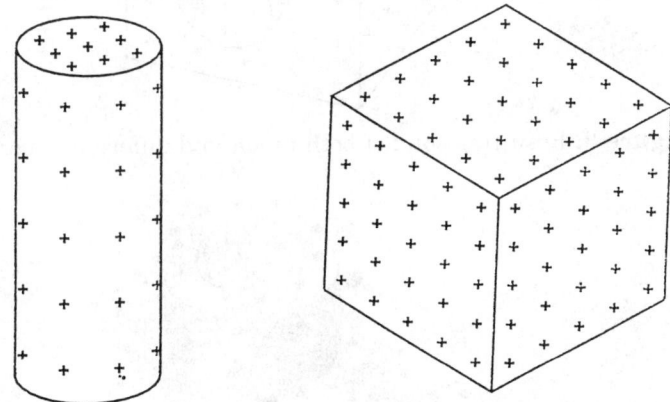

Figure 2. Sample–grid on specified surfaces of primitives.

The tolerance requirements, when translated into spherical deviation form, indicate that for measurement of straightness or angularity a sequence of points needs to be probed along a chosen straight line. This can extend into an array when area shapes are considered. For local straightness closely spaced points are required, whilst for global straightness of a feature, the points need to be spread across the full extent. For each of these measurements only three points are required but intermediate points may also be retained for secondary checking operations, should the test with the primary points fail. The grid of available points is reviewed for all tolerance requirements and those not required for either primary or secondary inspection are eliminated.

In order to carry out the inspection on the CMM, each surface reference point is replaced by its equivalent probing point centres, moved away from the surface, along the normal, by the chosen probe radius. The spherical tolerance deviation values can then be attributed to their specified locations, as shown on a real part in Figure 4. The order is chosen in which the sample–grid points will be probed. All measurements are made with respect to the global reference or to any previously measured point. For a feature, such as a hole, the first point measured on that feature becomes the absolute starting point for the internal geometry. Thus the group of points spaced around the hole are all referenced off the first point measured.

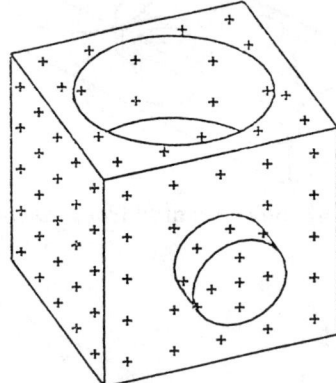

Figure 3. Resulting solid of both union and difference operations.

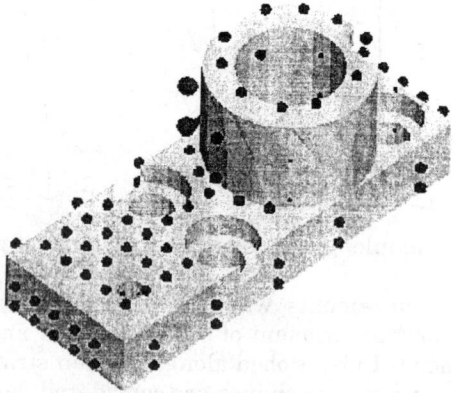

Figure 4. Inspection points and spherical deviations (shown enlarged).

Error detection

The strategy has been used to interpret manufacturing errors. A 'top–down' approach was adopted, as the method of manufacture indicated that errors were rare. The probing sequence was carried out as described. Differences between the theoretical and measured values were compared against the smallest deviation value attributed to each point. For errors less than their corresponding allowable deviations the 'true' state was declared. If a feature's points were true then the feature itself was deemed to be 'true'.

If all points on the body are true, no more investigation is required and the part accepted. If points are found to contain unacceptable errors, these are considered firstly within their lower feature groups to determine the type of error occurring. These may in their turn point to lower geometric errors attributable to tooling and setting errors. In order to provide advice on the likely causes of these errors, a code is built up giving the relationship between the errors detected. This is then compared with a list of known manufacturing and machining errors.

Conclusions

The research into an integrated approach to CAD modelling, inspection and error analysis, has shown the need to recast tolerancing conditions for use with CMMs. The chosen approach was to create sample–grids of points across the modelling features during their construction within the CAD system. This approach was successfully carried out in an experimental programme and showed that an automatic procedure could be created from the selection of the probing points through to the identification of errors and their likely causes.

Acknowledgements

The research here presented was carried out under a project supported by the ACME Directorate of the SERC and a group of collaborating companies. This support is gratefully acknowledged by the authors.

References

ACIS, 1992, *ACIS Interface Guide*, (Spatial Technology Inc. and Three-Space Ltd.).
BS308, 1986, *Engineering drawing practice*, (British Standards Institution, London).
Medland, A.J. and Mullineux, G. 1993, Strategies for the automatic path planning of coordinate measuring machines, *Manufacturing Systems*, **22**, 159–164.
Medland, A.J., Mullineux, G. and Rentoul, A.H. 1993, Integration of constraint and solid modellers, *Proc. 30th. Int. MATADOR Conf.*, (MacMillan) 565–571.
Medland, A.J., Mullineux, G. and Rentoul, A.H. 1994a, A strategy for the verification of tolerances by coordinate measuring machines', *J. of Engineering Manufacture*, submitted
Rentoul, A.H., Mullineux, G. and Medland, A.J. 1994b, Interpretation of errors from inspection results', *Computer Integrated Manufacturing Systems*, to appear.

OFF-LINE PROGRAMMING FOR COORDINATE MEASURING MACHINES

Dr K Sivayoganathan, Ms A Leibar and Dr V Balendran

Manufacturing Automation Research Group
Department of Manufacturing Engineering
The Nottingham Trent University
Burton Street
Nottingham, NG1 4BU
email:man3sivayk@uk.ac.ntu

Even though the advantages of off-line programming for CNC machines are known, such systems are not widely available for measuring machines. This paper presents a $2^1/_2$ D off-line programming system for Coordinate Measuring Machines (CMMs). The system has been developed on a PC system, takes IGES (Initial Graphics Exchange Specification) files or DMIS (Dimensional Measuring Interface Specification) files as input, and produces DMIS output to drive CNC CMMs. The system also produces the native language CNC program for the CMM in the Department, and IGES output to CAD systems. The system relies on user interaction in the creation of the CMM program and provides a simulation of the program generated. The salient features of the system and the future directions for such systems are discussed.

Introduction:

A Coordinate Measuring Machine (CMM) has a series of movable members, a sensing probe and a workpiece support member, which can be operated in such a way that the probe can be brought into a fixed and known relationship with points on the workpiece surface, and the coordinates of these points can be displayed or otherwise determined with respect to the origin of the coordinate system (BS6808 1987). CMMs are like milling machines in usage. The major difference between them is in the light weight construction of CMMs as they do not need to transmit heavy forces like the milling machines. CMMs can have contact ("touch probe") or non-contact sensor systems. The probe is usually mounted on a mechanism that permits it to move along one or more perpendicular axes. The mechanism contains electronic sensors which provide a digital readout of the position of the probe. When the probe "senses" the surface of a part being measured, the location of the probe in space is recorded and the data is manipulated to provide required results. When the Coordinate Measuring Machine (CMM) was introduced in 1960s, it brought about a subtle, yet significant, change in the manufacturing practices of machined parts. When machine tools performed individual operations, the machine operator performed the measuring function and made adjustments accordingly. When CNC machining with automatic tool changers entered the scene, the measurement task became beyond the capability of the machine tool operator. This change created the market demand for 3-D flexible measuring capability. The complexity of the measuring task and the need for a clean, controlled environment resulted in the CMM being located in the gauge room, or what has become the quality control area. Thus the feedback path was interrupted. The individual operator who had made the measurement to control the process was now operating automatic machine tools without the benefit of continuous measuring data (Bosch 1992). CMMs can provide the feedback needed, with improved productivity, if setup properly.

CMMs and Off-line Programming

A reason that more CMM programming is not done on off-line graphical programming systems (typical of CNC machine-tool programming) is due to CMM programming capability lagging CNC programming systems according to Mason (1992). Some manufacturing firms have been linking CAD systems to coordinate-measuring machines for a few years (Hahn, Roder, Michalowsky, Purucker, Hornung 1988). However, early attempts were not very successful. Since then, new software developments have made CAD to CMM links more productive. Over the past few years, some CAD software designers have begun to develop off-line programming facilities for CMMs. A software (preprocessor) is then necessary, for the integration of a CAD system and a CMM. Some CMM vendors have taken advantage of these developments and have provided post-processors for the different CMMs. Most CAD system vendors, however, are slow in supplying this link.

This paper describes the work done to provide a graphical off-line programming system for CMMs by extracting data from CAD systems and output DMIS (Dimensional Measuring Interface Specification) programs or the CMM specific programs (Leibar 1993). IGES (Initial Graphics Exchange Specification) 'neutral data' files are used to extract the data from CAD systems.

IGES, DMIS and CMES

Most of the work presented in this paper deals with the IGES, DMIS and CMES. Brief descriptions are given below.

IGES

IGES defines a file structure format, a language format, and the representation of geometric, topological, and non geometric product definition, in these formats [IGES 91]. Developers must write software to translate from their system to IGES format, or vice versa. The software that translates from a CAD system to an IGES file is called a pre-processor and the software that translates from IGES to a CAD system is called a post-processor. The combination of pre-processor and post-processor determines the success of an IGES translation. IGES is supported by most of the CAD vendors. It is a popular method attempted by developers trying to move CAD data from one system to another (Sivayoganathan, Balendran, Czerwinski, Keat, Leibar, Seiler, 1993). However it has not lived up to expectations, although many of the expectations were unrealistic. The primary expectation was that a user can take any CAD file, translate it into IGES, read the IGES file into another CAD system, and have all the data transferred, including resolution of system differences. In real life a spread of transfer success is witnessed (Mayer 1987). It is difficult to give an average figure, since the success rate depends on the IGES processors used and the types of entities in the files to be translated. The quality of IGES processors varies widely. Some processors handle surfaces and complicated structures whereas others handle little more than lines, points and circles. Obviously success depends in part on the type of systems employed. Even though the need for CAD data transfer is great, IGES has not been a perfect answer and a number of alternatives exist, each with its own strengths and weaknesses.

DMIS

The objective of the Dimensional Measuring Interface Specification (DMIS) is to provide a standard for the bidirectional communication between computer systems and inspection equipment. The specification is a vocabulary of terms which establishes a neutral format for inspection programs and inspection results data. While primarily designed for communication between automated equipment, DMIS is designed to be both man-readable and man writable, allowing inspection programs to be written (and inspection results to be analyzed) without the use of computer aids. Even before its approval as a national standard, DMIS was opening up a new industry, with DMIS products and DMIS supported equipment appearing regularly in the market. Vendors of measuring and CAD equipment see this as a boon to the industry, as these companies can now write a single interface to the standard (Anon 1990). DMIS like IGES is also based in the concept of neutral format. An equipment which interfaces to others through the DMIS vocabulary will have a pre-processor to convert its own internal data into the DMIS format, and a post-processor to convert the DMIS format into its own data structure. The implementation of DMIS is dependant on individual users. A section of a DMIS program may look as follows:-

    F(WallIn1)=FEAT/LINE,BND,CART,x1,y1,z1,x2,y2,z2
    F(Point1)=FEAT/POINT,CART,x1,y1,z1

CMES

CMES is the language and it forms an operating environment for the LK CMM machines. CMES allows for simple inspection programming or the 'High Level' programming (for constructs like 'IF', 'THEN', 'ELSE') if the option is available. Basically, points are taken on features, which are then used to find or calculate the necessary outcome. There are a good set of datum setting and manipulation commands, feature measurement commands along with file management commands. Most of the commands are of the 'two character type. For example, 'ID' is for internal diameter of a hole (LK 1993). To measure an inside distance in CMES two contact points are taken and saved in two different variables. Before using the command to measure the slot the CMM needs to know the number of the variables where the points were saved. The commands are:
UP,1,2
LI, X

Off-line Program Generator

For this generator, IGES is used to extract data from a CAD system. Then, using the mouse and the keyboard the user interacts with the system (Leibar 1993). Firstly, the master datum is defined on the part. The features to be inspected on the part are then selected. During the selection of features, sub-datum(s) can be defined, if needed. Once the selection is finished, the manner of execution of the measurement has to be defined. This can be performed either following the order of selection or taking the computer solution. The output of the inspection program will be in the international standard DMIS or in the CMES language.

The system has the following features:-

1. Set up master datum or sub-datum.
2. Twelve simple and complex feature measurements (hole / cylinder, arc inside / outside, length inside / outside, length between points, length between two features, pitch circle diameter etc.).
3. Feature manipulation (tolerances, delete, undelete) and probe manipulation (moves relation to features),
4. File(s) management for input & output of IGES, DMIS and CMES.
5. Simulation of inspection and to change the inspection method.

The system automatically adjusts the path of the probe to avoid collisions and also provides for prototype part measurement. The shortest path measurement is provided by looking at the nearest point the probe is required to move to after the last movement. The system handles all the normal measurements on the principal planes easily. This has to be extended to inspect on auxiliary planes.

Some Points to Note

When driving the probe, the working plane for which the inspection path is to be created, has to be considered. The coordinates when the feature was created are stored as x, y and z values. But when the CMM program is created, the coordinates must be related to the plane where the measurements are carried out. This relationship is different in CMES and DMIS.

- In CMES, before coordinates values are given, the axis to which these coordinates refer have to be set.
- In DMIS, the three coordinates (x, y, z) should always be defined. For instance, if a movement of the probe only in one axis direction is to take place, even when only one of the coordinates will change, the three coordinates of the target point must be given.

The inspection program was generated to take the minimum number of contact-points the CMM needs to calculate the actual feature values. But for some features (e.g, hole) more contact-points could be taken. One of the constraints is where the contact-points are taken in the feature. For the same type of feature the contact-point position is defined and cannot be altered, but the CMM allows the user to take a point in any part of the feature. If this is achieved, it could be applied to find the possible shortest-path. In this instance, a new problem would arise because the number of combinations would increase drastically as the number of possible combinations of points to consider would be enormous. In the solution proposed by the system as the contact-points are fixed, the number of combinations are lower. Even then some assumptions have to be made to calculate the shortest path, as otherwise it would be almost impossible to find a solution.

Future Development

Work currently undertaken by the International Standard Organization (ISO) for the Exchange of Product Data (STEP) project is to develop a three-layer model to represent the total information content of a product. This is achieved by creating a product model which contains all the information a product needs throughout its life time [Owen 93]. Formal methods are used in the development of models and as a result STEP avoids some of the problems that were encountered with IGES, such as ambiguities with respect to entity definitions. The tendency of present CAD systems is the move towards 'feature-based' systems. The idea is to create the designs based not only in the geometry but also in the functionality of the features. This will provide more relevant information about the part to the subsequent processes like manufacturing, inspection etc. The STEP specification is seen as an ideal standard output for CAD/CAM systems and so we have to wait until such an option is commercially available.

References

Anon 1990, DMIS approved as American National Standard, *Tooling & Production*, June 1990, 24.

Bosch, J.A. 1992, The Changing Roles of Coordinate Measuring Machines, *Industrial Engineering*, November 1992, 46-48.

BS6808, 1987, Coordinate Measuring Machines, British Standard Institution.

Hahn, D. Roder, J. Michalowsky, W. Purucker, H. Hornung, J. 1988, Linking Design and Test, *Computer Aided Design* report, July 1988, 1-9.

IGES 1991, Initial Graphics Exchange Specification V5.1, IGES/PDES Organization, September 1991.

Leibar, A. M. 1993, Intelligent Integration of Computer Aided Inspection with CAD/CAM Systems, MPhil Thesis, The Nottingham Trent University, Nottingham.

LK 1993, CMES Reference Manual, LK ltd., Castle Donington.

Mason, F. 1992, Program your CMM Off-Line, *American Machinist*, October 1992, 45-47.

Mayer, R. J. 1987, IGES: One Answer to the Problems of CAD, *Computer Aided Design*, June 1987, 209-214.

Owen, J. 1993, *STEP: An Introduction*, 1st edition, (Information Geometers, Winchester)

Sivayoganathan, K. Balendran, V. Czerwinski, A. Keat, J. Leibar, A. M. Seiler, A. 1993, 'CAD/CAM Data Exchange Application', *Advances in Manufacturing Technology VII*, Bath, Sept 1993, 306-310.

ERROR COMPENSATION ON CNC MACHINE TOOLS USING FUZZY LOGIC CONTROL

Erping Zhou, *D.K. Harrison, *D.Link and R. Healey

Bolton Institute of Higher Education
Bolton, Lancs, BL3 5AB

*Staffordshire University
Stafford, Staffs, ST18 0AD

In-process measuring or in-cycle gauging system using touch trigger probes have been widely applied in CNC machine tools. Within this system, the compensation scheme takes up a significant part. The machining and measuring processes are complex and non-linear since they are affected by geometric and thermal errors. It is impossible to determine each error item by a touch trigger probe system. This paper describes a new compensation technique which uses a fuzzy logic control system. The fuzzy control is a control strategy that can be used in ill-defined, complex, non-linear processes and is also suitable when a single precise measure of performance is not meaningful or practical. A compensation model employing a fuzzy control system has been presented and the technique has been justified using a CNC milling machine.

Introduction

In-process measuring (IPM) or gauging is a sensor-based measuring system and measuring is carried on the machine tool as part of the total machining cycle but not at the same time as cutting takes place. The main areas of application for this method are on CNC lathes and CNC machining centres where small-to-medium batch work is done. A typical IPM system normally consists of sensors, sensor interfaces, microprocessors or microcomputers, software measuring programs and compensation methodologies.

Measuring is used to determine which position or size of tool or fixture needs adjustment, and when the part program should be modified. In machining and /or measuring operations of the machine tool, the accuracy of the workpiece dimensions is affected by the error of the relative movement between the cutting tool or the probe and the workpiece. Factors that influence measuring accuracy may be described as inherent geometrical errors of the machine tool, probe errors, process dependent errors and environmental errors [Hermann, G. 1985]. The main factors among them that affect the relative position are the geometric errors of the machine tool and the thermal effects on these geometric errors [Fan. K.C. et al 1992]. Since there is a lack of three- dimensional comparison equipment, there are no direct dynamic methods to determine the volumetric errors of machine tools. Many mathematical models and different methodologies have been investigated and employed by numerous international bodies and researchers. However, it is an area that still needs to be researched since there is no one method which totally satisfies the requirements.

Review of compensation technologies

From the emergence of IPM, feedback methods have been utilized [Roe, J.M. 1984]. Roe described a compensation technique by comparing expected and measured values and using this error value as an increment for the relevant offset to correct the machining process. Although this feedback method ignored the thermal effects on the machine and could not compensate the inherent geometrical errors of the machine tool, it has still been widely accepted as a base for IPM software design. The total source error of adjustment (TSEA) model [Cochran, D.S. and Swinehart, K. 1991] compensates the error sources through defining static errors and dynamic errors. Static error refers to the time-independent errors which only occur at the time of setup. Dynamic error refers to the time-dependent error occurring throughout the manufacturing operation. This model is used to systematically detect and minimize the error sources including thermal effects on the machine and manufacturing process. The model provides a relatively standard compensation method compared with Roe's feedback technique, but it still does not compensate for the inherent geometrical errors of the machine tool and needs to use IPM repeatedly during each machining cycle. An error compensation model developed by Fan [Fan, K.C. et al 1992] deals with the geometrical errors of the machine tool. The error compensation scheme is based on an analysis of the volumetric error in 3D machines which have been considered in terms of 21 error components in machine linkages, namely three linear errors and three angular errors along each moving axis, and three squareness errors related to the three axes. The error calibration was carried out on a vertical milling machine table by using a HP5528 interferometer and an electronic level. Having calibrated all error items, the actual volumetric error of this machine could be determined. In practice, this data could be used for error compensation. The time-variant volumetric errors due to the dynamic effect of the machine tools are not concerned.

To summarize the compensation techniques mentioned above, the key issue involved is the design of the compensation system and how many error items could be found by a TTP system and compensated through the IPM. It is difficult to understand the detailed and precise structure knowledge of the machining and measuring processes, since there are vagueness and imprecision inherent in real practical systems. The fuzzy logic control attempts to solve complex control problems by relaxing the specification of the controlled response in a manner that reflects the uncertainty or complexity of the process under control. Fuzzy logic control theory gives some mathematical precision to imprecise statements and vagueness inherent in non-linear and complex processes. Therefore, fuzzy logic control has been applied to the error compensation system.

Fuzzy rules and fuzzy inference for error compensation

In the error compensation system, the deviation of the measured value from the desired value can be expressed by the relative error (e_n). The change of error (ce_n) is defined as the difference of two adjacent errors e_n and e_{n-1} as follows:

$$e_n = v_m(\text{measured value}) - v_d(\text{desired value}) \qquad \text{(Eq.1)}$$
$$ce_n = e_n - e_{n-1} \qquad \text{(Eq.2)}$$

In the following example, sets A, B and C are generic for linguistic terms as Positive Big (PB), Positive Small (PS), Negative Medium (NM), and Zero (Z0) etc. Where E, CE, and U denote error, change of error and compensation respectively. And $\mu_A(x)$, $\mu_B(y)$ and $\mu_C(z)$ are membership functions, which take values in the range 0 to 1. Suppose that the error compensation process could be expressed in the following linguistic terms:

if error is PB and change of error is PM then compensation is NB;
if error is NS and change of error is NM then compensation is PB;
......

In general, the process is written as:

$$\text{IF (E IS } A_i \text{ AND CE IS } B_j \text{) THEN U IS } C_k \qquad \text{(Eq.3)}$$

which links fuzzy sets A defined by $\mu_A(x)$ ($x \in X$) and B defined by $\mu_B(y)$ ($y \in Y$) with an output set C defined by $\mu_C(z)$ ($z \in Z$). In fact, linguistic modes and knowledge contained rules setting is based on vague human judgements regarding the operation of the process. Table 1 displays the error compensation process by using the fuzzy rule-base Eq.3.

Table 1 Fuzzy Rules For Error Compensation Process

A\B	PB	PM	PS	Z0	NS	NM	NB
PB	NB	NB	NB	NB	NM	O	O
PM	NB	NB	NB	NM	NS	O	O
PS	NB	NB	NM	NS	Z0	PS	PM
Z0	NB	NM	NS	Z0	PS	PM	PB
NS	NM	NS	Z0	PS	PM	PB	PB
NM	O	O	PS	PM	PB	PB	PB
NB	O	O	PM	PB	PB	PB	PB

In the form of the fuzzy relation that combines all the fuzzy rules:

$$R = \bigcup_{ij}(Ei \times CEj) \times Uij \qquad \text{(Eq.4)}$$

Thus the fuzzy relation R in Eq.4 has the membership function:

$$\mu_R(x, y, z) = \max_{i=1...n}\{ \min (\mu_{Ai}(x), \mu_{Bi}(y), \mu_{Ci}(z))\} \qquad \text{(Eq.5)}$$

If any inputs of the process, error E and change of error CE, have been measured, the fuzzy control is obtained via a compositional rule of inference:

$$U = (E \times CE) \circ R \qquad \text{(Eq.6)}$$

or

$$\mu_C(z) = \max_{xy}\{ \min (\mu_A(x), \mu_B(y), \mu_R(x, y, z))\} \qquad \text{(Eq.7)}$$

The output of the fuzzy control from Eq.7 is a fuzzy set. As a process usually requires a non-fuzzy value of control, a "de-fuzzification stage" is needed. There are several ways of tackling this problem, the max-procedure is chosen in this case. Here:

$$U_{max} = \max_{z \in Z} C(z) \qquad (Eq.8)$$

Fuzzy controller design

Considering that all input and output variables for the error compensation process are non-fuzzy variables, while the input and output of the fuzzy controller are all fuzzy variables. It is necessary to convert non-fuzzy variables to fuzzy variables. Assuming that the variable range of errors is within [a,b], the following equation could be used to convert non-fuzzy variable e (error) varied in [a,b] into fuzzy variable E varied in $[E^-, E^+]$[Zhang, Q. J. & Lin, Q.J. et al 1992]:

$$E = \frac{E^+ - E^-}{b - a}[e - \frac{a + b}{2}] \qquad (Eq.9)$$

Here $[E^-, E^+] = [-5, 5]$. Using similar equations, the change of error (ce) and the compensation (u) are converted to fuzzy variables CE and U correspondingly. The discrete universes E, CE and U are defined as:

$$\{E\} = \{-5, -4, -3, -2, -1, 0, 1, 2, 3, 4, 5\}$$
$$\{CE\} = \{-5, -4, -3, -2, -1, 0, 1, 2, 3, 4, 5\}$$
$$\{U\} = \{-5, -4, -3, -2, -1, 0, 1, 2, 3, 4, 5\}$$

In the real compensation situation, the inputs to be measured are non-fuzzy vectors. When converting these variables into fuzzy variables, only one element of a particular input vector has a membership grade of one, the rest being all zero. Therefore, with fuzzy variable inputs E and CE, fuzzy compensation variables are inferred by using Eq.4 and Eq.6. In practice, a decision table (Table 2) that includes all input variables has been stored in a related computer. A compensation value can then be directly obtained if E and CE have been determined.

Table 2 The Decision Table

E\CE	-5	-4	-3	-2	-1	0	1	2	3	4	5
-5	5	5	5	5	5	5	3	3	0	0	0
-4	5	5	5	5	5	4	2	2	0	0	0
-3	5	5	5	5	5	3	1	1	0	0	0
-2	5	5	5	4	4	2	0	0	-2	-2	-4
-1	5	5	4	3	3	1	0	0	-1	-2	-3
0	5	4	3	2	1	0	-1	-2	-3	-4	-5
1	3	2	1	0	0	-1	-3	-3	-4	-5	-5
2	4	2	2	0	0	-2	-4	-4	-5	-5	-5
3	0	0	0	-1	-1	-3	-5	-5	-5	-5	-5
4	0	0	0	-2	-2	-4	-5	-5	-5	-5	-5
5	0	0	0	-3	-3	-5	-5	-5	-5	-5	-5

Error compensation for in-process measurement

To compensate the machining process errors caused by inherent geometrical errors of the machine tool, process dependent errors and environmental errors, etc.., a comparison methodology has been applied here. Firstly, a testpiece is clamped onto the machine table (Bridgeport Series I CNC milling machine) and it is to be cut to the shape as shown in Figure 1. Measurements are taken on the machine after cutting and each co-ordinate of the grid is recorded. Then, the testpiece is moved to a CMM and the same measurement for each grid is taken. An error item ($e_{1n} = v_m$(measured value from milling machine) - v_{cmm}(measured value from CMM)) is defined as the error mentioned above. In machining, the process errors $e_{pn} = e_{1n} + e_{2n}$. Where $e_{2n} = v_m - v_d$ (see Eq.1). This e_{pn} together with ce_n are input variables for the fuzzy controller. In practice, the error item e_{1n} at any point in the plane can approximately be obtained by the linear interpolation method.

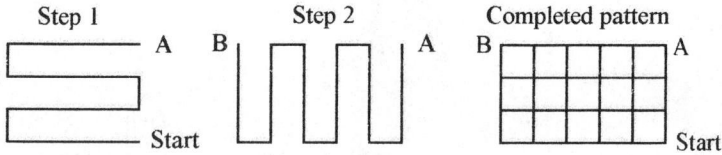

Figure 1. The testpiece inter-laced toolpath

Conclusions

This paper describes a compensation system using fuzzy logic control. The inherent geometrical errors of the machine tool, process dependent errors and environmental errors, etc. have been determined by comparing measured values on the testpiece with measured values on the CMM. Since fuzzy logic is employed the need for an exact mathematical relationship is avoided and thus the process evaluation is simplified.

References

Roe, J.M. 1984, In-Cycle Gauging, 1984, International Engineering Conference, Cleveland, Ohio, IQ84-263

Hermann, G. 1985, Process Intermittent Measurement of Tools and Workpieces, Journal of Manufacturing Systems 1985, Vol.4, part 1 pp 41-49

Cochran, D.S. & Swinehart, K., 1991, The Total Source Error of Adjustment model: A methodology for the elimination of setup and process adjustment, Int. J. Prod. Res. Vol. 29, No. 7, pp 1423-1435

Fan, K.C., Chen, L.C. & Lu, S.S. 1992, A CAD-Directed In-Cycle Gauging System with Accuracy Environment by Error Compensation Scheme, JAPAN/USA Symposium on Flexible Automation, Vol. 2, pp 1015-1021

Zhang, Q. J., Lin, Q. J. & Wu, H. 1992, Computer Aided Positional Error Compensation and Correction Using Fuzzy Inference, J. of Xi'An JiaoTong University, Vol. 26, Part 2, pp 55-62